Advances in
MICROBIAL ECOLOGY

Volume 11

ADVANCES IN MICROBIAL ECOLOGY

Advances in
MICROBIAL ECOLOGY

Volume 11

Edited by

K. C. Marshall
University of New South Wales
Kensington, New South Wales, Australia

PLENUM PRESS · NEW YORK AND LONDON

The Library of Congress cataloged the first volume of this title as follows:

Advances in microbial ecology. v. 1–
  New York, Plenum Press c1977–
  v. ill. 24 cm.
  Key title: Advances in microbial ecology, ISSN 0147-4863
  1. Microbial ecology — Collected works.
QR100.A36   576′.15   77-649698

ISBN 0-306-43340-0

© 1990 Plenum Press, New York
A Division of Plenum Publishing Corporation
233 Spring Street, New York, N.Y. 10013

Printed in the United States of America

Contributors

John Bauld, Division of Continental Geology, Bureau of Mineral Resources, Canberra, ACT 2601, Australia

T. E. Cloete, Department of Microbiology and Plant Pathology, University of Pretoria, Pretoria, South Africa

B. J. Finlay, Institute of Freshwater Ecology, Ambleside, Cumbria LA22 OLP, United Kingdom

Tim Ford, Laboratory of Microbial Ecology, Division of Applied Sciences, Harvard University, Cambridge, Massachusetts 02138

A. Gerber, Division of Water Technology, CSIR, Pretoria, South Africa

Graham W. Gooday, Department of Genetics and Microbiology, Marischal College, University of Aberdeen, Aberdeen AB9 1AS, Scotland

Don P. Kelly, Department of Biological Sciences, University of Warwick, Coventry CV4 7AL, England

Arthur L. Koch, Department of Biology, Indiana University, Bloomington, Indiana 47405

L. H. Lötter, City Health Department, Johannesburg, South Africa

Ralph Mitchell, Laboratory of Microbial Ecology, Division of Applied Sciences, Harvard University, Cambridge, Massachusetts 02138

M. O'Hara, Department of Microbiology and Genetics, Massey University, Palmerston North, New Zealand

Hans W. Paerl, Institute of Marine Sciences, University of North Carolina, Chapel Hill, Morehead City, North Carolina 28557

J. I. Prosser, Department of Genetics and Microbiology, Marischal College, University of Aberdeen, Aberdeen AB9 1AS, Scotland

Graham W. Skyring, CSIRO Division of Water Resources, Canberra, ACT 2601, Australia

Neil A. Smith, Department of Biological Sciences, University of Warwick, Coventry CV4 7AL, England

Gerald W. Tannock, Department of Microbiology, University of Otago, Dunedin, New Zealand

E. Terzaghi, Department of Microbiology and Genetics, Massey University, Palmerston North, New Zealand

D. F. Toerien, Division of Water Technology, CSIR, Pretoria, South Africa

David D. Wynn-Williams, British Antarctic Survey, Natural Environment Research Council, Cambridge CB3 OET, United Kingdom

Preface

The International Committee on Microbial Ecology (ICOME) sponsors both the International Symposium on Microbial Ecology, held in various parts of the world at three-year intervals, and the publication of *Advances in Microbial Ecology. Advances* was established to provide a vehicle for in-depth, critical, and even provocative reviews in microbial ecology and is now recognized as a major source of information for both practicing and prospective microbial ecologists. The Editorial Board of *Advances* normally solicits contributions from established workers in particular areas of microbial ecology, but individuals are encouraged to submit outlines of unsolicited contributions to any member of the Editorial Board for consideration for publication in *Advances*.

Chapters in Volume 11 of *Advances in Microbial Ecology* include those on microbial transformations of chitin by G. W. Gooday, organic sulfur compounds by D. P. Kelly and N. A. Smith, and phosphorus, including its removal in waste water treatment plants, by D. F. Toerien, A. Gerber, L. H. Lötter, and T. E. Cloete. The importance of diffusion processes in microbial ecology is discussed by A. L. Koch, and J. I. Prosser reviews the application of mathematical modeling to nitrification processes. Considerations of particular ecosystems include the Antarctic by D. D. Wynn-Williams and Australian coastal microbial mats by G. W. Skyring and J. Bauld. Other chapters include the regulation of N_2 fixation by H. W. Paerl, the role of microbial plasticity in ecology by E. Terzaghi and M. O'Hara, and the ecology of corrosion by T. Ford and R. Mitchell, of free-living protozoa by B. J. Finlay, and of lactobacilli in the gastrointestinal tract by G. W. Tannock.

K. C. Marshall, Editor
R. M. Atlas
J. G. Jones
B. B. Jørgensen

Contents

Chapter 1

Physiological Ecology of Free-Living Protozoa

B. J. Finlay

1. Introduction 1
2. Basic Requirements of Protozoa 2
3. Water, Food, and Polymorphism 7
   3.1. *Dictyostelium*—A Slime Mold 7
   3.2. *Sorogena*—A Polymorphic Ciliate 11
   3.3. Other Responses to Starvation 12
4. Microaerophily 13
   4.1. Interstitial Zone of Marine Sands 14
   4.2. *Loxodes*—A Microaerophilic Ciliate 15
   4.3. Zoochlorellae-Bearing Ciliates 18
5. Other Photosynthetic Symbionts 21
   5.1. Planktonic Foraminifera 22
   5.2. Sequestered Chloroplasts 24
6. Anaerobiosis and Methanogenic Symbionts 25
7. Postscript 28
   References 29

Chapter 2

Diffusion: The Crucial Process in Many Aspects of the Biology of Bacteria

Arthur L. Koch

1. Introduction 37
2. Microorganisms Are Small 38

3. An Outline of Diffusion 39
   3.1. Fick's Diffusion Law 39
   3.2. Steady State 40
   3.3. Diffusion through and around Obstacles 43
   3.4. Reaction along the Diffusion Path 46
   3.5. Dynamics of Diffusion 46
   3.6. Diffusion in a Stratified Ecosystem: Elementary Treatment 47
   3.7. Diffusion Limitation of Growth 47
   3.8. Literature on the Mathematics of Diffusion 49
4. Kinetics of Uptake from the Environment into the Metabolic Pool 50
   4.1. Transport in the Environment 51
   4.2. Transport through a Capsule 52
   4.3. Permeation through the S-layer or through the Outer Membrane 52
   4.4. Diffusion through the Peptidoglycan 54
   4.5. Role of Periplasmic Enzymes 54
   4.6. Uptake via Binding Proteins 55
   4.7. Passage through the Cytoplasmic Membrane 55
   4.8. The Role of Peripheral Enzymes 55
   4.9. Colimiting Transport 55
5. Diffusion through a Gel 57
   5.1. The Ogsten Theory 58
   5.2. Elaboration of the Renkin Theory 59
6. Two-Dimensional Cross-Diffusion 60
7. Efficiency of Uptake Systems 62
8. Conclusions 67
   References 68

Chapter 3

Ecological Aspects of Antarctic Microbiology

David D. Wynn-Williams

1. Introduction 71
2. Why Are Antarctic Habitats Distinctive? 74
   2.1. Continental Drift and the Isolation of Antarctica 74
   2.2. The Ice Habitats of Antarctica 75
   2.3. The Cold Deserts of Continental Antarctica 78
   2.4. Seasonal Changes in the Maritime Antarctic 81
3. Microbiology of Antarctic Cold Deserts 83
   3.1. Endolithic Communities 83
   3.2. Soil Microbial Xero- and Cryotolerance 90

4. Microbiology of Ice-Covered Water Bodies 98
   4.1. Permanently Ice-Covered Lakes 98
   4.2. Transient Rivers and Streams 107
   4.3. Temporarily Ice-Covered Freshwater Lakes 108
   4.4. Sea Ice Microbial Communities (SIMCO) 113
5. Microbiology of Patterned Ground 117
   5.1. Frost Tolerance in a Wet Habitat 117
   5.2. Microbial Stabilization of Fellfield Soils 119
   5.3. A Meeting of Ice and Steam 123
6. Terrestrial Nutrient Cycling—Peat and Penguins 125
   6.1. Moss–Peat Ecosystems 125
   6.2. Ornithogenic Soils ....:.............. 126
7. Environmental Impact 127
   7.1. Dry Valleys Drilling Project (DVDP) 127
   7.2. Microbial Aspects of Conservation in Coastal and Maritime
        Antarctica 130
8. Future Research Directions 131
   References 132

Chapter 4

The Microecology of Lactobacilli Inhabiting the Gastrointestinal Tract

Gerald W. Tannock

1. Introduction 147
2. The Microecology of Lactobacilli: Present Status 149
   2.1. Cell Walls of Lactobacilli and Adhesion Mechanisms 151
   2.2. Metabolism and Colonization 153
   2.3. Interactions with Other Microbes 155
   2.4. Testing the Colonization Abilities of Lactobacillus Strains 156
   2.5. Interactions with the Animal Host 158
3. The Microecology of Lactobacilli: The Future 160
   3.1. Plasmids 161
   3.2. Conjugation 162
   3.3. Transduction 162
   3.4. Transformation 163
   3.5. Transposons 164
4. Conclusion 165
   References 165

Chapter 5

Enhanced Biological Phosphorus Removal in Activated Sludge Systems

D. F. Toerien, A. Gerber, L. H. Lötter, and T. E. Cloete

1. Introduction 173
   1.1. Polyphosphates 174
   1.2. Activated Sludge Treatment 174
2. Background and Current Practices 175
   2.1. Evolution of Biological P-Removal Processes 175
   2.2. Design and Operational Aspects Pertinent to Biological P Removal . 184
3. Nutrient Dynamics in Activated Sludge Systems 189
   3.1. Factors Influencing P Release 190
   3.2. Factors Influencing P Uptake 195
4. Microbiology of P-Removal Activated Sludge Systems 196
   4.1. Introduction 196
   4.2. Bacteria Present in Activated Sludge Systems 196
   4.3. Numbers of Bacteria in Activated Sludge 197
   4.4. Selective Pressures in Activated Sludge Systems 199
5. Biochemical Model of Enhanced P Removal 201
   5.1. PolyP Metabolism 202
   5.2. Carbon Metabolism 206
   5.3. Transport of Metabolites 210
   5.4. Metabolic Control 213
   5.5. Current Biochemical Models of Enhanced P Removal 213
   5.6. Extended Model 215
6. Ecological Implications 218
   References 219

Chapter 6

The Ecology of Microbial Corrosion

Tim Ford and Ralph Mitchell

1. Introduction 231
2. The Role of the Surface Microbiota 232
   2.1. Aerobic Processes 232
   2.2. Anaerobic Processes 235
   2.3. Exopolymer–Metal Interactions 236
3. Acid Production 239
   3.1. Sulfur Oxidation 239
   3.2. The Role of Fungi 240

4. Influence of Iron and Manganese Deposition 242
    4.1. Tubercle Formation 242
    4.2. Colonization of Welds 243
5. Corrosion by Hydrogen-Consuming Bacteria 245
    5.1. Sulfate-Reducing Bacteria 245
    5.2. Activity of Methanogens 246
    5.3. Iron-Reducing Bacteria 246
6. Corrosion by Hydrogen-Producing Bacteria 246
7. Thermophilic Corrosion Processes 248
8. The Role of Consortia 251
9. Future Studies 252
    References 252

Chapter 7

Mathematical Modeling of Nitrification Processes

J. I. Prosser

1. Introduction 263
2. Nitrification 264
3. Mathematical Modeling 265
4. Pure-Culture Studies 266
    4.1. Basic Growth Kinetics 266
    4.2. Short-Term Measurements 266
    4.3. Measurement of Growth and Activity in Batch Culture 267
    4.4. Effect of Substrate Concentration 269
    4.5. Growth in Continuous Culture 270
    4.6. Effect of Temperature on Nitrification 272
    4.7. Effect of pH and Inhibitors 273
    4.8. Surface Growth 275
    4.9. Summary 277
5. Nitrification in Soil 277
    5.1. Measurement of Rates of Nitrification 277
    5.2. Effects of Environmental Factors 281
    5.3. Other Nitrogen Transformations 282
    5.4. Soil Columns 285
6. Nitrification in Aquatic Ecosystems 287
    6.1. Measurement of Rates of Nitrification 287
    6.2. System Models 287
7. Nitrification in Sewage and Waste Water Treatment Processes 289
    7.1. Basic Kinetics 290
    7.2. Measurement of Growth Constants 290

7.3. Effect of Biomass Concentration 293
7.4. Effects of Temperature, pH, and Inhibition 294
7.5. Surface Growth 296
7.6. Summary 298
8. Concluding Remarks 299
   References 300

Chapter 8

Physiological Ecology and Regulation of N_2 Fixation in Natural Waters

Hans W. Paerl

1. Historical and Current Perspectives 305
2. Aquatic N_2-Fixing Microorganisms: Their Diversity and Habitats 307
3. The Physiological Ecology of Aquatic N_2 Fixation 319
4. Environmental Constraints and Limitations on Aquatic N_2 Fixation 323
5. Roles of Organic Matter and Microzone Formation 327
6. Evolutionary and Ecological Considerations 333
7. Ecosystem-Level Regulation of N_2 Fixation: Are There Fundamental
   Differences between Freshwater and Marine Habitats? 334
8. Conclusions 336
   References 337

Chapter 9

Organic Sulfur Compounds in the Environment: Biogeochemistry, Microbiology, and Ecological Aspects

Don P. Kelly and Neil A. Smith

1. Introduction 345
   1.1. Background 345
   1.2. The Atmospheric Component of the Sulfur Cycle 346
2. Organic Sulfur Compounds in the Natural Environment 348
   2.1. Methylated Sulfides, Carbon Sulfide, and Carbonyl Sulfide 349
   2.2. Carbon Disulfide and Carbonyl Sulfide 358
   2.3. Sulfoxides and Sulfonates 360
   2.4. Higher Alkyl and Aromatic Sulfides and Polysulfur Compounds 361
   2.5. Naturally Occurring Aromatic and Heterocyclic Sulfur Compounds .. 362
   2.6. Mammalian Odors: Pheromones, Attractants, and Repellants 363
3. Microbiological Degradation of Organic Sulfur Compounds 368
   3.1. Methylated Sulfides 368
   3.2. Other Sulfides 369

3.3. Carbon Disulfide and Carbonyl Sulfide 370
3.4. Sulfonates 372
3.5. Thiophenes and Aromatic Sulfur Compounds 372
3.6. Miscellaneous Sulfur Compounds 374
4. Some Effects of Organic Sulfur Compounds on Global Ecology 374
References 375

Chapter 10

The Ecology of Chitin Degradation

Graham W. Gooday

1. Chitin and Its Occurrence 387
   1.1. Chitin Structure 387
   1.2. Occurrence of Chitin 388
   1.3. Fossil Chitin 389
   1.4. Annual Production of Chitin 390
   1.5. Amount of Chitin in the Biosphere 390
2. Pathways of Chitin Degradation 391
3. Chitin Digestion by Microbes 393
4. Chitin Degradation in the Sea 393
   4.1. Degradation in the Water Column 393
   4.2. Degradation in Sediments 394
   4.3. Degradation in the Deep Sea 395
   4.4. Rates of Degradation 397
5. Chitin Degradation in Estuaries 398
   5.1. Estuarine Chitinoclastic Microbes 398
   5.2. Rates of Degradation 400
6. Chitin Degradation in Freshwaters 402
   6.1. Freshwater Chitinoclastic Microbes 403
   6.2. Rates of Degradation 403
7. Chitin Degradation in Soil 405
   7.1. Microbial Degradation 405
   7.2. Soil Chitinase Activity 408
   7.3. Effects of Addition of Chitin to Soil 409
   7.4. Addition of Chitin to Soil for Biological Control of Disease 410
8. Adhesion of Microbes to Chitin 411
9. Chitin Digestion in Animals 416
10. Involvement of Chitin Degradation in Pathogenesis and Symbiosis 418
11. Summary 419
References 419

Chapter 11

Microbial Plasticity: The Relevance to Microbial Ecology

E. Terzaghi and M. O'Hara

1. Introduction 431
2. Interorganismal Exchange 433
3. Programmed Rearrangement 434
   3.1. Topology of Rearrangement 434
   3.2. Inversions 434
   3.3. Transposition and Deletion 437
4. Unprogrammed Rearrangement 438
   4.1. Agents of Change 438
   4.2. Evolution of Function 442
   4.3. Cryptic Functions 444
5. Evidence for Genomic Change 446
   5.1. Chromosomal 446
   5.2. Plasmid 447
6. Phenotypic Changes of Unknown Genetic Basis 449
7. Significance of Microbial Plasticity to Microbial Ecology 451
   References 453

Chapter 12

Microbial Mats in Australian Coastal Environments

Graham W. Skyring and John Bauld

1. Introduction 461
2. Occurrence and Characteristics of Australian Microbial Mats 463
   2.1. Location and Habitat 463
   2.2. Community Structure and Microbial Components 468
3. Phototrophic Activity and Environmental Constraints 471
   3.1. Light 471
   3.2. Photoautotrophic Activity and Primary Productivity 473
   3.3. Photoheterotrophic Activity 476
   3.4. Environmental Controls on Phototrophic Processes 477
   3.5. Interactions with Other Biota 478
   3.6. Nitrogen Fixation 480
4. Degradative Processes and Environmental Constraints 481
   4.1. Aerobic Processes 481
   4.2. Fermentation and Sulfate Reduction 482
   4.3. Methanogenesis 486

5. Conclusions 486
    5.1. Ecology 487
    5.2. Minerals, Petroleum, and Modeling 487
    5.3. Serendipity, Controversy, and Speculation 488
    5.4. Some Unsolved Problems and Unanswered Questions 489
    References 491

Index 499

1

Physiological Ecology of Free-Living Protozoa

B. J. FINLAY

1. Introduction

There are about 20,000 species of the single-celled animal-like organisms we call protozoa, and they are the most abundant phagotrophs in the biosphere. One milliliter of sea water contains about 1000 heterotrophic flagellates (Sherr and Sherr, 1984; Fenchel, 1988), freshwater sediments contain about 10,000 ciliates (Finlay, 1980, 1982), and organically rich habitats such as activated sludge plants support at least 10^5 ciliates and flagellates per milliliter (Curds, 1973). The calcareous and siliceous oozes that cover most of the marine benthos are largely composed of the sedimented shells and skeletons of planktonic foraminifera and radiolaria, and at least one wonder of the ancient world, the great pyramid of Cheops at Gizeh, consists almost entirely of the compacted shells of the fossil foraminiferan *Nummulites gizehensis* (see Haynes, 1981).

Much of the current interest in the ecology of protozoa is focused on the unique role played by small (<20 μm) planktonic heterotrophic flagellates, for these, the smallest of the free-living protozoa, seem to have a pivotal role in the so-called microbial loop (Azam *et al.*, 1983; Porter *et al.*, 1985; Sherr *et al.*, 1986; Fenchel, 1986a, 1988). Most ecological investigations of the protozoa start with a description of the species concerned and of their spatial distributions, but the emerging picture of the ecology of these flagellates has grown out of a different approach, in which much emphasis has been placed on the physiology of the organisms. They were virtually ignored until about 10 years ago, when the search began in earnest for the organisms that controlled the abundance of bacteria in the sea. Likely contenders seemed to be the small, ubiquitous, heterotrophic flagellates. They were obviously well adapted for feeding on bacteria, and

B. J. FINLAY • Institute of Freshwater Ecology, Ambleside, Cumbria LA22 OLP, United Kingdom.

laboratory experiments showed that they had the physiological capacity to rapidly clear bacteria from relatively large volumes of water (about 10^5 times their own cell volume per hour). And when these feeding rates were combined with counts of flagellates in the sea, it became obvious that they could easily control bacterial abundance close to the observed value of 10^6 cells per milliliter. Many of these flagellates have not yet acquired species names, although in many cases their gross morphology is well known and the physiology of their feeding is well documented (Fenchel, 1986c).

The capacity of these organisms to efficiently graze dilute bacterial suspensions secures their niche in the plankton, and the example illustrates the close link that always exists between physiology and ecology. In this review, I will not be concerned further with the physiological ecology of suspension feeding, for which an excellent review already exists (Fenchel, 1986b). Rather, I will address some of the other ways in which the physiology of protozoa equips them for life in the natural environment. But first, I must consider some basic requirements of free-living protozoa.

2. Basic Requirements of Protozoa

Over short periods, the protozoon will survive if it can generate enough ATP to sustain basic physiological functions and if it is immersed in water that is not too hot and contains some suitable inorganic ions. Over longer periods of time, the cell will need some mechanism for capturing sufficient food and the essential nutrients for renewal of cell components. And, if the protozoon is to persist and reproduce in a particular niche, it will need to perceive significant changes in the quality of its local environment and be able to coexist with competitors and potential predators.

The protozoon will deploy a vast range of physiological processes, some of which, like ionic and osmotic regulation, are vital to continued existence. Other processes may be considered useful but not necessarily vital, since they serve only to increase the fitness of the protozoon in a particular niche. Examples would be the ability to reduce the chances of being eaten by growing spines (Kuhlmann and Heckmann, 1985) or by escaping into anoxic water.

Some physiological processes become important or vital only when the protozoon develops certain gross physical properties. The force of gravity and the low rate of molecular diffusion through protoplasm are virtually irrelevant to a tiny planktonic flagellate, but both are important to some large ciliates, which respond physiologically, especially to gravity (see Section 4.2). One of the largest of protozoa, the naked ameba *Pelomyxa,* is very dense because of the inclusion of many mineral grains. Large amebae cannot swim, and the force of gravity keeps the protozoon in the sediment, where it apparently circumvents the limited diffusion of O_2 through its large and usually spherical mass by exploiting pathways of anaerobic metabolism (see Section 6). We could, of course, turn this argument around and claim that the adaptive significance of the great mass of *Pelomyxa* is to keep it in the sediment, because it is intolerant of oxygen (which is true). In either case, the close link between the physiology and the physical dimensions of *Pelomyxa* is obvious.

If we seek to define further basic requirements of protozoa in general by examining where they are found in the natural environment, we will encounter some difficulties. We might try to analyze the wealth of published data for various physical and chemical variables associated with protozoa but, as we can see in Table I, that will tell us very little about the requirements of protozoa in the natural environment. Almost all aquatic habitats that have been investigated, from acid mine drainage (Lackey, 1938) to hypersaline lagoons (Wilbert and Kahan, 1981) and sulfide-rich sediments (Fenchel, 1969), have been shown to contain protozoa. We might infer from this that protozoa are either extremely adaptable or very diverse. They are certainly diverse and, as discussed below, many are very adaptable. But before exploring the significance of the information in Table I, we might question the usefulness of these data.

The problem with this type of information is that environmental factors do not act in isolation from one another. The most useful data (and the most difficult to obtain) would be the combinations of factors that exclude protozoa from a habitat. To take just one example, at neutral pH most ammonia is present as the ammonium ion NH_4^+, which

Table I. Approximate Ranges of Some Physical and Chemical Factors in Aquatic Environments

Factor	Environmental range		Range probably tolerated by various protozoa		Units
	Min	Max	Min	Max	
Temperature	-2	~100	-2[a]	~50[b]	°C
pH	<2[c]	>11[c]	<2[d]	11[e]	$-\log_{10}a_{H^+}$
Salinity	0.12[f]	332[g]	<0.12	332[g]	g liter^{-1}
		>100[h]		(>100)[h]	
Oxygen	0	>30[i]	0	(>30)[i]	mg liter^{-1}
Hydrogen sulfide	0	700[j]	0	500[j]	mg liter^{-1}
Ammonia[k]	0	>200	0	>200[l]	mg liter^{-1}

[a] See Lee and Fenchel (1972).
[b] There are a few records of protozoa at $>50°C$, but there is no good evidence that they can grow and divide at these temperatures (see Kahan, 1972, and Nisbet, 1984).
[c] Values outside this range are exceptional in unpolluted waters. *Sphagnum* bogs, which are rich in protozoa, usually fall within the pH range of 3.3–4.5 (Wetzel, 1983).
[d] Lackey (1938) found various protozoa living in waters polluted with acid mine drainage at pH 1.8.
[e] Epilimnion of Lake Simbi, Kenya (Finlay *et al.*, 1987a), and Wadi Natrum, Egypt (Imhoff *et al.*, 1979).
[f] Mean salt content of world river water.
[g] In waters where chloride is the dominant anion [North Arm of Great Salt Lake (Post, 1977) and deep water of the Dead Sea (Nissenbaum, 1975)].
[h] In waters where carbonate or bicarbonate is the dominant anion. The maximum figure is an estimate for Lake Nakuru, which has a high but variable salt content (Talling and Talling, 1965; Vareschi, 1982). Finlay *et al.* (1987a) recorded many protozoa in this lake when the salt content was 25 g liter^{-1}.
[i] Melack (1979) recorded >30 mg O_2 liter^{-1} in the epilimnion of Lake Simbi, Kenya. Finlay *et al.* (1987a) recorded the protozoa living in the same zone of this lake when O_2 was ~20 mg liter^{-1}.
[j] See Fenchel (1969).
[k] Mainly NH_4^+ and NH_4OH at pH 6–9.
[l] Bick and Kunze (1971).

is quite innocuous, even at high concentrations, but its toxicity increases substantially as the pH increases, along with the proportion of ammonium hydroxide (NH_4OH). At pH 9.5 there is about 3000 times as much NH_4OH as at pH 6.

Moreover, some of the factors quoted in Table I prevent others from occurring. Hydrogen sulfide and O_2 at high concentrations exclude each other chemically, so it is not surprising that there probably does not exist a protozoon that has the biochemical capacity to tolerate high levels of both.

There is one factor that is probably guaranteed to exclude protozoa: high temperatures. It is very difficult to get any protozoon to grow at temperatures above 40°C, and there are virtually no records of growth above 50°C. There are isolated reports of the successful gradual adaptation of some protozoa to temperatures close to 70°C (see Noland and Gojdics, 1967), but the rule is that temperatures >50°C in the natural environment are devoid of protozoa and indeed all other eucaryotes. Procaryotes, on the other hand, are found at all temperatures up to 100°C and beyond.

Our only rather weak conclusion, then, is that protozoa are probably found in the aquatic environment wherever eucaryotes are found. Perhaps the basic requirements of protozoa can be better understood if we consider individual species and the conditions under which they are normally found. Data of this type for seven ciliate species have been taken from the literature and illustrated in Fig. 1. The figure includes so-called extreme tolerances (in this case, meaning the range within which each species has ever been found in the natural environment) of seven ciliate species to five factors: temperature, O_2, pH, ammonia, and hydrogen sulfide. The first impression is of different tolerances in the different species. *Tintinnidium* is apparently tolerant of only quite narrow ranges of all factors, *Glaucoma* occurs throughout broad ranges, and *Brachonella* is somewhere in between, at least for some factors. There are several reasons for believing that these data tell us much less than we might think about extreme tolerances of these ciliates.

First, *Tintinnidium* has been recorded by relatively few investigators, whereas *Glaucoma* has been recorded by many. This may be because *Tintinnidium* is quite rare or because protozoologists are often fascinated by sewage treatment plants and sites of organic pollution, where *Glaucoma* is often found. Only a tiny minority of aquatic habitats have been investigated for protozoa, so it is perhaps unwise to deduce extreme tolerances for ciliates such as *Tintinnidium* from the limited data available.

Second, we must cope with the perennial problem of discriminating and identifying ciliate species that are virtually indistinguishable from one another. This problem certainly exists for the sibling species of the *Tetrahymena pyriformis* complex (see Nanney, 1982). *Tetrahymena*, like *Glaucoma*, is credited with wide ranges of tolerance for many environmental factors. In the case of *T. pyriformis*, the broad tolerance almost certainly can be ascribed not to one species but to many species that cannot be differentiated by using a light microscope. And, as with *T. pyriformis*, there is no guarantee that other "species" may not also be species complexes whose representatives are phenotypically identical but members of genetically distinct species which, together, cover broad ranges of environmental factors.

Third, we may have been diverted from a true appreciation of the niches of the

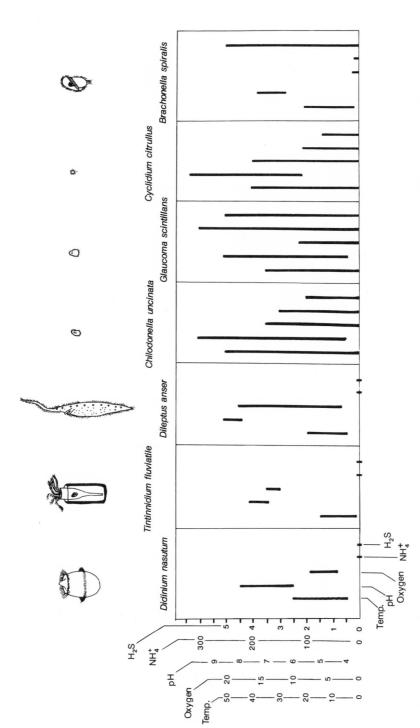

Figure 1. Ranges of "extreme tolerance" to five environmental factors reported for seven ciliate species. [Compiled from information in Bick (1972) and Bick and Kunze (1971).]

protozoa concerned. Notice, for example, that the smallest ciliates have the widest ranges, and they are all bacterial feeders. Two of them are filter feeders (*Glaucoma* and *Cyclidium*), and *Chilodonella* is a browser on surfaces. Most bacteria-feeding filter feeders are fairly nonselective for the quality of particles they ingest (Fenchel, 1986b). They will usually find enough food in sediments, where bacterial abundance is high and fairly constant, as well as in eutrophic lakes and sites affected by organic pollution. Perhaps they will be found wherever there is enough bacterial food.

The other three ciliates are quite different. *Didinium* is usually a specialized predator, especially of large ciliates like *Paramecium;* it also has the capacity to encyst. If it encysts because of the absence of suitable food and it is therefore recorded as being absent from a site, a physiological response to starvation might be misconstrued as an intolerance of ammonia, low temperatures, or some other factor.

Tintinnidium is usually planktonic in open waters, where it feeds on nanoplankton and, perhaps more important, makes a lorica which is embellished with diatom shells. We can appreciate how the seasonal occurrence of *Tintinnidium* might be at least as dependent on the seasonal cycle of algal productivity and seasonal occurrence of specific types of algae as on the temperature or chemistry of the water. Similarly, we can see how the food niche of the carnivorous ciliate *Dileptus* is more specialized than that of a filter-feeding bacterivore.

It is perhaps not surprising, then, that the more specialized ciliates seem to have narrower ranges of "tolerance." This may be because they, unlike the others, are not species complexes or because they are indeed less tolerant. Alternatively, the quality and abundance of suitable prey may be of paramount importance to ciliates such as *Dileptus* and *Didinium,* which have specialized organelles for feeding on relatively specialized diets. *Dileptus* and *Didinium* may or may not be tolerant of high temperatures and low oxygen tension, but that potential tolerance, if it exists, will remain redundant and not expressed if the food they require is not also found under the same conditions.

There is, however, one of these ciliates whose physiology is obviously linked to its chemical environment; that is the anaerobic ciliate *Brachonella,* which is tolerant of hydrogen sulfide and intolerant of oxygen. Even with *Brachonella,* however, the chemical environment is just a part of the niche of the ciliate. There is little point in having the physiological capacity for life in a hot anaerobic puddle if there is nothing suitable to eat, and, vice versa, the protozoon is unlikely to have a feast in an extreme environment unless it is also physiologically tolerant of that environment. It is surely desirable to consider the complete niche of a protozoon before pronouncing on the causes of presence in or absence from a habitat.

The rather pessimistic conclusion at this point might be that we know very little about the factors that are important in the ecology of free-living protozoa: we often do not know what the important factors are, we are invariably ignorant of the genetically determined limits of tolerance and, as a consequence, we are often unaware of the conditions required for survival, persistence, and growth in a habitat. But there is more to ecology than recording the co-occurrence of organisms and some easily measured parameters. There is, for example, some evidence that it may be more rewarding to take account of as much as possible of the protozoon's niche. In practice, this means becoming thoroughly familiar with the biology, and especially the physiology, of the organisms concerned. We must know how they feed and what they feed on, how they exploit the activities of any symbionts, how they cope with transient resources like dissolved oxygen and free water, how they respond to predators and a patchy food supply, and how their physiologies are adapted to the threat of starvation. In short, we must become familiar with the many ways in which protozoa function in relation to their environment or, to put it more grandly, we must attempt to understand their physiological ecology. In the account that follows, I have attempted to summarize much of what is known about how some free-living protozoa function in the natural environment.

3. Water, Food, and Polymorphism

This section might also have been entitled "soil protozoa," for fluctuations in the supply of water and food are the fundamental problems facing soil protozoa, and a polymorphic life cycle is the common solution. All protozoa need water to move and feed. If they are to persist in an environment with a transient supply of free water, they must have the capacity to respond physiologically to avoid dehydration, starvation, or both. Almost all soil protozoa share one basic strategy—the capacity to wrap themselves in a thick protective coat. This may be a shell, in the case of the testate amebae, or a cyst wall, in the case of many soil ciliates and amebae. Cysts can also be formed directly in response to a depleted food supply; again, this is a more common characteristic in soil protozoa.

Unlike the truly aquatic protozoa, the forms living in soil cannot rely on dispersal by water, so they exploit the only alternative available—in the encysted state, they are dispersed by air (see Corliss and Esser, 1974). This solution is also used by most other soil microorganisms and the larger fungi; just as mushrooms and toadstools show a bewildering variety of morphological and physiological adaptations geared to spore dispersal, so the soil protozoa show an impressive variety of strategies. The protozoan strategies are all the more remarkable because the potential for polymorphic life cycles, intercellular communication, cell aggregation, and differentiation are the product of the genome of a unicellular organism. This striking ability long ago convinced cell biologists that certain soil protozoa, and in particular the cellular slime mold *Dictyostelium,* could be a simple model for studies on intercellular communication and differentiation that would be relevant to the related but more intractable problems in higher organisms. The sophistication of the *Dictyostelium* life cycle, the ease with which the organism can be studied in the laboratory, and its justified status as a model organism are the principal reasons why we can now say something about the physiological ecology of the organism.

3.1. *Dictyostelium*—A Slime Mold

Dictyostelium is one of many slime molds that live in the surface layer of soil and the leaf litter of forests. Many species have been isolated and studied (see Raper, 1984), although almost all recent research on cell cultures has been confined to one species, *D.*

discoideum. The vegetative amebae of *Dictyostelium* thrive as independent cells in soil if they are provided with a high oxygen tension, moderate temperatures (~ 15°C), a near-saturated atmosphere, adequate bacterial food, and a film of soil moisture thick enough to support ameboid movement (Raper, 1984). The amebae are about 10 μm in diameter, they move sluggishly (0.2 μm sec^{-1}), they ingest bacteria and other microorganisms by phagocytosis, and they divide by binary fission, possibly as often as several times per day. But such a sluggish browser cannot continue to grow indefinitely on the patchy resources of clumps of soil bacteria, and the vegetative growth phase is halted when the colony has exhausted its food supply. The amebae then have three options: to migrate to another patch of bacteria [using their chemotaxis to folic acid secretions by bacteria (Pan *et al.,* 1975)], to encyst temporarily as microcysts, or to exploit their innate strategy for communal dispersal. The last process seems to be a general feature of the cellular slime molds, and it is now understood in some detail.

The process begins with cell aggregation as some cells in the population assert themselves as collection points by producing and secreting adenosine 3,5-cyclic monophosphate (AMP) [Mato and Konijn, 1979; Gerisch, 1982; Newell *et al.,* 1987; see Gerisch (1987) for a discussion of the many other roles of cAMP in *Dictyostelium*]. This binds to specific receptors in the plasma membrane of other amebae, and a chain reaction is set off in the latter, culminating in pseudopod extension and cell elongation in the direction of the source of attractant (Fig. 2). The receptive amebae also secrete cAMP and in so doing relay the stimulus for attraction to amebae lying further and further from the original collection point. The amebae closest to the center continue to secrete pulses of cAMP, producing radially propagated waves of attractant that sustain the process of population aggregation. Individual cells respond to each impulse with a movement of about two cell lengths (~ 20 μm) over a period of 50–100 sec (Raper, 1984).

The efficiency of this process is further increased by two factors. First, the threshold concentration of cAMP is at least 100 times lower in aggregating cells than in vegetative cells (Bonner *et al.,* 1969), and second, the concentration of attractant is kept within fairly narrow limits. The amebae do this themselves: elevated concentrations of attractant signal the production and secretion of phosphodiesterases, which hydrolyze the excess extracellular cAMP. Cell aggregation by such a method is particularly effective because the cells deploy a variety of physiological and biochemical processes to overcome two fundamental problems: the inefficiency of a biased random walk as a means of aggregation in organisms that move as slowly as *Dictyostelium,* and the inefficiency of relying on diffusion to transport an attractant from a point source over a relatively large area. The pattern of cell movement and aggregation is usually referred to as chemotaxis. Unlike the so-called chemotaxis in bacteria (which *is* a biased random walk), and like the few other recorded taxes in microorganisms, cell movement in *Dictyostelium* does appear to be a true chemotaxis insofar as both pseudopod extension and cell movement are invariably oriented in the direction of the source of attractant. This behavior implies that the cell knows where it is going and that it is capable of perceiving a spatial gradient, presumably through an excess binding of attractant to surface receptors on the cell surface closest to the source of attractant. This notion is

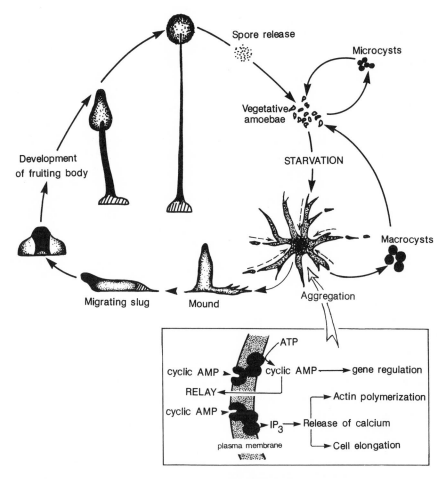

Figure 2. The complex life cycle of the cellular slime mold *Dictyostelium*. Boxed area shows a simplified picture (adapted from Newell *et al.*, 1987) of the biochemical and physiological processes underlying cell aggregation. IP$_3$, Inositol 1,4,5-triphosphate.

supported by some theoretical considerations (Berg and Purcell, 1977) and the observation that pseudopod retraction and the formation of new pseuopodia can be induced by the local application of high concentrations of cAMP from a micropipette (Gerisch, 1982). Cells also produce small "pilot" pseudopodia in a more or less random fashion, but these are too small ($\sim$0.5 μm) to play any role in sensing spatial gradients of attractant. The beneficial consequence for *Dictyostelium* is that it is able to move in an almost straight line toward the aggregation center of the population.

The problem of diffusion of the attractant has been overcome by the relay system, which periodically amplifies the signal of the attractant as it radiates out from the point source. The advantage is that the distance traveled by the attractant remains proportional to time and that the concentration of the signal remains virtually constant with distance. Neither of these benefits would obtain if aggregation depended on the secretion of an attractant from a single point source, in which case cell aggregation would be effective only over trivially small distances.

Aggregated cells adhere to each other and develop into a vertical wormlike mound about 1 mm high, which subsequently bends over to become a migrating slug [also referred to as a pseudoplasmodium or grex; see Vardy et al. (1986) for a discussion of the mode of locomotion]—a collection of about 10^5 amebae enclosed in a cellulose-rich sheath that commonly migrates at a speed of about $1-2$ mm hr^{-1}. It is at this stage that the differentiation into cell types takes place, with the anterior 15–20% of cells becoming pre-stalk cells and most of the remainder becoming pre-spore cells. The mechanism of differentiation is still not clear, but the fate of cells seems to depend on the stage in the division cycle at which they are when the differentiation-inducing starvation occurs. Cells that divide just before or just after starvation become pre-stalk cells, whereas those dividing at other times become pre-spore cells (Gomer and Firtel, 1987). The actual differentiation is probably effected by an inducing factor. One of these has recently been identified as 1-(3,5-dichloro-2,6-dihydroxy-4-methoxyphenyl)-1-hexanone (Morris *et al.*, 1987), a representative of a new class of effector molecule soluble in lipid and water and thus easily moved from cell to cell.

The final stage in the life cycle is the development of a fruiting body, when the differentiated cells begin to carry out different functions. The vacuolated and turgid stalk cells secrete a sheath composed largely of cellulose, which lengthens, carrying the developing sorus full of spore cells to the apex, often several millimeters from the substratum. Maximum spacing of the fruiting bodies is apparently ensured by their negative chemotaxis to ammonia, produced by the fruiting bodies themselves (Feit and Sollitto, 1987). The spore cells develop into spores, which are later released and dispersed to germinate in favorable conditions and to complete the cycle.

At all stages in this life cycle, *Dictyostelium* is sensitive to a variety of environmental factors that influence the rate of development and transfer to the next stage. Thus, aggregation is hastened in response to decreased humidity, increased temperature, and increased light, and migrating slugs are extremely sensitive to temperature (ceasing migration when the temperature is raised) and humidity (the slime sheath becomes tougher if the humidity falls). The light sensitivity of the migrating slug is particular interesting. It is mediated by a double photoreceptor pigment system (Poff *et al.*, 1974; Poff and Whitaker, 1979) with a maximum absorbance at 430 nm, and the behavioral response is phototaxis. Again, this does seem to be a real taxis: the slug will migrate in a straight line toward a light source that changes direction. Poff *et al.* (1974) concluded that a cylindrical slug could perceive light direction if unilateral light was focused on the distal surface, causing cells there to migrate faster and so bend the slug until it faced the light. If the slug is exposed to overhead light, it immediately stops horizontal migration and begins formation of the fruiting body (Newell *et al.*, 1969). Solitary vegetative amebae may also display a complex phototaxis (see Fisher *et al.*, 1985).

The slug plays a clear role in the life cycle of *Dictyostelium:* it brings a large number of cells to a site where a fruiting body can be erected for the efficient dispersal of spores. In so doing, it uses the environmental cues of humidity, temperature, and light to guide it to the soil surface. It must migrate quickly and efficiently because the slug does not feed on its travels, and if it takes several days to reach its destination for fruiting, it may be reduced to 10% of its original size. The fruit it forms will then be proportionately small (Newell *et al.*, 1969).

Polymorphic diversity in *Dictyostelium* is completed with the macrocyst, a true sexual stage and an alternative morphogenetic route following cell aggregation. Some cells become giants, which chemoattract and engulf other amebae of the same genotype. Nuclear fusion within the giant then transforms the cell into a zygote (Szabo *et al.*, 1982; Lewis and O'Day, 1986). The factors which induce this sexual stage are not entirely clear, although the presence of light in the visible spectrum favors the development of sorocarps and suppresses the formation of macrocysts (Chang *et al.*, 1983), and mating-type attraction involves the secretion of volatile sex hormones (Lewis and O'Day, 1977; Filosa, 1979).

3.2. *Sorogena*—A Polymorphic Ciliate

The cellular slime molds are not alone in the production of aerial sorocarps. *Sorogena* is a carnivorous ciliate that grows in the thin water film on dead attached plant parts [e.g., twigs, pods, and dried fleshy fruits; the type material was isolated from dry figs (Olive, 1978; Bradbury and Olive, 1980)]. It appears to feed only on the ciliate *Colpoda*, and it will select *Colpoda* from mixed assemblages of ciliate species (Bradbury and Olive, 1980). It feeds rapidly and, when the food supply is close to exhaustion, the cells begin to aggregate into a mound that extends upward to the water surface, becomes encapsulated in a protective sheath (at this stage it is called a sorogen), and secretes a stalk that raises the sorogen about 1 mm into the air (Olive and Blanton, 1980). When development of the stalk is complete, the cells within the sorogen encyst, producing a sorus that dries, fractures, and disperses the cysts, which subsequently germinate in favorable conditions.

There are several notable differences between the life cycles of *Dictyostelium* and *Sorogena*. Some are due to the fact that the two organisms are quite different types of protozoon. For example, the stalk of the fruiting body in *Dictyostelium* consists of a thickened cellulose wall, specialized but nonviable ameboid cells, and cellulose microfibrils, whereas the *Sorogena* stalk is a noncellular protein–polysaccharide complex secreted by all aggregated cells from subpellicular stalk material vesicles (Blanton and Olive, 1983a,b; Blanton *et al.*, 1983). The exact nature of the stalk material is unknown, but it does have the valuable property of hydrating and increasing in volume 100–1000 times, thus rapidly pushing the sorogen skyward. It then dehydrates to form a solid sheath.

The attractant for aggregation in *Dictyostelium* is certainly cyclic AMP, but this is unlikely in *Sorogena* (Olive and Blanton, 1980). The identity of the attractant in *Sorogena* is unknown.

The absence of a slug stage from the *Sorogena* life cycle is probably related to the nature of the habitat of the ciliate. The availability of water on aerial plant parts will change more quickly than in leaf litter and beneath the soil surface, the habitat of trophic *Dictyostelium*. Accordingly, it may be vitally important to *Sorogena* that it has the capacity to encyst as quickly as possible. *Sorogena* develops from aggregate to fully developed fruiting body in 30–45 min, but the process takes about 8 hr in *Dictyostelium*.

Another function of the slug in *Dictyostelium* is to bring many cells together into the open, where an aerial fruiting body can be formed. This function is, of course, redundant in an organism living on aerial plant parts.

Finally, *Sorogena* seems to feed on only one type of encysting ciliate. In so doing, it probably relies less on environmental cues (e.g., there is no phototaxis) than on the cues provided by *Colpoda;* when *Colpoda* encysts, *Sorogena* encysts. *Sorogena* is obviously aware of the presence of *Colpoda* in ways that are not understood (e.g., selective predation). One environmental cue that *Sorogena* does not ignore is the alternating cycle of night and day. Aggregation invariably begins before sunrise; sorocarps ascend into the air in early morning, when they can take advantage of the dew and the moist air.

Laying aside the differences, the polymorphic life cycles in both *Dictyostelium* and *Sorogena* are obviously the product of a great variety of physiological responses to environmental changes. The level of sophistication might convince us that at least some soil protozoa have gone a considerable way toward maximizing the chances of cells finding and exploiting the available food.

3.3. Other Responses to Starvation

The problem of periodic starvation and coping with a patchy food resource is not unique to the soil protozoa, although the strategy of encystment as a guard against dehydration is undoubtedly more common in soil. Many protozoa living in permanently aquatic environments may have lost the ability to encyst, but they must still cope with food resources that are patchy or temporally variable. Some do form cysts (see Reid, 1987), but a great variety of other adaptations have also evolved. Some species of the ciliates *Tetrahymena* and *Blepharisma* will, following exhaustion of their normal bacterial food, develop into "macrostomes" capable of ingesting much larger food particles (including their clone mates) (see Fenchel 1986a), and the size of the predatory ciliate *Didinium* varies in response to the changing size of its ciliate prey (Hewett, 1980).

Another consequence of a truly aquatic life style is that it does not confine cells to life in thin water films and narrow crevices, so it becomes possible in certain circumstances for species to become quite large. Some amebae and ciliates grow up to 1 mm or more, but none of these are found in the soil. Large protozoa can starve for relatively long periods and yet retain normal cell functions. When a large ameba is starved, it gradually gets smaller but its weight-specific respiration rate stays relatively constant. If a small flagellate is starved, it has only a few hours before it respires itself into extinction. It is not so surprising, then, that a small flagellate reduces its specific respiration rate by a factor of about 50 when it is deprived of food (Fenchel and Finlay, 1983) (Fig. 3).

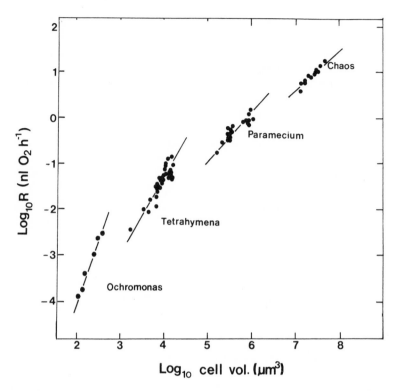

Figure 3. Relationships between respiration rate (R) and cell size in four types of protozoa in various physiological states. *Chaos* is a large naked ameba, *Paramecium* and *Tetrahymena* are ciliates, and *Ochromonas* is a heterotrophic flagellate. Weight-specific respiration rate is fairly constant in *Chaos* but falls by at least an order of magnitude in starving *Ochromonas*. [Adapted from Fenchel and Finlay (1983).]

There is also at least one documented case of polymorphism induced not by starvation but by the presence of a predator. The ciliate *Lembadion* has a large mouth, which it uses to ingest other ciliates such as *Euplotes*. But *Euplotes* is sensitive to the close presence of *Lembadion,* and it responds by modifying its cell shape: it develops prominent lateral wings and a dorsal ridge, thereby reducing its chances of being eaten (Kuhlmann and Heckmann, 1985). The nature of the morphogen is unknown.

4. Microaerophily

Although it may often be difficult to correlate the occurrence of protozoan species with specific environmental factors, it is nevertheless true that some species are found only within certain broad categories of habitat: anaerobic ciliates are not found living in the surface waters of lakes, and protozoa with functional sequestered chloroplasts are not found living in anaerobic sediments. Some habitats, and especially those that are vertically stratified, are in reality a series of microhabitats, providing distinctly different physical and chemical conditions which dictate a corresponding variety of physiological adaptations by the resident organisms. Two examples would be the interstitial habitat in marine sands and the water columns of stratified lakes. Both range from fully oxygenated to anaerobic conditions, and both support a diversity and abundance of protozoa with well-described vertical distributions. The physiological adaptations facilitating these vertical distributions are only beginning to be understood for a few species.

4.1. Interstitial Zone of Marine Sands

The nature and vertical distribution of the interstitial protozoan fauna of marine sands depends on such factors as the sizes of sand particles (which control the pore volume), the quantity of decomposable organic matter (which dictates the nature and abundance of the microbial community), and the extent of tidal flushing (which, together with the physical transport of O_2 by other processes and the extent of microbial O_2 consumption, determines the depth of the oxidized zone and the steepness of the oxygen and redox gradients). A typical profile might show an oxidized zone about 5 cm deep, a transition zone of decreasing O_2 tension about 10 cm thick, and an underlying anaerobic and reducing zone (Fenchel, 1969, 1986a). The depth and thickness of these zones do, of course, change seasonally. Ciliates are relatively abundant in each zone, and specific types tend to be associated with each layer (Fig. 4). Some, like the

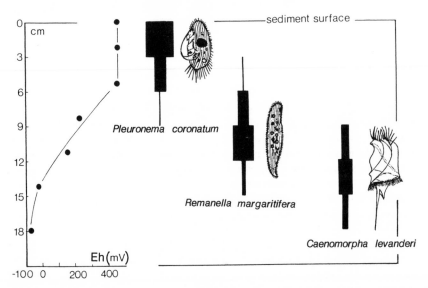

Figure 4. Vertical distribution of the abundance (relative scale) of three ciliate species in marine sand in relation to the redox (Eh) profile. [Adapted from Fenchel (1969, 1986a).]

oligotrichs and hypotrichs, tend to be restricted to the oxidized zone; some heterotrichs and several species of the karyorelictid genus *Remanella* aggregate around the oxic–anoxic boundary, while the underlying anaerobic zone is dominated by anaerobic ciliates. Because ciliates effectively partition the entire benthic habitable zone, many species can coexist as stratified populations in marine sands. How is this partition brought about? How do ciliates seek out and remain within their preferred habitats, and what types of physiological adaptations are critical to the maintenance of stratification? Some answers have recently been obtained, not from studies of ciliates in marine sands but from work on similar species, especially the microaerophilic ciliate *Loxodes,* living in stratified freshwater lakes. *Loxodes* is a close relative of the marine interstitial ciliate *Remanella* and, except for the salinities of their habitats, the two genera are quite similar ecologically.

4.2. *Loxodes*—A Microaerophilic Ciliate

Loxodes is common and often abundant in productive lakes and ponds, where it feeds on small algae and most other organic particles bigger than about 5 µm (Finlay and Berninger, 1984). When the water column is mixed and oxygenated, as it is during the winter in temperate latitudes, *Loxodes* lives exclusively in the sediment, but a remarkable change occurs during the summer, when the water column stratifies and the bottom water becomes anoxic: *Loxodes* migrates out of the sediment, becoming planktonic and reaching maximum abundance close to the oxic–anoxic boundary in the water column (Finlay, 1981; Finlay and Fenchel, 1986a). *Loxodes* is, apparently, microaerophilic (Fig. 5). What are the underlying physiological and behavioral processes?

We now know that *Loxodes* responds to three principal, interacting factors: oxygen, blue light, and the force of gravity (Finlay *et al.,* 1986; Fenchel and Finlay, 1986b; Finlay and Fenchel, 1986b). Cells left in the dark will seek out a low oxygen tension ($\sim 5\%$ of the saturation value), but if they are then exposed to light, they will swim into the dark or into anaerobic water: high O_2 tensions and O_2 and light together promote transient avoidance responses, an increased motility, and, if it is physically possible, positive geotaxis (Fenchel and Finlay, 1984, 1986a). Cells displaced into anaerobic water will also show transient avoidance responses and increased motility. They will also perform negative geotaxis, each cell turning its anterior end up and swimming vertically upward. The net result is that in a lake, *Loxodes* will quickly be transported by geotaxis to a zone that is neither anaerobic nor fully aerobic, where cells can then accumulate through a reduction in random motility induced by the "ideal" Po_2.

Loxodes is sensitive to quite low light levels (Finlay and Fenchel, 1986b)—about 10 W m^{-2} will cause cells to swim down from the oxic–anoxic boundary and into anaerobic water. This is probably the main reason why cells often reach peak abundance just inside the anaerobic zone in lakes. It also partly explains the maintenance of the ability of *Loxodes* to switch between aerobic and anaerobic respiration. *Loxodes* can use nitrate as a terminal electron acceptor, an unusual if not unique achievement for a eucaryote but a credible strategy for an organism that is periodically displaced into anaerobic water (Finlay *et al.,* 1983a; Finlay, 1985).

The ability of *Loxodes* to perceive O_2, light, and gravity is fundamental to the

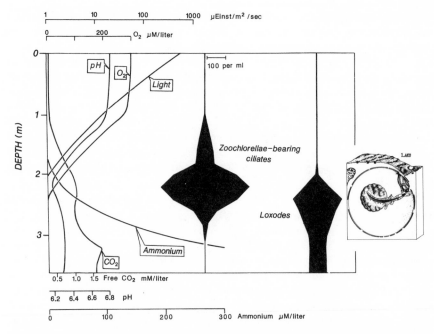

Figure 5. Vertical distribution of microaerophilic zoochlorellae-bearing ciliates (*Euplotes daidaleos, Disematostoma* spp., and *Prorodon* sp.) and the ciliate *Loxodes* (*L. magnus* + *L. striatus*) in a small productive pond [adapted from Berninger *et al.* (1986)]. The profiles of some relevant physical and chemical factors are also shown. Diagram at right shows Müller body structure in *Loxodes* [adapted from Fenchel and Finlay (1986a)].

microaerophilic behavior of this ciliate. Gravity perception is the easiest to comprehend because it is associated, in *Loxodes,* with a special organelle, the Müller body, and because it is quite easy to observe, even with a low-power microscope, that *Loxodes* is aware of what is up and what is down. The mechanoreceptor consists of a 3-μm, roughly spherical mineral body that is largely composed of barium sulfate (Rieder *et al.,* 1982; Finlay *et al.,* 1983b) and is located in a larger (7 μm) fluid-filled vacuole (Fig. 5). The mineral body is connected to the wall of the vacuole by a semirigid stalk that contains nine microtubules and is anchored in a nonciliated kinetosome (ciliary basal body); the neighboring kinetosome bears an emergent cilium (Fenchel and Finlay, 1986a). Gravity-induced movement of the mineral body will stretch the membranes between and around the kinetosomes, leading to a change in ion permeability and a change in the cell membrane potential. This changed potential will be propagated across the cell surface to modify the behavior and beating frequency of the cilia. This explanation has not been demonstrated empirically, but it is consistent with the postulated mechanism of similar mechanoreceptors in other invertebrates (see Fenchel and Finlay, 1986a) and is substantiated by the observation that the number of mechanoreceptors per cell is proportional to cell surface area— a relationship that must hold if large cells are not to be disadvantaged by the finite signal propagation from a few Müller bodies.

Our picture of the biochemistry and physiology of oxygen and light perception is less complete, but some key observations have been made. The most significant discovery is that the responses to oxygen and light are physiologically linked; cells living at low O_2 tensions react to light in the same way that cells living in the dark respond to an increase in the O_2 tension. Cells in anaerobic water are insensitive to light. Cyanide at micromolar concentrations blocks the responses to light and oxygen, and cells exposed to cyanide behave as they do (i.e., with increased motility) in anaerobic water (Fenchel and Finlay, 1986b).

It is known that the pigment mediating photosensitivity is a blue light receptor, possibly a flavin (Finlay and Fenchel, 1986b), and that one product of its excitation in the presence of oxygen is the toxic oxygen radical superoxide (O_2^-). We know that *Loxodes* has only low levels of superoxide dismutase (which catalyzes the removal of O_2^-) and catalase (Finlay *et al.*, 1986), and we can probably assume that the rate of intracellular O_2^- production in the dark increases at high oxygen tensions. Superoxide, or some product of its dismutation (e.g., H_2O_2) might then be the "common currency" for oxygen perception, either binding to the cytochrome oxidase or reducing some other component of the electron transport system (ETS), e.g., cytochrome *c*. The major failing of this explanation is that the ETS is located in the mitochondria, whereas the superoxide-producing pigment lies in granules within the cell membrane. Furthermore, we can imagine how the redox state of the ETS might influence the membrane potential of the mitochondria, but it is difficult to see how that signal might be transmitted to the cell membrane to effect a change in the rhythm and frequency of ciliary beat. The discovery of a cyanide-sensitive oxygen receptor in the cell membrane would, of course, clarify our picture of the interacting perceptions of light, oxygen, and gravity (Fig. 6).

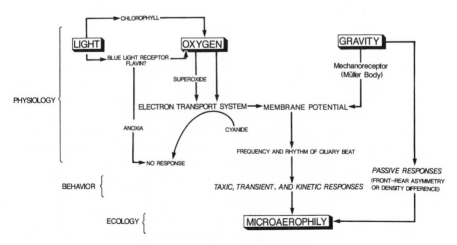

Figure 6. Some of the probable physiological and behavioral responses to the three cardinal factors (light, oxygen, and gravity) controlling microaerophily in ciliated protozoa.

Loxodes is clearly adept at using its perception of some key environmental factors to hold its position close to the oxic–anoxic boundary, but it is not so clear why it bothers to do this. One possibly relevant observation is that these behaviors, and a capacity for anaerobiosis, allow *Loxodes* to live in a zone that is virtually free of potential predators; crustacean zooplankton and the larvae of invertebrates and fish are almost all confined to the overlying oxygenated water, where they will be denied access to *Loxodes* and the other large microaerophilic ciliates. Many of these predators also swim deeper during the day (Stich and Lampert, 1981; Zaret and Suffern, 1976), so the daytime downward migration of *Loxodes* could be the fine tuning of its predator avoidance strategy.

4.3. Zoochlorellae-Bearing Ciliates

There has long been a widespread misconception that freshwater protozoa with algal symbionts are only found close to the water surface. Many of these protozoa are, however, microaerophilic (Berninger *et al.*, 1986; Finlay *et al.*, 1987b)—a behavior that might be regarded as curious in cells that need light for photosynthesis (Fig. 5). It is easier to understand this behavior if we consider some of the physiological interactions between host and symbiont.

Most biochemical and physiological research on zoochlorellae-bearing ciliates has been carried out with *Paramecium bursaria*. It is known that the symbionts are capable of sustaining the host paramecia in the absence of food (Karakashian, 1963), that the symbionts release a great deal (sometimes more than 80%) of their photosynthate as maltose (Muscatine *et al.*, 1967) and occasionally as other sugars (Reisser et al., 1984), and that these liberated sugars are used by the host paramecia (Brown and Nielsen, 1974). The nature and function of zoochlorellae seem to be similar, save for minor discrepancies, throughout a variety of ciliates and other invertebrates (Smith and Douglas, 1987; Rees, 1987), and I will assume that the zoochlorellae-bearing ciliates considered here contain "typical" zoochlorellae.

The vertical distribution of these ciliates in a pond resembles the distribution of *Loxodes* in that they both reach peak abundance close to the oxic–anoxic boundary (Fig. 5). There are, however, some differences: *Loxodes* peaks just beneath the zoochlorellae-bearing ciliates, and it also penetrates deeper into anoxic water. Also shown in Fig. 5 is the vertical distribution of some factors that are potentially important: the ciliates will require O_2 for respiration, and the symbionts will require light and CO_2 for photosynthesis and a source of nitrogen for growth. We can see at once how the vertical distribution of the ciliates might be a compromise solution, since the resources they require come from opposite directions—light and O_2 from above, and NH_4^+ and CO_2 by upward diffusion from the sediment. How do the ciliates find their way to this optimum zone? Do they perceive and react to some or all of these factors that we consider relevant, or are they guided by some master factor? The available evidence favors the latter. Cells stratified in glass chambers will also seek out a favorable environment, but they appear to be guided by only one factor, the oxygen tension. Experiments on two microaerophilic species (*Euplotes daidaleos* and *Frontonia vernalis*) show that they, like *Loxodes*, are governed in their migratory behavior by the O_2 tension and little else (this is in contrast to the situation for *P. bursaria*, which also has a well-developed step-down photophobic response; see Cronkite and van den Brink, 1981; Niess *et al.*, 1981; Finlay *et al.*, 1987b). The symbionts will produce O_2 in the light, and the ciliates apparently take account of this by integrating the extracellular and intracellular Po_2 (Finlay *et al.*, 1987b). The behavioral response of the ciliates to an increase in light level (with the concomitant stimulation of photosynthetic rate) is equivalent to their response to an increase in the extracellular oxygen tension in the dark: their motility increases when the Po_2 is not close to some "ideal," low value. We can conclude that these ciliates are "guided" to the microaerobic zone by the Po_2 because there they will benefit from an optimum supply of the other factors (light, CO_2, and NH_4^+) that they require. The strategy will invariably be successful because the chemical and microbial processes controlling the concentrations of CO_2 and NH_4^+ in the water are largely dependent on the O_2 concentration, which in turn is partly dependent on light penetration.

Another factor of some relevance to the ciliates is the concentration of food particles, especially bacteria and small algae. The autotrophic symbionts are essential when particulate food becomes depleted, but another advantage to the ciliate of remaining close to the oxic–anoxic boundary is that the concentrations of food are usually higher there than in the epilimnion. How does a microaerophilic zoochlorellae-bearing ciliate maintain a balance between the intracellular processes of heterotrophic and autotrophic nutrition?

If particulate food is abundant, the ciliates will exploit it and grow quickly. They may be able to increase the supply of nutrients to the zoochlorellae and so increase their growth rate; otherwise, growth of the ciliates will outstrip that of the zoochlorellae and the latter will be diluted out of the host population. When particulate food is absent or depleted, the ciliate will have a low growth rate, but the zoochlorellae will still receive their basic requirements for growth and increase in number in the cytoplasm of the host. The number of zoochlorellae in ciliates is known to vary within certain limits. What controls these limits?

The conventional wisdom concerning nutrient exchange in these associations is that protein catabolism by the host produces ammonia, which passes to the algae, where it is assimilated into amino acids and proteins, the amino acids possibly being released to the host. At first sight, our knowledge of the vertical distribution of these ciliates in lakes is consistent with this view; the elevated concentrations of NH_4^+ and CO_2 at the oxic–anoxic boundary will supplement sources from the metabolism of the ciliates. But the ciliates will be able to control the growth rate of the algae only if the latter are completely dependent on a nutrient supply that is controlled by the host. If the principal reason for microaerophily is to provide access to high levels of *extra*cellular nutrients, which will freely enter the cell, the ciliates will lose their control of symbiont growth rate. Zoochlorellae-bearing ciliates do not fill up and burst, which implies that they can control algal growth rate. The mechanism of control is unknown, but the results of some recent work on the zoochlorellae-bearing coelenterate *Hydra viridis* may also apply to these ciliates. Rees (1987) has shown that symbiont-bearing animals have high levels of the enzyme glutamine synthetase (GS), which catalyzes the assimilation of ammonia into glutamine, and low levels of the enzyme glutamate dehydrogenase (GDH), which catalyzes the oxidative deamination of glutamate, to release ammonia. Symbiotic animals also have 50% more GS than do aposymbiotic animals, whereas the latter have higher levels of GDH. The symbiotic animals release no ammonium in the light, whereas aposymbiotic animals release a lot. 3-(3,4-Dichlorophenyl)-1,1-dimethyl urea (DCMU), an inhibitor of photosynthesis, has no effect on ammonium release. The most complete explanation is that the nature of nitrogen metabolism in *Hydra* is profoundly affected by the presence or absence of zoochlorellae. When the symbionts are present, the host shows net assimilation of nitrogen and thereby controls the growth rate of the symbionts. In addition, the capacity of the host to control the supply of nitrogen ensures that the host will obtain maximum benefit from its maltose-excreting symbionts.

In *Hydra,* the rate of maltose release from symbionts is strongly dependent on the pH of the vacuole (Smith and Douglas, 1987). As the pH decreases, maltose synthesis and release increases, reaching a peak at about pH 4. When the pH increases, maltose release declines and the carbon that would have been lost from the algae is channeled into growth. Thus, if the host can control the pH of the perialgal vacuole, it can effectively control both the growth rate of its symbionts and the rate of supply of photosynthate. One method of controlling pH would be to deploy the ammonium ion, the presence of which causes an increase in pH. By restricting the supply of ammonium to the symbiont, the host can restrain growth of the symbiont and, by keeping the pH low, simultaneously maintain a high level of maltose excretion.

The zoochlorellae of ciliated protozoa are remarkably similar, both morphologically and physiologically, to those living in *Hydra viridis.* In particular, maltose excretion by the symbionts of the ciliates is also pH dependent (Reisser, 1981; Weis, 1983) (reaching a maximum at about pH 4), and in *P. bursaria,* the levels of GS and glutamine synthase (GOGAT) (enzymes responsible for ammonium assimilation) are higher in the ciliate fraction than in the algal fraction and GS activity is absent from aposymbiotic cells (Albers and Wiessner, 1985) (Fig. 7). DCMU fails to stimulate release of ammonium from illuminated zoochlorellae-bearing ciliates. Weis (1977) has shown that most of the zoochlorellae in a ciliate tend to be at the same stage in the life cycle—their growth and development are virtually synchronous. This synchrony is not due to exogenous factors, because different host cells in the same population contain algal populations at different stages in the life cycle. It appears that each ciliate controls the growth rate of its symbionts.

If this growth control is exercised by the host modulating the nitrogen supply to its symbionts, and if the "normal" state of the symbionts is one of nitrogen deficiency, it is difficult to believe that the ciliates seek out a habitat like the microaerobic zone because it is high in ammonium. Far more likely is the possibility that the cells are there because of the elevated CO_2 concentration, the lower pH, and the lower Po_2. The ciliates maintain their symbionts because they are efficient excretors of carbon compounds. They can be efficient only if they have a substantial capacity for CO_2 fixation. The zoochlorellae of *P. bursaria* have significantly higher activities of ribulose biphosphate carboxylase (RuBPc; the enzyme that catalyzes the incorporation of CO_2 into organic form within the Calvin cycle) than do other nonsymbiotic strains of chlorellae (Reisser

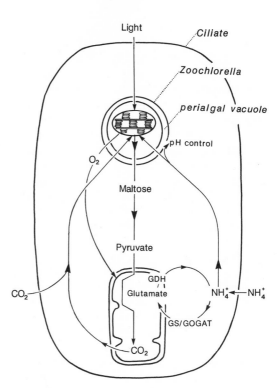

Figure 7. Probable metabolic interactions in a zoochlorellae-bearing ciliate. Only two inclusions are shown: a symbiont (the zoochlorella) and a host mitochondrion. GDH, Glutamate dehydrogenase (oxidative deamination of glutamate); GS/GOGAT, glutamine synthetase/glutamate synthase (assimilation of ammonia). [Based on Albers and Wiessner (1985), Rees (1987), and many other sources.]

and Benseler, 1981). These higher activities remain relatively constant irrespective of the external CO_2 tension. This is obviously useful to the ciliate if there is plenty of CO_2, but the potential drawback is that this enzyme can promote the reaction with either CO_2 or O_2. At elevated O_2 concentrations, RuBPc becomes an oxygenase, consuming the oxygen and producing glycolic acid, which is subsequently broken down to liberate CO_2. Protozoa are probably too small to effectively control an intraextracellular O_2 gradient, but they can respond behaviorally by migrating to the microaerobic zone. Microaerophily might simply be a means of obtaining maximum benefit from the sugar-producing machinery of photosynthetic symbionts.

5. Other Photosynthetic Symbionts

In addition to the microaerophilic ciliates, many other protozoa, especially planktonic forms, establish relationships with photosynthetic organisms. In some cases the success of the symbiosis manifests itself as massive growths, as in the "red tides" attributable to the cryptomonad-bearing ciliate *Myrionecta* (*Mesodinium*) *rubrum* (see Section 5.2) or the occasional high biomass of epilimnetic zoochlorellae-bearing ciliates, which may even exceed that of the phytoplankton (Hecky and Kling, 1981). Most planktonic foraminifera and many radiolaria possess photosynthetic symbionts (Anderson et al., 1979; Anderson, 1983; Taylor, 1982; Lee, 1983), as do many of the oligotrich ciliates in the surface waters of the sea (Stoecker et al., 1987; Jonsson, 1987) and in lakes (Rogerson et al., 1989).

The symbiont-bearing foraminifera are particularly common in nutrient-poor oceanic waters, and the efficient cycling of nutrients between the symbiotic partners has been described as an adaptive strategy to circumvent nutrient limitation (Hallock, 1981). If this is true, it indicates that there is a significant difference between symbiont-bearing protozoa in the sea and the freshwater forms we have already considered. The micro-aerophilic zoochlorellae-bearing ciliates deploy a variety of behavioral responses to trap themselves at the metalimnion, where they benefit from the upwelling of nutrients. This is an impossible strategy for protozoa living in the surface waters of the oceans, but these do, nevertheless, maintain photosynthetic symbionts. How do they manage this?

5.1. Planktonic Foraminifera

The question is most easily dealt with by considering specific symbiotic associations. The foraminiferan *Globigerinoides sacculifer* lives in the euphotic zone of tropical oceans. It grows to a diameter of 0.5–1 mm and is covered with spines 1–2 mm long. Like all planktonic foraminifera, it has a shell of calcium carbonate and contains many (about 500) photosynthetic dinoflagellate symbionts. During the day, the cytoplasm containing these symbionts migrates out of the shell aperture and along the spines. At night, the cytoplasm and symbionts are pulled back inside the shell; this, presumably, is when the photosynthate is released to the foraminiferan.

Jorgensen et al. (1985) placed oxygen microelectrodes in the symbiont "halo" and demonstrated a high photosynthetic rate: when illuminated, the Po_2 at the shell surface increased to 2.5 times the ambient seawater concentration, and an individual foraminiferan produced 14.9 nmol of O_2 hr^{-1}, equivalent to the provision of 10 times the quantity of organic carbon respired by the protozoon. The same authors went on to test the hypothesis of Hallock (1981) that the symbiotic association might be limited by the supply of nitrogen and phosphorus. It was, after all, well known (e.g., Bé et al., 1977; Anderson et al., 1979) that planktonic foraminifera would not grow in the laboratory unless they were also fed live zooplankton, which presumably ameliorated deficiencies in nitrogen and phosphorus. Calculations of the diffusive flux of inorganic P and N into *Globigerina* showed that the supply fell far short of what was needed. *Globigerina* has no means of filtering organic particles from the ocean, but its relatively large size and numerous "sticky" pseudopods mean that it is well adapted for "diffusion feeding" (see Fenchel, 1986b), whereby it periodically captures crustacean nauplii and other zooplankton. This external supply of nutrients would meet its requirements for P and N. The same conclusion about the need for occasional ingested zooplankton to satisfy nutrient deficiency can be inferred from the measurement by Swanberg and Harbison (1980) of photosynthetic rate in the giant (up to 3 m long) colonial radiolarian *Collozoum longiforme*. The colony consists of large numbers of radiolaria, each surrounded by symbiotic dinoflagellates and embedded in a secreted dense gelatinous matrix (see also Swanberg and Anderson, 1981). The remnants of digested prey remain for some time in this matrix, where they can be identified and counted. In *Collozoum*, there is a clear relationship between the photosynthetic rate per unit of chlorophyll *a* and the number of recently digested tintinnids. The most likely explanation is that the digested tintinnids ameliorate the nutrient deficiency in the algal symbionts.

One of the disadvantages of having a shell of $CaCO_3$ is that it causes the foraminifer to sink. It is not at all clear how they manage to stay in the euphotic zone, although the ability to modify the porosity and the thickness of shells (Boltovskoy and Wright, 1976) or produce "flotation bubbles" (Bé *et al.*, 1977) or an extensive vacuolar system (Anderson and Bé, 1976) probably helps. Buoyancy is an absolute requirement for a large planktonic organism with a calcite shell because the solubility of calcite increases as the hydrostatic pressure and partial pressure of CO_2 increase and the temperature decreases. Thus, dissolution increases with depth, and below the "compensation depth," the rate of dissolution exceeds that of precipitation. This depth (for calcite) is somewhere between 500 and 3000 m in the Pacific and slightly greater in the Atlantic (the Pco_2 is higher in the deep water of the Pacific). Therefore, the synthesis and maintenance of a calcite shell is possible only in surface waters. The traditional view (e.g., Muller, 1978, and references therein) has been that the formation of the shell is also related to the photosynthetic activity of the foraminifer, since photosynthetic uptake of CO_2 would drive the process of calcification:

$$Ca^{2+} + 2HCO_3^- \rightarrow CaCO_3 + CO_2 + H_2O$$
$$\downarrow$$
calcification

Other calcified protozoa (e.g., coccolithophorids) and other organisms (e.g., corals) also demonstrate a relationship between the rates of photosynthesis and calcification. However, the relationship between these two processes may not be simple. Erez (1983) demonstrated that DCMU at micromolar concentrations would completely inhibit photosynthesis but that the rate of calcification remained unaffected. The latter was, however, enhanced by light (the mechanism is unknown; see also Kuile and Erez, 1984).

But why should planktonic foraminifera produce calcite tests? One might suppose that the long spines provide an ideal support for the "halo" of photosynthetic symbionts (see photographs in Bé *et al.*, 1977), but the nonspinose species also bear symbionts. Another function has been suggested by Hallock (1981), who concluded, on purely theoretical grounds, that mixotrophy in nutrient-poor environments like the surface waters of oligotrophic oceans would be limited only by nutrient availability and the ability to cycle nutrients between the participating partners, with minimal leakage to the external environment. The growth of the symbiotic association would decline drastically as the loss rate of nutrients increased. The calcite test of the planktonic foraminifera might be regarded therefore as a barrier to nutrient loss and a means of accumulating nutrients needed for growth. The same might be said for the gelatinous matrix enclosing colonial radiolaria.

5.2. Sequestered Chloroplasts

There are two additional broad categories of association between protozoa and photosynthetic organisms, and they differ markedly from each other in their degree of functional integration. There is now abundant evidence that many oligotrich ciliates and some other protozoa, including heliozoans (Patterson and Dürrschmidt, 1987) can sequester the chloroplasts from ingested algae (Blackbourn et al., 1973; Laval-Peuto and Febvre, 1986; Laval-Peuto et al., 1986; McManus and Fuhrman, 1986; Jonsson, 1987; Stoecker et al., 1987; Rogerson et al., 1989). The chloroplasts continue to photosynthesize and fix carbon (Jonsson, 1987; Stoecker et al., 1987), which is presumably excreted to the host as a usable carbohydrate. The symbiosis is balanced heavily in favor of the host, which makes a minimal investment in maintaining its symbionts; the symbionts do not divide, and they eventually deteriorate and are probably digested. All protozoa that sequester chloroplasts can probably live without them, and they do not appear to be very selective in the types of algae they use. Nevertheless, because oligotrich ciliates are an important component of the planktonic microbial community, especially in the sea, chloroplast retention is probably a phenomenon with some significance in the functioning of planktonic food webs. Stoecker et al. (1987) counted more than 40% of planktonic ciliates with chloroplasts in the surface water in spring and summer.

The final type of association, representing extreme functional integration, is found in the relationship between the marine planktonic ciliate *Myrionecta* (*Mesodinium*) *rubrum* and its cryptophyte symbionts. The ciliate is, by virtue of its symbionts and the absence of a functional mouth, entirely autotrophic; it fixes CO_2 and P (Barber et al., 1969) and possesses an assimilatory nitrate reductase (Packard et al., 1978), a unique property for a ciliate. The symbionts color the ciliates reddish brown, and when the latter are abundant, which is usually in waters with nutrient-rich upwelling, the water can be colored red. Fenchel (1968) calculated the average biomass of *Myrionecta* in "clouds of red water" as 0.35 g liter^{-1}, and Smith and Barber (1979) measured the contribution of the ciliates to the chlorophyll *a* in the waters off the Peruvian coast as 100–125 mg m^{-3}, responsible for photosynthetic rates of 1–2 g m^{-3} hr^{-1}. The symbiont contains a nucleus, mitochondria, and many chloroplasts, and it is separated from the host cytoplasm by only a single membrane. It occupies much of the volume within the ciliate (Hibberd, 1977; Oakley and Taylor, 1978). The ciliate is extremely motile, with sustained swimming speeds of about 2 mm sec^{-1} and, possibly because of its phototaxis, it can perform vertical migrations over long distances (>20 m in 6 hr; see Lindholm, 1985). This behavior undoubtedly enables the ciliate to escape the nutrient depletion in the ambient water that would otherwise be caused by its remarkable photosynthetic ability.

Thus, protozoa seem to have established every conceivable type of association with photosynthetic symbionts. Their dependence on the symbionts varies among species, from casual exploitation to complete physiological integration. There are probably many ecological factors that select for and control the viability of different types of association, but the external nutrient supply is probably one of the more important. When nutrients are always limiting, as in the oligotrophic ocean, there are advantages to permanent symbiotic association if nutrient leakage from the symbionts is controlled. Alternatively, filter-feeding ciliates can supplement their nutrition by the periodic sequestering of chloroplasts, which are retained in a functional state. The alternative strategy, in stratified freshwater lakes where organic decomposition releases essential nutrients into illuminated water, is to migrate and take the photosynthetic symbionts to the source of nutrients. The problem then faced by the protozoon is a periodic excess of nutrients, and algal growth must be suppressed, probably by nitrogen limitation, to maintain maximum production of photosynthate.

Finally, and lest we become too impressed with the exploitative abilities of heterotrophic protozoa, I should mention that certain autotrophic flagellates perform the complementary feat of supplementing their nutrition by ingesting bacteria (e.g., Bird and Kalff, 1986).

6. Anaerobiosis and Methanogenic Symbionts

Most eucaryotes contain mitochondria, in which O_2 is used as a terminal electron acceptor to oxidize the reducing equivalents passing down electron transport chains (the cytochromes). This process is efficient in terms of energy production, but it is of course impossible in the absence of oxygen; therefore, the protozoa that live permanently in such habitats as the rumens of mammals and the anoxic layers of marine and freshwater sediments must rely on anaerobic metabolic pathways for the release of energy. None of these protozoa have mitochondria, but they all possess intracellular bacteria and most of them also have microbodies (Stumm and Zwart, 1986; Müller, 1988). In some protozoa, the function of these microbodies is known, and they are referred to as hydrogenosomes (Müller, 1980, and references therein; Yarlett et al., 1983).

The hydrogenosome functions like a mitochondrion in that it oxidizes pyruvate and can respire oxygen, but it does not contain cytochromes. Its most important and unusual function is that it can eliminate reducing equivalents in the form of H_2 gas. The other main products of pyruvate catabolism in the hydrogenosome are CO_2, acetate, and energy. The protozoon benefits from the additional energy provided by this process over cytosolic fermentation, but how can the process be maintained? How, for example, can a hydrogen-producing organelle be maintained inside a cell when so many other cell functions depend on the maintenance of a low hydrogen pressure, e.g., the reoxidation of reduced NAD in the cytosol—$NADH + H^+ \rightarrow NAD^+ + H_2$—which is feasible only at low H_2 pressures? This problem must be especially important in the hydrogenosome-containing protozoa because hydrogenosomes, unlike mitochondria, have only a low ability to reoxidize glycolytically formed NADH (Müller, 1980).

The answer began to be revealed when anaerobic rumen ciliates were investigated by epifluorescence microscopy. It had earlier been discovered that certain coenzymes such as the deazaflavin F_{420}, which are unique to methanogenic bacteria, would fluoresce if excited with the appropriate wavelengths (Doddema and Vogels, 1978). When rumen ciliates were excited with these wavelengths and examined microscopically, many fluorescing ectosymbiotic bacteria were observed (Vogels *et al.*, 1980). Fenchel *et al.* (1977) had already shown that many free-living marine sapropelic ciliates had ectosymbiotic bacteria and that some also had endosymbiotic bacteria. None of these ciliates had mitochondria, but they all appeared to have microbodies. It was subsequently shown, by isolation of the symbionts and by measurement of the release of methane from both anaerobic amebae (van Bruggen *et al.*, 1983) and ciliates (van Bruggen, 1986; Wagener and Pfennig, 1987), that the endosymbiotic bacteria were methanogens. *Methanobacterium formicicum* isolated from *Metopus striatus* was the first methanogen to be isolated from a protozoon (van Bruggen *et al.*, 1984). The same bacterium was isolated from the ciliate *Plagiopyla nasuta* (Goosen *et al.*, 1986) and from sapropelic amebae (van Bruggen *et al.*, 1985), and a new species of methanogen, *Methanoplanus endosymbiosus*, was isolated from the marine ciliate *Metopus contortus* (van Bruggen *et al.*, 1986).

The functional integration of hydrogenosome and methanogen would appear to benefit both partners in the symbiosis. The protozoon benefits from the higher yield accompanying the production of a more oxidized end product. It also benefits from the removal of waste H_2, which is used by the symbiont as a substrate. The symbiosis is no doubt strengthened by the close association between hydrogenosome and methanogen in the cell (Stumm and Zwart, 1986). However, it would be misleading to imply that the nature of this symbiosis is completely understood or that it is known to exist in the same form in all anaerobic protozoa. The function of hydrogenosomes has been described only in some anaerobic flagellates and rumen ciliates (see Müller, 1988). The presumption is that the microbodies in the sapropelic ciliates are hydrogenosomes and that the thick-rod bacteria in *Pelomyxa palustris* have the same function. Moreover, not all symbiotic bacteria are intracellular; some are located in special pellicular grooves or are embedded in mucilage (Fenchel *et al.*, 1977). The ectosymbionts of rumen ciliates are methanogens (Vogels *et al.*, 1980; Krumholz *et al.*, 1983), but the identity of the ectosymbionts of marine anaerobic ciliates is unknown.

It is also frequently observed that protozoa contain at least one type of bacterium in addition to the methanogens. These bacteria could also be methanogens, capable of using the acetate produced by the hydrogenosome (i.e., $CH_3COO^- + H_2O \rightarrow CH_4 + HCO_3^-$), but they have not been isolated, and they do not show the typical fluorescence of methanogens. Another possibility, alluded to by van Bruggen (1986), involves "interspecies hydrogen transfer" between three symbiotic partners (the hydrogenosome and the two types of symbiotic bacteria; Fig. 8). Hydrogen transfer between different bacterial types has been described, as in the coculture described by Zinder and Koch (1984) in which one species oxidizes acetate to CO_2 and H_2 while the accompanying species, a methanogen, reduces the CO_2 to CH_4. This cooperation succeeds if the methanogen keeps the H_2 pressure low enough (probably $<10^{-4}$ atm) for the oxidation of acetate to be thermodynamically feasible. This, however, creates problems for the methanogen, which, being forced to grow at such low H_2 pressures, may only grow very slowly. As a consequence, such an endosymbiotic coculture will be maintained only if the ciliate itself is growing very slowly. This was recently demonstrated by Wagener and Pfennig

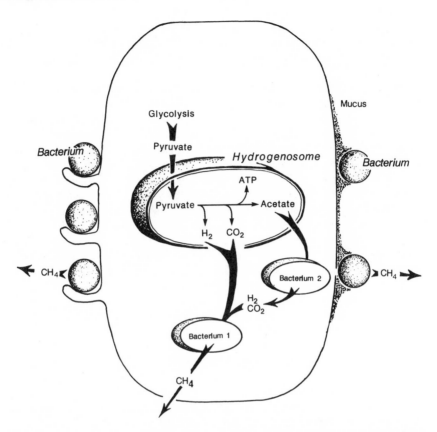

Figure 8. Some probable and possible metabolic interactions within an anaerobic H_2-producing protozoon containing two types of endosymbiotic bacteria (not to scale). Bacterium 1 is shown as a methanogen ($4H_2 + HCO_3^- \rightarrow CH_4 + 3H_2O$), and bacterium 2 is shown as an acetate oxidizer ($CH_3COO^- + 4H_2O \rightarrow 4H_2 + 2HCO_3^- + H^+$). The true identity of the latter is unknown; it is possible that it too is a methanogen ($CH_3COO^- + H_2O \rightarrow CH_4 + HCO_3^-$). The identity of the ectosymbiotic bacteria is unknown. Some show autofluorescence but there is no independent proof that they are methanogens. [Based on Fenchel *et al.* (1977), Van Bruggen *et al.* (1984), and Stumm and Zwart (1986).]

(1987), who managed to produce rapid growth in culture of the freshwater sapropelic ciliate *Trimyema compressum*. The growth of the bacterial symbionts could not match that of the ciliate, and the symbionts were gradually diluted out. The ciliates eventually became entirely free of symbionts. The persistence of this ciliate in the absence of oxygen, and without its endosymbionts, clearly demonstrates the facultative nature of the symbiosis. When there is an excess of substrate, the ciliate can obviously manage without the detoxification assistance of its endosymbionts. This observation, together with the finding that starving anaerobic protozoa tend to have much higher numbers of methanogens (Stumm *et al.*, 1982; Wagener and Pfennig, 1987), may also indicate that the bacteria supply something else to the protozoa. They may excrete carbohydrates or other metabolically useful compounds, or they may be farmed and digested by the protozoa.

The study of protozoan endosymbiosis is an expanding area, and the study of the symbionts of anaerobic protozoa is one of the more recent, vigorous branches. As usual, the exploration of a new branch of science gives rise to at least as many questions as it solves. What is the true nature and function of the ectosymbionts of sapropelic ciliates and their nonmethanogen symbionts? Are the thick bacteria in *Pelomyxa* equivalent in function to the hydrogenosomes in other anaerobic protozoa, and does that function include an ability to consume oxygen? Remarkably, both oxygen consumption (Chapman-Andresen and Hamburger, 1981; see also Whatley and Whatley, 1983) and methanogenesis have been demonstrated in *Pelomyxa*. Is it within the bounds of possibility that such a large (up to several millimeters in diameter) protozoon can keep these processes spatially separated within the cell and that peripheral oxygen consumption maintains an anaerobic core within the cell, where methanogenesis is possible? This would be an impressive performance for a single-celled organism.

The impact that *Pelomyxa* and other "methanogenic protozoa" make on the flow of carbon through anaerobic sediments might be equally impressive. We recently counted just over 100 *Pelomyxa* per ml of sediment in a productive pond (Finlay *et al.*, 1988). The number of endosymbionts in *Pelomyxa* varies with the size of the host, between about 10^5 and 10^9 per cell (Chapman-Andresen and Hamburger, 1981; van Bruggen *et al.*, 1983, 1985). The smallest cells of *Pelomyxa* contain about 5×10^4 methanogens (van Bruggen *et al.*, 1985), so the minimum estimate for the number of endosymbiotic methanogens in the pond would be 5×10^6 per ml of sediment, which is higher than most estimates of the abundance of viable free-living methanogens (e.g., Jones *et al.*, 1982). Perhaps the origin of sediment-derived methane should be reexamined.

7. Postscript

This review has not been exhaustive. Rather, the intention has been to select a variety of examples of the ways in which free-living protozoa respond and adapt to their natural environments. The conclusion to be drawn from the review is that a clear understanding of the factors controlling the occurrence and distribution of protozoa can be obtained only after a certain amount of information about the biology, and especially the physiology, of the organisms has been gathered. The ways in which this picture is built up vary greatly, and they depend on the different types of protozoa concerned. Laboratory investigations of the biochemistry and physiology of *Dictyostelium* are considerably more advanced and sophisticated than studies of the ecology of the organism in the soil. On the other hand, far more is known about the spatial distribution of certain freshwater ciliates (Fig. 9) than about their basic biology, and many of them have never been obtained in culture. We can be confident, however, that the imbalance in these complementary approaches will gradually disappear as the scientific community pursues

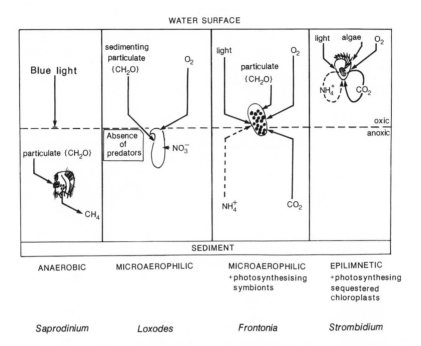

Figure 9. Vertical distribution and life styles of four distinct types of ciliated protozoa in a stratified freshwater lake. The ciliates are all dependent to some extent on particulate food. The anaerobes have no photosynthetic symbionts, but they have symbiotic methanogens which probably remove H_2 as CH_4. Microaerophilic ciliates such as *Loxodes* are apparently free of symbionts. They seek out water which is dark and microaerobic, and they escape into anaerobic water (where they respire NO_3^-) when the microaerobic zone is illuminated. *Frontonia vernalis* lives in illuminated water in the microaerobic zone, where it benefits from the upward diffusion of CO_2 that is fixed by its zoochlorellae. Extracellular NH_4^+ in the same zone is not vital to the symbionts if it is in the interest of the ciliates to keep the symbionts N-depleted. *Strombidium* is truly planktonic and it lives exclusively in the epilimnion. The sequestered chloroplasts are periodically renewed from ingested algae; they fix CO_2, but there is not yet any evidence that they assimilate nitrogen. The spatial distribution of each ciliate is largely controlled by (1) the biochemical potential for anaerobic life, (2) the activities and requirements of symbionts, and (3) physiological responses to nonbiological factors, especially light, oxygen, and gravity.

its current interest in defining the role played by "phagotrophic protists" in the natural environment.

References

Albers, D., and Wiessner, W., 1985, Nitrogen nutrition of endosymbiotic *Chlorella* spec., *Endocyt. Cell Res.* **2:**55–64.

Anderson, O. R., 1983, *Radiolaria*, Springer-Verlag, New York.

Anderson, O. R., and Bé, A. W. H., 1976, The ultrastructure of a planktonic foraminifer, *Globigerinoides sacculifer* (Brady), and its symbiotic dinoflagellates, *J. Foraminiferal Res.* **6:**1–21.

Anderson, O. R., Spindler, M., Bé, A. W. H., and Hemleben, C., 1979, Trophic activity of planktonic foraminifera, *J. Mar. Biol. Assoc. U. K.* **59:**791–799.

Azam, F., Fenchel, T., Field, J. G., Gray, J. S., Meyer-Reil, L. A., and Thingstad, F., 1983, The ecological role of water-column microbes in the sea, *Mar. Ecol. Prog. Ser.* **10:** 257–263.

Barber, R. T., White, A. W., and Siegelman, H. W., 1969, Evidence for a cryptomonad symbiont in the ciliate, *Cyclotrichium meunieri, J. Phycol.* **5:**86–88.

Bé, A. W. H., Hemleben, C., Anderson, O. R., Spindler, M., Hacunda, J., and Tuntivate-Choy, S., 1977, Laboratory and field observations of living planktonic foraminifera, *Micropaleontology* **23:** 155–179.

Berg, H. C., and Purcell, E. M., 1977, Physics of chemoreception, *Biophys. J.* **20:**193–219.

Berninger, U.-G., Finlay, B. J., and Canter, H. M., 1986, The spatial distribution and ecology of zoochlorellae-bearing ciliates in a productive pond, *J. Protozool.* **33:**557–563.

Bick, H., 1972, Ciliated Protozoa. An Illustrated Guide to the Species Used as Biological Indicators in Freshwater Biology, World Health Organization, Geneva.

Bick, H., and Kunze, S., 1971, Eine Zusammenstellung von autokologischen und saprobiologischen Befunden an Susswasserciliaten, *Int. Rev. Gesamten Hydrobiol.* **56:**337–384.

Bird, D. F., and Kalff, J., 1986, Bacterial grazing by planktonic lake algae, *Science* **231:**493–495.

Blackbourn, D. J., Taylor, F. J. R., and Blackbourn, J., 1973, Foreign organelle retention by ciliates, *J. Protozool.* **20:**286–288.

Blanton, R. L., and Olive, L. S., 1983a, Ultrastructure of aerial stalk formation by the ciliated protozoan *Sorogena stoianovitchae, Protoplasma* **116:**125–135.

Blanton, R. L., and Olive, L. S., 1983b, Stalk function during sorogenesis by the ciliated protozoan *Sorogena stoianovitchae, Protoplasma* **116:**136–144.

Blanton, R. L., Warner, S. A., and Olive, L. S., 1983, The structure and composition of the stalk of the ciliated protozoan *Sorogena stoianovitchae, J. Protozool.* **30:**617–624.

Boltovskoy, E., and Wright, R., 1976, *Recent Foraminifera,* Dr. W. Junk, The Hague.

Bonner, J. T., Barkley, D. S., Hall, E. M., Konijn, T. M., Mason, J. W., O'Keefe, G., and Wolfe, P. B., 1969, Acrasinase, and the sensitivity to acrasin in *Dictyostelium discoideum, Dev. Biol.* **20:**72–87.

Bradbury, P. C., and Olive, L. S., 1980, Fine structure of the feeding stage of a sorogenic ciliate, *Sorogena stoianovitchae* gen. n., sp. n., *J. Protozool.* **27:**267–277.

Brown, J. A., and Nielsen, P. J., 1974, Transfer of photosynthetically produced carbohydrate from endosymbiotic chlorellae to *Paramecium bursaria, J. Protozool.* **21:**569–570.

Chang, M. T., Raper, K. B., and Poff, K. L., 1983, The effect of light on morphogenesis of *Dictyostelium mucoroides, Exp. Cell Res.* **143:**335–341.

Chapman-Andresen, C., and Hamburger, K., 1981, Respiratory studies on the giant amoeba *Pelomyxa palustris, J. Protozool.* **28:**433–440.

Corliss, J. O., and Esser, S. C., 1974, Comments on the role of the cyst in the life cycle and survival of free-living protozoa, *Trans. Am. Microsc. Soc.* **93:**578–593.

Cronkite, D., and van den Brink, S., 1981, The role of oxygen and light in guiding "photoaccumulation" in the *Paramecium bursaria-Chlorella* symbiosis, *J. Exp. Zool.* **217:**171–177.

Curds, C. R., 1973, The role of protozoa in the activated-sludge process, *Am. Zool.* **13:**161–169.

Doddema, H. J., and Vogels, G. D., 1978, Improved identification of methanogenic bacteria by fluorescence microscopy, *Appl Environ. Microbiol.* **36:**752–754.

Erez, J., 1983, Calcification rates, photosynthesis and light in planktonic foraminifera, in: *Biomineralization and Biological Metal Accumulation* (P. Westbroek and E. W. de Jong, eds.), pp. 307–312, D. Reidel Publishing Co., Dordrecht, The Netherlands.

Feit, I. N., and Sollitto, R. B., 1987, Ammonia is the gas used for the spacing of fruiting bodies in the cellular slime mold, *Dictyostelium discoideum, Differentiation* **33:**193–196.

Fenchel, T., 1968, On "red water" in the Isefjord (inner Danish waters) caused by the ciliate *Mesodinium rubrum*, *Ophelia* **5**:245–253.

Fenchel, T., 1969, The ecology of marine microbenthos. IV. Structure and function of the benthic ecosystem, its chemical and physical factors and the microfauna communities with special reference to the ciliated protozoa, *Ophelia* **6**:1–182.

Fenchel, T., 1986a, *Ecology of Protozoa*, Springer-Verlag, Berlin.

Fenchel, T., 1986b, Protozoan filter feeding, *Prog. Protistol.* **1**:65–113.

Fenchel, T., 1986c, The ecology of heterotrophic flagellates, in: *Advances in Microbial Ecology*, Vol. 9 (K. C. Marshall, ed.), pp. 57–97, Plenum Press, New York.

Fenchel, T., 1988, Marine plankton food chains, *Annu. Rev. Ecol. Syst.* **19**:19–38.

Fenchel, T., and Finlay, B. J., 1983, Respiration rates in heterotrophic free-living protozoa, *Microb. Ecol.* **9**:99–122.

Fenchel, T., and Finlay, B. J., 1984, Geotaxis in the ciliated protozoon *Loxodes*, *J. Exp. Biol.* **110**:17–33.

Fenchel, T., and Finlay, B. J., 1986a, The structure and function of Müller vesicles in loxodid ciliates, *J. Protozool.* **33**:69–76.

Fenchel, T., and Finlay, B. J., 1986b, Photobehavior of the ciliated protozoon *Loxodes:* taxic, transient and kinetic responses in the presence and absence of oxygen, *J. Protozool.* **33**:139–145.

Fenchel, T., Perry, T., and Thane, A., 1977, Anaerobiosis and symbiosis with bacteria in free-living ciliates, *J. Protozool.* **24**:154–163.

Filosa, M. F., 1979, Macrocyst formation in the cellular slime mold *Dictyostelium mucoroides:* involvement of light and volatile morphogenetic substance(s), *J. Exp. Zool.* **207**:491–495.

Finlay, B. J., 1980, Temporal and vertical distribution of ciliophoran communities in the benthos of a small eutrophic loch with particular reference to the redox profile, *Freshwater Biol.* **10**:15–34.

Finlay, B. J., 1981, Oxygen availability and seasonal migrations of ciliated protozoa in a freshwater lake, *J. Gen. Microbiol.* **123**:173–178.

Finlay, B. J., 1982, Effects of seasonal anoxia on the community of benthic ciliated protozoa in a productive lake, *Arch. Protistenkd.* **125**:215–222.

Finlay, B. J., 1985, Nitrate respiration by protozoa (*Loxodes* spp.) in the hypolimnetic nitrite maximum of a productive freshwater pond, *Freshwater Biol.* **15**:333–346.

Finlay, B. J., and Berninger, U.-G., 1984, Coexistence of congeneric ciliates (Karyorelictida: *Loxodes*) in relation to food resources in two freshwater lakes, *J. Anim. Ecol.* **53**:929–943.

Finlay, B. J., and Fenchel, T., 1986a, Physiological ecology of the ciliated protozoon *Loxodes*. *Rep. Freshwater Biol. Assoc.* **54**:73–96.

Finlay, B. J., and Fenchel, T., 1986b, Photosensitivity in the ciliated protozoon *Loxodes:* pigment granules, absorption and action spectra, blue light perception, and ecological significance. *J. Protozool.* **33**:534–542.

Finlay, B. J., Span, A. S. W., and Harman, J. M. P., 1983a, Nitrate respiration in primitive eukaryotes. *Nature* (London) **303**:333–336.

Finlay, B. J., Hetherington, N. B., and Davison, W., 1983b, Active biological participation in lacustrine barium chemistry, *Geochim. Cosmochim. Acta* **47**:1325–1329.

Finlay, B. J., Fenchel, T., and Gardener, S., 1986, Oxygen perception and O_2 toxicity in the freshwater ciliated protozoon *Loxodes*, *J. Protozool.* **33**:157–165.

Finlay, B. J., Curds, C. R., Bamforth, S. S. and Bafort, J. M., 1987a, Ciliated protozoa and other microorganisms from two African soda lakes (Lake Nakuru and Lake Simbi, Kenya), *Arch. Protistenkd.* **133**:81–91.

Finlay, B. J., Berninger, U.-G., Stewart, L. J., Hindle, R. M., and Davison, W., 1987b, Some factors controlling the distribution of two pond-dwelling ciliates with algal symbionts (*Frontonia vernalis* and *Euplotes daidaleos*), *J. Protozool.* **34**:349–356.

Finlay, B. J., Berninger, U.-G., Clarke, K. J., Cowling, A. J., Hindle, R. M., and Rogerson, A., 1988,

On the abundance of protozoa and their food in a productive freshwater pond, *Eur. J. Protistol.* **23**:205–217.

Fisher, P. R., Haeder, D. P., and Williams, K. L., 1985, Multidirectional phototaxis by *Dictyostelium discoideum* amoebae, *FEMS Microbiol. Lett.* **29**:43–47.

Gerisch, G., 1982, Chemotaxis in *Dictyostelium, Annu. Rev. Physiol.* **44**:535–552.

Gerisch, G., 1987, Cyclic AMP and other signals controlling cell development and differentiation in *Dictyostelium, Annu. Rev. Biochem.* **56**:853–879.

Gomer, R. H., and Firtel, R. A., 1987, Cell-autonomous determination of cell-type choice in *Dictyostelium* development by cell-cycle phase, *Science* **237**:758–762.

Goosen, N. K., Horemans, A. M. C., Hillebrand, S. J. W., Stumm, C. K., and Vogels, G. D., 1988, Cultivation of the sapropelic ciliate *Plagiopyla nasuta* Stein and isolation of the endosymbiont *Methanobacterium formicicum, Arch. Microbiol.* **150**:165–170.

Hallock, P., 1981, Algal symbiosis: A mathematical analysis, *Mar. Biol.* **62**:249–255.

Haynes, J. R., 1981, *Foraminifera,* Macmillan, London.

Hecky, R. E., and Kling, H. J., 1981, The phytoplankton and protozooplankton of the euphotic zone of Lake Tanganyika: Species composition, biomass, chlorophyll content, and spatio-temporal distribution, *Limnol. Oceanogr.* **26**:548–564.

Hewett, S. W., 1980, Prey-dependent cell size in a protozoan predator, *J. Protozool.* **27**:311–313.

Hibberd, D. J., 1977, Observations on the ultrastructure of the cryptomonad endosymbiont of the red-water ciliate *Mesodinium rubrum, J. Mar. Biol. Assoc. U. K.* **57**:45–61.

Imhoff, J. F., Sahl, H. G., Soliman, G. S. H., and Truper, H. G., 1979, The Wadi Natrum: Chemical composition and microbial mass developments in alkaline brines of eutrophic desert lakes, *Geomicrobiol. J.* **1**:219–234.

Jones, J. G., Simon, B. M., and Gardener, S., 1982, Factors affecting methanogenesis and associated anaerobic processes in the sediments of a stratified eutrophic lake, *J. Gen. Microbiol.* **128**:1–11.

Jonsson, P. R., 1987, Photosynthetic assimilation of inorganic carbon in marine oligotrich ciliates (Ciliophora, Oligotrichina), *Mar. Microb. Food Webs* **2**:55–68.

Jorgensen, B. B., Erez, J., Revsbech, N. P., and Cohen, Y., 1985, Symbiotic photosynthesis in a planktonic foraminiferan, *Globigerinoides sacculifer* (Brady), studied with microelectrodes, *Limnol. Oceanogr.* **30**:1253–1267.

Kahan, D., 1972, *Cyclidium citrullus* Cohn, a ciliate from the hot springs of Tiberias (Israel), *J. Protozool.* **19**:593–597.

Karakashian, S. J., 1963, Growth of *Paramecium bursaria* as influenced by presence of algal symbionts, *Physiol. Zool.* **36**:52–68.

Krumholz, L. R., Forsberg, C. W., and Veira, D. M., 1983, Association of methanogenic bacteria with rumen protozoa, *Can. J. Microbiol.* **29**:676–680.

Kuhlmann, H. W., and Heckmann, K., 1985, Interspecific morphogens regulating prey-predator relationships in protozoa, *Science,* **227**:1347–1349.

Kuile, B. T., and Erez, J., 1984, In situ growth rate experiments on the symbiont-bearing foraminifera *Amphistegina lobifera* and *Amphisorus hemprichii, J. Foraminiferal Res.* **14**:262–276.

Lackey, J. B., 1938, The fauna and flora of surface waters polluted by acid mine drainage, *Public Health Rep.* **53**:1499–1507.

Laval-Peuto, M., and Febvre, M., 1986, On plastid symbiosis in *Tontonia appendiculariformis* (Ciliophora, Oligotrichina), *Biosystems* **19**:137–158.

Laval-Peuto, M., Salvano, P., Gayol, P., and Greuet, C., 1986, Mixotrophy in marine planktonic ciliates: ultrastructural study of *Tontonia appendiculariformis* (Ciliophora, Oligotrichina), *Mar. Microb. Food Webs* **1**:81–104.

Lee, C. C., and Fenchel, T., 1972, Studies on ciliates associated with sea ice from Antarctica. II. Temperature responses and tolerances in ciliates from Antarctic, temperate and tropical habitats, *Arch. Protistenkd.* **114**:237–244.

Lee, J. J., 1983, Perspective on algal endosymbionts in larger foraminifera, *Int. Rev. Cytol. Suppl.* **14:**49–77.

Lewis, K. E., and O'Day, D. H., 1977, Sex hormone of *Dictyostelium discoideum* is volatile, *Nature* (London) **268:**730–731.

Lewis, K. E., and O'Day, D. H., 1986, Phagocytic specificity during sexual development in *Dictyostelium discoideum, Can. J. Microbiol.* **32:**79–82.

Lindholm, T., 1985, *Mesodinium rubrum*—a unique photosynthetic ciliate, *Adv. Aquat. Microbiol.* **3:**1–48.

Mato, J. M., and Konijn, T. M., 1979, Chemosensory transduction in *Dictyostelium discoideum,* in: *Biochemistry and Physiology of Protozoa* (M. Levandowsky and S. H. Hutner, eds.), Vol. 2, 2nd ed., pp. 181–219, Academic Press, New York.

McManus, G. B., and Fuhrman, J. A., 1986, Photosynthetic pigments in the ciliate *Laboea strobila* from Long Island Sound, U.S.A. *J. Plank. Res.* **8:**317–327.

Melack, J. M., 1979, Photosynthesis and growth of *Spirulina platensis* (Cyanophyta) in an equatorial lake (Lake Simbi, Kenya), *Limnol. Oceanogr.* **24:**760–767.

Morris, H. R., Taylor, G. W., Masento, M. S., Jermyn, K. A., and Kay, R. R., 1987, Chemical structure of the morphogen differentiation inducing factor from *Dictyostelium discoideum, Nature* (London) **328:**811–814.

Müller, M., 1980, The hydrogenosome, in: *The Eukaryotic Microbial Cell* (G. W. Gooday, D. Lloyd, and A. P. J. Trinci, eds.), pp. 127–142, Cambridge University Press, Cambridge.

Müller, M., 1988, Energy metabolism of protozoa without mitochondria, *Annu. Rev. Microbiol.* **42:**465–488.

Muller, P. H., 1978, Carbon fixation and loss in a foraminiferal-algal symbiont system, *J. Foraminiferal Res.* **8:**35–41.

Muscatine, L., Karakashian, S. J., and Karakashian, M. W., 1967, Soluble extracellular products of algae symbiotic with a ciliate, a sponge and a mutant *Hydra, Comp. Biochem. Physiol.* **20:**1–12.

Nanney, D. L., 1982, Genes and phenes in *Tetrahymena, Bioscience* **32:**783–788.

Newell, P. C., Telser, A., and Sussman, M., 1969, Alternative developmental pathways determined by environmental conditions in the cellular slime mold *Dictyostelium discoideum, J. Bacteriol.* **100:**763–768.

Newell, P. C., Europe-Finner, G. N., and Small, N. V., 1987, Signal transduction during amoebal chemotaxis of *Dictyostelium discoideum, Microbiol. Sci.* **4:**5–11.

Niess, D., Reisser, W., and Wiessner, W., 1981, The role of endosymbiotic algae in photoaccumulation of green *Paramecium bursaria, Planta* **152:**268–271.

Nisbet, B., 1984, *Nutrition and Feeding Strategies in Protozoa,* Croom Helm, London.

Nissenbaum, A., 1975, The microbiology and biogeochemistry of the Dead Sea, *Microb. Ecol.* **2:**139–161.

Noland, L. E., and Gojdics, M., 1967, Ecology of free-living protozoa, in: *Research in Protozoology* (T.-T. Chen, ed.), Vol. 2, pp. 215–266, Pergamon Press, New York.

Oakley, B. R., and Taylor, F. J. R., 1978, Evidence for a new type of endosymbiotic organization in a population of the ciliate *Mesodinium rubrum* from British Columbia, *Biosystems* **10:**361–369.

Olive, L. S., 1978, Sorocarp development by a newly-discovered ciliate, *Science* **202:**530–532.

Olive, L. S., and Blanton, R. L., 1980, Aerial sorocarp development by the aggregative ciliate, *Sorogena stoianovitchae, J. Protozool.* **27:**293–299.

Packard, T. T., Blasco, D., and Barber, R. T., 1978, *Mesodinium rubrum* in the Baja California upwelling system, in: *Upwelling Ecosystems* (R. Boje and M. Tomczak, eds.), pp. 73–89, Springer-Verlag, Berlin.

Pan, P., Hall, E. M., and Bonner, J. T., 1975, Determination of the active portion of the folic acid molecule in cellular slime mold chemotaxis, *J. Bacteriol.* **122:**185–191.

Patterson, D. J., and Dürrschmidt, M., 1987, Selective retention of chloroplasts by algivorous heliozoa: fortuitous chloroplast symbiosis? *Eur. J. Protistol.* **23**:51–55.

Poff, K. L., and Whitaker, B. D., 1979, Movement of slime molds, in: *Encyclopedia of Plant Physiology* (W. Haupt and M. E. Feinleib, eds.), pp. 355–382, Springer-Verlag, Berlin.

Poff, K. L., Loomis, W. F., and Butler, W. L., 1974, Isolation and purification of the photoreceptor pigment associated with phototaxis in *Dictyostelium discoideum, J. Biol. Chem.* **249**:2164–2167.

Porter, K. G., Sherr, E. B., Sherr, B. F., Pace, M., and Sanders, R. W., 1985, Protozoa in planktonic food webs, *J. Protozool.* **32**:409–415.

Post, F., 1977, The microbial ecology of the Great Salt Lake, *Microb. Ecol.* **3**:143–165.

Raper, K. B., 1984, *The Dictyostelids*, Princeton University Press, Princeton, N.J.

Rees, T. A. V., 1987, The green *Hydra* symbiosis and ammonium. I. The role of the host in ammonium assimilation and its possible regulatory significance, *Proc. R. Soc. Lond. B* **229**:299–314.

Reid, P. C., 1987, Mass encystment of a planktonic oligotrich ciliate, *Mar. Biol.* **95**:221–230.

Reisser, W., 1981, The endosymbiotic unit of *Stentor polymorphus* and *Chlorella* sp. Morphological and physiological studies, *Protoplasma* **105**:273–284.

Reisser, W., and Benseler, W., 1981, Comparative studies on photosynthetic enzymes of the symbiotic *Chlorella* from *Paramecium bursaria* and other symbiotic and non-symbiotic *Chlorella* strains, *Arch. Microbiol.* **129**:178–180.

Reisser, W., Fischer-Defoy, D., Staudinger, J., Schilling, N., and Hausmann, K., 1984, The endosymbiotic unit of *Climacostomum virens* and *Chlorella* sp. I. Morphological and physiological studies on the algal partner and its localization in the host cell, *Protoplasma* **119**:93–99.

Rieder, N., Ott, H. A., Pfundstein, P., and Schoch, R., 1982, X-ray micro-analysis of the mineral contents of some protozoa, *J. Protozool.* **29**:15–18.

Rogerson, A., Finlay, B. J., and Berninger, U.-G., 1989, Sequestered chloroplasts in the freshwater ciliate *Strombidium viride* (Ciliophora, Oligotrichida), *Trans. Amer. Microsc. Soc.* **108**:117–126.

Sherr, B. F., and Sherr, E. B., 1984, Role of heterotrophic protozoa in carbon and energy flow in aquatic ecosystems, in: *Current Perspectives in Microbial Ecology* (M. J. Klug and C. A. Reddy, eds.), pp. 412–423, American Society for Microbiology, Washington, D.C.

Sherr, E. B., Sherr, B. F., and Paffenhöfer, G.-A., 1986, Phagotrophic protozoa as food for metazoans, a "missing" trophic link in marine pelagic food webs?, *Mar. Microb. Food Webs* **1**:61–80.

Smith, D. C., and Douglas, A. E., 1987, *The Biology of Symbiosis,* Edward Arnold, London.

Smith, W. O., and Barber, R. T., 1979, A carbon budget for the autotrophic ciliate *Mesodinium rubrum, J. Phycol.* **15**:27–33.

Stich, H.-B., and Lampert, W., 1981, Predator evasion as an explanation of diurnal vertical migration by zooplankton, *Nature* (London) **293**:396–398.

Stoecker, D. K., Michaels, A. E., and Davis, L. H., 1987, Large proportion of marine planktonic ciliates found to contain functional chloroplasts, *Nature* (London) **326**:790–792.

Stumm, C. K., and Zwart, K. B., 1986, Symbiosis of protozoa with hydrogen-utilizing methanogens, *Microbiol. Sci.* **3**:100–105.

Stumm, C. K., Gijzen, H. J., and Vogels, G. D., 1982, Association of methanogenic bacteria with ovine rumen ciliates, *Br. J. Nutr.* **47**:95–99.

Swanberg, N. R., and Anderson, O. R., 1981, *Collozoum caudatum* sp. nov.: A giant colonial radiolarian from equatorial and Gulf Stream waters, *Deep-Sea Res.* **28A**:1033–1047.

Swanberg, N. R., and Harbison, G. R., 1980, The ecology of *Collozoum longiforme*, sp. nov., a new colonial radiolarian from the equatorial Atlantic Ocean, *Deep-Sea Res.* **27**:715–732.

Szabo, S. P., O'Day, D. H., and Chagla, A. H., 1982, Cell fusion, nuclear fusion, and zygote differentiation during sexual development of *Dictyostelium discoideum, Dev. Biol.* **90**:375–382.

Talling, J. F., and Talling, I. B., 1965, The chemical composition of African lake waters, *Int. Rev. Gesamte Hydrobiol. Hydrogr.* **50**:421–463.

Taylor, F. J. R., 1982, Symbioses in marine microplankton, *Ann. Inst. Oceanogr. Paris* **58(S)**:61–90.

Van Bruggen, J. J. A., 1986, Methanogenic bacteria as endosymbionts of sapropelic protozoa, Ph.D. thesis, University of Nijmegen, Nijmegen, The Netherlands.

Van Bruggen, J. J. A., Stumm, C. K., and Vogels, G. D., 1983, Symbiosis of methanogenic bacteria and sapropelic protozoa, *Arch. Microbiol.* **136:**89–95.

Van Bruggen, J. J. A., Zwart, K. B., Van Assema, R. M., Stumm, C. K., and Vogels, G. D., 1984, *Methanobacterium formicicum,* an endosymbiont of the anaerobic ciliate *Metopus striatus* McMurrich, *Arch. Microbiol.* **139:**1–7.

Van Bruggen, J. J. A., Stumm, C. K., Zwart, K. B., and Vogels, G. D., 1985, Endosymbiotic methanogenic bacteria of the sapropelic amoeba *Mastigella, FEMS Microbiol. Ecol.* **31:**187–192.

Van Bruggen, J. J. A., Zwart, K. B., Hermans, J. G. F., Van Hove, E. M., Stumm, C. K., and Vogels, G. D., 1986, Isolation and characterization of *Methanoplanus endosymbiosus* sp. nov., and endosymbiont of the marine sapropelic ciliate *Metopus contortus* Quennerstedt, *Arch. Microbiol.* **144:**367–374.

Vardy, P. H., Fisher, L. R., Smith, E., and Williams, K. L., 1986, Traction proteins in the extracellular matrix of *Dictyostelium discoideum* slugs, *Nature* (London) **320:**526–529.

Vareschi, E., 1982, The ecology of Lake Nakuru (Kenya). III. Abiotic factors and primary production, *Oecologia* **55:**31–101.

Vogels, G. D., Hoppe, W. F., and Stumm, C. K., 1980, Association of methanogenic bacteria with rumen ciliates, *Appl. Environ. Microbiol.* **40:**608–612.

Wagener, S., and Pfennig, N., 1987, Monoxenic culture of the anaerobic ciliate *Trimyema compressum* Lackey, *Arch. Microbiol.* **149:**4–11.

Weis, D., 1977, Synchronous development of symbiotic chlorellae within *Paramecium bursaria, Trans. Am. Microsc. Soc.* **96:**82–86.

Weis, D., 1983, Infection in *Paramecium bursaria* as an inductive process, in: *Endocytobiology,* Vol. II (H. E. A. Schenk and W. Schwemmler, eds.), pp. 523–532, Walter de Gruyter & Co., Berlin.

Wetzel, R., 1983, *Limnology,* 2nd ed., W. B. Saunders, Philadelphia.

Whatley, F. R., and Whatley, J. M., 1983, *Pelomyxa palustris,* in: *Endocytobiology,* Vol. II (H. E. A. Schenk and W. Schwemmler, eds.) pp. 413–426, Walter de Gruyter & Co., Berlin.

Wilbert, N., and Kahan, D., 1981, Ciliates of Solar Lake on the Red Sea shore, *Arch. Protistenkd.* **124:**70–95.

Yarlett, N., Hann, A. C., Lloyd, D., and Williams, A. G., 1983, Hydrogenosomes in a mixed isolate of *Isotricha prostoma* and *Isotricha intestinalis* from ovine rumen contents, *Comp. Biochem. Physiol.* **74B:**357–364.

Zaret, T. M., Suffern, J. S., 1976, Vertical migration in zooplankton as a predator avoidance mechanism, *Limnol. Oceanogr.* **21:**804–813.

Zinder, S. H., and Koch, M., 1984, Non-aceticlastic methanogenesis from acetate: Acetate oxidation by a thermophilic syntrophic coculture, *Arch. Microbiol.* **138:**263–272.

2

Diffusion
The Crucial Process in Many Aspects of the Biology of Bacteria

ARTHUR L. KOCH

1. Introduction

The basis of all motion in biology is diffusion. The movement may be as simple as the diffusion of a precursor from the point of formation to an enzyme that processes it; it may be as complex as the mechanism of chemotaxis or muscle contraction. The movements of nutrients up to a cell or a collection of cells is probably the first thing that comes to mind when one considers bacterial ecosystems. Diffusion is also relevant to the movement of microorganisms and nutrients through slimes and gels in natural ecosystems. Of course, in microbiology there are practical applications of diffusion, such as the assay of antibiotic concentrations, in which diffusion, among other factors, controls the size of the zone of inhibition.

The strategy used by microorganisms of remaining small and simple makes their dependence on diffusion more obvious. In this review, I will discuss the general problem of being small and then present an outline of the mathematics of diffusion appropriate to certain microbiological problems. This discussion will include some interesting and not intuitively obvious results. I will consider primarily the roles of diffusion that are relevant to uptake into individual cells and, at the macroscopic level, into systems of cells. These examples include diffusion through the environment, passage through the porins and outer membrane of Gram-negative organisms, passage through the thick and thin layers of peptidoglycan, passage through the membrane, passive transport, and active transport. I will consider diffusion into a colony and diffusion within microbial mats and other two-dimensional ecosystems.

ARTHUR L. KOCH • Department of Biology, Indiana University, Bloomington, Indiana 47405.

2. Microorganisms Are Small

Microorganisms are so named because they are small. The cells of most procaryotic organisms (eu- and archaebacteria) are usually much smaller than those of eucaryotic organisms. They are small because of their evolutionary development, and they are abundant because their niches in the world are broad and are suitable for organisms of small size and simple lifestyle. The strategy of microbes as they survive and propagate is to do certain things more simply and better than their eucaryotic neighbors and to do certain things that the eucaryotes cannot do.

The advantage of being small is that simple mechanisms, even though they are primitive, can work well. In many cases, no mechanism need be constructed; the default laws of physics and chemistry suffice. No doubt the key factor involves the simplicity of uptake of nutrients and energy sources from the environment and the excretion of waste products. Simple diffusion in the environment and within the cell is adequate if the cell is small. A bacterium such as *Escherichia coli* or *Bacillus subtilis* does need to construct barriers (e.g., the cytoplasmic membrane and outer membrane) to diffusion and then modify those barriers to carry out needed transports via special mechanisms. The cytoplasmic membrane smoothly covers and is pressed against the inside of the taut peptidoglycan wall by osmotic forces. The outer membrane (when present) has special permeability properties. In contrast, a typical eucaryotic cell has a complicated system of plumbing consisting of the endoplasmic reticulum, the Golgi apparatus, vesicles, and membranous systems connecting the outside of the cell with the double nuclear membrane. The contents of all of those organelles, although mechanically inside the cell, are logically and topologically outside of it because the mechanism of membrane fusion in principle allows them to be ejected from the cell by exocytosis. For much larger, multicellular organisms, transport becomes still more elaborate: there is a forced circulation system to bathe mammalian cells with blood and lymph, there are a respiratory system, a gastrointestinal alimentary system, and a system for motility in animals that permit them to search out selected-for environments. Plants in their own way are as complicated.

A bacterium can do much less of these transport processes, but thanks to its small size also needs to do much less. Its facility for transport follows from the simple argument that the surface-to-biomass ratio decreases inversely in proportion to the linear dimension or directly with the surface-to-volume ratio. Therefore, a small cell can generally do without complex plumbing. There are exceptions, such as the formation of extensive photosynthetic membrane systems inside of phototrophic bacteria, particularly when the cells are cultivated at low light intensities. There are other ways in which smallness can be an advantage. Certain forces, such as the earth's gravity, are of little influence, and hydrostatic pressure is easier to resist. However, surface tension and viscosity become much more important.

Small size means that the bacteria can *be* simple and can do without the more elaborate organization of the higher cells. Thus, they manage without cytoskeletons, actin filaments, and microtubules. They have no actin, dynein, or tubulins. They have no internal organelles and very limited external appendages.

The higher microorganisms are in some senses very large. Protozoa have very large cells, and they have plumbing systems, such as gullets and vacuoles. The cells of some algae are large, with large vacuoles. But some algae and fungi, though nominally microorganisms, are organisms of the most macro kind. Thus, a fairy ring of mushrooms can be considered a single organism (one can even argue that it is one continuous cytoplasm), and fairy rings with a diameter of 4 miles have been observed. Similarly, some kelps grow to be taller than the tallest trees. But these are special adaptations, and transport is carried out by them in the same way in these as in other eucaryotic microorganisms.

What are the problems of being small? One is that bacteria are unable, as individuals, to alter their environments. Another is that although it may be advantageous from the point of transport to be even smaller than they are, the necessary elements for biological replication and function cannot be crammed into a still smaller space. This limits how small bacteria can be and still function autonomously. The main disadvantage of this lifestyle is that the procaryotes face all of the problems of eucaryotic organisms but without the same tools. They must grow, move, divide, etc., with much more limited means.

The list of problems to be addressed is large, and I have dealt with some aspects of some of these problems before (Koch, 1971, 1985). In particular, I have considered the procaryotic strategy for solving the problem of growth and division in the presence of an internal (turgor) pressure (Koch, 1983). In this strategy, the stress-bearing part of the wall plays an integral role.

3. An Outline of Diffusion

3.1. Fick's Diffusion Law

Diffusion is a process that results from the random, chaotic, noncoherent motion of molecules due to their thermal energy. It also applies to larger objects that move in a nonsystematic way. Even plants, animals, and people, although they may move by various means, obey the laws of diffusion when their movements occur nonpurposefully.

Figure 1. Fick's diffusion law. Shown is a rectangular parallelepiped; material is diffusing through it from a concentration that is high on the left side to one that is low on the right. The symbols dq/dt, q', and v are used synonymously for the flux in the text. All symbols are defined in the text. Fick's law is Eq. (1) of the text and is repeated in the figure.

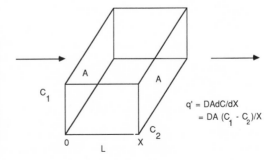

$$q' = DAdC/dX$$
$$= DA\,(C_1 - C_2)/X$$

Diffusion results in the net movement of material from levels of high to low concentration. If q is the amount of material, t is the time, D is the diffusion constant, A is the surface area, and dC/dX is the concentration gradient, then the flux of material is

$$dq/dt = -D \cdot A \cdot dC/dX \qquad (1)$$

(Fig. 1). The flux, dq/dt, will also be designated by v and q' below. This equation (Fick's first law) can be mathematically transformed for many particular cases, but it emphasizes the fact that increased diffusion transport requires a higher diffusion constant, an increased surface area, or an increased concentration gradient.

3.2. Steady State

When conditions have been constant for a long enough time in a system with sources and sinks of the substance in question, then dC/dX and dq/dt will become constant. Then the mathematics become simple. Sometimes a fraction of a second is sufficiently long; sometimes the system very slowly approaches a steady state. The time to achieve steady state can be important and must be always considered.

3.2.1. One Dimension: Linear Flow

When the actual system is the same in two dimensions but varies in the third, then at steady state dC/dx is constant and constant independent of the position between a source and sink (Fig. 2). In this hypothetical case, the concentrations at the left and right may be maintained high and low, by rapid stirring of large volumes, or by production and consumption. For this case, the flux, dq/dt, depends on the distance in between and the concentrations, C_1 and C_2, at the source and sink:

$$dq/dt = D \cdot A \cdot (C_1 - C_2)/X \qquad (2)$$

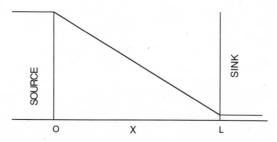

Figure 2. Linear diffusion (one-dimensional diffusion). The simplest diffusion system is outlined. Flow occurs from a reservoir of constant high concentration to one of constant low concentration. In the region in between, the flow is by diffusion through a rectangular parallelepiped shown in cross section. At the steady state, the concentration gradient becomes constant through this region, and Eq. (1) can be integrated to show that the concentration decreases linearly from source to sink.

3.2.2. Two Dimensions: Circular Symmetry

If the source of substance (e.g., a metabolic product) is inside a circular disk of radius, r_i, and the sink is outside a larger circle of radius, r_o, and the concentrations are maintained constant in the center and external environments (Fig. 3), then at steady state

$$dq/dt = D(2\pi r)\ wdC/dr \qquad (3)$$

applies with the same value of dq/dt for all values of r between the radius of the source and the sink. In this equation, w is the width or thickness of the system. Consequently, we can rearrange and integrate to obtain

$$\Delta C = [(dq/dt)/2\pi Dw]\ \ln\ (r_o/r_i) \qquad (4)$$

where ΔC is the concentration difference. In the steady state, dq/dt will adjust to satisfy this equation. (See curves in Fig. 3 that are arranged to be similar to those in Fig. 2. Also shown is the concentration dependence at which the source is external and the sink is internal.)

3.2.3. Three Dimensions: Spherical Symmetry

For spheres that absorbed all particles of a substance when they collided with it, von Smoluchowski (1918) showed 70 years ago the following relationships:

$$dq/dt = 4\pi DRC_2 \qquad (5)$$

$$C = C_2\ (1 - R/r) \qquad (6)$$

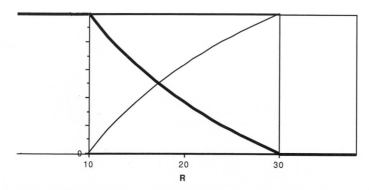

Figure 3. Diffusion through a circularly symmetrical barrier (two-dimensional diffusion). The figure's main purpose is for contrast with the one-dimensional case shown in Fig. 2 and three-dimensional case shown in Fig. 4. The cylindrical diffusion geometry would apply to objects like trichomes and filaments. Two cases are shown: one where the interior is the source, and one where it is the sink.

Here, C_2 is the concentration at large distances from the sphere where it is assumed constant; i.e., C_2 is uniform throughout the environment except in the neighborhood of the absorbing sphere (Fig. 4). The quantity, R, is the radius of the absorbing sphere. Equation (6) can be derived from Eq. (5) in the same way as shown above, i.e., from the fact that dq/dt is the same through any spherical surface surrounding the absorbing sphere. This equation applies only when the surface of the sphere is so effective at absorbing the material that the concentration of the surface is maintained at zero. When not all collisions of the particles with the sphere result in absorption, the same equation in which R is less than the radius of the actual sphere applies. This formulation has had much use in microbiology, serving as a basis for the kinetics of viral absorption and uptake through the environment linked together by lectins, etc., to free cells, aggregates of cells, and cells. It is discussed more thoroughly by Koch (1960, 1971). The example cited here is a unique case because the shape of the concentration curve is a hyperbola with a half distance equal to the radius of the sphere.

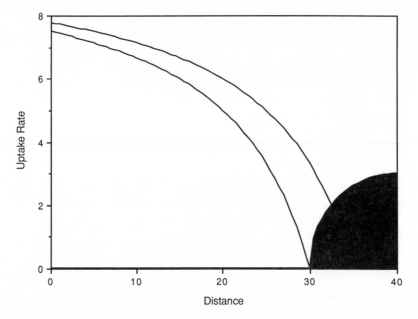

Figure 4. Absorption by a sphere (three-dimensional diffusion). A cross section of one-quarter of a sphere is shown. The steady-state concentration profile is shown for the case where the surface concentration is maintained at zero and where it is lowered by uptake, but the concentration on the sphere's surface is greater than zero.

3.3. Diffusion through and around Obstacles

3.3.1. The Principle of Diffusion

If objects that do not permit diffusion are present in the path, the rate of diffusion is reduced. The most simplistic treatment imagines that a certain fraction of the volume, ϕ, is occupied and the problem can be treated by simply pretending that normal diffusion occurs in the unoccupied fraction, $1 - \phi$. This is clearly wrong for many geometries. If the objects formed a solid wall across the area, diffusion would be totally blocked. With a few holes in the wall, some diffusion could occur. However the reduction in influx of a wall with a few holes is much less than might at first be suspected. Several cases are relevant in microbiology. One specific case is considered below, but the rule is quite general and applies to interposed porous plates and to gels and fibrous solids in a variety of geometries. The fluid paths, of course, must be large compared to the diffusing species size and then diffusion in a gel is minimally lower than the rate in an obstacle-free fluid.

3.3.2. Diffusion between Sources and Sinks
       through a Solid Slab Containing a Porous Plate

Imagine that a single thin, porous plate is interposed in a one-dimensional linear-flow diffusion path; assume that it has a thickness, X', and has small holes with a collective area of fA, where f is the fraction of the surface area that consists of holes (Fig. 5). It does not matter whether the slab is gradually curved or a plane or whether the

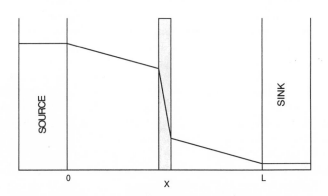

Figure 5. One-dimensional flow with an interposed barrier. This is the case depicted in Fig. 2 but with the addition of a thin porous barrier halfway between the source and sink. The analysis given in the text shows that the flux is decreased by factor of $1/[1 + (X'/X)(1/f - 1)]$, where X' is the thickness of the plate, X is the distance between source and sink, and f is the fraction of the area of the plate that is free for diffusion. In many situations, this factor is only slightly less than unity and the effects of the barrier can be neglected.

diffusion is into or out of the bathing medium. To the extent that this plate is a barrier to diffusion, a greater concentration gradient will develop across it. For arithmetical ease, let us assume that this plate is halfway between the source and the sink. There are now three regions of diffusion, and the same value of $dq/dt = q'$ applies to all in the steady state. Consequently, Fick's law can be applied to each of the three regions:

$$
\begin{aligned}
(q'/DA)(X - X')/2 &= C_2 - C_{2'} \\
(q'/DAf)(X') &= C_{2'} - C_{1'} \\
(q'/DA)(X - X')/2 &= C_{1'} - C_1
\end{aligned} \tag{7}
$$

$$
\overline{[q'/DA][X + X'(1/f - 1)] = C_2 - C_1}
$$

The last equation is the sum of the three above. By comparing Eq. (7) with Eq. (2) written in the form

$$
q'X/DA = C_2 - C_1 \tag{8}
$$

it can be seen that the flux is decreased by a factor of $1/[1 + (X'/X)(1/f - 1)]$. Clearly, as the thickness of the porous plate, X', becomes smaller, the impediment to diffusion becomes smaller unless the fractional area of the pore, f, is very small. As an example, if X'/X is 0.01 and f is 0.01, then the flow is decreased to 50.25% of the value with no interposed plate. If there were many such plates (n) and each plate was X'/n thick, the flux without obstacles must be multiplied by the same factor, $1/[1 = (X'/X)(1/f - 1)]$. Were the plates parallel to the diffusion path and completely solid, the diffusion would be decreased simply by the decrease in area. This would depend on different, but simple, aspects of the geometry.

In Berg's book *Random Walks in Biology* (1983), this same point is made for two cases. One is seemingly the same as that considered above, and Berg arrived at the same general conclusion. However, he does not arrive at the identical answer because of an important difference in detail. I assumed above that the concentration was exactly the same at all points at a given distance from the source or sink (Fig. 6A). That is, I tacitly assumed that there were very many holes but that they were all very small. Berg started with the presumption that there were circular holes in the thin plate that were big enough and few enough that each acted as a bottleneck. Moreover, he assumed that they were far enough apart that the diffusion current through one did not influence the diffusion through neighboring ones. Consequently, the diffusion current through a hole of radius, r, would look like that shown in Fig. 6B and not like that shown in Fig. 6A. For his case, the rate of diffusion is given by

$$
q' = 2Dr(C_1 - C_2) \tag{9}
$$

This differs from the standard form for diffusion [Eq. (2)] in that A/X has been replaced by $2r$. The two expressions have the same dimensions. Neither I nor Berg prove this relationship because that requires a great deal of mathematical analysis [a proof can be found on p. 42 of Crank (1975)]. Thus, a thin plate with not too many fairly large holes,

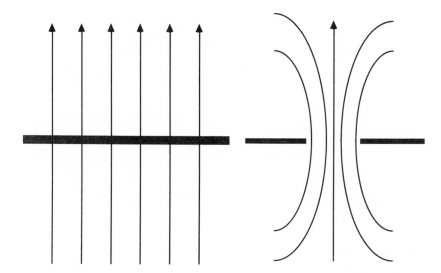

Figure 6. Effect of barrier geometry on pattern of diffusion flux. (A) Flows through a porous plate with many fine holes close together. This is the flow pattern assumed in the derivation of Eq. (7) shown in Fig. 5. (B) Flow pattern through a hole well separated from other holes. The flow through a single hole is given by Eq. (9).

when placed in a linear diffusion gradient, slows diffusion by a factor of $1/[1 + 1/(2nrX)]$, where n is the number of holes per unit area. Again, not much of the area need be devoted to holes to make the slowing negligible.

But the mathematics are not important; what is important is the understanding that the diffusion process will in large part automatically adjust to ameliorate the effect of partial blocks to diffusion. Thus, the surface of a leaf of a higher plant may be largely impermeable to water, CO_2, and O_2 because of the cuticle layer, but a few open stomata make this surface of the leaf a minor barrier to diffusion. Correspondingly, a soil system will be quite permeable to water, gases, etc., even though it is largely made of soil particles that are impermeable to these substances.

The second case in Berg's book which is of critical importance in biology is that of a spherical object, such as a cell, that is a sink for a substance present in the environment but whose surface is impermeable except for a number of circular disks. The analogy is to a sprinkling of uptake sites spread on the surface of the cell membrane. How much less efficient is the surface because only a fraction is absorptive? The flux for this case is

$$q = q'_0[1/(1 + \pi R/Na)] \tag{10}$$

where q'_0 is the rate of uptake if the whole surface is absorptive [$q'_0 = 4\pi DRC_2$, as given by Eq. (5)], R is the radius of the sphere, a is the radius of an uptake site, and N is the number of uptake sites per sphere.

Again, it is clear that the cell can approach the physical limit of diffusion in its environment *without making the entire surface into a perfect absorber.* This general concept would apply to microbial mats and "pelleted" consortial ecosystems. If some members of the consortium can consume a resource, the presence of other members that do not have this function impedes very little the diffusion to those that do.

3.4. Reaction along the Diffusion Path

When some reaction, either reversible or irreversible, takes place along the diffusion path, the calculation of concentrations is greatly complicated. The key point, however, is that diffusion is slowed. Again, I consider only one case: the diffusion of, say, detergent out of a dialysis bag into a surrounding bath, with good mixing both inside and outside the bag. If there is no ligand for the detergent, then $q' = DA (C_1 - C_2)/X$ as before, where X is the thickness of the dialysis tubing. With C_2 kept near zero, dialysis may be fast, but if the ligand is present at a concentration, L, and has a dissociation constant, K, then C_1 will be reduced by $(K + L)/L$, and so will the diffusion rate. Eventually, but slowly, the detergent would be depleted in the dialysis bag. This same principle applies generally to a number of diffusion geometries. Of course, in a system in which irreversible reactions take place, then the diffusing substance may be permanently trapped or lost.

3.5. Dynamics of Diffusion

The process of diffusion from a nonrenewable point source is the same as that generating a normal Gaussian distribution. Consequently, the concentration profile is a bell-shaped curve at any time after diffusion starts to occur. This rule applies for diffusion in any number of dimensions. With time, the curve becomes lower and broader. A case in which this dynamics of diffusion is important is the process of sporulation and dispersal, whether from a single fruit of a fungal fruiting body or from a sporangium of a fungus. When, however, there is synchronous germination from reproductive organs distributed in space, then the geometry of the source is different; the kinetic course will also be quite different but can be calculated. An important case in microbiology is the surface of a mat that consumes a substance initially available at the start of the season in a limited amount in the water column.

The key case is that of a spherical organism which, by some stirring or dispersal process, finds itself suddenly and discontinuously in an environment where a nutrient is uniformly distributed. With time, uptake becomes constant according to Eqs. (5) and (6). Thus, the surface concentration is maintained at zero at the surface; then, in the first instant of time after such a condition becomes applicable, there is an infinite flux. Eventually, a finite concentration gradient develops at the surface in the environment; thereafter, the flux will be finite and may change only slowly with time. In the transient, the mathematics is complex (see Koch, 1960, 1971), but the results are shown graphically in Koch (1971). It was from analysis of the speed of coming to a steady state of uptake that I was able to show that individual organisms could not swim fast enough to significantly increase their absorptive rate of small molecules (Koch, 1971). Later this conclusion was confirmed by Berg and Purcell (1977) and Purcell (1977). I was also able to show that a long filamentous organism in an unstirred environment would face a continuously decreasing rate of nutrient supply, in contrast to a spherical organism, which would achieve a constant steady state of uptake.

3.6. Diffusion in a Stratified Ecosystem: Elementary Treatment

Biomass is generated mainly at interfaces where different resources meet. A simple microbiological example is that of a still lake in which some oxidant diffuses downward (subscript 1), some reductant diffuses upward (subscript 2), and both are consumed by the microbiota. (This problem is discussed extensively in Section 6). In time, the consuming organisms increase in mass so that they totally eliminate the diffusing resources in some particular regions. Gradually, the organism forms a thinner but denser layer at the level at which the concentrations of substrates vanish. In this quasi-steady state, the following relations hold:

$$D_1 A dS_1/dX = dq_1/dt = dB/Y_1 dt$$
$$D_2 A dS_2/dX = dq_2/dt = dB/Y_2 dt \tag{11}$$

where D is the diffusion constant, S is the substrate, Y is the yield coefficient, and B is the bacterial biomass. As the system develops, the bulk of the organisms will be narrowly located at a position X_c, part way between the positions where S_1 and S_2 are maintained constant. Designating the source concentrations by the subscript o, we can express the concentration gradients by S_{1O}/X_c and $S_{2O}/(L - X_c)$, where L is the length between the two sources. Substituting these values and eliminating dB/dt, we obtain:

$$X_c = \frac{LD_2 S_{2O} Y_1}{D_1 S_{1O} Y_1 + D_2 S_{2O} Y_2} \tag{12}$$

In the simplest case, this equation defines the position of stratification of the microorganisms. If we think of the knallgas bacteria that exploit the reaction $2H_2 + O_2 \rightarrow 2H_2O$, and if we identify subscript 1 with hydrogen and subscript 2 with oxygen, then the position of the layer of bacterial growth in a lake is graphed in Fig. 7 for various concentrations of H_2 available at the bottom surface of the lake. The kinetics of the development of the layering are quite complex even in the simplest case. An example will be elaborated below in Section 6.

3.7. Diffusion Limitation of Growth

3.7.1. Solitary Cells

When some nutrient in the environment has been habitually present at low levels, microorganisms can be expected to evolve to become as efficient as physically possible

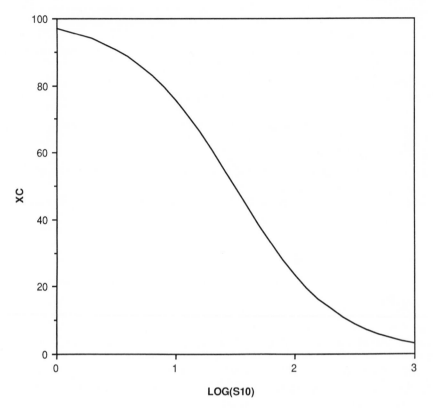

Figure 7. Position of layering in a still lake. Equation (12) is used to calculate the position of the steady-state growth layer in a hypothetical lake in which knallgas bacteria grow from hydrogen gas diffusing from the bottom and oxygen diffusing from the air above. X_c is the position of the biomass maximum in a column of length L. X_c as normalized is shown as a function of the H_2 concentration.

if given time periods that are long on the evolutionary scale. The limiting efficiency is given by $3D/R^2$, where R is the radius of a spherical organism and D is the diffusion constant of the substrate in the environment (Koch, 1971). Efficiency is defined as the number of cell volumes of medium from which nutrient is completely extracted by a cell per unit time. This formula for the limiting efficiency, illustrated in Fig. 8, can be derived from Eq. (6) (Koch, 1971). It has been shown that the cytoplasmic membrane of even a so-called copiotroph like *Escherichia coli* approaches this limit. However the porins, which a water-soluble nutrient must traverse to enter the periplasmic space in order to be consumed by a Gram-negative organism, are a significant impediment to growth when the environment substrate concentration is low. This work is reviewed in Section 4.9.

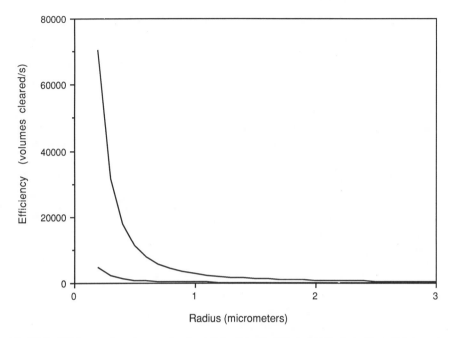

Figure 8. Efficiency of uptake as a function of the size of a spherical biological object. Efficiency is equal to $3D/R^2$. The upper line corresponds to a diffusion constant of 9.4×10^{-6} cm²/sec for glycine and typical of glucose or low-molecular-weight nutrients. The lower line is for $D = 6.3 \times 10^{-7}$ cm²/sec for serum albumin, which corresponds to a typical protein or dextran as the diffusing species. It is assumed that the environment is still so that there is no convection.

3.7.2. Groups of Cells

Many procaryotic cells adhere to form groups of cells in clusters, chains, filaments, microcolonies on surfaces. They may be mixed communities of different species. The larger such aggregates become, the more important is diffusion through the external medium (Fig. 8). In an environment in which there is convective mixing because of fluid flows, diffusion through an unstirred layer can still be important and dependent on a number of physical factors. When the aggregate can be considered a ball and when the unstirred layer is several diameters thick, then the limiting efficiency is still given by $3D/R^2$.

3.8. Literature on the Mathematics of Diffusion

Most of the key mathematics were derived by Fourier early in the last century. He was interested in the equivalent process of the diffusion of heat. Consequently, there are many handbooks and textbooks in engineering covering the material needed to predict heat flows and temperature distributions for a vast range of geometrical situations and heat input conditions. These equations could be located and applied to particular biological problems. Carslaw and Jaeger's *Conduction of Heat in Solids* (1959) is the standard but out-of-print reference on diffusion. Even though the equations contain many complicated functions, the availability of modern computing facilities renders them useful. The standard mathematical treatise on diffusion is that by Crank (1975). Rashevsky's work (1960) contains a number of applications of diffusion to problems in cellular physiology. An excellent introductory treatment is given by Berg (1983). I have written, in the contexts of microbiology and low-level mathematics, several relevant papers (Koch, 1959, 1960, 1982a,b) and reviews (Koch, 1971, 1979, 1985).

4. Kinetics of Uptake from the Environment into the Metabolic Pool

The main concern of living creatures is the exploitation of resources from their environments; that is what it is all about. Consequently, "How are nutrients sequestered by microbes?" must rank as the first question in microbial ecology. General microbiology has devoted itself to food chains and the special ways that substances in the environment can be used and brought into the core metabolic processes of cells, but generally this field has omitted the process of accumulation. I have addressed the accumulation question in general terms twice before (Koch, 1971, 1985), but is such an important topic that I will treat it yet again, with different details. Let us consider the most basic case: a hypothetical spherical organism alone in a large aqueous environment that contains an essential nutrient. Our hypothetical procaryote is a composite of real organisms and has a capsule, an S-layer, an outer membrane, a thick peptidoglycan layer, a periplasmic space with periplasmic enzymes, a sophisticated cytoplasmic membrane, and peripheral enzymes. When the nutrient (or its derivative) is delivered to the "acid-soluble pool" of the cell, as far as this review is concerned the story ends, except for the assumption that it is consumed for cell growth and by-product formation resulting in a very low internal level. Not every one of these listed steps is relevant for any one real organism or for any one actual nutrient.

The general constraint is that in the steady state as much nutrient must pass inward toward the cell through any one compartment as through any other. Consequently, the concentration of the nutrient at which transport is fast will change little in going from one phase to the next but change a great deal from input to output for stages at which the transport is slow. This must occur to generate the suitable steady-state concentration gradient that results in the same steady-state flux as for the other steps.

I am assuming that transport comes quickly to a steady state. But of course, on the evolutionary time scale, transport systems, like all biological processes, are not static: microorganisms may be able to adapt or may evolve to speed a step that otherwise would be rate limiting. These processes I will not consider here. In the following subsections, I will consider each of these stages as if it were the rate-limiting step. However, in Section 4.9 I will consider some cases in which two steps together limit growth.

4.1. Transport in the Environment

In an unstirred environment where uptake at the outermost aspect of the cell is fast, Eqs. (2) and (5) specify the rate limited by simple diffusion for flat and spherical surfaces, respectively. When the environment is stirred, then D becomes effectively bigger and the uptake rate larger. But the problem is not quite that simple if the consuming unit is small and buffeted about by the stirring as well. The details were worked out 30 years ago (Koch, 1960; Valentine and Allison, 1959), and the answer is that stirring hardly affects the rate of absorption of objects even as large as viruses onto objects as small as bacteria. Larger rate increases caused by mixing occur if the body absorbing the substance is large. Because much of the early work concerned tissues of eucaryotic systems, the concept of the unstirred layer evolved and dominated thinking (Fig. 9). This idea is that mixing in the environment is rapid up to a distance of about 10 μm, almost independently of the intensity and type of stirring mechanism, but it is negligible inside this distance. Thus, the external environment is uniform up to that distance, and the effect of diffusion need only be considered for that final few micrometers as a one-dimensional problem for large flat absorbers or for spherical organisms by the von Smoluchowski formula. In this region, the flux can then be treated, depending on shape, by Eq. (2) or (5). Evidently, this is an inappropriate approximation for organisms that are smaller than the "unstirred layer" but appropriate for mammalian cells and tissues and for ordinary measuring electrodes. The concept would be pertinent to assemblages of organisms such as biofilms, cell aggregates, particles of detritus

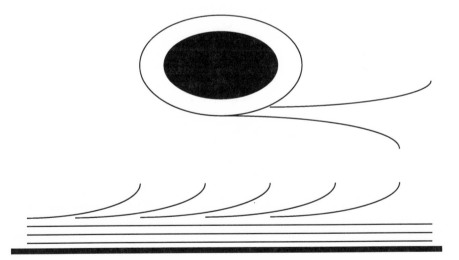

Figure 9. The concept of the unstirred layer. Shown are a flat surface and an ellipsoidal surface. According to this concept, they are covered with a fluid layer that is still no matter how violent the stirring of the bulk fluid. Outside of this layer, the model assumes that the stirring is totally effective whether mixing is due to slow laminar or violent turbulent flow.

colonized by bacteria, etc. To reiterate the concept: laminar flow and mild turbulent flow do mix the environment near to the surface (10 μm), but diffusion transports material to the absorptive surface layer. The concept is useful except for small, well-separated cells.

4.2. Transport through a Capsule

Many microorganisms secrete strands of a hydrophilic colloid that then crosslink to make a gel. This gel, which can be very dilute, remains around the cell and prevents turbulence in the environment from affecting the cell, thus creating a cell's private, unstirred layer. With respect to uptake of low-molecular-weight nutrients, this is largely irrelevant for a free-living cell; however, the gel may isolate the cell from the immune system of a host or from grazing by predators and may allow the cell to remain attached to a substrate material. In aggregates of cells, the gel keeps cells separated from each other, which improves the diffusion process in comparison with the circumstance in which cells are held close to one another. No stirring occurs in the capsule, and transport through it depends solely on the appropriate diffusion law for the geometry and for the case of nutrient molecules that are very small relative to the pores in the gel structure. In Section 5, I will deal with diffusion through a gel more critically; here it suffices to state that small molecules are somewhat impeded, whereas large molecules can still slowly (possibly very slowly) permeate the gel.

4.3. Permeation through the S-Layer or through the Outer Membrane

Both the S-layer and the outer membrane are obstacles that can be represented as thin layers corresponding to the one-dimensional problem whose solution is given by Eq. (2). However, Eq. (2) must be modified by the inclusion of a factor, f, on the right to correct either for the fraction of the surface that is not covered by the proteins of the S-layer, through which diffusion can occur in Gram-positive organisms, or for the fraction of the surface covered by the pores of the porin system, in the case of the outer membrane of Gram-negative bacteria.

For this geometry, transport was analyzed by Renkin (1954). He idealized the situation by imagining rigid spheres passing through right cylindrical holes in the membrane (Fig. 10A). He combined two considerations: the sphere (radius, R) could not pass unless it totally missed the edge of the cylinder (radius, r) on its initial approach, and a sphere would experience drag in diffusing through a cylindrical pore. Drag arises because the fluid in the pore must move around the spherical particle as it passes. The first consideration led to the inclusion of a factor of $[1 - (R/r)]^2$. The second consideration required the inclusion of the expanded form of Stokes' law. Stokes' law deals with the movement of a spherical body in fluid and serves as a basis for much of the hydrodynamic theory used in biophysical chemistry. The expanded version includes the drag as the sphere passes through a cylindrical tube. Combining these concepts, Renkin

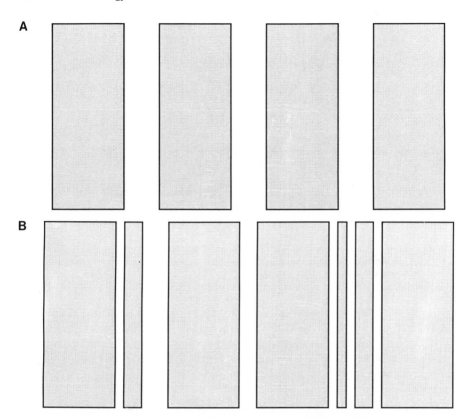

Figure 10. The Renkin model for passage through a wall. (A) A wall penetrated by right cylindrical pores. Renkin's analysis depends on the incorporation of two factors. One is the chance of the spherical molecule hitting the pore but missing the wall. The second factor relates to drag while passing through the hole. (B) Elaboration of the Renkin model to approach that of a porous wall or gel. It is imagined that many pores of various diameters penetrate the wall. They may vary in length as well, but it is shown in Fig. 14 that variation in path length is of minor significance.

advocated that the following additional factor to be introduced into the form of Fick's law that would apply if there were no wall:

$$F = [1 - (R/r)]^2 (1 - 2.104 (R/r) + 2.09(R/r)^3 - 0.95 (R/r)^5] \qquad (13)$$

This formula indicates that if the spheres are slightly smaller than the holes, the throughput will be small as a result of both factors (see Fig. 11). I shall use this same formulation below in considering diffusion through three-dimensional solids. Here it suffices to state that Renkin's treatment accounts well for the permeability characteristics of porins

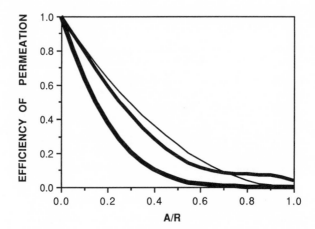

Figure 11. The Renkin function. The thick line was calculated from Eq. (12). The thin line shows the first factor, which is the geometry of entering the pore without hitting its lip. The intermediate line shows the effect of the second factor (Stokes' drag) alone.

(Nikaido, 1979; Nikaido and Rosenberg, 1981; Koch and Wang, 1982). The key point of the latter studies was that even though glucose is much smaller than the exclusion limit of 700 daltons (porin size), diffusion through the porins alters the growth rates of cells growing at very low levels of substrate because there is an interaction between the flux through the outer and inner membranes (see Section 4.9).

4.4. Diffusion through the Peptidoglycan

Recent studies by Demchick and Koch (unpublished) have shown that the wall of *Bacillus subtilis* is permeable to dextrans of about 35,000 daltons in size. This is consistent with earlier studies of Ou and Marquis (1970) with another Gram-positive organism; however, we measured the permeability properties of isolated intact walls, and Ou and Marquis measured the space accessible to different-size probe molecules in wall fragments. As such, the peptidoglycan represents a negligible barrier to low-molecular-weight nutrients. But in some cases, the rigid wall probably is a barrier to passage of protein.

4.5. Role of Periplasmic Enzymes

In Gram-negative bacteria, quite a number of hydrolytic enzymes are excreted into the periplasmic space. Some are constitutive and some are inducible. They serve the key role of decreasing the local concentration of nutrients being imported from the environment, thus speeding the diffusion-driven process into the periplasm by increasing the concentration gradient. Also, since all molecular species diffuse independently, they create a new substance for selective passage into the inside of the cell.

4.6. Uptake via Binding Proteins

Binding proteins are moderate size proteins (about 30,000 daltons) that bind to a substrate in the periplasmic space and then bind to a membrane-bound receptor to deliver the substrate. On the face of it, this seems silly. Diffusion of the complex of the binding protein and substrate must be much slower than that of the unfettered substrate. Work by Adams and Delbruck (1968) explained the paradox. If the protein had an increased probability for associating with the cytoplasmic membrane, then when the substrate bound to the binding protein, the diffusion problem would be changed from three dimensional to a two-dimensional problem of scanning the surface for the receptor. This decrease in dimensionality, in turn, would speed the uptake process.

4.7. Passage through the Cytoplasmic Membrane

It may be that the most sophisticated parts of the procaryotic cell are those proteins built into the membranes. They are so intricate and well machined that despite extensive work of many scientists, we do not understand how any integral membrane transport works. Seemingly, we are getting closer to an understanding. For example, we have the X-ray structure and directed-site mutagenesis and sequence data for bacteriorhodopsin and directed-site mutagenesis and sequence data for the galactoside permease (a symporter). It is understood which domains of the protein are inside and which outside, but we do not yet understand how the permease functions to pass exactly one galactoside molecule together with one proton through the membrane (in either direction) without allowing either the proton or the galactoside to pass by itself. Likewise, we have much information on the role of the binding protein in ATP-linked transport but only a poor understanding of mechanism. To further emphasize the point, we have had many elaborate physiological and genetic studies of PTS (phosphotransferase system) systems, as a representative of group translocation processes, yet still nothing is known of the mechanism at the membrane level of the transport event.

4.8. The Role of Peripheral Enzymes

It has become increasingly apparent in recent years that certain enzymes hover near the inner face of the cytoplasmic membrane. β-Galactosidase is the case known longest to me (Preiss and Pollard, 1960–1961). The value of this to the cell is the same as that of the periplasmic enzymes; i.e., they decrease the substrate concentration to near zero and thus favor influx in accordance with Fick's law or prevent product inhibition for facilitated or permease-directed mechanisms.

4.9. Colimiting Transport

Sections 4.1–4.8 dealt with the component transport processes as if each were the single step limiting transport. In these sections, I considered a process taking material from a source where the concentration is equal to that present in the environment across a single step and delivered it to a sink where the concentration is maintained very near zero.

For the full general treatment, equations for each step would be written in terms of the steady-state concentrations for the source and sink of that stage. This is akin to the method leading to Eq. (7) in Section 3.3.2. If there were n stages in the overall process, then there would be n equations and $n + 1$ concentrations. These would involve the bulk external concentration and the internal concentration (which we assume to be zero). Consequently, there are $n - 1$ unknown concentrations that are the sinks for one process and the sources for the next. These unmeasurable concentrations can be eliminated algebraically, and we are left with one equation in terms of known concentrations and the parameters of the transport processes.

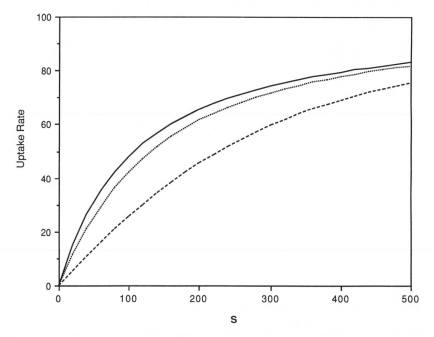

Figure 12. The Best equation. The velocity of a process that is the concatenation of a first-order reversible process and a subsequent irreversible simple enzymelike process is shown; illustrative values have been calculated from Eq. (14). This equation is appropriate in this review because it corresponds to the situation of passive diffusion through the outer membrane, followed by uptake via a permease at the cytoplasmic membrane. There are other microbiological situations where the equation applies for other reasons (see Koch, 1985). The equation depends on the parameter J. At one extreme (dashed line), the equation reduces to the familiar Monod hyperbolic expression. At the other extreme, it degenerates into the Blackman law of the minimum. The latter is probably a more apt description in most situations than the former. Of course, the Best treatment is more flexible but requires more accurate data. Solid line, $J = 16$; short dashed line, $J = 64$; dashed line, $J = 256$. Curves are calculated for $K = 100$.

The first microbiologist to consider the effect of colimiting steps was Powell (1967), who considered the case of a Fickian diffusion process preceding a enzymatic one. He was not the last to derive the same relationship. I have listed the papers concerning this in Koch (1985). Nor was Powell the first to derive the relationship. On this basis, I have called the relationship the Best equation, after Jay Boyd Best (1955). Since I have presented the derivation elsewhere, (Koch, 1967, 1985; Koch and Coffman, 1970), I will not do so here. The equation is

$$v = V(S + K_m + J) \{1 - [1 - 4SJ/(S + K_m + J)^2]^{1/2}\}/2J \quad (14)$$

were V is the maximum flux through the enzyme-like step in the system, K_m is the Michaelis–Menten or Monod constant, and J is parameter equal to VX/AD. If V is small, J is small. Then the square root can be expanded and the equation reduces to $v = VS/(K_m + S)$, thus becoming the normal Monod expression because the diffusion step is no longer limiting. Similarly, the equation reduces to either the first order Fick's law, if J is large and S is small, or to $v = V$, if both S and J are large. These latter two solutions correspond to the two branches of Blackman's law of the minimum (see Koch, 1982b).

Results are shown graphically in Fig. 12. The key point is that the Best equation ranges from the Monod enzymes hyperbola to the Blackman case, where the uptake rate versus concentration curve is composed of two straight-line segments.

The work of Powell (1967) and that of my colleagues (Koch, 1967; Koch and Coffman, 1970; Koch and Wang, 1982) show that the Blackman treatment is a better approximation in many cases than is the Monod, although the Best is better still. However, the latter has one more parameter, is much more difficult to calculate by computer, and makes analytical work virtually impossible. Many theoretical studies in microbial ecology are predicated on the Monod expression. I think that now that approach is inappropriate and that it would be more general (and no more work for the computer) to take the Blackman model as the default.

5. Diffusion through a Gel

Many gels are capable of appearing rigid even though the content of solids is very small. The solid structure is achieved because a system of long molecules composed of covalently linked monomers has formed. The system must form as a loose network with occasional interstrand bridges. The crosslinks may form either with hydrogen bonds or with covalent crosslinks. Agarose gels are examples of the former, and polyacrylamide gels exemplify the latter. The fraction of the volume actually occupied with solids, ϕ, may be very small; consequently, the average lacuna may be very large compared with the size of the permeant molecules. For this reason and for the reasoning associated with the prototype mathematics developed in Section 3.3.2, diffusion is impeded very little. But it is impeded to some degree because occasionally a molecule will diffuse into a cul-de-sac and must then diffuse backwards and around before forward motion is possible.

To date there is no realistic treatment of diffusion through a random three-dimensional network. Therefore, one can only appeal to approximations, as discussed in Sections 5.1 and 5.2.

5.1. The Ogsten Theory

To understand exclusion chromatography, Ogsten (1958) pictured a gel as a three-dimensional haystack of linear straws. "What is the probability," he asked, "that a sphere projected at random into the stack would *not* overlap one of the straws?" This is a problem related to one studied a century earlier by Buffon, whose question was, "What is the chance that a needle dropped at random on a system of equidistant parallel lines would not intersect one of them." This earlier mathematics was important in determining pi, and pi also enters the present problem. The solution for the three-dimensional problem is

$$P = \exp\left(-2\pi\nu LR^2 - 4\pi\nu R^2/3\right) \tag{15}$$

where P is the probability, R is the spherical radius of the ball, L is the length of the linear fiber, and ν is the average number of fibers per unit volume. This theory can be

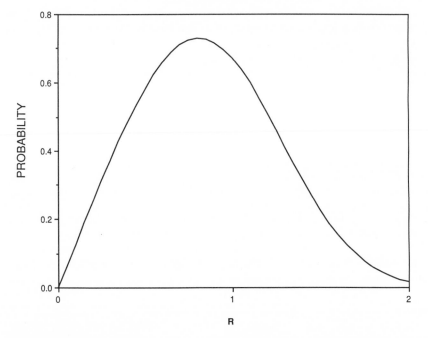

Figure 13. The Ogsten model. It is assumed that the diffusing species is a solid sphere. Notice that the distribution is positively skewed and very closely approaches the log normal distribution. It is in fact the chi-squared distribution for 2 degrees of freedom.

modified when appropriate to take into account fibers that are of appreciable diameter. The theory has been successfully used to model gel filtration and gel electrophoresis. In general, the model has been very useful (see Rodbard, 1974). A numerical example is given in Fig. 13. One additional point must be made: the distribution of pore spaces is a highly skewed one; that is, there are rare spaces that are very much larger than the mean. They are much more abundant than corresponding spaces that are the same amount smaller than the mean. Thus, the space size distribution is positively skewed and approaches the log normal distribution. The log normal distribution has applications in many areas of biology and indeed in many other fields (Koch, 1966, 1969). This fact is relevant to the present case because another treatment derived for gel filtration starts with the assumption that the pores are distributed according to a truncated log normal distribution proposed by Rodbard (1974) to succeed the normal distribution assumed by Acker (1967) and because I will use the concept again in the next section.

5.2. Elaboration of the Renkin Theory

From Renkin's model for diffusion through pores, reviewed in Section 4.3, it is a conceptually trivial generalization to imagine that a three-dimensional maze corresponds to many Renkin paths through the solid (see Fig. 10B). The individual paths would be of varied length and diameter, and composite behavior of all is required. Of course, it was a simple matter to generate a computer program that calculates the flux due to the one-dimensional diffusion through a single class of pore of constant dimensions and add this to a second class, and so on. Thus, the computer averages over the range of variation of the different parameters and combination of types of variation. To see all of the factors, let us rewrite Eq. (2) as

$$q' = N\pi r^2 \, FDC/X \tag{16}$$

where N is the number of pores, r is the radius of the pores, F is the Renkin factor, D is the diffusion constant, C is the external concentration, and X is the thickness of the barrier layer. As before, the concentration on the other side is assumed to be zero. A was defined as the total area of surface and f as the fraction of the surface that is open; in these terms, $Af = N \pi r^2$. The quantities that might vary in different paths through the gel include X, which could be longer or shorter, depending on how tortuous a channel was present and r, the radius of the pore, which enters directly in Eq. (16) and indirectly in the calculation of the F factor from Eq. (13).

Computer results for the initial rate of flux through a porous wall with either a homogeneous set of pores or pores that are variable in size or length are shown in Fig. 14. It can be seen that the same degree of variation affects the rate much more when it is the pore radius that varies than when the path length varies.

The program (which will be sent to anyone requesting it) is also of use when the permeant molecules have a range of sizes. This aspect would be of interest with respect to smaller random polysaccharide chains and partially degraded proteins diffusing in or out of biogels. The molecular weight changes both the value of a, the radius of the permeant molecule (in the F factor), and D where it is inversely proportional to a.

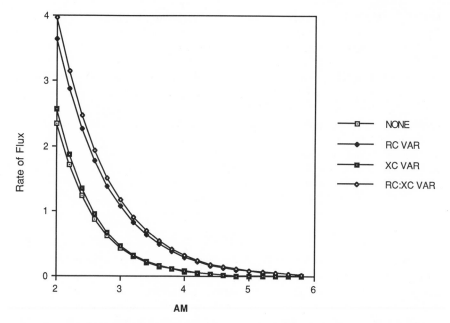

Figure 14. Variability of pore size and pore length. Lines correspond to a uniform set of pores each 5 nm in radius and 25 nm long (□), to a 30% coefficient of variation (cv) in pore diameter (♦), to a 30% cv in pore length (■), and to variability in both (◇).

6. Two-Dimensional Cross-Diffusion

A characteristic feature of life is that it occurs most abundantly at interfaces (see Marshall, 1976). This may be the surest fact in ecology: life is abundant at the earth–air interface, at land–water interfaces, at thermal vents under the sea, etc. Of course, the reason is that these are places where there is the continual mixing of resources from different locations. Had these resources been premixed, life would have already grown and the resources would no longer be available to exploit. Thus, in an estuary minerals pour into a semisaline world that is usually continually (with or without tides) irrigated and illuminated. This presents a pattern that occurs over and over again in larger and in smaller scale. Here we will consider systems that are essentially two-dimensional sheets where one resource diffuses in from one side and a second from the other side to be combined for the production of biomass (and products). The Precambrian microbial mats, the top few millimeters of mud on a tidal flat, the stratified lakes in which reducing power (in the form of H_2) diffuses upward from the silt on the bottom and O_2 diffuses downward from the air–water interface are all situations that are idealized in the model developed in this section.

The computer program that I have prepared is in FORTRAN and requires a number-crunching computer; the program (Fig. 15) is versatile enough for application to actual systems. I hope that interested readers will set this program up, in FORTRAN or in another high-level language, try it, and then modify it for their own uses. The program imagines that the habitat is composed of 101 layers in which diffusion of resources and growth of a single species of microorganism takes place. Outside of these layers, the resources remain at constant concentration of one resource, S10, at one end and of the other resource, S20, at the other. (To the extent possible, I have simulated the symbols used in Section 3.7. In that section, these two variables were designated S_{1O} and S_{2O}.) As the program simulates time, substrates diffuse, based on the concentration difference on either side of each layer (diffusion constants DA1 and DA2). I have assumed that bacterial growth is hyperbolic (Monod's law) in both substrates (although it could readily be rewritten for Blackman's law). Thus, I have assumed that the change in biomass and a proportionate change in both resources (the yield coefficients are Y1 and Y2) depend on the triple product of the concentrations of the biomass, B(I), the two resources, S1(I) and S2(I), divided by two terms, each of which is the sum of Michaelis–Menten-type constants, K1 and K2, and the appropriate one of the two resource concentrations, K1 + S1(I) and K2 + S2(I). I is an index that is 2 for the top layer and 102 for the bottom-most layer in the part of the water column. With this rule, the computer reports back every N = 1000 units of time for 20 periods. In the report, it lists the resources and the biomass at each tenth of the total distance through the column. It lists the biomasses over a finer spacial gradation in the region very near the middle of the column where the biomass becomes concentrated. There are some special features in the code that help prevent the program from getting into numerical difficulties, as well as other niceties. The output also includes statistical information about the distribution of the biomass; this includes mean, standard deviation, skewness measure, and kurtosis measure.

The simulation starts by using the adopted values of the constants shown in Fig. 15. Initially, the biomass is very low, the two resources are high, and all are uniform throughout the column. Then consumption and growth take place essentially exponentially until the resources in column are depleted (see Fig. 16A). From then on, the biomass growth is limited by the diffusion of both resources. By the last period simulated, the biomass is narrowly concentrated in the middle; there the two resources diffuse in from opposite directions and are quantitatively consumed. For our assumed geometry, each resource decreases in a linear way, starting from its source. The resources are almost absent in the region of intense growth and are totally absent on the far side. The distribution of biomass at the final time point has a slightly positively kurtose, very slightly negatively skewed but quite narrow (see Fig. 16B). The distribution becomes narrower and narrower when the program is allowed to work for a longer total time.

This exercise for a hypothetical, highly idealized case is, I hope, of value for more than focusing attention on similarities of the many two-dimensional microbial biospheres in which one-dimensional diffusion of two resources is the limiting process. The major point is the wide range of the variables parameters that lead to narrowly concentrated regions of intense biological activity.

```
C       MAT1 ONE DIMENSIONAL ECOSYSTEM
        PROGRAM MAT    (INPUT,OUTPUT,TAPE5=INPUT,TAPE6=OUTPUT)
        DIMENSION S1(103),S2(103),B(103),DB(103),SUM(5)
      1 ,DS1(103),DS2(103)
        REAL K1, K2
        N=1000
        V=0.002
        K1=.0001
        K2=.0001
        Y1=1.
        Y2=1.
        DA1=0.4
        DA2=0.4
        S10=.1
        S20=.1
        KK=20
1       FORMAT (11F11.6)
2       FORMAT (1I11,7F11.6)
3       FORMAT (1H1)
        WRITE (6,3)
        WRITE (6,1)((L-2)/100.,L=2,102,10)
        DO 50 I=2,102
        B(I)=.000001
        S1(I)=S10
50      S2(I)=S20
        S1(1)=S10
        S1(103)=0
        S2(1)=0
        S2(103)=S20
        B(1)=0
        B(103)=0
        DO 80 K=1,KK
        DO 90 J=1,N
        DO 100 I=2,102
        DEL=V*B(I)*S1(I)*S2(I)/((K1+S1(I))*(K2+S2(I)))
        IF (Y1*S1(I).GT.Y2*S2(I)) GO TO 102
        IF (DEL.GT..1*Y1*S1(I)) GO TO 103
        DB(I)=DEL
        GO TO 109
103     IF (DEL.GT.15.*Y1*S1(I)) GO TO 104
        DB(I)=Y1*S1(I)*(1.-EXP(-DEL/(Y1*S1(I))))
        GO TO 109
104     DB(I)=Y1*S1(I)
        GO TO 109
102     IF (DEL.GT..1*Y2*S2(I)) GO TO 105
```

Figure 15. FORTRAN program to simulate a one-dimensional ecosystem.

7. Efficiency of Uptake Systems

In discussing the limiting efficiency of microorganisms sequestering resources and in order to pinpoint key questions one at a time, I will choose my cases. First, I will neglect the question of motility and ask, "What can an organism do to grow in environments where the nutrient concentration is 'vanishingly small'?" Many organisms need to survive under conditions of feast alternating with conditions of famine (Koch, 1971); they have learned to achieve resting states (Morita, 1982, 1985) and are able to make rapid changes in metabolism. But that is quite different from the case of an organism

```
          DB(I)=DEL
          GO TO 109
105       IF (DEL.GT.15.*Y2*S2(I)) GO TO 107
          DB(I)=Y2*S2(I)*(1.-EXP(-DEL/(Y2*S2(I))))
          GO TO 109
107       DB(I)=Y2*S2(I)
109       DS1(I)=-(DB(I)/Y1)+DA1*(S1(I-1)-2*S1(I)+S1(I+1))
100       DS2(I)=-(DB(I)/Y2)+DA2*(S2(I-1)-2*S2(I)+S2(I+1))
          DS1(102)=DS1(102)+DA1*(S1(102)-S1(103))
          DS2(2)=DS2(2)-DA2*(S2(1)-S2(2))
          DO 110 I=2,102
          B(I)=B(I)+DB(I)
          S1(I)=S1(I)+DS1(I)
          S2(I)=S2(I)+DS2(I)
          IF (S1(I).LT.0.) S1(I)=0.
110       IF (S2(I).LT.0.) S2(I)=0.
90        CONTINUE
          DO 151 J=1,5
151       SUM(J)=0.
          DO 152 J=1,5
          DO 152 M=2,102
152       SUM(J)=SUM(J)+B((M-1))*((M-1)/100.)**(J-1)
          X=SUM(1)
          DO 153 J=2,5
          SUM(J-1)=SUM(J)/X
153       VAR=SUM(2)-SUM(1)**2
          SKEW=(SUM(3)-3*SUM(2)*SUM(1)+2.*SUM(1)**3)/VAR**1.5
          XKURT=(SUM(4)-4.*SUM(3)*SUM(1)+6.*SUM(2)*SUM(1)**2-3.*SUM(1)**4)/
      XVAR**2
          SD=VAR**0.5
          XL=LOG10(X)
          CV=SD/SUM(1)
          WRITE (6,2) K*N,X,SUM(1),SD,SKEW,XKURT,XL,CV
          WRITE (6,1) (B(L),L=2,102,10),(S1(L),L=2,102,10),
      1 (S2(L),L=2,102,10),(LOG10(B(L)),L=2,102,10)
          WRITE (6,2) (K-1)*N
          WRITE (6,1)((L-2)/100.,L=47,57)
          WRITE (6,1) (B(L),L=47,57),(S1(L),L=47,57),(S2(L),L=47,57)
      1 ,(LOG10(B(L)),L=47,57)
80        WRITE (6,3)
          STOP
          END
```

Figure 15. (*continued*)

growing under chemostat-type (steady-state) conditions for a very long time. These are conditions in which an organism grows in free suspension under constant conditions.

Although that condition is the goal of the experimenter, it is rarely the outcome because of the inevitability of wall growth. With evolutionary time available, mutation and selection would lead, and have led, to organisms capable of adhering to surfaces. An especially favored site is close to the inlet where the nutrient enters in the system. But, again, let us not consider this case at this point either; rather, let us consider an organism ideally suited to a free-living habit of growth in an environment with a continuous but infinitesimal level of a needed nutrient (say, a carbon and energy source).

It is the common conjecture that such an environment would produce oligotrophs. The *a priori* property of such organisms would be that they somehow achieve a high

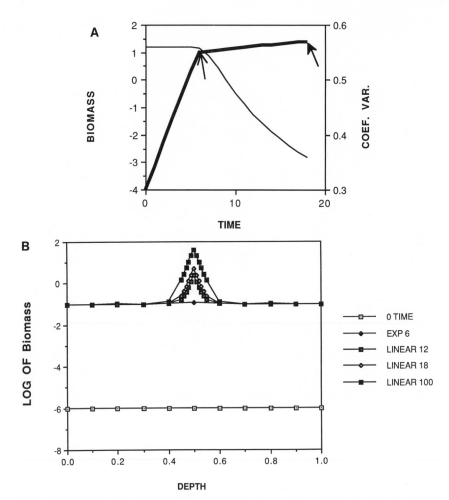

Figure 16. Development of a zone of bacterial growth. Part of the output of the program shown in Fig. 15 is represented. (A) Increase in biomass of a lake when temperature conditions become favorable. For simplicity, it is assumed that the maximum growth rate is constant. It is also assumed that at the initial time, the substrates are available at a constant concentration throughout the lake and that, similarly, residual bacteria are uniformly distributed throughout the water column. This is to simulate the growing season of a lake that has "turned over" during the previous winter. Also shown is the coefficient of variation of the spacial distribution of bacteria in the water column. (B) Biomass distribution initially (0 time), at the end of the exponential phase (exp 6), and into the linear phase where diffusion of the substrates is limiting to growth (linear 12, linear 18, and linear 100). Numbers correspond to growth times. See text for definitions and explanation of the parameters used in the simulation. The numerical values of the parameters are given by the negatively indented steps in the program listed in Fig. 15. Most parameters that have been adopted are stated in lines 6 to 16 of the program can be defined as follows: N, number of time steps per reported cycle; V, maximum growth rate; K_1, K_m of substrate 1; K_2, K_m of substrate 2; Y_1, yield of substrate 1; Y_2, yield of diffusion constant of substrate 2; S_{10}, source concentra-

surface-to-volume ratio. They would either be very small or, alternatively, be of complex shape, such as the prosthecate bacteria, which have one or more appendages that provide extra absorptive surface. I assume that such structures would require extra genetic and metabolic capabilities and thus increase the basic size of such organisms unless some other metabolic capabilities, or potential capabilities, present in other types of bacteria were deleted. Such appendages serve the same role as does the basic structure of plant cells. Many plant cells are formed of a thin layer of protoplasm covering a large vacuole. It is noteworthy that many eucaryotic algae also employ this strategy to achieve a high surface-to-cytoplasmic-volume ratio. With both the prosthecate and the large central vacuole strategies, the cells "look" bigger from a diffusion point of view than they physically are. But again, that is not the question at hand.

There is a more direct approach to increase growth rate at low substrate concentration: increase the uptake capacity. The number of uptake sites per unit area of membrane may in principle be increased simply by mutating a regulatory gene or duplicating a structural gene, genes, or entire operon. Since such mutations are quite common in many systems, this is a reasonable general expectation. On the negative side, too many permease systems of one kind may crowd out other essential membrane components. There may be other negatively selective aspects of having too many uptake sites; e.g., a momentary excess of nutrient in the environment may be toxic because the uptake system leads to a surfeit of substance internally.

Still, one might expect that microorganisms from nature, particularly from clean aquatic environments, would be highly evolved for uptake from water containing very low levels of nutrient. But there is a point beyond which more and better premeases would not aid the cell even if the extra permeases were not detrimental. This point is determined by the maximum rate at which nutrients can diffuse up of a cell from the bulk environment. We can view the cell as spherical because whether it is actually spherical or not. A single cell, without or without prosthecae, in aqueous suspension would be periodically subjected to rotation by Brownian motion and/or by chemotaxis mechanisms. This would give the effect of a spherical organism of a somewhat larger surface area. It is to be clearly noted that in an open environment, motility is not useful in increasing the rate of uptake. No organism can swim fast enough to beat the speed of diffusion of low-molecular-weight substances (Koch, 1971; Berg and Purcell, 1977; Kelly et al., 1988).

Ten years ago, I considered the question of diffusion limitation of growth and designed an experiment and a theory to test the question formulated as follows: How close to the physical limit of diffusion is any real organism? The approach (Koch, 1979,

---

tion of substrate 1; S_{20}, source concentration of substrate 2; and KK, the number of cycles followed. Steps 21 through 25 of the program establish the initial concentrations of bacteria, B, and the substrates in the water column. For the case chosen here, the bacterial concentration was uniform (given in line 23 of Fig. 15) and the substrate concentrations of both were the same as for the continuing sources. In the earliest phase, growth occurs throughout the column and is exponential; later growth occurs only in the middle of the column, limited by diffusion of the substrates into this region.

1982a,b; Koch and Wang, 1982) was as follows: (1) growth was measured very accurately and sensitively by a computer-linked, very stable, double-beam spectrophotometer; (2) the information was analyzed by a sophisticated computer program that statistically tested and compared many kinetic patterns for the uptake of nutrient at low concentration; and (3) the data were fitted to the physical theory for the diffusion-limited uptake of nutrient, given the morphometric aspects of the cells.

The apparent deficiency of my experiments was that I used a strain of *Escherichia coli* because I was familiar with its properties. Of course, I should have used a true oligotroph, as any "born-again" microbial ecologist would. Therefore, I viewed my experiments as prototype experiments. Part way through these experiments, I came across a paper by Matin and Veldkamp (1978), who had isolated two organisms from nature selected to be those that grew best in chemostat culture at low and at very low dilution rates. They had measured various morphometric and uptake parameters. From these measurements, I found that my laboratory strain of *E. coli*, ML308, tacitly assumed to be a "copiotroph," was more efficient than either their *Spirillum* sp. or *Pseudomonas* sp. So I continued my prototype experiments. [Only years later did I see the pond from which the isolation was made. It was sufficiently eutrophic that I became, and am, discouraged. Only when my experiments are repeated with a *bona fide* oligotroph (Poindexter, 1981) removed from nature for only a few generations will we know the essential nature of oligotrophy and even the reasons for being an oligophile.]

We found that if *E. coli* was grown with a low level of glucose as the carbon and energy source and was not allowed to go into stationary or lag phase for many generations, it exhibited uptake kinetics following the Best equation (see Section 4.9) and did not fit as well the other plausible mechanisms (Koch and Wang, 1982). The Best equation applies to the case of combination of a straight Fickian diffusion barrier followed by a first-order, one-way, saturable uptake process.

I was able to establish that the Fickian process was due to the diffusion of glucose through the holes in the outer membrane formed by the porins. A kinetic analysis showed that the subsequent uptake (presumably through the cytoplasmic membrane) was slower than the diffusion-limited process, where the efficiencies would be $3D/R^2$. At this point, a carbon-limited chemostat was set up with an effective doubling time of 11.7 hr and operated for 34 days. The organisms present at the end of this run had evolved in the sense that the maximum growth rate in an excess of glucose was smaller.

[The reduction in growth rate is not unexpected. Previously, I had observed similar phenomenon with shorter-term chemostats (Koch, 1959). In both cases, the rate of growth became maximal in less than a minute when a sample from the chemostat was diluted into rich medium, but took many hours to increase to the growth rate that a glucose-grown batch culture of the same organism exhibited. I dubbed this effect the New Yorker phenomenon, after the well-known fact that if one stands on the corner of 42nd Street and Times Square and offers people free five-dollar bills, no native will take them. Similarly, these organisms failed for a long time to exploit an unlimited glucose supply at the rate they could. Anthropomorphically, it seems that they assume that the excess supply is only temporary and thus a fraud, whereas when supplied with a variety of nutrients supporting fast growth, they very rapidly adopt a mode of rapid growth.]

On the other hand, the initial dependence of growth rate on glucose concentration was much stronger. The analysis of the kinetics indicated that the outer membrane had become more permeable and that the cytoplasmic membrane had become more efficient. The former finding was probably accounted for by a mutational shift of the type of porin (the *ompC* product, for example, instead of the *ompF* product), although at that time I was unable to pinpoint precisely the change. It is also possible that the number of porins per cell was increased.

The second implication of these studies, that the cytoplasmic membrane had become more efficient at glucose uptake, implies that the cytoplasmic membrane had more or better PTS assemblies and/or additional transport systems for glucose. The final level was such that had it been possible to grow *E. coli* without its outer membrane, its uptake systems would have been so efficient that it would have been a "black hole" for glucose. Then any molecule of glucose getting within a few cellular radii from the cell would have been inevitably consumed and converted into protoplasm and products. These studies have not been extended and, as far as I know, *E. coli* ML308 after chemostat selection is the only proven "oligotroph" among microorganisms.

The important conclusion that derives from these experiments is that the outer membrane is a detriment to the cell under truly nutrient-limiting conditions. This implies that either such conditions are not of biological significance or, more likely, that the advantages of an outer membrane outweigh its disadvantages. Of course, the major advantage of the outer membrane is that it allows the cell to keep out unwanted antibiotics and detergents (such as bile acids) while permitting nutrients and waste products to pass.

8. Conclusions

The above text is very deficient; it does not cite any work in the field of microbial ecology! I hope this is acceptable to the reader, because my purpose has been to bring concepts of physics to the physiology of microbial growth. This review's job is to give people who already are facing ecological problems some perspective on limitation due to diffusion. Most have not heretofore needed to delve into what appears to be esoterica. The key points that come out of the analysis are as follows. (1) Diffusion suffices for transport in procaryotic ecosystems because of the geometry and scale of free-living cells and aggregates of cells. (2) Obstacles may be quite unimportant in impeding diffusion in certain circumstances because of the detailed consequences of the diffusion law in special geometries. (3) Geometry can be quite important in diffusion and microbial morphology is the result of evolutionary adaptation to optimize microbial shapes. (4) Diffusion of nutrients in the aqueous environment of a free-living cell or aggregate can be growth limiting, and aspects of cell physiology, particularly the structure and function of elements in the wall, is the result of adaptation to such limitation. (5) It is important to consider the total path for nutrient utilization, not just the cytoplasmic membrane-bound permeases.

But there is an even broader generalization. Geological surfaces define the regions of the world that provide excellent habitats for procaryotes. Consider a primitive pond or shallow sea at a time of the great evolutionary radiation: oxygen-producing photosynthesis has just been perfected, and tremendous blooms of organisms had started for the first time in the sunlit water column. The biomass and fossil biomass of the world were then sharply increasing. The oxidizing potential of the atmosphere was increasing. Sooner or later, organisms died and sank to the bottom, carrying with them essential trace minerals, rich supplies of carbon, nitrogen, and phosphorus, and reducing potential. This growth-and-death process created oligotrophic conditions in the water column but opened a new, potentially rich habitat on the bottom surface. Now the evolutionary forces were directed toward different goals in the column than on the surface at the bottom. On the latter, evolution led to the perfection of organisms able to adhere, extract nutrients, and live in the surface microbial ecosystem. In the clear waters, overhead light could pass and allow the surface ecosystem to be powered directly by photosynthesis. Lower regions in the mat liberated reducing potential, and upper layers carried out oxidizing processes. This division led to new niches in the new subhabitats. Such evolutionary strides resulted in two-dimensional microbial mats with complex and sophisticated interrelationships among members of the consortia. There were other situations that would also lead to mats, and there would be circumstances in which the organisms in mats would have changed to create the typical mushroom-shaped stromatolites. But the essential concept is that the planar geometry of the habitat with resources coming from above and below led to organisms well adapted to living in a thin layer where movement of cells and resources was important only on the vertical axis. This is the prototype of the systems of biofilm development (Characklis and Cooksey, 1983; Characklis, 1984), and it is part of the reason that many microbes have effective ways of adhering to inanimate and living surfaces (Caldwell and Lawrence, 1986). The more effective the constituent microbe inhabitants become in exploiting their parts of the mat ecosystem, the sharper become the layers where different organisms predominate. This leads to an ecosystem of tightly banded structure, which can be epitomized by the several millimeters of surface in tidal flats. To study such ecosystems, it has been necessary to develop a new technology of microelectrodes to have the necessary fine spatial resolution (Revsbech and Jorgensen, 1986). To reach an understanding of this or any type of microbial ecosystem, we must integrate diffusion of nutrients and the capabilities of enzyme processes into the biology to rationalize what the various organisms are doing.

ACKNOWLEDGMENT. This review was written because of a biofilm that caused me to lose my grip and be carried down a small waterfall. The result was adequate to immobilize me sufficiently and for a long enough period to organize the diverse body of material presented here. I hope that it is lucid enough to ensure that the subject of diffusion under different geometrical conditions becomes a subject incorporated into the teaching, philosophy, and research in microbial ecology.

References

Acker, G. K., 1964, Molecular exclusion and restricted diffusion process in molecular-sieve chromatography, *Biochemistry* **3**;723–730.

Adams, G., and Delbruck, M., 1968, Reduction of dimensionality in biological diffusion processes, in: *Structural Chemistry and Molecular Biology* (A. Rich and N. Davidson, eds.), pp. 198–215, W. H. Freeman and Co., San Francisco.

Berg, H. C., 1983, *Random Walks in Biology*, Princeton University Press, Princeton, N.J.

Berg, H. C., and Purcell, E. M., 1977, Physics of chemoreception, *Biophys. J.* **20:**193–219.

Best, J., 1955, The inference on intracellular enzymatic properties for kinetics data obtained from living cells, *J. Cell Comp. Physiol.* **46:**1–27.

Caldwell, D. E., and Lawrence, J. R., 1986, Growth kinetics of *Pseudomonas fluorescens* microcolonies within the hydrodynamic boundary layers of a surface microenvironment, *Microbial Ecol.* **12:**299–312.

Carslaw, H. S., and Jaeger, J. C., 1959, *Conduction of Heat in Solids*, Oxford University Press, Oxford.

Characklis, W. G., 1984, Biofilm development: A process analysis, in: *Microbial Adhesion and Aggregation* (K. C. Marshall, ed.), pp. 137–157, Springer-Verlag, Berlin.

Characklis, W. G., and Cooksey, K. E., 1983, Biofilms and microbial fouling, *Adv. Appl. Microbiol.* **29:**93–138.

Crank, J., 1975, *The Mathematics of Diffusion*, 2nd ed., Oxford University Press, Oxford.

Kelly, F. X., Dapsis, K. J., and Lauffenburger, D. A., 1988, Effect of bacterial chemotaxis on dynamics of microbial competition, *Microb. Ecol.* **14:**115–131.

Koch, A. L., 1959, The dynamics of coliphage plaque formation. I. Macroplaque experiments, *Virology* **8:**273–292.

Koch, A. L., 1960, Encounter efficiency of coliphage-bacterium interaction, *Biochim. Biophys. Acta* **39:**311–318.

Koch, A. L., 1966, The logarithm in biology. I. Mechanisms generating the log-normal distribution exactly, *J. Theor. Biol.* **12:**276–290.

Koch, A. L., 1967, Kinetics of permease catalyzed transport, *J. Theor. Biol.* **14:**103–130.

Koch, A. L., 1969, The logarithm in biology. II. Distributions simulating the log-normal, *J. Theor. Biol.* **23:**251–268.

Koch, A. L., 1971, The adaptive responses of *Escherichia coli* to a feast and famine existence, *Adv. Microb. Physiol.* **6:**147–217.

Koch, A. L., 1979, Microbial growth in low concentrations of nutrients, in: *Strategies of Microbial Life in Extreme Environments* (M. Shilo, ed.), pp. 261–279, Springer-Verlag, Berlin.

Koch, A. L., 1982a, Diffusion limit and bacterial growth, in: *Overproduction of Microbial Products* (V. Krumphanzl, B. Sikyta, and Z. Vanek, eds.), pp. 571–580, Academic Press, London.

Koch, A. L., 1982b, Multistep kinetics: Choice of models for the growth of bacteria, *J. Theor. Biol.* **98:**401–417.

Koch, A. L., 1983, The surface stress theory of microbial morphogenesis, *Adv. Microbiol. Physiol.* **24:**301–366.

Koch, A. L., 1985, The macroeconomics of bacterial growth, in: *Bacteria in Their Natural Environments* (M. M. Fletcher and G. D. Floodgate, eds.), pp. 1–42, Academic Press, London.

Koch, A. L., and Coffman, R., 1970, Diffusion permeations or enzyme limitation: A probe for the kinetics of enzyme induction. *Biotech. Bioeng.* **XII:**651–677.

Koch, A. L., and Wang, C. H., 1982, How close to the theoretical diffusion limit do bacterial uptake systems function? *Arch. Microbiol.* **131:**36–42.

Marshall, K. C., 1976, *Interfaces in Microbial Ecology*, Harvard University Press, Cambridge, Mass.

Matin, A., and Veldkamp, H., 1978, Physiological basis of the selective advantage of a *Spirillum* sp. in a carbon-limited environment, *J. Gen. Microbiol.* **105:**187–197.

Morita, R. Y., 1982, Starvation-survival of heterotrophs in the marine environment, in: *Advances in Microbial Ecology, Vol. 6* (K. C. Marshall, ed.), pp. 171–198, Plenum Press, New York.

Morita, R. Y., 1985, Starvation and miniaturisation of heterotrophs, with special emphasis on maintenance of the starved viable state, in: *Bacteria in Their Natural Environments* (M. M. Fletcher and G. D. Floodgate, eds.), pp. 111–130, Academic Press, London.

Nikaido, H., 1979, Nonspecific transport through the outer membrane, in: *Bacterial Outer Membrane* (M. Inouye, ed.), pp. 361–407, John Wiley & Sons, New York.

Nikaido, H., and Rosenberg, E. Y., 1981, Effect of solute size on the diffusion rates through the transmembrane pores of the outer membrane of *Escherichia coli, J. Gen. Physiol.* **11**:121–135.

Ogsten, A. G., 1958, The spaces in a uniform random suspension of fibres, *Trans. Faraday Soc.* **54**:1754–1757.

Ou, L.-T., and Marquis, R. E., 1970, Electromechanical interaction on cell walls of gram-positive cocci, *J. Bacteriol.* **101**:92–101.

Poindexter, J., 1981, Oligotrophy: Fast and famine existence, in: *Advances in Microbial Ecology,* Vol. 5 (M. Alexander, ed.), pp. 63–89, Plenum Press, New York.

Powell, E. O., 1967, The growth rate of microorganisms as a function of substrate concentration, in: *Microbial Physiology and Continuous Culture* (E. O. Powell, C. Evans, R. E. Strange, and D. W. Tempest, eds.), pp. 34–55, Her Majesty's Stationery Office, London.

Preiss, J. W., and E. Pollard, 1960–61, Localization of β-galactosidase in cells of *Escherichia coli* by low voltage electron bombardment. *Biophys. J.* **1**:429–435.

Purcell, E. M., 1977, Life at low Reynolds number, *Am. J. Phys.* **45**:3–12.

Rashevsky, N., 1960, *Mathematical biophysics,* Vol. 1, 3rd rev. ed., pp. 1–148, Dover, New York.

Renkin, E. M., 1954, Filtration, diffusion, and molecular sieving through porous cellulose membranes, *J. Gen. Physiol.* **38**:225–243.

Revsbech, N. P., and Jorgensen, B. B., 1986, Microelectrodes: Their use in microbial ecology, in: *Advances in Microbial Ecology,* Vol. 9 (K. C. Marshall, ed.), pp. 293–252, Plenum Press, New York.

Rodbard, D., 1974, Estimation of molecular weight by gel filtration and gel electrophoresis, in: *Methods of Protein Separation,* Vol. 2 (N. Catsimpoolas, ed.), pp. 145–180, Plenum Press, New York.

Valentine, R. C., and Allison, A. C., 1959, Virus particle adsorption. I. theory of absorption and experiments on the attachment of particles to nonbiological surfaces, *Biochim. Biophys. Acta* **34**:10–23.

von Smoluchowski, M., 1918, Versuch einer mathematischen Theorie der Koagulationskinetik kolloider Losungen, *Zeitschr. Phys. Chem.* **92**:129–168.

3

Ecological Aspects of Antarctic Microbiology

DAVID D. WYNN-WILLIAMS

1. Introduction

If the science of microbiology is approaching maturity, then Antarctic microbiology is only just emerging from its infancy. The early expeditions of the 20th century used classical medical methodology to isolate and identify bacteria, yeasts, and fungi from sea water, soil, snow, air, and animals (Ekelöf, 1908; Tsiklinsky, 1908; Gazert, 1912; McLean, 1918, 1919). The initial emphasis was on survey and taxonomy, although Gazert (1912) noted the influence of marine bacteria on nutrient cycling during the German Antarctic Expedition of 1901–03. However, it is Ekelöf of the Swedish National Antarctic Expedition 1901–03 who may be regarded as the father of Antarctic microbial ecology. Between February 1902 and November 1903, he made a seasonal study of the soil and air microbiota at Snow Hill Island (64° 30′S) off the east coast of the Antarctic Peninsula (Fig. 1). Using rich medical media, he monitored viable bacteria, yeasts, and other microfungi but made no mention of the organisms resembling cyanobacteria and microalgae which are frequently the dominant primary producers in terrestrial Antarctic ecosystems (Ekelöf, 1908).

Studies of marine cyanobacteria and microalgae predate and parallel those of the heterotrophs (Hooker, 1847; West and West, 1911; Fritsch, 1912). Consideration of the Antarctic marine phytoplankton is outside the scope of this review, but the cyanobacteria and microalgae are a feature of the Antarctic microbiota, occurring in virtually all habitats where there is free water. A greater predominance of cyanobacteria in the Antarctic relative to Arctic benthic and soil ecosystems was first recorded by McLean (1918) of the 1911–14 Australasian expedition. He also noted that heterotrophic soil

DAVID D. WYNN-WILLIAMS • British Antarctic Survey, Natural Environment Research Council, Cambridge CB3 OET, United Kingdom.

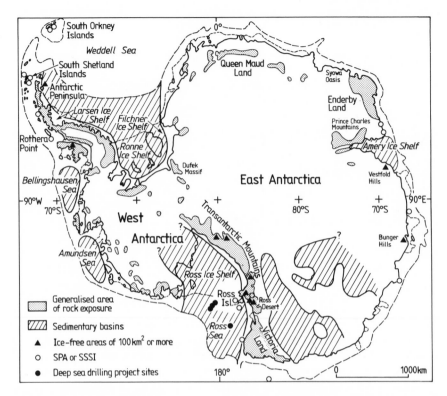

Figure 1. Ice-free land and sedimentary basins of Antarctica, showing SPA and SSSI locations and Deep Sea Drilling Project sites. [Redrawn from Elliot (1985), with permission.]

bacteria could not only survive but multiply at the low temperatures (1 to 2°C) prevailing at Commonwealth Bay (67°S). However, Ekelöf (1908) had earlier found an optimal temperature of 17.5°C for bacteria isolated from Antarctic soils that are warmed by insolation. These early expeditions laid the foundations for ecological studies of psychrotrophy, colonization, and nutrient cycling in Antarctic microbial ecosystems which are still of relevance.

After the "heroic age," there was a lull in Antarctic microbiological research until more coordinated expeditions were instigated (Roberts, 1958). However, it was not until the 1939–41 United States expedition to the Bay of Whales on the Ross Ice Shelf that the next microbial monitoring of air, snow, and soil was carried out (Darling and Siple, 1941). Antarctic air transport was by then more reliable, enabling collections from as far south as the Ford Range (78°S) and Scott Glacier (86°S) in the Transantarctic Mountains. Despite the instigation of a comprehensive biological research program in the South Orkney Islands, the Falkland Islands Dependencies Survey (later to become the British Antarctic Survey) did not include microbiological projects until 1963 (Heal *et al.*, 1967; Latter and Heal, 1971; Bailey and Wynn-Williams, 1982).

The microbiological activities of other nations during this period were similarly sporadic (Bryant, 1945; Prévot and Moureau, 1952; Hasle, 1956, 1969). Impetus for coordinated research was provided by the International Geophysical Year (IGY) in 1957–58, which led to the construction of many permanent Antarctic stations (Roberts, 1958), some with facilities suitable for microbiological studies. The United States McMurdo Station on Ross Island (77° 55'S) provided air support for work in the climatically and edaphically extreme ecosystems of Ross Island and the McMurdo Dry Valleys region (henceforth referred to as the Ross Desert) of southern Victoria Land. Soil samples from this region soon revealed bacteria and yeasts tolerant of low temperatures, extreme drought, high salinity, and oligotrophic conditions (Flint and Stout, 1960; di Menna, 1960; Straka and Stokes, 1960).

As more became known about the unique meromictic lakes of the Ross Desert and Enderby Land (Armitage and House, 1962; Burton, 1981), their microbiota received increasing attention. However, the work of this period was mainly a taxonomic survey of bacteria (Boyd, 1962; Meyer, 1962; Margni and Castrelos, 1963; Sieburth, 1963), yeasts (Soneda, 1961; Meyer, 1962), microfungi (Tubaki, 1961; Harder and Persiel, 1962; Corte and Daglio, 1963), and algae plus cyanobacteria (Drouet, 1961, 1962). Ecological studies of processes such as bacterial S cycling (Barghoorn and Nichols, 1961) and N_2 fixation by bacteria (Boyd and Boyd, 1962) and cyanobacteria (Holm-Hansen, 1963) were still not integrated into coordinated programs. Microbiological research up to this period has been reviewed by Sieburth (1963, 1965).

As the species lists became more detailed, the emphasis changed to microbial ecology (Boyd et al., 1966; Baker, 1970a,b) and ultimately to ecophysiology, biochemistry, and biophysics. In recent years, microbiological studies have become integrated into ecological programs such as the 10-year moss–peat ecosystem study at Signy Island, South Orkney Islands, in the maritime Antarctic (Davis, 1981) and the Antarctic Cryptoendolithic Microbial Ecosystems (ACME) research program in the Ross Desert (Friedmann, 1982). Long-term programs permit analysis of seasonal trends, including wintering strategies, and interactions between microbial groups and their grazers. In such terrestrial and freshwater microbiological programs, data have been obtained by overwintering scientists (Wynn-Williams, 1980; Ellis-Evans, 1982) and by satellite-linked automatic microclimate-monitoring stations in remote regions of the Ross Desert (McKay and Friedmann, 1984). A comprehensive overview of Antarctic microbial ecosystems has been prepared by Vincent (1988).

Five major influences have promoted studies of microbial ecology during the last 25 years:

1. The development of life detection systems and quarantine parameters for the U.S. Viking Mars probe, using the Ross Desert as a testing ground (Horowitz et al., 1972; Vishniac and Mainzer, 1973; Cameron et al., 1976).
2. Environmental impact on pristine Antarctic terrestrial and freshwater (or saline) habitats during activities such as the Dry Valleys Drilling Project (DVDP) (Cameron, 1972a; Parker, 1978), the construction of stations (Wynn-Williams, 1985a), and the possible exploitation of oil resources (Konlechner, 1985).

3. Interest in the fundamental, unifying physiological principles of freezing tolerance, cryopreservation, anhydrobiosis, and psychrophily in microorganisms
(Morita *et al.*, 1977; Vishniac and Hempfling, 1979a; Herbert, 1986).
4. The discovery of the ecophysiological significance of sea–ice microbial communities for generating biomass and inoculating inshore and oceanic waters of
the Southern Ocean (Palmisano and Sullivan, 1983, 1985a,b; Laws, 1985).
5. Increasing awareness of the value of Antarctica for testing fundamental ecological principles in a relatively unpolluted "natural laboratory" with uniquely
simple ecosystems. Extreme examples include the endolithic microbial ecosystems of the Ross Desert and the terrestrial food pyramid at Signy Island, at
whose apex the mesostigmatid mite *Gamasellus racovitzai* is the sole predator
(Lister, 1984).

In this review, I have been selective and have focused on aspects of microbial
ecology considered to be somewhat unusual and distinctive to the Antarctic region.
Vincent (1988) comprehensively reviews the microbial ecosystems of the region, while
microbial ecology is included in reviews of Antarctic terrestrial (Boyd *et al.*, 1966;
Cameron, 1971; Block, 1984), freshwater (Heywood, 1977, 1984, 1987), saline (Wright
and Burton, 1981), and marine habitats (Sieburth, 1965; Heywood and Whitaker, 1984;
Horner, 1985a,b; Garrison *et al.*, 1986). Advances in the study of microbial ecosystems
of the land, freshwater, and sea (excluding the oceanic phytoplankton) will be discussed
here with reference mainly to bacteria, cyanobacteria, microalgae, yeasts, and microfungi. However, as with all natural systems, there is an overlap of habitats, strategies,
and populations which contributes to the dynamics and resilience of the Antarctic
ecosystem.

2. Why Are Antarctic Habitats Distinctive?

2.1. Continental Drift and the Isolation of Antarctica

The drifting apart of the tectonic plates bearing the continental land masses comprising the supercontinent of Gondwana in the Southern Hemisphere (Fig. 2) opened up
the vast expanse of the Southern Ocean. Elliot (1985) has summarized a probable
sequence of these events, although this interpretation is not universally accepted (Mercer, 1983): As the continent drifted southwards from the tropics, a major cooling of
Antarctic surface waters occurred about 38 million years ago (38 Ma), resulting in the
first formation of sea ice in the region. The Drake Passage opened up between South
America and the Antarctic Peninsula at about 23 Ma, permitting free flow of the wind-
driven circumpolar surface currents of the Southern Ocean. These are easterly at high
latitudes and westerly at lower latitudes ca. 60°S (Fig. 3). One tentative model indicates
that ice began to accumulate on the land mass by about 30 Ma and to calve into icebergs
by 25 Ma, but it was not until about 15 Ma that the Antarctic ice sheet reached its
maximum size. By 7 Ma it had become cold enough for massive floating ice shelves,
such as the Ross Ice Shelf, to form over parts of the ocean. The increased albedo

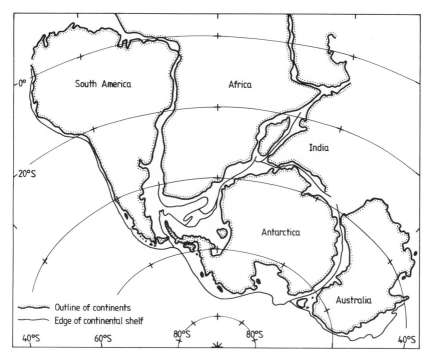

Figure 2. The break-up of Gondwana in about 180 Ma and the translocation of Antarctica. [From Elliot (1985), with permission.]

(reflectivity) caused by the ice mass perpetuated the cold climate and expansion of the ice cover, and it was only 10,000 years ago that recession of the ice started to reexpose the peri-Antarctic islands in the Southern Ocean and the periphery of the land mass. This continental movement results in the incongruity of a fossil tree on Mt. Fleming (77° 30'S) at the edge of the polar plateau. The biota inhabiting the region today is therefore the result of relatively recent long-range recolonization, which undoubtedly accounts for some of the distinctive species distribution. The present frigid climatic regime belies the tropical origin of the continental land mass. A legacy of this geological disruption is the occurrence of more than three major volcanoes in Antarctica: Mt. Erebus on Ross Island, Mt. Melbourne in northern Victoria Land, and Deception Island in the South Shetland Islands. All three support an unusual combination of thermophilic and psychrotolerant microorganisms and provide invaluable testing grounds for hypotheses of long- and short-range dispersal.

2.2. The Ice Habitats of Antarctica

Unlike the Arctic, which has a relatively small, floating ice cover and is therefore essentially a marine environment, Antarctica is a rocky continent of ca. 12.38 million km^2 nearly covered by a massive ice sheet 11.97 million km^2 in extent. It has the highest

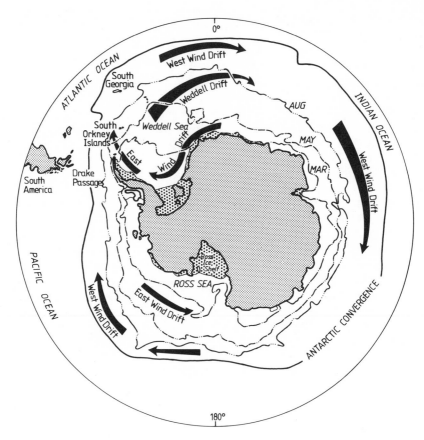

Figure 3. Currents of the Southern Ocean and the annual expansion of sea ice cover. [Redrawn from Foster (1984), with permission.]

mean elevation (2,300 m) of any continent and the lowest temperature ($-89.6°C$, recorded in July 1983). The vast area of ice reflects 40–90% of the incident solar radiation, thereby sustaining the low-temperature regime and causing a mass of cold, dense air to accumulate on the polar plateau and flow outward toward the sea as strong katabatic winds, occasionally reaching >160 km/h. These winds tend to isolate Antarctica from many of the influences of the more northerly continents. This isolation is accentuated by the cold waters of the Southern Ocean, which are cooled by the peripheral ice mass, the katabatic winds, and the fluctuating sea–ice. The net effect of this heat sink is to preserve a distinct easterly circulating circumpolar ocean of 35 million km², the dynamics of which are described by Foster (1984). These cold seas meet the warmer waters of the more northerly oceans at latitude ca. 50°S at a relatively well defined front called the Antarctic Convergence (Fig. 3) (Deacon, 1964).

The isolating effect of the Antarctic Convergence is further accentuated by the large distances to the nearest major land masses—1,000 km to South America and over 4,000 km to South Africa and New Zealand. Moreover, the low temperatures of the katabatic winds, the high albedo of the ice sheet, and the small amount of insolation in winter (none at all for several months at locations significantly south of the Antarctic Circle) combine to result in the freezing of the sea at ca. $-1.8°C$ to form fast ice, which may then become thickened and distorted to become mobile pack ice. This annual, self-accelerating freezing cycle expands the area of the sea ice from ca. 4 million km^2 in summer to ca. 20 million km^2 in winter. This expansion, in turn, radically alters the climate of the Antarctic islands from wet maritime to dry continental as they become surrounded by sea ice. It results in the most extreme example of seasonality in the world, both in the sea and on land (Phillpot, 1985).

The pack ice dampens the oceanic swell, resulting in the settling out of particulate material, both living and nonliving. This, in turn, improves the clarity of the water, but at the same time the thick ice cover greatly decreases the penetration of photosynthetically active radiation (PAR; wavelengths of light usable by plants, 400–700 nm) into the water column. The ice layer formed on freshwater and meromictic saline lakes has a similar stabilizing effect, accentuated by the elimination of turbulent inflow by the freeze-up of meltstreams in winter.

Water bodies have a thermal buffering capacity by virtue of their latent heat. However, the interface of the water and land with the cold winter air at temperatures significantly below the freezing point of water ($-30°C$, even in the maritime Antarctic) is a zone of great thermal stress. This, in turn, may result in great osmotic stress as ice freezes out, leaving microsites of concentrated salt solutions at temperatures well below freezing. This phenomenon is most noticeable in sea ice (Palmisano and Sullivan, 1983) but also occurs on a smaller scale in freshwater ice and in soil (Chambers, 1967). Thus, transient new microhabitats are created seasonally—ice crystal surfaces, which can be colonized by microorganisms, and supercooled brine channels or pockets that select for psychrotolerant, halotolerant species.

In winter, the surface of the ice itself on land, lakes, and sea is essentially devoid of the free water necessary to support a microbiota. Therefore, paradoxically, the surface of sea ice has the characteristics of a cold desert until the spring thaw. However, a phototroph-dominated sea ice microbial community (SIMCO) (Palmisano and Sullivan, 1983) develops. A similar community develops in Antarctic freshwater lakes, although the ice is more compact because of the lower concentration of salts in the surface water of all but the most saline lakes. On land, a greenhouse effect caused by black-body radiation occurs at the ground–ice interface. However, this occurs only in late winter because of the lower heat retention capacity of rock or even wet soil relative to that of a water body. The ice sheet of the polar plateau is permanently too cold to permit the occurrence of free water in all but the most sheltered sites adjacent to protruding rocky nunataks. However, deep within the ice sheet and permafrost, viable microorganisms have been detected, preserved essentially anhydrobiotically (Cameron et al., 1974; Abyzov et al., 1982). In contrast, the glaciers flowing from the plateau in sheltered valleys and at lower altitude to the sea occasionally bear localized pockets of meltwater, cryoconite pools (Wharton et al., 1981, 1985), produced by the black-body radiation of superficial windblown debris.

The ice on land, lakes, and sea therefore provides a potential habitat for a variety of microorganisms, which in turn act as the basis for a food web initiated primarily during the spring thaw, when the large ice resource melts to provide free water. There are, however, large areas of exposed rock and soil on nunataks, inland mountain ranges, and certain coastal valleys of the continent where the snow and ice rarely melts but sublimes because of the dryness of the katabatic winds. Free water is either scarce or localized to streams, ponds, or lakes. These areas are the Antarctic cold deserts, the best researched of which is the Ross Desert region of the McMurdo Dry Valleys (Fig. 1).

2.3. The Cold Deserts of Continental Antarctica

Despite ca. 90% of the world's water constituting the south polar ice sheet, cold desert regions, such as the Ross Desert (McCraw, 1967) and the Vestfold Hills (Pickard, 1986), constitute extensive ice-free areas of valleys and ridges fringing the polar plateau (Fig. 1). The relative humidity of the Ross Desert is often <10% in winter, when the katabatic winds are strongest (Thompson et al., 1971). In summer, gentler, moister onshore winds prevail, but precipitation is scarce and never falls as rain. The snowfall, averaging 5–10 cm yr^{-1}, rarely accumulates on the valley floors because the potential sublimation rate is ca. 50 cm yr^{-1} (Harris and Cartwright, 1981). Air temperatures range from −55°C in winter to +10°C in summer, with an annual mean of ca. −20°C and a ca. 2-month growing season for the cryptogamic flora. However, snow accumulates on the surrounding mountain tops, supplying melt streams that feed the seasonal rivers of the region.

The desert surface biota is subjected to prolonged periods of desiccation, hypersalinity, transient and diurnal freeze–thaw cycles, and prolonged low temperatures during winter. Moreover, the katabatic winds disrupt the desert soil and sandblast the rocks. Anhydrobiotic conditions occur on the nanoscale due not only to an absolute shortage of water but also to the freezing out of ice and the accumulation of salts. All three of these influences act on the cell membranes as osmotic stress and ionic imbalance in addition to the dehydration of biologically vital molecules and phase changes in membrane lipids. Such rigorous conditions severely limit the ability of microorganisms to colonize the region (Wynn-Williams, 1986). Microbial community development is mainly restricted to three types of habitat: endolithic communities inside rocks (Friedmann and Ocampo, 1976); freshwater communities in transient pools, streams, and rivers and their peripheries (Vincent and Howard-Williams, 1986a), including epiphytes on cryptogamic plants; and meromictic, hypersaline, ice-covered lakes (Heywood, 1977; Burton, 1981; Parker and Simmons, 1985). Each of these habitats supports a microbiota that requires distinctive ecophysiological strategies for survival.

The endolithic habitat was described originally for porous rocks in hot deserts such as the Negev (Friedmann, 1971, 1980). Its communities have been defined by Golubic et al. (1981) as chasmoendolithic if found in fissures and cryptoendolithic if located in the structural cavities within the rock. In the Ross Desert it is mainly represented by light-colored porous Beacon sandstone (Friedmann and Ocampo-Friedmann, 1984a). Chasmoendoliths are more frequent in moister regions such as coasts and Ross Desert valleys oriented at right angles to the katabatic winds. The characteristics of the Antarctic endolithic habitat have been summarized by Friedmann and Ocampo-Friedmann (1984a) and represent a habitat affording conservation of moisture, adequate illumination for primary production, thermal buffering against freeze–thaw cycles, and protection from abrasion. Exfoliation of the community may be regarded as a microbial inoculum for the surrounding soil. The exposure of originally protected endolithic cells requires their rapid penetration of the rock pores if they are to survive. The chasmolithic habitat comprises colonized fissures in sandstone and other translucent rocks such as frost-shattered marble. Associated sublithic habitats beneath translucent stones are based on similar protective principles (Broady, 1981a).

The rigors of physical and chemical weathering processes in cold desert soils are analogous to, although less extreme than, weathering processes detected on Mars (Gibson *et al.*, 1983), which had earlier encouraged the testing in the Ross Desert of microbiological concepts, methods, and equipment for the Viking Mars program of 1976 (Horowitz *et al.*, 1972). The cyclic heating and cooling of the Ross Desert soil results in patterned ground delimited by sand wedges (Berg and Black, 1966). Despite the presence of permafrost, frost polygons, which are defined by ice wedges and occur extensively in the moister maritime Antarctic (Chambers, 1967), are not evident in the Ross Desert.

As a result of the upward translocation of salts from the substratum by capillary action (Ugolini and Anderson, 1973) and the gradual transfer of seaspray carried inland by onshore winds over hundreds of years (Nakaya *et al.*, 1979), salts accumulate at the surface of cold desert soils. This process is accentuated by the desiccating effect of katabatic winds, frequently resulting in dense white accretions on rock and soil surfaces. Unlike the case in the maritime Antarctic, the absence of rainfall precludes leaching.

Although not strictly part of a cold desert system, the volcanic ash of regions such as the high slopes (>3,500 m) of Mt. Erebus on Ross Island and Mt. Melbourne in northern Victoria Land, and low-lying areas (5–500 m) of Deception Island (South Shetland Islands) in the maritime Antarctic, may also be locally dry despite steam moistening. Volcanic ash is very porous and requires a plentiful supply of steam, precipitation, meltwater, or seaspray to sustain a microbiota (Cameron and Benoit, 1970).

It is evident that it is not only the low temperature that regulates terrestrial microbial survival and growth in this region but also restricted availability of water, associated osmotic stress, and substratum instability. Littoral communities may be moistened for only a few days per season, depending on the input of meltwater (Wynn-Williams, 1985b), necessitating a rapid physiological response to available water. However, there are many transient streams derived from ice and snow melt and some substantial season-al rivers (Vincent and Howard-Williams, 1986b) such as the Onyx River flowing into Lake Vanda in the Ross Desert (Chinn, 1981), the Talg and Tierney Rivers in the Vestfold Hills (Adamson and Pickard, 1986), and the river complex flowing into Brandy Bay (64°S 58′W) on James Ross Island, Antarctic Peninsula (M. R. A. Thomson, personal communication). The seasonal ebb and flow of meltwater is a feature of continental Antarctic desert rivers, which ultimately drain, evaporate, or freeze solid and sublime each year. The Onyx River is particularly unusual since it flows inland from the coast into Lake Vanda, which does not fill up because of continuous ablation of ice from its frozen surface at a rate of ca. 30 cm yr^{-1} (McKay et al., 1985) by the dry katabatic winds. Maritime Antarctic rivers simply freeze.

The cold desert lakes of the Ross Desert and Vestfold Hills are usually ice covered unless exceptionally hypersaline. The ice minimizes turbulence, resulting in stratification (meromixis). They are frequently saline in at least part of the profile, depending on the surrounding topography and throughput of meltwater. The chemical balance of lakes near the coast, such as Lake Fryxell in Taylor Valley, Lakes Nurume and Hunazoko in the Syowa Oasis (Watanuki et al., 1977), and Ace Lake in the Vestfold Hills (Burton and Barker, 1979), is similar to that of seawater. However, others further inland in the Ross Desert have disproportionate accumulation of ions such as sodium, potassium, and chloride (Lake Bonney) or calcium and chloride, as in Lake Vanda and Don Juan Pond (Watanuki et al., 1977). Deep Lake and Don Juan Pond have such high total salts concentrations (28 and ca. 33%, respectively) that their freezing points are -18 and $-53°C$, respectively (Cameron et al., 1972; Torii et al., 1989). The chemistry, physics, and origin of Antarctic inland waters have been comprehensively reviewed by Heywood (1977, 1984, 1987), Burton (1981), and Torii et al. (1989).

The salinity of certain inland lakes distant from the coast is enigmatic. Nakaya et al. (1979) suggested that the salt accumulation occurred over a time scale of thousands of years. Sea salt particles were sprayed by sporadic onshore winds and precipitated onto the snow and ice of glaciers and lakes. The resulting meltwater poured into the dry valleys to produce very dilute lake water with an ionic ratio very similar to that of seawater. Evaporation and sublimation of this water resulted in accumulation of salts to concentrations at which ion exchange occurred between the water and sediment. At high salinities, insoluble calcium and magnesium in the rock dissolved as equivalent amounts of sodium and potassium became immobilized. After this phase of concentration, calcium sulfate, calcium carbonate, and then chlorides of sodium, magnesium, and potassium were deposited at increasingly lower temperatures, culminating in the saturated calcium chloride solution of Don Juan Pond (Craig et al., 1974; Watanuki et al., 1977). Finally, fresh meltwater from surrounding glaciers accumulated on top of the saline water and froze to form an ice cover that now helps to sustain stratification of salinity, temperature, and anoxia (Wharton et al., 1986).

The ice cover of freshwater and saline cold desert lakes regulates the percentage transmission of PAR, which governs the abundance and composition of the cyanobacterial and algal microbiota (Palmisano and Simmons, 1987). Solar heating of benthic sediments results in temperature inversions, with warm water (up to 25°C) at the bottom of lakes such as Lake Vanda. Moreover, stratification frequently results in gradients of dissolved oxygen, ranging in Lake Vanda from 400% supersaturation at 0°C immediately under the ice to anoxic conditions near the bottom sediment at ca. 70 m. It also results in gradients of inorganic and organic nutrients. Within such profiles there are distinctive cycles of nitrogen, phosphorus, and sulfur which are sustained in part by the microbiota.

The ice on large Ross Desert lakes is perennial except for a moat that melts round the shore during the summer. The ablation rate on some lakes is high, e.g., 1.5 m yr^{-1} from ice ca. 3.3 m thick on Lake Hoare, Taylor Valley. Blown sand accumulates on the clean ice and melts into it by local heating caused by black-body radiation. This probably continues as subsurface melting, resulting in surface ponding and translocation of the sand in meltwater. From the foregoing, it is clear that the extreme climatic and edaphic conditions of Antarctic cold desert regions provide a variety of unusual niches for both terrestrial and freshwater microorganisms.

2.4. Seasonal Changes in the Maritime Antarctic

Although seasonal changes affect the biota of the continental Antarctic, it is in the coastal continental and maritime Antarctic where the microbiota is most influenced by the onset of spring and winter because of the sudden availability of meltwater. The maritime Antarctic has been defined by Holdgate (1964) as having at least one summer month, with a mean air temperature above freezing point at sea level, and mean monthly winter temperatures that rarely fall below $-10°C$. Its growing season is ca. 5 months. The rapid transition from winter to spring is to a large extent initiated by the breakout of the sea ice, with a resulting drop in albedo, which accelerates a rise in air temperature and a wetter climatic regime with precipitation frequently falling as rain. As the snow cover melts, the greenhouse effect initiates the ground thaw, which is accelerated by rainfall and meltwater so that the transition from winter frigidity to summer microbial activity may occur within a single day (Wynn-Williams, 1980). Solutes in the soil water may result in a local thaw in microenvironments at temperatures below freezing point in advance of the main thaw, thereby extending the growing season. It is the point of freeze–thaw transition rather than the temperature itself which is critical for the onset of microbial activity. Such transitions paradoxically release organic nutrients from the microbiota itself and from the cryptogamic lichen, moss and liverwort flora, and invertebrate microfaunal communities. These communities develop especially well in the milder climate of the maritime Antarctic (Collins et al., 1975; Block, 1984; Smith, 1984b).

In coastal continental regions such as at Cape Bird, Ross Island, the scant moss flora is near the limits of its viability because of the shortness (ca. 2 months) of the growing season, the low mean air temperature, the frequency of deleterious freeze–thaw conditions, and the additional stress of desiccation and wet–dry cycles. Additional nutrients are often available from the coastal macrofauna, which includes penguins, other seabirds, and seals. Their large-scale activities create ornithogenic soils that accumulate large amounts of N and P excreted by the animals. On volcanic ash substrata, penguin rookeries are frequently the only sites of large microbial communities, due partly to increased nutrient availability and partly to reduced drainage (Speir and Cowling, 1984).

On some maritime Antarctic coasts and islands, the extensive area of land exposed supports substantial lakes devoid of ice cover in summer. Despite their proximity and consequent similarity of their climates, the temperature regimes of their marine, terrestrial, and freshwater environments differ markedly. Ground temperatures at Signy Island range seasonally from ca. $-20°C$ in winter to occasionally $>30°C$ in summer, with a midsummer mean of 3 to 4°C (Walton, 1982). The corresponding ranges for freshwater and coastal seawater are $-0.5°C$ to $+6°C$ (Heywood, 1984) and $-1.8°C$ to $+2°C$ (Tanner and Herbert, 1981), respectively.

The recession of ice in geological time has resulted in lakes at Signy Island that are thought to represent an evolutionary sequence (Priddle and Heywood, 1980). Proglacial lakes are still being formed by damming and ice recession. Lakes are oligotrophic where the substratum and catchment are poor in organic and inorganic nutrients. However, the influence of marine birds and mammals results in extensive eutrophication (Ellis-Evans, 1981b; Heywood, 1984). Such effects are very localized, and lakes with very different levels of eutrophication occur within 2 km of each other, permitting comparative nutritional and interactive studies within similar climatic regimes (Ellis-Evans, 1981a, 1985a,b; Ellis-Evans and Wynn-Williams, 1985).

In terrestrial maritime ecosystems, freeze–thaw cycles may occur throughout the summer, but the latent heat of wet ground makes their periodicity much longer than on the dry soils and rocks of the Ross Desert. Maritime lakes freeze over in winter, but most become ice-free in summer; as a result, winter stratification is disrupted annually by the input of meltwater and wind-induced turbulence (Ellis-Evans, 1982). The transition, often from anoxia to full aeration, permits a study of the response of the microbiota to extremes of E_h and nutrient availability, including the influence of the epontic phototrophic microbiota on the underside of the ice (Parker et al., 1982b; Vincent and Howard-Williams, 1985). The equivalent habitat in and under the sea ice has additional supercooled brine channels running vertically through the ice and into the underlying seawater (Palmisano and Sullivan, 1985b).

The bare mineral soil with a discontinuous cryptogamic vegetation, termed fellfield, is the dominant terrestrial environment. Continental fellfields are dry, rocky pavements or sand dunes (Miotke, 1985) or gritty volcanic ash, whereas maritime fellfields frequently have a high silt content and remain moist for much of the summer. The frost polygons and stripes of wetter ground are caused by the growth of ice needles in the soil profile. This results in upward movement of the soil surface and gravitational sorting of soil particles, with the larger components migrating to the periphery and dimensionally homogeneous fines in the center. Frost polygons and stripes are defined by ice wedges (Chambers, 1967) and may become bounded by cryptogamic vegetation.

Under favorable conditions of moisture, temperature, shelter, and nutrients, the vegetation becomes continuous to form carpets or turves of mosses and liverworts over peat, occasionally with two flowering plants: a grass, Deschampsia antarctica, and a herb, Colobanthus quitensis. In very sheltered areas, Deschampsia forms swards over a primitive brown earth soil as far south as Terra Firma Islands at 69°S (Smith and Poncet, 1987), with a correspondingly larger and more diverse microbiota (Wynn-Williams, 1982) and a better-developed heterotrophic population (Davis, 1981, 1986; Wynn-Williams, 1984). The plants support an epiphytic microbial population, and the microbiota supports an invertebrate fauna that includes protozoa (Smith, 1978), nematodes (Caldwell, 1981), and springtails and mites (Cannon and Block, (1988). Indigenous land vertebrates are absent from Antarctica, making the absence of macroherbivores a unique distinction from the Arctic. At the summit of the food pyramid at Signy Island is the mite Gamasellus racovitzai, whose prey diversity in the field is so restricted that its diet can be characterized in fresh specimens by gel electrophoresis of their respective esterases (Lister et al., 1987).

The relative simplicity of these ecosystems and their low species diversity relative to temperate soils makes them potentially vulnerable to environmental impact from humans (Parker, 1972, 1978; Walton, 1987) and other animals such as fur seals, whose population continues to escalate rapidly with destructive results after the cessation of depredations by sealers in the last century (Smith, 1988).

Five aspects of Antarctic microbial ecology will now be considered in more detail. These reflect many of the advances in approach and methodology which have been a feature of Antarctic microbiology in recent years.

3. Microbiology of Antarctic Cold Deserts

3.1. Endolithic Communities

Phototrophic and heterotrophic microbes have been found living as endolithic communities in the fabric of porous Beacon sandstone in the Ross Desert (Friedmann and Ocampo, 1976; Friedmann, 1977, 1982). The composition of these communities shows a remarkable interaction between the morphological and physiological characteristics of the organisms and the spatial and physicochemical properties of the habitat (Fig. 4) (Friedmann and Ocampo-Friedmann, 1984a).

Chasmoendolithic (or chasmolithic) communities are widespread in Antarctica (Friedmann, 1977, 1982; Broady, 1981b,c), and Friedmann and Ocampo-Friedmann (1984a) found cyanobacteria, green algae, and Xanthophyceae to be the dominant components. In addition to sandstone, their substrata include granite, granodiorite, and marble outcrops in regions ranging from southern Victoria Land (Friedmann, 1978, 1982) to Signy Island in the maritime Antarctic. Species diversity is generally higher than that of cryptoendolithic communities and often includes filamentous green algae. Studies by Broady (1981b, 1986a) of chasmolithic algae and cyanobacteria in coastal areas of Princess Elizabeth Land and Mac.Robertson Land, respectively east and west of the Amery Ice Shelf (Fig. 1), showed zonation within fissures related to moisture availability, salt concentrations, and rates of erosion of the rock surface. The area of rock covered ranged from 2 to 21%, which was greater than any other form of vegetation in the areas studied.

Cryptoendolithic communities consist of diverse combinations of cyanobacteria, green algae, filamentous fungi, yeasts and lichens (Friedmann, 1982), and associated heterotrophic bacteria (Hirsch et al., 1985; Hirsch, 1986; Siebert and Hirsch, 1988). They are oriented in layers parallel to the rock surface defined by physical and physiological constraints. Some communities, usually in moister areas, contain cyanobacteria only, whereas others contain lichens and free-living algae and cyanobacteria. Only cyanobacteria occur in hot desert endolithic communities, where they depend on liquid water from early morning dew and mist, absorbed through a mucilagenous sheath (Friedmann, 1971, 1980). In cold deserts, water rarely occurs in liquid form but

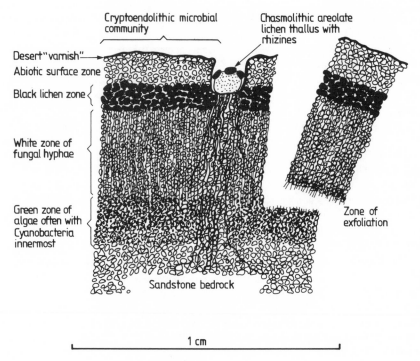

Figure 4. Cryptoendolithic and chasmoendolithic microbial communities in exfoliating sandstone. [Redrawn from Friedmann (1982), with permission.]

rather occurs as water vapor, which is utilized more readily by green algae, which are therefore the more common cryptoendolithic group. However, the addition of liquid water to Antarctic communities under optimal light conditions stimulates photosynthesis nearly 20-fold and is about six times more effective than water vapor under similar conditions (Vestal, 1988b).

In the lichenized community (Fig. 4), the black lichen zone (colored predominantly by fungal pigment) overlies a white layer of the lichen mycobiont, which in turn overlies a green layer of algae and occasionally cyanobacteria. This layer does not necessarily contain the lichen phycobiont and frequently consists of nonlichenized phototrophs (Tschermak-Woess and Friedmann, 1984). The species composition of the community is paradoxical, since the six cryptoendolithic lichen species identified to date (Hale, 1987) are also found in the maritime Antarctic, where they develop characteristic lichen morphologies, in contrast to the diffuse intergrain growth of their pseudo-tissue in sandstone (Friedmann, 1982). Moreover, a comparison of CO_2 gas exchange by cryptoendolithic lichens of the genus *Buellia* from southern Victoria Land (Kappen and Friedmann, 1983) with that of crustose epilithic forms of *Buellia frigida* in the coastal Antarctic at Cape Hallett, northern Victoria Land (Lange and Kappen, 1972) showed similar temperature ranges for photosynthesis in both groups.

The visual appearance of endolithic communities *in vivo* suggests considerable spatial variation in biomass. However, samples of phospholipids extracted as a measure of viable biomass give mean values varying within only an order of magnitude (McKay *et al.*, 1983). Determining absolute values in rocks is technically difficult. Estimates of viable organic matter derived from analyses of ATP in crushed rock unexpectedly suggested a highly productive endolithic community, but comparison with the total organic C value converted from Kjeldahl analyses indicated that only 2–5% of the biomass was viable (Friedmann *et al.*, 1980). This range was confirmed by Tuovila and LaRock (1987), who eliminated the possibility of overestimation due to extracellular accumulation of ATP in the rock pores by adding an apyrase to the crushed community under conditions that permitted only hydrolysis of free ATP. They, too, found variation in biomass C, from 10 to 2400 mg m^{-2} (assuming uniform thickness of the community in the top 5 mm of rock). Nevertheless, by the same method, 1g of endolithic rock contains at least as much living biomass as is found in at least 1 liter of seawater (Karl *et al.*, 1977). In a comparative study, Vestal (1988a) used a lipid phosphate extraction determination to estimate a biomass of 1.92–3.26 g of C m^{-2}). This represented between 7 and 54% (dry weight) of the biotic zone of the rock and comprised 0.3–9.6% of its total C content, the residue probably being mainly inactive cells. A contrasting approach of direct counts from the crushed outer 15 mm of rock containing endolithic lichens (Greenfield, 1989) indicated a biomass (dry weight per square meter) of 9.0 g of fungal hyphae, 0.5 g of algal cells, and 0.014 g of bacterial cells. These diverse determinations indicate a relatively small but potentially active endolithic microbiota (Vestal, 1988b).

Why is it that an endolithic microbial community composed of physiologically nonadapted organisms survives in such a biologically hostile environment? The answer may lie in the conservation of moisture and nutrients, the buffering of temperature extremes, the penetration of light as an energy source to drive anabolism, and protection from abrasion. Snow is the only source of water for the habitat (Friedmann, 1978; Friedmann and McKay, 1985), occasionally melting on the rock surface, to be imbibed through the crust. Using a satellite-linked automatic weather station, Friedmann *et al.* (1987) monitored the endolithic microclimate and "nanoclimate" (on the scale of the organisms themselves) continuously for 2 years (1984–86). This revealed snowfalls at least three times a month which correlated with internal rock humidity at 10-mm depth for all but the most transient of events. The major snowfalls, which technically represent available water, and elevated humidity both correlated with the elevation of air and rock temperatures to ca. 0°C.

The water- and gas-permeable "desert varnish" of fine mineral material which lodges between the grains at the surface (Friedmann, 1971; Friedmann and Ocampo, 1976) helps to retain the sporadic input of moisture for extended periods of time despite the extremely low ambient relative humidities (Kappen *et al.*, 1981). Moreover, Friedmann and Weed (1987) have found that the siliceous varnish stabilizes the surface to prevent erosion by wind, frost, and salt, thereby prolonging the duration of biologically favorable conditions underneath. Friedmann (1980) concluded that despite the irregularity of moisture input from snow, the endolithic communities of cold deserts are limited by low temperature, whereas those of hot deserts are limited more by low relative humidity within rocks because of their high temperatures. As eukaryotic algae and fungi in the hydrated state are unable to withstand a rapid rise in temperature and sudden water loss, this may account for the sole occurrence of cyanobacteria in hot desert endolithic communities. Conversely, at low ambient Antarctic temperatures, Kappen et al. (1981) found that the relative humidity within sandstone was generally high, ranging from 16–47% on clear days to over 70% on overcast days. After snowfall it rose to ca. 80%, which was sustained for 5 subsequent days. Corresponding rock temperature was consistently below 0°C on south-facing rock faces, but on north faces it rose to 7°C on clear days while remaining below 0°C on overcast days. Such conditions favor eukaryotic algae.

Insolation has a direct heating effect on the rock surface, and the obscuration of the radiation by shading results in rapid freezing. This incurs not only the loss of free water by ice formation but also the physical and osmotic disruption of cell membranes. The frequency of short-period (few minutes), low-amplitude (few degrees) freeze–thaw cycles resulting from the passage of clouds in front of the sun or gusts of wind (Friedmann and McKay, 1985; Friedmann et al., 1987) may stress cells on the rock surface beyond their limits of viability, thereby partly accounting for the sterility of the rock surface. The short-period oscillations do not affect endolithic communities a few millimeters below the surface because of thermal buffering by the rock. However, the short-period freeze–thaw cycles are superimposed on low-frequency (diurnal), large-amplitude (20°C) cycles that cause cessation of metabolism in prolonged shadow, which may be continuous for part of the winter. The influence of insolation is therefore profound, and the microbiota is efficient at rapidly switching its metabolism on and off according to the prevailing conditions. Using data obtained from an automatic weather station, Friedmann et al. (1987) estimated that metabolism was possible in the endolithic community for <1000hr yr^{-1} during a growing season from mid-November to mid-February. However, temperatures above 0°C, representing more optimal conditions for metabolism, occurred only for 50–550hr yr^{-1}, depending on the orientation of the surface (Friedmann et al., 1987). The lower temperature limit for photosynthesis is ca. $-6°C$, which may be exceeded in sunlit rock for up to 13 hr per day in midsummer (Kappen and Friedmann, 1983).

Kappen and Friedmann (1983) and Vestal (1988b) showed the minimum temperature for endolithic metabolism to lie between -6 and $-8°C$, with an optimum at ca. $+15°C$ under natural light conditions. This temperature is very rarely attained even in north-facing rock in summer (Friedmann et al., 1987). Nevertheless, the temperature range for photosynthesis in cryptoendolithic lichens is similar to that of crustose epilithic lichens living in less extreme Antarctic climates (Lange and Kappen, 1972).

CO_2 fixation is primarily photosynthetic (Vestal, 1988b). Dark fixation accounts for only 2.8% of the total, and chemolithotrophic energy sources, including thiosulfate, ferrous sulfate, and sodium nitrite, and ammonium, are not stimulatory. CO_2 fixation is dependent on the amount of incident PAR available. At the rock surface, this reaches 1500 μE m^{-1} sec^{-1}, similar to insolation in hot deserts, but there is a steep gradient through the rock profile. In dry rocks, ca. 0.1% of ambient light reaches the upper level of the lichen zone at 2-mm depth, whereas only 0.01% reaches the green algal layer at ca. 5-mm depth. Cyanobacteria, if present, occur below the green algal layer and are able to use even lower levels of PAR. Wetting the rock increases the penetration of PAR by as much as 10-fold; as a result, the availability of water coincides with increased illumination. Although the stratification of the phototrophs may be partially dependent on their photosynthetic compensation point, Kappen and Friedmann (1983) showed that light is unlikely to be the rate-limiting factor.

It seems paradoxical that the black lichenized layer present in many endolithic communities occurs nearest the surface, apparently screening underlying algae and cyanobacteria from the PAR. This is probably an energetic compromise whereby the phototrophs are sufficiently photoadapted to exploit the available PAR, while the black pigment may absorb heat to overcome temperature limitation and may also screen out potentially damaging UV radiation (Kappen et al., 1981).

The endolithic nitrogen cycle and budget is unique in that it is dependent entirely on dry deposition and precipitation from abiotic sources of ammonia and nitrate such as electric discharges (aurorae) for its supply of fixed nitrogen. Greenfield (1989) showed that 87% of the total N of endolithic rock was organic N which amounted to 2864 mg m^{-2}. Of this, 70–80% consisted of amino acids and hexosamines, but only 35% of this was live biomass N. Colonized rock contained 64 and 80 mg of N m^{-2} as exchangeable ammonium and nitrate, respectively, the residue being nonexchangeable ammonium in the substratum. Fresh and old snow contained 118 and 437 µg of mineral N liter^{-1}, respectively, mainly as ammonium. No N_2-fixing cyanobacteria have been detected to date, and there are probably insufficient energy substrates to support significant heterotrophic fixation. Neither nitrification nor denitrification has been detected. The community is not N limited (Friedmann and Kibler, 1980; Greenfield, 1989) but the N cycle is incomplete. Nitrate and ammonia are assimilated by microorganisms, which then decompose to release ammonia that is reassimilated. Fixed nitrogen is ultimately lost through leaching, ammonia volatilization, and disintegration of the community by biological and abiological weathering (Friedmann and Kibler, 1980).

Endolithic organisms obtain inorganic nutrients not only from precipitation but also by solubilizing the substratum. This is evident from the leaching of brown iron-bearing minerals in the colonized zone, which appears lighter than the outer crust and the inner substratum (Friedmann, 1982). Although providing essential minerals, this activity is potentially self-destructive since it weakens the crucial structure of its rock habitat. The rock grains are also loosened by the abiotic stress of steep thermal gradients and, more rarely, ice crystal formation (Hall, 1986a,b). Moreover, they are probably pushed apart by hydrostatic pressures exerted by growing cells, especially cyanobacteria such as Gloeocapsa, whose sheaths may expand 20-fold during imbibition of moisture (Friedmann, 1971). This action results in exfoliation (Fig. 4), whereupon the protective structure and its microbial community are lost. The residual cells on the parent rock must grow into its fabric if they are not to be abraded off. Exfoliative weathering is evident from the mosaic of light patches of rock leached by former endolithic communities in the iron-stained sandstone (Friedmann, 1982). The newly exposed sites become silicified to form a protective crust for subsequent community development, and the cycle is repeated (Friedmann and Weed, 1987). The biological and geological processes occur on comparable and overlapping time scales, consistent with the very short growing season (Friedmann et al., 1987).

These findings demonstrate an unusual aspect of microbial ecology—an avoidance strategy. Cryptoendolithic microbes are not especially adapted physiologically to their environments but are adapted morphologically (Friedmann et al., 1982). They are dependent for survival on the structures and the physical and chemical properties of their microhabitats. They live in an almost closed system of finite space and resources. Since CO_2 exchange through the rock crust is very slow (Kappen and Friedmann, 1983), accumulation of CO_2 in the endolithic airspace may serve to regulate photosynthesis to avoid outgrowing the niche.

When an endolithic community dies, its former presence is revealed by its leaching pattern, termed a trace fossil (Friedmann, 1986; Friedmann and Weed, 1987). Fossil microbes are rare, although fossil endoliths as old as Precambrian have been described (Campbell, 1982). Such fossils can be aged by calibrating the quartz rinds, which indicate that endolithic communities have been present in the area for the last few million years and were in existence before its last glaciation. During the last glaciation there were nunataks (rocky outcrops protruding through the ice sheet) which remained ice-free under conditions similar to those that have occurred in the present Ross Desert region for the last 4 million years. Friedmann and Ocampo-Friedmann (1984b) therefore argue that remnants of the earlier life forms of Antarctica could have found refuge in the nunataks. Present cryptoendolithic communities may therefore be descendants of these survivors which have evolved during the cooling of the continent. Colwell et al. (1987) have presented conflicting evidence, based on sequence analysis of bulk 5S rRNA extracted from endolithic communities, suggesting that the heterotrophic bacterial microbiota was dominated by Vibrio spp. of marine origin which were presumably carried inland by onshore winds in summer. However, these findings have been questioned by Hirsch and co-workers, who found the heterotrophic microbiota of the same site to be dominated by Gram-positive bacteria such as coryneforms, Micrococcus spp., and Deinococcus spp. (Hirsch et al., 1985; Siebert and Hirsch, 1988).

Much of the impetus for studies of Antarctic desert soil and subsequently endolithic communities sprang from the search for terrestrial analogs of the Martian biome (Fig. 5) during development of life detection systems for the Viking Mars Program (Cameron et al., 1970b,c, 1976; Horowitz et al., 1972; Siegel et al., 1983). The Ross Desert test sites were originally thought to be largely abiotic (Horowitz et al., 1969), and indeed some contain levels of ions, such as borate, toxic to microbes (Cameron et al., 1968). However, direct observation and growth experiments in situ revealed a sparse soil microbiota (Vishniac and Mainzer, 1973), and the final Viking lander life detection system was based on metabolic analysis of sieved soil. Of four experiments, three gave negative indications of life and the fourth was ambiguous (Klein, 1977). It was after these exobiological missions that endolithic microbes were discovered in the Ross Desert (Friedmann and Ocampo, 1976), and it was recommended that future experiments include analysis of rock samples. Subsequent information collected on the Mar-

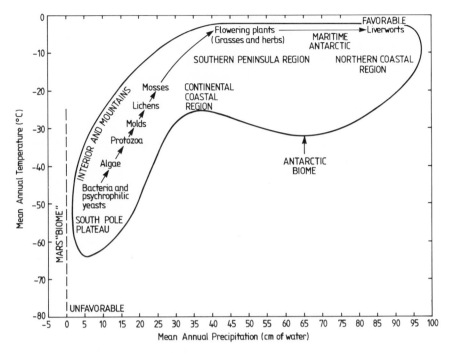

Figure 5. Comparison of the Mars "biome" with Antarctic terrestrial biomes in relation to mean annual water precipitation. [Redrawn from Cameron *et al.* (1976), with permission.]

tian environment indicates that even endolithic communities would be unlikely to remain viable under present conditions (Klein, 1979). Nevertheless, Huguenin *et al.* (1983) believe that there are areas less hostile than the regions sampled by the Viking lander where moisture may accumulate within the upper 1 m of the soil profile and be potentially life supporting. They are therefore concerned about the chances of contamination of Mars by terrestrial microbes, which is an escalation of the concern for maintenance of "pristine" conditions in Antarctica (Cameron, 1972b). Since the primordial Martian environment was probably conducive to life before its atmospheric pressure dropped (McKay, 1986), the emphasis for life detection on Mars, and research into its Antarctic analogs, has now changed to collecting evidence of earlier life forms. The remarkable scarcity of organic matter in Martian soil may be due to highly reactive soils causing abiotic decomposition so that fossil C cannot be relied on (Klein, 1979). The discovery of trace fossils in the Ross Desert may, however, have parallels on Mars, where endolithic microbiota may have withdrawn into porous rocks as the environment deteriorated (Friedmann, 1986; Friedmann and Weed, 1987). Recent microbial ecological studies in Antarctic cold deserts may therefore be of value for future paleontological and exobiological research (McKay, 1986).

3.2. Soil Microbial Xero- and Cryotolerance

Since moisture is central to the basic requirements for survival listed in Table I (Horowitz, 1979), the ability to tolerate long periods of desiccation, phase changes, and osmotic stress is crucial, particularly in cold deserts. This is illustrated by the undecomposed state of mummified seals centuries old, found frequently in Ross Desert Dry Valleys (Dort, 1981).

Table I. Ecological Factors Governing the Distribution of
Terrestrial Microbial Life in Antarctic Cold Deserts[a]

Favorable conditions	Unfavorable conditions
North–south orientation of valley (across katabatic winds)	East–west orientation of valley (along katabatic winds)
Gentle north-facing slope	Flat or south-facing slope
High, constant solar radiaton	Low, sporadic solar radiation
Low albedo of substratum	High albedo of sustratum
Microclimate above freezing	Microclimate below freezing
Thermal buffering by substratum	Low thermal capacity of substratum
Infrequent freeze–thaw cycles	Frequent freeze–thaw cycles
Absence of wind (low cooling, evaporation, and abrasion)	Strong, cold, dry winds laden with sand or ice grains
Northerly winds (warmer, moister)	Southerly katabatic winds
High precipitation	Low, transient precipitation
High humidity	Low humidity
Available water (meltwater, streams, ponds, lakes)	No available water, distant from water bodies
Slow or impeded drainage	Rapid drainage, porous soil
Infrequent wet–dry cycles	Frequent wet–dry cycles
Translucent stones (sublithic greenhouse effect)	Opaque stones
Porous or fissured translucent rock (endolithic and chasmolithic growth)	Smooth, dense, opaque rock
Stable substratum (soil crust, desert varnish)	Unstable substratum (rock or sand dunes with no crust)
Low salinity of soil or water	Highly saline soil or water
Balanced ionic composition	Unbalanced ionic composition
Nontoxic soils	Toxic soils (excess boron, lead, nitrite, sulfide, chromate)
Approximately neutral pH	Acid or alkaline pH
Organic nutrients (guano, dead cells, leachates, secretions, excretions)	No organic nutrients, carbon and nitrogen limited
Autotrophic metabolism	Heterotrophic metabolism
Adaptive strategies (physiological or morphogenetic)	Slow metabolic response to change, inflexibility
Constant, abundant aerial inoculum	Sporadic, sparse aerial inoculum
Animal vectors	No animal vectors

[a]After Cameron et al. (1970b).

Microbes in unsheltered soils of dry valleys of the Ross Desert are more vulnerable to climatic and edaphic change than their endolithic and coastal counterparts (Fig. 5). Not only is there an absolute shortage of water, but the accumulation of salts in the soil makes imbibition of liquid water energetically demanding. The dry valley floors of the Ross Desert receive significantly less precipitation than the uplands, and the funneled katabatic winds disrupt intergrain microsites of elevated humidity. The desiccation gradient is so steep that even the reservoir of ice in the permafrost cannot furnish adequate liquid water for the consistent microbial growth necessary to stabilize the soil surface (Cameron and Devaney, 1970; Wynn-Williams, 1986, 1989). Algae and cyanobacteria are restricted to soils that are moistened during at least part of the year. *Nostoc* and *Phormidium,* which are to a significant extent tolerant of wet–dry cycles, are the most abundant genera in the Ross Desert and other desert areas such as the Vestfold Hills (Broady, 1986b) and Mawson Rock, Enderby Land (Broady, 1982). As primary producers, they generate soil microbial crusts (Cameron and Devaney, 1970) which occasionally develop into substantial mats (Wynn-Williams, 1985b).

A survey of all published physical and chemical analyses of Ross Desert soils has revealed not only a diversity of analytical methods but also substantial variation in results from reportedly the same sites (Boyd and Boyd, 1963; Cameron, 1969; Allen *et al.,* 1974; Vishniac and Hempfling, 1979b; Parker *et al.,* 1982a). This makes overall comparisons invalid. However, the results for selected Ross Desert soils given in Table II, all obtained by the same methods, illustrate the heterogeneity of habitat composition and associated microbial composition. The use of relatively rich media to obtain the viable counts (full-strength tryptone soya agar for bacteria) is not commensurate with the organic content of the soil, making the counts of relative value only. Despite their proximity to the Polar Plateau, the orientation of Wheeler and Balham Valleys across the katabatic winds makes them conspicuously moister and more microbiologically productive than adjacent wind-desiccated valleys such as Victoria Valley. An ornithogenic soil (enriched with guano) on Dunlop Island, included to show the influence of coastal climate, gave results similar to those obtained for the moister dry valleys. There is a conspicuous correlation between the presence of algae and large bacterial populations, consistent not only with the common response to moisture but also with the C limitation of Ross Desert microbial metabolism.

The dubious validity of using mean edaphic data to analyze local microbial responses on earlier ecological surveys of the region (Boyd *et al.,* 1966; Cameron, 1971, 1972b, 1974; Vishniac and Hempfling, 1979a) was confirmed by Klingler and Vishniac (1989), who demonstrated a 10-fold difference in mineral content of soil samples from the same area. They consequently recommended that water potentials be determined on the same samples that are used for microbiological investigations. It is now appreciated that steep physical and chemical gradients and local heterogeneity on a macro-, micro-, and nanoscales are a fundamental characteristic of Ross Desert habitats, in marked contrast to more temperate regions, where the free flow of soil water minimizes such gradients. Because the typically low clay content of Ross Desert soils, matric forces exert only a minor influence on their water potential while osmotic forces play the major role (Klingler and Vishniac, 1989). Thus, when the water availability necessary to

Table II. Ross Desert Soil Edaphic and Microbial Characteristics[a]

Characteristic	Finding at given location					
	Victoria Valley	Coalsack Bluff	Barwick Valley	Wheeler Valley	Balham Valley	Dunlop Island
Position	77° 20'S	84° 14'S	77° 20'S	77° 12'S	77° 25'S	77° 15'S
Sample number	634	742	625	615	522	740
Substratum	Dune sand	Sand on Berg Moraine	Pattern ground sand	Gray-brown sand	Wet salty sand	Ornithogenic coastal
Moisture (% dry wt)	0.14	1.58	1.74	4.30	13.60	34.50
pH	8.0	6.5	8.3	8.1	7.7	7.4
Organic C (wt %)	0.02	0.38	0.02	0.17	0.61	4.62
Organic N (wt %)	0.003	0.024	0.004	0.024	0.114	0.589
Inorganic ions (ppm dry wt)						
Ca	140	50	1	20	500	150
NH_4	1.0	0.1	0.8	ND	0.0	26
NO_3	2	742	1	7	900	50
Cl	835	110	5	41	8,875	4,260
PO_4	0.05	0.00	0.15	0.20	0.06	0.4
SO_4	6	120	3	15	30	520
CFU g (dry wt)$^{-1}$						
Bacteria						
2–5°C	50	29,000	52,000	120,000	92,000	24,000
10–22°C	12,000	400	85,000	1.3×10^6	190,000	160,000
Yeasts	0	5,100	300	2	0	0
Microfungi	0	0	0	200	0	0
Algae	0	0	100	6.4×10^6	0	100,000

[a]After Cameron (1969, 1971) and Cameron et al. (1970b).

support microbial life in soils is marginal, the local variation in water potential effected by heterogeneous accumulation of salts can determine the distribution of viable cells. An inverse correlation has been found between the frequency of isolation of the psychrophilic yeast *Cryptococcus vishniacii* and the total cationic equivalents in the soils sampled (Vishniac and Hempfling, 1979a). The heterogeneous distribution of this yeast had hitherto been unexplained (Vishniac and Klingler, 1986; Klingler and Vishniac, 1989). These authors calculated that in the most salt-laden soil tested (>60.5 meq g^{-1} dry soil), the water content would have to be ca. 4.5% to permit yeast growth. At <19.4 meq, as little as 1% water would support growth. Yeast growth was not inhibited by water potentials down to ca. -1.75 MPa, and its function continued at limited rates down to ca. -3.5 MPa. To maintain a physiologically favorable equilibrium of solute concentration across the cell membrane in microhabitats of such low water potentials, microbial cells must accumulate solutes. However, the solutes must not be inhibitory to enzyme function or deleterious to membrane structure at high equilibrium concentrations. Suitable solutes are termed compatible (Brown, 1978).

Under conditions of osmotic stress, xerotolerant yeasts can switch their metabolism to accumulate polyols or trehalose as compatible solutes (Brown, 1978) while entering the anhydrobiotic state. Cold desert strains show a ruderal growth strategy (Pugh, 1980) by growing opportunistically in sporadic traces of water. A recurring feature of Antarctic desert microbes, probably related to this survival strategy, is the production of mucilaginous capsules by yeasts (Aksenov *et al.*, 1973) and heterotrophic bacteria (Uydess and Vishniac, 1976) and of sheaths by cyanobacteria.

Sun *et al.* (1978) noted that two yeasts that were widely distributed throughout the Antarctic, including the Ross Desert, were rich in lipids, which probably help to protect membranes against cold and desiccating conditions. Polyols acting as compatible solutes also confer cryotolerance, not only in yeasts but also in filamentous fungi (Lewis and Smith, 1967), algae (Brown, 1978), nematodes (Womersley and Smith, 1981), and Antarctic cryptogams and invertebrates (Block and Sømme, 1982; Sømme and Block, 1982; Tearle, 1987). The metabolic strategy of polyol accumulation for cryo- and xeroprotection and the synthesis of mucilagenous sheaths for water budgeting may constitute "unifying hypotheses" linking many Antarctic terrestrial organisms.

Cameron (1971) observed that microbes in disrupted soil crumbs are more sensitive to freeze–thaw cycles than are those experiencing the same cycles in their undisturbed habitats. Shelter is crucial for organisms in such a delicate balance of survival. Factors affecting the lethality of freeze–thaw cycles include the rate of cooling and warming, the range of temperature change [$-15°C$ to $+ 27.5°C$ in 3 h in Ross Desert soil on one occasion (Cameron, 1974)], the state of hydration of the cells, their metabolic activity, their compatible solute content, their ability to switch metabolism to synthesize cryoprotectants, their intrinsic membrane characteristics, and their degree of association with soil particles. Many microbes are unable to meet the minimal requirements for viability, resulting in a very sparse soil surface microbiota (Cameron and Conrow, 1969) and the concentration of populations in sublithic and chasmolithic communities (Broady, 1981a–c) or in association with coalesced algae (Cameron, 1972c).

Hypersaline soils are common in the Ross Desert, but they reach an extreme at the edge of Don Juan Pond, Wright Valley, where the calcium chloride concentration reach-

94   David D. Wynn-Williams

es ca. 33%. Despite the steep osmotic gradient and ionic imbalance, the soil is occupied by a mineral and cyanobacterial mat composed of *Oscillatoria* and other phototrophic species cemented by organic material (Siegel *et al.*, 1983). The total organic content of the mat averages 8.7% dry weight, of which chlorophyll determinations indicated 3.6–9.0% to be photosynthetic biomass. Although bacteria are rare, fungal filaments (*Penicillium*) occur and, surprisingly, tardigrades are also present. The mat contain a range of active enzymes, and 0.5–5.0 ppm ATP has been found. This remarkably diverse, trophically complete ecosystem is an example of the hardiness of the Antarctic microbiota.

In soils, psychotrophic bacteria prevail over psychrophiles, which are rare outside the permafrost, probably because of occasional high ground temperatures. They are found more often in consistently cold environments such as the Southern Ocean (Morita *et al.*, 1977). The only truly psychrophilic [as defined by Morita (1975)] terrestrial microbes in the Ross Desert are potentially indigenous yeasts (Vishniac and Hempfling, 1979a). On the basis of studies of yeast isolates and known physicochemical and nutritional characteristics of Ross Desert soils, Vishniac and Hempfling (1979b) defined a model Antarctic indigene as a "unique species with random distribution in endolithic habitats, psychrophilic, xeroduric, urease and protease negative, nitrate positive and growth factor independent." The demanding requirements for survival, combined with the scarcity of the microbiota (and consequently restricted interspecies interactions), makes this ecosystem valuable for characterizing the fundamental nature of indigeneity. The chances of survival of allochthonous species settling out from the airspora (Fig. 6) (Cameron, 1972a; Johnson *et al.*, 1978) in soil conditions at the limit of life support are very low, and they are found localized in greatest diversity to more favorable sites. However, sites at more hostile locations are recolonized only by emigrants from the very small populations of the few indigenous taxa and are therefore distributed randomly (Vishniac and Hempfling, 1979a). *Cryptococcus vishniacii* was shown to have a random distribution and grew at −3°C. Some biotypes grew at >20°C and were therefore psychotrophic, but others did not grow above 20°C and were therefore considered psychrophilic. Their growth at very low water potentials meets the xeroduric requirement, and their organic nutrient requirement was several orders of magnitude below the organic content of the soil.

The organic content of the soil is probably replenished partly by cells, debris, and leachates from disintegrated endolithic communities but more substantially from the input of the airspora itself. Vishniac and Hempfling (1979a) calculated that for an hourly deposition of 11 fungal spores, each containing 14% trehalose (Sussman and Halvorson, 1966), the sugar supply alone would permit a maximum doubling rate for yeast cells at 10°C. Since Gregory (1966) reported 137 fungal spores m^{-3} in an air mass of polar origin and Horowitz *et al.* (1972) isolated nonadapted microbes from the Ross Desert region), it is probable that the incident airspora supplies a significant proportion of the energy substrate requirement of the heterotrophic soil microbiota. Characterization of yeast isolates (Vishniac and Hempfling, 1979b) showed that the enzymic and growth requirements were also met, and thus the indigeneity (as defined by Vishniac and Hempfling, 1979a) of *C. vishniacii* was confirmed.

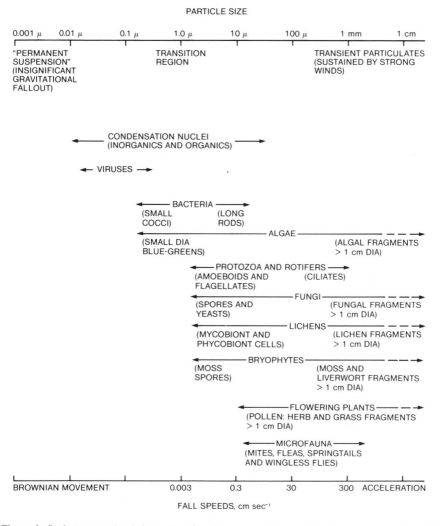

Figure 6. Settlement speeds relative to size of particles, cells, and propagules arriving as potential aerial colonizers of Antarctic fellfield soils. [From Cameron *et al.* (1976), with permission.]

As in the endolithic ecosystem, the soil nitrogen cycle is incomplete. Nitrate plus minimal amounts of nitrite, both probably largely of atmospheric origin from electrical discharges, accumulate in desert soils and are utilized by yeasts. The resulting organic C:N ratios are typically narrow, between 4:1 and 9:1, indicating C rather than N limitation (Cameron, 1971).

Ross Desert permafrost isolates take longer to regenerate than the surface isolates and may represent a sheltered relict indigenous microbiota in a quiescent state (Cameron *et al.*, 1970c) (cf. the endolithic microbiota). They are unlikely to be surface microbes leached into the profile because of the net upward movement of water in the dry valleys (McCraw, 1967). Viable microbes observed in cores drilled down to 380 m (representing the Pleistocene era) in the permafrost of Taylor Valley (Ross Desert) included yeasts and streptomycetes (Cameron and Morelli, 1974). The dearth of water in their original habitat may have improved their viability by freeze–drying the cells before burial. Preserving conditions included low redox potential, a slow rate of cooling, and a relatively low temperature (-8 to $-14°C$). Such isolates must always be treated with caution because of the difficulty of avoiding contamination due to penetration of drilling fluid. However, they have relevance to Earth's climatic evolution and to exobiological life detection studies because similar permafrost may exist near the poles of Mars or beneath the floors of its evident water-created valleys (McKay, 1986).

Streptomycetes and coryneforms are dominant in certain Antarctic regions (Benoit and Hall, 1970) and have been for a long time, as indicated by the permafrost record. They grow well on glycerol, which may be one of the more abundant nutrients released into the environments since it is produced by various organisms, including yeasts, algae, and invertebrates, as a common compatible solute. Other such nutrients potentially available from the extant biota include ribitol from algae, arabitol from fungi, and trehalose from yeasts (Brown, 1978; Tearle, 1987). Siebert and Hirsch (1988) demonstrated the assimilation of ribitol and mannitol by a group of Gram-positive cocci including *Micrococcus, Deinococcus,* and *Brevibacterium,* from the Ross Desert. However, Hirsch (1986) pointed out that the concentration of substrate required by many terrestrial microbes, including Antarctic strains, is often very low because they are frequently oligotrophic. He isolated typically oligotrophic strains, including *Caulobacter* and *Blastobacter,* from Antarctic rock material.

Intrinsic protection from harmful UV radiation afforded by their pigmentation may partly account for the frequency of *Micrococcus* and *Deinococcus* isolates at the surface of soils. This property assumes increasing importance in view of recent findings that UV radiation is increasing with the expansion of the ozone "hole" over Antarctica (Tuck, 1987). Drier soils are dominated by Gram-positive bacteria, predominantly coryneforms (Johnson and Bellinoff, 1981; Johnson *et al.*, 1981) but also including *Bacillus* species (Madden *et al.*, 1979). Various workers have monitored the diversity of bacteria (Cameron, 1971), yeasts (Atlas *et al.*, 1978), and filamentous microfungi (Sun *et al.*, 1978) of the region. The evident relatively low species diversity detected makes interpretation of ecological interactions easier in the Antarctic than in more temperate regions (Cameron, 1972a). Coryneforms are found at all locations, including the maritime Antarctic, while Gram-negative rods such as *Pseudomonas* are conspicuously absent (Table III). The latter are found in more favorable conditions such as moss–peat communities and fellfield soils bearing cryptogamic vegetation at maritime Antarctic locations such as Signy Island. It is noticeable that although Deception Island is also maritime, its volcanic ash substratum supports a microbiota similar to that of the Ross Desert (Cameron and Benoit, 1970).

Table III. Distribution of Microbial Genera by Antarctic Location and General Environment as Indicated by Numbers of Isolates

			Number of isolates				
		Ross Desert region					South Shetlands
Organism	Barwick Valley (SSSI)	Balham Valley (moist)	Wright Valley[a]	Marble Point (coast)	McMurdo Station soil	Air samples (sum of 4 yr)	(Deception Island, Maritime Antarctic)
Corynebacterium	12	7	12	7	5	19	11
Arthrobacter	5	11	2	0	11	17	15
Brevibacterium	5	2	3	0	8	21	1
Micrococcus	2	6	5	3	3	36	3
Streptomyces	1	2	1	2	1	4	0
Nocardia	2	1	0	2	0	1	1
Bacillus	6	0	1	0	0	23	0
Achromobacter	0	0	4	0	2	3	0
Flavobacterium	1	0	0	0	3	3	0
Pseudomonas	0	0	0	0	4	3	0
Staphylococcus	0	0	0	0	0	12	0
Sarcina frigia	0	0	7	1	0	0	0
Yeast	2	0	0	1	3	3	1

[a]Includes Lake Vanda, Lake Vida, and Don Juan Pond. After Johnson *et al.* (1978).

4. Microbiology of Ice-Covered Water Bodies

4.1. Permanently Ice-Covered Lakes

4.1.1. Strata as Microbial Niches

Ice cover on a lake has a profound influence on the water column and benthic sediments under it. In a situation analogous to the endolithic habitat, it tends to isolate the biota from the external environment and, by excluding wind-mediated turbulence in the water column, helps to maintain well-defined stratified microbial niches. The permanently ice-covered lakes of the Ross Desert and Vestfold Hills range from freshwater to hypersaline and have been more extensively studied than any of the other continental lake systems. As most of the lakes characteristically have little or no outflow, the ingress of water into the peripheral moat which melts in summer frequently results in steep salinity (chlorinity) gradients, e.g., from freshwater immediately under the ice to brine about four times the salinity of seawater at the bottom of Lake Vanda (Table IV). It is this salinity rather than ice cover which is the primary stabilizer of stratification, as is shown by the stratification of ice-free hypersaline Deep Lake in the Vestfold Hills (Hand, 1980). The potential niches provided by such stratification will be considered with respect to the profile.

The nature of the ice cover affects the characteristics of the water column. The clean, clear, thin ice of Lake Vanda transmits 18% of incident PAR, whereas the 4.5-m-thick rock-strewn ice of Lake Hoare transmits <1%. Because of its great clarity (extinction coefficient of 0.055), the water of Lake Vanda is heated mainly by solar radiation absorbed by the benthic sediments. In contrast, the cloudy water of Lake Fryxell (extinction coefficient of 0.773) restricts penetration of light below 10 m, which mainly accounts for the water temperature of only 4°C at 11 m.

Homogeneity of the ice affects gaseous diffusion and particle penetration. Lake Vanda ice has cracks permitting permeation of gases, whereas most other lakes have relatively continuous, thicker ice (up to 6 m on Lake Hoare), which provides a more effective seal (Parker and Wharton, 1985). The accumulation of O_2 to supersaturation close to the undersurface of the ice cover (Parker et al., 1982a) can reach 51.4 mg m^{-3} in Lake Hoare, whereas saturation occurs at 14.5 mg m^{-3} (Wharton et al., 1986). Perennially high dissolved oxygen (HDO) is rare in nature, and such water therefore represents an unusual microbiological niche. The excess is mainly due to the inflow of fresh, aerated meltwater that freezes on the underside of the ice, excluding O_2 from the new crystals into a progressively concentrated solution. An additional source of O_2 is photosynthesis by plankton and cyanobacterial mats, despite respiration at rates unknown. Because the perennial HDO in Lake Hoare is relatively unchanging, the microbiota must adapt or expire by oxygen toxicity. Mikell et al. (1984) found in the Lake Hoare HDO zone an oligotrophic, ultramicroplanktonic community that did not show O_2 inhibition of assimilation or respiration. This contrasted with experimental inhibition under identical conditions of cells from normally oxygenated oligotrophic Mountain Lake, Virginia. All isolates from the Lake Hoare HDO zone were catalase positive, an infrequent characteristic in oligotrophic bacteria that possibly protects the cells against

Table IV. Physical and Chemical Characteristics of Lake Vanda[a]

Characteristic	Value at			
	Ice–water interface (3.1 m)	Immediately above oxycline (58 m)	Immediately below oxycline (60 m)	Sediment–water interface (72 m)
Ice thickness	3.1	NA	NA	NA
PAR transmission (%)	18.6	1.3	1.0	0.02
Temperature (°C)	0	20	22	25
Specific gravity	1.00	1.04	1.05	1.09
Chlorinity (g kg^{-1})	0	31	44	72
Conductivity (mS cm^{-1})	0.8	56	61	81
Oxygen (% saturation)	130	55	0	0
Nitrate-N (mg m^{-3})	45	812	84	0
Ammonium-N (mg m^{-3})	7.9	6,622	10,696	19,124
Soluble P (mg m^{-3})	0.6	3.1	25	161
Alkalinity (mg CaCO$_3$ liter^{-1})	36	NA	NA	NA

[a]Data from Parker and Wharton (1985), Love et al. (1983), and Vincent and Howard-Williams (1985). NA, Not applicable.

toxic peroxide accumulation. Moreover, some also were superoxide dismutase positive. In contrast, the benthic heterotrophic microbiota was inhibited by HDO, suggesting the stratification of a layer of HDO-tolerant plankton. Reduced metabolic activity during oligotrophic conditions may itself reduce O_2 uptake and consequent generation of toxic peroxides. The exocellular polymers frequently produced by HDO-tolerant microbes may restrict O_2 diffusion so that nanoscale gradients may afford local protection.

The lower surface of the ice itself does not appear to be a site of active colonization by algae or cyanobacteria (cf. sea ice microbial communities), although there is an upper euphotic phytoplankton community in the 0- to 15-m water column. However, in lakes such as Lake Hoare, pieces of benthic mat within the HDO zone accumulate at the ice–water interface, buoyed by O_2-enriched bubbles of photosynthetic origin or N_2 in other niches. They migrate through the ice profile by melting and freezing cycles driven by black-body radiation. The bubbles become elongated vertically within the ice, and strata probably representing annual layers, are visible. The mat finally emerges from the upper surface and is lost to the system by wind abrasion but may act as a potential source of inoculum or nutrient elsewhere. It is estimated that of the total benthic mat, the percentage capable of "lifting off" by such buoyancy is 10% in Lake Hoare and 25% in Lake Fryxell. Of this portion, ca. 10% lifts off per annum, but only 5–10% of this freezes into the ice. The total annual loss is therefore only a small fraction of the benthic biomass. It is doubtful whether the solid freshwater ice itself can be regarded as a microbial niche, as there is no evidence of active community development, unlike the SIMCO, which develops in porous sea ice. However, there may be HDO-tolerant bacteria associated with freshwater cyanobacterial mat debris.

In Lake Vanda, the HDO zone (Table IV) is less marked because of O_2 leakage through cracked ice (Parker and Wharton, 1985). In the 15- to 38-m zone, the temperature remains relatively constant at 10°C, with the HDO concentration at 150% saturation. This large thermohaline convection cell supports a mideuphotic algal community. The next stratal niche down the profile, at peak HDO concentration (175% saturation), is a zone of elevated concentration of gaseous forms of mineral N. Nitrous oxide concentrations are 10–20 times higher than have been reported in temperate lakes, even those where intense nitrification occurs. The N_2O–N maximum at 54-m depth (4.3 μg-atom liter^{-1}) nearly coincides with the nitrate-N peak at 55 m (233 μg-atom liter^{-1}) and nitrite-N maximum at 56 m (1.9 μg-atom liter^{-1}). This 54- to 56-m zone correlates with a deep band of nitrifying bacteria (Fig. 7) (Vincent et al., 1981). The spatial separation of peak algal activity from the nitrous oxide and nitrate peaks indicates that bacteria and not algae or cyanobacteria are responsible for these accumulations. At the region of highest nitrifier activity, denitrification is undetectable and nitrous oxide accumulates. The lower zone of denitrification spanning the oxycline coincides with a sharp decline in nitrous oxide concentration. Nitrifiers might have been expected to peak in the oxycline zone, but a concentration of ammonium-N increasing from 144 μg-atom liter^{-1} at 55 m through 600 μg at 59 m to 1747 μg at the anoxic sediment–water interface, probably associated with high sulfate reduction, may have inhibited them. This clearly displayed profile of an N-exchange niche indicates that nitrification is the source and denitrification the sink for dissolved inorganic N.

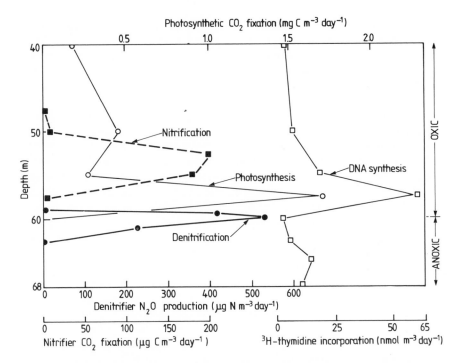

Figure 7. Microbial metabolic activity relative to depth and the oxycline in stratified Lake Vanda, Wright Valley, Ross Desert with reference to its nitrogen cycle. [From Vincent *et al.* (1981), with permission of Macmillan Journals Ltd.].

Photosynthesis at low light intensities and considerable depths is common in cold desert lakes. In Lake Vanda, maximal photosynthesis just above the oxycline at 59–60 m coincides with peak DNA synthesis, indicating total microbial activity (Vincent *et al.*, 1981). At this depth, because of the exceptionally low extinction coefficient, the PAR is still 1.3% of the incident level (Table IV), thus providing yet another ecophysiological niche for C fixation. However, the phytoplankton biomass of the whole water column is extremely small and has low activity and assimilation efficiency. Nevertheless, phototrophs in this deep zone share a common ecophysiological characteristic with organisms in several diverse Antarctic habitats—exceptionally well-developed photoadaptation (Priddle *et al.*, 1986). For example, the lower range of light intensity (μE m^{-2} sec^{-1}) reported to be usable for photosynthesis is 0.5–4.0 at 9-m depth in Lake Fryxell (Vincent, 1981), 0.2–20.0 in the lichen zone (2-mm depth) of cryptoendolithic sandstone communities at Linnaeus Terrace (Vestal, 1988b), and 2.5–20.0 at the ice–water interface of the sea ice microbial community in McMurdo Sound off Ross Island (Fig. 1) (Palmisano and Sullivan, 1985b). Vincent and Howard-Williams (1985) have shown a higher 480:663-nm light absorbance ratio in Lake Fryxell at 9 m (3.2) than immediately under the ice (2.5), suggesting a higher carotenoid accessory pigment content in the deeper population. Moreover, phycoerythrin, which is a common accessory pigment in a variety of Antarctic cyanobacterial communities (Johnson and Sieburth, 1979; Wynn-Williams, 1989) was more abundant in the Lake Fryxell population at 9-m depth. The ability to operate efficiently at low light levels may therefore be associated partly with the accumulation of accessory pigments, combined with the high efficiency of conversion of absorbed light into photosynthate (Vincent and Howard-Williams, 1985).

The stratal niches of Lake Vanda have been described as an example of extreme gradients in a saline lake with a relatively shallow anoxic layer. In contrast, Ace Lake in the Vestfold Hills (Fig. 1) is a stratified lake with an ice cover of 2 m and a depth of 23 m, of which the bottom 14 m are anoxic (Fig. 8). In this anoxic zone an active microbial S cycle (Fig. 9) is interlinked with C cycling, resulting in substantial methane accumulation despite low rates of methanogenesis (Hand and Burton, 1981). There is a 2-m-deep oxycline at the bottom of the euphotic zone in which there is a distinct band of photosynthetic bacteria of the genera *Chromatium* and *Rhodospirillum* together with the sulfate reducer *Desulfovibrio*. These phototrophs use sulfide as an electron donor, oxidizing it to sulfate, which *Desulfovibrio* reduces back to sulfide, resulting in a tightly linked S cycle within the water column (Hand, 1980). The depletion of sulfate by microbial sulfate reduction with increasing depth precludes competition between sulfate reducers and methanogens. Methanogens consequently become dominant in the saline bottom water zone, which is derived from seawater. Particulate organic debris in and near the sediment is converted to methane and CO_2, which diffuse upward to be utilized by methylotrophic bacteria. Because of the minimal input from the catchment, little new carbon is introduced into the lake and its oxycline-based stratal niche, which merely recycles C fixed by the sparse phytoplankton and benthic cyanobacterial mats within the upper euphotic zone.

Because of the stable stratification maintained by the salinity gradient, halotolerance is a requirement for some Antarctic aquatic microbes in the bottom lake water zone of marine origin. The salinity of Ace Lake increases uniformly from ca. 28 g liter^{-1} (80% that of seawater) at the surface to ca. 40 g liter^{-1} at the bottom. However, nearby Deep Lake is so saline (280 g liter^{-1}) that, despite experiencing the same climatic conditions as Ace Lake, it remains ice-free throughout the year since its freezing point is $-27°C$. However, Hand (1980) has suggested that many of the 10^5 bacteria ml^{-1} enumerated in Deep Lake water by direct counting may be unable to grow. There is no evidence of bacterial activity in the lake, probably because of a combination of nutrient deficiency, hypersalinity, low temperature, and disruption of potentially protective stratal niches by wind-induced mixing. Wright and Burton (1981) suggest that the bacteria are not viable because they use intracellular potassium chloride as a compatible solute to maintain osmotic equilibrium, and at such high concentrations their enzymes cease to function unless modified. However, modification makes them cold sensitive, thereby precluding metabolic activity. Nevertheless, McMeekin and Franzmann (1988) have used growth models to predict that the dominant *Halobacterium* sp., having temperature optima 4–5°C lower than that of the temperate counterparts and a potential for growth at $-5.4°C$, could have attained the observed population density within 8 years. Although unicellular green algae are found in the water column, no reproducible evi-

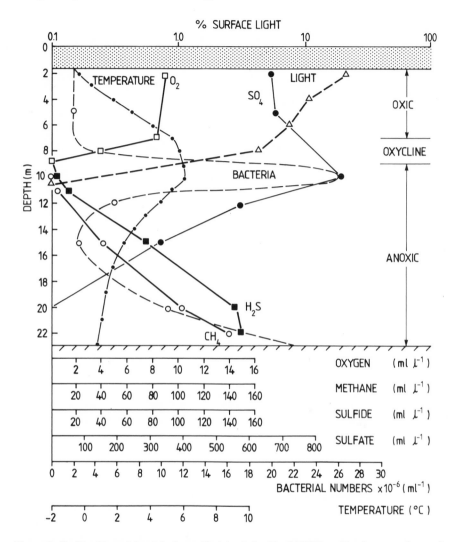

Figure 8. Profile of bacterial activity in stratified Ace Lake, Vestfold Hills, with reference to the anoxic zone and sulfur cycling. [Combined from Hand (1980) and Hand and Burton (1981), with acknowledgment to A.N.A.R.E.]

dence of primary production has been detected (Campbell, 1978). It is possible that they survive by synthesizing glycerol as a compatible solute as does the green alga *Dunaliella* (Brown, 1978), which occurs in saline lakes of the Vestfold Hills (Wright and Burton, 1981). This again draws parallels with terrestrial osmo- and cryoprotection mechanisms.

Although the exceptionally low temperatures of Deep Lake and Don Juan Pond have a limiting effect on microbial metabolism, temperature is not the limiting factor in

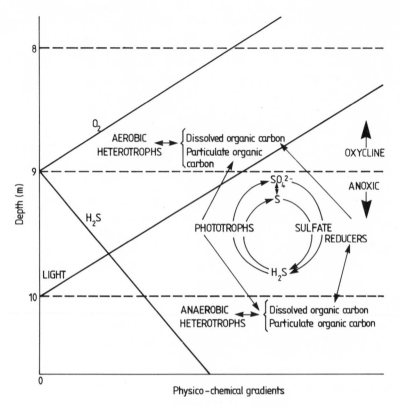

Figure 9. The sulfur cycle of Ace Lake, Vestfold Hills. [From Hand (1980), with acknowledgment to A.N.A.R.E.]

most lakes of Antarctic cold deserts. Despite extreme photoadaptation, it is shortage of energy from light, and consequent restriction of C fixation, that is the ultimate growth-limiting factor. Moreover, the position of the deep chlorophyll maximum (DCM) at the oxycline of lakes such as Vanda suggests that nutrients from the N- and P-rich anoxic zone are a more dominant influence than temperature. This conclusion is supported by the similarity between production in the DCM of Lakes Vanda and Fryxell. Despite the latter being 15°C colder than Lake Vanda, phosphorus diffuses upward from its anoxic zone at a rate of 18 µg P m^{-2} day^{-1}, compared with only 1.2 µg P m^{-2} day^{-1} in Lake Vanda (Vincent and Howard-Williams, 1985). The values summarized in Table IV suggest that Lake Vanda may be classified as one of the most oligotrophic waters in the world (Vincent and Vincent, 1982).

4.1.2. Stromatolites, Past and Present

Stromatolites are organo-sedimentary structures produced by the trapping, binding, and precipitation of sediment by living microorganisms, particularly cyanobacteria.

They are evident in the fossil record from the Precambrian to the Holocene, but declined in diversity in the late Precambrian and in abundance from the middle Ordovician. They illustrate the important role played by microorganisms in the Precambrian ecosystem, but they have been subsequently outcompeted by eukaryotic algae, excluded noncompetitively by metazoan browsers and burrowers, and possibly inhibited by UV in shallow waters, thereby being restricted to deep waters or shaded habitats (Parker et al., 1981). They currently occur infrequently in a variety of habitats, including cold fresh and saline Antarctic lakes beneath thick permanent ice of poor light transmittance. They are adapted to extremely low light intensities, a broad range of salinity, and a broad range of dissolved O_2 from supersaturation to anoxia. Benthic studies have been focused on Lakes Fryxell and Hoare, Ross Desert, but stromatolites from five other Victoria Land lakes have also been described (Parker and Wharton, 1985).

Stromatolitic mats are formed by cyanobacterial filaments gliding together and intermeshing to make a cohesive mucilaginous film (cf. cyanobacterial "rafts" colonizing fellfield soils). Cyanobacterial sheaths provide cementation and rigidity. The mesh forms knots on upward protrusions from the sand which become foci for tufts growing toward the light. Calcite is precipitated within the mat concurrently (Wharton et al., 1983).

Five major types of benthic cyanobacterial mats have been recognized by Parker and Wharton (1985), not all of which are stromatolitic—moat, ice cover, lift-off, pinnacle, and prostrate. Moat mats and ice cover mats are littoral mats and superficial cryoconite communities (Wharton et al., 1981), respectively. The littoral mats are widespread in freshwater lakes. Lift-off mats are of two types, those that break loose and those that remain in place as columnar stromatolites. Both are made buoyant by gas bubbles of photosynthetic origin.

A pinnacle mat, modified to show how O_2 bubbles cause the buoyancy of columnar and lift-off mats, is illustrated in Fig. 10. All stromatolites contain calcite crystals to some degree, especially in the organic layers beneath the mat itself. The sand is of allochthonous origin, probably arriving partly from meltstream input and partly by cyclic dumping after melting through the ice (Simmons et al., 1987). The converse of the lift-off mode of C loss from the lake ecosystem occurs in deeper, dimly lit areas of lakes where bubbles are not formed and prostrate mats result. During austral summer, glacial meltstreams deposit sand-size sediment on top of the mat surface. The cyanobacteria grow through this layer, leaving a layer of "dead" cells under the sediment. As there is no disturbance of this layering, the organic matter becomes buried and lost to the ecosystem while remaining within the habitat (Parker et al., 1982c). Prostrate mats are found extensively in aerated waters of Lakes Fryxell and Vanda and in anoxic water in Lake Hoare, which has yielded a stromatolite containing ca. 400 alternating layers (Wharton et al., 1983). These layers, like the sediment layers in the ice cover, may indicate climatic or seasonal changes and may reflect the recent history of a lake.

Filaments of the cyanobacterium *Phormidium frigidum,* common in soils, lakes, streams, and cryoconite holes on glaciers, consolidate the matrix in all stromatolites. It is usually admixed with *Lyngbya martensiana,* particularly in pinnacle mats of Lake Vanda, which also have the unique feature of a moss, *Bryum* cf. *algens.* Pennate diatoms are also common to all except anaerobic mats. Species diversity in various types of

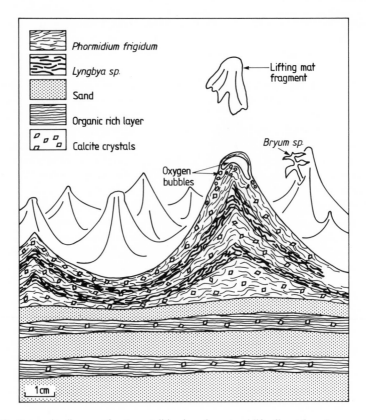

Figure 10. Composite diagram of a stromatolitic pinnacle mat and lift-off mat from ice-covered stratified lakes of the Ross Desert. [Redrawn from Love *et al.* (1983), with permission.]

benthic mats from five representative lakes has been summarized in Wharton *et al.* (1983). Heterocystous cyanobacteria are restricted to moat and cryoconite communities. Their N_2 fixation is unlikely to be significant since NH_4-N and NO_3-N are probably the most readily used sources of N for aquatic mats, especially on anoxic sediments which permit N mobilization. Moreover, the light intensity under ice is probably too low to generate sufficient ATP for the nitrogenase system, and the HDO would be inhibitory (Allnutt *et al.*, 1981).

Stromatolitic cyanobacteria are exceptionally well photoadapted for low levels of PAR. The pinnacle mat of Lake Vanda, dominant from the under-ice surface down to ca. 50 m, receives 18–2.5% of incident PAR, equivalent to 270–37.5 μE m^{-2} sec^{-1}. The columnar lift-off mat of Lake Hoare between 4.6 and 12 m receives even less light (1.0–0.1% of incident PAR, equivalent to 15–1.5 μE m^{-2} sec^{-1}), similar to the levels penetrating endolithic and SIMCO habitats. The amount of PAR reaching the mat regulates O_2 production and hence the formation of lift-off mats in shallower depths. Mats fix over 12 times more C per square meter per day (up to 940 mg C) than the plankton per cubic meter (up to 7.4 mg C). This difference is reduced to a factor of 2 on correcting for cell biomass but still shows the importance of mats for primary production (Parker and Wharton, 1985). On comparing productivity with mat morphology, the order of yield is moat > pinnacle > columnar lift-off > aerobic prostrate > anaerobic prostrate. All mats contain a large biomass of bacteria and yeasts supported by leachates and decomposition products, but food chains are either absent or poorly developed (Parker and Simmons, 1985). The food web is best developed in moat mats that have a low diversity of primarily bacteria-feeding protozoa (Cathey *et al.*, 1981) and three groups of microbe-feeding metazoa: rotifers, tardigrades, and nematodes, in decreasing order of abundance.

A prominent feature of mat communities is the ecophysiological flexibility of the dominant cyanobacterium *Phormidium frigidum*. It tolerates high levels of PAR (up to 1500 μE m^{-2} sec^{-1} in full sunlight) and of UV radiation in moat mats but can photosynthesize with as little as 1.5 μE m^{-2} sec^{-1}. It can tolerate freezing and metabolizes at ca. 0°C but also occurs in relatively warm water at 7°C in Lake Vanda. It grows not only in lift-off mats in Lake Hoare at an HDO concentration of up to 50 mg O$_2$ liter^{-1} but also in its anaerobic prostrate mats. Its salt tolerance ranges from freshwater to saline, and it tolerates a wide range of alkalinity. It also represents an eco-physiological link not only between soils, streams, and lakes but also between the continental and maritime Antarctic and with habitats as diverse as hot springs in Yellow-stone National Park, U.S.A., and hypersaline habitats in Shark Bay, Australia (Parker *et al.*, 1981). Moreover, pinnacle mats containing *P. frigidum* closely resemble the paleolithic stromatolite *Conophyton*, providing a link with the paleohistory of Antarctica (Love *et al.*, 1983).

4.2. Transient Rivers and Streams

Antarctic cold desert rivers and streams are more common than might be expected. Some, such as the Onyx and Alph Rivers in Victoria Land and the Talg and Tierney Rivers in the Vestfold Hills, are relatively substantial, with flow rates in excess of 2 million m^{-3} yr^{-1}. Their common factor is transience. For as little as 2 months per year, the Onyx River in Wright Valley flows inland for 30 km from the Wilson Piedmont Glacier into Lake Vanda, which does not, however, fill up because the input is balanced by water loss through evaporation and ablation. This is also the fate of the river itself. As the winter approaches, its flow dwindles as the supply of glacial meltwater diminishes and eventually either dries up or freezes and sublimes. It is recognizable in late winter only as a dry river bed with dehydrated cyanobacterial mats. The microbiota of such habitats, whether on the large scale of a river or the small scale of a meltstream, must therefore have ecophysiological resilience similar to that of soil microbes but with additional requirements for life in flowing water with its associated disturbance of the substrate by turbulence and sediment deposition. River discharge depends on meltwater input, which in turn depends on air temperature and incident radiation acting on snow and ice. Water flow therefore varies topographically, climatically, diurnally, and season-ally (Howard-Williams and Vincent, 1985).

The prolonged cloudless conditions of the continental climate result in potentially inhibitory levels of solar radiation. Protection is obtained either by synthesis of extra protective pigments such as carotenoids (Vincent and Howard-Williams, 1986b) or occupation of sublithic habitats (Howard-Williams *et al.*, 1986a). Epilithic communities are rare on exposed Ross Desert soils but are found in streams, being dominated usually by *Gloeocapsa* and *Schizothrix* (Vincent and Howard-Williams, 1986b). However, exposure to winds resulting from fluctuations in stream level make the habitat transitory and the residents hardy (Howard-Williams *et al.*, 1986b).

The residents of streams depend mainly on photoautotrophs for primary production of fixed C for growth and metabolism, although bacteria are present and active in this community. The cyanobacteria *Phormidium* and *Nostoc* are the dominant colonizers, forming dense, cohesive mats up to 8 mm thick. Whereas *Nostoc* mats (almost exclusively *N. commune*) are superficial in growth habit and have an organic C content of >20% (dry weight) (Vincent and Howard-Williams, 1986a), *Phormidium* mats become more integrated with the mineral substratum and contain 6% or less organic C. This indicates potential for *Phormidium* to stabilize the stream bed substratum. However, large amounts of mat material are lost from streams by disruptive turbulence during rapid glacial melt.

A large proportion of the epilithion survives the cold ($-55°C$), dark winter months to act as an inoculum in spring. The Onyx river bed has a 42% cover of cyanobacterial mats over winter, which increases to 80% in the slow-flowing reaches in late summer. The response to rehydration includes rapid, light-dependent CO_2 fixation. This is detectable in *Phormidium* mats within 20 min of rehydration, but full metabolic recovery, which is dependent on enzyme synthesis, takes about 2 days (Vincent and Howard-Williams, 1986a). Ammonium uptake begins immediately on wetting, and nitrate and phosphate are absorbed after only a brief delay. Photosynthetic rates are very slow, indicating C-turnover times of ca. 800 days for *Nostoc* and 200 days for *Phormidium* communities, which is an order of magnitude slower than in temperate streams (Vincent and Howard-Williams, 1986a,b). The stream mat communities in the Ross Desert must therefore take several seasons to establish, due mainly to low summer temperatures, usually 0–2°C, as light and nutrients are unlikely to be limiting.

Stream cyanobacteria are psychrotolerant rather than psychrophilic, having a temperature optimum of ca. 18°C, and are therefore metabolizing suboptimally in the field (Seaburg *et al.*, 1981). Low temperatures reduce DNA synthesis by bacteria, nitrogenase activity in *Nostoc,* microbial lipid biosynthesis, glucose catabolism, and community respiration and photosynthesis (Vincent and Howard-Williams, 1986b). High CO_2 production is usually associated with heterotrophic bacteria which are seen embedded in the mucilaginous sheaths of the cyanobacterial matrix. Bacterial rods, filaments, and cocci and filamentous fungi probably intercept photosynthetic leachate amounting to 1–3 g of dissolved organic carbon (DOC) m^{-3} (Downes *et al.*, 1986). The microbiota of Antarctic streams illustrate another example of an opportunistic survival strategy.

4.3. Temporarily Ice-Covered Freshwater Lakes

In contrast to the permanent stratification of Ross Desert lakes, those of the maritime Antarctic are subjected to extensive seasonal disruption by meltwater influx and usually wind-mediated turbulence. This occurs when the ice cover melts in spring, although ice may persist for more than 8 months and occasionally not melt at all. Moreover, the absence of salinity gradients makes stratification unstable and transient. Maritime Antarctic freshwater microbial populations are more diverse than those in continental lakes. Nutrient input is generally greater, and anoxia, if occurring in the water column at all, is confined to a few months per year.

The topography and catchment age of lakes at a typical maritime location such as Signy Island, South Orkney Islands (Fig. 1), have a prime influence on lake characteristics. The variety of lakes experiencing the same climatic regime, ranging from proglacial through oligotrophic to eutrophic, permits analysis of the evolution of maritime Antarctic lakes in relation to those of the continent (Priddle and Heywood, 1980).

Supraglacial lakes on ice or proglacial lakes in rock basins are frequently ephemeral but may develop a microbiota and even a simple food web derived from local inocula and global immigration. The most common lake type is oligotrophic, evolving from proglacial lakes as deglaciation proceeds and less of the catchment is ice covered. They are dependent mainly on the mineral substratum and aerial input for nutrients (Fig. 11). They vary in size, depth, and clarity, which is dependent on sediment loading and phytoplankton biomass and can have an extinction of $<10\%$ m^{-1}. The phytoplankton is mainly composed of pico-cyanobacteria and small chlorophytes and chrysophytes; planktonic diatoms are rare.

In winter, less translucent ice and thicker snow cover than on continental lakes preclude phytoplankton growth under the ice at Signy Island, although microbial decomposer activity continues (Ellis-Evans, 1985b). This differential microbial response results in the development of anoxic conditions, the extent of which depends on the degree of eutrophication of the lake (Ellis-Evans, 1981a,b). In most oligotrophic lakes, the oxycline remains within the sediment profile, whereas in highly eutrophic lakes it can extend to the ice–water interface (Ellis-Evans, 1982). Eutrophication can result from nutrient input from surrounding vegetation and pollution by animals, especially seals and penguins, which are abundant at Signy Island (Fig. 11B). Consequent increased O_2 uptake leads to an extended anoxic zone, and winter dissolved-O_2 concentrations range from 0.3 to 3.2 g m^{-3} in enriched lakes, compared with 3.1–8.5 g m^{-3} in oligotrophic lakes (Priddle and Heywood, 1980). The reducing conditions release phosphorus and DOC from the sediments and also change the composition of the nitrogen pool. Regenerated available N and P accumulate within the anoxic zone but to a lesser extent in the oxic water column. Clarity of the water may transmit adequate PAR to support a deep phytoplankton maximum at the oxycline, if present, where N and P become available from the anoxic zone. The phytoplankton may be highly efficient at utilizing these nutrients, such as in shallow Heywood Lake, where they incorporate 40% of available N (preferably as ammonium) and 12% of available P during the spring bloom (Hawes, 1983).

As in cold desert lakes, cyanobacteria of maritime Antarctic lakes form benthic mats which are photoadapted to lower light levels than are the phytoplankton (Priddle et al., 1986). Unlike cold deserts, where snowfall on lake ice is minimal, in the maritime zone it can be substantial, reducing PAR penetration from >20 to $<1\%$ of incident

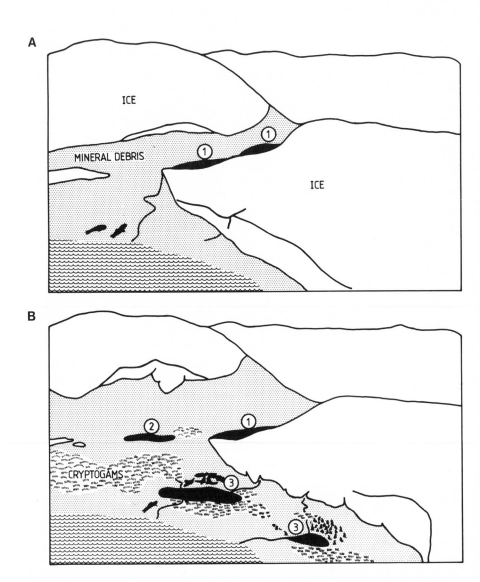

Figure 11. Evolution of lakes on Signy Island, maritime Antarctica. A. Proglacial lakes (1) dammed by ice. B. Oligotrophic lake (2) resulting from ice melt and recession, and eutrophic lakes (3) enriched by cryptogamic vegetation, penguins, and especially seals. [From Priddle and Heywood (1980), with permission.]

radiation (Hawes, 1985). At shallow depths in oligotrophic lakes at Signy Island, the availability of N and, more critically, P is growth limiting for phytoplankton, but at lower depths light limits the rate at which nutrients can be taken up. Temperature, in the range 0–4°C, is also limiting since resident bacteria are psychrotolerant rather than psychrophilic, indicating a closer link with the terrestrial microbiota than with the nearby marine community (Ellis-Evans and Wynn-Williams, 1985).

Cyanobacterial mats in oligotrophic lakes at Signy Island are seasonal in shaded or turbid waters but perennial in clearer shelf waters. Deep water mats are dominated by *Tolypothrix* in lakes with low sediment loading and *Trebonema* in lakes with higher sediment loading. As in Lake Vanda, they occasionally have aquatic mosses growing through them (Heywood *et al.*, 1980). Despite their genetic, structural, and metabolic disparities, these two phototrophic groups are similarly photoadapted (Priddle, 1980b). Although growth occurs throughout most of the year, C fixation rates are low, ranging typically from 11 to 45 mg C m^{-2}, with a recorded maximum of 320 mg C m^{-2} (Priddle, 1980a). Mats in shallow shelf areas are subject to freezing and have a different cyanobacterial community dominated by *Phormidium*. This has a high carotenoid content, as is also found in coastal Antarctic pond populations and continental Antarctic streams (Priddle and Belcher, 1981). Shelf zone mats are highly productive during their short growing season, e.g., 1480 mg C m^{-2} day^{-1} fixed in a shelf community in Lake Kamome, Syowa Oasis (Tominaga, 1977).

In younger lakes, the heterotrophic bacterial community resembles that of fellfield soil, but oligotrophic conditions in lakes are selective for strains with higher substrate affinities. The distinction between the terrestrial and freshwater microbiota is shown by the transience of catchment strains of *Janthinobacterium (Chromobacterium) lividum* in the oligotrophic water of Heywood Lake, Signy Island, relative to the persistence of the probably indigenous strain of *Chromobacterium fluviatile* under the same conditions (Ellis-Evans and Wynn-Williams, 1985). The distinction is emphasized concurrently by the flushing of the yeasts prevalent in the relatively nutrient-rich environment of the moss phyllosphere into the relatively yeast-free lake water in spring (Table V). The relatively low microbial species diversity of these habitats simplifies monitoring such indicator microbes.

The decomposer activity of Signy Island lake bacteria is mainly aerobic in the turbulent summer water and aerated sediments but can have a substantial anaerobic component in winter (Ellis-Evans, 1985b). Benthic mats appear to stabilize the redox potential of the underlying sediments (Ellis-Evans, 1984). Meanwhile, in some lakes the heterotrophic bacterioplankton population density closely follows the phytoplankton population, whose leachates and necromass appear to be growth limiting under the prevailing low-temperature regime. Although the sediment bacterial population is several orders of magnitude larger than that of the bacterioplankton, microautoradiography indicates that only a small proportion is actively metabolizing (Ellis-Evans, 1985b). After the mixing of the water in summer and consequent disruption of redox gradients in the water column and sediment, carbon sediments out in the autumn. As aerobic microbial activity removes oxygen, methanogenesis progresses up to the surface, accounting for 13% of the decomposition of organic C in Sombre Lake (Ellis-Evans, 1984). As

Table V. Seasonal Changes in Counts of *Chromobacterium* and Yeasts in Heywood Lake (Signy Island) and Its Catchment[a]

| | CFU (mean ± SE g^{-1} fresh weight or ml) | | | | |
| | *Chromobacterium* | | | Yeasts | |
Season	Moss	Lake water	Lake sediment	Moss	Lake water
Early winter	<100	130 ± 28	0	ND	ND
Late winter	<100	0	0	800 ± 200	0
Spring thaw	<100	0	900 ± 57	10,700 ± 500	100
Late spring	ND	ND	ND	1,000 ± 250	1,200 ± 200
Summer (thaw + 30 days)	522 ± 176	400 ± 29	5,000 ± 420	4,500 ± 700	1,600 ± 200
Autumn	2,783 ± 2,420	1,350 ± 194	12,000 ± 422	ND	ND

[a]After Ellis-Evans and Wynn-Williams (1985). ND, Not determined.

sulfate concentrations in all Signy Island lakes are very low, competition between sulfate reducers and methanogens is generally biased in favor of methanogens.

As N_2 fixation does not appear to be significant in Signy Island lakes, N conservation must occur to prevent loss of nitrogen from the ecosystem. Some bacterial isolates will reduce nitrate to nitrite, but few continue the reduction further. The majority anaerobically reduce nitrate to ammonia, which is assimilated and therefore recycled. Ammonia is also made available by hydrolysis of urea, which is an important component of phytoplankton debris (Ellis-Evans, 1985b). Experiments on bacterial isolates indicate that nitrate reduction and urea hydrolysis are unimportant in oligotrophic lake water but are of considerable significance in algal-cyanobacterial mats. The incomplete N cycle parallels those of cold desert endolithic ecosystems.

The luxuriant benthic vegetation in oligotrophic Moss Lake (Signy Island) supports a variety of protozoa and other invertebrate grazers. Eutrophication in nearby Heywood Lake has impoverished the benthic vegetation and consequently reduced the species diversity of grazers. Both lake types have relatively well developed food chains (by comparison with Ross Desert lakes) containing a few nektobenthic herbivores, e.g., the anostrachan *Branchinecta poppei* and the copepod *Pseudoboeckella gaini*, and one large predator (of *P. gaini*), the copepod *Parabroteas sarsi*. The bulk of the fauna are microherbivores such as protozoa, rotifers, tardigrades, nematodes, and enchytraeids that graze the epiphytic community of the mats. This simple food pyramid resembles that of the terrestrial ecosystem (Section 6.1) and contrasts with the paucity of nutrient cycling and food webs in Ross Desert lakes (Parker and Simmons, 1985).

4.4. Sea Ice Microbial Communities (SIMCO)

During the annual advance and retreat of sea ice in the Southern Ocean, ca. 17 million km^2 of sea ice habitat forms and melts (Gordon, 1981). This transient habitat supports a diverse assemblage of autotrophic and heterotrophic microorganisms which are probably to a large extent interactive and may therefore be regarded as a community. This concept is questioned by Horner *et al.* (1988), who propose a standardization of terminology based on assemblages in the surface, interior, and bottom zones of sea ice. As it contains a substantial microalgal and bacterial community, the SIMCO represents a vast potential inoculum and enrichment for the underlying water column.

The ocean is a turbulent environment, and microhabitats developed by the microbiota of the upper water column are rapidly disrupted. Sea ice provides a stable environment in which strata analogous to those of endolithic habitats and the strata of ice-covered lakes are established. Based on observations of unexpectedly high algal concentrations in newly formed Weddell Sea ice, Clarke and Ackley (1984) suggested that when seawater is in a state of incipient freezing, ice crystals may be nucleated by algal cells in the water column. They reported observing algae within initial 3- to 4-mm ice needles (termed frazil ice) and proposed that algae were further concentrated by the scavenging of more cells from the water column as frazil crystals rose to the surface. These two mechanisms for cell concentration may be unique to the sea ice of the Weddell Sea, which is derived mainly from frazil ice.

In McMurdo Sound, where a different temperature regime prevails, sea ice is formed predominantly at the ice–water interface by conductive heat losses along a temperature gradient within the ice. This results in the downward growth of columnar crystals termed congelation ice (Garrison et al., 1983) and is not dependent on nucleation by algae. This is consistent with the nondetection of microbes within crystals in McMurdo Sound sea ice by Sullivan and Palmisano (1984). Nevertheless, considering the large number of bacteria associated with the entire profile of land-fast and floating park ice, Sullivan (1985) suggested that bacteria rather than algae might be biological nucleators for sea ice. No Antarctic sea ice diatoms tested to date in pure culture has been shown to be nucleation active, but several ice bacteria from McMurdo Sound were active at $-10°C$. One psychrophilic strain was active with an ice nucleation temperature of -2.0 to $-2.8°C$.

When the frazil ice fraction of surface water exceeds 30–40%, sintering of crystals results in a continuous cover resembling congelation ice (Gow et al., 1981). Cells frozen in at this stage will later experience the lowest winter temperatures, down to ca. $-50°C$. As the ice thickens, salt is excluded and streamers of cold dense brine sink and are replaced by warmer, less concentrated seawater rising toward the ice face. The ice forms as a vertically orientated interlocking matrix of crystals which entraps marine microbes in brine pockets and channels as the matrix develops. In waters off Enderby Land and the Amery Ice Shelf region (Fig. 1), the bottom of the ice cover is hard, flat congelation ice (Fig. 12) (McConville and Wetherbee, 1983). In McMurdo Sound, however, positively buoyant "platelet" crystals up to 3 cm in diameter and 2–3 mm thick accumulate under the congelation ice to form an additional tightly packed layer ca. 0.5 to 4.0 cm thick (Horner, 1985a; Makyut, 1985). The congelation ice content of the Weddell Sea is only 47% of the total, the rest being frazil ice, often in alternating layers indicative of episodic freezing. The microhabitats for sea ice microbes are therefore significantly different in three sections of the Southern Ocean.

The ice of the Amery Ice Shelf region bears two distinct but interconnected microbial communities (Fig. 12A–C). One is a uniform matlike covering that is suspended into the underlying water column as colonial strands up to 15 cm long. These are dominated by a mucilaginous tube-dwelling pennate diatom, two species of ribbonlike Entomoneis (formerly Amphiprora), and Nitzschia frigida. They are colonized by epiphytic pennate diatoms. The strands anastomose and are held together by charged polysaccharides. The mucilaginous sheaths not only constitute attachments and stalks but also trap bubbles to aid buoyancy and help with selective iron retention (McConville, 1985). Since bubble-free strands are negatively buoyant, they are loosely secured to the ice by the penetration of a few diatom chains 2–3 mm into brine channels. The other type of community penetrates up to 5 cm into brine channels in the hard congelation ice without impeding the flow of brine. In contrast to this selective community, the microbiota of frazil-based Weddell Sea ice more closely resembles that of the underlying water but at a density 50 times greater (Garrison et al., 1982), probably due to its origin.

The microbes of McMurdo Sound are incorporated in the basic SIMCO interstitially between platelets (Bunt and Lee, 1970) and in brine pockets which later interconnect to become brine channels ca. 4 mm in diameter. Depending on the ice

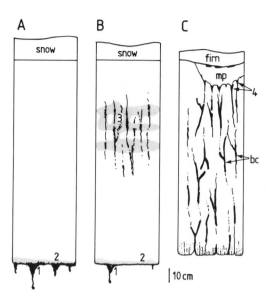

Figure 12. The development of sea-ice microbial communities from waters off the Amery Ice Shelf. A: Suspended colonial and prostrate mat communities (1) and colonial communities penetrating fine brine-channels (2) on bottom ice. B: Algae developing in interior ice (3) later in the season as brine channels interconnect. C. Surface melt pool algal communities (4) penetrating interconnected brine channels (bc) draining freely into the sea. [From McConville and Wetherbee (1983), with permission.]

thickness, there may be as many as 50–300 channels per m². The SIMCO is dominated by two groups of diatoms: attached mucilaginous forms which penetrate brine channels and hang into the water column as off Amery Ice Shelf (e.g., *Nitzschia, Amphiprora,* and *Fragilaria*), and free-living forms such as *Biddulphia* and *Coscinodiscus*. Interior ice (Fig. 12B) often contains a colored band of algae that is probably residual from the autumn bloom (Hoshiai, 1977; Garrison *et al.*, 1982). Surface ice communities in melt pools (Fig. 12C) expand their habitats by black-body radiation. The pools are dominated by small diatoms and flagellates such as *Pyramimonas* and *Phaeocystis pouchetii*, which is dominant in the oceanic spring bloom as the sea ice disperses (Bunt, 1964). The SIMCO *Phaeocystis* population may act as an inoculum for the bloom, although Mc-Conville *et al.* (1985) concluded that ice algae were in poor physiological condition at the time of their release and were probably of minor importance (cf. terrestrial snow algae). Nevertheless, their input of organic nutrient into the water column for heterotrophic microbial activity is probably substantial.

Despite their abundance in the saline water of Ross Desert lakes, cyanobacteria have not been reported in any Antarctic SIMCO after direct observation (Sullivan *et al.*, 1985) or taxonomic collation (Horner, 1985b). Heterotrophic bacteria are abundant in the SIMCO, but their vertical distribution has been described only for McMurdo Sound and the Weddell Sea. Sullivan and Palmisano (1984) described two groups: free-living bacteria, which made up 70% of the total, and attached or epibacteria. Free-living forms frequently produce fibrillar exocellular polymeric substances. Symbiotic relationships were suggested by the close physical association of epibacteria with healthy diatoms, especially *Amphiprora* spp., which constituted 13% of the diatom population. *Amphiprora* may leak photosynthate that is made available to the epibacteria or, conversely, may not produce antibiotic material that would inhibit bacteria. The stalked bacterium *Prosthecobacter* is conspicuous among the epibacterial microbiota.

To survive and flourish in the SIMCO, microbial colonizers must interact with the physically stable substratum and exploit the abundant inorganic nutrients of the Southern Ocean. Holm-Hansen *et al.* (1977) reported the availability in the water column of 40–90 μM silicate, 10–20 μM nitrate, and 1–2 μM phosphate, all of which are concentrated further in the SIMCO by salt exclusion during freezing. However, environmental temperatures are low, ranging from ca. $-50°C$ at the upper surface in winter to $-1.8°C$ at the ice–water interface.

The melting of surface ice and the concentration of brine during freezing results in salinities ranging from <0.5% in meltwater (Whitaker, 1977) to ca. 16% in interstitial brine from 1-year ice collected in late winter (Palmisano and Sullivan, 1985a). This brine has a freezing point depression of $-15.2°C$. At least some sea ice algae have cryoprotectant mechanisms: Burch and Marchant (1983) have shown that two flagellated ice algae, *Chlamydomonas* sp. and *Pyramimonas gelidicola,* sustain their peripheral microtubule integrity (usually unstable below ca. 4°C) and flagellar function down to -14 and $-18°C$, respectively. Both are highly tolerant of osmotic stress and can grow at salinities ranging from 5 to 14%, almost exactly the range reported for the SIMCO. *Pyramimonas, Chlamydomonas,* and *Dunaliella* are all found in hypersaline lakes in the nearby Vestfold Hills, where they tolerate low temperatures and high salinities by accumulating polyols. As in other Antarctic habitats, polyols may therefore be used as compatible solutes by residents of SIMCO brine channels. Moreover, algal mucilages also provide freezing protection, as they prevent adhesion of ice crystals to the cell wall and lower the freezing point of the microenvironment of the cell (McConville, 1985).

The PAR gradient of the SIMCO ranges from ca. 1500 μE m^{-2} sec^{-1} in full sunlight to c. 6 μE m^{-2} sec^{-1} in the platelet layer under ice with a snow cover (Palmisano and Sullivan, 1985b) and total darkness for several months in winter. However, ice algae such as *Fragilariopsis sublinearis* have been shown to photoadapt by increasing their chlorophyll *a* content logarithmically and have an absorption spectrum appropriate for the transmitted light (Bunt, 1968). The compensation point for these organisms was remarkably low at ca. 1 μE m^{-2} sec^{-1} and even as low as 0.5 μE m^{-2} sec^{-1} (Bunt, 1964). However, they assimilate only ca. 0.6 mg C mg chlorophyll^{-1} hr^{-1} (assuming ca. 33% loss of extracellular leachate). This value is considerably lower than that for temperate phytoplankton (2 to 10 mg C). It is notable that there is a similar extreme degree of photoadaptation to very low light levels in several distinct terrestrial, freshwater and marine habitats. This probably indicates a limit for photosynthetic efficiency for opportunistic prokaryotes and eukaryotes in microenvironments where other factors are not consistently rate limiting.

Despite the prevalence of psychrotolerant and psychrophilic microbes among the ice algae and bacteria (Bunt, 1968; Kottmeier and Sullivan, 1988), Palmisano and Sullivan (1983) concluded that temperature variation could not explain seasonal changes in McMurdo SIMCO standing crop, population density, or spatial distribution. During the brief ca. 3-month growing season, estimates of algal and bacterial growth rates *in situ* indicate doubling times of 7 and 14 days, respectively (Sullivan *et al.*, 1985). The annual production of the McMurdo SIMCO was estimated as 4.9 g C m^{-2}, based the relationship 38 g C g chlorophyll a^{-1}. However, such values are difficult to interpret without knowing the variation in ice thickness (Horner *et al.*, 1988). Errors likely to make this an underestimate of productivity include incompleteness of the bloom, losses due to grazing, and secondary production by ice bacteria. Bacterial heterotrophy does not add C to the system but can change its availability to other organisms. Losses due to respiration are likely, but this process helps to reduce potentially toxic levels of O_2 resulting from its freezing out as in the HDO zone of Ross Desert lakes.

The variety of sea ice conformations provides diverse habitats for microorganisms (Table VI). Populations can achieve extraordinary densities, such as 5-cm-thick folds of an aggregation of the diatom *Biddulphia punctata,* on an underwater shelf of a grounded iceberg amounting to a standing crop of ca. 40 g m^{-2} (Whitaker, 1977). Moreover, the presence of the SIMCO extends the marine growing season. Primary production occurs at a rate of 81 μg C liter^{-1} hr^{-1} in early ice during November, when it is negligible in the water column. Production may reach as high as 363 μg C liter^{-1} hr^{-1} in localized melt pools near the time of ice breakout (McConville and Wetherbee, 1983). On a larger scale, Ackley *et al.* (1979) estimated that up to 21% of primary production in the Weddell Sea was due to the SIMCO. This order of contribution is of considerable significance for the energy and nutrient budget of the Southern Ocean.

5. Microbiology of Patterned Ground

5.1. Frost Tolerance in a Wet Habitat

The maritime Antarctic fellfield environment is characterized by patterned ground resulting from freeze–thaw cycles in the presence of free water. Unlike continental cold deserts, fellfield soils are not generally saline because of leaching by plentiful precipitation. Conditions for microbial growth on mineral and cryptogamic substrata differ nutritionally and edaphically, and there are gradients between the two microhabitats at the periphery of frost polygons. Cryptogams release soluble organic matter, predominantly polyols, into the mineral soil (Tearle, 1987). These potential compatible solutes (Section 3.2) confer a degree of freezing and desiccation tolerance to cryptogams, but repeated freeze–thaw transitions damage a proportion of the cryptogamic cell membranes, resulting in the release of soluble cell contents which are utilized by the microbiota. Tearle (1987) has shown that lichens at Signy Island are more resistant than mosses to frost damage. However, net losses from mixed cryptogamic vegetation result in polyol-plus-sugar concentrations of up to 1.5% (dry weight) in mineral soil. The polyol content (mainly arabitol and ribitol) of lichen tissue is seasonally constant at 97%

Table VI. Sea Ice Habitats Present in Winter at the South Orkney Islands[a]

Habitat	Occupants and comments
Coastal tide crack zone	
Snow melt pools	Terrestrial and snow algal groups after flood
Infiltration ice–slush	Abundant diatoms, mainly *Navicula glacei*
Vertical walls of tide cracks	Abundant algae, including *Navicula glacei, Nitzschia curta, Nitzschia lineata,* and *Thalassiosira tumida*
Ice foot	43 species; *Phaeocystis antarctica* replaced by *Navicula glacei* and *Nitzschia curta* later
Anchor ice	Similar to ice foot
Fast ice sheet	
Snow melt puddles	No microalgae found
Infiltration ice–slush	*Navicula glacei* common; other microbes later
Congelation layers and columnar sea ice	Abundant diatoms later; *Navicula glacei, Chaetoceros schimperianum,* and *Nitzschia curta* dominant
Lower surface of fast ice sheet	*Nitzschia curta* dominant; others sparse
Loose congelation ice crystals (platelets)	*Amphiprora kjelmani* and two *Nitzschia* spp. very abundant in October–November if ice persists
Simple pack ice	
Snow	No microalgae found
Infiltration ice zone	Dense concentration of *Nitzschia closterium*
Underwater ice mass	Sparse growth of *Nitzschia* (=*Fragilariopsis*) spp., including *Nitzschia curta*
Offshore tide cracks	
Dense snow slush on seawater	Dense growth of *Thalassiosira tumida; Nitzschia* (=*Fragilariopsis*) also present
Icebergs	
Ice foot	Unexamined dense microalgal growth visible
Shelf area	Massive growth of about 50 diatom spp.; *Biddulphia punctata* extremely abundant
Vertical walls	Abundant microalgal growth visible down to 17 m

[a]After Whitaker (1977).

of the total polyol-plus-sugar pool, which amounts to 24% of cell dry weight. However, the polyol content of mosses varies from 8% in summer to 56% in winter, about three times that of equivalent temperate species. It is possible that this substantial supply of polyols not only acts as an energy source for the heterotrophic microbiota but may also provide cryoprotection. This effect may escalate during the freezing process, since solutes become more concentrated in the soil pore microhabitats of the microbiota as ice crystallizes out. Metabolic adaptation or selection for available substrate is suggested by atypical utilization of trehalose, another leachate and potential cryoprotectant, by *Janthinobacterium (Chromobacterium) lividum* from Antarctic soils (Wynn-Williams, 1983).

Cryoprotection is especially important in wet soils because of the vulnerability of fully hydrated microbial cells. However, it may not be energetically feasible to accumulate an adequately protective concentration of compatible solutes against a steep osmotic gradient. Nevertheless, if freezing is delayed by supercooling, as in Antarctic mites and insects (Cannon and Block, 1988), the growing season may be extended significantly. The ability of yeasts to accumulate cryoprotectants and supercool may partly account for their large numbers in moss communities at Signy Island (Baker, 1970a) and their exponential increase under conditions that damage the moss tissue on which they are resident (Wynn-Williams, 1980; Tearle, 1987). Lyakh *et al.* (1984) showed that after periodic freezing to $-13°C$ and thawing of the Antarctic yeast *Nadsoniella nigra* var. *hesuelica,* 33% of the populations was viable after one cycle and 10% was still viable after ten cycles. The opportunistic metabolism of yeasts at low metabolic temperatures, combined with frequent capsulation which acts as a moisture-regulating mechanism (Aksenov *et al.*, 1973) and the ability to store potential compatible solutes as food reserves (Brown, 1978), conveys ecophysiological advantages. This may account for the widespread geographical distribution of yeasts in diverse Antarctic habitats (Castrelos *et al.*, 1977; Atlas *et al.*, 1978; Wynn-Williams, 1980, 1982; Fletcher *et al.*, 1985).

Algae and cyanobacteria, which are frequently abundant on frost-sorted ground of the coastal and maritime Antarctic, also show frost tolerance. The cyanobacterium *Nostoc commune,* a frequent colonizer of moist fellfield soils, fixes CO_2 at temperatures down to $-5°C$, whereas the green alga *Prasiola crispa,* common on soils under a direct marine influence, respires at as low as $-15°C$ (Becker, 1982). It has been suggested that their ability to accumulate sugar monophosphates (up to 50% of the water-soluble metabolites from photosynthesis in *Nostoc commune*) may offer cryoprotection. The synthesis of polyols and sugars as compatible solutes may be a common factor in the cryptogamic flora and invertebrate microfauna of the maritime Antarctic.

The microbiota of maritime Antarctic soils not only seems tolerant of freeze–thaw cycles but is efficient at activating its respiratory and growth metabolism at near-freezing temperatures. Water availability, rather than temperature itself, appears to be the limiting factor. Gross microbial respiratory activity, 83–88% of total respiration in organic substrata (Wynn-Williams, 1984), was detectable at $-2°C$ and escalated rapidly during the freeze–thaw transition. Moreover, it was maximal at only 0.5 to 1.0°C both in field-fresh material and under simulated field conditions (Wynn-Williams, 1980, 1982).

Some maritime Antarctic microbes benefit from the action of freeze–thaw cycles on wet rocks such as quartz-micaschist and marble at Signy Island, both of which split along cleavage planes (Hall, 1986a,b). This process creates chasmolithic microhabitats which are colonized by cyanobacteria and algae, especially in translucent marble. The hydrostatic action of their growth further weakens the rock, which is also weathered biologically by crustose lichens (Walton, 1985).

5.2. Microbial Stabilization of Fellfield Soils

The process of microbial colonization and stabilization in the structureless Antarctic fellfield soils is dependent on a sequence of interacting physical, chemical, and biological requirements, a failure in any one of which can upset the delicate balance of such a vulnerable system (Fig. 13). The success of colonization also depends on the number and characteristics of incident microbial propagules (Broady, 1979a) and their ability to settle out from the airstream before they are blown away from a potentially suitable habitat (Fig. 6). This is particularly important for small Antarctic islands, which

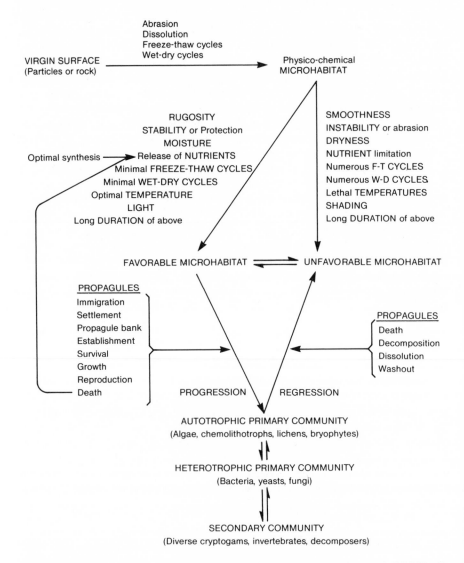

Figure 13. Physical, edaphic, and biological processes in the colonization of Antarctic fellfield soils. [From Wynn-Williams (1986).]

are typically swept by strong winds (Walton, 1982). Animal vectors may be contributory, as Schlichting *et al.* (1978) found evidence of a probable transfer of a cyanobacterial species endemic to a pond on Ellesmere Island (Arctic) to the Antarctic by Arctic terns. The overlying winter snow and ice cover, which accumulates propagules in spring and for several subsequent months, melts to provide plentiful free water (Walton, 1984). As in lake and sea ice, it frequently develops an extensive algal and cyanobacterial community which may color the snow red, green, or yellow as the melt proceeds. The meltwater, laden with live and dead algal cells and leachates, percolates down to the soil surface and into meltstreams, which enrich and inoculate the soil, albeit with a senescing population.

Once microbes have immigrated and settled, conditions may not be favorable for progress; therefore, they become part of a propagule "bank" in the soil. P. V. Tearle (personal communication) noted that despite the infrequent occurrence of purple pigmented *Janthinobacterium lividum* in fresh samples of fellfield fines at Signy Island, it became the dominant viable bacterium isolated from soil crusts incubated at constant 15°C and constant saturated relative humidity *in vitro*. If propagules meet the necessary ecophysiological requirements (Table VII), they will survive to colonize and stabilize their new environment (Wynn-Williams, 1986).

To be effective in stabilizing the soil, microbes must have at least one of two prime attributes: first, filamentous morphology to provide a fibrous framework that enmeshes soil particles and other microbes, and second, mucilaginous capsules or sheaths that cement the particles and microbes together. Filamentous cyanobacteria frequently produce polysaccharide sheaths, combining both attributes with an independent energy source, photoautotrophy (Broady *et al.*, 1984). Cameron and Devaney (1970) concluded that capsulated bacteria preceded cyanobacteria and algae as colonizers of deserts because the phototrophs had a greater moisture requirement. Bacterial capsules not only store moisture for the individual but can form aggregates that not only conserve water more efficiently but also improve nutrient exchange and soil stability. Such aggregates, predominantly cyanobacterial and algal, are also apparent in moister maritime Antarctic soils (Wynn-Williams, 1986, 1989).

Soil fines in the center of frost-sorted stripes or polygons are eminently suitable for studying microbial colonization processes because of their relatively homogeneous particle size distribution and flat surface. Natural microbial assemblages can be examined directly with minimal disruption, using epifluorescence microscopy (EFM) (Wynn-Williams, 1985c, 1986). Among the primary colonizers of the substratum are cyanobacteria and green algae which autofluoresce. Although the relatively low species diversity of the fellfield soil crust microbiota makes marker species such as the large filaments of *Phormidium* spp. recognizable (Broady *et al.*, 1984), the density of the population and the diversity of its cell morphology and orientation preclude visual quantification *in situ*. However, pseudo-color television image analysis (TVIA) permits quantification of this community. TVIA has hitherto been applied to ecological studies of freshwater and marine planktonic bacteria and cyanobacteria (Sieracki *et al.*, 1985; Bjørnsen, 1986; Estep *et al.*, 1986). Its use for the determination of microbial biomass in ecosystems has been reviewed by Fry (1988). The EFM–TVIA combination permits not only the quan-

Table VII. Requirements for Potential Microbial Colonizers in a Cold Desert or Fellfield Environment

Requirement	Attribute or strategy	References
Translocation to a favorable site, e.g., moist soil, translucent porous rock	Aerial transport Flowing meltwater Migration into the substratum	Cameron (1972b) Wynn-Williams (1982) Friedmann and Ocampo-Friedmann (1984a)
Compatibility with substratum structure	Diverse or variable propagule morphology	Cameron and Devaney (1970), Friedmann (1982)
Tolerance of water stress Desiccation and wet–dry cycles	Comulative compatible solutes, e.g., polyols Ability to supercool	Brown (1978) Block (1980)
Freezing and freeze–thaw cycles	Freezing resistance Ability to freeze–dry Rapid rehydration Endolithic thermal buffering	Meryman (1966) Cameron and Blank (1967) Aksenov et al. (1973) McKay and Friedmann (1985)
Osmotolerance in saline soils	Cumulative compatible solutes, e.g., polyols	Miller et al. (1983), Brown (1978)
Tolerance of low mean ground temperatures	Psychrotolerant enzymes or membranes Membrane adaptation High Q_{10} at low temperatures	Herbert (1986) Bhakoo and Herbert (1980) Svensson (1980)
Tolerance of variable, low PAR intensity	Shade adaptation Spectral phootoadaptation	Vincent and Vincent (1982) Van Liere and Walsby (1982)
Nutritional independence	Autotrophy Lichenization	Friedmann and Ocampo (1976) Friedmann et al. (1982)
Tolerance of starvation	Dwarf cells Dormancy, quiescence Polyol–sugar storage Adhesion to particles Fast metabolic response Low activation energy Ruderal strategy	Novitsky and Morita (1976) Poindexter (1981) Lewis and Smith (1967) Dawson et al. (1981) Wynn-Williams (1982) Poindexter (1981) Pugh and Allsopp (1982)
Avoidance of unstable or unfavorable substrata	Sublithic, chasmolithic, and endolithic growth.	Broady (1981a,b, 1986b), Friedmann (1980)
Substratum stabilization	Filamentous morphology Mucilage cementation	Friedmann (1982) Cameron and Devaney (1970)

tification of cell images but also differentiation between microbial groups in mixed populations with respect to pigmentation, dimensions, and morphology (Wynn-Williams, 1989).

Visual distinction between cyanobacteria and algae is possible by differences in the emission spectra of their photosynthetic pigments (Cohen-Bazire and Bryant, 1982;

Wood *et al.*, 1985). All cyanobacteria contain orange-fluorescing phycocyanin, and many also contain yellow-fluorescing phycoerythrin (Wynn-Williams, 1989). Cyanobacteria and eucaryotic algae both contain red-fluorescing chlorophyll *a*. Optical filters, particularly narrow-band interference filters, such as 640 nm for phycocyanin (Schreiber, 1979), have been used to separate cyanobacteria from green algae in fellfield soil crusts and to monitor phycoerythrin-rich species (Wynn-Williams, 1989). The method has also been applied to heterotrophic colonizers such as bacteria, yeasts, and microfungi stained with Acridine Orange and a photofading retardant (Wynn-Williams, 1985c).

EFM–TVIA shows that three distinct groups of cyanobacterial filaments (probably *Phormidium* spp.) are dominant colonizers of fellfield soils in frost-sorted polygons at Signy Island. A dense mesh of fine filaments (ca. 2 μm in diameter) is usually overlain by a coarser network of cells 5 μm in diameter and a patchy surface distribution of either broad filaments (>7 μm in diameter) or gelatinous aggregates of cells. The area of colonization by phototrophs is closely correlated with filament length per unit area. The distribution of these colonizers is related to the occurrence of "rafts" of cells which have been detected as preformed colonizing units settling from meltwater onto artificial substrata (Wynn-Williams, 1986). In midsummer, water is not limiting for colonization but P and energy substrates probably are. Although glucose stimulates microbial colonization, natural polyol-rich moss extract is more beneficial for aggregation processes (Tearle, 1987; Wynn-Williams, 1986). Heterotrophs apparently promote colonization, possibly forming biofilms over particle surfaces (Fletcher and Marshall, 1982). Observations of fellfield fines at Signy Island suggest that the cyanobacterial-algal-bacterial colonizer crust disintegrates into rafts of a size that survives the physical upheaval of freezing and coalesces again on thawing to reestablish a stabilizing microbial crust.

5.3. A Meeting of Ice and Steam

The occurrence of steam-warmed ground on the slopes of Mt. Erebus on Ross Island, Mt. Melbourne in northern Victoria Land, and Deception Island in the South Shetland Islands (Fig. 1) provides an incongruous Antarctic habitat for microbial colonization. Not only is the ground exceptionally hot by Antarctic standards, but the isolation of the habitats, particularly the summits of the mountains (altitudes of 3794 and 2733 m, respectively), provides a unique measure of long-range aerial dispersal since some taxa found are not of local origin. The sites surveyed by Broady (1984) for the occurrence of algae were at an altitude of between 3350 and 3793 m, certainly not under any direct marine influence. Four species of cyanobacteria and twelve species of eukaryotic green algae were identified, of which one cyanobacterium could grow at 50°C and three could grow at 40°C, and all chlorophycean algae grew at 22°C. Only cyanobacteria were found on soils at >39°C (up to a maximum of 59°C), while 86% of the algae occurred at below 30°C. As in cold soils of the Ross Desert, the phototrophic microbiota is prevalent where there is moisture—in this case, condensed steam which freezes on cold ground but may thaw with solar heating. The detection of seven taxa not previously recorded in Antarctica is probably indicative of the isolation of the site. It is stimulating to conjecture how a thermophilic *Lyngbya* sp. arrived. It has several characteristics favorable for long-range dispersal and colonization: resistance to desiccation, to deep freezing, and to storage in the dark at 14–15°C for over a year, autotrophic energy metabolism, a broad temperature range for growth, and an independent source of nitrogen (N_2 fixation).

The vigorous activity of Mt. Erebus a few hundred years ago almost certainly annihilated the vegetation (Giggenbach et al., 1973), so the population is not a relict biota. The potential for recolonization may not yet have been realized, since the statistical chances of the microbiota capable of filling all the available niches are so low. The nearest volcano capable of exchanging thermophilic propagules is Mt. Melbourne, 400 km to the north. Deception Island and the geothermal soils of New Zealand are approximately equidistant at ca. 4000 km. The prevailing winds are contrary to propagule input from both of these sources. Prolonged winter darkness (16 weeks) may be partly responsible for the poor representation of cyanobacteria, particularly if the soil remains unfrozen and respiration is sustained. Moreover, the acid volcanic soil favors algae rather than cyanobacteria, and rapid desiccation of porous soil is an additional stress. This unusual localized environment may result in the evolution of new strains.

In a similar study of geothermal soils on Mt. Melbourne (Broady et al., 1987), significant N_2 fixation was detected. This was probably due mainly to the heterocystous cyanobacterium *Mastigocladus laminosus* but possibly also to the bacterial genera *Bacillus* and *Klebsiella,* particularly in cyanobacteria-free subsurface soil. Thermophilic fungi and actinomycetes were isolated, but many fungal and all yeast and bacterial isolates were mesophilic or even psychrotolerant, consistent with a cosmopolitan origin. Of the total of 19 species of algae plus cyanobacteria, seven were common to both mountains, whereas four were restricted to Mt. Erebus and eight others were restricted to Mt. Melbourne. This may be due to heterogeneous deposition of the airspora or to the inability of later potential colonizers to compete with the extant thermotolerant residents. The discovery of the testate amoeba *Corythion dubium* on Mt. Melbourne is a remarkable example of global species distribution, since it is common at Signy Island (Smith, 1978) and has a bipolar distribution. However, it was not detected in any of 78 samples collected on Mt. Erebus. These inconsistencies present valuable ecological challenges in a region that to date has received minimal human disruption or contamination.

Deception Island is also extremely active, resulting in the creation of a new island within the caldera in 1967. The microbiota colonizing the island was surveyed a year later by Cameron and Benoit (1970). The microbiota is under a strong marine influence and is in an airstream that probably has a South American influence. After a year, the volcanic ash contained up to 1100 bacterial colony-forming units (CFU) g (dry weight)$^{-1}$, a large percentage of which grew at 2°C. There were also up to 37 mold CFU and ca. 10 yeast CFU g^{-1}, but the distribution was very heterogeneous. At one site sampled after 1 and 12 years, the mesophilic bacterial population had increased from 5 CFU to 1.9 million CFU, yeasts from 0 to 1220 CFU, and molds from 2 to 60,000 CFU g dry soil^{-1} (D. D. Wynn-Williams, unpublished observations). Cosmopolitan molds and the yeast *Cryptococcus* were detected within 1 year. Cyanobacteria were apparently sparse, possibly because of the methodology used but also because of the porosity and acidity of the ash. In this maritime region, mosses colonize the soil relatively rapidly (Smith, 1984a) and support a microbial population of their own. The protozoan population likewise recolonizes the moister sites much more rapidly than in the isolated sites on the Ross Dependency volcanoes (Smith, 1985).

6. Terrestrial Nutrient Cycling—Peat and Penguins

Two distinctive ecosystems will be considered: moss–peat communities of the maritime Antarctic and the ornithogenic soils of coastal penguin rookeries.

6.1. Moss–Peat Ecosystems

Study of the microbiology of moss–peat communities has focused on the decomposition of soluble and insoluble organic material and the consumption of the resulting biomass by microfaunal grazers. A moss peat habitat was the basis of the Signy Island terrestrial sites (SIRS) ecosystem program, which covered a variety of trophic levels and nutrient interactions over a 10-year period. The findings were synthesized by Davis (1981) and included microbial aspects of aerobic respiration, decomposition, and later N cycling, anaerobic microbiology, and methanogenesis.

The major regulatory mechanism of the system is climatic change from winter to summer, accentuated to a large extent by the formation and departure of winter sea ice, which stabilizes the macroclimate (Collins *et al.*, 1975). Changes in the macroclimate, in turn, affect the microclimate (Walton, 1982), the most profound effect being on the freeze–thaw transition in spring. Freeze–thaw cycles can occur at any time of the year, but the initial melt of winter ice releases a pulse of nutrients, including polyols and sugars (Tearle, 1987). The microbiota responds almost immediately by a rapid increase in respiration and evident exponential yeast growth even though the temperature remains near 0°C because of buffering by latent heat exchange (Wynn-Williams, 1980, 1982). Psychrotolerant opportunistic yeasts, which are probably epiphytic in origin rather than part of the underlying peat microbiota (Baker, 1970a), exploit soluble nutrients released from moss tissue by frost damage. Respiration rates in moss carpet communities were remarkably similar wherever in the maritime Antarctic the community was sampled. This was probably a measure of the relatively stable water regimes of these communities, unlike moss turfs, which were more variable in both water content and respiration rate (Wynn-Williams, 1984). The microfloral population was characterized by ruderal (opportunistic) species that exploit the disruptive effects of freeze–thaw cycles (Pugh and Allsopp, 1982).

As the substratum of a moss–peat habitat has a total organic content of ca. 98%, the rate of cellulose decomposition is a major factor influencing the existence of the habitat itself (Wynn-Williams, 1980, 1988). The balance between moss growth and decomposition defines the formation of peat banks or carpets (Davis, 1986). Temperature and moisture are the prime regulatory factors, although the ground-heating effect of solar radiation has been suggested to account for rapid cellulose decomposition at Rothera

Point on the Antarctic Peninsula (Wynn-Williams, 1988). Excess moisture results in waterlogging and anaerobiosis in moss carpets. In addition to CO_2 release, methanogenesis is significant but heterogeneous in time and location in the moss carpet ecosystem (Yarrington and Wynn-Williams, 1985), accounting for some of the discrepancies in the Davis (1981) ecosystem model, which was based solely on aerobic pathways. Drier moss turf communities have a higher cellulose content and higher C:N ratio than do moss carpets (Davis, 1986), which have smaller bacterial and yeast populations (Wynn-Williams, 1985a; Christie, 1987b).

Davis (1981) concluded that the efficiencies and pathways of organic matter transfer were similar in moss turf and carpet sites despite differences in redox conditions, water regimes, mean ground temperature, and nutrient levels. However, the higher microbial decomposer activity in the carpet led to a smaller accumulation of dead organic matter and, consequently, no formation of peat banks.

Substantial numbers of algae and cyanobacteria (especially *Nostoc*) were found in SIRS communities (Broady, 1979b; Christie, 1987a), especially in the wet site. The diverse microbiota developing in the moss–peat matrix is the basis for the food pyramid which culminates in the sole predatory mite, *Gamasellus racovitzai* (Davis, 1981; Burn, 1984; Lister *et al.*, 1987).

6.2. Ornithogenic Soils

The scale of nutrient enrichment, physical disturbance, and local warming has a major ecological impact on otherwise cold, oligotrophic coastal soils. Guano deposition by penguins is a major source of N, phosphate, and other nutrients, but their decomposition and recycling is dependent on moisture, which is probably the major regulatory factor. In rookeries at Capes Royds and Bird on Ross Island, the limited precipitation of the continental weather regime at 77°S results in only 5% of guano-C being decomposed annually (Orchard and Corderoy, 1983). Ramsay (1983) showed correspondingly that <8% of the bacterial population of rookery soils was metabolically active, even under summer conditions. It is probable that decomposer bacteria from the penguin gut either may not be able to metabolize at the much lower temperatures of rookery soils or may need time to adapt to the new environment. Moreover, the bacteria may be inhibited by accumulations of antibiotic substances such as acrylic acid, derived from the dominant marine phytoplankton *Phaeocystis* consumed by krill (*Euphausia superba*), which constitutes a major portion of a penguin's diet (Sieburth, 1961).

Because water is plentiful in the maritime Antarctic, uric acid in penguin rookery guano in the South Shetland Islands is rapidly decomposed (Myrcha *et al.*, 1985). About 50% of the C and N in fresh guano is volatilized in the first 3 weeks. Most of the dissolved ammonia is oxidized to nitrate in the subsoil, and urea and chitin are degraded totally in the second stage of mineralization. Conversely, microorganisms are responsible for immobilizing at least 11% of the phosphate brought from the sea by penguins. This ecosystem represents a cycle that is closely linked to the productivity of the marine ecosystem but which gradually enriches the terrestrial environment. Strong onshore winds are common in the maritime Antarctic, so volatile nutrients are blown inland directly or dissolved in spray and precipitation. Ornithogenic nutrients are also redistributed in the form of dust and are translocated to a lesser extent by animals. In the Ross Desert, where the atmosphere is the prime source of fixed N, enrichment by rookery volatiles, aerosols, and dust may therefore be an important source of N and P.

7. Environmental Impact

7.1. Dry Valleys Drilling Project (DVDP)

Environmental impact is traditionally a controversial issue because of the difficulties of defining the initial baseline, the parameters crucial to the detection of impact, and the biological significance of the impact. Moreover, the environment is not static but is in a state of flux, responding to physical and chemical changes of overlapping periods of time, which in turn affect the biota. There will also be microbiological "feedback" with respect to evolution, adaptive strategies, and biological modification of the environment. An assessment of impact must therefore consider rates and cycles as well as standing crops and short-term change. There is an assumption that monitoring the microbiota provides a measurement of biological impact. This basic concept has been put in perspective by McGinnis (1978), who challenged the techniques of impact detection used for the DVDP from the drilling engineer's viewpoint. These included microbial aerosol sampling, microbial soil and water sampling, and sediment core culturing for microorganisms. He questioned why an alteration in microbiological populations of soils at a site is considered detrimental to the environment and queried the initial status of "pristine" conditions, the level of background "noise" in the microbial flux, and the significance of the changes relative to this level.

We now know more about the early history of Antarctica and Gondwana, so what we currently consider to be the typical Antarctic microbiota may have been tropical in origin (Boyd *et al.*, 1966; Friedmann and Ocampo-Friedmann, 1984b). We now also know about the interchange of bacterial plasmids in the natural Antarctic population (Kobori *et al.*, 1984; Siebert and Hirsch, 1988): although the microbiota may outwardly appear unchanged, its genetic pool may be modified by contact with "alien" organisms. Modern genetic hybridization techniques now enable the phylogenetic categorization of Antarctic yeasts (Baharaeen *et al.*, 1983; Baharaeen and Vishniac, 1984) and bacteria (Colwell *et al.*, 1987); therefore, the origins of isolates, indigenous or adventitious, can be established more accurately. Such considerations draw attention to the composition of the incident Antarctic airspora (Fig. 6) relative to the global propagule "rain." The association of an increased density of airspora with human activity has been shown by Cameron *et al.* (1974), who found 400 times more bacteria and molds in air inside a tent at Lake Vida than at the inactive drilling site nearby. However, aerial contamination is localized and unlikely to have a major effect on the ecosystem.

The value of Antarctica for recognition of environmental impact lies in the relative species paucity of its endemic microbiota. Whether or not the invasion of alien species has a deleterious effect on the ecosystem is open to discussion, but the establishment of pathogenic species, such as the pathogenic fungi *Phialophora dermatidis* and *Phialophora gougerotii* near the McMurdo research station, is conspicuous (Sun *et al.*, 1978).

The detection of alien species depends on a prior knowledge of the endemic biota. Baseline studies were conducted in relatively undisturbed Barwick Valley (Parker *et al.*, 1978b), which is a Site of Special Scientific Interest (SSSI) for microbiology, and further inland in the Transantarctic Mountains (Cameron *et al.*, 1970a; Cameron and Ford, 1974; Parker *et al.*, 1977, 1982a) as far south as 87° 21'S (Cameron, 1972c), where *Arthrobacter* sp. was present in the soil. However, the total viable bacterial population isolated on relatively rich tryptone soya agar was ca. 100 cells g^{-1} of soil. No spore-forming bacteria or yeasts, molds, or protozoa were isolated. However, Vishniac (1983) has subsequently shown that nonclassical methods are often required to isolate microbes from such extreme environments.

The DVDP was the first comprehensive drilling program to investigate the subsurface geology and biology of Antarctica, concentrating on the dry valleys of the Ross Desert (Mudrey *et al.*, 1978). Its potential influence on the microbial ecology of the region was part of an environmental appraisal that also implemented a monitoring plan for evaluating the appraisal's effectiveness (Parker *et al.*, 1978a). From the outset, the value of conserving the region for microbial ecological studies was recognized. Already a gradient of microbial abundance is apparent, decreasing from the coast of south Victoria Land to Mt. Howe at 87°S (Cameron *et al.*, 1977). This is partly due to decreasing favorability of habitat, but disruption and chemical and microbiological contamination brought about by human activity is a contributory factor. The summary of microbiological aspects of environmental impact during the DVDP presented by Parker *et al.* (1978b) is based mainly on total viable counts, which must be interpreted cautiously because the method is selective for organisms that can grow under the artificial conditions of an agar plate.

Environmental impact affecting the microbiota may have several effects: disruption of the substratum, toxification of the soil, enrichment of the soil, and local interference with thermal balance and light regimes. The ecological response of the microbiota falls into three categories (Cameron, 1972b): (1) the diversion of the natural evolutionary sequence alters the delicate balance of niche occupation and competition; (2) the microbiota is enhanced in both number and diversity by nutrient enrichment, inhibitor dilution (e.g., salt leaching), inoculation (e.g., sewage input), and physical disruption, which not only releases nutrients but may stir up quiescent cells from lower depths and permafrost (Cameron and Morelli, 1974); and (3) the microbiota may alternatively be inhibited or eliminated by disruption of habitats, toxic pollution, or competition from alien species. Growth rates of endolithic communities, lichens, and mosses are so slow and the growing season is so short in the Ross Desert that material disturbed or removed from a site may take decades to regenerate, if it ever does so.

Drilling fluids are a hazard to the environment since they are difficult to contain if spillages or breakages occur. Three fluids are commonly used: local meltwater or lake water (at Don Juan Pond), Antarctic-grade diesel fuel, DFA (at Lake Vida), or saturated calcium chloride muds (at Lake Fryxell). Each of these fluids potentially jeopardizes microbiological studies. Water from Don Juan Pond was used as a circulation fluid and was returned at the end of drilling. Don Juan Pond is probably unique in the Ross Desert in that it is virtually saturated with calcium chloride solution and has a water activity of 0.45. It has a single dominant, if sparse, bacterial species, *Achromobacter parvalus*, which is of ecological value for monitoring microbial response to environmental change (Cameron *et al.*, 1972, 1973, 1974). Its extreme edaphic conditions have led to study of its algal-cyanobacterial mat and associated heterotrophs as an extraterrestrial analog (Siegel *et al.*, 1983). However, it is not now possible to regard the Don Juan Pond ecosystem as "pristine."

Spillages of DFA at Lake Vida resulted in a decreased heterotrophic microbiota but no evidence of development of a hydrocarbon-degrading community (Morelli *et al.*, 1972; Cameron *et al.*, 1973, 1974). However, natural abundance of oil-degrading bacteria accounted for the recovery of soil subjected to a small DFA spillage at Commonwealth Glacier site. The heterogeneity of Ross Desert habitats and microbiota again precludes generalization about the ability for recovery. A calcium chloride spillage on soil at Lake Fryxell not only reduced the total viable microbial count but also decreased its diversity (Cameron *et al.*, 1973, 1974).

Cameron (1972a) presented a vivid illustrated account of the types of microbiological pollution that were causing concern in the Ross Desert region. Fast-growing *Bacillus* species (growing at 37°C) were found in air samples downwind of a camp on the Beardmore Glacier, and *Staphylococcus* strains of unequivocal human origin were found near a hut at Lake Bonney, Taylor Valley. Fortunately, human coliforms do not survive in such conditions, so a population decreases 6–8 orders of magnitude in 10–20 days according to strain (Boyd and Boyd, 1963). However, *Clostridium* species persist in the soil and have been found in the vicinity of Syowa Station in Enderby Land (Miwa, 1975). Despite attempts to contain such pollution, as research groups increase in size, so the risk of environmental contamination escalates and vigilance is necessary.

Materials introduced by the "heroic" expeditions of the period 1908–14 at Capes Royds (Shackleton) and Evans (Scott) provide a valuable opportunity to monitor long-term responses of alien microbes to Antarctic conditions with minimal subsequent interference (Meyer *et al.*, 1962, 1963). After 50 years, a sealed bottle of baker's yeast still contained viable cells, including the yeasts *Saccharomyces cerevisiae* and *Rhodotorula pallida*, the molds *Absidia corymbifera* and *Rhizopus arhizus*, and bacteria of the genera *Bacillus*, *Pseudomonas*, and *Micrococcus*. A tin of "pearl barley" contained strains of *Pseudomonas*, *Bacillus*, *Streptomyces*, and *Mucor*, while hay contained *Pseudomonas*, *Cryptococcus*, *Rhodotorula*, *Mucor*, and *Penicillium*. Of perhaps more significance from the safety viewpoint was the survival in feces of strains of *Staphylococcus*, *Mycobacterium*, *Alcaligenes*, and *Bacillus*, anaerobic strains of *Clostridium*, *Bacteroides*, and *Lactobacillus*, and yeasts of the genera *Rhodotorula*, *Candida*, *Phoma*, and *Botrytis*. The survival of these mesophilic organisms emphasizes the need for minimizing human pollution, which would apparently be cumulative, in contrast to the gradual dissipation of such inocula in temperate conditions.

7.2. Microbial Aspects of Conservation in Coastal and Maritime Antarctica

Disruption, pollution, and microbial contamination have a greater environmental impact on the resident microbiota of maritime relative to continental Antarctica because their metabolically active state makes them more sensitive to changes in the environment. They may die, proliferate, or diversify, depending on the nature of the impact, and an alien population, including pathogens, may become established and outcompete the endemic microbiota. For example, large numbers of coliforms were detected in a sewage-contaminated street at coastal McMurdo Station (Boyd and Boyd, 1963), where the warmth emanating from buildings is likely to sustain liquid water for proliferation.

Lipps (1978) presented a graphic account of the impact of humans along the Antarctic Peninsula, and many sites described have now been cleaned up. The remains of food and other biodegradable materials such as timber harbor microbes alien to the Antarctic, but few microbiological impact studies have been conducted in this region. Wynn-Williams (1985a) noted that after the construction of a new research station at Rothera Point, the viable heterotrophic bacterial count had increased by a factor of up to 330 at a site ca. 200 m distant. This may have been due either to microbial contamination or enrichment of the moss–peat substrate with dust or volatiles.

Research stations require various oils and fuels for logistic support, and Fig. 1 shows the extent of potentially oil-bearing sedimentary basins that might become economically viable as sites for prospecting if the world financial climate changes. The effects of oil spillages have received little attention to date. However, Konlechner (1985) showed that most of the loss in 13 months from two small experimental spillages (100 ml of alkane-based crude oil in plots 20 × 20 cm) at Cape Bird, Ross Island, was due to volatilization of short-chain hydrocarbons up to n-14 rather than microbial degradation. As the oil soaked into the soil profile, a hydrocarbon-degrading population was enriched *in situ*, but it was estimated that a substantial amount of the oil would remain for several centuries.

Intensive sampling pressure also disrupts natural edaphic conditions for the microbiota of soils and peats (Walton, 1987). There is also an inheritance of our interference with the biota in the last century, when the delicate balance of whale, krill, and fur seal populations was destabilized. Fur seals displaced by overcrowding at South Georgia are now migrating to Signy Island, where their population, having been less than 100 up to 1975, has risen sharply from 2000 in 1977 to >16,500 in 1988 (Smith, 1988). They have now destroyed the moss research sites described in Section 6.1. These were the sites for studies of aerobic and anaerobic decomposition, methanogenesis, and N cycling (Yarrington and Wynn-Williams, 1985; Christie, 1987a) which now cannot be repeated. They have also had a major influence in the eutrophication of Amos Lake, resulting in a dense phytoplankton and a diverse heterotrophic microbiota (Ellis-Evans and Sanders, 1988). The phytoplankton have a very high assimilation efficiency and a high tolerance of anoxia and sulfide accumulation. As many of the lakes at Signy Island are within reach of fur seals, measures such as electric fences have been found necessary to protect certain ongoing microbiological research programs.

Concern for the future of various aspects of Antarctic research has led to the designation of Specially Protected Areas (SPA) and SSSIs under the auspices of the Antarctic Treaty (Heap, 1987) (Fig. 1). SPAs are fully protected, and SSSIs may be sampled and monitored but are closed to tourists. Of these, eight SPAs and nine SSSIs are of relevance to terrestrial and freshwater microbiological research (Broady, 1987). It is to be hoped that these sites will encourage microbial ecological research while respecting the delicacy and value of the ecosystems in the spirit of the Antarctic Treaty.

8. Future Research Directions

Distinctive, well-defined habitats in the Antarctic permit the study of fundamental processes in microbial ecology in the relative absence of disturbance, pollution, and introduced species. The environmental extremes of these habitats emphasize the limits of biological structures and metabolic mechanisms and the flexibility of their bio-chemicals, pathways, and strategies. Findings to date draw attention to several potentially unifying themes for further research:

1. The involvement of polyols as compatible solutes in cryo-, xero-, and halotol-erance.
2. The nature and common principles of photoadaptation in prokaryotic and eukaryotic phototrophs.
3. The unifying concept of strata as ecological niches in endolithic habitats, ice-covered stratified lakes, and sea ice microbial communities.
4. Convergent ecophysiology of endolithic microbial communities in desert climates, potentially including Mars.
5. The validity of the concept of closed ecosystems in endolithic habitats and ice-covered cold desert lakes.
6. The characteristics and validity of indigenous Antarctic microbes in soils and water bodies and adaptability to change in environment.
7. The physicochemical principles of microbial filament–mucilage meshes for stabilizing soil crusts, stream felts, lake benthic mats, and sea ice microbial communities.
8. The significance of the sea ice microbiota for seeding and enriching the Southern Ocean in spring.
9. Biogeographical distribution of microbial phylogenetic groups (5S and 16S tRNA techniques) with reference to their origins (relict microbiota).
10. The spread of genetic information by plasmids into the Antarctic microbiota and within it.
11. The rate of spread of pathogens in the global airspora and by the activities of human beings.

The advent of diverse new techniques such as satellite-linked automatic microclimate monitoring, 5S and 16S tRNA analyses for phylogenetic groupings, and television image analysis for quantifying microbes *in situ* has greatly broadened the scope of

Antarctic microbial ecology. In combination with time-honored classical approaches, these advances in microbial ecology will permit further elucidation of Antarctic ecosystems.

ACKNOWLEDGMENTS. I would like to thank personnel at Signy Island and other Antarctic stations and at British Antarctic Survey headquarters, Cambridge, for their help with research reported in this review. I gratefully acknowledge valuable discussions with Drs. W. Block, J. C. Ellis-Evans, and L. Greenfield. The research was supported by the British Antarctic Survey.

References

Abyzov, S. S., Bobin, N. E., and Kudryashov, B. B., 1982, Quantitative analysis of microorganisms during the microbiological investigation of Antarctic glaciers, *Biol. Bull. Acad. Sci. USSR* **9**:558–564.

Ackley, S. F., Buck, K. R., and Taguchi, S., 1979, Standing crop of algae in the sea ice of the Weddell Sea region, *Deep Sea Res.* **26A**:269–281.

Adamson, D. A., and Pickard, J., 1986, Cainozoic history of the Vestfold Hills, in: *Antarctic Oasis. Terrestrial Environments and History of the Vestfold Hills* (J. Pickard, ed.), pp. 63–98, Academic Press, North Ryde, Australia.

Aksenov, S. I., Babyeva, I. P., and Golubev, V. I., 1973, On the mechanism of adaptation of microorganisms to conditions of extreme low humidity, *Life Sci. Space Res.* **11**:55–61.

Allen, S. E., Grimshaw, M., Parkinson, J. A., and Quarmby, C., 1974, *Chemical Analysis of Ecological Methods*, Blackwell Scientific Publications, London.

Allnutt, F. C. T., Parker, B. C., Seaburg, K. G., and Simmons, G. M. Jr., 1981, In situ nitrogen (C_2H_2) fixation in lakes of southern Victoria Land, Antarctica, *Hydrol. Bull.* **15**:99–109.

Armitage, K. B., and House, H. B., 1962, A limnological reconnaissance in the area of McMurdo Sound, Antarctica, *Limnol. Oceanogr.* **7**:36–41.

Atlas, R. M., Di Menna, M. E., and Cameron, R. E., 1978, Ecological investigations of yeasts in Antarctic soils, *Antarct. Res. Ser. Wash.* **30**:27–34.

Baharaeen, S., Melcher, U., and Vishniac, H. S., 1983. Complementary DNA-25S ribosomal RNA hybridization: An improved method for phylogenetic studies, *Can. J. Microbiol.* **29**:546–551.

Baharaeen, S., and Vishniac, H. S., 1984, 25S ribosomal RNA homologies of basidiomycetous yeasts: taxonomic and phylogenetic implications, *Can. J. Microbiol.* **30**:613–621.

Bailey, A. D., and Wynn-Williams, D. D., 1982, Soil microbiological studies at Signy Island, South Orkney Islands, *Br. Antarct. Surv. Bull.* **51**:167–191.

Baker, J. H., 1970a, Quantitative study of yeasts and bacteria in a Signy Island peat, *Br. Antarct. Surv. Bull.* **23**:51–55.

Baker, J. H., 1970b, Yeasts, moulds and bacteria from an acid peat on Signy Island, in: *Antarctic Ecology*, Vol. 2 (R. M. Laws, ed.), pp. 717–722, Academic Press, London.

Barghoorn, E. S., and Nichols, R. L., 1961, Sulfate-reducing bacteria and pyritic sediments in Antarctica, *Science* **134**:190.

Becker, E. W., 1982, Physiological studies on Antarctic *Prasiola crispa* and *Nostoc commune* at low temperatures, *Polar Biol.* **1**:99–104.

Benoit, R. E., and Hall, C. L., 1970, The microbiology of some Dry Valley soils of Victoria Land, Antarctica, in: *Antarctic Ecology* (M. Holdgate, ed.), pp. 697–701, Academic Press, London.

Berg, T. E., and Black, R. F., 1966, Preliminary measurements of growth of nonsorted polygons, Victoria Land, Antarctica, *Antarct. Res. Ser. Wash.* **8**:61–108.

Bhakoo, M., and Herbert, R. A., 1980, Fatty acid and phospholipid composition of five psychotrophic *Pseudomonas* spp. grown at different temperatures, *Arch. Microbiol.* **126:**51–56.

Bjørnsen, P. K., 1986, Automatic determination of bacterioplankton biomass by image analysis, *Appl. Environ. Microbiol.* **51:**1199–1204.

Block, W., 1980, Survival strategies in polar terrestrial arthropods, *Biol. J. Linn. Soc.* **14:**29–38.

Block, W., 1984, Terrestrial microbiology, invertebrates and ecosystems, in: *Antarctic Ecology,* Vol. 1 (R. M. Laws, ed.), pp. 163–236, Academic Press, London.

Block, W., and Sømme, L., 1982, Cold hardiness of terrestrial mites at Signy Island, maritime Antarctic, *Oikos* **38:**157–167.

Boyd, W. L., 1962, Comparison of soil bacteria and their metabolic activities in Arctic and Antarctic regions, *Polar Rec.* **11:**319.

Boyd, W. L., and Boyd, J. W., 1962, Presence of *Azotobacter* species in polar regions, *J. Bacteriol.* **83:**429–430.

Boyd, W. L., and Boyd, J. W., 1963, Viability of coliform bacteria in Antarctic soils, *J. Bacteriol.* **85:**1121–1123.

Boyd, W. L., Staley, J. T., and Boyd, J. W., 1966, Ecology of soil microorganisms of Antarctica, *Antarct. Res. Ser. Wash.* **8:**125–159.

Broady, P. A., 1979a, Wind dispersal of terrestrial algae at Signy Island, South Orkney Islands, *Br. Antarct. Surv. Bull.* **48:**99–102.

Broady, P. A., 1979b, The terrestrial algae of Signy Island, South Orkney Islands, *Br. Antarct. Surv. Sci. Rep.* **98:**1–117.

Broady, P. A., 1981a, The ecology of sublithic terrestrial algae at the Vestfold Hills, Antarctica, *Br. Phycol. J.* **16:**231–240.

Broady, P. A., 1981b, The ecology of chasmolithic algae at coastal locations of Antarctica, *Phycologia* **20:**259–272.

Broady, P. A., 1981c, Ecological and taxonomic observations on subaerial epilithic algae from Princess Elizabeth Land and MacRobertson Land, Antarctica, *Br. Phycol. J.* **16:**257–266.

Broady, P. A., 1982, Ecology of non-marine algae at Mawson Rock, Antarctica, *Nova Hedw.* **36:**209–229.

Broady, P. A., 1984, Taxonomic and ecological investigations of algae on steam-warmed soil on Mt. Erebus, Ross Island, Antarctica, *Phycologia* **23:**257–271.

Broady, P. A., 1986a, A floristic survey of algae at four locations in northern Victoria Land, *N.Z. Antarct. Rec.* **7:**8–19.

Broady, P. A., 1986b, Ecology and taxonomy of the terrestrial algae of the Vestfold Hills, in: *The Vestfold Hills: An Antarctic Oasis* (J. Pickard, ed.), pp. 165–202, Academic Press, Sydney.

Broady, P. A., 1987, Protection of terrestrial plants and animals in the Ross Sea regions, *N. Z. Antarct Rec.* **8:**18–41.

Broady, P. A., Garrick, R., and Anderson, G., 1984, Culture studies on the morphology of ten strains of Antarctic Oscillatoriaceae (Cyanobacteria), *Polar Biol.* **2:**233–244.

Broady, P. A., Given, D., Greenfield, L. G., and Thompson, K., 1987, The biota and environment of fumaroles on Mount Melbourne, northern Victoria Land, *Polar Biol.* **7:**97–113.

Brown, A. D., 1978, Compatible solutes and extreme water stress in eukaryotic micro-organisms, *Adv. Microb. Physiol.* **17:**181–242.

Bryant, H. M., 1945, Biology at East Base, Antarctic Peninsula, Antarctica, *Proc. Am. Phil. Soc.* **89:**256–269.

Bunt, J. S., 1964, Primary productivity under sea-ice in Antarctic waters. 1. Concentrations and photosynthetic activities of microalgae in the waters of McMurdo Sound, Antarctica, *Antarct. Res. Ser. Wash.* **1:**13–26.

Bunt, J. S., 1968, Some characteristics of microalgae isolated from Antarctic sea-ice, *Antarct. Res. Ser. Wash.* **11:**1–13.

Bunt, J. S., and Lee, C. C., 1970, Seasonal primary production in Antarctic sea ice at McMurdo Sound in 1967, *J. Mar. Res.* **28**:304–320.

Burch, M. D., and Marchant, H. J., 1983, Motility and microtubule stability of Antarctic algae at subzero temperatures, *Protoplasma* **115**:240–250.

Burn, A. J., 1984, Energy partitioning in the Antarctic collembolan *Cryptopygus antarcticus*, *Ecol. Entomol.* **9**:11–21.

Burton, H. R., 1981, Chemistry, physics and evolution of Antarctic saline lakes—a review, *Hydrobiologia* **83**:339–362.

Burton, H. R., and Barker, R. J., 1979, Sulfur chemistry and microbiological fractionation of sulfur isotopes in a saline Antarctic lake, *Geomicrobiol. J.* **1**:329–340.

Caldwell, J. R., 1981, Biomass and respiration of nematode populations in two moss communities at Signy Island, *Oikos* **37**:160–166.

Cameron, R. E., 1969, Cold desert characteristics and problems relevant to other arid lands, in: *Arid Lands in Perspective* (W. G. McGinnies and B. J. Goldman, eds.), pp. 167–205, American Association for the Advancement of Science, Washington, D.C.

Cameron, R. E., 1971, Antarctic soil microbial investigations, in: *Research in the Antarctic* (L. O. Quam and H. D. Porter, eds.), pp. 137–189, American Association for the Advancement of Science, Washington, D.C.

Cameron, R. E., 1972a, Pollution and conservation of the Antarctic terrestrial ecosystem, in: *Proceedings, Colloquium on Conservation Problems in Antarctica* (B. C. Parker, ed.), pp. 267–308, Allen Press, Lawrence, Kans.

Cameron, R. E., 1972b, Microbial and ecological investigations in Victoria Dry Valley, Southern Victoria Land, Antarctica, *Antarct. Res. Ser. Wash.* **20**:195–260.

Cameron, R. E., 1972c, Farthest South algae and associated bacteria, *Phycologia* **11**:133–139.

Cameron, R. E., 1974, Application of low latitude microbial ecology to high latitude deserts, in: *Polar Deserts and Modern Man* (T. L. Smiley and J. H. Zumberge, eds.), pp. 71–90, University of Arizona Press, Tucson.

Cameron, R. E., and Benoit, R. E., 1970, Microbial and ecological investigations of recent cinder cones, Deception Island—A preliminary report, *Ecology* **51**:802–809.

Cameron, R. E., and Blank, G. B., 1967, Desert soil algae survival at extremely low temperatures. *Cryogenic Technol.* **3**:151–156.

Cameron, R. E., and Conrow, H. P., 1969, Soil moisture, relative humidity, and microbial abundance in dry valleys of Southern Victoria Land, *Antarct. J. U.S.* **4**:23–28.

Cameron, R. E., and Devaney, J. R., 1970, Antarctic soil algal crusts. A scanning electron and optical microscope study, *Trans. Am. Microsc. Soc.* **80**:264–273.

Cameron, R. E., and Ford, A. B., 1974, Baseline analyses of soils from the Pensacola Mountains, *Antarct. J. U.S.* **9**:116–119.

Cameron, R. E., and Morelli, F. A., 1974, Viable microorganisms from ancient Ross Island and Taylor Valley drill cores, *Antarct. J. U.S.* **9**:113–115.

Cameron, R. E., David, C. N., and King, J., 1968, Soil toxicity in Antarctic dry valleys, *Antarct. J. U.S.* **3**:164–166.

Cameron, R. E., Hanson, R. B., Lacy, G. L., and Morelli, F. A., 1970a, Soil microbial and ecological investigations in the Antarctic interior, *Antarct. J. U.S.* **5**:87–88.

Cameron, R. E., King, J., and David, C. N., 1970b, Microbial ecology and micro-climatology of soil sites in Dry Valleys of southern Victoria Land, Antarctica, in: *Antarctica Ecology*, Vol. 1 (M. W. Holdgate, ed.), pp. 702–716, Academic Press, London.

Cameron, R. E., King, J., and David, C. N., 1970c, Soil microbial ecology of Wheeler Valley, Antarctica, *Soil Sci.* **109**:110–120.

Cameron, R. E., Morelli, F. A., and Randall, L. P., 1972, Aerial aquatic and soil microbiology of Don Juan Pond, Antarctica, *Antarct. J. U.S.* **7**:254–258.

Cameron, R. E., Morelli, F. A., and Johnson, R. M., 1973, Aerobiological monitoring of dry valley drilling sites, *Antarct. J. U.S.* **8:**211–214.

Cameron, R. E., Morelli, F. A., Donlan, R., Guilfoyle, J., Markley, B., and Smith, R., 1974, Dry Valley Drilling Project environmental monitoring, *Antarct. J. U.S.* **9:**141–144.

Cameron, R. E., Honour, R. C., and Morelli, F. A., 1976, Antarctic microbiology—preparation for Mars life detection, quarantine, and back contamination, in: *Extreme Environments; Mechanisms of Microbial Adaptation* (M. R. Heinrich, ed.), pp. 57–82, Academic Press, New York.

Cameron, R. E., Honour, R. C., and Morelli, F. A., 1977, Environmental impact studies of Antarctic sites, in: *Adaptations within Antarctic Ecosystems*, Proc. 3rd SCAR Symp. Antarct. Biol. (G. A. Llano, ed.), pp. 1157–1176, Gulf Publishing Co., Houston.

Campbell, P. J., 1978, Primary productivity of a hypersaline Antarctic lake, *Aust. J. Mar. Freshwater Res.* **29:**717–724.

Campbell, S. E., 1982, Precambrian endoliths discovered, *Nature* (London) **299:**429–431.

Cannon, R. J. C., and Block, W., 1988, Cold tolerance of microarthropods, *Biol. Rev.* **63:**23–77.

Castrelos, O. D., Ikonicoff, S. I., Del Prete, L., Milano, O. C., and Margni, R. A., 1977, Microbiologia de la Antartida, *Inst. Antart. Argent. Publ.* **209:**4–25.

Cathey, D. D., Parker, B. C., Simmons, G. M. Jr., and Yongue, W. H., Jr., 1981, Artificial substrates in southern Victoria Land lakes of Antarctica, *Hydrobiologia* **85:**3–16.

Chambers, M. J. G., 1967, Investigations of patterned ground at Signy Island, South Orkney Islands. III. Miniature patterns, frost heaving and general conclusions, *Br. Antarct. Surv. Bull.* **12:**1–22.

Chinn, T. J. H., 1981, Hydrology and climate of the Ross Sea area, *J. R. Soc. N.Z.* **11:**373–386.

Christie, P., 1987a, Nitrogen in two contrasting Antarctic bryophyte communities, *J. Ecol.* **75:**73–94.

Christie, P., 1987b, C-to-N ratios in two contrasting Antarctic peat profiles, *Soil Biol. Biochem.* **19:**777–778.

Clarke, D. B., and Ackley, S. F., 1984, Sea ice structure and biological activity in the Antarctic marginal ice zone, *J. Geophys. Res.* **89:**2087–2097.

Cohen-Bazire, G., and Bryant, D. A., 1982, Phycobilisomes: Composition and structure, in: *The Biology of Cyanobacteria* (N. G. Carr and B. A. Whitton, eds.), pp. 143–190, Blackwell Scientific Publications, Oxford.

Collins, N. J., Baker, J. H., and Tilbrook, P. J., 1975, Signy Island, maritime Antarctic, in: *Structure and Function of Tundra Ecosystems* (T. Rosswall and O. W. Heal, eds.), *Ecol. Bull.* (Stockholm) **20:**345–374.

Colwell, R. R., MacDonell, M. T., Friedman, I., and Vestal, J. R., 1987, Identification of Antarctic endolithic microorganisms by 5S rRNA sequence analysis, Abstracts, Modern Approaches in the Biology of Terrestrial Microorganisms and Plants in the Antarctic, Institut für Polarokologie, Kiel, September 1987.

Corte, A., and Daglio, C. A. N., 1963, Micromicetes aislados en el Antartico, *Inst. Antart. Argent. Publ.* **74:**1–27.

Craig, K. R., Fortner, R. D., and Weand, B. L., 1974, Halite and hydrohalite from Lake Bonney, Taylor Valley, Antarctica, *Geology* **2:**389–390.

Darling, C. A., and Siple, P. A., 1941, Bacteria of Antarctica, *J. Bacteriol.* **42:**83–98.

Davis, R. C., 1981, Structure and function of two Antarctic terrestrial moss communities, *Ecol. Monogr.* **5:**125–143.

Davis, R. C., 1986, Environmental factors influencing decomposition rates in two Antarctic moss communities, *Polar Biol.* **5:**95–104.

Dawson, M. P., Humphrey, B., and Marshall, K. C., 1981, Adhesion: A tactic in the survival strategy of a marine vibrio during starvation, *Curr. Microbiol.* **6:**195–198.

Deacon, G. E. R., 1964, A discussion on the physical and biological changes across the Antarctic Convergence. Introduction, *Proc. R. Soc. Ser. A* **281:**1–6.

di Menna, M. E., 1960, Yeasts from Antarctica, *J. Gen. Microbiol.* **23:**295–300.

Dort, W., 1981, The mummified seals of Southern Victoria Land, Antarctica, *Antarct. Res. Ser. Wash.* **30:**123-154.

Downes, M. T., Howard-Williams, C., and Vincent, W. F., 1986, Sources of organic nitrogen, phosphorus and carbon in Antarctic streams, *Hydrobiologia* **134:**215-225.

Drouet, F., 1961, A brief review of the freshwater algae of Antarctica, in: *Science in Antarctica*, Part 1, Report by the Committee on Polar Research, pp. 10-12, National Academy of Sciences, Washington, D.C.

Drouet, F., 1962, The Oscillatoriaceae and their distribution in Antarctica, *Polar Rec.* **11:**320-321.

Ekelöf, E., 1908, Bakteriologische Studien während der Schwedischen Südpolar-Expedition 1901-1903, in: *Wissenschaftliche Ergebnisse der Schwedischen Südpolar-Expedition 1901-1903* (O. Nordenskjöld, ed.), Lithogr. Inst. Generalstabs, Stockholm.

Elliot, D. H., 1985, Physical geography—Geological evolution, in: *Key Environments—Antarctica* (W. N. Bonner and D. W. H. Walton, eds.), pp. 39-61, Pergamon Press, Oxford.

Ellis-Evans, J. C., 1981a, Freshwater microbiology in the Antarctic—I. Microbial numbers and activity in oligotrophic Moss Lake, *Br. Antarct. Surv. Bull.* **54:**85-104.

Ellis-Evans, J. C., 1981b, Freshwater microbiology in the Antarctic—II. Microbial numbers and activity in mesotrophic Heywood Lake, *Br. Antarct. Surv. Bull.* **54:**105-121.

Ellis-Evans, J. C., 1982, Seasonal microbial activity in Antarctic freshwater lake sediments, *Polar Biol.* **1:**129-140.

Ellis-Evans, J. C., 1984, Methane in maritime Antarctic freshwater lakes, *Polar Biol.* **3:**63-72.

Ellis-Evans, J. C., 1985a, Interactions of bacterio- and phytoplankton in nutrient cycling with eutrophic Heywood Lake, Signy Island, in: *Antarctic Nutrient Cycling and Food Webs* (W. R. Siegfried, P. R. Condy, and R. M. Laws, eds.), pp. 261-264, Springer-Verlag, Berlin.

Ellis-Evans, J. C., 1985b, Decomposition processes in maritime Antarctic lakes, in: *Antarctic Nutrient Cycling and Food Webs* (W. R. Siegfried, P. R. Condy, and R. M. Laws, eds.), pp. 253-260, Springer-Verlag, Berlin.

Ellis-Evans, J. C., and Sanders, M. W., 1988, Observations on microbial activity in a seasonally anoxic, nutrient enriched maritime Antarctic lake, *Polar Biol.* **8:**311-318.

Ellis-Evans, J. C., and Wynn-Williams, D. D., 1985, The interaction of soil and lake microflora at Signy Island, in: *Antarctic Nutrient Cycling and Food Webs* (W. R. Siegfried, P. R. Condy, and R. M. Laws, eds.), pp. 662-668, Springer-Verlag, Berlin.

Estep, K. W., MacIntyre, F., Hjörleifsson, E., and Sieburth, J. McN., 1986, MacImage: A user-friendly image-analysis system for accurate mensuration of marine organisms, *Mar. Ecol. Prog. Ser.* **33:**243-253.

Fletcher, L. D., Kerry, E. J., and Weste, G. M., 1985, Microfungi of Mac.Robertson and Enderby Lands, Antarctica, *Polar Biol.* **4:**81-88.

Fletcher, M., and Marshall, K. C., 1982, Are solid surfaces of ecological significance to aquatic bacteria?, In: *Advances in Microbial Ecology,* Vol. 6 (K. C. Marshall, ed.), pp. 199-236, Plenum Press, New York.

Flint, E. A., and Stout, J. D., 1960, Microbiology of some soils of Antarctica, *Nature* (London) **188:**767-768.

Foster, T. D., 1984, The marine environment, in: *Antarctic Ecology* (R. M. Laws, ed.), pp. 345-372, Academic Press, London.

Friedmann, E. I., 1971, Light and scanning electron microscopy of the endolithic desert habitat, *Phycologia* **10:**411-428.

Friedmann, E. I., 1977, Microorganisms in Antarctic desert rocks from Dry Valleys and Dufek Massif, *Antarct. J. U.S.* **12:**26-29.

Friedmann, E. I., 1978, Melting snow in the Dry Valleys is a source of water for endolithic microorganisms, *Antarct. J. U.S.* **13:**162-163.

Friedmann, E. I., 1980, Endolithic microbial life in hot and cold deserts, *Orig. Life* **10:**233-245.

Friedmann, E. I., 1982, Endolithic microorganisms in the Antarctic cold desert, *Science* **215:**1045–1053.

Friedmann, E. I., 1986, The Antarctic cold desert and the search for life on Mars, *Adv. Space Res.* **6:**12:265–268.

Friedmann, E. I., and Kibler, A. P., 1980, Nitrogen economy of endolithic microbial communities in hot and cold deserts, *Microb. Ecol.* **6:**95–108.

Friedmann, E. I., and McKay, C. P., 1985, Methods for the continuous monitoring of snow: Application to the cryptoendolithic microbial community of Antarctica, *Antarct. J. U.S.* **20:**179–181.

Friedmann, E. I., and Ocampo, R., 1976, Endolithic blue-green algae in the Dry Valleys. Primary producers in the Antarctic desert ecosystem, *Science* **193:**1247–1249.

Friedmann, E. I., and Ocampo-Friedmann, R., 1984a, Endolithic microooorganisms in extreme dry environments: Analysis of a lithobiontic microbial habitat, in: *Current Perspectives in Microbiology* (M. J. Klug and C. A. Reddy, eds.), pp. 177–185, American Society for Microbiology, Washington, D.C.

Friedmann, E. I., and Ocampo-Friedmann, R., 1984b, The Antarctic cryptoendolithic ecosystem. Relevance to exobiology, *Orig. Life* **14:**771–776.

Friedmann, E. I., and Weed, R., 1987, Trace-fossil formation in modern microbial communities: Biogenous and abiotic weathering in the Antarctic cold desert, *Science* **236:**703–705.

Friedmann, E. I., La Rock, P., and Brunson, J. O., 1980, Adenosine triphosphate (ATP), chlorophyll, and organic nitrogen in endolithic microbial communities and adjacent soils in the Dry Valleys of S. Victoria Land, *Antarct. J. U.S.* **15:**164–166.

Friedmann, E. I., Friedmann, R. O., and McKay, C. P., 1982, Adaptation of cryptoendolithic lichens in the Antarctic desert, *Comm. Natl. Francais Res. Antarct. Rep.* **51:**65–72.

Friedmann, E. I., McKay, C. P., and Nienow, J. A., 1987, The cryptoendolithic microbial environment in the Ross Desert of Antarctica: Continuous nanoclimate data, 1984 to 1986, *Polar Biol.* **7:**273–287.

Fritsch, F. E., 1912, *Natural History,* Vol. 6, "Freshwater Algae, National Antarctic Expedition 1901–1904, British Museum (Natural History), London.

Fry, J. C., 1988, Determination of biomass, in: *Methods in Aquatic Bacteriology* (B. Austin, ed.), pp. 27–72, Wiley, London.

Garrison, D. L., Buck, K. R., and Silver, M.W., 1982, Ice algal communities in the Weddell Sea, *Antarct. J. U.S.* **17:**157–159.

Garrison, D. L., Buck, K. R., and Silver, M. W., 1983, Studies of ice-algal communities in the Weddell Sea, *Antarct. J. U.S.* **18:**179–181.

Garrison, D. L., Sullivan, C. W., and Ackley, S. F., 1986, Sea ice microbial communities in Antarctica, *BioScience* **36:**243–250.

Gazert, H., 1912, Untersuchungen über Meeresbakterien und ihren Einfluss auf den Stoffwechsel in Meere, in: *Deutsche Südpolar Expedition, 1901–1903,* G. Reimes, Berlin, Vol. 7, pp. 268–296.

Gibson, E. K., Wentworth, S. J., and McKay, D. S., 1983, Chemical weathering and diagenesis of a cold desert soil from Wright Valley: An analog of Martian Weathering processes, *J. Geophys. Res.* **88** (Suppl.)**:**A912–A928.

Giggenbach, W. F., Kyle, P. R., and Lyon, G. G., 1973, Present volcanic activity on Mount Erebus, Ross Island, Antarctica, *Geology* **1:**135–136.

Golubic, S., Friedmann, E. I., and Schneider, J., 1981, Lithobiontic ecological niche, with special reference to microorganisms, *J. Sedim. Petrol.* **51:**475–478.

Gordon, A. L., 1981, Seasonality of Southern Ocean sea-ice, *J. Geophys. Res.* **86:**4193–4197.

Gow, A. J., Weeks, W. F., Goroni, J. W., and Ackley, S. F., 1981, Physical and structural characteristics of sea-ice in McMurdo Sound, *Antarct. J. U.S.* **16:**94–95.

Greenfield, L. G., 1989, Forms of nitrogen in Beacon sandstone rocks containing endolithic microbial communities in Southern Victoria Land, Antarctica, *Polarforschung* **58:**211–218.

Gregory, P. H., 1966, Dispersal, in: *The Fungi, an Advanced Treatise*, Vol. II (G. C. Ainsworth and A. S. Sussman, eds.), pp. 709–732, Academic Press, New York.

Hale, M., 1987, Epilithic lichens in the Beacon sandstone formation, Victoria Land, Antarctica, *Lichenologist* **19**:269–287.

Hall, K., 1986a, Rock moisture content in the field and the laboratory and its relationship to mechanical weathering studies, *Earth Surface Processes Landforms* **11**:131–142.

Hall, K., 1986b, Freeze-thaw simulations on quartz-micaschist and their implications for weathering studies on Signy Island, Antarctica, *Br. Antarct. Surv. Bull.* **73**:19–30.

Hand, R. M., 1980, Bacterial population of two saline Antarctic lakes, *Proc. 4th Int. Symp. Environ. Biogeochem.*, pp. 123–129, Springer-Verlag, Berlin.

Hand, R. M., and Burton, H. R., 1981, Microbial ecology of an Antarctic saline meromictic lake, *Hydrobiologia* **81/82**:363–374.

Harder, R., and Persiel, I., 1962, The occurrence of lower soil Phycomycetes in the Antarctic, *Arch. Mikrobiol.* **41**:44–50.

Harris, H. J. H., and Cartwright, K., 1981, Hydrogeology of the Don Juan Basin, Wright Valley, Antarctica, *Antarct Res. Ser. Wash.* **33**:161–184.

Hasle, G. R., 1956, Phytoplankton and hydrography of the Pacific part of the Atlantic Ocean, *Nature* (London) **177**:616–617.

Hasle, G. R., 1969, An analysis of the phytoplankton of the Pacific Southern Ocean: Abundance, composition and distribution during the Brategg Expedition, 1947–48, *Hvalradets Skr.* **52**:1–168.

Hawes, I., 1983, Turbulence and its consequences for phytoplankton development in ice covered Antarctic lakes, *Br. Antarct. Surv. Bull.* **60**:69–82.

Hawes, I., 1985, Factors controlling phytoplankton population in maritime Antarctic lakes, in: *Antarctic Nutrient Cycling and Food Webs* (W. R. Siegfried, P. R. Condy, and R. M. Laws, eds.), pp. 245–252, Springer-Verlag, Berlin.

Heal, O. W., Bailey, A. D., and Latter, P. M., 1967, Bacteria, fungi and protozoa in Signy Island soils compared with those from a temperate moorland, *Phil. Trans. R. Soc. Lond. B* **252**:191–197.

Heap, J. (ed.), 1987, *Handbook of the Antarctic Treaty System*, 5th ed., Scientific Committee on Antarctic Research, Cambridge.

Herbert, R. A., 1986, The ecology and physiology of psychrophilic microorganisms, in: *Microbes in Extreme Environments* (R. A. Herbert and G. A. Codd, eds.), pp. 1–25, Academic Press, London.

Heywood, R. B., 1977, Antarctic freshwater ecosystems: review and synthesis, in: *Adaptations within Antarctic Ecosystems*, Proc. 3rd SCAR Symp. Antarct. Biol. (G. A. Llano, ed.), pp. 801–828, Gulf Publishing Co., Houston.

Heywood, R. B., 1984, Antarctic Inland waters, in: *Antarctic Ecology* (R. M. Laws, ed.), pp. 279–344, Academic Press, London.

Heywood, R. B., 1987, Limnological studies in the Antarctica Peninsula region, in: *Antarctic Aquatic Biology*, BIOMASS Sci. Ser., Vol. 7 (S. Z. El-Sayed, ed.), pp. 157–173, SCAR, Cambridge.

Heywood, R. B., and Whitaker, T. M., 1984, The Antarctic marine flora, in: *Antarctic Ecology* (R. M. Laws, ed.), pp. 373–420, Academic Press, London.

Heywood, R. B., Dartnall, H. J. G., and Priddle, J., 1980, Characteristics and classification of the lakes of Signy Island, South Orkney Islands, Antarctica, *Freshwater Biol.* **10**:47–59.

Hirsch, P., 1986, Microbial life at extremely low nutrient levels, *Adv. Space Res.* **6**:12:287–298.

Hirsch, P., Gallikowski, C. A., and Friedmann, E. I., 1985, Microorganisms in soil samples from Linnaeus Terrace, southern Victoria Land: preliminary observations, *Antarct. J. U.S.* **20**:183–186.

Holdgate, M. W., 1964, Terrestrial ecology in the maritime Antarctic, in: *Biologie Antarctique* (R. Carrick, M. Holdgate, and J. Prévost, eds.), pp. 181–194, Hermann, Paris.

Holm-Hansen, O., 1963, Algae: Nitrogen fixation by Antarctic species, *Science* **139**:1059–1060.

Holm-Hansen, O., Azam, F., Carlucci, A. F., Hodson, R. E., and Karl, D. M., 1977, Microbial distribution and activity in and around McMurdo Sound, *Antarct. J. U.S.* **12**:29–32.

Hooker, J. D., 1847, Flora Antarctica, part 55, Algae, in: *The Botany of the Antarctic Voyage,* Vol. II, pp. 454–519, Reeve Bros., London.

Horner, R. A., 1985a, Ecology of sea ice microalgae, in: *Sea Ice Biota* (R. A. Horner, ed.), pp. 83–103, CRC Press, Boca Raton.

Horner, R. A., 1985b, Taxonomy of sea ice microalgae, in: *Sea Ice Biota* (R. A. Horner, ed.), pp. 147–157, CRC Press, Boca Raton.

Horner, R. A., Syvertsen, E. E., Thomas, D. P., and Lange, C., 1988, Proposed terminology and reporting units for sea ice algal assemblages, *Polar Biol.* **8:**249–253.

Horowitz, N. H., 1979, Biological water requirements, in: *Strategies of Microbial Life in Extreme Environments* (M. Shilo ed.), pp. 15–27, Dahlem Konferenzen, Berlin.

Horowitz, N. H., Bauman, A. J., Cameron, R. E., Geiger, P. J., Hubbard, J. S., Shulman, G. P., Simmonds, P. G., and Westberg, K., 1969, Sterile soil from Antarctica: Organic analysis, *Science* **164:**1054–1056.

Horowitz, N. H., Cameron, R. E., and Hubbard, J. S., 1972, Microbiology of the Dry Valleys of Antarctica, *Science* **176:**242–245.

Hoshiai, T., 1977, Seasonal changes of ice communities in the sea-ice near Syowa Station, Antarctica, in: *Polar Oceans* (M. J. Dunbar, ed.), pp. 301–317, Arctic Institute of North America, Calgary, Alberta.

Howard-Williams, C., and Vincent, W. F., 1985, Ecosystem properties of Antarctic streams, *N. Z. Antarct. Rec.* (Special Issue) **6:**21–31.

Howard-Williams, C., Vincent, C. L., Broady, P. A., and Vincent, W. F., 1986a, Antarctic stream ecosystems: Variability in environmental properties and algal community structure, *Int. Rev. Gesamte Hydrobiol.* **71:**511–544.

Howard-Williams, C., Vincent, W. F., and Wratt, G. S., 1986b, The Alph River ecosystem: A major freshwater environment in southern Victoria Land, *N.Z. Antarct. Rec.* **7:**21–33.

Huguenin, R. L., Miller, K. J., and Leschine, S. B., 1983, Mars: A contamination potential?, *Adv. Space Res.* **38:**35–38.

Johnson, P. W., and Sieburth, J. McN., 1979, Chroococcoid cyanobacteria in the sea: A ubiquitous and diverse phototrophic biomass, *Limnol. Oceanogr.* **24:**928–935.

Johnson, R. M., and Bellinoff, R. D., 1981, A taxonomic study of a dominant coryneform bacterial type found in Antarctic soils, *Antarct. Res. Ser. Wash.* **30:**169–184.

Johnson, R. M., Madden, J. M., and Swafford, J. R., 1978, Taxonomy of Antarctic bacteria from soils and air, primarily of the McMurdo Station and Dry Valleys region, *Antarct. Res. Ser. Wash.* **30:**35–64.

Johnson, R. M., Inai, M., and McCarthy, S., 1981, Characteristics of cold desert Antarctic coryneform bacteria, *J. Ariz. Nev. Acad. Sci.* **16:**51–60.

Kappen, L., and Friedmann, E. I., 1983, Ecophysiology of lichens in the Dry Valleys of Southern Victoria Land, Antarctica. II. CO_2 gas exchange in cryptoendolithic lichens, *Polar Biol.* **1:**227–232.

Kappen, L., Friedmann, E. I., and Garty, J., 1981, Ecophysiology of lichens in the Dry Valleys of Southern Victoria Land, Antarctica. I. Microclimate of the cryptoendolithic lichen habitat, *Flora* **171:**216–235.

Karl, D. M., LaRock, J. W., and Schultz, D. J., 1977, Adenosine triphosphate and organic carbon in the Cariaco Trench, *Deep Sea Res.* **24:**105–113.

Klein, H. P., 1977, The Viking biological investigations: General aspects, *J. Geophys. Res.* **82:**4677–4680.

Klein, H. P., 1979, The Viking mission and the search for life on Mars, *Rev. Geophys. Space Phys.* **17:**1655–1662.

Klingler, J. M., and Vishniac, H. S., 1989, Water potential of Antarctic soils, *Polarforschung* **58:**231–238.

Kobori, H., Sullivan, C. W., and Shizuya, H., 1984, Bacterial plasmids in Antarctic natural microbial assemblages, *Appl. Environ. Microbiol.* **48:**515–518.

Konlechner, J. C., 1985, Investigation of the fate and effects of a paraffin-based crude oil in an Antarctic terrestrial ecosystem, *N.Z. Antarct. Rec.* **6:**40–46.

Kottmeier, S. T., and Sullivan, C. W., 1988, Sea ice microbial community (SIMCO). 9. Effects of temperature and salinity on rates of metabolism and growth in autotrophs and heterotrophs. *Polar Biol.* **8:**293–304.

Lange, O. L., and Kappen, L., 1972, Photosynthesis of lichens from Antarctica, *Antarct. Res. Ser. Wash.* **20:**83–95.

Latter, P. M., and Heal, O. W., 1971, A preliminary study of the growth of fungi and bacteria from temperate and Antarctic soils in relation to temperature, *Soil Biol. Biochem.* **3:**365–379.

Laws, R. M., 1985, Ecology of the Southern Ocean, *Am. Sci.* **73:**26–40.

Lewis, D. H., and Smith, D. C., 1967, Sugar alcohols (polyols) in fungi and green plants, *New Phytol.* **66:**143–184.

Lipps, J., 1978, Man's impact along the Antarctic Peninsula, in: *Environmental Impact in Antarctica* (B. C. Parker, ed.), pp. 333–372, Virginia Polytechnic Institute, Blacksburg.

Lister, A., 1984, Predation in and Antarctic micro-arthropod community, *Acarology* **6:**886–892.

Lister, A., Usher, M. B., and Block, W., 1987, Description and quantification of field attack rates by predatory mites: An example using an electrophoresis method with a species of Antarctic mite, *Oecologia* **72:**185–191.

Love, F. G., Simmons, G. M. Jr., Parker, B. C., Wharton, R. A. Jr., and Seaburg, K. G., 1983, Modern Conophyton-like microbial mats discovered in Lake Vanda, Antarctica, *Geomicrobiol. J.* **3:**33–48.

Lyakh, S. P., Kozlova, T. M., and Salivonik, S. M., 1984, Effect of periodic freezing and thawing of cells of the Antarctic black yeast *Nadsoniella nigra* var. hesuelica, *Microbiology* **52:**486–491.

Madden, J. M., Siegel, S. K., and Johnson, R. M., 1979, Taxonomy of some Antarctic *Bacillus* and *Corynebacterium* species, *Antarct. Res. Ser. Wash.* **30:**77–103.

Makyut, G. A., 1985, The ice environment, in: *Sea Ice Biota* (R. A. Horner, ed.), pp. 21–82, CRC Press, Boca Raton.

Margni, R. A., and Castrelos, O. D., 1963, Exámenes bacteriológicos de aire, nieve y suelo de Cabo Primavera y Estación Científica Ellsworth, *Inst. Antart. Argent. Publ.* **76:**1–15.

McConville, M. J., 1985, Chemical composition and biochemistry of sea ice microalgae, in: *Sea Ice Biota* (R. A. Horner, ed.), pp. 105–209, CRC Press, Boca Raton.

McConville, M. J., and Wetherbee, R., 1983, The bottom-ice microalgal community from annual ice in inshore waters of East Antarctica, *J. Phycol.* **19:**431–439.

McConville, M. J., Mitchell, C., and Wetherbee, R., 1985, Patterns of carbon assimilation in a microalgal community from annual sea ice, *Polar Biol.* **4:**135–142.

McCraw, J. D., 1967, Soils of Taylor dry valley, Victoria Land, Antarctica, with notes on soils from other localities in Victoria Land, *N.Z. J. Geol. Geophys.* **10:**498–539.

McGinnis, L. D., 1978, appendix: Letter and critique, in: *Environmental Impact in Antarctica* (B. C. Parker, ed.), pp. 253–254, Virginia Polytechnic Institute, Blacksburg.

McKay, C. P., 1986, Exobiology and future Mars missions: The search for Mars' earliest biosphere, *Adv. Space Res.* **6:**12:269–285.

McKay, C. P., and Friedmann, E. I., 1984, Continuous temperature measurements in the cryptoendolithic microbial habitat by satellite-relay data acquisition system, *Antarct. J. U.S.* **19:**170–172.

McKay, C. P., and Friedmann, E. I., 1985, Temperature variations in the cryptoendolithic microbial environment in the Antarctic Dry Valleys, *Polar Biol.* **4:**19–25.

McKay, C. P., Weed, R., Tyler, D. A., Vestal, J. R., and Friedmann, E. I., 1983, Studies of cryptoendolithic communities in the Antarctic cold desert, *Antarct. J. U.S.* **18:**227–228.

McKay, C. P., Clow, G., Wharton, R. A., Jr., and Squyres, S., 1985, The thickness of ice on perennially frozen lakes. *Nature* (London) **313:**561–562.

McLean, A. L., 1918, Bacteria of ice and snow in Antarctica, *Nature* (London) **102**:35–39.

McLean, A. L., 1919, Bacteriological and other researches, Australasian Antarctic Expedition 1911–1914, Sci. Rep. C, Vol. 7, pp. 1–128.

McMeekin, T. A., and Franzmann, P. D., 1988, Effect of temperature on the growth rates of halotolerant and halophilic bacteria isolated from Antarctic saline lakes. *Polar Biol.* **8**:281–285.

Mercer, J. H., 1983, Cenozoic glaciation in the Southern Hemisphere, *Ann. Rev. Earth Planet Sci.* **11**:99–132.

Meryman, H. T., 1966, Review of biological freezing, in: *Cryobiology* (H. T. Meryman, ed.), pp. 1–106, Academic Press, London.

Meyer, G. H., 1962, Microbiological populations of Antarctic air, soil, snow and melt pools, *Polar Rec.* **11**:317–318.

Meyer, G. H., Morrow, M. B., and Wyss, O., 1962, Viable microorganisms in a fifty year old yeast preparation in Antarctica, *Nature* (London) **196**:598–599.

Meyer, G. H., Morrow, M. B., and Wyss, O., 1963, Viable microorganisms from faeces and foodstuffs from early Antarctic expeditions, *Can. J. Microbiol.* **9**:163–167.

Mikell, A. T. Jr., Parker, B. C., and Simmons, G. M. Jr., 1984, Response of an Antarctic lake heterotrophic community to high dissolved oxygen, *Appl. Environ. Microbiol.* **47**:1062–1066.

Miller, K. J., Leschine, S. B., and Huguenin, R. L., 1983, Halotolerance of micro-organisms isolated from saline Antarctic Dry Valley soils, *Antarct. J. U.S.* **18**:222–223.

Miotke, F., 1985, Die Dünen in Victoria Valley, Victoria Land, Antarktis. Ein Beitrag zur äolischen Formung im extrem kalten Klima, *Polarforschung* **55**:79–125.

Miwa, T., 1975, Clostridia isolated from the soil in the east coast of Lützow-Holm Bay, East Antarctica, *Antarct. Rec.* **53**:89–99.

Morelli, F. A., Cameron, R. E., Gensel, D. R., and Randall, L. P., 1972, Monitoring of Antarctic Dry Valley drilling sites, *Antarct. J. U.S.* **7**:92–94.

Morita, R. Y., 1975, Psychrophilic bacteria, *Bacterio. Rev.* **39**:146–167.

Morita, R. Y., Griffiths, R. P., and Hyasaka, S. S., 1977, Heterotrophic activity of microorganisms in Antarctic waters, in: *Adaptations within Antarctic Ecosystems* (G. A. Llano, ed.), pp. 99–113, Smithsonian Institution, Washington, D.C.

Mudrey, M. G., Jr., McGinnis, L. D., and Treves, S. B., 1978, Summary of field activities of the Dry Valley Drilling Project, 1972–73 and 1973–74, in: *Environmental Impact in Antarctica* (B. C. Parker, ed.), pp. 179–210, Virginia Polytechnic Institute, Blacksburg.

Myrcha, A., Pietr, S. J., and Tatur, A., 1985, Role of pygoscelid penguin rookeries in nutrient cycles at Admiralty Bay, King George Island, in: *Antarctic Nutrient Cycles and Food Webs* (W. R. Siegfried, P. Condy, and R. M. Laws, eds.), pp. 156–162, Springer-Verlag, Berlin.

Nakaya, S., Motoori, Y., and Nishimura, M., 1979, One aspect of the evolution of saline lakes in the Dry Valleys of south Victoria Land, *Mem. Natl. Inst. Polar Res. (Tokyo)* (Special Issue) **13**:49–52.

Novitsky, J. A., and Morita, R. Y., 1976, Morphological characterization of small cells resulting from nutrient starvation of a psychrophilic marine vibrio, *Appl. Environ. Microbiol.* **32**:617–622.

Orchard, V. A., and Corderoy, D. M., 1983, Influence of environmental factors on the decomposition of penguin guano in Antarctica, *Polar Biol.* **1**:199–204.

Palmisano, A. C., and Simmons, G. M., Jr., 1987, Spectral downwelling irradiance in an Antarctic lake, *Polar Biol.* **7**:145–151.

Palmisano, A. C., and Sullivan, C. W., 1983, Sea-ice microbial communities (SIMCO). 1. Distribution, abundance, and primary production of ice microalgae in McMurdo Sound, Antarctica, in 1980, *Polar Biol.* **2**:171–178.

Palmisano, A. C., and Sullivan, C. W., 1985a, Physiological response of micro-algae in the ice platelet layer to low light conditions, in: *Antarctic Nutrient Cycles and Food Webs* (W. R. Siegfried, P. R. Condy, and R. M. Laws, eds.), pp. 84–88, Springer-Verlag, Berlin.

Palmisano, A. C., and Sullivan, C. W., 1985b, Growth, metabolism and dark survival in sea ice microalgae, in: *Sea Ice Biota* (R. A. Horner, ed.), pp. 131–146, CRC Press, Boca Raton.

Parker, B. C., 1972, Conservation of freshwater habitats on the Antarctic Peninsula, in: *Proceedings of the Colloquium on Conservation Problems in Antarctica, 1971, Blacksburg, Virginia* (B. C. Parker, ed.), pp. 143–162, Allen Press, Lawrence, Kans.

Parker, B. C., 1978, Potential impact on Lake Bonney of activities associated with modelling freshwater Antarctic ecosystems, in: *Environmental Impact in Antarctica* (B. C. Parker, ed.), pp. 255–278, Virginia Polytechnic Institute, Blacksburg.

Parker, B. C., and Simmons, G. M., Jr., 1985, Paucity of nutrient cycling and absence of food chains in the unique lakes of southern Victoria Land, in: *Antarctic Nutrient Cycling and Food Webs* (W. R. Siegfried, P. R. Condy, and R. M. Laws, eds.), pp. 238–244, Springer-Verlag, Berlin.

Parker, B. C., and Wharton, R. A., Jr., 1985, Physiological ecology of bluegreen algal mats (modern stromatolites) in Antarctic oasis lakes, *Arch. Hydrobiol. Alg. Stud.* **38/39:**331–348.

Parker, B. C., Ford, A. B., Allnutt, T., Bishop, B., and Wendt, S., 1977, Baseline microbiological data for soils of the Dufek Massif, *Antarct. J. U.S.* **12:**24–26.

Parker, B. C., Mudrey, M. G., Jr., Cartwright, K., and McGinnis, L. D., 1978a, Environmental appraisal for the Dry Valley Drilling Project, Phases III, IV, V (1973–74, 1974–75, 1975–76), in: *Environmental Impact in Antarctica* (B. C. Parker, ed.), pp. 37–144, Virginia Polytechnic Institute, Blacksburg.

Parker, B. C., Howard, R. V., and Allnutt, F. C. T., 1978b, Summary of environmental monitoring and impact assessment of the DVDP, in: *Environmental Impact in Antarctica* (B. C. Parker, ed.), pp. 211–251, Virginia Polytechnic Institute, Blacksburg.

Parker, B. C., Simmons, G. M., Jr., Love, F. G., Wharton, R. A., Jr., and Seaburg, K. G., 1981, Modern stromatolites in Antarctic Dry Valley lakes, *BioScience* **31:**656–661.

Parker, B. C., Boyer, S., Allnutt, F. C. T., Seaburg, K. G., Wharton, R. A. Jr., and Simmons, G. M. Jr., 1982a, Soils from the Pensacola Mountains, Antarctica: Physical, chemical and biological characteristics, *Soil. Biol. Biochem.* **14:**265–271.

Parker, B. C., Simmons, G. M., Jr., Kaspar, M., Mikell, A., Love, F. G., Seaburg, K. G., and Wharton, R. A., Jr., 1982b, Physiological adaptations of biota in Antarctic oasis lake—year 2, *Antarct. J. U.S.* **17:**191–193.

Parker, B. C., Simmons, G. M., Jr., Wharton, R. A., Jr., Seaburg, K.G., and Love, F. G., 1982c, Removal of organic and inorganic material from Antarctic lakes by aerial escape of blue-green algal mats, *J. Phycol.* **18:**72–78.

Phillpot, H. R., 1985, Physical geography—climate, in: *Key Environments—Antarctica* (W. N. Bonner and D. W. H. Walton, eds.), pp. 23–38, Pergamon Press, Oxford.

Pickard, J. (ed.), 1986, *Antarctic Oasis. Terrestrial Environments and History of the Vestfold Hills*, Academic Press, North Ryde, Australia.

Poindexter, J. S., 1981, Oligotrophy: Fast and famine existence, in: *Advances in Microbial Ecology,* Vol. 5 (M. Alexander, ed.), pp. 63–89, Plenum Press, New York.

Prévot, A. R., and Moureau, M., 1952, Recherches sur les bactéries anaerobies de la Terre Adélie (prelevées par la première expedition antarctique française), *Ann. Inst. Pasteur* **82:**13–19.

Priddle, J., 1980a, The production ecology of benthic plants in some Antarctic lakes. I. *In situ* production studies, *J. Ecol.* **68:**141–153.

Priddle, J., 1980b, The production ecology of benthic plants in some Antarctic lakes. II. Laboratory physiology studies, *J. Ecol.* **68:**155–166.

Priddle, J., and Belcher, J. H., 1981, Freshwater biology at Rothera Point, Adelaide Island. 2. Algae, *Br. Antarct.Surv. Bull.* **53:**1–10.

Priddle, J., and Heywood, R. B., 1980, Evolution of Antarctic lake ecosystems, *Biol. J. Linn. Soc.* **14:**51–66.

Priddle, J., Hawes, I., and Ellis-Evans, J. C., 1986, Antarctic aquatic ecosystems as habitats for phytoplankton, *Biol. Rev.* **61:**199–238.

Pugh, G. J. F., 1980, Strategies in fungal ecology, *Trans. Br. Mycol. Soc.* **75**:1–14.
Pugh, G. J. F., and Allsopp, D., 1982, Micro-fungi on Signy Island, South Orkney Islands, South Atlantic Ocean, *Br. Antarct. Surv. Bull.* **57**:55–68.
Ramsay, A. J., 1983, Bacterial biomass in ornithogenic soils of Antarctica, *Polar Biol.* **1**:221–225.
Roberts, B., 1958, Chronological list of Antarctic expeditions, *Polar Rec.* **9**:191–239.
Schlichting, H. E., Jr., Speziale, B. J., and Zink, R. M., 1978, Dispersal of algae and protozoa by Antarctic flying birds, *Antarct. J. U.S.* **13**:147–149.
Schreiber, U., 1979, Cold-induced uncoupling of energy transfer between phycobilins and chlorophyll in *Anacystis nidulans*, *FEBS Lett.* **107**:4–9.
Seaburg, K. G., Parker, B. C., Wharton, R. A., Jr., and Simmons, G. M., Jr., 1981, Temperature-growth responses of algal isolates from Antarctic oasis lakes, *J. Phycol.* **17**:353–360.
Siebert, J., and Hirsch, P., 1988, Characterization of 15 selected coccal bacteria isolated from Antarctic rock and soil samples from the McMurdo Dry Valleys (South Victoria Land), *Polar Biol.* **9**:37–44.
Sieburth, J. McN., 1961, Antibiotic properties of acrylic acid, a factor in the gastrointestinal antibiosis of polar marine animals, *J. Bacteriol.* **82**:72–79.
Sieburth, J. McN., 1963, Bacterial habitats in the Antarctic environment, in: *Symposium on Marine Micro-biology* (C. H. Oppenheimer, ed.), pp. 533–548, Charles C. Thomas, Springfield, Ill.
Sieburth, J. McN., 1965, Microbiology of Antarctica, in: *Biogeography and Ecology in Antarctica* (J. van Mieghem and van Oye, eds.), pp. 267–295, Dr. W. Junk, The Hague.
Siegel, B. Z., Siegel, S. M., Chen, J., and La Rock, P., 1983, Extraterrestrial habitat on earth: The algal mat of Don Juan Pond, *Adv. Space Res.* **3**:39–42.
Sieracki, M. E., Johnson, P. W., and Sieburth, J. McN., 1985, Detection, enumeration, and sizing of planktonic bacteria by image-analyzed epifluorescence microscopy, *Appl. Environ. Microbiol.* **49**:799–810.
Simmons, G. M., Wharton, R. A., Jr., McKay, P., Nedell, S., and Clow, G., 1987, Sand/ice interactions and sediment deposition in perennially ice-covered Antarctic lakes. *Antarct. J. U.S.* **22**:237–240.
Smith, H. G., 1978, The distribution and ecology of terrestrial Protozoa of Subantarctic and Antarctic islands, *Br. Antarct. Surv. Sci. Rep.* **95**:1–104.
Smith, H. G., 1985, The colonization of volcanic tephra on Deception Island by Protozoa: Long-term trends, *Br. Antarct. Surv. Bull.* **66**:19–33.
Smith, R. I. L., 1984a, Colonization and recovery by cryptogams following recent volcanic activity on Deception Island, South Shetland Islands, *Br. Antarct. Surv. Bull.* **62**:25–51.
Smith, R. I. L., 1984b, Terrestrial plant biology of the sub-Antarctic and Antarctic, in: *Antarctic Ecology* (R. M. Laws, ed.), pp. 61–162, Academic Press, London.
Smith, R. I. L., 1988, Destruction of Antarctic terrestrial ecosystems by a rapidly increasing fur seal population. *Biol. Conserv.* **45**:55–72.
Smith, R. I. L., and Poncet, S., 1987, *Deschampsia antarctica* and *Colobanthus quitensis* in the Terra Firma Islands, *Br. Antarct. Surv. Bull.* **74**:31–35.
Sømme, L., and Block, W., 1982, Cold hardiness of Collembola at Signy Island, Maritime Antarctica, *Oikos* **39**:168–176.
Soneda, M., 1961, On some yeasts from the Antarctic region, *Biol. Results Jpn. Res. Exp.* **15**:3–10.
Speir, T. W., and Cowling, J. C., 1984, Ornithogenic soils of the Cape Bird Adelie penguin rookeries, Antarctica. 1. Chemical properties, *Polar Biol.* **2**:199–206.
Straka, R. P., and Stokes, J. L., 1960, Psychrophilic bacteria from Antarctica, *J. Bacteriol.* **80**:622–625.
Sullivan, C. W., 1985, Sea ice bacteria: Reciprocal interactions of the organisms and their environment, in: *Sea Ice Biota* (R. A. Horner, ed.), pp. 159–171, CRC Press, Boca Raton.
Sullivan, C. W., and Palmisano, A. C. 1984, Sea-ice microbial communities: Distribution, abundance and diversity of ice bacteria in McMurdo Sound, Antarctica, *Appl. Environ. Microbiol.* **47**:788–795.
Sullivan, C. W., Palmisano, A. C., Kottmeier, S., McGrath Grossi, D., and Moe, R., 1985, Influence of

light on growth and development of the sea-ice microbial community of McMurdo Sound, in: *Antarctic Nutrient Cycles and Food Webs* (W. R. Siegfried, P. R. Condy, and R. M. Laws, eds.), pp. 78–83, Springer-Verlag, Berlin.

Sun, S. H., Huppert, M., and Cameron, R. E., 1978, Identification of some fungi from soil and air of Antarctica, *Antarct. Res. Ser. Wash.* **30:**1–26.

Sussman, A. S., and Halvorson, H. O., 1966, *Spores,* Harper and Row, New York and London.

Svensson, B. H., 1980, Carbon dioxide and methane fluxes from the ombrotrophic parts of a Subarctic mire, *Ecol. Bull.* (Stockholm) **30:**235–250.

Tanner, A. C., and Herbert, R. A., 1981, Nutrient regeneration in maritime Antarctic sediments, *Kiel Meeresforsch.* **5:**390–395.

Tearle, P. V., 1987, Cryptogamic carbohydrate release and microbial response during spring freeze-thaw cycles in Antarctic fellfield fines, *Soil Biol. Biochem.* **19:**381–390.

Thompson, D. C., Bromley, A. M., and Craig, R. M. F., 1971, Ground temperatures in an Antarctic dry valley, *N.Z. J. Geol. Geophys.* **14:**477–483.

Tominaga, H., 1977, Photosynthetic nature and primary productivity of Antarctic freshwater phytoplankton, *Jpn. J. Limnol. Oceanogr.* **11:**596–607.

Torii, T., Matsumoto, G. I., and Nakaya, S., 1989, The chemical characteristics of Antarctic lakes and ponds, with special emphasis on the distribution of nutrients. *Polarforschung* **58:**219–230.

Tschermak-Woess, E., and Friedmann, E. I., 1984, *Hemichloris antarctica,* new genus, new species, Chlorococcales, Chlorophyta. A cryptoendolithic alga from Antarctica, *Phycologia* **23:**443–454.

Tsiklinsky, Mlle., 1908, La flore microbienne dans les régions du Pole Sud, in: *Expedition Antarctique Française 1903–1905* (J. Charcot, ed.), pp. 1–33, Masson et Cie., Paris.

Tubaki, K., 1961, On some fungi isolated from the Antarctic materials, *Biol. Results Jpn. Antarctic Res. Exp.* (Special Publ.) **14:**3–9.

Tuck, A. F., 1987, *Stratospheric Ozone, United Kingdom Stratospheric Ozone Group, First Report,* Her Majesty's Stationery Office, London.

Tuovila, B. J., and LaRock, P. A., 1987, Occurrence and preservation of ATP in Antarctic rocks and its implications in biomass determinations, *Antarct. J. U.S.* **19:**181–182.

Ugolini, F. C., and Anderson, D. M., 1973, Ionic migration and weathering in frozen Antarctic soils, *Soil Sci.* **115:**461–470.

Uydess, I. L., and Vishniac, W. V., 1976, Electron microscopy of Antarctic soil bacteria, in: *Extreme Environments; Mechanisms of Microbial Adaptation* (M. R. Heinrich, ed.), pp. 29–56, Academic Press, New York.

Van Liere, L., and Walsby, A. E., 1982, Interactions of cyanobacteria with light, in: *The Biology of Cyanobacteria* (N. G. Carr and B. A. Whitton, eds.), pp. 9–45, Blackwell Scientific Publications, Oxford.

Vestal, J. R., 1988a, Biomass of the cryptoendolithic microbiota from the Antarctic desert, *Appl. Environ. Microbiol.* **54:**957–959.

Vestal, J. R., 1988b, Carbon metabolism in the cryptoendolithic microbiota from the Antarctic desert, *Appl. Environ. Microbiol.* **54:**960–965.

Vestal, J. R., Federle, T. W., and Friedmann, E. I., 1984, The effects of light and temperature on the Antarctic cryptoendolithic microbiota in vitro, *Antarct. J. U.S.* **19:**173–174.

Vincent, W. F., 1981, Production strategies in Antarctic inland waters: Phytoplankton eco-physiology in a permanently ice-covered lake, *Ecology* **62:**1215–1224.

Vincent, W. F., 1985, Factors controlling phytoplankton production in Lake Vanda (77 deg S), *Can. J. Fish. Aquat. Sci.* **39:**1602–1609.

Vincent, W. F., 1988, *Microbial Ecosystems of Antarctica,* Cambridge University Press, Cambridge.

Vincent, W. F., and Howard-Williams, C., 1985, Ecosystem properties of Dry Valley lakes, *N.Z. Antarct. Rec.* (Special Issue) **6:**11–20.

Vincent, W. F., and Howard-Williams, C., 1986a, Antarctic stream ecosystems: Physiological ecology of a blue-green algal epilithion, *Freshwater Biol.* **16:**219–234.

Vincent, W. F., and Howard-Williams, C., 1986b, Microbial ecology of Antarctic streams, in: *Perspectives in Microbial Ecology* (F. Megusar and M. Cantar, eds.), pp. 201–206, Slovene Society for Microbiology, Ljubljana, Yugoslavia.

Vincent, W. F., and Vincent, C. L., 1982, Factors controlling phytoplankton production in Lake Vanda (77 deg S), *Can. J. Fish. Aquat. Sci.* **39:**1602–1609.

Vincent, W. F., Downes, M. T., and Vincent, C. L., 1981, Nitrous oxide cycling in Lake Vanda, Antarctica, *Nature* (London) **292:**618–620.

Vishniac, H. S., 1983, An enation system for the isolation of Antarctic yeasts inhibited by conventional media, *Can. J. Microbiol.* **29:**90–95.

Vishniac, H. S., and Hempfling, W. P., 1979a, Evidence of an indigenous microbiota (yeast) in the Dry Valleys of Antarctica, *J. Gen. Microbiol.* **112:**301–314.

Vishniac, H. S., and Hempfling, W. P., 1979b, *Cryptococcus vishniacii* sp. nov., an Antarctic yeast, *Int. J. Syst. Bacteriol.* **29:**153–158.

Vishniac, H. S., and Klingler, J. M., 1986, Yeasts in the Antarctic deserts, in: *Perspectives in Microbial Ecology* (F. Megusar and M. Gantar, eds.), pp. 46–51, Slovene Society for Microbiology, Ljubljana, Yugoslavia.

Vishniac, W. V., and Mainzer, S. E., 1973, Antarctica as a Martian model, *Life Sci. Space Res.* **11:**25–31.

Walton, D. W. H., 1982, The Signy Island terrestrial reference sites: XV. Microclimate monitoring, 1972–74, *Br. Antarct. Surv. Bull.* **55:**111–126.

Walton, D. W. H., 1984, The terrestrial environment, in: *Antarctic Ecology* (R. M. Laws, ed.), pp. 1–60, Academic Press, London.

Walton, D. W. H., 1985, Preliminary study of the action of crustose lichens on rock surfaces in Antarctica, in: *Antarctic Nutrient Cycles and Food Webs* (W. R. Siegfried, P. Condy, and R. M. Laws, eds.), pp. 180–185, Springer-Verlag, Berlin.

Walton, D. W. H., 1987, Antarctic terrestrial ecosystems, *Environ. Int.* **13:**83–93.

Watanuki, K., Torii, T., Murayama, H., Hirabayashi, J., Sano, M., and Abiko, T., 1977, Geochemical features of Antarctic lakes, *Antarct. Rec.* **59:**18–25.

West, W., and West, G. S., 1911, Freshwater algae, in: *British Antarctic Expedition, 1907–09. Reports of the Scientific Investigations. Biology,* Vol. 1 (J. Murray, ed.), pp. 263–298, Heinemann, London.

Wharton, R. A., Vinyard, W. C., Parker, B. C., Simmons, G. M., and Seaburg, K. G., 1981, Algae in cryoconite holes in Canada Glacier in southern Victoria Land, Antarctica, *Phycologia* **20:**208–211.

Wharton, R. A., Jr., Parker, B. C., and Simmons, G. M., Jr., 1983, Distribution, species composition and morphology of algal mats in Antarctic Dry Valley lakes, *Phycologia* **22:**355–366.

Wharton, R. A., Jr., McKay, C. P., Simmons, G. M., Jr., and Parker, B. C., 1985, Cryoconite holes on glaciers, *BioScience* **35:**499–503.

Wharton, R. A., Jr., McKay, G. M., Simmons, G. M., Jr., and Parker, B. C., 1986, Oxygen budget of a perennially ice-covered Antarctic dry valley lake, *Limnol. Oceanogr.* **31:**437–443.

Whitaker, T. M., 1977, Sea ice habitats of Signy Island (South Orkneys) and their primary productivity, in: *Adaptations within Antarctic Ecosystems* (G. A. Llano, ed.), pp. 75–82, Smithsonian Institution, Washington, D.C.

Womersley, C., and Smith, L., 1981, Anhydrobiosis in nematodes. I: the role of glycerol, myoinositol and trehalose during desiccation, *Comp. Biochem. Physiol.* **70B:**579–586.

Wood, A. M., Horan, P. K., Muirhead, K., Phinney, D. A., Yentsch, C. M., and Waterbury, J. B., 1985, Discrimination between types of pigments in marine *Synechococcus* spp. by scanning spectroscopy, epifluorescence microscopy and flow cytometry, *Limnol. Oceanogr.* **30:**1303–1315.

Wright, S. W., and Burton, H. R., 1981, The biology of Antarctic saline lakes, *Hydrobiologia* **81/82:**319–338.

Wynn-Williams, D. D., 1980, Seasonal fluctuations in microbial activity in Antarctic moss peat, *Biol. J. Linn. Soc.* **14:**11–28.

Wynn-Williams, D. D., 1982, Simulation of seasonal changes in microbial activity of maritime Antarctic peat, *Soil. Biol. Biochem.* **14:**1–12.

Wynn-Williams, D. D., 1983, Distribution and characteristics of *Chromobacterium* in the maritime and sub-Antarctic, *Polar Biol.* **2:**101–108.

Wynn-Williams, D. D., 1984, Comparative respirometry of peat decomposition on a latitudinal transect in the maritime Antarctic, *Polar Biol.* **3:**173–181.

Wynn-Williams, D. D., 1985a, Comparative microbiology of moss-peat decomposition on the Scotia Arc and Antarctica Peninsula, in: *Antarctic Nutrient Cycles and Food Webs* (W. R. Siegfried, P. R. Condy, and R. M. Laws, eds.), pp. 204–210, Springer-Verlag, Berlin.

Wynn-Williams, D. D., 1985b, The biota of a lateral moraine and hinterland of the Blue Glacier, South Victoria Land, Antarctica, *Br. Antarct. Surv. Bull.* **66:**1–5.

Wynn-Williams, D. D., 1985c, Photofading retardant for epifluorescence microscopy in soil micro-ecological studies, *Soil Biol. Biochem.* **17:**739–746.

Wynn-Williams, D. D., 1986, Microbial colonisation of Antarctic fellfield soils, in: *Perspectives in Microbial Ecology* (F. Megusar and M. Cantar, eds.), pp. 191–200, Slovene Society for Microbiology, Ljubljana, Yugoslavia. Microbial Ecology, Ljubljana, August 1986.

Wynn-Williams, D. D., 1988, Cotton strip decomposition relative to environmental factors in the Maritime Antarctic, in: *Cotton Strip Assay: An Index of Decomposition in Soils* (A. F. Harrison, P. M. Latter, and D. W. H. Walton, eds.), ITE Symp., Vol. 24, pp. 126–133. Institute of Terrestrial Ecology, Grange-over-Sands.

Wynn-Williams, D. D., 1989, TV image analysis of microbial communities in Antarctic fellfields, *Polarforschung* **58:**239–250.

Yarrington, M. R., and Wynn-Williams, D. D., 1985, Methanogenesis and the anaerobic microbiology of a wet moss community at Signy Island, in: *Antarctic Nutrient Cycles and Food Webs* (W. R. Siegfried, P. R. Condy, and R. M. Laws, eds.), pp. 229–233, Springer-Verlag, Berlin.

4

The Microecology of Lactobacilli Inhabiting the Gastrointestinal Tract

GERALD W. TANNOCK

1. Introduction

Microbiological interest in lactobacilli inhabiting the gastrointestinal tract of vertebrate animals dates, traditionally, from the publications of Metchnikoff originating more than 80 years ago. One can, at least, trace to that epoch the concept that lactobacilli, ingested as a dietary supplement in fermented milk products, have health-promoting effects. Metchnikoff, convinced that the human colon acted as a reservoir of proteolytic bacteria that generated substances toxic to the host's tissues, advocated the consumption of milk products that had undergone a lactic fermentation. Milk fermented by lactic acid-producing bacteria, it had been observed, did not provide a suitable medium for proteolytic (putrefactive) microbes. Colonization of the colon by bacteria capable of producing a lactic acid fermentation, it was reasoned, would inhibit the proliferation of putrefactive microbes in that site, thus protecting the host from diseases caused by toxins generated by proteolytic bacteria. Fermented milk products which had long been a part of the diets of eastern Europeans became popular in the West (Metchnikoff, 1907; Tannock, 1981; Tannock, 1984).

The lactic acid-producing bacteria (*Streptococcus thermophilus, Lactobacillus delbrueckii* subsp. *bulgaricus*) used in the production of fermented milk products such as yogurt do not colonize the gastrointestinal tract of humans. Therefore, Rettger and colleagues, some 50 years ago, promoted the daily consumption of milk containing an inhabitant of the human intestinal tract, *Lactobacillus acidophilus,* as a means of colonizing the digestive tract with beneficial bacteria. "Acidophilus milk" was reported by Rettger *et al.* to ameliorate the condition of subjects suffering from a rather cosmopoli-

GERALD W. TANNOCK • Department of Microbiology, University of Otago, Dunedin, New Zealand.

tan collection of diseases: simple constipation, constipation with biliary symptoms, chronic diarrhea following bacillary dysentery, colitis, sprue, or eczema (Rettger *et al.*, 1935; Tannock, 1981).

Today, fermented milk products containing viable bacterial cells used in the preparation of the food are popular in many regions of the world. While all have their advocates regarding health-promoting properties, lactobacillus-containing foods are best enjoyed for their organoleptic qualities rather than from the point of view of colonization of the intestinal tract with lactic acid-producing bacteria. This statement can be made for three reasons.

First, it is extremely difficult for allochthonous microbes to establish themselves in an already colonized ecosystem. From soon after birth, the gastrointestinal tract of animals harbors a complex collection of microbial types, most of which are obligate anaerobes that inhabit the large bowel. About 400 species of bacteria have been isolated from human feces, although 30 to 40 species are commonly encountered (Draser and Barrow, 1985; Lee, 1985). Microbes such as these, colonizing the intestinal ecosystem, occupy all available niches; consequently, newly introduced bacteria cannot establish in the intestinal tract. This phenomenon, known as microbial interference or colonization resistance, has been well studied in relation to nonspecific resistance to infectious diseases (Tannock, 1981, 1984). It is unlikely, therefore, that lactic acid-producing bacteria introduced into the digestive tract with food will colonize the gastrointestinal tract of adult animals.

Second, lactobacilli are not the numerically dominant bacterial type in the gastrointestinal tract. This fact is particularly pertinent with respect to the human intestinal tract, where, although *Lactobacillus* species are harbored by about 70% of subjects consuming a "Western diet," they are not among the 25 most prevalent species (Finegold *et al.*, 1974, 1983). Curiously, lactobacilli are sometimes regarded as the numerically dominant organisms in the intestinal tract of infants suckled at the breast. How this erroneous impression arose is not known, but it is clear from the work of Tissier, at the turn of the century, and research by modern researchers that *Bifidobacterium* species are the most common lactic acid-producing bacteria in the intestinal tract of breast-fed infants (Moreau *et al.*, 1986; Stark and Lee, 1982; Tissier, 1905).

Finally, the belief that the consumption of lactobacillus-containing preparations promotes health is based largely on anecdotal information. The scientific literature is distressingly sparse concerning controlled, scientifically valid investigations of fermented milk products and their influence on health (Tannock, 1981).

It is clear, therefore, that interest in the microecology of lactobacilli inhabiting the gastrointestinal tract should not be based on, or influenced by, pseudo-scientific concepts of a bygone era. Lactobacilli should be studied because they are bacteria encountered commonly in gastrointestinal ecosystems, where they participate in interesting ecological phenomena. Whether the lactobacilli contribute to the well-being of their host is a question that can be answered only after the microbial ecology of lactobacilli in the gastrointestinal tract has been studied comprehensively and the intimate interactions of the bacteria with their host elucidated.

2. The Microecology of Lactobacilli: Present Status

Of the 44 species of lactobacilli listed in *Bergey's Manual of Systematic Bacteriology* (Kandler and Weiss, 1986), 16 have been detected in gastrointestinal or fecal material from vertebrate animals (Table I). Until recently, however, the taxonomy of the genus *Lactobacillus* has been extremely confusing, and it is likely that many strains have been misidentified. An accurate list of gastrointestinal species awaits the isolation and identification of many more gastrointestinal isolates of lactobacilli by modern methods of classification. Several species (for example, *L. fermentum* and *L. reuteri; L. acidophilus* and *L. gasseri*) cannot be distinguished from each other by simple physiological tests. The mol% G + C of DNA, cell wall components, and electrophoretic mobilities of lactate dehydrogenase and soluble cellular proteins are important additional distinguishing tests (Kandler and Weiss, 1986). Perhaps a modern collaborative study involving microbial ecologists and a taxonomist would provide more reliable information on the spectrum of *Lactobacillus* species inhabiting the gastrointestinal tract of vertebrate animals.

Some strains of lactobacilli inhabiting the gastrointestinal tract of animals exhibit a particularly intimate association with their hosts. These lactobacilli colonize stratified, squamous epithelial surfaces lining a proximal region of the host's digestive tract. In

Table I. Gastrointestinal Species of Lactobacilli

Species	Association with epithelia (where known)
L. delbrueckii (*leichmannii*)	+ (Lin and Savage, 1984; Roach *et al.*, 1977; Wesney and Tannock, 1979)
L. acidophilus	+ (Lin and Savage, 1984)
L. animalis	
L. crispatus	
L. gasseri	+ (Lin and Savage, 1984; Roach *et al.*, 1977; Wesney and Tannock, 1979)
L. ruminis	
L. salivarius	+ (Barrow *et al.*, 1980; Fuller, 1973; Tannock *et al.*, 1982)
L. vitulinus	
L. casei	
L. coryniformis	
L. curvatus	
L. murinus	+ (Lin and Savage, 1984)
L. plantarum	
L. brevis	
L. fermentum	+ (Barrow *et al.*, 1980; Lin and Savage, 1984)
L. reuteri	+ (Lin and Savage, 1984; Wesney and Tannock, 1979; Tannock *et al.*, 1982)

mice, for example, lactobacilli inhabit the tissue surface lining the esophagus and forestomach; in rats, lactobacilli colonize the forestomach; in fowl, they inhabit the epithelial surface of the crop; in pigs, lactobacilli colonize the epithelial surface of the esophagus and of a small area of tissue in the stomach (pars esophagea) (Barrow *et al.*, 1980; Fuller and Turvey, 1971; Fuller *et al.*, 1978; Savage, 1977; Tannock and Smith, 1970; Tannock *et al.*, 1982). Seven species of lactobacilli have been described as capable of colonizing epithelial surfaces in animal hosts (Table I), but it is by no means sure that all strains of a given species have tissue-associating ability. Under natural conditions, lactobacilli colonize epithelial surfaces in mice, rats, fowl, and pigs soon after birth (hatching). Lactobacilli attain high population levels on the epithelial surface (of the order of 10^8 bacteria per gram of sample) and a layer of lactobacilli, sometimes several cells thick, is present on the tissue surface throughout the host's life. The lactobacilli are shed from the colonized surface and can be detected throughout the remainder of the animal's digestive tract. Colonization of epithelial surfaces by lactobacilli is easily demonstrated by microscope observation of stained histological sections of digestive tract samples prepared with a microtome–cryostat, by phase contrast microscopy of epithelial cells brushed gently from tissue surfaces, or by electron microscopy (Fuller, 1973; Fuller and Brooker, 1974; Savage *et al.*, 1968; Savage and Blumershine, 1974; Savage, 1983; Tannock, 1988a).

Lactobacillus strains have been reported to adhere to human fetal intestinal cells and to intestinal cells from pigs and calves under *in vitro* conditions (Conway *et al.*, 1987; Kleeman and Klaenhammer, 1982; Mayra-Makinen *et al.*, 1983). These observations presumably have little significance, since association of lactobacilli with intestinal epithelia *in vivo* has not been demonstrated.

Strains of lactobacilli that associate with epithelial surfaces exhibit host specificity. This means that epithelium-associating strains isolated from mice or rats will not colonize the crop epithelium of chickens. Conversely, isolates from poultry will not associate with rodent or porcine epithelia. The host specificity of epithelium-associating strains has been demonstrated in both *in vitro* and *in vivo* experiments (Fuller, 1973; Suegara *et al.*, 1975; Lin and Savage, 1984; Tannock *et al.*, 1982; Wesney and Tannock, 1979). Several factors are likely to be involved in this phenomenon.

(1) Lactobacilli isolated from fowl adhere only to crop epithelial cells, not to cells from the rat forestomach in *in vitro* experiments in which the bacteria have been cultivated in a standard laboratory medium. Rat isolates of lactobacilli cultured in the same standard medium adhere only to epithelial cells from rats (Fuller, 1973; Suegara *et al.*, 1975). Bacterial strains from different hosts cultured under the same physiological conditions thus exhibit host specificity, which suggests that interactions occur between specific adhesins and receptors on bacterial and host cells.

(2) The stratified, squamous epithelia lining the proximal digestive tracts of fowl, pigs, and rodents differ in chemical composition. The murine forestomach, for example, has a keratinized epithelium, whereas the chicken crop does not (Kaplan *et al.*, 1983). Lactobacilli may have to be nutritionally adapted to inhabit certain types of epithelial surfaces.

(3) Physiological conditions, including the nature of the diet, may influence the

colonization of epithelial surfaces by lactobacilli. Although it is clear that marked differences in diet affect the composition of the normal microbiota of the gastrointestinal tract, investigation of subtle physiological differences on intestinal microbes has proved technologically difficult (Brockett and Tannock, 1981, 1982; Tannock, 1983a,b).

Although many strains of lactobacilli do not associate with epithelial surfaces and many animal species do not demonstrate the phenomenon of epithelial colonization by microbes, the tissue-associating lactobacilli provide a useful focus for further discussion of the microecology of gastrointestinal lactobacilli. In considering the epithelium-associating strains, we can pose a number of questions concerning their colonization of the gastrointestinal tract: (1) What are the cell surface properties of lactobacilli that permit them to adhere to mammalian cells? What is the adhesion mechanism? (2) What metabolic properties of lactobacilli permit them to colonize gastrointestinal habitats? (3) What ecological advantage is gained by lactobacilli that can colonize epithelial surfaces? (4) In what ways do lactobacilli interact with other microbial types in the ecosystem? (5) What influence do lactobacillus populations inhabiting the gastrointestinal tract have on their animal hosts? Attempts to answer these questions will be made in subsequent sections of this review.

2.1. Cell Walls of Lactobacilli and Adhesion Mechanisms

Electron microscope examination of thin sections of lactobacilli show that they have a typical Gram-positive cell wall structure. The cell wall contains peptidoglycan of various chemotypes of the cross-linkage group A. This means that cross-linkage of peptide moieties in the peptidoglycan extends from the w-amino group of the diamino acid in position three of one peptide subunit to the carboxyl group of D-alanine in position four of another adjacent peptide subunit. The L-lysine (position 3)–D-aspartic acid (position 4) peptide variation is the most widespread peptidoglycan type in lactobacilli, cross-linkage occurring through single D-isoasparaginyl residues. Some species have m-diaminopimelic acid-type peptidoglycan, with direct cross-linkage between peptide subunits. Others contain L-ornithine–D-aspartic acid-type peptidoglycan, and some have interpeptide bridges composed of L-serine and/or L-alanine (Kandler and Weiss, 1986; Schleifer and Kandler, 1972). The cell wall also contains polysaccharide (for example, polymers of glucose, galactose, and rhamnose in equimolar amounts in the case of L. acidophilus) attached to peptidoglycan by phosphodiester bonds. Membrane teichoic acid is present in all species; cell wall-bound teichoic acid is found only in some species. Extracellular slime in large amounts is produced from sucrose by L. confusus and particular strains of some other heterofermentative species (Kandler and Weiss, 1986).

Apart from the interest of taxonomists in the peptidoglycan structure, lipoteichoic acids appear to be the best-studied cell wall polymers of lactobacilli. Lipoteichoic acids are "amphiphiles," since they are molecules containing both hydrophilic and hydrophobic regions (Wicken and Knox, 1980). The lipoteichoic acids are typically linear polymers of glycerophosphate (25–30 residues linked 1–3 by phosphodiester bonds) sometimes substituted with sugars or D-alanine. The phosphomonoester end of the polymer is covalently linked to a glycolipid. The lipid moiety of the molecule is intercalated with the cell membrane. Lipoteichoic acids can be detected at the surface of bacterial cells and in the cell wall material as well as in association with cell membranes. Passage of lipoteichoic acids from the cell membrane location, through the cell wall and into the external milieu of the cell apparently occurs. Two forms of lipoteichoic acid can be detected in both cellular and extracellular locations: a high-molecular-weight micellar form and a lower-weight deacylated form that cannot produce hydrophobic aggregates (Wicken and Knox, 1980).

Acylated lipoteichoic acids adsorb to the surface of red blood cells from a number of animal species, and such "sensitized" erythrocytes are agglutinated by specific antibodies (Knox and Wicken, 1973). Lipoteichoic acid antigens have been used in serological grouping of Gram-positive bacteria (Wicken and Knox, 1975). The adsorption of lipoteichoic acids to mammalian cells, however, is of more interest in relation to the adhesion of lactobacilli to epithelial surfaces in the gastrointestinal tract. Lipoteichoic acids have been detected in extracts of gastrointestinal strains of lactobacilli, including some known to associate with epithelial surfaces (Sherman and Savage, 1986). The exact mechanism by which lactobacilli adhere to mammalian cell surfaces is not known, but electron microscopy of lactobacillus cells associated with epithelia and stained with ruthenium red (a carbohydrate-specific stain) demonstrates the presence of acidic carbohydrate-containing material (including fibrillae) between attached bacteria and epithelium (Brooker and Fuller, 1975). In vitro experiments show that attachment of lactobacilli to epithelial cells can be blocked by carbohydrate-specific molecules (lectins) such as monovalent concanavalin A (Fuller and Brooker, 1974).

Lipoteichoic acids may participate in attachment of lactobacilli to mammalian cells in a manner similar to that described for Streptococcus pyogenes. According to Beachey and colleagues, lipoteichoic acid synthesized by Lancefield group A streptococci forms a complex with protein molecules (such as M-protein molecules) at the streptococcal surface to produce fibrillae that bridge the gap between bacterial cell and epithelial cell (Christensen et al., 1985). The lipoteichoic acid–protein bridging ligands are anchored in the cell membrane of the streptococcal cells at one end and to fibronectin (a glycoprotein with a molecular weight of about 450,000) receptors in the epithelial cell membrane at the other via the hydrophobic ends of lipoteichoic acid molecules. It is not recorded whether lipoteichoic acids from gastrointestinal strains of lactobacilli interact with fibronectin, nor does effort appear to have been devoted to detecting and characterizing proteins on the surface of such lactobacilli. This is surprising considering the evidence provided by Suegara et al. (1975) that heat treatment of lactobacilli (60°C for 60 min) prevents adhesion of the bacteria to mammalian cells under in vitro conditions and the observation by Conway (1986) that an extracellular protein from a strain of L. fermentum appears to influence lactobacillus adhesion.

If bridging ligands composed of lipoteichoic acid and protein can be demonstrated in lactobacillus–epithelial cell interactions, the lack of adhesion to epithelial cells by certain lactobacillus strains could be explained as follows: (1) lipoteichoic acid, although synthesized, may not be present at the cell surface of some lactobacilli, or (2) protein molecules able to bind to the lipoteichoic acid backbone are not synthesized in sufficient quantity by the bacteria to form bridging ligands. The specificity of adhesion exhibited by gastrointestinal lactobacilli might be due to as yet unknown differences in fibronectin or glycolipid receptors in cell membranes from different animal species.

Although the group A streptococcus model of adhesion is an attractive one to pursue for the lactobacilli, an alternative mechanism should also be considered. This alternative adhesion mechanism involves protein molecules (lectins) which bind specific carbohydrates. Indeed, the majority of mechanisms that permit bacteria to adhere to eucaryotic or microbial surfaces, and that have been studied in detail, appear to involve lectin–glycolipid or lectin–glycoprotein interactions. Lectins may be synthesized by bacterial cells (e.g., fimbriae of enterobacteria), be derived from the host diet (e.g., wheat germ lectin), or be synthesized by eucaryotic cells (e.g., liver cells) (Ofek and Perry, 1985). This means that lactobacillus adhesion could be mediated by (1) lectins synthesized by lactobacilli or by host epithelial cells or (2) lectins that are derived from the animal's diet and may coat epithelial surfaces. These lectins would bind to specific carbohydrates on epithelial cells or to lactobacillus cells, as appropriate, which would explain the mechanism of specificity by certain lactobacilli for a particular animal host.

Speculation on adhesion mechanisms may be entertaining, but it is purposeless unless it stimulates research aimed at gaining information concerning the characteristics of carbohydrates and proteins on lactobacillus cell surfaces and of glycoproteins and glycolipids of epithelial cell membranes in the proximal regions of the gastrointestinal tracts of animals.

Adhesion of lactobacilli to epithelial surfaces is the beginning of the colonization process for some strains of lactobacilli in the gastrointestinal tract. Anchored to the surface of an epithelium, these lactobacilli are able to resist the flushing action of the relatively rapid flow of digesta passing through the proximal digestive tract. Successful colonization by any strain of lactobacillus, however, depends on the metabolic ability of the microbe to function in the gastrointestinal ecosystem.

2.2. Metabolism and Colonization

Lactobacilli are demanding in their nutritional requirements since they require, in addition to carbohydrates as an energy and carbon source, a variety of nucleotides, amino acids, and vitamins for growth (Kandler and Weiss, 1986). Based on the observation that mutant strains of lactobacilli that have lost their requirement for an exogenous source of amino acids can be obtained it can be postulated that the multiple nutritional requirements of the lactobacilli existing in nature today reflect a stepwise natural selection of deficient strains out of a population with a full component of biosynthetic pathways (Morishita *et al.*, 1974, 1981). The lactobacilli inhabiting the gastrointestinal tract can no doubt satisfy their nutritional requirements by utilizing substances in the diet or secretions (exfoliated epithelial cells, digestive secretions, and mucus) of the host. Epithelium-associated lactobacilli are well placed to utilize nutrients diffusing from epithelial cells and underlying tissues (Eyssen *et al.*, 1965). Lactobacilli are not noted for the production of extracellular degradative enzymes, so other members of the normal microbiota may act on complex molecules within the digestive tract before appropriate nutrients become available to the lactobacilli (Thomas and Pritchard, 1987).

Mention of the keratinized epithelium of the mouse forestomach has already been made (Kaplan *et al.*, 1983), and it has been tempting to imagine that lactobacilli colonizing this site might have keratinolytic activity. Examination of 17 gastrointestinal strains of lactobacilli (*L. delbrueckii, L. reuteri, L. gasseri, L. fermentum, L. salivarius,* and *L. acidophilus* isolates), including some epithelium-associating strains from mice, by a sensitive assay that uses fluorescein isothiocyanate-labeled keratin as substrate has failed to detect keratinolytic activity in these bacteria (Mikx and De Jong, 1987; F. H. M. Mikx and G. W. Tannock, unpublished data).

Strains of lactobacilli isolated from the gastrointestinal tract can ferment a wide range of carbohydrates, and strains inhabiting the proximal digestive tract may well have access to any such molecules in the host's diet (Roach *et al.*, 1977). Certainly, starch is fermented by lactobacilli in the chicken crop, and the lactobacillus population of the rodent stomach can be markedly influenced by the composition of the food ingested by the host animal (Brockett and Tannock, 1981; Champ *et al.*, 1981).

Lactobacilli are aciduric or acidophilic bacteria, growth ceasing in laboratory media only when a pH of about 4 is reached (Kandler and Weiss, 1986; Kashket, 1987). The acid tolerance of lactobacilli presumably permits their successful transit through the acid barrier of the human stomach (Drasar and Barrow, 1985) and their ability to inhabit the stomach of pigs. Colonization of the pig stomach by lactobacilli, however, involves bacterial association with the epithelial surface of the pars esophagea: the tissue surface is probably much less acid than the stomach contents. Similarly, the rodent forestomach and the crop of poultry are not sites of hydrochloric acid secretion. Some lactobacilli can "ferment" arginine with the release of ammonia, influencing their environmental pH as well as generating energy (Kandler and Weiss, 1986; Kashket, 1987).

The lactobacilli are generally aerotolerant bacteria, but optimal growth is achieved under microaerophilic or anaerobic conditions. They do not possess cytochrome systems to carry out oxidative phosphorylations, but oxygen can be the terminal electron acceptor in reactions involving the reoxidation of pyrimidine nucleotides mediated by flavine-containing oxidases and peroxidases (Condon, 1987; Kandler and Weiss, 1986). Fermentation of hexoses can occur by either the Embden-Meyerhof pathway (homolactic fermentation) or the 6-phosphogluconate pathway (heterolactic fermentation), depending on the species of lactobacillus (Kandler and Weiss, 1986). Thus, lactobacilli are well equipped to generate energy when inhabiting the anaerobic regions of the distal intestinal tract. The proximal digestive tract, where lactobacilli reside in some animal hosts, probably contains some oxygen since a yeast, *Candida pintolopesii,* that inhabits the acid-secreting part of the murine stomach in some animal colonies harbors mitochondria (McCarthy *et al.*, 1987). This suggests that the ability to utilize oxygen has survival value for an organism inhabiting the proximal digestive tract. The ability of lactobacilli to tolerate the presence of oxygen and even to utilize it, to a limited extent, as a final electron acceptor while remaining essentially fermentative enables these bacteria to inhabit a variety of gastrointestinal sites.

Although lactobacilli are well adapted in metabolic terms for life in various regions of the gastrointestinal tract, one can imagine that these bacteria must face severe competition in the ecosystem from the numerous other microbial types inhabiting the gastrointestinal tract. Competition in ecosystems is, in general, poorly understood at the molecular level. Competition for nutrients is possibly the major growth-limiting factor for microbes in most habitats. Enzyme affinity for particular substrates does not appear to have been compared between lactobacilli and other gastrointestinal inhabitants. Antagonistic interactions between lactobacilli and other microbes have been studied, but only at a superficial level.

2.3. Interactions with Other Microbes

Lactobacilli, apparently because of their ability to produce lactic acid, are able to inhibit the growth of other microbial types in the gastrointestinal tract as evidenced by the facts that (1) the presence of lactobacilli in the chicken crop results in lower numbers of coliforms in that organ (Fuller, 1977), and (2) lactobacilli inhibit a *Ristella* strain (classification of Prévot; a *Bacteroides* or *Fusobacterium*) in the gastrointestinal tract of lactose-fed gnotobiotic mice (Ducluzeau *et al.*, 1971).

Other antagonistic interactions involving lactobacilli have been described, but the antagonistic mechanism(s) have not been elucidated. First, *Lactobacillus fermentum* strains appear to suppress *Escherichia coli, Staphylococcus epidermidis,* and *Enterococcus faecalis* var. *liquefaciens* in the proximal gastrointestinal tract of gnotobiotic rats (Watanabe *et al.*, 1977). Also, mice or rats treated orally with penicillin no longer harbor lactobacilli on the epithelial surface of the forestomach. Under these circumstances, a yeast (*Candida pintolopesii*) that colonizes the secretory epithelium of the stomach in the members of some animal colonies now also spreads onto the forestomach epithelium. Lactobacilli recolonize the forestomach when penicillin treatment is discontinued, displacing the yeast population, which is once again confined to the secretory part of the stomach (Savage, 1969).

Antagonistic interactions occurring in cultures under laboratory conditions between lactobacilli and other microbial types and between lactobacillus strains have been reported on numerous occasions (De Klerk and Coetzee, 1961; Barefoot and Klaenhammer, 1983, 1984; Joerger and Klaenhammer, 1986; McCormick and Savage, 1983; Muriana and Klaenhammer, 1987; Reddy *et al.*, 1984; Shahani *et al.*, 1977; Tagg *et al.*, 1976). The significance of these interactions in ecological terms is not known, however, since the interactions between strains have never been examined under conditions pertaining to the gastrointestinal tract. Some of these *in vitro* interactions appear to involve the production of lactic acid or hydrogen peroxide; others are due to inhibitory substances known as bacteriocins.

There is no universally accepted definition of a bacteriocin, but Tagg *et al.* (1976) have listed six criteria that could be used to define these substances: bacteriocins have a narrow inhibitory spectrum of activity centered about the producer species, they have an essential, biologically active protein moiety, they have a bactericidal mode of action, they attach to specific receptors on susceptible cells, the genetic determinants of bacteriocin production and of immunity to bacteriocins are plasmid borne, and production of bacteriocin may be lethal to the bacteriocin-producing cell. Some inhibitory substances produced by lactobacilli have a relatively wide spectrum of activity, however, and are thus more like conventional antibiotics. Some examples of proteinaceous inhibitory agents produced by lactobacilli are described below.

(1) An unidentified species of *Lactobacillus* produces a heat-stable, acid-pH-stable bacteriocin. The bacteriocin is inactivated by trypsin and papain, is precipitable by ammonium sulfate, and inhibits the growth *in vitro* of an obligate anaerobe, *Clostridium ramosum* (McCormick and Savage, 1983).

(2) An acid-stable, heat-stable inhibitory substance (bulgarican) produced by *L. delbrueckii* subsp. *bulgaricus* DDS in milk cultures inhibits a wide range of bacterial species of the genera *Bacillus, Clostridium, Escherichia, Lactobacillus, Proteus, Pseudomonas, Salmonella, Sarcina, Serratia, Staphylococcus,* and *Streptococcus* (Reddy *et al.,* 1984).

(3) *Lactobacillus acidophilus* DDS1 cultured in milk produces an inhibitory substance (acidophilin) that has a wide spectrum of activity similar to that of bulgarican (Shahani *et al.,* 1977).

(4) Lactacin B is produced by *L. acidophilus* N2 and inhibits strains of *L. delbrueckii (leichmannii), L. bulgaricus, L. helveticus,* and *L. lactis.* Lactacin B is a heat-stable peptide, is sensitive to the action of proteolytic enzymes, and has a molecular weight of about 6000 (Barefoot and Klaenhammer, 1983, 1984).

(5) Lactacin F is a bacteriocin produced by *L. acidophilus* strain 88 that is inhibitory toward *L. acidophilus* 6032, *L. lactis* 970, *L. helveticus* 87, *L. delbrueckii* subsp. *bulgaricus* 1489, *L. delbrueckii (leichmanni)* 4797, *L. fermentum* 1750, and *Enterococcus faecalis* 19433. The bacteriocin is heat stable and sensitive to proteolytic enzymes. The genetic determinants for lactacin F production can be transferred conjugally between *L. acidophilus* strains, apparently due to transiently occurring plasmids [102 and 78 kilobases (kb)] that are usually integrated into the chromosome (Muriana and Klaenhammer, 1987).

(6) *Lactobacillus helveticus* 481 produces a bacteriocin (helveticin J) active against strains of *L. helveticus, L. delbrueckii* subsp. *bulgaricus,* and *L. lactis.* The inhibitor is sensitive to heat (30 min at 100°C) and proteolytic enzymes and has a molecular weight of about 37,000 (Joerger and Klaenhammer, 1986).

A distressing feature of bacteriocin research involving Gram-positive bacteria is that once having been detected and purified *in vitro,* the substances are never heard of again. Surely some *in vivo* testing of these bacteriocin producers for improved colonization ability compared with nonproducing, bacteriocin-sensitive strains is desirable?

2.4. Testing the Colonization Abilities of *Lactobacillus* Strains

Ecological studies invariably require an experimental model with which hypotheses conceived on the basis of the observation of complex, natural ecosystems can be tested under controlled conditions. Studies of the microecology of gastrointestinal lactobacilli are no exception. Experimental models for this work fall into two groups: *in vitro* cell adhesion tests and animal experimentation. *In vitro* adhesion tests with lactobacilli were pioneered by Fuller (1973), who demonstrated that epithelial cells could be brushed from the surface of stratified, squamous epithelia. The cells, suspended in a standard buffer solution, could be mixed with cultured, washed lactobacillus cells and, after a suitable incubation period, observed by light microscopy for evidence of bacterial adhesion to the epithelial cells. Sources of epithelial cells suitable for this test include germfree animals, fasted fowl (food deprivation results in reduced crop populations of lactobacilli), and newborn piglets that have been prevented from suckling the dam. This type of adhesion test has proved useful in screening lactobacillus isolates for epithelium-associating ability. Isolates eventually, however, must be checked by *in vivo* experimentation. Problems with the *in vitro* test include difficulties in harvesting enough of epithelial cells from the organs of small animals (chickens, piglet, and rodents) and harvesting only epithelial cells from the outermost stratum of tissue (the cells that lactobacilli would encounter *in vivo*).

Kotarski and Savage (1979) investigated an *in vitro* method of measuring lactobacillus adhesion capability. They injected [3H]thymidine-labeled lactobacilli into stomachs removed from germfree mice. Radioactivity associated with the nonsecretory (forestomach) and secretory stomach regions was then measured. In concurrent experiments, germfree mice were monoassociated with lactobacilli used in the *in vitro* tests. Evidence of forestomach colonization by lactobacilli was determined by microscope examination of histological sections of the stomach prepared with a microtome–cryostat. Of two strains tested, one colonized the epithelium of the forestomach of mice in the *in vivo* experiments and the other strain did not. Radioactive measurements from the *in vitro* experiments showed adhesion of both strains of lactobacilli to both nonsecretory and secretory parts of the stomach. It must be concluded, therefore, that *in vitro* adhesion tests, because of the impossibility of reproducing in a test tube all of the factors inherent in the natural ecosystem, can only be considered screening methods as far as predicting colonization abilities of lactobacilli are concerned. Adhesion to epithelia under *in vitro* conditions does not guarantee that epithelial association and colonization will occur in the natural ecosystem.

Germfree animals have been used extensively in testing the ability of various lactobacillus strains to colonize the gastrointestinal tract (Lin and Savage, 1984; Schaedler *et al.*, 1965; Roach *et al.*, 1977; Tannock *et al.*, 1982; Wesney and Tannock, 1979). Such animals, housed within isolators, can be inoculated with a single type of lactobacillus and maintained under microbiologically constant conditions. It is probable, however, that any lactobacillus strain will be able to colonize the gastrointestinal lumen of monoassociated animals. The populations of indigenous bacteria normally inhabiting the digestive tract of conventional animals are absent; so too are the regulatory forces generated by the indigenous populations. Although useful for recognition of epithelium-associating strains of lactobacilli, germfree animals do not provide a realistic model with which to study the general colonization characteristics of these bacteria (Tannock and Archibald, 1984).

A more realistic animal model has been developed recently that uses mice housed within isolators using gnotobiotic technology but harboring a complex intestinal microbiota free of lactobacilli (Tannock and Archibald, 1984). Experiments using these lactobacillus-free (LF) mice have shown that the presence of a complex normal microbiota in the digestive tract influences the colonization of the gastrointestinal tract by lactobacilli. Specifically, a strain of lactobacillus isolated from domestic fowl will colonize the gastrointestinal contents of germfree mice. When used to inoculate LF mice, however, the lactobacillus strain is eliminated from the gastrointestinal tract. Lactobacillus strains isolated from rodents colonize the gastrointestinal lumen of both germfree and LF mice, with certain strains associating with the forestomach epithelium in both types of animal. LF mice, therefore, offer an experimental model in which important regulatory aspects of the gastrointestinal ecosystem are retained and with which experiments can be carried out under the microbiologically constant conditions of a gnotobiotic isolator.

In principle, colonization experiments using newborn conventional animals that have not yet been colonized by lactobacilli should be possible. In practice, however, such experiments are difficult because lactobacilli are generally present in the newborn digestive tract once the young animal has suckled for the first time. Lactobacillus strains for use in experiments with conventional animals must be labeled in some manner (e.g., by antibiotic resistance) so that they can be selectively enumerated in laboratory cultures of gastrointestinal specimens. Conventional animals, by definition, are not held under gnotobiotic conditions. There is the added complication, therefore, that lactobacillus types originating from sources other than laboratory cultures will enter the digestive tracts of the experimental animals. The new lactobacillus strains may supplant the experimental strains, making evaluation of experimental results difficult. There is little doubt, though, that as our knowledge of the microecology of the lactobacilli increases, experiments involving conventional animals in open environments will be necessary. In other words, it will be necessary to gradually work toward colonization experiments involving the natural ecosystem in which all competitive, regulatory, and potentially disruptive forces are functioning; otherwise we will not be able to obtain a full explanation of the colonizing abilities and versatility of the gastrointestinal lactobacilli.

2.5. Interactions with the Animal Host

The presence of large populations of lactobacilli inhabiting the proximal regions of the digestive tract of some animal species might be expected to have major influences on the host. Metabolic products of the bacteria could permeate the digesta and be carried into the intestinal tract. Absorption of microbial products from the intestinal tract could influence host tissues remote from the sites of bacterial colonization. Thus, lactobacilli colonizing the digestive tract might produce systemic as well as localized effects of the animal host. Several areas of research support this view.

(1) Lactobacilli inhabiting the crop of fowl produce lactic acid, which lowers the pH of the crop contents (Fuller, 1977). Similarly, the "acid barrier" of the piglet stomach during the first week of life is due mostly to organic acids produced by bacteria, including lactobacilli (Ratcliffe, 1985). Acidic pH is known to minimize the extent to which the proximal digestive tract of animals is colonized by microbes and acts as a barrier through which potential pathogens must pass to reach the intestinal tract (Drasar and Barrow, 1985). The metabolism of lactobacilli in the digestive tract thus constitutes an important part of the nonspecific disease resistance mechanisms of fowl and piglets.

(2) Lactic acid produced by lactobacilli can serve as a source of energy for the host animal. Pigs can absorb and utilize both the L(+) and D(−) isomers of lactic acid (Ratcliffe, 1985).

(3) The utilization of lactose by lactobacilli in the intestinal tract has been suggested to benefit lactose-intolerant human beings. A large proportion of adult humans do not produce sufficient lactase to metabolize a 50-g peroral dose of lactose. Such individuals cannot ingest more than 300 ml of milk without suffering the symptoms of lactose intolerance (abdominal discomfort and diarrhea). It has been proposed that lactobacilli ingested in fermented milk products could lower the lactose concentration of milk, not only during preparation of dairy-based foods but also after the entry of the bacteria into the intestinal tract (Alm, 1982). Some aspects of this interesting concept should be rigorously examined. Are, for example, lactobacilli cultured in milk metabolically active, even for a short period, after their introduction into the gastrointestinal tract?

(4) A metabolite of an *L. delbrueckii* subsp. *bulgaricus* strain has been shown to inactivate *Escherichia coli* enterotoxin in the intestine of pigs. The specific nature of the toxin inhibitor (molecular weight of $<10^3$) has unfortunately not been elucidated (Mitchell and Kenworthy, 1976).

(5) Acute bovine pulmonary emphysema is a respiratory disease of cattle that occurs soon after a change from a poor-quality diet to one of lush green forage. The disease can be induced experimentally by infusing the rumen of cattle with the L-isomer of tryptophan. Infusion with tryptophan results in high concentrations of skatole in the rumen contents and blood of the animals. Experimental administration of skatole into the rumen or blood circulation of cattle also produces the disease. Acute bovine pulmonary emphysema therefore results from the conversion of tryptophan to skatole in the rumen and absorption of the skatole into the blood. A search for skatole-producing bacteria in the rumen has led to the isolation of lactobacillus strains that can decarboxylate indoleacetic acid (a metabolite of tryptophan) to form skatole (Yokoyama and Carlson, 1979, 1981). The specific conditions that promote the production of toxic levels of skatole in the rumen by lactobacilli have not yet been described.

(6) Amine synthesis in lactobacilli has been described by Rodwell (1953) and Nugon-Baudon *et al.* (1985). Monoassociation of germfree chickens with an amine-producing lactobacillus strain capable of colonizing the crop epithelial surface results in higher levels of tyramine, putrescine, and cadaverine in the crop and liver than in untreated chickens. The higher concentrations of amines may be responsible for the excitability, growth disorders, and cannibalism observed in the inoculated birds (Nugon-Baudon *et al.*, 1985).

(7) The normal microbiota of the animal body provides a major source of antigens to which the immunological tissues of the host respond. As a result, the immunological mechanisms of the host are "primed" by the presence of the normal microbiota: conventional animals have a larger stock of immunocompetent cells than do their germfree counterparts (Tannock, 1981). Lactobacilli inhabiting the gastrointestinal tract play a part in this priming phenomenon, since cells producing antibodies that react with lactobacilli can be detected in the spleen of mice from 7 days after birth (Berg and Savage, 1975). Certain strains of lactobacilli promote nonspecific resistance to the establishment of *Salmonella* in murine tissues, as has been demonstrated by the "vaccination" of mice

with lactobacilli and the monoassociation of germfree mice with lactobacillus strains before intravenous challenge with the pathogen (Roach and Tannock, 1980). The decreased susceptibility of mice to the implantation of tumor cells after inoculation with lactobacillus preparations is presumably due to a similar nonspecific stimulus of the immunological tissues by lactobacilli (Kato et al., 1981, 1983).

(8) Numerous articles (summarized by Whitt and Savage, 1987) have appeared relating to the influence of the normal microbiota on enterocyte migration, cell mass, extractable protein, and enzyme concentrations in enterocytes in the proximal small bowel of rodents. Lactobacilli inhabiting the forestomach of rodents have been invoked as important influences on small bowel physiology because they are strategically located to affect duodenal characteristics. Some studies, indeed, have reported lower alkaline phosphatase activity in the duodenum of lactobacillus-associated rodents than in that of germfree or other gnotobiotic animals (Kawai and Morotomi, 1978; Yolton and Savage, 1976). Sampling times and the nature of samples used in this type of study can markedly affect the conclusions drawn. Thus, recent studies suggest that lactobacilli do not directly influence cell mass or alkaline phosphatase, phosphodiesterase, or thymidine kinase activities in the small bowel (Whitt and Savage, 1987).

(9) The growth-stimulating properties of subtherapeutic concentrations of antibiotics in animal feeds have never been explained in terms of a molecular mechanism. A growth-stimulating effect is not observed in germfree animals fed antibiotics, however, which suggests that bacteria, perhaps belonging to the normal microbiota, are involved in growth depression of conventional animals (Visek, 1978). Enterococci and clostridia may be the mediators of growth depression in poultry, but the biochemical mechanism by which this effect is achieved has not been reported. Bile salt hydrolase activity in the small bowel seems to be related to growth depression: antibiotic regimens that produce a decrease in bile salt hydrolase activity in the small bowel of fowl have a growth-promoting effect (Feighner and Dashkevicz, 1987). Lactobacilli inhabiting the gastrointestinal tract of mice commonly have bile salt hydrolase activity (Tannock, 1983a, unpublished data). If this is true of lactobacilli harbored by other animal species, investigation of the growth-depressing effect of bile salt-hydrolyzing strains is warranted. Deconjugation of bile salts in the small bowel by such strains could lead to decreased micelle formation and hence to reduced lipid absorption by the host animal. In addition, elevated levels of deconjugated bile salts can be toxic to enterocytes so that carbohydrate and protein absorption are impaired (Gracey, 1983).

It appears, therefore, that colonization of the gastrointestinal tract by lactobacilli may be a mixed blessing. Whereas certain activities of the bacteria may promote animal health, other activities may be detrimental. Further research is obviously required to clarify many of the interactions occurring between lactobacilli and their hosts.

3. The Microecology of Lactobacilli: The Future

Studies of the microecology of the gastrointestinal tract have thus far relied heavily on methods and techniques derived from traditional medical microbiology. Most of the useful information that can be derived from "dipping and counting" methods has probably been obtained during the past 20 years. Future studies of the normal microbiota will need to use the technologies of molecular genetics and molecular biology if new knowledge is to be acquired (Tannock, 1988b). As far as lactobacilli are concerned, there is a need to derive isogenic strains that differ in a single characteristic. These isogenic strains can then be compared as to their colonization capabilities, for example, and the importance of specific bacterial phenotypic traits in colonization of the gastrointestinal tract can be identified. The major obstacle to such studies is the need to develop methods by which lactobacilli can be genetically manipulated. Largely because of industrial interest in lactic acid-producing bacteria, rather than because they are common inhabitants of the digestive tract, some progress in developing these methods has been made in recent years. In general, methods for the introduction of DNA into lactobacillus cells are required. Thus, plasmid vectors, transformation methods, transduction, and conjugation in lactobacilli need to be investigated. In addition, the use of transposons to generate mutant strains of lactobacilli would be of great assistance in microoecological investigations. Unlike chemical mutagens or irradiation, transposons produce lesions at specific insertion sites in DNA molecules and at the same time provide a phenotypic marker (antibiotic resistance) that enables mutants to be easily detected during screening procedures (Tannock, 1988b).

3.1. Plasmids

Plasmids were first demonstrated in lactobacilli by Chassy et al. (1976). They appear to be relatively common in lactobacilli, including strains isolated from the gastrointestinal tract (Chassy, 1987; Klaenhammer and Sutherland, 1980; Klaenhammer, 1984; Nes, 1984; Lin and Savage, 1985; Tannock and Savage, 1985). The ability of some strains of lactobacilli to adhere to and colonize epithelial surfaces does not appear to be plasmid associated: a strain of L. gasseri known to colonize the forestomach epithelium of mice does not harbor plasmids (Tannock and Savage, 1985). In addition, a strain of L. reuteri able to adhere to crop epithelial cells in vitro has been reported by Sarra et al. (1986) to harbor six plasmids. A culture cured of all plasmids was obtained by culturing the bacteria at 49°C for a prolonged period. The cured strain was able to adhere to crop cells well, as did the wild-type culture.

Most of the plasmids thus far described in lactobacilli are cryptic, but a few have associated phenotypes. (1) A 13-kb plasmid is associated with N-acetyl-D-glucosamine fermentation, proteolysis, and lactic acid production in L. helveticus subsp. jugurti S36-2 (Smiley and Fryder, 1978a,b). (2) A 34.5-kb plasmid is associated with lactose metabolism in L. casei 64H. The first two steps of lactose metabolism are affected: phosphotransferase uptake and β-D-galactoside galactohydrolase (Chassy et al., 1978). The galactohydrolase gene is located on a 7.9-kb fragment of DNA (Lee et al., 1982). (3) Tetracycline and erythromycin resistance is associated with 15- and 56.2-kb plasmids in L. fermentum LF601 (Ishiwa and Iwata, 1980). Other antibiotic resistances have been reported by Vescovo et al. (1982) to be associated with the presence of various plasmids in strains of L. reuteri and L. acidophilus.

Lactobacillus plasmids encoding an easily recognizable phenotypic characteristic are of interest because it may be possible to derive plasmid vectors from them. Some small cryptic plasmids from *L. casei* have been restriction endonuclease mapped and cloned in *Escherichia coli* and in *Streptococcus sanguis* as preliminary steps in screening plasmids as potential vector molecules (Lee-Wickner and Chassy, 1985; Damiani *et al.*, 1987). A plasmid with potential as a shuttle vector has been derived by combining a 4.3-kb fragment from pBR322, a 1.4-kb fragment encoding chloramphenicol resistance from pC194, and a 1.8-kb *Bcl*I fragment from *L. plantarum*. This plasmid replicates in a Gram-positive host, *Bacillus subtilis* (Chassy, 1987). A number of streptococcal vectors are available and may prove useful in introducing cloned DNA fragments into lactobacilli (Chassy, 1987).

3.2. Conjugation

Conjugal transfer of the broad-host-range plasmid pAMβ1 (encoding erythromycin resistance) occurs between *Streptococcus lactis, Enterococcus faecalis, Streptococcus sanguis,* and lactobacilli (Gibson *et al.*, 1979; Shrago *et al.*, 1986; Tannock, 1987; Vescovo *et al.*, 1983). The plasmid also transfers between *L. plantarum* strains and between strains of *L. reuteri* (Shrago *et al.*, 1986; Tannock, 1987). Plasmid pIP501 (encoding erythromycin and chloramphenicol resistance) transfers from *Enterococcus faecalis* to *L. plantarum* strains (West and Warner, 1985). A 54-kb *L. casei* plasmid that encodes enzymes involved in lactose metabolism (phosphoenol pyruvate-dependent phosphotransferase; phospho-β-D-galactoside galactohydrolase) has been found to transfer intraspecifically to non-lactose-fermenting strains (Chassy and Rokaw, 1981). The plasmid mobilizes cryptic, non-self-transmissible plasmids and therefore may be useful in mobilizing recombinant plasmids, which could thus be transferred into lactobacillus recipients during conjugation (Chassy, 1987). DNA encoding the ability to produce lactacin F can be transferred conjugally between strains of *L. acidophilus*. Lactacin F production may be associated with one or both of two plasmids (pPM52 and pPM68) that appear to have a transient existence in lactobacillus transconjugants. They may usually be integrated into the bacterial chromosome, their autonomous state during conjugation being a mobilization event (Muriana and Klaenhammer, 1987).

Although conjugative plasmids have not yet been used to introduce cloned DNA into lactobacillus cells, the potential exists for their use in the genetic manipulation of lactobacilli. Broad-host-range plasmids such as pAMβ1 could be particularly useful in this respect: recombinant plasmids could be introduced into streptococcal cells by transformation and then transferred to lactobacillus recipients by conjugation. Success may best be achieved by using deletion forms of conjugative plasmids that retain their *tra* and replication regions yet are sufficiently small molecules to transform streptococcal recipient cells at a high frequency.

3.3. Transduction

Both virulent and lysogenic bacteriophages occur in lactobacilli, often showing a narrow host range (Yokokura *et al.*, 1974; Stetter, 1977; Stetter *et al.*, 1978; Sozzi *et al.*,

1981). Transduction has not yet been demonstrated in lactobacilli (Kandler and Weiss, 1986).

3.4. Transformation

A natural competency for transformation has not been observed in lactobacilli. Since transformation of cells with plasmid DNA has proved so useful in the genetic manipulation and study of other bacterial genera, however, much effort is now being directed toward the development of methods for transforming lactobacilli (Gilmore, 1985; Chassy, 1987). Several avenues of study are available.

(1) The complete removal of the bacterial cell wall (protoplasting) permits the introduction of DNA into the cells in the presence of polyethylene glycol. The transformed protoplasts must be cultured under conditions that permit synthesis of new cell wall material so that vegetative bacterial cells are regenerated. Protoplast transformation and regeneration can be achieved easily, and at a high frequency, with *Bacillus subtilis* strains (Chang and Cohen, 1979). Osmotically fragile forms of lactobacilli can be obtained by treating cells with muralytic enzymes. In general, lysozyme and mutanolysin are required in combination to produce lactobacillus protoplasts, and regeneration to vegetative forms occurs somewhat unreliably and at low frequencies, if at all (Lee-Wickner and Chassy, 1984; Vescovo *et al.*, 1984). Spheroplasts of lactobacilli (osmotically fragile bacillary forms) can be obtained by treating cells with low concentrations of lysozyme alone. The spheroplasts can be reliably regenerated to osmotically stable forms (Connell *et al.*, 1988).

Polyethylene glycol-mediated transformation of osmotically sensitive forms of a few specific strains of lactobacilli, including gastrointestinal strains, has been reported using either plasmid or chromosomal DNA (Lin and Savage, 1986; Morelli *et al.*, 1987). Whether this approach is suitable for the genetic manipulation of a wide range of lactobacillus species and can be developed into a general method for the transformation of lactobacilli remains to be seen.

(2) Transfection and transformation of lactobacillus protoplasts and spheroplasts by polytheylene glycol-induced fusion with liposomes containing DNA has been reported. Liposomes prepared from egg yolk lecithin have been used to encapsulate bacteriophage DNA or plasmid DNA to protect the nucleic acid molecules from the large amounts of DNase secreted by lactobacillus strains. Liposomes can encapsulate DNA molecules as large as 36.6 kb (Shimizu-Kadota and Kudo, 1984; Chassy, 1987).

(3) Fusion of protoplasts (mediated by polyethylene glycol) belonging to two different strains of bacteria is another method with potential for the genetic manipulation of lactobacilli. The broad-host-range plasmid pAMβ1 and trehalose-fermenting ability have been transferred from *Streptococcus lactis* SH4174 protoplasts to *L. reuteri* DSM20016 protoplasts by this method (Cocconcelli *et al.*, 1986). Protoplast fusion has also been reported for two strains of *L. fermentum* harboring plasmids encoding erythromycin or tetracycline resistance (Iwata *et al.*, 1986).

(4) The poor efficiency of the transformation methods listed above when used with lactobacilli has stimulated interest in a relatively new technique for the introduction of DNA into cells. Electroporation has been used successfully to transform mammalian and

plant cells as well as bacteria and yeasts (Karube *et al.*, 1985; Chassy and Flickinger, 1987). In electroporation, a high-voltage electric discharge is passed through a suspension of cells. Transient "pores" are formed in the cell membrane in an electric field above a certain threshold level, permitting the entry of DNA molecules. Bacterial strains differ in the voltage and time constant that must be used to obtain transformation. Electroporation has been used to transfect/transform *L. casei* strains with bacteriophage DNA (40 kb) and plasmid DNA (4–28 kb) at efficiencies of 5.0 × 10³–8.5 × 10⁴ transformants per μg of DNA (Chassy and Flickinger, 1987). This method has the advantage of producing good transformation efficiencies while requiring neither the preparation of osmotically sensitive cells or their regeneration nor the use of polyethylene glycol.

3.5. Transposons

Transposons are segments of DNA that are endowed with the ability to transpose from site to site in DNA of chromosomal, plasmid, or bacteriophage origin. They are composed of insertion sequences that flank additional DNA that sometimes encodes antibiotic resistance. Transposons can induce mutations when they become inserted into genes, and bacterial strains that have acquired a transposon can be easily recognized and selected if the transposon confers an antibiotic resistance phenotype on the bacterial cells. Labeled probes prepared from transposons can then be used to pinpoint mutations in cloned fragments of bacterial DNA. The introduction of transposons into bacterial cells can be achieved by using plasmids as vectors; in some Gram-positive bacteria, however, self-transmitting conjugative transposons occur (Clewell *et al.*, 1985). Although transposon mutagenesis in lactobacilli has not yet been described, the conjugative transposon Tn*919* transfers conjugally from *Enterococcus faecalis* to *L. plantarum* (Hill *et al.*, 1985). The potential for using transposons to obtain mutant strains of lactobacilli therefore exists. This technique will be preferable to the use of chemical mutagens because of the propensity of the latter agents to produce mutations, usually clustered together, in many genes (Tannock, 1988b).

The creation of isogenic strains of lactobacilli by using one or a combination of the above techniques will greatly enhance studies on the colonization characteristics of these bacteria. Isogenic strains can be studied in suitable animal models to observe the effects of altered phenotype upon colonization ability. These studies will, in turn, lead to characterization of the molecular mechanisms by which lactobacilli colonize the digestive tracts of vertebrate animals. Some suitable phenotypes that could be compared initially include (1) colonization of the gastrointestinal tract of suckling animals by lactose-fermenting and non-lactose-fermenting isogenic strains (influence of specific dietary component), (2) population levels of bacteriocin and non-bacteriocin-producing isotypes in the digestive tract when these isotypes are used to inoculate animals harboring other lactobacillus strains (influence of antagonistic substances in vivo), and (3) epithelium-associating ability of isotypes differing in lectin agglutination patterns (influence of cell surface carbohydrates).

4. Conclusion

The lactobacilli inhabiting the gastrointestinal tract of vertebrate animals provides an excellent tool with which to begin to determine the molecular mechanisms involved in colonization of the digestive tract by bacterial members of the normal microbiota. Already well studied in terms of ultrastructure and metabolism, relatively easily cultured and manipulated in the laboratory, and with the potential to be genetically modified, lactobacillus strains can be used to study microecological phenomena more easily than would be the case with obligately anaerobic members of the normal microbiota. Appropriate lactobacilli can be chosen to study the colonization of all regions of the digestive tract since, in various animal species, areas ranging from the esophagus to the colon are inhabited by these bacteria. Lactobacilli colonizing epithelial surfaces provide tools to study molecular interactions between the normal microbiota and animal tissues; lactobacilli also serve as tools for studying the ecology of microbes inhabiting the contents of the digestive tract.

Historically, it has been common for teachers of microbiology to ascribe altruistic motives to members of the normal microbiota: the microbes nobly aid the host in numerous ways while seeking little in return. In reality, a precarious relationship between the normal microbiota and vertebrate animals has evolved in which some of the metabolic activities of the normal microbiota appear to be beneficial to the animal host. The normal microbiota is, however, only tenuously regulated and confined to certain body regions by mechanisms deployed by the host as the result of the evolution of intimate relationships between microbes and vertebrates. As with other forms of life, members of the normal microbiota selfishly exploit their habitat to its fullest potential in response to the biological urge to continue the existence of their species. All too easily, the commensal/mutualistic balance between the normal microbiota and the host can be destroyed resulting in disease processes involving members of the microbiota (e.g., anaerobic infections, dental caries, endocarditis, urinary tract infection, malabsorption syndromes, and vaginitis). The role of lactobacilli in the gastrointestinal ecosystem is by no means understood. While some lactobacillus strains appear to be beneficial to the host under certain circumstances, the potentially detrimental effects of the metabolic activity of lactobacilli has hardly been investigated. Future studies of the microecology of the gastrointestinal lactobacilli will permit analysis of the historically held belief that lactobacilli inhabiting the digestive tract have solely beneficial effects on the host animal.

References

Alm, L., 1982, Effect of fermentation on lactose, glucose, and galactose content in milk and suitability of fermented milk products for lactose intolerant individuals, *J. Dairy Sci.* **65:**346–352.

Barefoot, S. F., and Klaenhammer, T. R., 1983, Detection and activity of lactacin B, a bacteriocin produced by *Lactobacillus acidophilus*, *Appl. Environ. Microbiol.* **45:**1808–1815.

Barefoot, S. F., and Klaenhammer, T. R., 1984, Purification and characterization of the *Lactobacillus acidophilus* bacteriocin lactacin B, *Antimicrob. Agents Chemother.* **26:**328–334.

Barrow, P. A., Brooker, B. E., Fuller, R., and Newport, M. J., 1980, The attachment of bacteria to the gastric epithelium of the pig and its importance in the microecology of the intestine, *J. Appl. Bacteriol.* **48:**147–154.

Berg, R. D., and Savage, D. C., 1975, Immune responses of specific pathogen-free and gnotobiotic mice to antigens of indigenous and nonindigenous microorganisms, *Infect. Immun.* **11:**320–329.

Brockett, M., and Tannock, G. W., 1981, Dietary components influence tissue-associated lactobacilli in the mouse stomach, *Can. J. Microbiol.* **27:**452–455.

Brockett, M., and Tannock, G. W., 1982, Dietary influence on microbial activities in the caecum of mice, *Can. J. Microbiol.* **28:**493–499.

Brooker, B. E., and Fuller, R., 1975, Adhesion of lactobacilli to the chicken crop epithelium, *J. Ultrastruct. Res.* **52:**21–31.

Champ, M., Szylit, O., and Gallant, D. J., 1981, The influence of microflora on the breakdown of maize starch granules in the digestive tract of chicken, *Poultry Sci.* **60:**179–187.

Chang, S., and Cohen, S. N., 1979, High frequency transformation of *Bacillus subtilis* protoplasts by plasmid DNA, *Mol. Gen. Genet.* **168:**111–115.

Chassy, B. M., 1987, Prospects for the genetic manipulation of lactobacilli, *FEMS Microbiol. Rev.* **46:**297–312.

Chassy, B. M., and Flickinger, J. L., 1987, Transformation of *Lactobacillus casei* by electroporation, *FEMS Microbiol. Lett.*, **44:**173–177.

Chassy, B. M., and Rokaw, E., 1981, Conjugal transfer of lactose plasmids in *Lactobacillus casei*, in: *Molecular Biology, Pathogenesis and Ecology of Bacterial Plasmids* (S. Levy, R. Clowes, and E. Koenig, eds.), pp. 590, Plenum Press, New York.

Chassy, B. M., Gibson, E. M., and Giuffrida, A., 1976, Evidence for extrachromosomal elements in *Lactobacillus*, *J. Bacteriol.* **127:**1576–1578.

Chassy, B. M., Gibson, E. M., and Giuffrida, A., 1978, Evidence for plasmid-associated lactose metabolism in *Lactobacillus casei* subsp. *casei*, *Curr. Microbiol.* **1:**141–144.

Christensen, G. D., Simpson, W. A., and Beachey, E. H., 1985, Adhesion of bacteria to animal tissues: Complex mechanisms, in: *Bacterial Adhesion, Mechanisms and Physiological Significance* (D. C. Savage and M. Fletcher, eds.), pp. 279–305, Plenum Publishing Corp., New York.

Clewell, D. B., Fitzgerald, G. F., Dempsey, L., Pearce, L. E., An, F. Y., White, B. A., Yagi, Y., and Gawron-Burke, C., 1985, Streptococcal conjugation: Plasmids, sex pheromones, and conjugative transposons, in: *Molecular Basis of Oral Microbial Adhesion* (S. E. Mergenhagen and B. Rosan, eds.), pp. 194–203, American Society for Microbiology, Washington, D.C.

Cocconcelli, P. S., Morelli, L., Vescovo, M., and Botazzi, V., 1986, Intergeneric protoplast fusion in lactic acid bacteria, *FEMS Microbiol. Lett.* **35:**211–214.

Condon, S., 1987, Responses of lactic acid bacteria to oxygen, *FEMS Microbiol. Rev.* **46:**269–280.

Connell, H., Lemmon, J., and Tannock, G. W., 1988, Formation and regeneration of protoplasts and spheroplasts of gastrointestinal strains of lactobacilli, *Appl. Environ. Microbiol.* **54:**1615–1618.

Conway, P. L., 1986, Some dietary effects on bacterial adhesion in the alimentary tract with emphasis on *Lactobacillus*, Ph.D. Thesis, University of New South Wales, Kensington, Australia.

Conway, P. L., Gorbach, S. L., and Goldin, B. R., 1987, Survival of lactic acid bacteria in the human stomach and adhesion to intestinal cells, *J. Dairy Sci.* **70:**1–12.

Damiani, G., Romagnoli, S., Ferretti, L., Morelli, L., Bottazzi, V., and Sgaramella, V., 1987, Sequence and functional analysis of a divergent promoter from a cryptic plasmid of *Lactobacillus acidophilus* 168 S, *Plasmid* **17:**69–72.

De Klerk, H. C., and Coetzee, J. N., 1961, Antibiosis among lactobacilli, *Nature* (London) **192:**340–341.

Drasar, B. S., and Barrow, P. A., 1985, *Intestinal Microbiology*, American Society for Microbiology, Washington, D.C.

Ducluzeau, R., Dubos, F., and Raibaud, P., 1971, Effet antagoniste d'une souche de *Lactobacillus* sur une souche de *Ristella* sp. dans le tube digestif de souris "gnotoxeniques" absorbant du lactose, *Ann. Inst. Pasteur* (Paris) **121:**777–794.

Eyssen, H., Swaelen, E., Kowszyk-Gindifer, Z., and Parmentier, G., 1965, Nucleotide requirements of *Lactobacillus acidophilus* variants isolated from the crops of chicks, *Antonie van Leeuwenhoek J. Microbiol. Serol.* **31:**241–248.

Feighner, S. D., and Dashkevicz, M. P., 1987, Subtherapeutic levels of antibiotics in poultry feeds and their effects on weight gain, feed efficiency, and bacterial cholytaurine hydrolase activity, *Appl. Environ. Microbiol.* **53:**331–336.

Finegold, S. M., Attebery, H. R., and Sutter, V. L., 1974, Effect of diet on human fecal flora: Comparison of Japanese and American diets, *Am. J. Clin. Nutr.* **27:**1456–1469.

Finegold, S. M., Sutter, V. L., and Mathisen, G. E., 1983, Normal indigenous intestinal flora, in: *Human Intestinal Microflora in Health and Disease* (D. J. Hentges, ed.), pp. 3–31, Academic Press, New York.

Fuller, R., 1973, Ecological studies on the lactobacillus flora associated with the crop epithelium of the fowl, *J. Appl. Bacteriol.* **36:**131–139.

Fuller, R., 1977, The importance of lactobacilli in maintaining normal microbial balance in the crop, *Br. Poult. Sci.* **18:**85–94.

Fuller, R., and Brooker, B. E., 1974, Lactobacilli which attach to the crop epithelium of the fowl, *Am. J. Clin. Nutr.* **27:**1305–1312.

Fuller, R., and Turvey, A., 1971, Bacteria associated with the intestinal wall of the fowl (*Gallus domesticus*), *J. Appl. Bacteriol.* **34:**617–622.

Fuller, R., Barrow, P. A., and Brooker, B. E., 1978, Bacteria associated with the gastric epithelium of neonatal pigs, *Appl. Environ. Microbiol.* **35:**582–591.

Gibson, E. M., Chace, N. M., London, S. B., and London, J., 1979, Transfer of plasmid-mediated antibiotic resistance from streptococci to lactobacilli, *J. Bacteriol.* **137:**614–619.

Gilmore, M. S., 1985, Molecular cloning of genes encoding Gram-positive virulence factors, *Curr. Top. Microbiol. Immunol.* **118:**219–234.

Gracey, M., 1983, The contaminated small bowel syndrome, in: *Human Intestinal Microflora in Health and Disease* (D. J. Hentges, ed.), pp. 495–515, Academic Press, New York.

Hill, C., Daly, C., and Fitzgerald, G. F., 1985, Conjugative transfer of the transposon Tn*919* to lactic acid bacteria, *FEMS Microbiol. Lett.*, **30:**115–119.

Ishiwa, H., and Iwata, S., 1980, Drug resistance plasmids in *Lactobacillus fermentum*, *J. Gen. Appl. Microbiol.*, **26:**71–74.

Iwata, M., Mada, M., and Ishiwa, H., 1986, Protoplast fusion of *Lactobacillus fermentum*, *Appl. Environ. Microbiol.* **52:**392–393.

Joerger, M. C., and Klaenhammer, T. R., 1986, Characterization and purification of helveticin J and evidence for a chromosomally determined bacteriocin produced by *Lactobacillus helveticus* 481, *J. Bacteriol.* **167:**439–446.

Kandler, O., and Weiss, N., 1986, Regular, nonsporing Gram-positive rods, in: *Bergey's Manual of Systematic Bacteriology*, Vol. 2 (P. H. A. Sneath, ed.), pp. 1208–1234, Williams and Wilkins, Baltimore.

Kaplan, H. M., Brewer, N. R., and Blair, W. H., 1983, Physiology, in: *The Mouse in Biomedical Research* (H. L. Foster, J. D. Small, and J. G. Fox, eds.), pp. 247–292, Academic Press, New York.

Karube, I., Tamiya, E., and Matsuoka, H., 1985, Transformation of *Saccharomyces cerevisiae* spheroplasts by high electric pulse, *FEBS Lett.* **182:**90–94.

Kashket, E. R., 1987, Bioenergetics of lactic acid bacteria: Cytoplasmic pH and osmotolerance, *FEMS Microbiol. Rev.* **46:**233–244.

Kato, I., Kobayashi, S., Yokokura, T., and Mutai, M., 1981, Antitumor activity of *Lactobacillus casei* in mice, *Gann* **72:**517–523.

Kato, I., Yokokura, T., and Mutai, M., 1983, Macrophage activation by *Lactobacillus casei* in mice, *Microbiol. Immunol.* **27**:611–618.

Kawai, Y., and Morotomi, M., 1978, Intestinal enzyme activities in germfree, conventional, and gnotobiotic rats associated with indigenous microorganisms, *Infect. Immun.* **19**:771–778.

Klaenhammer, T. R., 1984, A general method for plasmid isolation in lactobacilli, *Curr. Microbiol* **10**:23–28.

Klaenhammer, T. R., and Sutherland, S. M., 1980, Detection of plasmid deoxyribonucleic acid in an isolate of *Lactobacillus acidophilus, Appl. Environ. Microbiol.* **39**:671–674.

Kleeman, E. G., and Klaenhammer, T. R., 1982, Adherence of *Lactobacillus* species to human fetal intestinal cells, *J. Dairy Sci.* **65**:2063–2069.

Knox, K. W., and Wicken, A. J., 1973, Immunological properties of teichoic acids, *Bacteriol. Rev.* **37**:215–257.

Kotarski, S. F., and Savage, D. C., 1979, Models for study of the specificity by which indigenous lactobacilli adhere to murine gastric epithelia, *Infect. Immun.* **26**:966–975.

Lee, A., 1985, Neglected niches. The microbial ecology of the gastrointestinal tract, in: *Advances in Microbial Ecology,* Vol. 8 (K. C. Marshall, ed.), pp. 115–162, Plenum Press, New York.

Lee, L.-J., Hansen, J. B., Jagusztyn-Krynicka, E. K., and Chassy, B. M., 1982, Cloning and expression of the β-D-phosphogalactoside galactohydrolase gene of *Lactobacillus casei* in *Escherichia coli* K-12, *J. Bacteriol.* **152**:1138–1146.

Lee-Wickner, L.-J., and Chassy, B. M., 1984, The production and regeneration of *Lactobacillus casei* protoplasts, *Appl. Environ. Microbiol.* **48**:994–1000.

Lee-Wickner, L.-J., and Chassy, B. M., 1985, Characterization and molecular cloning of cryptic plasmids isolated from *Lactobacillus casei, Appl. Environ. Microbiol.* **49**:1154–1161.

Lin, J. H.-C., and Savage, D. C., 1984, Host specificity of the colonization of murine gastric epithelium by lactobacilli, *FEMS Microbiol. Lett.* **24**:67–71.

Lin, J. H.-C., and Savage, D. C., 1985, Cryptic plasmids in *Lactobacillus* strains isolated from the murine gastrointestinal tract, *Appl. Environ. Microbiol.* **49**:1004–1006.

Lin, J. H.-C., and Savage, D. C., 1986, Genetic transformation of rifampicin resistance in *Lactobacillus acidophilus, J. Gen. Microbiol.* **132**:2107–2111.

Mayra-Makinen, A., Manninen, M., and Gyllenberg, H., 1983, The adherence of lactic acid bacteria to the columnar epithelial cells of pigs and calves, *J. Appl. Bacteriol.* **55**:241–245.

McCarthy, D. M., Jenq, W., and Savage, D. C., 1987, Mitochondrial DNA in *Candida pintolopesii,* a yeast indigenous to the surface of the secreting epithelium of the murine stomach, *Appl. Environ. Microbiol.* **53**:345–351.

McCormick, E. L., and Savage, D. C., 1983, Characterization of *Lactobacillus* sp. strain 100-37 from the murine gastrointestinal tract: Ecology, plasmid content, and antagonistic activity toward *Clostridium ramosum* H1, *Appl. Environ. Microbiol.* **46**:1103–1112.

Metchnikoff, E., 1907, *The Prolongation of Life. Optimistic Studies,* William Heinemann, London.

Mikx, F. H. M., and De Jong, M. H., 1987, Keratinolytic activity of cutaneous and oral bacteria, *Infect. Immun.* **55**:621–625.

Mitchell, I. De G., and Kenworthy, R., 1976, Investigations on a metabolite from *Lactobacillus bulgaricus* which neutralizes the effect of enterotoxin from *Escherichia coli* pathogenic for pigs, *J. Appl. Bacteriol.* **41**:163–174.

Moreau, M.-C., Thomasson, M., Ducluzeau, R., and Raibaud, P., 1986, Cinetique d'etablissement de la microflore digestive chez le nouveau-ne humain en fonction de la nature du lait, *Reprod. Nutr. Dev.* **26**:745–753.

Morelli, L., Cocconcelli, P. S., Bottazzi, V., Damiani, G., Ferretti, L., and Sgaramella, V., 1987, *Lactobacillus* protoplast transformation, *Plasmid* **17**:73–75.

Morishita, T., Fukada, T., Shirota, M., and Yura, T., 1974, Genetic basis of nutritional requirements in *Lactobacillus casei, J. Bacteriol.* **120**:1078–1084.

Morishita, T., Deguchi, Y., Yajima, M., Sakurai, T., and Yura, T., 1981, Multiple nutritional requirements of lactobacilli: Genetic lesions affecting amino acid biosynthetic pathways, *J. Bacteriol.* **148:**64–71.

Muriana, P. M., and Klaenhammer, T. R., 1987, Conjugal transfer of plasmid-encoded determinants for bacteriocin production and immunity in *Lactobacillus acidophilus* 88, *Appl. Environ. Microbiol.* **53:**553–560.

Nes, I. F., 1984, Plasmid profiles of ten strains of *Lactobacillus plantarum*, *FEMS Microbiol. Lett.* **21:**359–361.

Nugon-Baudon, L., Szylit, O., Chaigneau, M., Dierick, N., and Raibaud, P., 1985, Production of amines in monoxenic chicken inoculated with a lactobacillus strain isolated from holoxenic (conventional) cock crop, *Prog. Clin. Biol. Res.* **181:**119–122.

Ofek, I., and Perry, A., 1985, Molecular basis of bacterial adherence to tissues, in: *Molecular Basis of Oral Microbial Adhesion* (S. E. Mergenhagen and B. Rosan, eds.), pp. 7–13, American Society for Microbiology, Washington, D.C.

Ratcliffe, B., 1985, The influence of the gut microflora on the digestive processes, in: *Digestive Physiology in the Pig* (A. Just, H. Jorgensen, and J. A. Fernandez, eds.), pp. 245–267, National Institute of Animal Science, Copenhagen.

Reddy, G. V., Shahani, K. M., Friend, B. A., and Chandan, R. C., 1984, Natural antibiotic activity of *Lactobacillus acidophilus* and *bulgaricus*. III. Production and partial purification of bulgarican from *Lactobacillus bulgaricus*, *Cultured Dairy Prod. J.* **19:**7–11.

Rettger, L. F., Levy, M. N., Weinstein, L., and Weiss, J. E., 1935, *Lactobacillus acidophilus and Its Therapeutic Application*, Yale University Press, New Haven.

Roach, S., and Tannock, G. W., 1980, Indigenous bacteria that influence the number of *Salmonella typhimurium* in the spleen of intravenously challenged mice, *Can. J. Microbiol.* **26:**408–411.

Roach, S., Savage, D. C., and Tannock, G. W., 1977, Lactobacilli isolated from the stomach of conventional mice, *Appl. Environ. Microbiol.* **33:**1197–1203.

Rodwell, A. W., 1953, The occurrence and distribution of amino-acid decarboxylases within the genus *Lactobacillus*, *J. Gen. Microbiol.* **8:**224–232.

Sarra, P. G., Vescovo, M., and Fulgoni, M., 1986, Study on crop adhesion genetic determinant in *Lactobacillus reuteri*, *Microbiologica* **9:**279–285.

Savage, D. C., 1969, Microbial interference between indigenous yeast and lactobacilli in the rodent stomach, *J. Bacteriol.* **98:**1278–1283.

Savage, D. C., 1977, Microbial ecology of the gastrointestinal tract, *Annu. Rev. Microbiol.* **31:** 107–133.

Savage, D. C., 1983, Morphological diversity among members of the gastrointestinal microflora, *Int. Rev. Cytol.* **82:**305–334.

Savage, D. C., and Blumershine, R. V. H., 1974, Surface-surface associations in microbial communities populating epithelial habitats in the murine gastrointestinal ecosystem: Scanning electron microscopy, *Infect. Immun.* **10:**240–250.

Savage, D. C., Dubos, R., and Schaedler, R. W., 1968, The gastrointestinal epithelium and its autochthonous bacterial flora, *J. Exp. Med.* **127:**67–76.

Schaedler, R. W., Dubos, R., and Costello, R., 1965, Association of germfree mice with bacteria isolated from normal mice, *J. Exp. Med.* **122:**77–83.

Schleifer, K. H., and Kandler, O., 1972, Peptidoglycan types of bacterial cell walls and their taxonomic implications, *Bacteriol. Rev.* **36:**407–477.

Shahani, K. M., Vakil, J. R., and Kilara, A., 1977, Natural antibiotic activity of *Lactobacillus acidophilus* and *bulgaricus*. II. Isolation of acidophilin from *L. acidophilus*, *Cultured Dairy Prod. J.* **12:**8–11.

Sherman, L. A., and Savage, D. C., 1986, Lipoteichoic acids in *Lactobacillus* strains that colonize the mouse gastric epithelium, *Appl. Environ. Microbiol.* **52:**302–304.

Shimizu-Kadota, M., and Kudo, S., 1984, Liposome-mediated transfection of *Lactobacillus casei* spheroplasts, *Agr. Biol. Chem.* **48:**1105–1107.

Shrago, A. W., Chassy, B. M., and Dobrogosz, W. J., 1986, Conjugal plasmid transfer (pAMβ1) in *Lactobacillus plantarum*, *Appl. Environ. Microbiol.* **52:**574–576.

Smiley, M., Maret, R., and Fryder, V., 1978a, Proteolytic activity, lactic acid production, N-acetyl-D-glucosamine fermentation and plasmids in *Lactobacillus helveticus* subsp. *jugurti, Experientia* **34:**955.

Smiley, M. B., and Fryder, V., 1978b, Plasmids, lactic acid production, and *N*-acetyl-D-glucosamine fermentation in *Lactobacillus helveticus* subsp. *jugurti, Appl. Environ. Microbiol.* **35:**777–781.

Stark, P. L., and Lee, A., 1982, The microbial ecology of the large bowel of breast-fed and formula-fed infants during the first year of life, *J. Med. Microbiol.* **15:**189–203.

Stetter, K. O., 1977, Evidence for frequent lysogeny in lactobacilli: Temperate bacteriophages within the subgenus *Streptobacterium, J. Virol.* **24:**685–689.

Stetter, K. O., Priess, H., and Delius, H., 1978, *Lactobacillus casei* phage PL-1. Molecular properties and first transcription studies *in vivo* and *in vitro, Virology* **87:**1–12.

Sozzi, T., Watanabe, K., Stetter, K., and Smiley, M., 1981, Bacteriophages of the genus *Lactobacillus, Intervirology* **16:**129–135.

Suegara, N., Morotomi, M., Watanabe, T., Kawai, Y., and Mutai, M., 1975, Behavior of microflora in the rat stomach: Adhesion of lactobacilli to the keratinized epithelial cells of the rat stomach in vitro, *Infect Immun.* **12:**173–179.

Tagg, J. R., Dajani, A. S., and Wannamaker, L. W., 1976, Bacteriocins of gram-positive bacteria, *Bacteriol. Rev.* **40:**722–756.

Tannock, G. W., 1981, Microbial interference in the gastrointestinal tract, *ASEAN J. Clin. Sci.* **2:**2–34.

Tannock, G. W., 1983a, Effect of dietary and environmental stress on the gastrointestinal microbiota, in: *Human Intestinal Microflora in Health and Disease* (D. J. Hentges, ed.), pp. 517–539, Academic Press, New York.

Tannock, G. W., 1983b, Influence of host diet on gastrointestinal microbes, in: *Fibre in Human and Animal Nutrition* (G. Wallace and L. Bell, eds.), pp. 131–134, Royal Society of New Zealand, Wellington.

Tannock, G. W., 1984, Control of gastrointestinal pathogens by normal flora, in: *Current Perspectives in Microbial Ecology* (M. J. Klug and C. A. Reddy, eds.), pp. 374–382, American Society for Microbiology, Washington, D.C.

Tannock, G. W., 1987, Conjugal transfer of plasmid pAMβ1 in *Lactobacillus reuteri* and between lactobacilli and *Enterococcus faecalis, Appl. Environ. Microbiol.* **53:**2693–2695.

Tannock, G. W., 1988a, The normal microflora: New concepts in health promotion, *Microbiol. Sci.* **5:**4–8.

Tannock, G. W., 1988b, Molecular genetics: A new tool for investigating the microbial ecology of the gastrointestinal tract? *Microb. Ecol.* **15:**239–256.

Tannock, G. W., and Archibald, R. D., 1984, The derivation and use of mice which do not harbour lactobacilli in the gastrointestinal tract, *Can. J. Microbiol.* **30:**849–853.

Tannock, G., and Savage, D., 1985, Detection of plasmids in gastrointestinal strains of lactobacilli, *Proc. Univ. Otago Med. Sch.* **63:**29–30.

Tannock, G. W., and Smith, J. M. B., 1970, The microflora of the pig stomach and its possible relationship to ulceration of the pars oesophagea, *J. Comp. Pathol.* **80:**359–367.

Tannock, G. W., Szylit, O., Duval, Y., and Raibaud, P., 1982, Colonization of tissue surfaces in the gastrointestinal tract of gnotobiotic animals by lactobacillis strains, *Can. J. Microbiol.* **28:**1196–1198.

Tannock, G., Blumershine, R., and Archibald, R., 1987, Demonstration of epithelium-associated microbes in the oesophagus of pigs, cattle, rats and deer, *FEMS Microbiol. Ecol.* **45:**199–203.

Thomas, T. D., and Pritchard, G. G., 1987, Proteolytic enzymes of dairy starter cultures, *FEMS Microbiol. Rev.* **46**:245–268.

Tissier, H., 1905, Repartition des microbes dans l'intestin du nourisson, *Ann. Inst. Pasteur* (Paris) **19**:109–123.

Vescovo, M., Morelli, L., and Bottazzi, V., 1982, Drug resistance plasmids in *Lactobacillus acidophilus* and *Lactobacillus reuteri, Appl. Environ. Microbiol.* **43**:50–56.

Vescovo, M., Morelli, L., Bottazzi, V., and Gasson, M. J., 1983, Conjugal transfer of broad-host-range plasmid pAMβ1 into enteric species of lactic acid bacteria, *Appl. Environ. Microbiol.* **46**:753–755.

Vescovo, M., Morelli, L., Cocconcelli, P. S., and Bottazzi, V., 1984, Protoplast formation, regeneration and plasmid curing in *Lactobacillus reuteri, FEMS Microbiol. Lett.* **23**:333–334.

Visek, W. J., 1978, The mode of growth promotion by antibiotics, *J. Anim. Sci.* **46**:1447–1469.

Watanabe, T., Morotomi, M., Kawai, Y., and Mutai, M., 1977, Reduction of population levels of some indigenous bacteria by lactobacilli in the gastrointestinal tract of gnotobiotic rats, *Microbiol. Immunol.* **21**:495–503.

Wesney, E., and Tannock, G. W., 1979, Association of rat, pig and fowl biotypes of lactobacilli with the stomach of gnotobiotic mice, *Microb. Ecol.* **5**:35–42.

West, C. A., and Warner, P. J., 1985, Plasmid profiles and transfer of plasmid-encoded antibiotic resistance in *Lactobacilus plantarum, Appl. Environ. Microbiol.* **50**:1319–1321.

Whitt, D. D., and Savage, D. C., 1987, Lactobacilli as effectors of host functions: No influence on the activities of enzymes in enterocytes of mice, *Appl. Environ. Microbiol.* **53**:325–330.

Wicken, A. J., and Knox, K. W., 1975, Lipoteichoic acids: A new class of bacterial antigen, *Science* **187**:1161–1167.

Wicken, A. J., and Knox, K. W., 1980, Bacterial cell surface amphiphiles, *Biochim. Biophys. Acta* **604**:1–26.

Yokokura, T., Kodaira, S., Ishiwa, H., and Sakurai, T., 1974, Lysogeny in lactobacilli, *J. Gen. Microbiol.* **84**:277–284.

Yokoyama, M. T., and Carlson, J. R., 1979, Microbial metabolites of tryptophan in the intestinal tract with special reference to skatole, *Am. J. Clin. Nutr.* **32**:173–178.

Yokoyama, M. T., and Carlson, J. R., 1981, Production of skatole and *para*-cresol by a rumen *Lactobacillus* sp., *Appl. Environ. Microbiol.* **41**:71–76.

Yolton, D. P., and Savage, D. C., 1976, Influence of certain indigenous gastrointestinal microorganisms on duodenal alkaline phosphatase in mice, *Appl. Environ. Microbiol.* **31**:880–888.

5

Enhanced Biological Phosphorus Removal in Activated Sludge Systems

D. F. TOERIEN, A. GERBER, L. H. LÖTTER, and T. E. CLOETE

1. Introduction

Eutrophication is a worldwide water pollution problem which results in the overabundant growth of algae and/or macrophytes (Wetzel, 1983). Control of the access of phosphates (P) to the aquatic environment is widely used as a eutrophication control strategy (e.g., Lee *et al.*, 1978), thus requiring its removal from effluents by chemical and/or biological means.

Activated sludge systems, modified for enhanced P removal during effluent treatment, are now operative in at least eight countries spanning the globe (Australia, Brazil, Canada, France, New Zealand, South Africa, United States, and Zimbabwe). At present, more than 40 full-scale treatment plants have been constructed at a cost of several hundred million U.S. dollars. Substantial savings are achieved through biological rather than chemical P removal. For instance, we estimate savings of US$5–6 million per annum for some 32 nutrient removal treatment plants operating in South Africa.

Enhanced P removal in these activated sludge systems appears to be a result of microbial action. Fuhs and Chen (1975) first isolated bacteria, identified as *Acinetobacter* spp., from a P-removing activated sludge plant. These bacteria accumulated polyphosphates (polyP). The presence of *Acinetobacter* and other polyP bacteria has repeatedly been shown in activated sludge systems (Brodisch, 1985; Cloete, 1985; Cloete *et al.*, 1985; Brodisch and Joyner, 1983; Buchan, 1981, 1983). There is little doubt that

D. F. TOERIEN and A. GERBER • Division of Water Technology, CSIR, Pretoria, South Africa. L. H. LÖTTER • City Health Department, Johannesburg, South Africa. T. E. CLOETE • Department of Microbiology and Plant Pathology, University of Pretoria, Pretoria, South Africa.

polyP bacteria are responsible for enhanced biological P removal and that optimization of the process depends on a thorough understanding of the physiology and ecology of these bacteria.

1.1. Polyphosphates

Phosphorus occurs in procaryotic and eucaryotic cells in important cellular structural components such as nucleic acids, phospholipids, and proteins, as well as in the nucleotides involved in cellular bioenergetics (Kulaev, 1985). The metabolism of polyP in microorganisms has been reviewed extensively (Kulaev, 1975, 1985; Kulaev and Vagabov, 1983; Dawes and Senior, 1973; Harold, 1966). Kulaev (1985) pointed out that polyP is important in several cellular functions, e.g., (1) the accumulation of energy-rich ATP phosphorylic residues in an osmotically inert reserve material containing "activated phosphate," (2) making cells more independent of environmental conditions through the accumulation of polyP reserves, (3) regulation of ATP and other nucleotide levels in cells, (4) fulfilling the function of ATP in some cases by direct participation in phosphorylation reactions, (5) linking cation (e.g., K^+, Mg^{2+}, and Mn^{2+}) metabolism with that of polyP, and (6) contributing to cellular homeostasis and osmotic regulation.

1.2. Activated Sludge Treatment

Enhanced biological P removal from domestic wastewaters in full-scale activated sludge plants is currently perceived to hinge on the provision of alternate stages in which the activated sludge is subjected to anaerobic and aerobic conditions, respectively (e.g., Barnard, 1976). A characteristic feature of such plants is that P, after being released from the biomass in an anaerobic stage, is reincorporated in the biomass during aeration, together with part or all of the influent P (Gerber *et al.*, 1986).

Although there are many successful applications of enhanced P removal, the role of particular polyP organisms, the mechanisms involved, and indeed the functions of the anaerobic stage itself remain unclear (Gerber *et al.*, 1986). Nicholls and Osborn (1979) postulated that the anaerobic zone caused stress in the microbial population, which resulted in enzyme reactions similar to those occurring during certain nutrient imbalances and eventually led to the production of polyP granules. Fuhs and Chen (1975) were the first to implicate bacteria of the *Acinetobacter-Moraxella-Mima* group in biological P removal. They postulated that the anaerobic zone served as a fermentation zone for acidogenic bacteria to transform organic matter by fermentation. The fatty acids produced may be taken up by the polyP bacteria and stored in the form of poly-β-hydroxybutyrate (PHB) (Deinema *et al.*, 1980; Nicholls, 1978). The energy required for this uptake and storage is supplied by the hydrolysis of polyP reserves (Rensink, 1981), which explains P release in the anaerobic stage. During the aerobic phase, PHB is metabolized and the energy produced is partially used to reconstitute polyP reserves.

The mechanism of enhanced P removal in activated sludge systems must therefore depend on a group of organisms (the polyP organisms) which in nature are favored by fluctuating conditions of aerobiosis–anaerobiosis. Their selective advantage requires (1) their presence, (2) alternating aerobiosis–anaerobiosis, (3) degradable organic matter which can be fermented by acidogenic bacteria, and (4) the presence of a sufficient quantity of P to allow uptake of fatty acids during anaerobiosis and P uptake during aerobiosis.

When activated sludge systems were first modified to include a nonaerated stage for nitrogen removal, these systems fortuitously met the above conditions; hence, the first observation of and eventual utilization of enhanced P removal. However, the precise mechanism of enhanced P removal in activated sludge systems has not yet been clarified (Gerber *et al.*, 1987a; Comeau *et al.*, 1986). In this chapter, we examine the state of knowledge of enhanced biological P removal in activated sludge systems and its practical application.

2. Background and Current Practices

2.1. Evolution of Biological P-Removal Processes

Conventional activated sludge systems (Fig. 1) have been used to purify domestic wastewater for many decades (Ardern and Lockett, 1914). Microbes are retained in the system as an activated sludge. In the early 1970s, research in the United States and South Africa suggested the feasibility of enhanced P removal in modified activated sludge systems (Barnard, 1973; Levin *et al.*, 1972). For instance, Barnard (1973) found that high nitrogen (N) and P removal could be obtained in the Bardenpho process, with its four stages (Fig. 2). Some of the nitrate (NO_3^- – N) formed in the mixed liquor of the aerobic (or nitrification) stage was recycled together with the mixed liquor to the first anoxic (denitrification) stage and mixed with raw or settled sewage. Apart from serving as the principal zone for conversion of NO_3^- – N to gaseous end products, the primary anoxic stage also brought about significant degradation of the carbonaceous waste components, thus reducing subsequent oxygen requirements (1 mg NO_3^- – N ≡ 2.86 mg oxygen). Mixed liquor, which was not recycled, was passed to a second anoxic stage, where most of the remaining NO_3^- – N was reduced to the gaseous state by endogenous respiration. The secondary aeration stage served, among other purposes, to strip micro-bubbles adhering to floc particles as a result of secondary denitrification, before mixed

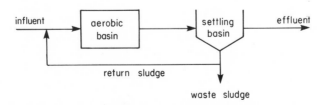

Figure 1. Flow schematic of the conventional activated sludge process.

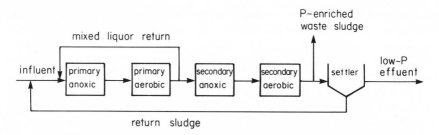

Figure 2. Flow schematic of the Bardenpho process for carbon and nitrogen removal.

liquor entered the clarifier, and to ensure positive dissolved oxygen (DO) concentrations in the clarifier, not only to forestall denitrification and rising sludge but also to prevent possible P release. In addition to high levels of N removal (90%), excellent P removal was observed (Barnard, 1974–76, 1982).

Most of the processes for enhanced biological P removal also incorporate stages for biological N removal. Although separate reactor–clarifier combinations can be used in series for carbon removal, nitrification, and denitrification, respectively (Fig. 3), the current trend is to have these processes, along with biological P removal, occur in so-called single-sludge systems, in which the microbial biomass responsible for these reactions is present as one fully mixed assemblage. Since the occurrence of N compounds such as $NO_3^- - N$ can have a severe impact on biological P elimination, N removal must also be considered when discussing biological P removal.

2.1.1. Single-Sludge N-Removal Systems

Wuhrmann (1957, 1960) and Ludzack and Ettinger (1962) were among the earlier researchers who proposed specific activated sludge process configurations for N removal. Wuhrmann (1957) proposed a single-sludge two-stage process, with the sewage fed into an aeration stage together with the recycled sludge from the clarifier. The nitrified mixed liquor passed from the aeration stage, where most of the carbonaceous compounds were removed, to an anoxic stage (no measurable dissolved oxygen), where denitrification occurred at a very slow rate due to low residual carbonaceous energy.

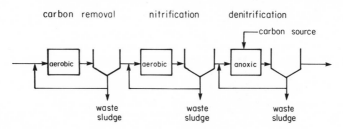

Figure 3. Flow schematic of the reactor/settler sequence constituting the three-sludge system for carbon and nitrogen removal.

This problem was overcome by dosing an N-free source of biodegradable organic carbon, such as methanol, into this zone, thereby allowing the volume of the denitrification stage to be reduced.

Ludzack and Ettinger (1962) placed an anoxic stage in front of the aeration stage and created an internal recycle from the aeration stage into the anoxic stage by means of hydraulic back-mixing. The organic-laden feed sewage was introduced into the anoxic stage while the recycled sludge was returned to the aerated stage. The NO_3^- $-$ N entering the anoxic stage was rapidly reduced in the presence of the biodegradable materials present in the incoming feed, and the NO_3^- $-$ N free flow leaving this stage was then introduced into the aeration basin, where nitrification occurred. By increasing the recycle rate, the NO_3^- $-$ N concentration in the effluent could be reduced. This configuration resulted in high rates of denitrification without requiring addition of an external carbon source. Sharma and Ahlert (1977) and Christensen and Harremoës (1977), among others, have published extensively on nitrification and denitrification systems.

2.1.2. Systems for Enhanced Biological P Removal

P removal by an activated sludge process was first investigated by Jenkins and Lockett (1943). They noted that only 54% of the P entering a treatment works was discharged in the effluent. Harris (1957) reported that certain microbial strains can accumulate P in excess of metabolic requirements, and Srinath *et al.* (1959) carried out the first detailed laboratory analysis of enhanced P removal in an activated sludge system.

Levin and Shapiro (1965) introduced the term "luxury uptake" to describe the ability of activated sludge to remove more P than that required for growth. They obtained patent rights (Levin, 1964) for the Phostrip process, which has been successful in pilot plant (Levin *et al.*, 1972) and full-scale (Levin *et al.*, 1975, Levin and Sala, 1987) applications. This process (Fig. 4) consists of an activated sludge reactor into

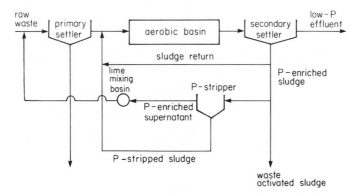

Figure 4. Flow schematic of the (sidestream) Phostrip process for carbon and phosphorus removal.

which recycled sludge, stripped of P, is introduced with an influent feed. In the earlier versions no provision was made for N removal, but a subsequent patent (Levin, 1972) also covering N removal was granted.

The Phostrip process is designed to handle carbonaceous oxidation, nitrification, and denitrification according to the proposal of Wuhrmann (1957). Aerated mixed liquor enters a settler, where activated sludge solids are separated and the supernatant is discharged. The solids are passed to an anaerobic clarifier where, in the absence of both dissolved oxygen and nitrate, P is released from the organisms into the supernatant. The P-stripped sludge is returned to the head of the activated sludge reactor, while the P-enriched secondary supernatant is treated separately with lime to precipitate P. The process can produce an effluent with a very low P concentration but achieves relatively poor denitrification unless an external carbon source is introduced into the anoxic stage.

Enhanced biological P removal in a full-scale plant was reported by Vacker *et al.* (1967). This plant was not specifically designed to achieve enhanced P removal and consisted of three modules operated in parallel and fed with the same influent feed. By varying the nature and degree of treatment, they were able to observe which parameters were correlated with effective P removal. Under conditions of limited nitrification, "luxury" P uptake occurred.

Barnard (1973) reviewed the available technology for biological nitrification and denitrification and proposed a four-stage process, the so-called Bardenpho process (Fig. 2). This process, for which the South African Council for Scientific and Industrial Research was granted patent rights (South African Inventions Development Corporation, 1973), consists of four stages—primary anoxic, primary aerated, secondary anoxic, and secondary aerated—followed by a clarifier. Carbon removal and nitrification take place in the main aeration basin. Nitrified mixed liquor is recycled from this basin to the primary anoxic basin where, in the absence of free dissolved oxygen, denitrification occurs, using the organic compounds in the influent wastewater as the carbon source. Mixed liquor not recycled to the primary anoxic basin passes on to the secondary anoxic basin, where additional denitrification takes place at a slow rate under conditions of endogenous respiration. Before entering the secondary clarifier, the mixed liquor from the secondary anoxic basin passes through a small reaeration basin, the function of which is to ensure that (1) NH_3 − N formed during endogenous respiration in the secondary anoxic basin is converted to NO_3^- − N, (2) aerobic conditions exist in the secondary clarifier, as any denitrification that occurs there under anoxic conditions would produce nitrogen gas that could cause rising sludge, and (3) aerobic conditions exist in the secondary clarifier to prevent P release from the sludge into the effluent.

After pilot plant trials with settled domestic wastewater, Barnard (1974) reported N and P removals of up to 95 and 97%, respectively, without the addition of chemicals. The results obtained were consistent with the observation by Milbury *et al.* (1971) that full-scale activated sludge plants in the United States, with unusually high levels of P removal, released P from the sludge microorganisms while the sludge was near the inlet end of the aeration basin, followed by P uptake along the length of the aeration basin. Barnard (1974) postulated that key requirements for enhanced biological P removal are exposure of sludge microorganisms to anaerobic conditions (i.e., an environment free of dissolved oxygen and NO_3^- — N), under which P release could occur. When subsequently exposed to an aerated or aerobic environment, the P that was previously released into solution and the P entering in the influent wastewater could either be taken up by the microorganisms in quantities greater than their normal metabolic requirements or be precipitated from solution as a result of changes in redox potential caused by exposure to different oxygen regimes.

McLaren and Wood (1976) and Davelaar et al. (1978) modified the Bardenpho process to include five stages, i.e., the Bardenpho sequence with an additional anaerobic stage placed at the head of the process (Fig. 5). The anaerobic stage received the influent sewage feed and the sludge recycled from the clarifier underflow. The outflow from the anaerobic stage was passed into the primary anoxic stage, which also received the mixed liquor recycle. The mixed liquor recycle was drawn either from a point within the primary aeration stage, which is the third stage of the process, or from the outflow from this stage. The remainder of the outflow entered a secondary anoxic stage, which was followed by a secondary aeration stage and finally a clarifier, from which the supernatant liquor was discharged and the settled solids were returned to the head of the process.

In some applications, the last two basins have been deleted from the five-stage process because inclusion of the secondary anoxic basin can result in only slightly greater total N removal because of the low rate of denitrification under endogenous respiration (Simpkins and McLaren, 1978). Such systems are known in South Africa as the three-stage modified Bardenpho process.

Barnard (1976) also postulated that enhanced biological P removal could be achieved in a simple activated sludge plant not designed for N removal by adding an anaerobic basin to the head end of the plant (Fig. 6). In this so-called Phoredox configuration, nitrification in the aeration basin would have to be minimized so that NO_3^- — N is not discharged into the anaerobic basin via the return sludge.

The UCT process (Fig. 7) was developed by Ekama et al. (1984). It is similar to the three-stage modified Bardenpho process except that the return sludge is discharged to the anoxic reactor rather than the anaerobic reactor, and mixed liquor is recycled from the anoxic reactor to the anaerobic reactor. Its principal advantage over the modified Bardenpho process is that there is an opportunity for NO_3^- — N in the return sludge to be

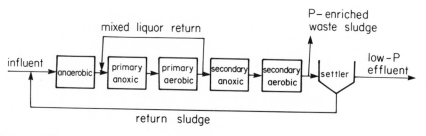

Figure 5. Flow schematic of the five-stage Bardenpho process for carbon, nitrogen, and phosphorus removal.

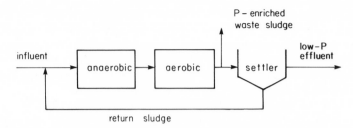

Figure 6. Flow schematic of the Phoredox process for carbon and phosphorus removal.

denitrified in the anoxic basin before entering the anaerobic basin. The modified UCT process contains an anoxic reactor divided into two separate compartments so that denitrification of the return sludge and mixed liquor recycle streams can be controlled separately.

The A/O process for biological P removal was developed in the United States (Hong *et al.*, 1982). It is similar to the Phoredox process (Barnard, 1976) except that the anaerobic and aerobic stages are divided into a number of compartments. This configuration can be designed with or without nitrification (Hong *et al.*, 1982). If total N removal is required, a three-stage A/O process incorporating an anoxic zone for denitrification is used (Hong *et al.*, 1982).

Arvin and Kristensen (1985) developed the Biodenipho process (Fig. 8). It is based on another biological N-removal process (the Biodenitro process), which was also developed in Denmark (Bundgaard *et al.*, 1983) but does not include an initial anaerobic basin. Both processes are continuous batch processes with an alternating mode of operation. Nitrification takes place during aeration in a simple basin, followed by a period of denitrification under unaerated conditions in the same basin. The advantages are (1) elimination of the need for mixed liquor recycle pumps and a recycle pipe or channel and (2) the ability to vary the detention times for nitrification and denitrification.

The Biodenipho process is based on four operational phases (Fig. 8), each with a

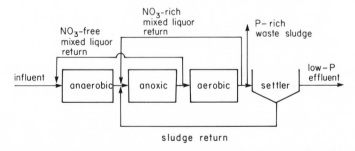

Figure 7. Flow schematic of the UCT process for carbon, nitrogen, and phosphorus removal.

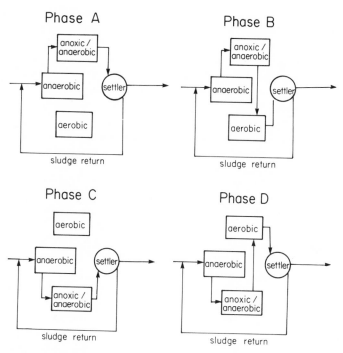

Phase A   Phase B

Phase C   Phase D

Figure 8. Flow schematic and operational sequence of the continuous-batch Biodenipho process for carbon, nitrogen, and phosphorus removal.

typical duration of 1 hr. Nitrification occurs in the aerated basin for 2 hr during phases A and B, followed by denitrification in the same basin during the 2-hr duration of the subsequent phases C and D. Similarly, nitrification takes place in the aerated basin associated with phases C and D, followed by denitrification during phases A and B of the next cycle. During phases A and C, wastewater does not pass through the aerobic basins indicated.

2.1.3. Operating Conditions of Full-Scale Enhanced P-Removal Plants

Extensive descriptions of the operation of full-scale biological nutrient removal plants exist (e.g., Osborn *et al.*, 1986; Pitman, 1984). Of central importance is that in the conventional activated sludge process, which is not designed or operated to achieve biological excess P removal, the bacteria use P only in quantities that satisfy their basic metabolic requirements. Because of nutrient imbalances in sewage, only a limited quantity of the feed P will be removed in such plants. In plants designed for enhanced biological P removal, however, an environment is created for the proliferation of bacteria that will accumulate P in excess of normal metabolic requirements. These bacteria can accumulate comparatively large quantities of P. Furthermore, certain chemical precipitates containing P may be adsorbed onto organic matter in activated sludge. Plants modified for enhanced P removal can, for example, produce effluents containing less than 1 mg liter^{-1} P from a feed sewage containing 10 mg P liter^{-1}.

2.1.3a. Anaerobic Conditioning of Sludge. Under anaerobic conditions, activated sludge that has accumulated excess P will tend to release P into the surrounding liquid (Fig. 9). The release rate depends on the activity of the sludge in terms of enhanced P removal and the length of time of exposure to anaerobiosis. Within the relatively short anaerobic detention times used in full-scale plants (0.5 to 3 hr), between 10 and 20% of the accumulated P will be released (Pitman, 1984). Such release is an important prerequisite for biological P removal and indicates that an activated sludge has been conditioned to remove further quantities of P.

Anaerobic conditioning is achieved by passing the activated sludge repeatedly through a zone free of NO_3^- − N and DO so that at any one time a fraction of the total activated sludge is subjected to an anaerobic environment. In certain plants, on/off operation is used to anaerobically condition the sludge. This may be achieved by stopping all aerators for a period of time. The off period is repeated after some time, and the plant is then operated in an aerobic state for a period. If the release capacity in the anaerobic zone is lost, e.g., due to the presence of NO_3^- − N, P removal will fail. In most plants this usually manifests itself shortly after the P-release capacity is lost (Okada *et al.*, 1987; Pitman, 1984). If sludge has been anaerobically conditioned on a batch or on/off basis, the conditioned sludge will continue to exhibit excess P uptake for a number of days or even weeks before the effect of the conditioning is lost (Pitman, 1984).

2.1.3b. Limitation of NO_3^- − N Entering the Anaerobic Stage. In plants with an anaerobic zone at the inlet end, NO_3^- − N feedback to this zone via the return sludge causes problems. When NO_3^- − N concentrations in the return sludge are high, an excessive denitrification load can be placed on this zone, and the anaerobic conditions required for P removal may be lost. As soon as NO_3^- − N is detectable in an anaerobic zone (even at 0.5 to 1.0 mg liter^{-1} as N) of large-scale plants, impending failure of P removal by the plant is indicated. Ideally, the return sludge should contain no NO_3^- −

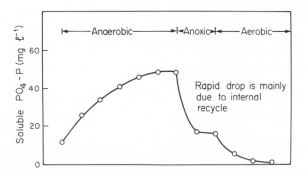

Figure 9. Spatial distribution of phosphorus concentration in biological nutrient removal systems.

N, but this is not always possible in practice. The degree of $NO_3^- - N$ feedback that can be tolerated depends on the strength of the sewage fed to the anaerobic zone and, in particular, its readily biodegradable chemical oxygen demand (COD) concentration. For a high-strength sewage with COD around 600 mg liter^{-1} and with a high readily biodegradable content, a return sludge $NO_3^- - N$ level of 7 mg liter^{-1} could be tolerated. Alternatively, for a low strength feed with a COD of about 200 mg liter^{-1}, a return sludge $NO_3^- - N$ level of 1.5 mg liter^{-1} could be undesirable. A low return sludge $NO_3^- - N$ concentration is ensured if the effluent from the biological reactor has a low $NO_3^- - N$ concentration. This depends on denitrification efficiency, which in turn depends on many factors. $NO_3^- - N$ feedback to the anaerobic zone can also occur via back-mixing from adjacent anoxic zones through ports in the baffles separating anaerobic and anoxic zones.

The UCT and modified UCT processes have been conceived specifically to prevent the uncontrolled discharge of $NO_3^- - N$ to the anaerobic stage via the underflow recycle. Recently, Pitman *et al.* (1987) and Nicholls *et al.* (1987) successfully incorporated anoxic reactors on the return sludge line of a three-stage modified Bardenpho plant (see Section 2.1.2) to achieve the same objective. Although the $NO_3^- - N$ removal rate in these reactors is slow because of the absence or limited availability of extracellular energy sources, the concentrated mass of organisms results in negligible NO_3 levels entering the anaerobic reactor.

2.1.3c. Oxygen and Denitrification. The ingress of excessive amounts of molecular oxygen (O_2) into an anaerobic reactor will ultimately, like $NO_3^- - N$, cause P release to cease. Excessive turbulence in streams entering an anaerobic zone or in the zone itself can introduce unnecessary O_2. For instance, cascades, screw pumps, and aerated grit channels can oxygenate the feed sewage, and screw pumps in sludge recycle lines can have the same effect. Excessive mixing of anaerobic zones could also be detrimental. The denitrification stage is of extreme importance in full-scale nutrient removal activated sludge plants. If nitrification of the effluent is required, there are technical and economic advantages in incorporating denitrification with enhanced P removal. For example, the O_2 requirement for nitrification of domestic wastewater is typically 25–35% of the total O_2 requirement (Ekama and Marais,1984a). A large part (63%) of the O_2 demand for nitrification can be recovered when nitrification is followed by denitrification. By siting a denitrification reactor upstream of the aerobic basin, an equivalent amount of influent biodegradable organic matter can be oxidized, thus reducing the required oxygen input requirements by way of aeration. According to the stoichiometry of nitrification, H^+ is released during the process. For every 1 mg of $NH_3 - N$ oxidized, 7.14 mg of alkalinity (as $CaCO_3$) is destroyed. Many low-alkalinity wastewaters thus cannot support full nitrification unless alkalinity is supplemented by lime addition or other means. If $NO_3^- - N$ is completely denitrified, however, 50% of the alkalinity removed during nitrification is recovered, often being sufficient for nitrification requirements.

The optimum mixed-liquor recycle rate would be such that $NO_3^- - N$ is fully consumed in the anoxic zone. Because of variations in the strength and volume of the feed sewage and the usual lack of online monitoring means, this ideal is normally not practically attainable. Therefore, a fixed recycle ratio of, say, 4 : 1 is usually set. In typical nutrient removal plants, P and $NH_3 - N$ concentrations in the mixed-liquor solution will decline between anaerobic and anoxic zones. This is due to a dilution effect caused by the recycled mixed liquor (Fig. 9). If, however, $NH_3 - N$ decreases to a greater extent than can be accounted for by dilution, the O_2 input to this zone is too great, leading to nitrification in the anoxic zone. Oxygen input should then be restricted because of possible failure to remove P.

In the aerobic zones, organic matter is oxidized and $NH_3 - N$ is nitrified. Reactor size and DO levels should be such that the above processes can operate effectively. Practical experience has shown that an average DO in the aerobic zone should be about 2–3 mg liter^{-1} and should not be allowed to decline below 1 mg liter^{-1} at any one point. A DO profile starting at about 1 mg liter^{-1} near the inlet and rising to about 3 mg liter^{-1} at the outlet is suitable. Filamentous bacterial growth is enhanced and nitrification is suppressed by DO values below 0.5 mg liter^{-1}. Excessively high DO levels (4–5 mg liter^{-1}) should be avoided, because "pinpoint" floc formation can be caused and power will be wasted. Trial-and-error allow optimum DO levels to be determined for each plant. Often the monitoring of $NH_3 - N$ and $NO_3^- - N$ levels in filtered mixed-liquor samples from the outflow of the aerobic zone gives a good indication of the sufficiency of aeration levels. For example, high $NH_3 - N$ levels and low $NO_3^- - N$ levels suggest inadequate aeration or *vice versa*.

The secondary anoxic zone, when present, serves to achieve additional denitrification. Experience in many plants has shown that only about 2 to 3 mg $NO_3^- - N$ liter^{-1} is removed in this zone because of a low denitrification rate (Pitman *et al.*, 1987). Therefore, inclusion of a secondary anoxic zone is desirable only when very low effluent $NO_3^- - N$ levels are desired. With low-strength sewage, containing low amounts of biodegradable solids, the secondary anoxic zone will remove minimal amounts of $NO_3^- - N$. Most of the biodegradable substrates will have been metabolized in the aerobic zone, resulting in a sludge with a low activity entering this zone. For high-strength sewage which contains higher quantities of biodegradable compounds, the latter compounds may not be totally consumed in the first aerobic zone. Then higher activities and denitrification rates occur in the secondary anoxic zones. If high-strength sewage with a sufficiently high proportion of biodegradable compounds is treated, activities in this zone might be so high that $NO_3^- - N$ is depleted, the zone goes anaerobic, and P is released. This P may not be removed in the final aerobic zone, and effluent P levels could be increased. A method of avoiding P release in the secondary anoxic zone is to allow DO or some $NO_3^- - N$ from the aerobic zone to enter this zone.

2.2. Design and Operational Aspects Pertinent to Biological P Removal

The process design for a given biological P-removal process should, among other things, specify aspects such as reactor volumes, sludge concentrations and wastage rates, recycle flows where applicable, oxygen requirements in the aeration tanks, sludge handling and disposal, and primary and secondary treatment facilities. The design selected is strongly dependent on the quality and hydraulic characteristics of the wastewater to be treated as well as on a host of local factors such as the availability of materials, land, and trained operators. In this section, no attempt has been made to present detailed design procedures for biological P removal, particulars of which are available in other publications (e.g., Water Research Commission, 1984). The major design and operational principles, which almost inevitably have an ecological impact on activated sludge plants or are affected by microbial considerations, are outlined below.

2.2.1. Flow Regime

Enhanced biological P-removal processes are either mainstream or sidestream systems, examples of which are the modified Bardenpho (Fig. 5) and Phostrip (Fig. 4) configurations, respectively. Both process types may be operated either with or without concomitant biological N removal. The common feature in both mainstream and sidestream processes is that the mixed sludge passes through an anaerobic–aerobic sequence in a cyclic mode, an operation which has proven to be of fundamental importance to establish a capacity to accumulate excess P. The microbiological and biochemical consequences of this process are discussed in Sections 4 and 5.

The feature distinguishing between mainstream and sidestream processes is that the sink for P removal in the former is the surplus or waste-activated sludge. If activated sludge is not wasted, enhanced P removal cannot occur and eventually all P entering the plant will have to exit via the effluent. In sidestream processes, partial P removal occurs in the sidestream through anaerobic stripping of P from the sludge to the liquid phase and subsequent chemical precipitation of the P. The P-depleted sludge is returned to the aeration stage(s), where P uptake occurs. Sidestream systems are, in fact, combined biological–chemical P-removal processes.

2.2.2. Sludge Age

Selection of the sludge age (solids retention time) is the most fundamental and important decision in the design of an activated sludge process. The choice depends on many factors, including consideration of the effluent quality required (Ekama and Marais, 1984b). A sludge age ranging from 1 to 3 days is generally sufficient if the primary objective is to remove organics (as measured by the COD). When nitrification is obligatory, sludge ages five to eight times longer than those for only COD removal may be required. In temperate regions where water temperatures may be below 12°C, the required sludge age is likely to be in the range 10–15 days, depending on the maximum specific growth rate of the nitrifying organisms. If denitrification is required in addition to nitrification, a fraction of the sludge mass must be kept unaerated to create a zone in which recycled NO_3^- − N can be denitrified. As nitrifiers are obligate aerobes, nitrification–denitrification plants should have sludge ages in excess of those where only nitrification is required to compensate for the presence of substantial unaerated mass fractions. The overall sludge age in such systems typically varies from 15 to 20 days.

Where P removal without obligatory N removal is required, the sludge age can be selected to prevent nitrification. In multiple-reactor-type plants, only anaerobic and aerobic reactors are theoretically needed (i.e., no anoxic reactor is present). The sludge age can then be reduced substantially, probably to less than 8 days.

Apart from its significance as a basic factor in activated sludge design, the sludge age is also important in other ways. Given a particular wastewater, the mass of sludge produced per unit influent COD decreases with increasing sludge age. Under steady-state operation, the wastage rate of excess sludge is just equal to the sludge production rate; hence, the mass of P removed per unit influent COD decreases accordingly. The requirement of increased sludge ages to ensure nitrification in anaerobic–anoxic–aerobic systems is therefore in conflict with the need to reduce sludge age for maximum P removal. The degree to which this may become important in practical systems would be influenced by factors such as the possible enrichment of the sludge with organisms capable of enhanced P uptake at a particular sludge age or the achievement of higher or lower incorporation levels of P into the biomass. Estimates of the degree of enhanced P storage in cells vary through wide limits, from about 5 to 35%. Whatever the actual upper limit, the P pool in the biomass is of finite extent, and to deal with a given P load, a finite sludge wastage rate is required. This is of particular significance in mainstream processes, in which the waste sludge is the only P sink.

2.2.3. Influent Wastewater Composition

Discharge of NO_3^- – N into the anaerobic stage delays or prevents the onset of anaerobiosis, the condition empirically found to be a prerequisite for enhanced P uptake. In addition, the generally limited concentration of organic substrates at the head of the plant is further depleted by being partially or exclusively utilized for denitrification rather than to effect P release (Mostert *et al.*, 1987). The COD consumption during denitrification is approximately 8.6 mg COD mg^{-1} NO_3^- – N reduced (Ekama and Marais, 1984a).

Because the nitrification capacity and denitrification potential for a given wastewater are approximately proportional to the influent total Kjeldahl nitrogen (TKN) and COD, respectively, the influent TKN/COD ratio is often used as a relative measure of the amount of NO_3^- – N generation in the process and the amount of NO_3^- – N that can be removed by denitrification. The maximum influent TKN/COD ratio that can be tolerated is 0.1 mg N mg^{-1} COD (Barnard, 1976). Complete denitrification can be achieved only for TKN/COD ratios of less than 0.08 mg N mg^{-1} COD (Ekama *et al.*, 1983). For ratios exceeding 0.14 mg N mg^{-1} COD, it is unlikely that excess biological P removal can be obtained when achievement of complete nitrification is obligatory This is due to the inability to effect complete denitrification under such conditions unless an additional external carbon source is added to the wastewater.

Enhanced P removal is crucially dependent on the exposure of the activated sludge to short-chain organic compounds in the anaerobic stage (Nicholls *et al.*, 1985, 1986; Osborn and Nicholls, 1985; Fuhs and Chen, 1975). Various measures have been used to ensure the supply of desirable, but not yet fully defined, carbon compounds (in cases where the influent lacks these) to the biomass. Venter *et al.* (1978) used intermittent operation of aerators near the inlet end of an extended aeration plant to promote *in situ* volatile fatty acid production in sludge settled to the floor, while Osborn and Nicholls (1978) and Pitman *et al.* (1987) added acid-fermented primary sludge to the anaerobic zone. The current trend is to introduce into the anaerobic stage soluble fermentation products elutriated from either acid-digested sludge or sludge fermented in primary sedimentation tanks or thickeners (Nicholls *et al.*, 1987; Pitman *et al.*, 1983, 1987; Rabinowitz *et al.*, 1987; Osborn *et al.*, 1986; Oldham, 1985; Barnard, 1984).

The amount of desirable biodegradable compounds available to P-accumulating organisms in the anaerobic stage depends largely on factors such as the mass of NO_3^- $-$ N in the sludge recycle and dissolved oxygen or NO_3^- $-$ N in the wastewater influent. Ekama *et al.* (1984) suggested that if NO_3^- $-$ N is positively excluded from the anaerobic reactor, excess P removal can be achieved if the readily biodegradable fraction is in excess of 50 mg COD liter^{-1}. The establishment of conditions to stimulate excess P removal becomes increasingly more difficult as the influent COD strength of a wastewater decreases. Unless special measures are implemented to increase the readily biodegradable fraction, such as by fermenting part or all of the influent wastewater, effluents with COD values lower than about 250 mg liter^{-1} are hardly capable of supporting excess P accumulation.

2.2.4. Sludge Management

The separation of excess sludge (biomass) from the water phase in enhanced P-removal plants is, in most respects, as vital as in conventional systems, and in others it is even more critical. This solid–liquid separation phase is usually achieved by gravity sedimentation, although flotation of solids may also be feasible. In the former case, sludge settlability governs the daily flow and load that can be treated in an activated sludge plant. The settling tank (secondary) that follows the activated sludge system combines the functions of a clarifier and a thickener to produce a clear final effluent and a continuous underflow of thickened sludge for return to the system. If the clarifier fails in either of these functions, sludge will escape with the effluent. Besides impaired effluent quality, loss of sludge represents an uncontrolled reduction of the sludge age, a factor that may affect plant performance (e.g., reduced nitrification). Additionally, P-accumulating sludges contain more P than do conventional sludges. The P content of sludge from conventional activated sludge plants typically varies from 1.5 to 3.0% on a dry-mass basis, compared with values ranging from about 4 to 18% in mainstream systems (Comeau *et al.*, 1987; Mino *et al.*, 1987; Okada *et al.*, 1987; Randall *et al.*, 1987; Tanaka *et al.*, 1987; Wentzel *et al.* 1987). It follows that when the suspended solids concentration of the effluent is about 20 mg liter^{-1} and the P content of the solids about 10%, the final effluent will contain 2.0 mg liter^{-1} of particulate P in addition to the dissolved P. Filtration of the final effluent for removal of suspended solids may be needed in such cases to achieve P standards for effluents.

Treatment and disposal of the P-rich waste mixed liquor or secondary sludge from a modified Bardenpho plant or any other biological P-removal process require special consideration. P may be released from such sludges back into the environment if the sludge is exposed to anaerobic conditions. Generally, the sludge should be kept aerobic during handling and treatment, and the site for final disposal of the sludge should be at a location where any liquid released from the sludge will not be directly discharged to a watercourse.

2.2.5. Scum Formation

Experience has shown (Pitman, 1984) that biological nutrient removal plants tend to favor the appearance of biological scums which accumulate on the surfaces of reactors and final clarifiers unless special design precautions are taken. The scums are usually caused by the excessive growth of filamentous organisms such as *Nocardia erythropolis* and *Microthrix parvicella* (Jenkins *et al.*, 1984; Osborn *et al.*, 1986), which produce a biosurfactant that enables them to float as a surface mat. Apart from being unsightly, these scums can produce offensive odors. If discharged in the final effluent, significant increases in COD and suspended solids content can be the result. The exact cause of these accumulations is not yet fully known, but these organisms are apparently favored by certain substrates in the influent sewage or specific environmental conditions in the plant, e.g., certain dissolved oxygen levels and temperatures (Jenkins *et al.*, 1984).

2.2.6. Sludge Bulking

Another major problem in the operation of activated sludge processes is the growth of filamentous organisms, which leads to sludge bulking. Apart from impairing P uptake (Rensink and Donker, 1987), the lower settling and compaction rates of such organisms make the separation of (bulking) sludge during the secondary clarification step very difficult. Blackbeard *et al.* (1986) surveyed 111 South African activated sludge plants (of which 26 were designed for nutrient removal) and found that filamentous organisms Type 0092, Type 0914, *Microthrix parvicella*, Type 1851, Type 0675, and Type 0041 (Eikelboom and van Buijsen, 1981), in order of descending frequency, dominated in the mixed liquor. These types are associated with low food-to-biomass (F/M) ratios typical of plants operated at long sludge ages. In a follow-up survey aimed specifically at nutrient removal plants, Blackbeard *et al.* (1987) found that among 33 such plants, filament Type 0092 occurred in 94% and dominated in 82% of the cases. In descending order of frequency of dominance, the next five most common filaments were Type 0675, Type 0041, *Microthrix parvicella*, Type 0914, and Type 1851, with frequencies of 45, 39, 33, 33, and 21%, respectively. Although ranked in a different order than in the first study (Blackbeard *et al.*, 1986), the six most frequently dominant filaments in nutrient removal plants were also the six most frequently dominant filaments among all activated sludge plants. The results suggest that the major filamentous organisms present in nutrient removal sludges are similar over a range of different wastewaters and system configurations. The problem of bulking sludge therefore does occur in enhanced P-removal activated sludge plants and has been described by Mulder and Rensink (1987) and Pitman (1984).

Exposing activated sludge to a substrate gradient under anoxic conditions has proven to be a successful way of suppressing the proliferation of filamentous micro-organisms such as *Sphaerotilus natans,* Type 021N, and *Thiothrix* (Hoffman, 1987; Price, 1982; Chambers, 1982; Tomlinson and Chambers, 1979). Growth suppression (and consequently improved sludge settlability) under these conditions can be ascribed to the lower denitrification rate associated with these organisms in comparison with floc-forming bacteria (Wanner *et al.*, 1987). Filamentous microorganisms such as *Microthrix parvicella,* on the other hand, do seem capable of utilizing $NO_3^- - N$ as final electron acceptor at a rate comparable to that of floc-forming bacteria and hence may occur (and even dominate) in systems with anoxic zones.

Researchers such as Watanabe *et al.* (1984) have indicated that anaerobic conditions, coupled with the imposition of a substrate gradient, are also unfavorable to activated sludge filamentous bulking organisms. In P-removal systems, with their sequential anaerobic–aerobic zones, the proliferation of polyP-accumulating organisms leads to the rapid sequestration of low-molecular-mass compounds, generated by acidogenesis and acetogenesis, in the anaerobic stage. Provided that the available substrate is largely utilized in this way, filamentous microorganism growth can be effectively suppressed in P-removal plants (Wanner *et al.*, 1987). However, if slowly hydrolyzable compounds pass through this reactor into the aerobic stage, then the relatively high rate of microbial utilization of hydrolysis products in this zone may result in low concentrations of organic compounds in the bulk liquid, a condition favoring the growth of filamentous microorganisms over floc-forming species (Ekama and Marais, 1985). These findings are in line with full-scale experience at nutrient removal plants in Johannesburg, South Africa, which generally supports the postulate that all organic compounds adsorbed on the floc must be oxidized before the sludge is returned to the process via the underflow recycle (Osborn *et al.*, 1986). Potential adsorption sites are thereby made available for the immediate uptake of readily assimilable organic material. Such compounds cannot then gain entry into the aerobic zone where, if present at low concentrations, the selective growth of filamentous organisms can be promoted.

The effect of anaerobiosis on filamentous bulking is less clear when the offensive organism is not solely dependent on heterotrophic metabolism under aerobic conditions (Wanner *et al.*, 1987). *Thiothrix* spp. are, for example, capable of using sulfate as terminal electron acceptor under anaerobic conditions and in addition are known to store polyP (Jenkins *et al.*, 1984; Brodisch and Joyner, 1983; Eikelboom and van Buijsen, 1981). Such situations are best dealt with by imposing staged anaerobic reactors, which approximate plug flow rather than completely mixed conditions.

3. Nutrient Dynamics in Activated Sludge Systems

The behavior of P-removal activated sludge systems is characterized by a P release in the anaerobic stage followed by excess P uptake in the aerobic stage (Fig. 9). A number of factors have been postulated or shown to play a role in P release and uptake phenomena.

3.1. Factors Influencing P Release

3.1.1. Anaerobiosis

Levin *et al.* (1972), Shapiro *et al.* (1967), Shapiro (1967), and Levin and Shapiro (1965) first investigated P release from activated sludge under anaerobic conditions and suggested that release was triggered by a lack of oxygen and/or a low redox potential. The P appeared to be released from the acid-soluble fraction of sludge cells and only to a minimal degree from nucleotides. The biochemical oxygen demand (BOD) of the liquid medium hardly changed during the anaerobic period. Many researchers have subsequently commented on P release in the anaerobic stage of activated sludge systems (e.g., Marais *et al.*, 1983; Barnard, 1975).

3.1.2. Redox Potential

Marais *et al.* (1983) suggested that redox potential as a parameter controlling P release has a deceptive persuasiveness about it. In samples from long-sludge-age plants, P is not immediately released when DO becomes undetectable, but release may only commence after a considerable time has elapsed. In samples spiked with influent, in contrast, P release is virtually immediate when DO becomes undetectable. As a consequence, it could be reasoned that it is not the DO *per se* but the redox potential that needs to be depressed sufficiently before P is released. Marais *et al.* (1983) pointed out that this hypothesis had not yet been disproved and remained open to question.

Gerber *et al.* (1987b) have shown conclusively that P release can be induced in P-removing activated sludges under anoxic and aerobic conditions when a sufficient quantity of acetate is added to the sludge (Fig. 10). These results allow the rejection of the hypotheses that a low redox potential and/or the presence of anaerobic conditions are prerequisites for P release by polyP bacteria.

3.1.3. Nitrate

Enhanced biological P removal in conjunction with biological N removal is detrimentally affected when denitrification is incomplete. A major factor adversely affecting P removal is, therefore, an excessive amount of NO_3^- − N present in the anoxic basin (Fukase *et al.*, 1985; Iwema and Meunier, 1985; Barnard, 1982; Rensink, 1981; Simpkins and McLaren, 1978; Venter *et al.* 1978; Davelaar *et al.*, 1978; Osborn and Nicholls, 1978; Barnard, 1976; Menar and Jenkins, 1969). No researcher has reported either adverse or beneficial effects on enhanced P accumulation of NO_3^- − N in the aerobic stage, and currently the consensus is that this ion plays a passive role under aerated conditions. For this reason, the subsequent discussion is focused on the effect of NO_3^- − N when introduced into the anaerobic zone. It stands to reason that NO_3^- − N is a potential problem only in actively nitrifying systems. Also, the introduction of O_2 into the anaerobic stage is equally detrimental, but because O_2 is introduced externally and

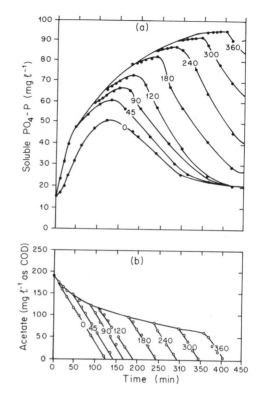

Figure 10. The time course of phosphate (a) and acetate (b) concentrations before and after aeration of an anaerobic mixture containing phosphate-releasing sludge. (Numbers on the curves represent the duration of the anaerobic period prior to aeration.) [Adapted with permission from Gerber *et al.* (1987b).]

hence can theoretically be controlled, it is not considered explicitly in the following discussion.

Comparatively few researchers have conducted formal studies into the role of NO_3^- − N during enhanced P removal, but several early explanations have been advanced to describe the observed deleterious effects. Among these are that the redox potential in the presence of NO_3^- − N would be too high to allow the P-removing bacteria to release P under anaerobic stress conditions (Venter *et al.*, 1978; Barnard, 1976) or too high to allow adequate organic acid formation in the anaerobic zone, organic acids being a preferential substrate for P-removing bacteria (Rensink, 1981). Alternatively, the easily biodegradable short-chain organic acids would be consumed by the denitrifying bacteria, leaving little or nothing for the P-removing bacteria (Osborn and Nicholls, 1978; Simpkins and McLaren, 1978). Implicit in the latter explanation is that the denitrifying organisms constitute a group distinct from the P accumulators.

It is only recently that the uptake and release kinetics of P in activated sludge, in the absence and presence of NO_3^- − N, respectively, have received explicit research attention. Using activated sludges with enhanced P-uptake capabilities, Hascoet and Florentz (1985), Iwema and Meunier (1985), and Malnou *et al.* (1984) were among the first to show that P release can be observed under anoxic conditions when the sludge is exposed to synthetic substrates containing meat extract and acetate, respectively. These results were at variance with previous findings that P release commenced only once strict anaerobiosis was achieved. Occasional ingress of NO_3^- − N into the anaerobic zone of continuous experimental systems operated by these investigators did not significantly affect subsequent P uptake, but the continuous influx of NO_3^- − N in time reduced or eliminated P-uptake ability, a finding that is in line with those of full-scale studies. P uptake occurred under anoxic conditions once the synthetic substrate was consumed, a result observed in many pilot plant and full-scale studies (e.g., McLaren and Wood, 1976).

In the studies of Malnou *et al.* (1984), Hascoet and Florentz (1985), Hascoet *et al.*, (1985), and Iwema and Meunier (1985), the actual NO_3^- − N concentration that could be tolerated in the anaerobic stage without halting enhanced P removal was variable and related to the concentration of organic material. Even high concentrations of NO_3^- − N had a negligible effect if the influent COD level was sufficiently high, a result confirming the conclusions of Davelaar *et al.* (1978). Once enhanced P-accumulating ability was lost, imposition of anaerobic conditions did not immediately lead to the appearance of enhanced P uptake, indicating that the biomass needs to adapt to the alternating anaerobic–aerobic phases or a polyP population must be selected for. Practical experience using continuous-flow systems at bench, pilot, or full scale has shown this period to vary from about 1 to 7 weeks (Okada *et al.,* 1987; Schönberger and Hegemann, 1987).

The findings of the above-mentioned researchers with respect to the role of NO_3^- − N in enhanced P removal were corroborated and extended by Gerber *et al.* (1986, 1987ab) and Mostert *et al.* (1987). By exposing activated sludge derived from full-scale plants with enhanced P accumulation to a variety of pure organic substrates, it was shown that the latter could be classified into one of three groups on the basis of P-release patterns under anoxic conditions.

The first group consists of the lower fatty acids (formic, acetic, and propionic), or their salts, each of which consistently proved capable of inducing P release from sludge under anaerobic, anoxic, and aerobic conditions. Typical results with this group of substrates in the presence of an initial concentration of 10 mg NO_3^- − N liter^{-1} are illustrated in Fig. 11. P release commenced immediately and remained practically linear with time up to the point where either all substrate was utilized or the biomass was depleted of its P-containing reserves participating in the release reaction.

The second group consists of compounds such as ethanol, citrate, methanol, butane diol, and glucose which invariably failed to induce P release unless strict anaerobiosis prevailed. Substrate was removed fairly rapidly during the anoxic phase, but P release commenced only once NO_3^- − N had decreased to negligible concentrations, and then at rates markedly lower than those observed during the early time period with compounds of group one. Transformation (fermentation) of compounds in group two apparently must occur before P release is triggered. Anaerobiosis provides a suitable environment for such reactions to take place.

The third group consists of substances such as butyrate, lactate, and succinate which proved capable of inducing P release under anoxic conditions in sludges from

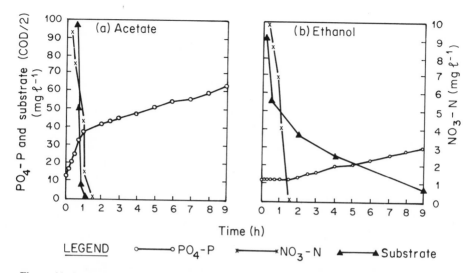

Time (h)

LEGEND   ○————○ PO₄-P   ×————×NO₃-N   ▲————▲ Substrate

Figure 11. Dependence of the time course of phosphate release during sequential anoxic/anaerobic periods on the nature of the organic substrate. Symbols: ○, PO_4-P; ×, $NO_3 - N$; ▲, substrate. [Reprinted with permission from Gerber *et al.* (1987a).]

some plants but not in others. P release, once started, occurred at rates comparable to those associated with compounds of group one. The results suggest that the sludge must undergo a period of conditioning or adaptation before these substrates are utilized under anoxic conditions.

The P-release phenomenon is therefore primarily dependent on the nature of the feed rather than the anaerobic state as such. This is in agreement with the observations reported by Malnou *et al.* (1984) and Hascoet and Florentz (1985). The time course of P concentration in the bulk liquid, and in the presence of $NO_3^- - N$ and substrate, is a multivariate function that depends on, among other factors, the nature of the substrate and the relative amounts of substrate and $NO_3^- - N$ in the mixture.

3.1.4. Sulfide

Although no formal study of the effect of sulfide on enhanced biological P removal has been undertaken, South African experience indicates that high concentrations are detrimental to the process. These concentrations have not been precisely quantified but are probably 25 mg liter⁻¹ (as S) or higher. Apart from being toxic, excessive sulfide may reduce the concentration of essential trace elements to subcritical levels, thus interfering directly with biochemical reactions occurring during P release. Once sulfate-reducing bacteria such as *Desulfovibrio desulfuricans* develop, the net availability of short-chain fatty acids declines because these bacteria utilize such compounds as carbon sources for growth (Marais, 1987). Furthermore, sulfide contributes to the phenomenon of bulking sludge due to the growth of filamentous organisms such as *Thiothrix* and is thus best avoided. It has, however, also been claimed that such organisms contribute to the process of enhanced P removal.

3.1.5. Substrate Composition

Readily bioassimilable compounds favor P removal in full-scale activated sludge systems (see Section 2.2.3). The P-release capacity of a P-removal activated sludge is limited (Toerien *et al.*, 1986). When increasing concentrations of acetate are added to such a sludge, the acetate removal reaches a limit (Fig. 12). This limit presumably corresponds to the polyP reserves available to effect acetate uptake by the polyP bacteria.

3.1.6. P Uptake and Release and Concurrent Anion/Cation Transport

Concurrent P and metal ion release or uptake has been reported (Somiya *et al.*, 1987; Toerien *et al.*, 1986; Comeau *et al.*, 1985; Arvin and Kristensen, 1985; Miyamoto-Mills *et al.*, 1983). P, Mg^{2+}, K^+, Mn^{2+}, and SO_4^{2-} concentrations increased under anaerobic conditions and diminished under aerobic conditions. Reported molar ratios of Mg^{2+}/P and K^+/P range from approximately 0.25 to 0.30 and 0.20 to 0.40, respectively, independent of the direction of transport. The molar ratio of Mn^{2+}/P is about 0.003. Toerien *et al.* (1986) also reported a cotransport of SO_4^{2-}, with a molar ratio of 0.19 (Fig. 13). Ca^{2+} and Fe^{3+} did not exhibit the same release/uptake patterns as P but diminished in the aerobic stage at approximate molar ratios of 0.05–0.15 and 0.0015, respectively.

Buchan (1983) indicated with electron microscopy and energy-dispersive X-ray

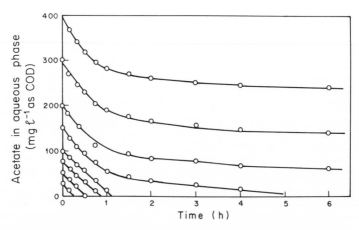

Figure 12. Time course of acetate utilization during anaerobic phosphate release after exposure of identical portions of phosphate-enriched activated sludge to increasing initial acetate concentrations.

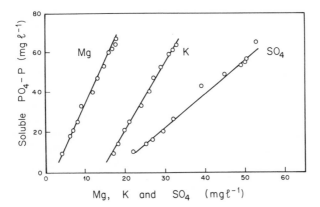

Figure 13. Exchange relationships between phosphate and other ions during phosphate uptake and release.

(EDAX) techniques that metal ions were present in the polyP granules of bacterial cells from enhanced P-removal activated sludge systems. Kulaev (1985) referred to the possible cotransport of metal ions and P across cellular membranes. He also suggested an important role for Mn^{2+} (or Mg^{2+}) in the biosynthesis of a certain polyP fraction. There is little doubt that metallic ions are associated with enhanced P removal in activated sludge systems. Whether these compounds contribute to process regulation still needs to be ascertained.

3.2. Factors Influencing P Uptake

3.2.1. Aerobiosis

Levin and Shapiro (1965) showed that mixed-liquor samples from an activated sludge system exhibited P uptake when aerated. If aeration was prolonged, P release to the bulk liquid occurred; a phenomenon ascribed to endogenous respiration reducing the active biomass and resulting in P release. They ascribed the P uptake under aerobic conditions as being necessary for ATP formation by oxidative phosphorylation. Barnard (1976) used two aerobic stages in his five-stage Bardenpho systems to achieve maximal nitrification–denitrification and to prevent anaerobiosis developing in the sludge of the final clarifier. Wells (1969), using batches of mixed liquor from the San Antonio plant (which exhibited P removal), showed that if a batch was sequentially aerated during the day and left unaerated during the night, P was taken up during the day and released during the night. However, the uptake decreases progressively in such experiments (Fig. 14).

Aerobic conditions therefore seem to be the major factor affecting P uptake in enhanced P-removal activated sludge systems.

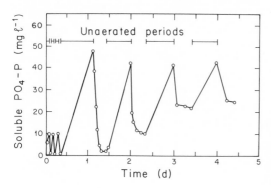

Figure 14. Time course of phosphate uptake and release in a batch test under a repeated sequence of alternating aerated and unaerated periods. [Adapted with permission from Wells (1969).]

3.2.2. Other Factors

The presence of polyP bacteria in a preconditioned sludge and certain metallic ions (see Section 3.1.6) are also necessary for successful P uptake. However, the precise influence of each of these factors is not yet fully understood.

4. Microbiology of P-Removal Activated Sludge Systems

4.1. Introduction

The microbiology of enhanced P-removal activated sludge systems has received some attention but is far from completely understood. Most studies on the population structure of activated sludge systems have used agar plates (Osborn *et al.*, 1986; Lötter and Murphy, 1985; Brodisch and Joyner, 1983; Buchan, 1983; Cloete *et al.*, 1985; Prakasam and Dondero, 1967a; McKinney and Weichlein, 1953; Allen, 1944). However, Prakasam and Dondero (1967b) reported higher numbers of cells with microscopic enumerations than with agar plate enumerations. Cloete and Steyn (1987) found that less than 10% of microscopic cell counts of activated sludges could be accounted for in agar plate enumerations. Prakasam and Dondero (1967a) stressed that a single culture medium for all activated sludge bacteria had not been developed. Toerien and Gerber (1986) cautioned about the interpretation of population studies based only on agar plating techniques.

We will briefly explore the types of bacteria encountered in activated sludge systems, their numbers, and the factors that contribute to the selection of bacteria in activated sludge systems.

4.2. Bacteria Present in Activated Sludge Systems

As a result of the classical work of Butterfield (1935), it was believed that *Zooglea ramigera* was the most predominant and active organism in activated sludge. Heukelekian and Littman (1939) confirmed the importance of zoogleal bacteria, but Dias and Bhat (1964), Van Gills (1964), McKinney and Weichlein (1953), and Allen (1944) reported the presence of other bacteria. Prakasam and Dondero (1967a, 1970) used approaches other than the normal taxonomic ones to characterize activated sludge bacterial populations.

Fuhs and Chen (1975) first focused attention on *Acinetobacter* spp. as being important in enhanced P-removal systems; this importance was subsequently supported by, among others, Osborn *et al.* (1986), Deinema *et al.* (1985), Gersberg and Allen (1985), Lötter and Murphy (1985), and Buchan (1983). Stephenson (1987) debated the role of *Acinetobacter* in biological P removal and concluded that further research was needed to clarify this issue.

The presence of other bacteria was also recorded: *Pseudomonas* (Brodisch and Joyner, 1983); *Klebsiella* (Gersberg and Allen, 1985); and *Moraxella, Aeromonas, Pseudomonas, Flavobacterium, Bordetella, Citrobacter, Shigella, Pasteurella, Klebsiella,* and *Yersinia* (Lötter and Murphy, 1985). The presence of Gram-positive bacteria was also reported (Osborn *et al.*, 1986).

Brodisch (1985) suggested that P removal was dependent on the correct balance between different bacterial types. She implicated *Aeromonas punctata* and *Acinetobacter calcoaceticus* in an effective combination. The former bacterium apparently ferments carbohydrates to end products such as acetate, which enhance the P-uptake ability of the latter.

4.3. Numbers of Bacteria in Activated Sludge

Prakasam and Dondero (1967a) evaluated various growth media for the enumeration of activated sludge bacteria and reported colony forming units (CFUs) ranging from 2×10^6 ml^{-1} to 6×10^8 ml^{-1}. Banks and Walker (1977) reported CFUs ranging from 1.6×10^9 ml^{-1} to 4.2×10^9 ml^{-1}. Isolates used in population studies are often isolated from 10^{-3} to 10^{-8} dilutions (e.g., Brodisch, 1985).

However, activated sludge systems are normally operated with a suspended solids content of several grams per liter. Typical values might range from 3 to 5 g liter^{-1}. If we assume that living cells constitute only 50% of the suspended solids and a typical bacterial cell mass (dry basis) is 10^{-13} g, then activated sludge systems contain at least 1.5×10^{10} to 2.5×10^{10} cells ml^{-1}. Cloete and Steyn (1987) reported an average of 1.54×10^9 bacteria ml^{-1} in four activated sludge samples when enumerated with Acridine Orange microscopic counts.

The viable (cultivable) bacteria in activated sludge, therefore, constitutes only a small fraction (<10%) of the total number of cells present (Cloete and Steyn, 1987). This is also the case for a number of other ecosystems, e.g., the rumen (Hungate, 1966). A question therefore arises as to whether those organisms that can be cultivated truly represent the entire population, which includes large numbers of apparently noncultivable cells. For activated sludge systems, this question is as yet unresolved, but observations made by Cloete (1985) caution against acceptance of such a hypothesis.

Cloete (1985) compared the total microscopic counts of two activated sludge samples that had been centrifuged onto density gradients with the number of *Acinetobacter* estimated with fluorescent antibodies and the numbers of volutin-containing bacteria revealed by metachromatic staining (Table I). *Acinetobacter* and volutin-containing bacteria occurred only in 3 of 10 fractions (8 of which are shown in Table I) and constituted only 2.4–20.5% of the total number of bacteria applied to the density gradients. Given the fact that the fluorescent antibody numbers of *Acinetobacter* corresponded closely to the CFUs of *Acinetobacter* of activated sludge samples (Cloete, 1985), it appears likely that the bacteria which were noncultivable were not *Acinetobacter*. The *Acinetobacter* (or cultivable bacteria) could therefore not represent the population structure of those bacterial groups that could not be cultivated.

For these reasons, Toerien and Gerber (1986) have cautioned against drawing firm conclusions about population structures from evidence obtained only with viable (or cultivable) isolates. This caveat pertains especially to the assumption that *Acinetobacter* is the main polyP bacterium in enhanced P-removal activated sludge systems (e.g., Osborn *et al.*, 1986). Stephenson (1987) also indicated the need for further microbiological studies.

We must therefore conclude that the roles of different organisms in activated sludge systems in general, and in enhanced P-removal systems specifically, require elucidation. Gaining better insight in the microbiology of activated sludge pressures operative in activated sludge systems because they create the conditions for specific populations to be present.

Table I. Total Microscopic Number of Bacteria, *Acinetobacter* Number, and Number of Volutin-Containing Bacteria in Different Activated Sludge Density Gradient Fractions[a]

Sample	Fraction	T	A	M	M/T%	M/A%	A/T%
1	1	184	0	0	0	0	0
	2	9	0	0	0	0	0
	3	77	70	59	77	91	85
	4	9	8	8	86	96	90
	5	8	6	6	75	93	80
	6	10	0	0	0	0	0
	7	45	0	0	0	0	0
	8	19	0	0	0	0	0
2	1	32	0	0	0	0	0
	2	108	0	0	0	0	0
	3	49	4	1	30	41	7
	4	10	8	8	75	93	81
	5	54	3	1	3	48	6
	6	68	0	0	0	0	0
	7	68	0	0	0	0	0
	8	17	0	0	0	0	0

[a] T, Total microscopic number of bacteria; A, *Acinetobacter* number; M, metachromatin number; expressed in units of 10^6 ml^{-1}. [Adapted from Cloete (1985).]

4.4. Selective Pressures in Activated Sludge Systems

4.4.1. Physical Factors

Physical factors are extremely important selective forces in any biological waste-water treatment system. For instance, in activated sludge systems the active biomass is returned to the reactors after settling out in a clarifier (Fig. 1). Bacteria, which become part of the flocs that settle out, will therefore be retained in the system (and be selected for), whereas bacteria that tend to stay in a planktonic form will not settle out and will leave the system with the effluent (be selected against).

In addition, alternating anaerobic, anoxic, and aerobic zones will select for organisms that can tolerate such conditions or, even more likely, be favored by such conditions. For instance, the inclusion of a zone devoid of oxygen but containing $NO_3^- - N$, which can act as electron acceptor, will favor the selection of denitrifiers because these organisms can utilize fermentation end products as energy sources in the presence of $NO_3^- - N$ (e.g., Gerber et al., 1986).

Scum-forming bacteria often reach problem levels in activated sludge systems (e.g., Osborn et al., 1986). Selection for these organisms is enhanced by the fact that most activated sludge systems do not have physical systems designed to remove scum layers. Scums are therefore retained in such systems and over a period of time can reach problem levels. For polyP bacteria to be present, they must be able to react positively to physical factors that will enhance their selection in enhanced P-removal activated sludge plants. The ability of Acinetobacter to form clusters that can easily become enmeshed in existing flocs (e.g., Osborn et al., 1986) and the greater density of polyP-containing cells compared with normal bacterial cells (Suresh et al., 1985) have been cited as reasons for the presence of polyP bacteria in modified activated sludge systems. In addition, the metabolic advantages for polyP bacteria created by alternating anaerobic and aerobic zones in activated sludge systems (see Section 5) also result from specific physical conditions.

4.4.2. Chemical Factors

Like other heterotrophic biological life forms, activated sludge bacteria require carbonaceous compounds and other nutrients for energy and synthesis. Broadly speaking, the requirements are for sources of energy, nitrogen, P, sulfur, minerals such as potassium, calcium, and magnesium, and trace elements such as iron, copper, zinc, and cobalt. Activated sludge bacteria may require some vitamins (or other factors) present in activated sludge (e.g., Prakasam and Dondero, 1967a).

Domestic sewage provides a nutritionally diverse substrate (e.g., Grady and Lim, 1980) that contains the compounds necessary for the growth of many bacteria. However, these compounds may not always be present in the quantities or ratios necessary to achieve specific goals, e.g., total denitrification or enhanced P removal (Osborn et al., 1986).

The readily degradable portion of sewage is extremely important for enhanced P removal (Siebritz *et al.*, 1983; Osborn *et al.*, 1986). PolyP bacteria such as *Acinetobacter.* spp are obligate aerobes requiring a fairly restricted spectrum of substrates for proliferation in the activated sludge process (Osborn *et al.*, 1986). These substrates include volatile fatty acids, the end products of anaerobic fermentation. Design of activated sludge systems sometimes includes supplementary acid production (Nicholls, 1975; Venter *et al.*, 1978).

The kinetics of P-removal processes have been investigated extensively (Wentzel *et al.*, 1985, 1987; Water Research Commission, 1984) and have been adapted for use by operators and managers of P-removal plants (Osborn *et al.*, 1986). The UCT model used in these applications can be broken up into a number of sections (Osborn *et al.*, 1986): (1) composition of the feed (in terms of COD), (2) degradation of carbonaceous material, (3) nitrification, (4) denitrification, (5) oxygen utilization, and (6) biological P removal. We shall not give more details about this model except to state that a number of chemical parameters of the feed are important determinants. These factors obviously play an important role in determining bacterial succession in P-removal activated sludge systems.

4.4.3. Biological Factors

Organisms strive to maximize their fitness for survival through r or K strategies (Roughgarden, 1971). In uncrowded environments, r strategists can attain maximum specific growth rates; in crowded environments, K strategists can attain maximum population densities. Roughgarden (1971) stated that high r and K values are mutually exclusive; a high r value occurs at the expense of a low K value, and *vice versa*.

Environments can be characterized on an r/K basis, determined by the extent and frequency of transient and stable conditions. The relative degree of a selective pressure is determined largely by environmental stability (MacArthur and Wilson, 1967), and it is useful to consider environments from an r/K perspective (Andrews and Harris, 1985).

Andrews and Harris (1985) indicated that the limitations imposed by r and K selection necessitate choices by organisms, each with inherent advantages and disadvantages. They give three alternative lifestyles: (1) survival of a microbial species as an r or K strategist, thriving in the corresponding environment but at the expense of barely surviving or becoming locally extinct elsewhere, (2) survival of r organisms in a K environment in a cryptic fashion as spores or other survival forms, and (3) life as a "diplomat," whereby the extremes are negotiated in favor of a middle course in which the microbe does neither very well nor very poorly under either condition.

The competitive ability of microbes in activated sludge systems must influence the importance of a specific microbe in such systems. Andrews and Harris (1985) stated that no organism is a "master of all trades." Tempest *et al.* (1983) rationalized from an ecophysiological perspective that for microorganisms in particular, the size of the genome must of necessity be kept to a minimum. It is therefore unlikely that a single organism will have the genetic capability to be both an r and a K strategist.

Factors in enhanced P-removal activated sludge systems tending to favor K strategists are: (1) high-density populations (achieved by physically returning microorganisms to the system after gravity settling in the secondary settling tank), (2) long sludge ages (up to 15 to 20 days) in such systems, and (3) low nutrient concentrations. Factors that favor r strategists include (1) alternating anaerobic and aerobic conditions, which stress aerobic and anaerobic organisms, respectively, (2) high nutrient concentrations at the point of entrance of influent, and (3) the presence of predators such as protozoa and rotifers (Madoni, 1986). It can be expected that both r and K strategists will occur in enhanced nutrient removal activated sludge systems and that keen competition between these organisms partly determines the microbial population composition in these systems.

5. Biochemical Model of Enhanced P Removal

Marais *et al.* (1983) stated that significant advances in the use of enhanced biological P removal in wastewater treatment will be contingent on a greater understanding of the biochemical mechanisms controlling the phenomenon. In the period prior to this statement, a few attempts had been made to explain observations made on activated sludge plants in terms of underlying biochemical processes. Fuhs and Chen (1975) provided the earliest microbial and biochemical concepts of enhanced P removal. After studies into the phenomena of P uptake and release, they concluded that a single microorganism, or closely related species, was responsible for the phenomenon. These bacteria accumulated both PHB and polyP, the former serving as a source of energy for polyP accumulation. In studying the nutritional requirements of the polyP-accumulating bacteria, they found that volatile fatty acids, in particular acetic acid, were the preferred substrates.

In contrast, Nicholls and Osborn (1979) proposed a conceptual model in which glucose is absorbed in the anaerobic zone and metabolized via the Embden-Meyerhof pathway to acetyl coenzyme A (CoA), which then acts as an electron sink by reduction to PHB. Hydrolysis of ATP for cell maintenance results in P release to the external medium. On subsequent aeration, the stored PHB is utilized for polyP synthesis.

Rensink (1981) modified the model of Nicholls and Osborn (1979) in a number of respects. They, in parallel with Fuhs and Chen (1975), proposed volatile fatty acids as substrates for polyP-accumulating oganisms which intracellularly store PHB while in the anaerobic zone. The energy requirement for PHB formation is supplied by hydrolysis of polyP, which results in P release. Marais *et al.* (1983) proposed that polyP accumulation in the polyP organisms serves as an energy source for cell maintenance and PHB synthesis under anaerobic conditions. In investigating possible pathways for PHB synthesis under anaerobic conditions, these authors could postulate only pathways with glucose as substrate and could not envisage acetate as a source of PHB.

Before proceeding to two more recent models which resolved many of the inconsistencies of previous hypotheses, it is necessary to consider PHB and polyP metabolism in greater detail. (In the rest of this section, frequent mention will be made of the dissociation products of orthophosphoric acid, PO_4^{3-}, HPO_4^{2-}, and $H_2PO_4^-$, respectively. Rather than referring to the specific anion that occurs, the notation "$PO_4 - P$" is used, with the understanding tha' any or all of the given anions may be involved in a given reaction. Abbreviated forms such as Ca and Mg for calcium and magnesium, respectively, are used similarly and may indicate both the free cation and the bound atom in a compound.)

5.1. PolyP Metabolism

5.1.1. Intracellular Localization

Volutin granules containing polyP have been demonstrated in bacteria by a number of researchers using cytochemical methods (Deinema *et al.*, 1980; Fuhs and Chen, 1975; Voelz *et al.*, 1966). The use of electron microscopy helped to establish that in heterotrophic procaryotes, polyP granules are associated with DNA fibrils and nucleoplasm (Deinema *et al.*, 1980; Friedberg and Avigad, 1968; Voelz *et al.*, 1966) or in sites of oxidation or reduction (Mudd *et al.*, 1958).

Electron microscopy was subsequently combined with EDAX analysis to reveal the chemical nature of the granules (Baxter and Jensen, 1980). These researchers showed that under normal growth conditions, appreciable amounts of K and comparatively low quantities of Ca and Mg are present in the polyP granules of *Plectonema boryanum* (Baxter and Jensen, 1980). When the medium contains an excess amount of a particular metal, such as Mg, Ba, or Mn, the metals accumulate in large quantities in the polyP granules. Using the same technique, Buchan (1983) showed that Ca was the chief cation associated with intracellular polyP in activated sludge bacteria. Van Groenestijn and Deinema (1985) detected the presence of Ca, Mg, and K in polyP granules in *Acinetobacter* and identified K as essential for polyP synthesis by this organism.

5.1.2. Effect of Growth Conditions on PolyP Accumulation

Mudd *et al.* (1958) showed a competitive relationship between nucleic acid synthesis and polyP accumulation in *Mycobacterium* spp. Glucose as substrate favored nucleic acid synthesis, whereas malate favored polyP accumulation, illustrating the link between carbon and polyP metabolism. In the same study, stored polyP was utilized for nucleic acid synthesis. The main form of polyP was acid insoluble (Mudd *et al.*, 1958). This type of polyP was also detected as the chief storage product in *Hydrogenomonas* spp. (Kaltwasser, 1962). In a complete growth medium, polyP was accumulated until the exogenous P was exhausted, after which the polyP was used for the continuing synthesis of organic P compounds. Under anaerobic conditions, no organophosphate was absorbed and no polyP was consumed (Kaltwasser, 1962). In the denitrifying organism *Micrococcus denitrificans*, $PO_4 - P$ was rapidly taken up by P-deficient cells under aerobic and anoxic conditions (Kaltwasser *et al.*, 1962).

The reciprocal relationship between nucleic acid synthesis and polyP accumulation (Mudd *et al.*, 1958) was subsequently confirmed by Harold (1963) in *Aerobacter aerogenes*. This relationship remained, however, a function of the growth medium. PolyP accumulation could be induced or suppressed at will by manipulation of the P and sulfur content of the medium (Harold and Sylvan, 1963). Addition of PO_4 $-$ P to *Aerobacter aerogenes* strains previously subjected to P starvation produced polyP accumulation (Harold, 1964).

In the nitrifying bacterium *Nitrosomonas europeae*, the steady-state concentration of polyP depends on the balance between the rates of ATP-yielding and ATP-utilizing reactions (Terry and Hooper, 1970), but polyP synthesis was favored when NH_3 $-$ N oxidation (ATP yielding) occurred in the absence of protein synthesis (ATP utilizing). In *Nitrobacter winogradskyi*, polyP synthesis commenced only after complete oxidation of NO_2^- $-$ N (Eigener and Bock, 1972).

Acinetobacter strains have been shown to accumulate polyP as a function of growth conditions (Deinema *et al.*, 1980). *Acinetobacter* also used stored polyP as a P source and responded with increased polyP synthesis after P starvation (van Groenestijn and Deinema, 1985). Induction of polyP accumulation by anaerobiosis was observed in the facultative anaerobe *Escherichia coli*. The chain lengths of the polyP molecules were dependent on the age of the culture (Rao *et al.*, 1985).

Subjection of the obligate aerobe *Acinetobacter calcoaceticus* to anaerobic conditions resulted in polyP degradation and release of PO_4 $-$ P to the external medium (Ohtake *et al.*, 1984; Murphy and Lötter, 1986).

In contrast to the utilization patterns of polyP described above, *Neisseria gonorrhoeae* does not respond with increased polyP synthesis when placed in a P-rich medium after P starvation. In addition, polyP is not utilized as a P source during P starvation (Noegel and Gotschlich, 1983).

5.1.3. Enzymatic Biosynthesis and Degradation

Polyp:ADP phosphotransferase (polyP kinase), which was first isolated from *E. coli* (Kornberg *et al.*, 1956), catalyzes the following reaction:

$$ATP + (Polyp)_n \leftrightarrow ADP + (Polyp)_{n+1}$$

Subsequently, the enzyme was isolated from a number of organisms accumulating polyP, including *Corynebacterium xerosis* (Muhammed, 1961), *Azotobacter vinelandii* (Zaitseva and Belozerskii, 1960), *Salmonella minnesota* (Mühlradt, 1971), *Arthrobacter atrocyaneus* (Levinson *et al.*, 1975), *Propionibacterium shermanii* (Robinson *et al.*, 1984), *Pseudomonas vesicularis* (Suresh *et al.*, 1985), and *Acinetobacter calcoaceticus* (Lötter and Dubery, 1987; T'seyen *et al.*, 1985).

In some procaryotic organisms this enzyme appears to be the key enzyme in polyP metabolism, as evidenced by the fact that mutants defective in poly P kinase of *Aerobacter aerogenes* (Harold and Harold, 1965) and *Anacystis nidulans* (Vaillancourt *et al.*, 1978) no longer produced polyP.

Kornberg *et al.* (1956) found ADP to be a potent inhibitor of polyP kinase. Subsequently, the apparent inhibition was identified as the reverse reaction:

$$ADP + (Polyp)_n \leftrightarrow ATP + (Polyp)_{n-1}$$

The reverse reaction could not be demonstrated in the enzyme from *Corynebacterium xerosis* (Muhammed, 1961), which showed an absolute requirement for Mg^{2+} and was inhibited by ADP and AMP. The same inability to catalyze the transfer of PO_4 − P to ADP was observed in the enzyme from *Azotobacter vinelandii*. ADP inhibited the enzyme. Non-adenine nucleotides did not replace ATP in polyP synthesis. The amount of ATP varied very little during synthesis, suggesting a consistent replenishment of the ATP used (Zaitseva and Belozerskii, 1960).

The enzyme from *Mycobacterium smegmatis* catalyzed the reverse reaction and had an absolute requirement for Mg^{2+}. The reaction was inhibited by AMP, ADP, and high concentrations of ATP (Suzuki *et al.*, 1972). The enzyme from *Arthrobacter atrocyaneus* showed the same inhibitory pattern, but the reaction proceeded with Mn^{2+} or Mg^{2+} (Levinson *et al.*, 1975).

P starvation stimulated polyP kinase activity in *Aerobacter aerogenes*, which resulted in a corresponding increase in the rate of polyP accumulation, when PO_4 − P was made available to the starved organism (Harold, 1964). Similarly, a rapid increase in polyP kinase activity occurred in *Arthrobacter atrocyaneus* when the organism was aerated in the absence of PO_4 − P (Levinson *et al.*, 1975). In contrast, the polyP kinase from *E. coli* reacted negligibly to an absence of PO_4 − P in the medium (Nesmeyanova *et al.*, 1973). A partially purified enzyme from *E. coli*, however, showed an absolute requirement for PO_4 − P for maximum activity (Li and Brown, 1973).

PolyP kinase activity was affected by the growth phase of *Pseudomonas vesicularis* (Suresh *et al.*, 1985) and *Acinetobacter calcoaceticus* (Osborn *et al.*, 1986). The activity of the *Acinetobacter calcoaceticus* enzyme was affected by the carbon source and stimulated by anaerobiosis (Lötter and Dubery, 1987). The polyP product produced by the action of the enzyme from *Propionibacterium shermanii* consisted of more than 200 PO_4 residues (Robinson *et al.*, 1984). Studies of the mechanism of this enzyme revealed that the elongation reaction occurs without dissociation of intermediate sizes of polymer from the enzyme. As a consequence, only high-molecular-weight polyP was synthesized (Robinson and Wood, 1986).

A second enzyme linking polyP with adenine nucleotide metabolism was detected in some microorganisms by Winder and Denneny (1957). This enzyme, polyP:AMP phosphotransferase, catalyzes the following reaction:

$$AMP + (polyP)_n \leftrightarrow ADP + (polyP)_{n-1}$$

Another enzyme linking polyP with energy metabolism has been detected in species of *Acetobacter*, *Achromobacter*, *Brevibacterium*, *Corynebacterium*, and *Micrococcus* (Murata *et al.*, 1980). This enzyme, polyP-dependent NAD+ kinase, catalyzes the following reaction:

$$NAD + (polyP)_n \leftrightarrow NADP + (polyP)_{n-1}$$

Szymona *et al.* (1967) detected an enzyme of polyP metabolism linked to hexose utilization in mycobacteria. This enzyme, polyP:glucose-6-phosphotransferase (polyP glucokinase), catalyzes the transfer of PO_4 to glucose:

$$\text{glucose} + (\text{polyP})_n \leftrightarrow \text{glucose-6-phosphate} + (\text{polyP})_{n-1}$$

Studies with partially purified enzyme suggested that ATP and polyP glucokinase activity occur in *Mycobacterium phlei* (Szymona and Ostrowski, 1964). Attempts to detect polyP glucokinase in *Azotobacter vinelandii* and *Rhodospirillium rubrum* were unsuccessful. A number of *Nocardia* species, however, contained both ATP and polyP glucokinase activity (Szymona *et al.*, 1967).

More recent studies of polyP:glucose phosphotransferase in *Nocardia minima* revealed three forms of the enzyme in this organism (Szymona and Szymona, 1979). ATP:- and polyP:phosphotransferase activities have been detected in *Propionibacterium shermanii* and *P. freudenreichii*. The rate of glucose phosphorylation was much more rapid with polyP than with ATP (Wood and Goss, 1985). Studies with the partially purified *Propionibacterium shermanii* enzyme suggested that ATP and polyP activities are catalyzed by the same protein (Pepin and Wood, 1986).

Another enzyme catalyzing the synthesis of polyP has been detected in some eubacteria (Kulaev *et al.*, 1971) and later in the bacterial parasite *Bdellovibrio bacteriovorus*. This enzyme, 1,3-diphosphoglycerate:polyphosphate phosphotransferase, participates in the following reaction:

$$\text{1,3-diphosphoglycerate} + (\text{polyP})_n \leftrightarrow \text{3-phosphoglycerate} + (\text{polyP})_{n+1}$$

In *Micrococcus lysodeikticus*, *Propionibacterium shermanii*, and *E. coli*, the activity of this enzyme was affected by culture age, being higher in the stationary phase. This enzyme contributed equally with polyP kinase to polyP synthesis in *E. coli* (Nesmeyanova *et al.*, 1973) and was not regulated to a significant degree by exogenous $PO_4 - P$ (Nesmeyanova *et al.*, 1974).

Degradation of polyP occurs via the participation of two enzymes. The enzyme polyP depolymerase splits the molecule into smaller fragments according to the following reaction:

$$(\text{polyP})_n + H_2O \rightarrow (\text{polyP})_{n-x} + (\text{polyP})_x$$

This enzyme has, however, been detected only in eucaryotes (Kulaev and Vagabov, 1983). The polyphosphatases split one terminal $PO_4 - P$ residue from the polyP molecule:

$$(\text{polyP})_n + H_2O \rightarrow (\text{polyP})_{n-1} + PO_4 - P$$

The occurrence of polyphosphatase is widespread in polyP-accumulating organisms (Kulaev and Vagabov, 1983). In *E. coli*, a significant proportion of the polyphosphatase is membrane bound and located on the outer side of the plasma membrane (Severin *et al.*, 1975; Nesmeyanova *et al.*, 1974).

The biosynthesis and secretion of polyphosphatase into the periplasm is repressed by exogenous PO_4 — P (Maraeva *et al.*, 1979; Nesmeyanova *et al.*, 1974; Harold, 1966) in some microorganisms, while a lack of PO_4 — P in the medium stimulates activity in others (Harold, 1964). The polyphosphatase enzyme in *Aerobacter aerogenes* mediates the main degradation pathway, suggesting that the main function of polyP is to regulate intracellular P levels rather than act directly as a phosphagen (Harold and Harold, 1965). Derepression of the *E. coli* polyphosphatase by orthophosphate starvation has also been observed (Yagil, 1975). The *Acinetobacter calcoaceticus* enzyme was initially stimulated by anaerobic conditions; after 4 h of anaerobiosis, the activity decreased to a level below that of the aerobic control (Ohtake *et al.*, 1984).

5.2. Carbon Metabolism

In obligate aerobes, the complete oxidation of carbon substrates to carbon dioxide and water is the most significant source of ATP (Mahler and Cordes, 1971). Oxidative phosphorylation to produce ATP is accomplished via the tricarboxylic acid cycle and the electron transfer chain, with oxidation of NADH to NAD. Carbohydrate substrates are degraded via the Embden-Meyerhof or Entner-Doudoroff pathway to pyruvate, which enters the tricarboxylic acid cycle via acetyl CoA.

Where acetic acid is the sole substrate, intervention of the so-called glyoxylic acid cycle is essential to keep the tricarboxylic acid cycle operational (Fig. 15). Control of the tricarboxylic acid cycle by variations in the NAD(P)H/NAD and ATP/ADP ratios is a well-established phenomenon (Mahler and Cordes, 1971). Specific aspects of this control have been observed with enzymes isolated from *Acinetobacter* spp. The enzymes citrate synthase (Weitzman and Dunmore, 1969; Weitzman and Jones, 1968) and α-ketoglutarate dehydrogenase (Weitzman, 1972) are inhibited by high NAD(P)H/NAD values. Citrate synthase and isocitrate dehydrogenase are inhibited by high ATP/ADP values (Parker and Weitzman, 1970; Weitzman and Dunmore, 1969).

In most organisms, the glyoxylic acid cycle is regulated by three-carbon (C3) compounds. In *Acinetobacter* spp., the four-carbon (C4) intermediates of the tricarboxylic acid cycle regulate the activity of isocitrate lyase (Herman and Bell, 1970; Bell and Herman, 1967) and thus the glyoxylic acid cycle. This cycle is therefore also indirectly controlled by the NAD(P)H/NAD(P) and ATP/ADP ratios.

5.2.1. Polyhydroxybutyrate Metabolism

A wide variety of microorganisms accumulate PHB (Dawes and Senior, 1973). PHB normally occurs in granules which intensely stain with Sudan Black in fixed bacterial preparations. The accumulation of PHB in batch and continuous culture has been reviewed by Dawes and Senior (1973) and will be dealt with here only in relation to the activated sludge process.

The biosynthesis and degradation of PHB have been intensively studied in *Azotobacter beijerinckii* and will be used to illustrate the enzymatic reactions essential to both

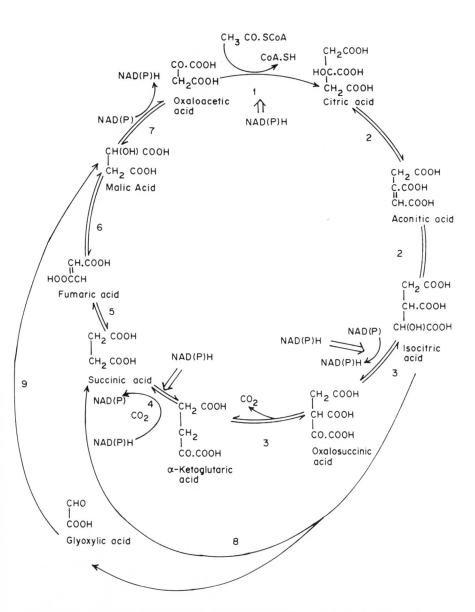

Figure 15. The tricarboxylic acid and glyoxylic acid cycles. Key to enzymes: 1, citrate synthase; 2, aconitase; 3, isocitrate dehydrogenase; 4, α-ketoglutarate dehydrogenase; 5, succinate dehydrogenase; 6, fumarase; 7, malate dehydrogenase; 8, isocitrate lyase; 9, malate synthase. Tricarboxylic acid cycle enzymes: 1–7; glyoxylic acid cycle enzymes: 1, 2, 8–9, 5, 6, 7. ⇏, Inhibition.

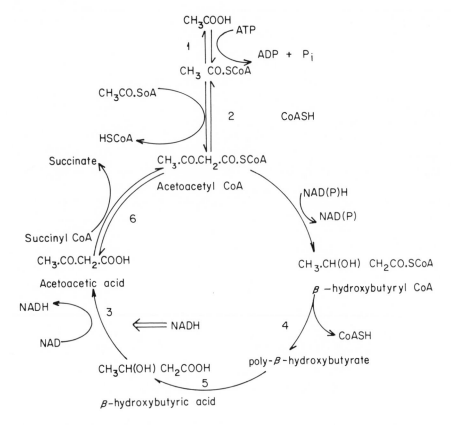

Figure 16. Synthesis and degradation of poly-β-hydroxybutyrate. Key to enzymes: 1, thiokinase; 2, β-ketothiolase; 3, β-hydroxybutyrate dehydrogenase; 4, β-hydroxybutyryl CoA polymerase; 5, poly-β-hydroxybutarate depolymerase; 6, acetoacetate succinyl CoA:CoA transferase. ⇒, Inhibition.

pathways (Fig. 16). The enzyme β-ketothiolase catalyzes the condensation of two molecules of acetyl CoA to produce acetoacetyl CoA. Although the equilibrium constant for this reaction does not favor the formation of acetoacetyl CoA, it can be argued that under conditions where acetyl CoA and $NAD(P)H_2$ are high and CoASH (the reduced form of CoA) low, the equilibrium will be displaced in favor of acetoacetyl CoA (Ritchie *et al.*, 1971). An acetoacetyl CoA reductase that catalyzes the reduction of acetoacetyl CoA to β-hydroxybutyryl CoA ˙has been detected (Ritchie *et al.*, 1971). Subsequent studies with this enzyme have shown that the K_m for acetoacetyl CoA is in the range of 1 to 2 μM and that at high concentrations of acetoacetyl CoA (>10 μM), inhibition occurs (Senior and Dawes, 1971). The final reaction is catalyzed by PHB granule-bound hydroxybutyryl CoA polymerase, which uses β-hydroxybutyryl CoA as substrate and releases CoASH (Senior and Dawes, 1971). The catabolism of intracellular PHB is probably not initiated until all available exogenous carbon is exhausted. There is no evidence to suggest that PHB undergoes turnover, particularly in view of its physical properties (Alper *et al.*, 1963). A requirement for the presence of oxygen for degradation has been demonstrated in some bacteria (Stockdale *et al.*, 1968; Macrae and Wilkinson, 1958).

The initial hydrolysis of PHB involves the production of β-hydroxybutyric acid, catalyzed by a PHB depolymerase. In contrast to the synthetic pathway, degradation appears to occur via the free acids (Wong and Evans, 1971; Sierra and Gibbons, 1963; Macrae and Wilkinson, 1958).

Every bacterium capable of accumulating PHB has been shown to possess a β-hydroxybutyrate dehydrogenase that catalyzes the oxidation of the β-hydroxybutyric acid to acetoacetic acid (Dawes and Senior, 1973). The presence of this enzyme has recently been confirmed in an *Acinetobacter calcoaceticus* isolate from an activated sludge plant (Lötter and Dubery, 1987). A partially purified form of this enzyme was inhibited by NADH, oxaloacetic acid, and, to a lesser degree, acetyl CoA (Lötter and Dubery, 1987). The final step in the degradation pathway in *A. beijerinckii* is conversion of acetoacetic acid to acetyl CoA by the action of acetoacetate: succinyl CoA transferase (Senior and Dawes, 1971).

5.2.2. Extracellular Polysaccharides

The extracellular polysaccharides can be conveniently divided into cell wall material and exoglycans, material exuded into the growth medium. This discussion is restricted to the latter group.

In contrast to extracellular polysaccharides, which function as food reserves, the exoglycans generally cannot be utilized as potential energy and carbon sources by the microorgansims that synthesize them. Most microbial exocellular polysaccharides are heteropolysaccharides comprising a variety of monosaccharide residues.

Morphological and biochemical studies of floc-forming bacteria indicate that exocellular polymers are responsible for the flocculant growth habit of these bacteria (Friedman *et al.*, 1969). Bacterial aggregation has generally been attributed to exocellular polysaccharides, although poly-amino acids could also play a functional role in this respect. These compounds are secreted at the bacterial surface under varying conditions. The availability of certain substrates and relative nutrient concentrations have also been shown to influence the composition and concentration of exocellular polymers (Harris and Mitchell, 1973).

Emulsan, the extracellular polymeric bioemulsifier produced by certain strains of *Acinetobacter calcoaceticus,* occurs in both cell-free and cell-associated forms (Goldman *et al.*, 1982). The major portion of the cell capsule consists of the cell-associated emulsan. A reduction of this form coincides with an increase in extracellular emulsifying activity (Pines *et al.*, 1983). This polymer appears to fulfill two roles: encapsulation of the cell and emulsification of hydrocarbon substrates. Emulsan production is not exclusive to strains growing on hydrocarbon substrates (Sar and Rosenberg, 1983). The flocculant ability of activated sludge bacteria is clearly an important factor in successful secondary settling prior to effluent discharge. In fact, the sedimentation properties of flocculating bacteria have been shown to improve with the concentration of exocellular polymer (Sheintuch *et al.*, 1986).

5.3. Transport of Metabolites

The proton motive force (pmf) is the most important energy source in bacterial active transport (Jain and Wagner, 1980; Wilson, 1978; Boyer, 1977; Brodie *et al.*, 1972). The so-called primary active transport systems mainly involve the generation of the pmf. The two most important energy sources for this purpose are ATP hydrolysis and substrate oxidation coupled to electron flow (Harold, 1974; Brodie *et al.*, 1972). The chemiosmotic theory of Mitchell (1977) is largely responsible for the concept of pmf. According to Mitchell (1968, 1977), the electron transport chain and ATPase complexes are incorporated into the membrane, where they operate as electrogenic pumps (Konings *et al.*, 1981; Harold, 1974). In the bacterial membrane, H^+ ions from the cytoplasm are translocated to the external medium via these proton pumps. The cytoplasmic membrane is practically impermeable to ions, particularly H^+ and OH^-. The translocation of H^+ results in the creation of two gradients. A pH gradient, or chemical H^+ gradient (ΔpH), is formed as a result of removal of H^+ from the cytoplasm, and an electrical potential (Ψ) is formed as a result of the loss of positive charges from the cytoplasm to the medium (Konings *et al.*, 1981; Harold, 1974). Both gradients exert an inward force on the H^+ which is known as the pmf:

$$\Delta\mu H^+ = \Delta\Psi - 2.3\ RT/F\ \Delta pH$$

(Mitchell, 1977).

Potassium transport has been widely studied in bacteria. Three distinct K^+ transport systems have been discovered in *E. coli*. One of the systems is a high-affinity, trace K^+-scavenging system that permits growth at very low external K^+ levels and is repressible by growth in high K^+ (Silver, 1978). This system is unusual in that it does not appear to have a periplasmic protein component and does not appear to perform K^+ uptake coupled to Na^+ extrusion, but may couple uptake to extrusion of another cation (Laimins *et al.*, 1978).

Harold (1977) earlier proposed an energy transduction reversible ATPase for *E. coli* which extruded H^+ in response to K^+ uptake. In contrast, *Streptococcus faecalis*, which possesses three systems similar to those described for *E. coli* (Silver, 1978), takes up K^+ with concurrent extrusion of equivalent amounts of H^+ and Na^+ (Harold *et al.*, 1970).

The major K^+ system is synthesized constitutively and dominates the K^+ economy at moderate or high external K^+ levels. The third system is also synthesized constitutively at low and high K^+ concentrations. Influx and efflux of K^+ appear to be independently regulated by feedback in response to osmotic pressure (Silver, 1978).

To perform its energy-providing function, the H^+ must be present in sufficient amounts to buffer the rate of fluctuations in the H^+- consuming and -generating processes (Skulachev, 1977). Skulachev (1978) went on to hypothesize a Na^+/K^+ gradient as a membrane link buffering system. Electrophoretic K^+ influx results in $\Delta\Psi$ being converted to pH, which increases the total $\mu\Delta H^+$ capacity of the system, rendering it more resistant to fluctuations. The ΔpH can be converted to ΔpNa^+ by means of a Na^+/H^+ antiporter. An increase in the intracellular concentration of K^+ is compensated to some degree by Na^+.

Mg^{2+} is the major intracellular divalent cation in all living cells. In bacterial cells, the Mg^{2+} content is equivalent to 20–40 mM Mg^{2+}, in comparison to 100–500 mM for K^+. Most of the intracellular Mg^{2+} is bound so that it is not osmotically active but relatively readily exchangeable. Extensive investigation of Mg^{2+} transport in *E. coli* and other microbial cells revealed the presence of two transport systems. The cobalt-resistant system is synthesized constitutively during growth in low or high Mg^{2+}. It transports magnesium as primary substrate but also accumulates Co^{2+}, Ni^{2+}, and Mn^{2+}. The other system is specific for Mg^{2+} and is repressible by growth in high Mg^{2+} (Park *et al.*, 1976).

Transport of Fe, especially under Fe starvation conditions, involves a series of high-affinity Fe-binding chelates that are excreted into a medium lacking Fe. After binding of Fe, the ferrichelate is taken up by highly specific transport systems (Silver, 1978). One of these systems is responsible for uptake of Fe chelated with hydroxyamino acids produced by other organisms.

High-affinity transport systems for scavenging traces of Mn^{2+} supply the low-level growth requirements for Mn^{2+} in most bacteria. Other elements which appear to be essential for the growth of some bacteria include Cu, Ni, Cr, and Co, for which evidence of specific transport systems does not exist (Silver, 1978). All bacteria appear to have outwardly oriented energy-dependent Ca transport systems (Silver, 1978). These transport systems are oriented in the cell membrane so as to reduce the intracellular Ca^{2+} level to a level below that in the outside medium.

Research with *E. coli* indicated that Ca is expelled by an antiport for protons, with the proton circulation as the sole energy source (Tsuchiya and Rosen, 1976; Rosen and McClees, 1974). In contrast, Kobayashi *et al.* (1978) proposed an ATP-linked Ca pump that expels Ca by electroneutral exchange for protons.

Bacterial cells excrete Na^+ as well as under physiologically active conditions to maintain a Na^+_{in}-to-Na^+_{out} gradient of between 1 : 3 and 1 : 50. In the case of *E. coli,* sodium egress occurs via a Na^+/H^+ antiport system that translocates Na^+ outward under the influence of an inwardly directed H^+ gradient (Schuldiner and Fishkes, 1978; West and Mitchell, 1974). Harold and Papineau (1972) proposed a somewhat different system for *Streptococcus faecalis,* namely, that the cells extrude Na^+ in exchange for H^+ via an ATP-linked Na^+ pump (Heefner and Harold, 1982) and H^+ is then extruded via the H^+ pump.

Four transport systems exist to transport $PO_4 - P$ across the cell membrane of *E. coli*. There are two constitutive systems (Pit and Pst), one of which (Pst) functions at five-times-higher levels after $PO_4 - P$ starvation (Rosenberg *et al.*, 1977), and two inducible organophosphate systems, which also transport $PO_4 - P$ as alternative substrate (Bennett and Malamey, 1970). The Pst system has the basic properties of an ATP-driven system, whereas the Pit system appears to be coupled to the proton motive gradient (Berger and Heppel, 1974). $PO_4 - P$ uptake and exchange take place via the Pit system, while the Pst system accomplishes only uptake (Rosenberg *et al.*, 1977). $PO_4 - P$ uptake in *E. coli* has been reported to require H^+ (for symport with $H_2PO_4^-$) and extracellular K^+ (Harold, 1977).

$PO_4 - P$ uptake in *S. faecalis* depends on the capacity of the cells to maintain a neutral or alkaline cytoplasm. The ATP-driven $PO_4 - P$ accumulation is probably an electroneutral exchange for OH^- (Harold and Spitz, 1975). P transport in *Paracoccus denitrificans* occurs in the same way (Burnell *et al.*, 1975).

The two constitutive $PO_4 - P$ accumulation systems appear to be sensitive to intracellular and extracellular $PO_4 - P$ levels in respect of the initial uptake rate. In PO_4-limited cells of *Corynebacterium bovis*, the initial rate was dependent on the external PO_4 concentration and was inversely related to the amount of intracellular PO_4 (Chen, 1974). In *Bacillus cereus*, the uptake rate is doubled in phosphate-starved cells due to the requirement of filling a primary PO_4 pool before metabolism can begin. Once the primary pool is filled, the rate of uptake falls to about half the initial value (Rosenberg *et al.*, 1969). The same pattern is observed in starved *E. coli* cells, although in this case two kinetically distinct low- and high-affinity systems have been recognized (Medveczky and Rosenberg, 1971).

As in the case of $PO_4 - P$, the SO_4^{2-} ion requires a specific binding protein for transport (Oxender, 1972). SO_4^{2-} transport in *Paracoccus denitrificans* has been shown to be uncoupler sensitive and can be driven by respiration or a transmembrane pH gradient. It has been proposed on the basis of experimental data that SO_4^{2-} transport operates by a mechanism of electroneutral symport in either direction across the plasma membrane.

Accumulative uptake occurs via respiration-driven H^+ expulsion (Burnell *et al.*, 1975). In some fungi, a synergistic relationship between Ca^{2+} and SO_4^{2-} uptake has been observed (Cuppoletti and Segel, 1975).

Studies with a variety of bacteria revealed that bacterial cells take up acetic acid in its undissociated form (Kell *et al.*, 1981; Konings *et al.*, 1981; Visser and Postma, 1973). The uptake may take place by facilitated diffusion. This possibility is supported by the fact that acetate and propionate share a common uptake system in *E. coli*, with K_m values of 36 and 220 μM, respectively (Kay, 1972). Whether uptake is by free diffusion, as suggested for butyrate and valerate (Salanitro and Wegener, 1971), or by facultative diffusion, the acids are taken up in the undissociated form.

Fumarate, malate, and succinate are taken up by *E. coli* via an active transport system which requires simultaneous uptake of two H^+ for each molecule of acid (Gutowski and Rosenberg, 1975). One of the most important transport systems for carbohydrate uptake by bacteria is the phosphoenol pyruvate phosphotransferase system. A membrane-bound enzyme system catalyzes phosphorylation and transfer of the sugar across the membrane (Postma and Roseman, 1976; Kaback, 1968).

5.4. Metabolic Control

Metabolic control is essential for the orderly continuation of life. Two main types of control mechanism have been identified in bacteria, in common with other cells. Enzyme activity can simply be altered by the presence of different substrate concentrations or product accumulation. Most known control mechanisms operate at the level of enzymatic substrate conversion. Metabolism can be controlled efficiently by regulating the flow of metabolites to the conversion site.

Transport systems can be controlled in the same way as enzymes, by internal or external metabolites. Most intracellular conversions are directly or indirectly dependent on the adenine nucleotides. These compounds play a unique role in the cell. In general, catabolic enzymes are activated by ADP or AMP and inhibited by ATP. All three nucleotides are present in the cell, and a change in one affects the other two. This phenomenon led to the formulation of the concept of adenylate charge, which varies from a value of) (all AMP) to 1 (all ATP) (Atkinson, 1968). A number of reaction rates appear to be regulated by the adenylate charge. In fact, the effect of the charge is greater than the effect of one of the nucleotides alone (Swedes et al., 1975).

Another important form of enzyme regulation is by reversible covalent modification. Although covalent protein modification is more common in eucaryotic cells, examples have been reported in bacteria (Krebs, 1985).

Regulation of enzyme activity by controlling the enzyme concentration can be effected at the level of transcription or translation.

5.5. Current Biochemical Models of Enhanced P Removal

Comeau et al. (1985) postulated a model in which polyP is proposed as a source of energy, both for the replenishment of the pmf and for substrate storage under anaerobic conditions. Under anaerobic conditions, diffusion of substrates (such as acetate) as neutral molecules into the cells will decrease the pH gradient of bacteria. PolyP bacteria utilize their polyP reserves to reestablish the pH gradient, possibly directly or via the production of ATP. PolyP can also provide energy for the formation of acetyl CoA. Storage (of acetyl CoA) as PHB requires NADH, which the tricarboxylic acid cycle can produce anaerobically. $PO_4 - P$ expulsion takes place because of the excess of $PO_4 - P$ molecules accumulating in the cell. A pH gradient-sensitive carrier could "sense" that PO_4 cannot be used for synthesis. Metallic cations are cotransported with $PO_4 - P$ (Fig. 17).

Consumption of external or stored substrate in the presence of oxygen (or oxidized nitrogen under anoxic conditions) will allow polyP bacteria to produce a pmf. The pmf can be used, in particular, for PO_4 transport and ATP production. ATP is then utilized for growth but also for polyP storage as a result of the availability of $PO_4 - P$ and of energy.

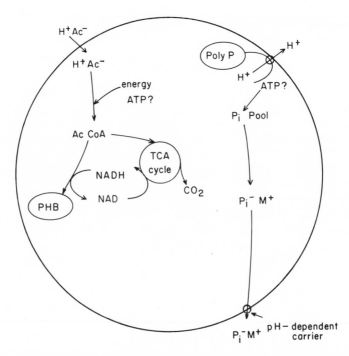

Figure 17. The Comeau model for anaerobic metabolism (Comeau *et al.*, 1985). TCA, Tricarboxylic acid; Ac CoA, acetyl CoA.

Metallic cations are cotransported with $PO_4 - P$ molecules (Comeau *et al.*, 1985) (Fig. 18).

The Comeau model differs from that of Wentzel *et al.* (1986) in that it does not address the control mechanisms which regulate the cellular response to different environmental conditions. Wentzel *et al.* (1986) postulated that ATP/ADP and NADPH/NAD ratios control polyP and PHB synthesis and degradation and therefore the metabolic behavior essential to enhanced P removal. These researchers, while recognizing the role of cations in metabolic transport, did not identify the relevant ions. The electroneutrality of the various transmembrane movements is, however, maintained in this model.

In the later model of Wentzel *et al.* (1986), the lack of a terminal electron acceptor (anaerobic conditions) results in an increase in the NADPH/NAD ratio, which inhibits the tricarboxylic acid cycle. The resultant increase in acetyl CoA levels stimulates PHB synthesis. The decrease in the ATP/ADP ratio in the absence of oxidative phosphorylation stimulates polyP degradation. The release of $PO_4 - P$ in apparent exchange for acetic acid is explained by the necessity to reinstate the pmf, which is dissipated by the removal of a hydrogen ion during transport of the acetic acid into the cell. These processes are depicted in Fig. 19.

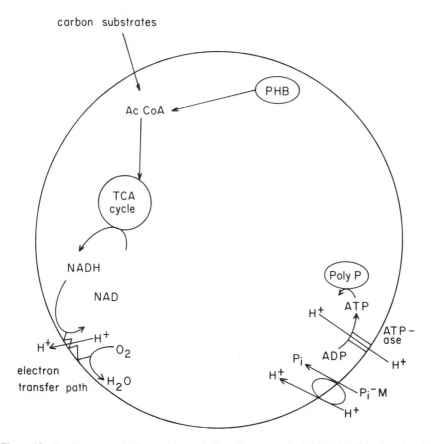

Figure 18. The Comeau model for aerobic metabolism (Comeau *et al.*, 1985). TCA, Tricarboxylic acid.

Under aerobic conditions, the NADPH/NAD ratio decreases, releasing the inhibition of the tricarboxylic acid cycle and permitting oxidation of stored substrate. Oxidative phosphorylation proceeds, ATP is produced, and $PO_4 - P$ is actively transported into the cell (Fig. 20).

5.6. Extended Model

The models of Comeau *et al.* (1985) and Wentzel *et al.* (1986) can now be extended and, in some aspects, confirmed by more recent research. Lack of oxidative phosphorylation under anaerobic conditions leads to rapid depletion of the cellular ATP. While substrate uptake by passive diffusion has been demonstrated, activation of acetic acid to acetyl CoA requires an energy source.

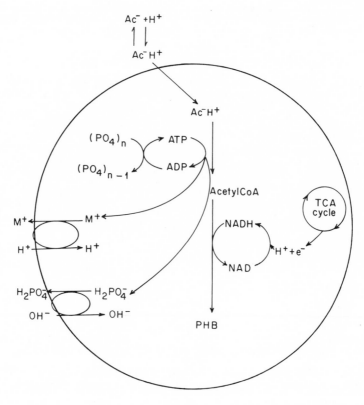

Figure 19. Anaerobic metabolism according to Wentzel *et al.* (1986). TCA, Tricarboxylic acid.

In microorganisms, the activation of acetic acid generally proceeds through the following steps (Mahler and Cordes, 1971):

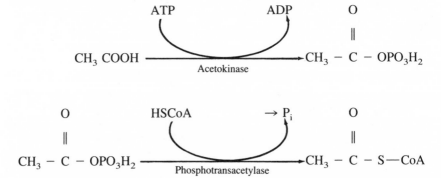

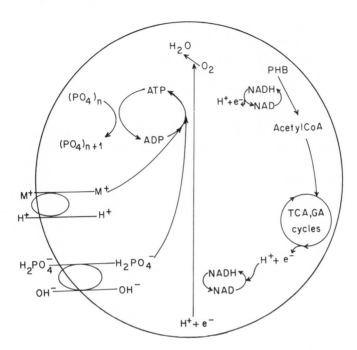

Figure 20. Aerobic metabolism according to Wentzel *et al.* (1986). TCA, Tricarboxylic acid; GA, glyoxylic acid.

Phosphotransacetylase activity has been observed in the anaerobic zones of P-removing plants, thus confirming this route for polyP bacteria (Lötter, 1985).

The PO_4 − P released in the above reaction sequence must contribute to the total intracellular PO_4 − P released under these conditions. PolyP degradation is stimulated by the reduction in ATP, thus replenishing intracellular ATP levels. The concomitant release of SO_4^{2-} with P under these conditions (Toerien *et al.*, 1986) still requires explanation.

The initial step in SO_4^{2-} utilization is the formation of a compound which also serves as a general agent for esterification of sulfate with polysaccharides, namely, 3'-phosphoadenosine-5'-phosphosulfate (PAPS). This compound is formed in a two-step process as follows:

$$SO_4^{2-} + ATP \rightarrow \text{adenosine-5'-phosphosulphate (APS)} + PP_1$$

Although the equilibrium of this equation catalyzed by ATP sulfurylase lies to the left, removal of PP_i by pyrophosphatase action drives APS formation. In the second step, APS is phosphorylated by adenylyl kinase:

$$APS + ATP \rightarrow PAPS + ADP$$

Both reactions require the presence of Mg^{2+} (White *et al.*, 1978). Under anaerobic conditions, the pyrophosphatase is inhibited by high intracellular $PO_4 - P$ levels. The build-up of pyrophosphate (PP_i) will drive the equilibrium of the ATP sulfurylase reaction to the left, resulting in a build-up of sulfate, which is eventually released by osmotic pressure. At the same time ATP is released, which can then be utilized as an energy source. The substrate uptake and phosphate release occurs in the manner described by Wentzel *et al.* (1986).

Mg^{2+} and K^+, which act as counterions in the polyP chains, are released into the intracellular medium as the polyP is hydrolyzed. These cations are expelled from the cell in order to maintain normal osmotic pressure.

Under aerobic conditions, the high NADH level maintained by the lack of a terminal electron acceptor is reduced; the inhibition of the tricarboxylic acid cycle and inhibition of β-hydroxybutyrate dehydrogenase (Lötter and Dubery, 1987) allow PHB to be metabolized, resulting in a flow of electrons through the electron transfer chain and a regeneration of ATP via oxidative phosphorylation. Energy becomes available for the active transport of $PO_4 - P$ and SO_4^{2-}. The cations Mg^{2+} and K^+, which are essential to the stabilization of the polyP chains, are taken up and polyP synthesis proceeds.

6. Ecological Implications

The process of enhanced P removal in modified activated sludge systems is now an accepted, and proven, alternative in eutrophication control strategies. It is ironic that this process was discovered by chance and that hundreds of millions of dollars have been invested in it without a full understanding of its basic mechanisms. However, this probably reflects the need to have lower-cost P-removal processes that can function without the addition of chemicals to effluents.

Optimization of enhanced P-removal processes depends on a complete understanding of the ecophysiology of polyP organisms. These organisms, probably ubiquitous in nature, are selected for (or are adapted) by (1) alternating anaerobic–aerobic conditions in activated sludge systems receiving their influent organic load in the anaerobic zone and (2) the chemical and physical properties of the sludge flocs of which they have become part and which are settled out in sedimentation tanks (clarifiers) before being returned to the influent end of the plants.

Our understanding of the ecophysiology of the polyP organisms has increased tremendously in the past few years. If *Acinetobacter* is used as a model of polyP organisms, several important aspects can be recorded:

1. This obligate aerobe proliferates in an environment with periodic anaerobic conditions by being physiologically active during the anaerobic stage. By using stored polyP as an energy source, fermentation end products such as volatile fatty acids are taken up and stored as PHB. However, the success of this activity is dependent on the outcome of competition with a host of organisms, particularly denitrifiers, which can take up fermentation end products under anaerobic conditions.

2. Under aerobic conditions, *Acinetobacter* is not dependent on external compounds as energy source but can use the stored PHB in the presence of oxygen to generate energy for growth and storage of polyP. In this way, *Acinetobacter* utilizes the opportunities afforded by alternating anaerobic–aerobic conditions to compete effectively with other aerobes in the aerobic stages of enhanced P-removal plants. However, it is now clear that other variables, such as the presence of specific cations, also play important roles in the metabolism of polyP bacteria. It also still remains to be proven beyond doubt that *Acinetobacter* is an acceptable model for all polyP bacteria in enhanced P-removal plants.

We accept intuitively that there is competition between different microbes in activated sludge systems. However, the possibility that denitrifiers and polyP bacteria are locked in severe competition for fermentation end products in the presence of nitrate, highlights the significance of competition as an important determinant in achieving P removal or other effluent treatment goals.

Another area that has as yet received insufficient attention is the precise role of the physical configuration of an activated sludge system as a selector (determinant) of microbial population composition. It can be argued that most activated sludge systems are unique because they differ in one or other dimensional aspect from other systems. Such uniqueness could be important in the physical selection of microbes present in these units. For instance, we accept that the clarifier of an activated sludge system is the prime mechanism for the retention of the active microbial population in the system. However, our understanding of how particular bacteria are included in the flocs and how the flocs behave in clarifiers is far from complete. In addition, we do not know how factors such as size and shape of reactors, aeration mechanisms (e.g., aeration through diffusers or by surface aerators), and type of pumps used influence the type or numbers of bacteria present.

Thorough microbiological studies of activated sludge systems, with particular reference to ecophysiological and physical conditions which select for or maintain specific microbial groups in such systems, must be an integral part of that research which will significantly advance our understanding of enhanced P-removal activated sludge systems in the next decade.

References

Ardern, E., and Lockett, W. T., 1914, Experiments on the oxidation of sewage without the aid of filters, *J. Soc. Chem. Ind.* **33**:523–539.

Allen, L. A., 1944, The bacteriology of activated sludge, *J. Hyg.* **43**:424–431.

Alper, R., Lundgren, D. G., Marchessault, R. H., and Cote, W. A., 1963, Properties of poly-β-hydroxybutyrate. 1. General considerations concerning the naturally occurring polymer. *Biopolymers* **1**:545–556.

Andrews, J. H., and Harris, R. F., 1985, *r*- and *K*-selection and microbial ecology, in: *Advances in Microbial Ecology*, Vol. 9 (K. C. Marshall, ed.), pp. 99–48, Plenum, New York.

Arvin, E., and Kristensen, G. H., 1985, Exchange of organics, phosphate and cations between sludge and water in biological phosphorus and nitrogen removal processes, *Water Sci. Technol.* **17**:147–162.

Atkinson, D. E., 1968, The energy charge of the adenylate pool as a regulatory parameter. Interaction with feedback modifiers, *Biochemistry* **7**:4030–4034.

Banks, C. J., and Walker, I., 1977, Sonication of activated sludge flocs and the recovery of their bacteria on solid media, *J. Gen. Microbiol.* **98**:363–368.

Barnard, J. L., 1973, Biological denitrification, *Water Pollut. Control* **72**:705–720.

Barnard, J. L., 1974, Cut P and N without chemicals, *Water Wastes Eng.* **11**:33–44.

Barnard, J. L., 1975, Nutrient removal in biological systems, *Water Pollut. Control* **74**:143–154.

Barnard, J. L., 1976, A review of biological phosphorus removal in the activated sludge process, *Water SA* **2**:136–144.

Barnard, J. L., 1982, The influence of nitrogen on phosphorus removal in activated sludge plants, *Water Sci. Technol.* **14**:31–45.

Barnard, J. L., 1984, Activated primary tanks for phosphate removal, *Water SA* **10**:121–126.

Baxter, M., and Jensen, T. H., 1980, Uptake of magnesium, strontium, barium and manganese by *Plectonema boryanum* (Cyanophyceae), *Protoplasm* **104**:81–89.

Bell, E. J., and Herman, N. J., 1967, Effect of succinate and isocitrate lyase synthesis in *Mima polymorpha, J. Bacteriol.* **93**:2020–2021.

Bennett, R. L., and Malamy, M. H., 1970, Arsenate resistant mutants of *Escherichia coli* and phosphate transport, *Biochem. Biophys. Res. Commun.* **40**:496–503.

Berger, E. A., and Heppel, L. A., 1974, Different mechanisms of energy coupling for the shock-sensitive and shock-resistant amino acid permeases of *Escherichia coli, J. Biol. Chem.* **249**:7747–7755.

Blackbeard, J. R., Ekama, G. A., and Marais, G. v. R., 1986, A survey of filamentous bulking and foaming in activated sludge plants in South Africa, *Water Pollut. Control* **85**:90–100.

Blackbeard, J. R., Gabb, D. M. D., Ekama, G. A., and Marais, G. v. R., 1987, Identification of filamentous organisms in nutrient removal activated sludge plants in South Africa, in: *Proceedings of the Institute of Water Pollution Control (S A Branch) Biennial Conference, Port Elizabeth,* Paper no. 10.

Boyer, P. D., 1977, Coupling mechanisms in capture, transmission and use of energy, *Annu. Rev. Biochem.* **46**:957–966.

Brodie, A. F., Hirata, H., Asano, A., Cohen, N. S., Hinds, T. R., Aithal, H. N., and Kalra, V. K., 1972, The relationship of bacterial membrane orientation to oxidative phosphorylation and active transport, in: *Membrane Research* (C. Fred Fox, ed.), pp. 445–472, Academic Press, New York.

Brodisch, K. E. U., 1985, Interaction of different groups of microorganisms in biological phosphate removal, *Water Sci. Technol.* **17**:89–97.

Brodisch, K. E. U., and Joyner, S. J., 1983, The role of microorganisms other than *Acinetobacter* in biological phosphate removal in the activated sludge process, *Water Sci. Technol.* **15**:117–125.

Buchan, L., 1981, The location and nature of accumulated phosphorus in seven sludges from activated sludge plants which exhibited enhanced phosphorus removal, *Water SA* **7**:1–7.

Buchan, L., 1983, Possible biological mechanism of phosphorus removal, *Water Sci. Technol.* **15**:87–103.

Bundgaard, E., Kristensen, G. H., and Arvin, E., 1983, Full-scale experience with phosphorus removal in an alternating system, *Water Sci. Technol.* **15**:197–217.

Burnell, J. N., John, P., and Whatley, F. R., 1975, Phosphate transport in membrane vesicles of *Paracoccus denitrificans, FEBS Lett.* **58**:215–218.

Butterfield, C. T., 1935, Studies of sewage purification. II. A Zooglea-forming bacterium isolated from activated sludge, *Public Health Rep.* **50**:671–684.

Chambers, B., 1982, Effect of longitudinal mixing and anoxic zones on settleability of activated sludge, in: *Bulking of Activated Sludge: Preventative and Remedial Methods* (B. Chambers and E. J. Tomlinson, eds.), pp. 166–186, Ellis Horwood Ltd., Chichester.

Chen, M., 1974, Kinetics of phosphorus absorption by *Corynebacterium bovis, Microb. Ecol.* **1**:164–175.

Christensen, M. H., and Harremoës, P., 1977, Biological denitrification of sewage: A literature review, *Prog. Water Technol.* **8**:509–555.

Cloete, T. E., 1985, The detection of Acinetobacter in activated sludge and its possible role in biological phosphorus removal, D.Sc. thesis, University of Pretoria, Pretoria, South Africa.

Cloete, T. E., and Steyn, P. L., 1987, A combined fluorescent antibody-membrane filter technique for enumerating *Acinetobacter* in activated sludge, in: *Advances in Water Pollution Control, Biological Phosphate Removal from Wastewaters* (R. Ramadori, ed.), pp. 335–338, Pergamon Press, Oxford.

Cloete, T. E., Steyn, P. L., and Buchan, L., 1985, An aut-ecological study of *Acinetobacter* in activated sludge, *Water Sci. Technol.* **17**:139–146.

Comeau, Y., Hall, K. J., Hancock, R. E. W., and Oldham, W. K., 1985, Biochemical model for enhanced biological phosphorus removal, in: *Proceedings of University of British Columbia Conference on New Directions and Research in Waste Treatment and Residuals Management,* pp. 324–346, University of British Columbia, Vancouver, Canada.

Comeau, Y., Hall, K. J., Hancock, R. E. W., and Oldham, W. K., 1986, Biochemical model for enhanced biological phosphorus removal, *Water Res.* **20**:1511–1521.

Comeau, Y., Oldham, W. K., and Hall, K. J., 1987, Dynamics of carbon reserves in biological dephosphatation of wastewater, in: *Advances in Water Pollution Control. Biological Phosphate Removal from Wastewaters* (R. Ramadori, ed.), pp. 39–55, Pergamon Press, Oxford.

Cuppoletti, J., and Segal, I. H., 1975, Kinetics of sulphate transport by *Penicillium notatum.* Interactions of sulphate, protons, and calcium, *Biochemistry* **14**:4712–4718.

Davelaar, D., Davis, T. R., and Wiechers, S. G., 1978, The significance of an anaerobic zone for the biological removal of phosphate from wastewaters, *Water SA* **4**:54–60.

Dawes, E. A., and Senior, P. J., 1973, The role and regulation of energy reserve polymers in microorganisms, *Adv. Microb. Physiol.* **10**:135–266.

Deinema, M. H., Habets, L. H. A., Scholten, J., Turkstra, E., and Webers, H. A., 1980, The accumulation of polyphosphate in *Acinetobacter* spp., *FEMS Microbiol. Lett.* **9**:275–279.

Deinema, M. H., Van Loosdrecht, M., and Scholten, A., 1985, Some physiological characteristics of *Acinetobacter* spp. accumulating large amounts of phosphate, *Water Sci. Technol.* **17**:119–125.

Dias, F. F., and Bhat, J. V., 1964, Microbial ecology of activated sludge. I. Dominant bacteria, *Appl. Microbiol.* **12**:412–417.

Dold, P. L., and Marais, G. v. R., 1986, Evaluation of the general activated sludge model proposed by the IAWPRC task group, *Water Sci. Technol.* **18**:63–89.

Eigener, U., and Bock, E., 1972, Auf-und Abbau der Polyphosphat-fraktion in Zellen von *Nitrobacter winogradskyi* (Buch), *Arch. Mikrobiol.* **81**:367–378.

Eikelboom, D. H., and van Buijsen, H. J., 1981, *Microscopic Sludge Investigation Manual,* Report No. A94A of the TNO Instituut voor Milieu-hygiene en -gesondheidstechniek, Delft, The Netherlands.

Ekama, G. A., and Marais, G. v. R., 1984a, Biological nitrogen removal, in: *Theory, Design and Operation of Nutrient Removal Activated Sludge Processes,* pp. 6-1–6-26, Water Research Commission, Pretoria, South Africa.

Ekama, G. A., and Marais, G. v. R., 1984b, Carbonaceous material removal, in: *Theory, Design and Operation of Nutrient Removal Activated Sludge Processes,* pp. 4-1–4-20, Water Research Commission, Pretoria, South Africa.

Ekama, G. A., and Marais, G. v. R., 1985, The implications of the IAWPRC hydrolysis hypothesis on low F/M bulking, *Water Sci. Technol.* **18**:11–19.

Ekama, G. A., Siebritz, I. P., and Marais, G. v. R., 1983, Considerations in the process design of nutrient removal activated sludge processes, *Water Sci. Technol.* **15**:283–318.

Ekama, G. A., Marais, G. v. R., and Siebritz, I. P., 1984, Biological excess phosphorus removal, in: *Theory, Design and Operation of Nutrient Removal Activated Sludge Processes,* pp. 7-1–7-32, Water Research Commission, Pretoria, South Africa.

Friedberg, I., and Avigad, G. 1968, Structures containing polyphosphate in *Micrococcus lysodeikticus*, *J. Bacteriol.* **96**:544–553.

Friedman, B. A., Dugan, P. R., Pfister, R. M., and Remsen, C. C., 1969, Structure of exocellular polymers and their relationship to bacterial flocculation, *J. Bacteriol.* **98**:1328–1334.

Fuhs, G. W., and Chen, M., 1975, Microbiological basis of phosphate removal in the activated sludge process for the treatment of wastewater, *Microb. Ecol.* **2**:119–138.

Fukase, T., Shibata, M., and Miyaji, Y., 1985, Factors affecting biological removal of phosphorus, *Water Sci. Technol.* **17**:187–198.

Gerber, A., Mostert, E. S., Winter, C. T., and de Villiers, R. H., 1986, The effect of acetate and other short-chain carbon compounds on the kinetics of biological nutrient removal, *Water SA* **12**:7–12.

Gerber, A., Mostert, E. S., Winter, C. T., and de Villiers, R. H., 1987a, Interactions between phosphate, nitrate and organic substrate in biological nutrient removal processes, *Water Sci. Technol.* **19**:183–194.

Gerber, A., de Villiers, R. H., Mostert, E. S., and van Riet, C. J. J., 1987b, The phenomenon of simultaneous phosphate uptake and release, and its importance in biological nutrient removal, in: *Advances in Water Pollution Control. Biological Phosphate Removal from Wastewaters* (R. Ramadori, ed.), pp. 123–134, Pergamon Press, Oxford.

Gersberg, R. M., and Allen, D. W., 1985, Phosphorus uptake by *Klebsiella pneumoniae* and *Acinetobacter calcoaceticus*, *Water Sci. Technol.* **17**:113–118.

Goldman, S., Shabtai, Y., Rubinovitz, C., Rosenberg, E., and Gutnick, D. L., 1982, Emulsan in *Acinetobacter calcoaceticus* RAG-1: Distribution of cell free and cell associated cross reacting material. *Appl. Environ. Microbiol.* **44**:165–170.

Grady, C. P. L., and Lim, H. C., 1980, *Biological Wastewater Treatment: Theory and Applications,* Marcel Dekker, Inc., New York.

Gutowski, S. J., and Rosenberg, H., 1975, Succinate uptake and related proton movements in *Escherichia coli* K12, *Biochem. J.* **152**:647–654.

Harold, F. M., 1963, Accumulation of inorganic polyphosphate in *Aerobacter aerogenes*, *J. Bacteriol.* **86**:216–221.

Harold, F. M., 1964, Enzymic and genetic control of polyphosphate accumulation in *Aerobacter aerogenes*, *J. Gen. Microbiol.* **35**:81–90.

Harold, F. M., 1966, Inorganic polyphosphates in biology: Structure, metabolism, and function, *Bacteriol. Rev.* **30**:772–794.

Harold, F. M., 1974, Chemiosmotic interpretation of active transport in bacteria, *Ann. N.Y. Acad. Sci.* **227**:297–311.

Harold, F. M., 1977, Membranes and energy transduction in bacteria, *Curr. Top. Bioenerg.* **6**:83–149.

Harold, F. M., and Harold, R. L., 1965, Degradation of inorganic polyphosphate in mutants of *Aerobacter aerogenes*, *J. Bacteriol.* **89**:1262–1270.

Harold, F. M., and Papineau, D., 1972, Cation transport and electrogenesis by *Streptococcus faecalis, J. Membr. Biol.* **8**:45–62.

Harold, F. M., and Spitz, E., 1975, Accumulation of arsenate, phosphate and aspartate by *Streptococcus faecalis, J. Bacteriol.* **122**:266–277.

Harold, F. M., and Sylvan, S., 1963, Accumulation of inorganic polyphosphate in *Aerobacter aerogenes, J. Bacteriol.* **86**:222–231.

Harold, F. M., Baarda, J. R., and Pavlasova, E., 1970, Extrusion of sodium and hydrogen ions as the primary process in potassium ion accumulation by *Streptococcus faecalis, J. Bacteriol.* **101**:152–159.

Harris, E., 1957, Radiophosphorus metabolism in zooplankton and micro-organisms, *Can. J. Zool.* **35**:769–782.

Harris, R. H., and Mitchell, R., 1973, The role of polymers in microbial aggregation, *Annu. Rev. Microbiol.* **27**:27–50.

Hascoet, M. C., and Florentz, M., 1985, Influence of nitrates on biological phosphorus removal from wastewater, *Water SA* **11**:1–8.

Hascoet, M. C., Florentz, M., and Granger, P., 1985, Biochemical aspects of enhanced biological phosphorus removal from wastewater, *Water Sci. Technol.* **17**:23–41.

Heefner, D. L., and Harold, F. M., 1982, ATP-driven sodium pump in *Streptococcus faecalis*, *Proc. Natl. Acad. Sci. USA* **79**:2798–2802.

Herman, N. J., and Bell, E. J., 1970, Metabolic control in *Acinetobacter* sp. I: Effect of C_4 versus C_2 and C_3 substrates on isocitrate lyase synthesis, *Can. J. Microbiol.* **16**:769–774.

Heukelekian, H., and Littman, M. L., 1939, Carbon and nitrogen transformations in the purification of sewage by activated sludge process. II. Morphological and biochemical studies of zoogleal organisms, *Sewage Works J.* **11**:752–763.

Hoffman, H., 1987, Influence of oxic and anoxic mixing zones in compartment systems on substrate removal and sludge characteristics in activated sludge plants, *Water Sci. Technol.* **19**:897–910.

Hong, S.-N., Krichten, D. J., Kisenbauer, K. S., and Sell, R. L., 1982, A Biological Wastewater Treatment System for Nutrient Removal, presented at the EPA Workshop on Biological Phosphorus Removal in Municipal Wastewater Treatment, Annapolis, Md.

Hungate, R. E., 1966, *The Rumen and Its Microbes*, Academic Press, New York.

Iwema, A., and Meunier, A., 1985, Influence of nitrate on acetic acid induced biological phosphate removal, *Water Sci. Technol.* **17**:289–294.

Jain, M., and Wagner, R. C., 1980, Passive facilitated diffusion, in: *Introduction to Biological Membranes* (M. Jain and R. C. Wagner, eds.), pp. 232–247, John Wiley and Sons, New York.

Jenkins, D., Richard, M. G., and Neethling, J. B., 1984, Causes and control of activated sludge bulking, *J. Water Pollut. Control Fed.* **83**:455–472.

Jenkins, S. H., and Lockett, W. T., 1943, Loss of phosphorus during sewage purification, *Nature* (London) **151**:306–307.

Kaback, H. R., 1968, The role of the phosphoenol pyruvate-phosphotransferase system in the transport of sugars by isolated membrane preparations of *Escherichia coli*, *J. Biol. Chem.* **143**:3711–3724.

Kaltwasser, H., 1962, Die Rolle der Polyphosphate im Phosphat-stoffwechsel eines Knallgasbakteriums (*Hydrogenomonas* Stamm 20), *Arch. Mikrobiol.* **41**:282–306.

Kaltwasser, H., Vogt, G., and Schlegel, H. G., 1962, Polyphosphat-synthese während der Nitrat-Atmung von *Micrococcus denitrificans*, Stamm 11, *Arch. Mikrobiol.* **44**:259–265.

Kay, W. W., 1972, Genetic control of the metabolism of propionate by *Escherichia coli*, K12. *Biochim. Biophys. Acta* **264**:508–521.

Kell, D. B., Peck, M. W., Rodger, G., and Morris, J. G., 1981, On the permeability to weak acids and bases of the cytoplasmic membrane of *Clostridium pasteurianum*, *Biochim. Biophys. Res. Commun.* **99**:81–88.

Kobayashi, H., van Brunt, J., and Harold, F. M., 1978, ATP-linked calcium transport in cells and membrane vesicles of *Streptococcus faecalis*, *J. Biol. Chem.* **253**:2085–2092.

Konings, W. N., Hellingwerf, K. J., and Robellard, G. T., 1981, Transport across bacterial membranes, in: *Membrane Transport* (S. L. Bonting and J. J. H. de Pont, eds.), pp. 257–283, Elsevier/North Holland Biomedical Press, Amsterdam.

Kornberg, A., Kornberg, S. R., and Simms, E. S., 1956, Metaphosphate synthesis by an enzyme from *Escherichia coli*, *Biochim. Biophys. Acta* **20**:215–227.

Krebs, E. G., 1985, The phosphorylation of proteins: A major mechanism for biological regulation, *Biochem. Soc. Trans.* **13**:813–820.

Kulaev, I. S., 1975, Biochemistry of inorganic polyphosphates, *Rev. Physiol. Biochem. Pharmacol.* **73**:131–158.

Kulaev, I. S., 1985, Some aspects of environmental regulation of microbial phosphorus metabolism, *FEMS Symp.* **23**:1–25.

Kulaev, I. S., and Vagabov, V. M., 1983, Polyphosphate metabolism in micro-organisms, *Adv. Microb. Physiol.* **24**:83–171.
Kulaev, I. S., Bobyk, M. A., Nikolaev, N. N., Sergeev, N. S., and Uryson, S. O., 1971, Polyphosphate synthesizing enzymes in some fungi and bacteria, *Biokhimiya* **36**:943–949.
Laimins, L. A., Rhoads, D. B., Altendorf, K., and Epstein, W., 1978, Identification of the structured proteins of ATP-driven potassium transport system in *Escherichia coli, Proc. Natl. Acad. Sci. USA* **75**:3216–3219.
Lee, G. F., Rast, W., and Jones, R. A., 1978, Eutrophication of water bodies: Insights for an age-old problem, *Environ. Sci. Technol.* **12**:900–908.
Levin, G. V., 1964, Sewage Treatment Process, U.S. patent 3236766, applied for 31 March 1964.
Levin, G. V., 1972, Nitrate Removal from Sewage, U.S. patent 3654147, applied for 16 March 1971, granted 4 April 1972.
Levin, G. V., and Sala, U. D., 1987, Phostrip process—a viable answer to eutrophication of lakes and coastal sea waters in Italy, in: *Advances in Water Pollution Control. Biological Phosphate Removal from Wastewaters* (R. Ramadori, ed.), pp. 249–259, Pergamon Press, Oxford.
Levin, G. V., and Shapiro, J., 1965, Metabolic uptake of phosphorus by wastewater organisms, *J. Water Pollut. Control Fed.* **37**:800–821.
Levin, G. V., Topol, G. J., Tarnay, A. G., and Samworth, R. B., 1972, Pilot-plant tests of a phosphate removal process, *J. Water Pollut. Control Fed.* **44**:1940–1954.
Levin, G. V., Topol, G. J., and Tarnay, A. G., 1975, Operation of full-scale biological phosphorus removal plant, *J. Water Pollut. Control Fed.* **47**:577–590.
Levinson, S. L., Jacobs, L. H., Krulwich, T. A., and Li, H. C., 1975, Purification and characterization of a polyphosphate kinase from *Arthrobacter atrocyaneus, J. Gen. Microbiol.* **88**:65–74.
Li, H. C., and Brown, G. G., 1973, Orthophosphate and histone dependent polyphosphate kinase from *E. coli, Biochim. Biophys. Res. Commun.* **53**:875–881.
Lötter, L. H., 1985, The role of bacterial phosphate metabolism in enhanced phosphorus removal from the activated sludge process, *Water Sci. Technol.* **17**:127–138.
Lötter, L. H., and Dubery, I. A., 1987, Metabolic control in polyphosphate accumulating bacteria and its role in enhanced biological phosphate removal, in: *Advances Water Pollution Control. Biological Phosphate Removal from Wastewaters* (R. Ramadori, ed.), pp. 7–14, Pergamon Press, Oxford.
Lötter, L. H., and Murphy, M., 1985, The identification of heterotrophic bacteria in an activated sludge plant with particular reference to polyphosphate accumulation, *Water SA* **11**:179–184.
Ludzack, F. J., and Ettinger, M. B., 1962, Controlling operation to minimize activated sludge effluent nitrogen, *J. Water Pollut. Control Fed.* **34**:920–931.
MacArthur, R. H., and Wilson, E. O., 1967, *The Theory of Island Biogeography,* Princeton University Press, Princeton, N.J.
Macrae, R. M., and Wilkinson, J. F., 1958, Poly-β-hydroxybutyrate metabolism in washed suspensions of *Bacillus cereus* and *Bacillus megaterium, J. Gen. Microbiol.* **19**:210–222.
Madoni, P., 1986, Protozoa in Waste Treatment Systems, presented at the 4th International Symposium on Microbial Ecology, August 1986, Ljubljana, Yugoslavia.
Mahler, H. R., and Cordes, E. H., 1971, *Biological Chemistry,* 2nd ed., Harper and Row, New York.
Malnou, D., Meganck, M., Faup, G. M., and du Rostu, M., 1984, Biological phosphorus removal: Study of the main parameters, *Water Sci. Technol.* **16**:173–185.
Maraeva, O. B., Kolot, M. N., Nesmeyanova, M. A., and Kulaev, I. S., 1979, Interrelationships between metabolic and genetic regulation of alkaline phosphatase and poly- and pyrophosphate, *Biokhimiya* **44**:715–719.
Marais, G. v. R., 1987, The future of biological removal of phosphorus from wastewater, in: *Proceedings of the Australian Water and Wastewater Association, 1987 International Convention, Adelaide,* K.18-K.27.

Marais, G. v. R., Loewenthal, R. E., and Siebritz, I. P., 1983, Observations supporting phosphate removal by biological excess uptake—A review, *Water Sci. Technol.* **15**:15–41.

McKinney, R. E., and Weichlein, R. G., 1953, Isolation of floc-producing bacteria from activated sludge, *Appl. Microbiol.* **1**:259–261.

McLaren, A. R., and Wood, R. J., 1976, Effective phosphorus removal from sewage by biological means, *Water SA* **2**:47–50.

Medveczky, N., and Rosenberg, H., 1971, Phosphate transport in *Escherichia coli, Biochim. Biophys. Acta* **241**:494–506.

Menar, A. B., and Jenkins, D., 1969, Fate of phosphorus in waste treatment processes: The enhanced removal of phosphate by activated sludge, in: *24th Industrial Waste Treatment Conference*, p. 655–674, Purdue University, Lafayette, Ind.

Milbury, W. F., McCauley, D., and Hawthorne, C. H., 1971, Operation of conventional activated sludge for maximum phosphorus removal, *J. Water Pollut. Control. Fed.* **43**:1890–1901.

Mino, T., Arun, V., Tsuzuki, Y., and Matsuo, T., 1987, Effect of phosphorus accumulation on acetate metabolism in the biological phosphorus removal process, in: *Advances in Water Pollution Control. Biological Phosphate Removal from Wastewaters* (R. Ramadori, ed.), pp. 27–38, Pergamon Press, Oxford.

Mitchell, P., 1968, *Chemiosmotic Coupling and Energy Transduction,* Glyn Research Ltd., Bodmin, England.

Mitchell, P., 1977, A commentary on alternative hypotheses of protonic coupling in the membrane systems catalysing oxidative and photosynthetic phosphorylation, *FEBS Lett.* **78**:1–20.

Miyamoto-Mills, J., Larson, J., Jenkins, D., and Owen, W., 1983, Design and operation of a pilot-scale biological phosphate removal plant at the Central Contra Costa Sanitary District, *Wat. Sci. Tech.* **15**:153–179.

Mostert, E. S., Gerber, A., and van Riet, C. J. J., 1987, Fatty acid utilization by sludge from full-scale nutrient removal plants, with special reference to the role of nitrate, in: *Proceedings of the Institute for Water Pollution Control (Southern African Branch) Biennial Conference, Port Elizabeth,* Paper no. 23.

Mudd, S., Yoshida, A., and Koike, M., 1958, Polyphosphate as accumulator of phosphorus and energy, *J. Bacteriol.* **75**:224–235.

Muhammed, A., 1961, Studies on biosynthesis of polymetaphosphate by an enzyme from *Corynebacterium xerosis, Biochim. Biophys. Acta* **54**:121–132.

Mühlradt, P. F., 1971, Synthesis of high molecular weight polyphosphate with a partially purified enzyme from *Salmonella, J. Gen. Microbiol.* **68**:115–122.

Mulder, J. W., and Rensink, J. H., 1987, Introduction of biological phosphorus removal to an activated sludge plant with practical limitations, in: *Advances in Water Pollution Control. Biological Phosphate Removal from Wastewaters* (R. Ramadori, ed.), pp. 213–223, Pergamon Press, Oxford.

Murata, K., Uchida, T., Tani, K., Kato, J., and Chibata, I., 1980, Metaphosphate: A new phosphoryl donor for NAD phosphorylation, *Agr. Biol. Chem.* **44**:61–68.

Murphy, M., and Lötter, L. H., 1986, The effect of acetate and succinate on polyphosphate formation and degradation in activated sludge with particular reference to *Acinetobacter calcoaceticus, Appl. Microbiol. Biotechnol.* **24**:512–517.

Nesmeyanova, M. A., Dmitriev, A. D., and Kulaev, I. S., 1973, High molecular weight polyphosphates and enzymes of polyphosphate metabolism in the process of *E. coli* growth, *Mikrobiologiya* **42**:213–219.

Nesmeyanova, M. A., Dmitriev, A. D., and Kulaev, I. S., 1974, Regulation of the enzymes of phosphorus metabolism and the level of polyphosphate in *E. coli* K-12 by exogenous o-PO$_4$, *Mikrobiologiya* **43**:227–234.

Nicholls, H. A., 1975, Full scale experimentation on the new Johannesburg extended aeration plants, *Water SA* **1**:121–132.

Nicholls, H. A., 1978, Kinetics of phosphorus transformations in aerobic and anaerobic environments, *Prog. Water Technol.* **10**(Suppl. 1):89–102.

Nicholls, H. A., and Osborn, D. W., 1979, Bacterial stress: A prerequisite for biological removal of phosphorus, *J. Water Pollut. Control Fed.* **51**:557–569.

Nicholls, H. A., Pitman, A. R., and Osborn, D. W., 1985, The readily biodegradable fraction of sewage: Its influence on phosphorus removal and measurement, *Water Sci. Technol.* **17**:73–87.

Nicholls, H. A., Osborn, D. W., and Pitman, A. R., 1986, Biological phosphorus removal at the Johannesburg Northern and Goudkoppies Wastewater Purification Plants, *Water SA* **12**:13–18.

Nicholls, H. A., Osborn, D. W., and Pitman, A. R., 1987, Improvement to the stability of the biological phosphate removal process at the Johannesburg Northern Works, in: *Advances in Water Pollution Control: Biological Phosphate Removal from Wastewaters* (R. Ramadori, ed.), pp. 261–272, Pergamon Press, Oxford.

Noegel, A., and Gotschlich, E. C., 1983, Isolation of a high molecular weight polyphosphate from *Neisseria gonorrhoeae*, *J. Exp. Med.* **157**:2049–2060.

Ohtake, H., Takahashi, K., Tsuzuki, Y., and Toda, K., 1984, Phosphorus release from a pure culture of *Acinetobacter calcoaceticus* under anaerobic conditions, *Environ. Technol. Lett.* **5**:417–424.

Okada, M., Murakami, A., and Sudo, R., 1987, Ecological selection of phosphorus-accumulating bacteria in sequencing batch reactor activated sludge processes for simultaneous removal of phosphorus, nitrogen and organic substances, in: *Advances in Water Pollution Control: Biological Phosphate Removal from Wastewaters* (R. Ramadori, ed.), pp. 147–154, Pergamon Press, Oxford.

Oldham, W. K., 1985, Full-scale optimization of biological phosphorus removal at Kelowna, Canada, *Water Sci. Technol.* **17**:243–257.

Osborn, D. W., and Nicholls, H. A., 1978, Optimization of the activated sludge process for the biological removal of phosphorus, *Prog. Water Technol.* **10**:261–277.

Osborn, D. W., and Nicholls, H. A., 1985, Biological nutrient removal in South Africa, *Water SA* **12**:10–13.

Osborn, D. W., Lötter, L. H., Pitman, A. R., and Nicholls, H. A., 1986, *Enhancement of Biological Phosphate Removal by Altering Process Feed Composition*, Report No. 137/1/86, Water Research Commission, Pretoria, South Africa.

Oxender, D. L., 1972, Membrane transport, *Annu. Rev. Biochem.* **41**:777–814.

Park, M. H., Wong, B. B., and Lusk, J. E., 1976, Mutants in three genes affecting transport of magnesium in *Escherichia coli:* genetics and physiology, *J. Bacteriol.* **126**:1096–1103.

Parker, M. G., and Weitzman, P. D. J., 1970, Regulation of NADP-linked isocitrate dehydrogenase activity in *Acinetobacter*, *FEBS Lett.* **7**:324–326.

Pepin, C. A., and Wood, H. G., 1986, Polyphosphate glucokinase from *Propionibacterium shermanii*. Kinetics and demonstration that the mechanism involves both processive and nonprocessive type reactions. *J. Biol. Chem.* **261**:4476–4480.

Pines, O., Bayer, E. A., and Gutnick, D. L., 1983, Localization of emulsan-like polymers associated with the cell surface of *Acinetobacter calcoaceticus*, *J. Bacteriol.* **154**:893–905.

Pitman, A. R., 1984, Operation of biological nutrient removal plants, in: *Theory, Design and Operation of Nutrient Removal Activated Sludge Processes*, pp. 11-1–11-16, Water Research Commission, Pretoria, South Africa.

Pitman, A. R., Venter, S. L. V., and Nicholls, H. A., 1983, Practical experience with biological phosphorus removal plants in Johannesburg, *Water Sci. Technol.* **15**:233–259.

Pitman, A. R., Trim, B. C., and van Dalsen, L., 1988, Operating experience with biological nutrient removal at the Johannesburg Bushkoppie Works, *Water Sci. Techol.* **20**:51–62.

Postma, P. W., and Roseman, S., 1976, The bacterial phosphoenolpyruvate:sugar phosphotransferase system, *Biochim. Biophys. Acta* **457**:213–257.

Prakasam, T. B. S., and Dondero, N. C., 1967a, Aerobic heterotrophic bacterial populations of sewage and activated sludge. I. Enumeration, *Appl. Microbiol.* **15**:461–467.

Prakasam, T. B. S., and Dondero, N. C., 1967b, Aerobic heterotrophic bacterial populations of sewage and activated sludge. II. Method of characterization of activated sludge bacteria, *Appl. Microbiol.* **15**:1122–1127.

Prakasam, T. B. S., and Dondero, N. C., 1970, Aerobic heterotrophic bacterial populations of sewage and activated sludge. V. Analysis of population structure and activity, *Appl. Microbiol.* **19**:671–680.

Price, G. J., 1982, Use of an anoxic zone to improve activated sludge settleability, in: *Bulking of Activated Sludge: Preventative and Remedial Methods* (B. Chambers and E. J. Tomlinson, eds.), pp. 259–260, Ellis Horwood Ltd., Chichester.

Rabinowitz, B., Koch, F. A., Vassos, T. D., and Oldham, W. K., 1987, A novel operational mode for a primary sludge fermenter for use with the enhanced biological phosphorus removal process, in: *Advances in Water Pollution Control: Biological Phosphate Removal from Wastewaters* (R. Ramadori, ed.), pp. 349–352, Pergamon Press, Oxford.

Randall, C. W., Daigger, G. T., Morales, L., Waltrip, G. D., and Romm, E. D., 1987, High-rate economical biological removal of nitrogen and phosphorus, in: *Advances in Water Pollution Control. Biological Phosphate Removal from Wastewaters* (R. Ramadori, ed.), pp. 373–376, Pergamon Press, Oxford.

Rao, N. N., Roberts, M. F., and Torriani, A., 1985, Amount and chain length of polyphosphates in *Escherichia coli* depend on cell growth conditions, *J. Bacteriol.* **62**:242–247.

Rensink, J. H., 1981, Biologische Defosfatering en procesbepalende Factoren, presented at the NVA Symposium, Amersfoort, The Netherlands.

Rensink, J. H., and Donker, H. J. G. W., 1987, The influence of bulking sludge on enhanced biological phosphorus removal, in: *Advances in Water Pollution Control. Biological Phosphate Removal from Wastewaters* (R. Ramadori, ed.), pp. 369–372, Pergamon Press, Oxford.

Ritchie, G. A. F., Senior, P. J., and Dawes, E. A., 1971, The purification and characterization of acetoacetyl-coenzyme A reductase from *Azotobacter beijerinckii*, *Biochem. J.* **121**:309–316.

Robinson, N. A., and Wood, H. G., 1986, Polyphosphate kinase from *Propionibacterium shermanii*. Demonstration that the synthesis and utilization of polyphosphate is by a processive mechanism, *J. Biol. Chem.* **261**:4481–4485.

Robinson, N. A., Goss, N. H., and Wood, H. C., 1984, Polyphosphate kinase from *Propionibacterium shermanii*: Formation of an enzymatically active insoluble complex with basic proteins and characterisation of synthesized polyphosphate, *Biochem. Int.* **8**:757–769.

Rosen, B. P., and McClees, J. S., 1974, Active transport of calcium in inverted membrane vesicles of *Escherichia coli*, *Proc. Natl. Acad. Sci. USA* **71**:5042–5046.

Rosenberg, H., Medveczky, N., and La Nauze, J. M., 1969, Phosphate transport in *Bacillus cereus*, *Biochim. Biophys. Acta* **193**:159–167.

Rosenberg, H., Gerdes, R. G., and Chegwidden, K., 1977, Two systems for the uptake of phosphate in *Escherichia coli*, *J. Bacteriol.* **131**:505–511.

Roughgarden, J., 1971, Density-dependent natural selection, *Ecology* **52**:453–468.

Salanitro, J. P., and Wegener, W. S., 1971, Growth of *Escherichia coli* on short-chain fatty acids: Nature of the uptake system, *J. Bacteriol.* **108**:893–901.

Sar, N., and Rosenberg, E., 1983, Emulsifier production by *Acinetobacter calcoaceticus* strains, *Curr. Microbiol.* **9**:309–313.

Schönberger, R., and Hegemann, W., 1987, Biological phosphorus removal with and without sidestream precipitation, in: *Advances in Water Pollution Control. Biological Phosphate Removal from Wastewaters* (R. Ramadori, ed.), pp. 165–176, Pergamon Press, Oxford.

Schuldiner, S., and Fishkes, H., 1978, Sodium-proton antiport in isolated membrane vesicles of *Escherichia coli*, *Biochemistry* **17**:706–711.

Senior, P. J., and Dawes, E. A., 1971, Poly-β-hydroxybutyrate and the regulation of glucose metabolism in *Azotobacter beijerinckii, Biochem. J.* **125**:55–66.

Severin, A. I., Lusta, K. I., Nesmeyanova, M. A., and Kulaev, I. S., 1975, Membrane bound polyphosphatase of *Escherichia coli, Biokhimiya* **41**:357–362.

Shapiro, J., 1967, Induced rapid release and uptake of phosphate by microorganisms, *Science* **155**:1269–1271.

Shapiro, J., Levin, G. V. and Zea, G. H., 1967, Anoxically induced release of phosphate in wastewater treatment, *J. Water Pollut. Control Fed.* **39**:1810–1818.

Sharma, B., and Ahlert, R. C., 1977, Nitrification and nitrogen removal, *Water Res.* **11**:897–925.

Sheintuch, M., Lev, O., Einav, P., and Rubin, E., 1986, Role of exocellular polymer in the design of activated sludge, *Biotechnol. Bioeng.* **28**:1564–1576.

Siebritz, I. P., Ekama, G. A., and Marais, G. v. R., 1983, A parametric model for biological excess phosphorus removal, *Water Sci. Technol.* **15**:127–152.

Sierra, G., and Gibbons, N. E., 1963, Production of poly-β-hydroxybutyric acid granules in *Micrococcus halodenitrificans, Can. J. Microbiol.* **8**:249–253.

Silver, S., 1978, Transport of cations and anions, in: *Bacterial Transport* (B. P. Rosen, ed.), Microbiology Ser. Vol. 4, pp. 221–324, Marcel Dekker Inc., New York.

Simpkins, M. J., and McLaren, A. R., 1978, Consistent biological phosphate and nitrate removal in an activated sludge plant, *Prog. Water Technol.* **10**:433–441.

Skulachev, V. P., 1977, Transmembrane electrochemical H^+-potential as a convertible energy source for the living cell, *FEBS Lett.* **74**:1–9.

Skulachev, V. P., 1978, Membrane-linked energy buffering as the biological function of Na^+/K^+ gradient, *FEBS Lett.* **87**:171–179.

Somiya, I., Tsuno, H., and Nishikawa, M., 1987, Behaviour of phosphorus and metals in the anaerobic-oxic activated sludge process, in: *Advances in Water Pollution Control. Biological Phosphate Removal from Wastewaters* (R. Ramadori, ed.), pp. 321–324, Pergamon Press, Oxford.

South African Inventions Development Corporation, 1973, S.A. patent 72/5371, filed 27 June 1973.

Srinath, E. G., Sastry, C. A., and Pillai, S. C., 1959, Rapid removal of phosphorus from sewage by activated sludge, *Water Waste Treat.* **11**:410–415.

Stephenson, T., 1987, *Acinetobacter:* Its role in biological phosphate removal, in: *Advances in Water Pollution Control. Biological Phosphate Removal from Wastewaters* (R. Ramadori, ed.), pp. 313–316, Pergamon Press, Oxford.

Stockdale, H., Ribbons, D. W., and Dawes, E. A., 1968, Occurrence of poly-β-hydroxybutyrate in the *Azotobacteriaceae, J. Bacteriol.* **95**:1798–1803.

Suresh, N., Warburg, R., Timmerman, M., Wells, J., Coccia, M., Roberts, M. F., and Halvorson, H. O., 1985, New strategies for the isolation of microorganisms responsible for phosphate accumulation, *Water Sci. Technol.* **17**:99–111.

Suzuki, H., Kauko, T., and Ikeda, Y., 1972, Properties of polyphosphate kinase prepared from *Mycobacterium smegmatis, Biochim. Biophys. Acta* **268**:381–390.

Swedes, J. S., Sedo, R. J., and Atkinson, D. E., 1975, Relation of growth and protein synthesis to the adenylate energy charge in an adenine-requiring mutant of *Escherichia coli, J. Biol. Chem.* **250**:6930–6938.

Szymona, M., and Ostrowski, W., 1964, Inorganic polyphosphate glucokinase of *Mycobacterium phlei, Biochim. Biophys. Acta* **85**:283–295.

Szymona, O., and Syzmona, M., 1979, Polyphosphate and ATP-glucose phosphotransferase activities in *Nocardia minima, Acta Microbiol. Pol.* **28**:153–160.

Szymona, O., Uryson, S. O., and Kulaev, I. S., 1967, Detection of polyphosphate glucokinase in various microorganisms, *Biokhimiya* **32**:408–415.

Tanaka, T., Kawakami, A., Yoneyama, Y., and Kobayashi, S., 1987, Study on the reduction of returned phosphorus from a sludge treatment process, in: *Advances in Water Pollution Control. Biological Phosphate Removal from Wastewaters* (R. Ramadori, ed.), pp. 201–211, Pergamon Press, Oxford.

Tempest, D. W., Neijssel, O. M., and Zevenboom, W., 1983, Properties and performance of micro-organisms in laboratory culture; their relevance to growth in natural ecosystems, in: *Microbes in Their Natural Environment* (J. H. Slater, R. Whittenbury, and J. W. T. Wimpenny, eds.), pp. 119–149, Cambridge University Press, Cambridge.

Terry, K. R., and Hooper, A. B., 1970, Polyphosphate and orthophosphate content of *Nitrosomonas europaea* as a function of growth, *J. Bacteriol.* **103:**199–206.

Toerien, D. F., and Gerber, A., 1986, Bacterial population structure of activated sludge systems, Letter to the editor, *Water SA* **12:**239.

Toerien, D. F., Gerber, A., and Brodisch, K. E. U., 1986, Biological phosphate removal in activated sludge systems, in: *Perspectives in Microbial Ecology* (F. Megušar and M. Ganter, eds.), pp. 66–73, Slovene Society for Microbiology, Ljubljana, Yugoslavia.

Tomlinson, E. J., and Chambers, B., 1979, Methods for prevention of bulking in activated sludge, *Water Pollut. Control* **78:**524–538.

T'Seyen, J., Malnou, D., Block, J. C., and Faup, G., 1985, Polyphosphate kinase activity during phosphate uptake by bacteria, *Water Sci. Technol.* **17:**43–56.

Tsuchiya, T., and Rosen, B. P., 1976, Characterization of an active transport system for calcium in inverted membrane vesicles of *Escherichia coli*, *J. Biol. Chem.* **250:**7687–7692.

Vacker, D., Connell, C. H., and Wells, W. N., 1967, Phosphate removal through municipal wastewater treatment at San Antonio, Texas, *J. Water Pollut. Control Fed.* **39:**750–771.

Vaillancourt, S., Beauchemin-Newhouse, N., and Cedergren, R. J., 1978, Polyphosphate-deficient mutants of *Anacystis nidulans*, *Can. J. Microbiol.* **24:**112–116.

Van Gills, H. W., 1964, Bacteriology of Activated Sludge, Research Institute of Public Health Engineering, T.N.O., Delft, The Netherlands, Publ. 32.

Van Groenestijn, J. W., and Deinema, M. H., 1985, Effects of cultural conditions on phosphate accumulation and release by *Acinetobacter* strain 210A, in: *Proceedings of the International Conference on Management Strategies for Phosphorus in the Environment*, pp. 405–410, Seeper Ltd., London.

Venter, S. L. V., Halliday, J., and Pitman, A. R., 1978, Optimization of the Johannesburg Olifantsvlei extended aeration plant for phosphorus removal, *Prog. Water Technol.* **10:**279–292.

Visser, A. S., and Postma, P. W., 1973, Permeability of *Azotobacter vinelandii* to cations and anions, *Biochim. Biophys. Acta* **298:**333–340.

Voelz, H., Voelz, U., and Ortigoza, R. O., 1966, The polyphosphate overplus phenomenon in *Myxococcus xanthus* and its influence on the architecture of the cell, *Arch. Mikrobiol.* **53:**371–388.

Wanner, J., Ottova, V., and Grau, P., 1987, Effect of an anaerobic zone on settleability of activated sludge, in: *Advances in Water Pollution Control. Biological Phosphate Removal from Wastewaters* (R. Ramadori, ed.), pp. 155–164, Pergamon Press, Oxford.

Watanabe, A., Miya, A., and Matsuo, Y., 1984, Laboratory scale study on biological phosphate removal using synthetic waste water, *Newsl. IAWPRC Study Group on Phosphate Removal in Biological Sewage Treatment Processes* **2:**40–43.

Water Research Commission, 1984, *Theory, Design and Operation of Nutrient Removal Activated Sludge Processes*, Water Research Commission, Pretoria, South Africa.

Weitzman, P. D. J., 1972, Regulation of α-ketoglutarate dehydrogenase activity in *Acinetobacter*, *FEBS Lett.* **22:**323–326.

Weitzman, P. D. J., and Dunmore, P., 1969, Citrate synthases: Allosteric regulation and molecular size, *Biochim. Biophys. Acta* **171:**198–200.

Weitzman, P. D. J., and Jones, D., 1968, Regulation of citrate synthase and microbial taxonomy, *Nature* (London) **219:**270–272.

Wells, W. N., 1969, Differences in phosphate uptake rates exhibited by activated sludges, *J. Water Pollut. Control Fed.* **41:**765–771.

Wentzel, M. C., Dold, P. L., Ekama, G. A., and Marais, G. v. R., 1985, Kinetics of biological phosphorus release, *Water Sci. Technol.* **17:**57–71.

D. F. Toerien *et al.*

Wentzel, M. C., Lötter, L. H., Loewenthal, R. E., and Marais, G. v. R., 1986, Metabolic behaviour of *Acinetobacter* spp in enhanced biological phosphorus removal: A biochemical model, *Water SA* **12**:209–224.

Wentzel, M. C., Dold, P. L., Loewenthal, R. E., Ekama, G. A., and Marais, G. v. R., 1987, Experiments towards establishing the kinetics of biological excess phosphorus removal, in: *Advances in Water Pollution Control. Biological Phosphate Removal from Wastewaters* (R. Ramadori, ed.), pp. 79–97, Pergamon Press, Oxford.

West, I. C., and Mitchell, P., 1974, Proton/sodium ion antiport in *Escherichia coli, Biochem. J.* **144**:87–90.

Wetzel, R. C., 1983, *Limnology,* 2nd ed., W. B. Saunders, Philadelphia.

White, A., Handler, P., Smith, E. L., Hill, R. L., and Lehman, I. R., 1978, *Principles of Biochemistry,* McGraw-Hill, Kogakusha Ltd., Tokyo.

Wilson, D. B., 1978, Cellular transport mechanisms, *Annu. Rev. Biochem.* **47**:933–965.

Winder, F. G., and Denneny, J. M., 1957, The metabolism of inorganic polyphosphate in *Mycobacteria, J. Gen. Microbiol.* **17**:573–585.

Wong, P. P., and Evans, H. J., 1971, Poly-β-hydroxybutyrate utilization by soybean (*Glycine* max. mer) nodules and assessment of its role in maintenance of nitrogenase activity, *Plant Physiol.* **47**:750–755.

Wood, H. G., and Goss, N. H., 1985, Phosphorylation enzymes of the propionic acid bacteria and the roles of ATP, inorganic pyrophosphate and polyphosphates, *Proc. Natl. Acad. Sci. USA* **82**:312–315.

Wuhrmann, K., 1957, Die dritte Reinigungsstufe: Wege und bisherige Erfolge in der Eliminierung eutrophierender Stoffe, *Schweiz Z. Hydrol.* **19**:409–427.

Wuhrmann, K., 1960, Effects of oxygen tension on biochemical reactions in sewage treatment plants, in: *Advances in Biological Waste Treatment. Proceedings of the 3rd Conference on Biological Waste Treatment* (W. W. Eckenfelder and J. McCabe, eds.), Pergamon Press, New York, pp. 27–38.

Yagil, E., 1975, Derepression of polyphosphatase in *Escherichia coli* by starvation for inorganic phosphate, *FEBS Lett.* **55**:124–127.

Zaitseva, G. N., and Belozerskii, A. N., 1960, Formation and utilisation of polyphosphates catalyzed by an enzyme isolated from *Azotobacter vinelandii, Dokl. Akad. Nauk. SSR* **132**:950–953.

6

The Ecology of Microbial Corrosion

TIM FORD and RALPH MITCHELL

1. Introduction

Corrosion reactions may be induced or enhanced by microbial activity. The classic corrosion reaction is electrochemical, resulting in the dissolution of metal from anodic sites with subsequent electron acceptance at cathodic sites. Consumption of electrons varies, depending on the redox potential of the surface. In an aerobic environment, oxygen is the electron acceptor, forming metal oxides and hydroxides. At low redox potentials, protons become the electron acceptors, yielding hydrogen gas and other highly reduced products. The process of corrosion is accelerated by removal of the end products of the chemical reactions.

The involvement of microorganisms in this process has been difficult to evaluate. Analytic methods are designed to measure generalized corrosion, based on the Wagner-Traud theory for metal dissolution (Wagner and Traud, 1938). According to this hypothesis, corrosion is described as formation of random and dynamic cathodic and anodic sites on the metal surface (Little *et al.*, 1984). The heterogeneity of microbial communities on metal surfaces results in the formation of anodic and cathodic areas that are distinct in space and time.

Investigation of microbially induced corrosion processes involves the study of adhesion of microorganisms to metal substrata, metabolism within the surface colony, and bacteria–metal interactions. Because corrosion rates are invariably accelerated beneath natural surface films comprising complex microbial communities, study of microbial interactions within these communities has become central to this research. Corrosion by sulfate-reducing bacteria has been intensively studied during the past 40 years. The literature has been extensively reviewed (Crombie *et al.*, 1980; Iverson, 1981; Tiller, 1982; Gragnolino and Tuovinen, 1984; Hamilton, 1985; Pankhania, 1988). In this

TIM FORD and RALPH MITCHELL • Laboratory of Microbial Ecology, Division of Applied Sciences, Harvard University, Cambridge, Massachusetts 02138.

review, we intend to focus on aspects of microbial corrosion that have received little attention despite an increasing awareness of their importance. In the section describing the sulfate-reducing bacteria, we will summarize the current knowledge of the ecology of these microorganisms in corroding systems. Oxidative processes, biochemical reactions related to microbial surface film formation, and interactions within communities will be discussed in greater detail.

Traditionally, the microorganisms involved in corrosion have been categorized into four distinct groups: (1) The sulfate reducers, (2) "slime-forming" bacteria, (3) iron-oxidizing bacteria, and (4) a miscellaneous group containing sulfur-oxidizing bacteria, fungi, and algae (Donham et al., 1976; Tatnall, 1981a). While this approach may have some functional utility for the nonmicrobiologist, it provides limited insight to the microbial ecologist searching for the fundamental microbial processes responsible for the biodeterioration of metals.

Both sulfate-reducing and iron-oxidizing bacteria may produce copious quantities of exopolymer (slime), and the traditional "slime-forming" pseudomonads may oxidize iron. In this review, we will separate the ecological processes involved in microbial corrosion into sections for ease of discussion. However, it is important to be aware that all processes described may be occurring simultaneously within the same surface community. The following topics will be discussed: the role of the surface microbiota, acid production, metal deposition and tubercle formation, hydrogen-consuming reactions, hydrogen-producing reactions, and thermophilic corrosion. In the concluding section, we will discuss current knowledge of microbial consortia and their interactions relating to corrosion.

2. The Role of the Surface Microbiota

Information about the complex ecology of microbial communities adhering to and growing on solid surfaces is essential if we are to understand the function of microorganisms in corrosion processes. In natural ecosystems, the surface microbiota is controlled to a large degree by the nature of the substratum and the available nutrients in the water column. The microbial community on metal surfaces is dependent on the chemistry of the solid and the liquid phases with which it is in contact. Hence, extremely specific microbial communities can be associated with individual metals, e.g., copper alloys and titanium (Marszalek et al., 1979; Berk et al., 1981; Walch, 1986; Ford et al., 1989).

No metal substratum appears to be totally immune from microbial colonization, although the rate of colonization may be strongly affected by a sloughing oxide layer (aluminum bronze), toxic ions (copper nickel), or inclusion of biocides. The nature of the biofilm is such that the glycocalyx, the extracellular microbial material generally consisting of extracellular polysaccharides (Costerton et al., 1981), can coat or bind toxic ions and biocides, rendering them ineffective. Briefly, the sequence of events in the adhesion of bacteria to surfaces is as follows (Maki and Mitchell, 1988):

1. Adsorption of macromolecules to surfaces (concentration of nutrients, altered molecular activity)
2. Transport of bacteria to surfaces (fluid dynamics, sedimentation)
3. Physiochemical and biological mechanisms of bacteria in close proximity to the surface (chemotaxis of motile bacteria, Brownian motion)
4. Reversible adhesion of bacteria (long-range forces)
5. Irreversible adhesion of bacteria (polymeric bridging, short-range forces)

Adhesion of microorganisms and subsequent development of surface colonies has been the subject of extensive research [see reviews by Costerton et al. (1978), Absolom et al. (1983), Baier (1984), Baier et al. (1984), Marshall (1984), Savage and Fletcher (1985) and Hamilton (1987)].

There are a number of ways in which surface films can be responsible for accelerated corrosion: (1) heterogeneity (formation of oxygen concentration cells), (2) exopolymer–metal interactions (formation of ion concentration cells), (3) influence of iron- and manganese-oxidizing bacteria (direct oxidation of metal, formation of tubercles), (d) organic and inorganic acid production by microorganisms, and (e) creation of anaerobic niches.

Processes described in this review are as applicable to a municipal water distribution system as to a marine surface coated with invertebrates. Water piping systems are often covered with microorganisms and their exopolymers and inorganic deposits. There are a number of case histories (Tatnall, 1981b) in which water pipes are layered with the concretions, tubercles, and corrosion products of iron-oxidizing bacteria, with anaerobic communities present beneath the deposits. These concretions frequently block pipelines in a matter of weeks (Schmitt, 1986). An example from marine habitats is the association between corrosion and the edges of barnacle attachment sites, probably resulting from oxygen concentration cells formed by small areas not covered by cement (Griffin et al., 1988).

Surface deposits on metal pipes can be directly related to corrosion by (1) restriction of flow through pipelines, leading to development of anaerobic conditions, and (2) entrapment of debris, compounding the fouling problem and increasing the risk of oxygen concentration cell formation. Bacterial colonies only 8–10 cells thick can rapidly create conditions suitable for anaerobic growth (Costerton et al., 1987). Furthermore, both aerobic and anaerobic electrochemical reactions occur simultaneously within the surface film.

2.1. Aerobic Processes

In aerobic corrosion processes, oxygen acts as the electron acceptor at the cathodic sites to form hydroxides. Information about the influence of microorganisms on mechanisms of aerobic corrosion is limited because of the assumption that abiotic oxidative processes are much more rapid. However, the importance of microorganisms in aerobic degradation of metals may be significantly underrated. The most apparent influence of an aerobic community on a metal surface is the creation of differential aeration cells

(Iverson, 1974) (Fig. 1). The natural heterogeneity of a microbial film on a metal surface results in oxygen depletion under thick films relative to sparsely colonized areas. In this reaction, an area of oxygen depletion becomes anodic to more oxygenated areas. As a result, a susceptible metal goes into solution at the anode, while electrons combine with water and oxygen at the cathode.

Other aerobic microbial processes that may accelerate corrosion include formation of ion concentration cells, bacterial polymer–metal interactions, activities of metal-transforming and acid-producing bacteria, and thermophilic reactions. The latter are of increasing importance as the safety of high-temperature water systems (e.g., in electric power generation and industrial heat exchangers) becomes an increasing problem. Differentiation between microbial and chemical corrosion is difficult, particularly in aerobic environments, because the two processes enhance each other. However, application of newly developed nondestructive analytic methods, including measurement of galvanic currents (Little *et al.*, 1986a) and electrochemical impedance techniques (Kasahara and Kajiyama, 1986; Moosavi *et al.*, 1986; Dowling *et al.*, 1988), should permit separation of abiotic and microbial oxidative corrosion processes.

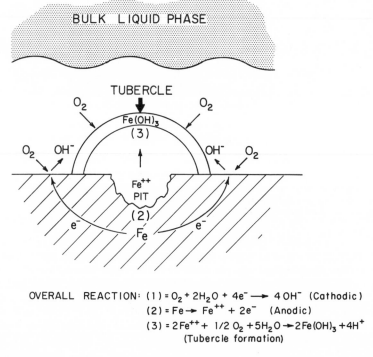

OVERALL REACTION: $(1) = O_2 + 2H_2O + 4e^- \rightarrow 4 OH^-$ (Cathodic)
$(2) = Fe \rightarrow Fe^{++} + 2e^-$ (Anodic)
$(3) = 2Fe^{++} + 1/2\ O_2 + 5H_2O \rightarrow 2Fe(OH)_3 + 4H^+$
(Tubercle formation)

Figure 1. Differential aeration cell formed by oxygen depletion under a microbial surface film. [After Iverson (1974).]

There are specific oxidative processes that are mediated by microorganisms, e.g., iron and manganese deposition. Although there is little evidence for enzymatic oxidation of Fe^{2+} to Fe^{3+} (Ghiorse, 1984), deposition of Fe^{3+} bound to extracellular polymers of Fe-depositing bacteria may accelerate the autooxidation of Fe^{2+}, particularly in slightly acidic conditions. There is, however, considerable evidence for direct enzymatic oxidation of Mn^{2+} [see Ghiorse (1984) for review]. These activities will be discussed in detail in Section 4.

2.2. Anaerobic Processes

Corrosion of metals in anaerobic environments has been linked to the activity of microorganisms since the work of Gaines (1910). Emphasis has consistently been placed on the sulfate-reducing bacteria (Hamilton, 1985). The mechanism of anaerobic corrosion in the presence of these bacteria is most clearly illustrated in Fig 2. The classic mechanism describing the anaerobic corrosion of iron can be summarized in the following equations:

$$4Fe^0 \rightarrow 4Fe^{2+} + 8e^- \quad \text{anodic process}$$
$$8H_2O \rightarrow 8H^+ + 8OH^- \quad \text{dissociation of } H_2O$$
$$8H^+ + 8e^- \rightarrow 4H_2 \quad \text{cathodic process}$$
$$4H_2 + SO_4^{2-} \rightarrow S^{2-} + 4H_2O \quad \text{depolarization}$$

The corrosion products from these reactions are ferrous sulfide (FeS) and ferrous hydroxide [Fe $(OH)_2$]. The overall reaction is driven by cathodic depolarization (von Wolzogen Kuhr and van der Vlugt, 1934). The rate-limiting reaction is the removal of hydrogen from cathodic sites. Because sulfate-reducing bacteria readily oxidize hydrogen, the consumption of hydrogen by the action of hydrogenase has become established as the major process in cathodic depolarization. This explanation, however, is a simplification of the corrosion reactions. King and Miller (1971) proposed that ferrous

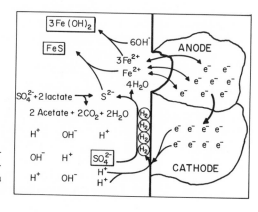

Figure 2. Proposed mechanism of anaerobic corrosion of iron in the presence of sulfate-reducing bacteria. [From Pankhania (1988).]

sulfide first acts to adsorb molecular hydrogen. The ferrous sulfide cathode is regenerated by the action of bacterial hydrogenase. Costello (1974) proposed that hydrogen sulfide, rather than the hydrogen ion, was in fact the cathodic reactant ($2H_2S + 2e^- \rightarrow 2HS^- + H_2$).

There has been considerable controversy surrounding the mechanism of anaerobic microbial corrosion. Studies by Iverson and co-workers (Iverson, 1981, 1984; Iverson and Olson, 1983, 1984; Iverson et al., 1986) suggest an even more complex mechanism involving both sulfide and phosphides. Although there is some recent support for the involvement of phosphides, the mechanisms are still far from clear (Weimer et al., 1988). Starkey (1986) suggested that several processes related to hydrogenase activity are involved in anaerobic corrosion and need to be considered. These processes concern the effects of ferrous sulfide, sulfur, ferrous hydrate, phosphides, and other products of the corrosion process, and may be further combined with formation of differential aeration and ion concentration cells.

Surprisingly, recent research strongly favors the original hypothesis of von Wolzogen Kuhr and van der Vlugt (1934). A number of studies have conclusively shown that sulfate reduction can occur with cathodically formed hydrogen from a steel surface (Hardy, 1983; Cord-Ruwisch and Widdel, 1986; Pankhania et al., 1986a). Pankhania et al. (1986a) were able to grow Desulfovibrio vulgaris in acetate medium in which the only energy source was hydrogen from a mild steel electrode.

Although anaerobic corrosion has traditionally been associated with sulfate-reducing bacteria, it should be noted that many other anaerobic reactions consume hydrogen and may act to cathodically depolarize a metal (Tomei and Mitchell, 1986; Pankhania, 1988). These include reduction of thiosulfate, sulfite, sulfur, fumerate and nitrate, as well as methanogenesis and acetogenesis. These processes are discussed in more detail in Section 5.

2.3. Exopolymer–Metal Interactions

In recent years, the ecological function of macromolecular products in surface biofilms has attracted increasing interest. Extracellular polymers produced by bacteria are usually acidic and contain functional groups that readily bind metal ions [e.g., uronic acid, cis-hydroxyl, and phosphoric acid groups (Martin, 1971)]. There is an extensive literature describing the complexation of bacterial exopolymers with metals (Friedman and Dugan, 1968; Martin et al., 1972; Bitton and Freihofer, 1978; Brown and Lester, 1979, 1982a,b; Stoveland and Lester, 1980; Lester et al., 1984; Rudd et al., 1984; Mittelman and Geesey, 1985). Not only are these polymers central to the structural integrity of the surface microbiota, but they may also be directly involved in metal dissolution from the corroding metal (Costerton et al., 1987).

Recently, the question of involvement of metal-binding exopolymers in corrosion has been addressed (Geesey et al., 1986, 1987, 1988; Ford et al., 1987a, 1988; Black et al., 1988; Jolley et al., 1988). The principle behind these investigations is derived from the probability that binding of metal ions within a biofilm–polysaccharide matrix will influence the stability of metals in the surface lattice. Initial investigations involved measurement of binding of metals to bacterial exopolymers (Geesey *et al.*, 1986; Ford *et al.*, 1987a, 1988; Black *et al.*, 1988). These studies have established that there are considerable differences in binding capacity between individual metal ions and specific exopolymers. Furthermore, the extent of the binding is dependent on environmental factors, such as pH, redox, and competing ions.

The role of metal binding is difficult to directly relate to corrosion. Ford *et al.* (1987a, 1988) speculate that differential binding abilities help to establish ion concentration cells. Concentration of a metal from the aqueous phase within a microbial film could affect the surface metal in two ways: (1) if the concentrated metal is more noble than the ions in the surface lattice, then dissolution of ions from the surface would be expected, and (2) if the surface ions are more noble than the ions concentrated by the exopolymer, then that metal would oxidize in preference to the surface ions, thus preventing corrosion (Ford *et al.*, 1988).

Use of surface analytical techniques may increase our understanding of metal-binding phenomena. Geesey and co-workers (1987) have used Fourier transform infrared spectroscopy to characterize binding of an acidic polysaccharide to thin copper films. Their measurements suggest a cupric ion interaction with carboxyl groups on the polymers tested. These interactions promote ionization of metallic copper (Geesey *et al.*, 1988). This group has also used Auger electron spectroscopy and X-ray photoelectron spectroscopy to study corrosion of copper in the presence of exopolymers (Jolley *et al.*, 1988). This technique showed removal of copper from the surface and incorporation into the polymer matrix. This approach to biocorrosion research is at an early stage. It is likely to provide important information about the biochemical function of bacterial exopolymers in metal dissolution.

Metal binding by microbial exopolymers is dependent on the chemical structure of the exopolymer. Chemical characterization of bacterial exopolymers has focused on bacteria that are readily cultured in the laboratory, especially the *Enterobacteriaceae* (Sutherland, 1985). In general, the most common sugar residues in these polymers are the hexoses. However, exopolymers also contain both amino sugars (e.g., D-glucosamine) and sugar acids (e.g., D-glucuronic acid). The most widespread non-sugar organic substituents are the O-acetyl and O-methyl residues and the pyruvate ketal groups. Some exopolymers also resemble the teichoic acids, generally associated with cell walls of Gram-positive bacteria, in that phosphorylated monosaccharides are present.

Polysaccharides produced by a number of bacteria isolated from different habitats have been partially characterized; these include activated sludge bacteria (Farrah and Unz, 1976; Tago and Aida, 1977), marine bacteria (Boyle and Reade, 1983; Wrangstadh *et al.*, 1986), soil bacteria (Beyer *et al.*, 1983; Congregado *et al.*, 1985), and a freshwater sediment bacterium (Platt *et al.*, 1985). The picture that has emerged is that bacterial exopolymers are chemically diverse. Culture conditions and growth phase directly affect the quantity (Sutherland, 1972; Williams and Wimpenny, 1977) and composition (Uhlinger and White, 1983; Christensen *et al.*, 1986) of exopolymer produced. Sutherland (1972) points out that when more than one exopolysaccharide is formed, as in the case of a consortium of microorganisms, variations in proportions of the different polymers may occur as a result of changing growth conditions. Considerably more information is necessary before the chemistry of metal binding can be accurately modeled.

One recent approach that may be useful is specific enzyme hydrolysis of carbohydrate or protein substrates (Sutherland, 1984). Flatau *et al.* (1985) used enzyme hydrolysis to determine cadmium-binding sites on the cell wall of a marine pseudomonad. Cadmium was found to bind mainly to ionic material such as polygalacturonic acids and, to a lesser extent, anionic material such as cellulose and dextrans.

The complexity of metal-binding phenomena is further complicated by the other components of the microbial exopolymer, especially the proteins. Preliminary findings suggest that the protein component of an exopolymer preparation can contribute significantly to metal binding (Mittelman and Geesey, 1985). They calculated a maximum binding ability (MBA) for the protein component of a freshwater sediment bacterium as 193 nmol of Cu/mg of protein. Conversely, the MBA for the carbohydrate component was 253 nmol of Cu/mg of carbohydrate.

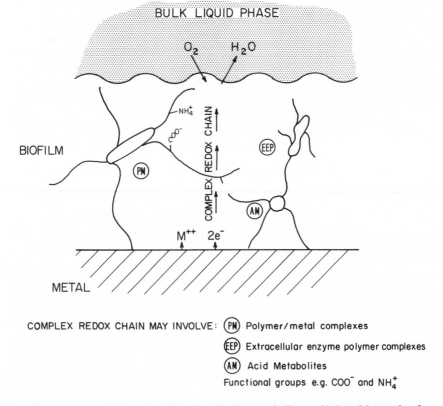

COMPLEX REDOX CHAIN MAY INVOLVE: (PM) Polymer/metal complexes

(EEP) Extracellular enzyme polymer complexes

(AM) Acid Metabolites

Functional groups e.g. COO^- and NH_4^+

Figure 3. Interactions within a microbial surface film that may facilitate oxidation of the metal surface.

Much of the preceding discussion may also apply to the microbial cell wall components, particularly the lipopolysaccharides (LPS). Unpublished research on LPS–metal interactions suggests an important role for LPS in initial colonization processes and therefore at least an indirect role in corrosion (I. Beech and C. Gaylarde, personal communication).

In addition, involvement of extracellular enzymes in metal-binding processes, and indeed in electron transfer within the microbial film, has yet to be investigated. The subject of bioelectrocatalysis, the acceleration of electrochemical reactions in the presence of enzymes, has been discussed by Tarasevich (1979). Reduction of molecular oxygen could be activated by direct electron exchange between an electrode and an enzyme. Extracellular enzymes in a microbial surface colony have the potential to affect corrosion of a metal by catalysis of electron transport reactions. Potential relationships between a microbial film and oxidation at a metal surface are illustrated diagrammatically in Fig 3. Further research, perhaps using immobilized enzymes with an artificial polymeric film, may enable us to quantify the importance of this process further.

3. Acid Production

Inorganic and organic acids produced by microorganisms tend to be the most directly corrosive metabolites (Iverson, 1974). The sulfur oxidizers (e.g., *Thiobacillus thiooxidans*) have been directly implicated in corrosion through acid production. Organic acids produced by *Lactobacillus delbrueckii* are involved in corrosion of steel piping (Allen *et al.*, 1948). Acetic and butyric acids produced by cellulose-degrading bacteria have been reported to etch metals buried in soil (Coles and Davies, 1956). More recently, Little *et al.* (1986b) have implicated production of isobutyric and isovaleric acids by an obligately thermophilic bacterium in the failure of welded nickel used in a hot water system. Pope *et al.* (1988) have reproduced in the laboratory a unique form of tunneling corrosion, seen on carbon steel in the field, by exposing the metal to organic acids typically produced by *Clostridium* species.

There is very little direct evidence available for corrosion by microbially produced organic acids. However the argument is convincing, considering the corrosive nature of these acids. Furthermore, Little *et al.* (1986a,b) have suggested that concentrations of organic acids, measured in culture media, may underestimate by orders of magnitude the quantity accumulated beneath a microbial surface film.

3.1. Sulfur Oxidation

In contrast to organic acids, there is considerable evidence for corrosion by inorganic acid production. Municipal wastewater systems suffer from corrosion of concrete infrastructures. Wet walls, grit chambers, sewer lines, and aeration basins are particularly susceptible. Since the work of Parker (1945a,b), the sulfur-oxidizing thiobacilli were believed to be the causative agents of concrete deterioration. The basic scenario was described by Rigdon and Beardsley (1958), Forrester (1959), and Hueck-van der Plas (1968).

The process of microbial corrosion of concrete illustrates how interactions between microorganisms and the environment can create an extreme corrosion problem. The generally accepted process is as follows: carbon dioxide and hydrogen sulfide (products of sewage decomposition) react with the damp concrete surfaces to form carbonates and calcium sulfate. As a result, the surface pH of the concrete drops from approximately 12.5 to about 8.5. Hydrogen sulfide is partially oxidized in air to form thiosulfuric and polythionic acids. These in turn lower the pH further, to approximately 7.5. Below pH 10, a number of different bacteria oxidize thiosulfates and polythionates to sulfate. Below pH 9, *Thiobacillus thioparus* becomes active and oxidizes thiosulfate to elemental sulfur. The pH continues to decline, and at about pH 5 *Thiobacillus thiooxidans* oxidizes elemental sulfur to sulfuric acid. The final pH can reach 0.6. The corrosion of concrete is therefore a result of direct attack by sulfuric acid. Although in this description only oxidizing bacteria are discussed, communities of anaerobic bacteria contribute to hydrogen sulfide production and are therefore indirectly involved in the degradation of concrete. Steel reinforcing bars in concrete structures, if exposed to sulfate-reducing bacteria, are also attacked (Moosavi *et al.*, 1986). A complete biochemical process has recently been described for concrete corrosion in the Hamburg sewer system (Sand and Bock, 1984). The activity of the thiobacilli is not solely confined to the degradation of concrete. *Thiobacillus ferrooxidans* oxidizes ferrous iron to ferric iron and has been implicated in corrosion of iron pipes and of equipment used for microbial mining of metal ores (Miller, 1970).

The presence of acid-producing thiobacilli can now be detected in environmental samples by analysis of signature fatty acids (Kerger *et al.*, 1987). These biomarker polar lipid fatty acids (PLFA) have been used to study diverse microbial consortia (White, 1983, 1984). A number of organisms have sufficiently unique PLFA patterns to be distinguishable within a consortium (Edlund *et al.*, 1985; Dowling *et al.*, 1986; Nichols *et al.*, 1986). The acid-producing thiobacilli have distinct PLFA patterns, permitting species identification. This chemical approach to characterization of consortia in a corroding system provides a powerful tool for the study of corrosion caused by the thiobacilli.

3.2. The Role of Fungi

Although fungi are frequently involved in biodeterioration of materials such as polymers, paints, organic coatings, food, etc. (Barry and Houghton, 1986), they are infrequently implicated in corrosion processes. Most work has concentrated on the corrosive effects of a contaminant of aviation fuels, *Cladosporium resinae* (Hendey, 1964; Hansen *et al.*, 1981; Salvarezza *et al.*, 1983), since reclassified as *Hormoconis resinae* (Videla *et al.*, 1988). Hendey (1964) reported that metabolites of *H. resinae* were capable of attacking aluminum. Acidic metabolites produced by this organism apparently facilitate the breakdown of the passive oxide film on aluminum alloys by chloride ions (Salvarezza et al., 1983).

The ecological niche occupied by fungi on corroding surfaces is unique to the survival and proliferation of the fungi. Damp storage tank walls and interfaces between fuel and water are particularly favorable habitats for growth of fungi, which are usually outcompeted by bacteria in aqueous environments. Videla et al. (1988) list the various mechanisms proposed to interpret fungal corrosion of aluminum in fuel–water systems: (1) production of organic acids from degradation of hydrocarbon chains; (2) establishment of concentration cells at attachment sites; (3) formation of tubercles by the symbiotic association of bacteria and fungi; and (4) removal of metallic atoms from the metal surface by extracellular enzyme activity. Organic acid production by H. resinae is a probable mechanism. The fungus has been found to produce dodecanoic, acetic, glycolic, and glyoxylic acids (Siporin and Cooney, 1975). Hansen et al. (1981) implicated localized acidification, enhanced many times at the metal–mycelium interface, in major pitting corrosion of aluminum.

H. resinae is not the only fungus isolated from fuel storage and distribution systems. However, it appears to be the most directly corrosive. Videla et al. (1988) report preliminary data to suggest that Penicillium, Trichosporon, Fusarium, and Aspergillus isolates from fuel are less corrosive than H. resinae, as judged by their capacity to metabolize the aliphatic fraction of the fuel.

Although fungal contaminants of fuel tanks have received the greatest attention, there has been some research on fungus-induced corrosion in other systems. Brown and Pabst (1975) studied degradative effects of microorganisms on the metals being tested as structural elements for the U.S. Orbiting Space Station. They found that fungi, including Aspergillus, Trichoderma, and Rhizopus, caused considerably more corrosion than bacteria on aluminum alloys. Although they did not evaluate mechanisms of corrosion, higher oxygen concentrations increased the corrosion rate. They suggested that differential aeration cell formation may be involved.

A severe case of fungal corrosion in the hold of a ship was reported by Stranger-Johannessen (1986). Considerable blistering and debonding of the protective paint film occurred, resulting in severe pitting of the steel beneath the blisters. Scopulariopsis brevicaulis, Penicillium sp., and two unidentified fungi were isolated from a sample of corroded steel plate. Subsequent laboratory experiments with these isolates showed that S. brevicaulis and one of the unidentified fungi produced the same blistering effect and severe pitting corrosion observed in the ship's hold. These two organisms, considered to have been introduced into the hold by a cargo of soybeans, grew on the protective paint, causing debonding and flaking.

The mechanism of the corrosion is unclear. The unidentified fungus produced acid, lowering the pH of growth media to 4.5. S. brevicaulis, on the other hand, did not produce acidity, suggesting that this fungus acts by a different mechanism. Considerable aeration differences occur on blistered, painted surfaces, which may account for some metal deterioration. Stranger-Johannessen (1986) found an even more aggressive effect with Cladosporium herbarum, isolated on another occasion from black spots on the paint in a ship's hold. She suggests that this form of metal deterioration may be common.

4. Influence of Iron and Manganese Deposition

Iron and manganese-depositing or -oxidizing bacteria deserve to be treated as a special case because of their direct involvement in metal transformations and subsequent key role in the ecology of microbial corrosion. Excellent reviews have recently been written on the biology of these organisms (Nealson, 1983a,b; Ghiorse, 1984; Jones, 1986). They discuss in detail the physiological basis for iron and manganese transformations. We will restrict our discussion to the importance of these organisms in corrosion reactions.

Rapid degradation occurs when stainless steel piping and vessels are hydraulically tested and allowed to stand. Kobrin (1976) identified *Gallionella* as the predominant organism present in microcolonies associated with the corroding pits. These colonies were rich in iron, manganese, and chlorides, and the type of corrosion was typical of ferric chloride attack. Two questions (raised at the T3J committee meeting on microbially induced corrosion, National Association of Corrosion Engineers, St. Louis, Mo., 1988) result from this analysis: why are manganese and chlorides concentrated in the colony when *Gallionella* does not react with either ion?, and what is the association between these bacteria and the welded seams with which they are generally associated? These questions remain unanswered. To the microbiologist, the potential for contamination of the system by iron- and manganese-depositing bacteria is clear, considering the source of test water. These organisms are widely distributed and readily isolated from environments where iron hydroxide and ferromanganese oxide deposits are found (Ghiorse, 1984).

Jones (1986) discussed the role of iron deposition in the ecology of the iron-oxidizing bacteria. There is evidence that mechanisms exist for iron deposition both intracellularly and extracellularly (Caldwell and Caldwell, 1980; Cowen and Silver, 1984). However, the evolutionary advantage to the organism is not clear (Jones, 1986). Extracellularly, the iron may have a structural role in colony morphology or in the coprecipitation of toxic ions. It may also protect the cell from oxygen. Intracellularly, iron deposition may prevent accumulation of toxic levels of iron. It has been suggested that in some bacteria, a high intracellular concentration of magnetite ensures migration of the bacterium toward the sediments and away from high oxygen concentrations (Frankel and Blakemore, 1984).

4.1. Tubercle Formation

The typical development of iron tubercles, commonly associated with failed water pipes, presents a unique ecological habitat (Lutey, 1980; Walch, 1986). Tubercles are initially formed by microbiological deposition of iron and manganese oxides. This deposition yields a large, localized mass of cells and minerals. The bacteria produce exopolymers that bind the mineral deposits together and limit diffusion of oxygen and nutrients in and out of the developing tubercle. Any oxygen that does penetrate is consumed, either abiotically or biotically, by oxidation of ferrous ions. Metabolic activity of microorganisms reduces the pH inside the tubercle by production of CO_2 and organic acids. Steep chemical gradients are formed. The outer layers of the tubercle are aerobic with a near neutral pH, while inner regions are highly reduced and acidic. A diverse microbial community can therefore coexist within the tubercle ecosystem.

Analysis of these deposits suggests the presence of sulfate- and nitrate-reducing bacteria, nitrite oxidizers, and various unidentified heterotrophs (Tuovinen *et al.*, 1980). They also include large populations of metal depositors, many of which are extremely difficult to isolate. Isolates include *Hyphomicrobium*, a manganese-depositing bacterium (Tyler and Marshall, 1967a,b), and *Gallionella*, an iron-depositing bacterium (Kobrin, 1976; Ridgway and Olson, 1981; Borenstein and Lindsay, 1988). Less frequently, *Sphaerotilis*, *Crenothrix*, *Leptothrix*, and *Siderocapsa* have been isolated from water distribution systems (Iverson, 1974; Kobrin, 1976; Tatnall, 1981b). However, there are many more genera of metal-transforming bacteria that may be present within the tubercle community. The wide range of bacteria known to deposit iron or manganous oxides is shown in Table I. We know very little about the involvement of most of these bacteria with corrosion.

Extensive elemental analysis of tubercles was undertaken by Tuovinen *et al.* (1980). Iron oxides were the main constituents. Many other elements, most notably calcium, magnesium, aluminum, and manganese, were present in varying quantities. A generalized picture of the distribution of iron oxides within the tubercle reflects the redox gradient. In the interior, green rust, a pyroaurite structure of varying Fe(II)/Fe(III) ratio and composition, predominates (Tuovinen and Mair, 1985). This is surrounded by magnetite (Fe_3O_4) with an outer crust of geothite (FeOH).

Activity of these metal-depositing bacteria provides a unique microhabitat for a corrosive consortium of bacteria. The problems associated with isolation of these microorganisms complicate a study of their ecology within a corroding system. Their direct effect on a metal is frequently masked by the action of the anaerobic community within the tubercle. However, there are cases in which deposits rich in iron and manganese bacteria show no evidence of sulfate-reducing bacteria. Kobrin (1986) describes one such case in which these deposits were observed along weld seams on stainless steel tanks and piping. Deep pits were found at the edges of the welds underneath the deposits. Corrosion was caused by concentration of iron and manganese chlorides, which typically severely pit stainless steel. Microbial corrosion in this case is indirect but extremely effective. More frequently, these deposits are associated with anaerobic consortia that can also induce severe pitting. Corrosion by these consortia is discussed in Section 5.

4.2. Colonization of Welds

One aspect of iron and manganese oxidation that has gained increasing interest is an association with the heat-affected areas of welds (Kobrin, 1986; Stoecker and Pope,

**Table I. Organisms Associated with Oxidation and
Deposition of Iron and/or Manganese[a]**

Iron-depositing bacteria	Manganese-depositing bacteria
Acholeplasma	*Aeromonas*
Actinomyces spp.	*Bacillus*
Aureobasidium	*Caulobacter*
Caulococcus	*Caulococcus*
Clonothrix	*Citrobacter*
Coniothyrium	*Clonothrix*
Crenothrix	*Cytophaga*
Cryptococcus	*Enterobacter*
Ferrobacillus	*Flavobacterium*
Gallionella	*Hyphomicrobium*
Hyphomicrobium	*Kuznetsovia*
Leptospirillum	*Leptothrix*
Leptothrix	*Metallogenium*
Lieskeella	*Micrococcus*
Metallogenium	*Nocardia*
Naumaniella	*Oceanospirillum*
Ochrobium	*Pedomicrobium*
Papilospora	*Pseudomonas*
Peloploca	*Siderocapsa* (*Arthrobacter*)
Pedomicrobium	Streptomyces
Planctomyces (*Blastocaulis*)	*Vibrio*
Seliberia	
Siderocapsa (*Arthrobacter*)	
Siderococcus	
Sphaerotilus	
Sulfolobus	
Toxothrix	
Thiobacillus	
Thiopedia	

[a]Compiled from tables by Nealson (1983a,b), Ghiorse (1984), and Jones (1986).

1986; Borenstein, 1988; Borenstein and Lindsay, 1988). There is no adequate explanation for the preferential colonization of welds. However, there are a number of factors that influence bacterial attachment to specific substrata (Little *et al.*, 1986c); the wettability of the surface (Baier, 1973; Dexter *et al.*, 1975) and the surface charge (Fletcher and Loeb, 1979) are important factors. Little *et al.* (1988) have analyzed welded copper–nickel piping systems and found metal segregation both in the weld and throughout the heat-affected zone. As a result, the surface characteristics are different from those of the other material. Irregularities at weld joints or leaching of stimulatory ions may also account for preferential colonization by microorganisms.

5. Corrosion by Hydrogen-Consuming Bacteria

5.1. Sulfate-Reducing Bacteria

Until recently, anaerobic corrosion was thought to be mediated by only two genera of hydrogenase-producing sulfate-reducing bacteria, *Desulfovibrio* and *Clostridium* (Donham *et al.*, 1976). A number of other hydrogenase-producing, hydrogen-utilizing sulfate reducers have recently been isolated (Widdel and Pfennig, 1981, 1982; Widdel *et al.*, 1983; Widdel, 1987). Typical species are shown in Table II. According to Hamilton (1985), growth of sulfate-reducing bacteria is possible on CO_2, on a range of organic compounds including benzoate but excluding sugars and hydrocarbons, and on fatty acids from acetate to stearate. An interesting phenomenon is the ability of a number of sulfate-reducing bacteria to grow on products of their own metabolism as environmental conditions change (Laanbroek *et al.*, 1982; Pankhania *et al.*, 1986b).

The relationship between sulfate-reducing bacteria and anaerobic corrosion has been extensively reviewed (Iverson, 1981; Hamilton, 1985; Pankhania, 1988). The large range of organic substrates and metabolic versatility of the organisms, including limited oxygen tolerance, explain the ubiquity of these bacteria. Hamilton (1985) points out that sulfate-reducing bacteria are likely to be found in all consortia in sulfate-containing environments. They often form syntrophic associations with other sulfur bacteria (Postgate, 1984). These associations have also been noted between sulfate-reducing bacteria and methanogens (Bryant *et al.*, 1977; Tomei *et al.*, 1985). They are based on nutritional interdependence and, frequently, thermodynamic considerations, resulting in increased activity causing acceleration of corrosion reactions. The subject of syntrophic associations, and particularly interspecific hydrogen transfer, is discussed in greater detail in Section 8.

Accelerated corrosion rates have also been observed when iron, corroded by sulfate-reducing bacteria, is exposed to intermittent aerobic–anaerobic conditions (Crombie *et al.*, 1980). Hardy and Bown (1984) observed a similar effect of aeration on mild

Table II. Some Typical Hydrogen-Utilizing Sulfur Bacteria[a]

Species	Carbon source	Electron acceptor
Desulfovibrio sapovorans	Fatty acids	Sulfate
Desulfobulbus propionicus	Propionate	Sulfate, nitrate
Desulfotomaculum acetoxidans	Acetate	Sulfate
Desulfuromonas acetoxidans	Acetate	Elemental sulfur
Desulfobacter postgatei	Acetate	Sulfate
Desulfosarcina variabilis	Organic and CO_2	Sulfate
Desulfonema magnum	Organic	Sulfate

[a] Adapted from Hamilton (1985).

steel. They proposed that the biogenic sulfide film differentiated into anodic and cathodic sites as a result of the aeration.

Recently, Pankhania (1988) has suggested that sulfide- oxidizing bacteria are present within the anaerobic microbial community. Aeration may cause these bacteria to become active, producing acidic metabolites that considerably enhance the corrosion reactions.

5.2. Activity of Methanogens

Since methanogenic bacteria also utilize molecular hydrogen, it has been suggested that they may promote corrosion in an anaerobic environment through cathodic depolarization (Tomei and Mitchell, 1986). This mechanism of corrosion has not yet been rigorously demonstrated. However, Daniels et al. (1987) have induced *Methanosarcina barkeri* to produce methane on metallic iron as the sole electron source. These authors conclude that chemical oxidation of the iron occurs and is accelerated by the removal of molecular hydrogen by the methanogen.

5.3. Iron-Reducing Bacteria

Anodic depolarization by iron-reducing bacteria has been the subject of extensive investigation by Westlake and his colleagues (Obuekwe *et al.*, 1981a–d, 1983; Obuekwe and Westlake, 1982; Westlake *et al.*, 1986; Semple and Westlake, 1987). From oil production systems, they isolated facultative anaerobes that were capable of using ferric ions as terminal electron acceptors. These isolates were initially characterized as *Pseudomonas* sp. (Obuekwe *et al.*, 1981a) but have since been reclassified as *Alteromonas putrefaciens* strains (Semple and Westlake, 1987). Their mode of action was to attach to mild steel, removing the passive Fe_2O_3 film and forming a dense fibrous mat (Westlake *et al.*, 1986). The surface of the metal was pitted beneath the mat as a result of anodic depolarization. The suggested mechanism is through reduction of the ferric ion, in this case an insoluble protective layer, to the soluble ferrous ion (Obuekwe *et al.*, 1981a). These strains were isolated from large populations of both aerobic and anaerobic bacteria present in oil production systems. Obuekwe *et al.* (1981a) suggest that the combination of anodic and cathodic depolarization, the latter by sulfate-reducing bacteria, accounts for the severe corrosion of these pipelines.

6. Corrosion by Hydrogen-Producing Bacteria

Damage to metals results from entry of atomic hydrogen into the lattice. This may result in embrittlement, with loss of ductility and tensile strength, crack propagation, or stress corrosion cracking (Smith and Miller, 1975; Staehle *et al.*, 1977). Hydrogen-induced failure of metals has been recognized as a serious problem because of the potential for catastrophic failure. Hydrogen embrittlement is a cause of corrosion-related failures in chemical and process plants, the oil industry, buried pipelines, and structural steels (Staehle *et al.*, 1977). Embrittlement constitutes a serious hazard, since catastrophic failures can occur without warning.

Problems can arise from the absorption of atomic hydrogen in fabrication or in service in a hydrogen-rich environment. Sources can be hydrogen gas, if dissociation into atomic hydrogen (H°) occurs, or electrolytic hydrogen, if recombination of H$^+$ into H$_2$ is prevented by sulfides or other hydrogen evolution poisons. Corrosion resulting from sorption of hydrogen is especially difficult to detect and control because variations in environmental and surface conditions affect hydrogen entry (Troiano, 1974).

The role of bacteria and other microorganisms in the hydrogen embrittlement process is not understood. Because the production of atomic hydrogen has been identified as a major factor in hydrogen damage of metals, it can be assumed that microbial transformations involving the production of hydrogen would be relevant. Furthermore, the presence of microbially produced sulfide increases significantly the probability that metals will absorb hydrogen (Walch, 1986).

It is well established that hydrogen gas can cause embrittlement in a variety of materials (Johnson, 1974). Even low concentrations can be effective provided that dissociation occurs, permitting absorption of atomic hydrogen into the metal lattice. A wide range of bacteria produce molecular hydrogen as an end product from the fermentation of carbohydrates. However, the potential contribution to embrittlement from hydrogen-producing microorganisms has received little attention. Given the ability of many bacteria growing on metal surfaces to produce hydrogen, it is likely that bacterial hydrogen is an important factor (Walch, 1986).

Under different nutrient regimes, sulfate reducers, which are traditionally thought to consume hydrogen and hence cathodically depolarize a metal, will in fact generate hydrogen. Many sulfate reducers possess two hydrogenases, one located in the periplasm of the cell, which oxidizes hydrogen, and another located in the cytoplasm, which is involved in hydrogen production (Odom and Peck, 1981). For a more extensive discussion of the hydrogenase system in sulfate-reducing bacteria, the reader is referred to Pankhania (1988). It is interesting to note that investigations of sulfate reducers utilizing lactate have demonstrated corrosion rates considerably lower than those obtained in the field. It is probable that under these conditions, the bacteria produce rather than consume hydrogen.

Recently, Walch and Mitchell (1986) have measured the microbially induced permeation of hydrogen into metals, using a modified Devanathan cell (Devanathan and Stachurski, 1962). The method is based on the ability to detect hydrogen that has passed through a fine metal membrane. An oxidation current is established that is proportional to the hydrogen flux through the metal. Hydrogen permeation kinetics are complex [see Bernstein and Thompson (1974) for a complete discussion]. Walch (1986) grew films of *Clostridium acetobutylicum*, a hydrogen producer, on the mild steel surface of a miniature version of the Devanathan cell and compared the permeation current with that produced by a film of *Clostridium limosum*, which produces minimal hydrogen. A typical hydrogen permeation current, together with the corresponding growth curve for *C. acetobutylicum*, is shown in Fig. 4. The hydrogen producer clearly induces hydrogen

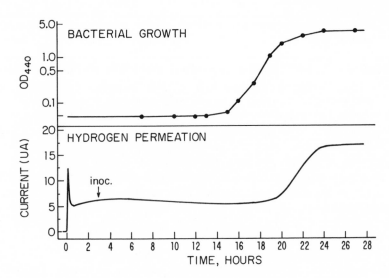

Figure 4. Relationship between hydrogen permeation of a metal surface and growth of a hydrogen-producing bacterium.

permeation through the membrane, and the currents obtained are sufficient to cause damage to sensitive metals (Nakai *et al.*, 1982).

The quantity of hydrogen absorbed by a metal is likely to be different from the total hydrogen produced within the microbial film (Walch, 1988). The most important factor is the likelihood of hydrogen consumption by other bacteria growing on the metal surface. Hydrogen embrittlement may be determined by the outcome of competition for hydrogen between the metal and hydrogen-consuming bacteria (Walch and Mitchell, 1986). Current research in our laboratory is directed toward quantifying these relationships, using highly defined metal surfaces. Potential competition for hydrogen between hydrogen-consuming and hydrogen-producing bacteria and hydrogen-absorbing metals is illustrated diagrammatically in Fig. 5.

7. Thermophilic Corrosion Processes

Microbial corrosion in thermophilic environments is worthy of separate discussion. Thermophiles are receiving increasing attention in metal reclamation (Brierly and Brierly, 1986), waste treatment (Zinder, 1986), and biotechnology (Beguin and Millet, 1986; Imanaka and Aiba, 1986; Ng and Kenealy, 1986; Weimer, 1986). The acceleration of biochemical and electrochemical reactions observed at thermophilic growth temperatures is of utmost importance in corrosion. The diversity of microorganisms in high-temperature, stressed environments is usually low (Kushner, 1978; Shilo, 1979). Cases of corrosion caused by thermophiles can, therefore, provide excellent models for mecha-

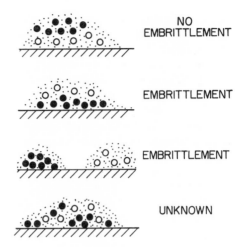

NO
EMBRITTLEMENT

EMBRITTLEMENT

EMBRITTLEMENT

UNKNOWN

Figure 5. Competition between hydrogen-con-
suming, hydrogen-producing bacteria and hy-   ● HYDROGEN PRODUCERS
drogen-absorbing metals. Symbols: ●, hydro-
gen producers; ○, hydrogen consumers.   ○ HYDROGEN CONSUMERS

nistic studies, since laboratory studies may be representative of natural environments
(Walch, 1986).

The ubiquity of thermophiles in hot water distribution systems has been apparent
for a number of years. Bacteria resembling *Thermus aquaticus* can easily be isolated
from commercial and domestic hot water heaters, where it appears that they grow
attached to the tank walls (Ramaley and Hixson, 1970; Brock and Boylen, 1973). A
large number of physiological types of bacteria, both aerobic and anaerobic, have
thermophilic or thermotolerant representatives, many of which grow in extremely low-
nutrient environments (Brock and Freeze, 1969; Brock and Boylen, 1973).

There are many anaerobic thermophiles capable of hydrogen oxidation that may be
involved in corrosion. Thermophilic species of sulfate-reducing bacteria have been
identified, including *Desulfovibrio thermophilus* (Rosanova and Khudyakova, 1974)
and *Thermodesulfobacterium commune* (Zeikus *et al.*, 1983). It has been suggested
(Belkin *et al.*, 1985) that sulfur respiration, although not energetically competitive at
mesophilic temperatures, is favored at high temperatures due to abiotic transformation of
elemental sulfur to sulfide and thiosulfate. The thermoacidophiles *Pyrodictium* sp. and
Thermoproteus sp., isolated from environments at temperatures as high as 105°C (Stetter
et al., 1983), can respire elemental sulfur. The majority of these isolates form H_2S and
CO_2 from sulfur and organic substrates (Zillig *et al.*, 1981, 1982). However, a number
of species have been found to grow chemolithoautotrophically with CO_2 as the sole
carbon source, obtaining energy from oxidation of hydrogen by sulfur (Fischer *et al.*,
1983; Stetter, 1986). In combination with iron, these chemolithoautotrophs may be able
to directly utilize cathodically produced hydrogen and therefore promote corrosion by

the same mechanisms described previously (see Section 2.2). Thermophilic methanogens may also be able to cathodically depolarize iron. A number of species that grow autotrophically on H_2 and CO_2, e.g., *Methanobacterium* sp. and *Methanthermus* sp. (Zeikus and Wolfe, 1972; Stetter *et al.*, 1981; Ahring and Westermann, 1984; Winter *et al.*, 1984), have been isolated from high-temperature environments.

The involvement of these bacteria in microbial corrosion is at present speculative. However, most recently isolated thermophilic isolates (85–105°C) have been anaerobic and involved in sulfur respiration, methanogenesis, or fermentation reactions. Involvement of these bacteria in corrosion reactions is likely to provide a fascinating field of study.

Aerobic thermophiles include potentially corrosive *Sulfolobus* species. In aerobic conditions, they are capable of acid production through oxidation of sulfur and can also directly oxidize ferrous iron. They can reduce ferric iron in microaerophilic habitats, and some species can also reduce sulfur anaerobically (Stetter, 1986). These bacteria, together with moderately thermophilic thiobacilli, are becoming increasingly important in metal recovery (Brierley and Brierley, 1986). Leaching operations are dependent on the kinetics of biological and chemical reactions, which are enhanced at elevated temperatures. The two genera have the characteristics of thermophily, acidophily, and chemolithotrophy, oxidizing ferrous and other metal sulfides. They also have the ability to grow in the presence of high concentrations of metal ions (Brierly and Brierly, 1986).

Thermophilic bacteria have been directly implicated in corrosion processes. Stoecker and Pope (1986) have demonstrated that the localized deterioration of a stainless steel tank containing water at 75–90°C was caused by a *Bacillus* species. Little *et al.* (1984, 1986b) isolated a bacterium, with characteristics similar to those of a *Thermus*, from a test designed to evaluate corrosion failure of brazed nickel joints at temperatures between 60 and 80°C. The failed brazed joint was coated with filamentous bacteria. The mechanisms of corrosion have been investigated (Little *et al.*, 1984, 1986b; Ford *et al.*, 1987b, 1988). Differential aeration, generally associated with microbial film formation, contributed significantly to corrosion of the metal at 60°C (Little *et al.*, 1986b; Wagner and Little, 1986). Isobutyric and isovaleric acids, produced by the thermophilic isolate, accelerated metal deterioration (Little *et al.*, 1986b).

Observations by Little *et al.* (1984) of copper deposits associated with microbial colonies on the failed brazed nickel provide evidence that copper–nickel galvanic couples may also be involved in the corrosion process. A significant decline in the metal concentration of the flowthrough water suggested that the microbial film was concentrating metals (Little *et al.*, 1984). Investigations have recently been directed toward metal binding by exopolymers from thermophilic isolates as a possible mechanism for metal concentration (see Section 2.3; Ford *et al.*, 1988).

Thermophilic microorganisms may present a much greater corrosion hazard than has previously been realized. Metal tolerance explains the ubiquity of thermophiles in industrial, commercial, and domestic hot water systems. This tolerance is probably not plasmid mediated (Chiong *et al.*, 1988) and may be associated with chelation by bacterial exopolymers (Geesey and Mittelman, 1985; Ford *et al.*, 1988). Lowered activation energy of enzymes at high temperatures probably accelerates corrosion reactions, explaining the rapid deterioration of metals in contact with high-temperature aquatic environments.

8. The Role of Consortia

The most discussed yet least addressed issue in biocorrosion research is the question that pervades all aspects of microbial ecology. How do laboratory experiments relate to field studies? The complexity of microbial interactions in the field cannot be realistically modeled in the laboratory. It is primarily for this reason that corrosion rates measured in the laboratory are invariably far lower than those encountered in the field. One of the primary reasons is the presence of consortia of microorganisms in surface biofilms. The study of consortia is complicated by our inability to isolate specific populations from synergistic communities. Laboratory studies of surface consortia have involved relatively simple interactions between two or three different species (e.g., Gaylarde and Johnston, 1986).

Sulfate reducers frequently form consortia with other anaerobes in which their metabolic products are utilized as substrates for other members of the consortium. Interspecific hydrogen transfer is an important process within anaerobic consortia. Where methanogenesis is the main terminal dissimilatory pathway, volatile acids are important intermediates. They are dissimilated by obligate proton reducers (Bryant, 1979; Zehnder, 1978). Reduction of protons to molecular hydrogen permits these bacteria to regenerate their coenzymes. However, this reaction is energetically unfavorable (e.g., $\Delta G^{0'}$, = $+11.5$ and $+18.2$ kcal per mole for butyrate and propionate, respectively) unless the end products are removed and maintained at concentrations below a critical level (Kaspar and Wuhrmann, 1978). Proton reducers can live synergistically with hydrogen oxidizers such as methanogens or sulfate reducers. Because of the difficulty of isolation of these bacteria, the literature describes few obligate proton reducers. The description of "*Methanobacillus omelianskii*" provides an early example of such a consortium (Bryant *et al.*, 1967). The bacterium was found to be a syntrophic association between two organisms. One bacterium degrades ethanol to acetate and hydrogen. The other, a methanogen, consumes the hydrogen produced, thus maintaining low partial pressures of hydrogen and allowing hydrogen reduction to occur.

Other proton reducers described include *Syntrophomonas wolfei*, which oxidizes even-numbered fatty acids to acetate and H_2 and uneven-numbered fatty acids to propionate, acetate, and H_2 (McInerney *et al.*, 1979, 1981a,b), and *S. wolinii*, which oxidizes propionate to acetate, CO_2, and H_2 (Boone and Bryant, 1980). An obligate proton reducer, *Syntrophus buswellii*, which degrades benzoate in consortia with methanogens, has also been identified (Mountfort and Bryant, 1982; Mountfort *et al.*, 1984). Shelton and Tiedje (1984) isolated and partially characterized bacteria that mineralized 3-chlorobenzoic acid in an anaerobic consortium. Stieb and Schink (1985) described a sporeforming obligately syntrophic bacterium, *Clostridium bryantii*. Tomei *et al.* (1985) isolated two sporeforming, butyrate-degrading obligate proton reducers in syntrophic association with methanogens. They were able to separate the sporeformers by pasteurization and reestablish them in coculture with hydrogen-oxidizing bacteria. Ahring and Westermann (1987a,b) describe a thermophilic, anaerobic, butyrate-degrading triculture and some of the kinetics of its complex metabolism.

Consortia appear to play an important role in anaerobic corrosion processes. Niches exist in which interspecific hydrogen transfer and fatty acid utilization by sulfate-reducing bacteria occur. Similar complex interactions develop between populations of aerobic microorganisms on metal surfaces. However, very little is known either about these consortia or their importance in metal corrosion.

9. Future Studies

Continued investigation of surface processes, including interspecific hydrogen transfer, extracellular enzyme activity, and macromolecular interaction with metal ions, is essential to our understanding of the microbial ecology of surfaces. However, development or adaptation of advance surface analytical techniques is necessary to directly correlate these biofilm activities with chemical transformations of metal surfaces.

Molecular techniques are already in use for detection of known corrosive organisms in the field. These methods need to be applied to specific surface communities. We know very little about interrelationships between organisms within a complex biofilm. Investigation of these questions in relation to corrosion of metals may provide information central to our understanding of a wide range of ecological processes occurring on surfaces.

ACKNOWLEDGMENTS. We would like to thank J. S. Maki for helpful discussion. Original research was supported by Office of Naval Research contract N00014-81-K-0624 and by National Oceanic and Atmospheric Administration sea grant NA87AA-D-SG020 to Harvard University.

References

Absolom, D. R., Lamberti, F. W., Policova, Z., Zingg, W., Van Oss, D. J., and Neumann, A. W., 1983, Surface thermodynamics of bacterial adhesion, *Appl. Environ. Microbiol.* **46**:90–97.

Ahring, B. K., and Westermann, P., 1984, Isolation and characterization of a thermophilic, acetate-utilizing methanogenic bacterium, *FEMS Microbiol. Lett.* **25**:47–52.

Ahring, B. K., and Westermann, P., 1987a, Thermophilic anaerobic degradation of butyrate by a butyrate-utilizing bacterium in coculture and triculture with methanogenic bacteria, *Appl. Environ. Microbiol.* **53**:429–433.

Ahring, B. K., and Westermann, P., 1987b, Kinetics of butyrate, acetate, and hydrogen metabolism in a thermophilic, anaerobic, butyrate-degrading triculture, *Appl. Environ. Microbiol.* **53**:434–439.

Allen, L. A., Cairns, A., Eden, G. E., Wheatland, A. B., Wormwell, F., and Nurse, T. J., 1948, Microbiological problems in the manufacture of sugar from beet: Part 1. Corrosion in the diffusion battery and in the recirculation system, *J. Soc. Chem. Ind.* **67**:70–77.

Baier, R. E., 1973, Influence of the initial surface condition of material on bioadhesion, in: *Proceedings of the 3rd International Congress on Marine Corrosion and Fouling* (R. F. Acker, ed.), pp. 15–48, Northwestern University Press, Evanston, Ill.

Baier, R. E., 1984, Adhesion in the biologic environment, *Biomater. Med.* **12**:133–160.

Baier, R. E., Meyer, A. E., Natiella, J. R., Natiella, R. R., and Carter, J. M., 1984, Surface properties determine bioadhesive outcomes: methods and results, *J. Biomed. Mater. Res.* **18**:337–355.

Barry, S., and Houghton, D. R. (eds.), 1986, *Biodeterioration 6*, C.A.B. International, Slough, United Kingdom.

Beguin, P., and Millet, J., 1986, Applied genetics of anaerobic thermophiles, in: *Thermophiles: General, Molecular and Applied Microbiology* (T. D. Brock, ed.), pp. 179–195, John Wiley & Sons, Inc., New York.

Belkin, S., Wirsen, C. O., and Jannasch, H. W., 1985, Biological and abiological sulfur reduction at high temperatures, *Appl. Environ. Microbiol.* **49**:1057–1061.

Berk, S., Mitchell, R., Bobbie, R., Nickels, J., and White, D. C., 1981, Microfouling on metal surfaces exposed to seawater, *Int. Biodeterior. Bull.* **17**:29–37.

Bernstein, I. M., and Thompson, A. W. (eds.), 1974, *Hydrogen in Metals*, Am. Society for Metals, Metals Park, Ohio.

Beyer, R., Melton, L. D., and Kennedy, L. D., 1983, The structure of the neutral polysaccharide gum secreted by *Rhizobium* strain CB744, *Carbohydr. Res.* **122**:155–163.

Bitton, G., and Freihofer, V., 1978, Influence of extracellular polysaccharides on the toxicity of copper and cadmium toward *Klebsiella aerogenes, Microb. Ecol.* **4**:119–125.

Black, J. P., Ford, T. E., and Mitchell, R., 1988, Corrosion behaviour of metal-binding exopolymers from iron- and manganese-depositing bacteria, CORROSION/88, Paper 94, National Association of Corrosion Engineers, Houston, Tex.

Boone, D. R., and Bryant, M. P., 1980, Propionate-degrading bacterium, *Syntrophobacter wolinni* sp. nov. gen. nov., from methanogenic ecosystems, *Appl. Environ. Microbiol.* **40**:626–632.

Borenstein, S. W., 1988, Microbiologically influence corrosion failures of austenitic stainless steel welds, CORROSION/88, Paper 78, National Association of Corrosion Engineers, Houston, Tex.

Borenstein, S. W., and Lindsay, P. B., 1988, Microbiologically influenced corrosion failure analyses, *Mater. Perform.* **27**:51–54.

Boyle, C. D., and Reade, A. E., 1983, Characterization of two extracellular polysaccharides from marine bacteria, *Appl. Environ. Microbiol.* **46**:392–399.

Brierley, J. A., and Brierley, C. L., 1986, Microbial mining using thermophilic microorganisms, in: *Thermophiles: General, Molecular and Applied Microbiology* (T. D. Brock, ed.), pp. 279–305, John Wiley & Sons, Inc., New York.

Brock, T. D., and Boylen, K. L., 1973, Presence of thermophilic bacteria in laundry and domestic hot-water heaters, *Appl. Microbiol.* **25**:72–76.

Brock, T. D., and Freeze, H., 1969, *Thermus aquaticus* gen. n. and sp. n., a non-sporulating extreme thermophile, *J. Bacteriol.* **98**:289–297.

Brown, M. J., and Lester, J. N., 1979, Metal removal in activated sludge: the role of bacterial extracellular polymers, *Water Res.* **13**:817–837.

Brown, M. J., and Lester, J. N., 1982a, Role of bacterial extracellular polymers in metal uptake in pure bacterial culture and activated sludge—I. Effects of metal concentration, *Water Res.* **16**:1539–1548.

Brown, M. J., and Lester, J. N., 1982b, Role of bacterial extracellular polymers in metal uptake in pure bacterial culture and activated sludge—II. Effects of mean cell retention time, *Water Res.* **16**:1549–1560.

Brown, L. R., and Pabst, G. S., 1975, Biodeterioration of stainless steel and aluminum alloys, in: *Proceedings of the 3rd International Biodegradation Symposium* (J. M. Sharpley and A. M. Kaplan, eds.), Applied Science Publishers Ltd., London.

Bryant, M. P., 1979, Microbial methane production—theoretical aspects, *J. Anim. Sci.* **48**:193–201.

Bryant, M. P., Wolin, E. A., Wolin, M. J., and Wolfe, R. S., 1967, *Methanobacillus omelianskii*, a symbiotic association of two species of bacteria, *Arch. Mikrobiol.* **59:**20–31.

Bryant, M. P., Campbell, L. L., Reddy, C. A., and Crabill, M. R., 1977, Growth of *Desulfovibrio* in lactate or ethanol media low in sulfate in association with H₂-utilizing methanogenic bacteria, *Appl. Environ. Microbiol.* **33:**1162–1169.

Caldwell, D. E., and Caldwell, S. J., 1980, Fine structure of *in situ* microbial iron deposits, *Geomicrobiol. J.* **2:**39–53.

Chiong, M., Barra, R., Gonzalez, E., and Vasquez, C., 1988, Resistance of *Thermus* spp. to potassium tellurite, *Appl. Environ. Microbiol.* **54:**610–612.

Christensen, B. E., Kjosbakken, J., and Smidsrod, O., 1985, Partial chemical and physical characterization of two extracellular polysaccharides produced by marine, periphytic *Pseudomonas* sp. strain NCMB 2021, *Appl. Environ. Microbiol.* **50:**837–845.

Coles, E. L., and Davies, R. L., 1956, The protection of cable sheathing: The "phenol corrosion" of lead, *Chem. Ind.* **39:**1030–1035.

Congregado, F., Estanol, I., Espuny, M. J., Fuste, M. C., Manresa, M. A., Marques, A. M., Guinea, J., and Simon-Pujol, M. D., 1985, Preliminary studies on the production and composition of the extracellular polysaccharide synthesized by *Pseudomonas* sp. EPS-5028, *Biotechnol. Lett.* **7:**883–888.

Cord-Ruwisch, R., and Widdel, F., 1986, Corroding iron as a hydrogen source for sulphate reduction in growing cultures of sulphate-reducing bacteria, *Appl. Microbiol. Biotechnol.* **25:**169–174.

Costello, J. A., 1974, Cathodic depolarization by the sulphate-reducing bacteria, *S. Afr. J. Sci.* **70:**202–204.

Costerton, J. W., Geesey, G. G., and Cheng, K.-J., 1978, How bacteria stick, *Sci. Am.* **238:**86–95.

Costerton, J. W., Irvin, R. T., and Cheng, K.-J., 1981, The bacterial glycocalyx in nature and disease, *Annu. Rev. Microbiol.* **35:**299–324.

Costerton, J. W., Cheng, K.-J., Geesey, G. G., Ladd, T. I., Nickel, J. C., Dasgupta, M., and Marrie, T. J., 1987, Bacterial biofilms in nature and disease, *Annu. Rev. Microbiol.* **41:**435–464.

Costerton, J. W., Geesey, G. G., and Jones, P. A., 1988, Bacterial biofilms in relation to internal corrosion monitoring and biocide strategies, *Mater. Perform* **27:**49–53.

Cowen, J. P., and Silver, M. W., 1984, The association of iron and manganese with bacteria on marine macroparticulate material, *Science* **224:**1340–1342.

Crombie, D. J., Moody, G. J., and Thomas, J. D. R., 1980, Corrosion of iron by sulphate-reducing bacteria, *Chem. Ind.* **21:**500–504.

Daniels, L., Belay, N., Rajagopal, B. S., and Weimer, P. J., 1987, Bacterial methanogenesis and growth from CO₂ with elemental iron as the sole source of electrons, *Science* **237:**509–511.

Devanathan, M. A. V., and Stachurski, Z., 1962, Adsorption and diffusion of electrolytic hydrogen in palladium, *Proc. R. Soc. Lond. A* **270:**90–110.

Dexter, S. C., Sullivan, J. D., Williams, J., and Watson, S. W., 1975, Influence of substrate wettability on the attachment of marine bacteria to various surfaces, *Appl. Microbiol.* **30:**298–308.

Donham, J. E., Farquhar, G., Johnston, D., Junkin, E., Lane, D., Edwards, D., and Magnon, L., 1976, The Role of Bacteria in the Corrosion of Oil Field Equipment, TPC Publication 3, National Association of Corrosion Engineers, Houston, Tex.

Dowling, N. J. E., Widdel, F., and White, D. C., 1986, Phospholipid ester-linked fatty acid biomarkers of acetate-oxidizing sulfate reducers and other sulfide forming bacteria, *J. Gen. Microbiol.* **132:**1815–1825.

Dowling, N. J. E., Guezennec, J., Lemoine, M. L., Tunlid, A., and White, D. C., 1988, Analysis of carbon steels affected by bacteria using electrochemical impedance and direct current techniques, *Corrosion* **44:**869–874.

Edlund, A., Nichols, P. D., Roffey, R., and White, D. C., 1985, Extractable and lipopolysaccharide fatty acid and hydroxy acid profiles from *Desulfovibrio* species, *J. Lipid Res.* **26:**982–988.

Farrah, S. R., and Unz, R. F., 1976, Isolation of exocellular polymer from *Zoogloea* strains MP6 and 106 and from activated sludge, *Appl. Environ. Microbiol.* **32**:33–37.

Fischer, F., Zillig, W., Stetter, K. O., and Schreiber, G., 1983, Chemolithoautotrophic metabolism of anaerobic extremely thermophilic archaebacteria, *Nature* (London) **301**:511–513.

Flatau, G. N., Clement, R. L., and Gauthier, M. J., 1985, Cadmium binding sites on cells of a marine pseudomonad, *Chemosphere* **14**:1409–1412.

Fletcher, M., and Loeb, G. I., 1979, Influence of substratum characteristics on the attachment of a marine pseudomonad to solid surfaces, *Appl. Environ. Microbiol.* **37**:67–72.

Ford, T. E., Maki, J. S., and Mitchell, R., 1987a, The role of metal-binding bacterial exopolymers in corrosion processes, CORROSION/87, Paper 380, National Association of Corrosion Engineers, Houston, Tex.

Ford, T. E., Walch, M., and Mitchell, R., 1987b, Corrosion of metals by thermophilic microorganisms, *Mater. Perform.* **26**:35–39.

Ford, T. E., Maki, J. S., and Mitchell, R., 1988, Involvement of bacterial exopolymers in biodeterioration of metals, in: *Biodeterioration 7*, (D. R. Houghton, R. N. Smith, and H. O. W. Eggins, eds.) pp. 378–384, Elsevier Applied Science, Barking, United Kingdom.

Ford, T. E., Walch, M., Mitchell, R., Kaufman, M. J., Vestal, J. R., Ditner, S. A., and Lock, M. A., 1989, Microbial film formation on metals in an enriched arctic river, *Biofouling* **1**:301–311.

Forrester, J. A., 1959, Destruction of concrete caused by sulfur bacteria in a purification plant, *Surveyor* **118**:881–884.

Frankel, R. B., and Blakemore, R. P., 1984, Precipitation of Fe_3O_4 in magnetotactic bacteria, *Phil. Trans. R. Soc. Lond. B* **304**:567–574.

Friedman, B. A., and Dugan, P. R., 1968, Concentration and accumulation of metallic ions by the bacterium *Zoogloea*, *Dev. Ind. Microbiol.* **9**:381–388.

Gaines, R. H., 1910, Bacterial activity as a corrosive influence in the soil, *Ind. J. Eng. Ind. Chem.* **2**:128–135.

Gaylarde, C., Johnston, J., 1986, Anaerobic metal corrosion in cultures of bacteria from estuarine sediments, in: *Biologically Induced Corrosion* (S. C. Dexter, ed.), pp. 137–143, National Association of Corrosion Engineers, Houston, Tex.

Geesey, G. G., and Mittelman, M. W., 1985, The role of high-affinity, metal-binding exopolymers of adherent bacteria in microbial-enhanced corrosion, CORROSION/85, Paper 297, National Association of Corrosion Engineers, Houston, Tex.

Geesey, G. G., Mittelman, M. W., Iwaoka, T., and Griffiths, P. R., 1986, Role of bacterial exopolymers in the deterioration of metallic copper surfaces, *Mater. Perform.* **25**:37–40.

Geesey, G. G., Iwaoka, T., and Griffiths, P. R., 1987, Characterization of interfacial phenomena occurring during exposure of a thin copper film to an aqueous suspension of an acidic polysaccharide, *J. Colloid Interface Sci.* **120**:370–376.

Geesey, G. G., Jang, L., Jolley, J. G., Hankins, M. R., Iwaoka, T., and Griffiths, P. R., 1988, Binding of metal ions by extracellular polymers of biofilm bacteria, *Wat. Sci. Tech.* **20**:161–165.

Ghiorse, W. C., 1984, Biology of iron- and manganese-depositing bacteria, *Annu. Rev. Microbiol.* **38**:515–550.

Gragnolino, G., and Tuovinen, O. H., 1984, The role of sulphate-reducing bacteria and sulphur oxidizing bacteria in the localized corrosion of iron-base alloy: A review, *Int. Biodeterior. Bull.* **20**:9–26.

Griffin, R. B., Cornwell, L. R., Seitz, W., and Estes, E., 1988, Localized corrosion under biofouling, CORROSION/88, Paper 400, National Association of Corrosion Engineers, Houston, Tex.

Hamilton, W. A., 1985, Sulphate-reducing bacteria and anaerobic corrosion, *Annu. Rev. Microbiol.* **39**:195–217.

Hamilton, W. A., 1987, Biofilms: Microbial interactions and metabolic activities, in: *Ecology of Microbial Communities*, Society for General Microbiology Symposium 41 (M. Fletcher, T. R. G. Gray, and J. G. Jones, eds.), pp. 361–387, Cambridge University Press, Cambridge.

Hansen, D. J., Tighe-Ford, D. J., and George, G. C., 1981, Role of the mycelium in the corrosive activity of *Cladosporium resinae* in a dieso/water system, *Int. Biodeterior. Bull.* **17:**103–112.

Hardy, J. A., 1983, Utilization of cathodic hydrogen by sulphate-reducing bacteria, *Br. Corros. J.* **18:**190–193.

Hardy, J. A., and Bown, J. L., 1984, The corrosion of mild steel by biogenic sulfide films exposed to air, *Corrosion* **40:**650–654.

Hendey, N. I., 1964, Some observations on *Cladosporium resinae* as a fuel contaminant and its possible role in the corrosion of aluminium alloy fuel tanks, *Trans. Br. Mycol. Soc.* **47:**467–475.

Hueck-van der Plas, 1968, The micro-biological deterioration of porous building materials, *Int. Bioderior. Bull.* **4:**11–28.

Imanaka, T., and Aiba, S., 1986, Applied genetics of aerobic thermophiles, in: *Thermophiles: General, Molecular and Applied Microbiology* (T. D. Brock, ed.), pp. 159–178, John Wiley & Sons, Inc., New York.

Iverson, W. P., 1974, Microbial corrosion of iron, in: *Microbial Iron Metabolism* (J. B. Neilands, ed.), pp. 475–512, Academic Press, New York.

Iverson, W. P., 1981, An overview of the anaerobic corrosion of underground metallic structures, evidence for a new mechanism, in: *Underground Corrosion* (E. Escalante, ed.), pp. 33–52, Technical Publication 741, American Society for Testing and Materials, Philadelphia.

Iverson, W. P., 1984, Mechanism of anaerobic corrosion of steel by sulfate reducing bacteria, *Mater. Perform.* **23:**28–30.

Iverson, W. P., and Olson, G. J., 1983, Anaerobic corrosion by sulphate-reducing bacteria due to highly reactive volatile phosphorus compound, in: *Microbial Corrosion*, pp. 46–53, The Metal Society, London.

Iverson, W. P., and Olson, G. J., 1984, Problems relating to sulphate-reducing bacteria in the petroleum industry, in: *Petroleum Microbiology* (R. M. Atlas, ed.), pp. 619–641, Macmillan, New York.

Iverson, W. P., Olson, G. J., and Heverly, L. F., 1986, The role of phosphorus and hydrogen sulfide in the anaerobic corrosion of iron and the possible detection of this corrosion by an electrochemical noise technique, in: *Biologically Induced Corrosion* (S. C. Dexter, ed.), pp. 154–161, National Association of Corrosion Engineers, Houston, Tex.

Johnson, H. H., 1974, Hydrogen gas embrittlement, in *Hydrogen in Metals* (I. M. Bernstein and A. W. Thompson, eds.), pp. 35–49, American Society for Metals, Metals Park, Ohio.

Jolley, J. G., Geesey, G. G., Hankins, M. R., Wright, R. B., and Wichlacz, P. L., 1988, Auger electron spectroscopy and X-ray photoelectron spectroscopy of the biocorrosion of copper by gum arabic, bacterial culture supernatant and *Pseudomonas atlantica* exopolymer, *J. Surf. Interface Anal.* **11:**371–376.

Jones, J. G., 1986, Iron transformations by freshwater bacteria, in: *Advances in Microbial Ecology*, Vol. 9 (K. C. Marshall, ed.), 149–185, Plenum Press, New York.

Kasahara, K., and Kajiyama, F., 1986, Role of sulfate reducing bacteria in the localized corrosion of buried pipes, in: *Biologically Induced Corrosion* (S. C. Dexter, ed.), pp. 171–183, National Association of Corrosion Engineers, Houston, Tex.

Kaspar, H. F., and Wurhmann, K., 1978, product inhibition in sludge digestion, *Microb. Ecol.* **4:**241–248.

Kerger, B. D., Nichols, P. D., Sand, W., Bock, E., and White, D. C., 1987, Association of acid-producing thiobacilli with degradation of concrete: analysis by 'signature' fatty acids from the polar lipids and lipopolysaccharide, *J. Ind. Microbiol.* **2:**63–69.

King, R. A., and Miller, J. D. A., 1971, Corrosion by the sulphate-reducing bacteria, *Nature* (London) **233:**491–492.

Kobrin, G., 1976, Corrosion by microbiological organisms in natural waters, *Mater. Perform.* **15:**38–43.

Kushner, D. J. (ed.), 1978, *Microbial Life in Extreme Environments*, Academic Press, New York.

Laanbroek, H., Abee, T., and Voogd, I. L., 1982, Alcohol conversions by *Desulfobulbus propionicus* Lindhorst in the presence and absence of sulphate and hydrogen, *Arch. Microbiol.* **133:**178–184.

Lester, J. N., Sterrett, R. M., Rudd, T., and Brown, M. J., 1984, Assessment of the role of bacterial extracellular polymers in controlling metal removal in biological waste water treatment, in: *Microbiological Methods for Environmental Biotechnology* (J. M. Grainger and J. M. Lynch, eds.), pp. 197–217, Academic Press, London and Orlando.

Little, B. J., Walch, M., Wagner, P., Gerchakov, S. M., and Mitchell, R., 1984, The impact of extreme obligate thermophilic bacteria on corrosion processes, in *Proceedings of the 6th International Congress on Marine Corrosion and Fouling,* pp. 511–520.

Little, B., Wagner, P., and Gerchakov, S. M., 1986a, A quantitative investigation of mechanisms for microbial corrosion, in: *Biologically Induced Corrosion* (S. C. Dexter, ed.), pp. 209–214, National Association of Corrosion Engineers, Houston, Tex.

Little, B., Wagner, P., S. M. Gerchakov, S. M., Walch, M., and Mitchell, R., 1986b, The involvement of a thermophilic bacterium in corrosion processes, *Corrosion* **42:**533–536.

Little, B. J., Wagner, P., Maki, J. S., Walch, M., and Mitchell, R., 1986c, Factors influencing the adhesion of microorganisms to surfaces, *J. Adhes.* **20:**187–210.

Little, B., Wagner, P., and Jacobus, J., 1988, The impact of sulfate-reducing bacteria on welded copper-nickel seawater piping systems, CORROSION/88, Paper 81, National Association of Corrosion Engineers, Houston, Tex.

Lutey, R., 1980, Microbiological corrosion, CORROSION/80, Paper 39, National Association of Corrosion Engineers, Houston, Tex.

Maki, J. S., and Mitchell, R., 1988, L'adhesion microbienne aux surfaces et ses consequences, in: *Microorganisms dans les Ecosystemes Oceaniques* (A. Bianchi, D. Marty, J.-C. Bertrand, C. Caumette, and M. Gauthier, eds.), pp. 387–409, Masson, Paris.

Marshall, K. C. (ed.), 1984, *Microbial Adhesion and Aggregation,* Dahlem Konferenzen, Springer-Verlag, Berlin.

Marszalek, D. S., Gerchakov, S. M., and Udey, L. R., 1979, Influence of substrate composition on marine microfouling, *Appl. Environ. Microbiol.* **38:**987–995.

Martin, J. P., 1971, Decomposition and binding action of polysaccharides in soil, *Soil Biol. Biochem.* **3:**33–41.

Martin, J. P., Ervin, J. O., and Richards, S. J., 1972, Decomposition and binding action in soil of some mannose-containing microbial polysaccharides and their Fe, Al, Zn, and Cu complexes, *Soil Sci.* **113:**322–327.

McInerney, M. J., Bryant, M. P., and Pfennig, N., 1979, Anaerobic bacterium that degrades fatty acids in syntrophic association with methanogens, *Arch. Microbiol.* **122:**129–135.

McInerney, M. J., Mackie, R. I., and Bryant, M. P., 1981a, Syntrophic association of a butyrate-degrading bacterium and *Methanosarcina* enriched from bovine rumen fluid, *Appl. Environ. Microbiol.* **41:**826–828.

McInerney, M. J., Bryant, M. P., Hespell, R. B., and Costerton, J. W., 1981b, *Syntrophomonas wolfei* gen. nov. sp. nov., an anaerobic, syntrophic, fatty acid-oxidizing bacterium, *Appl. Environ. Microbiol.* **41:**1029–1039.

Miller, J. D. A. (ed.), 1970, *Microbial Aspects of Metallurgy,* American Elsevier, New York.

Mittelman, M. W., and Geesey, G. G., 1985, Copper-binding characteristics of exopolymers from a freshwater sediment bacterium, *Appl. Environ. Microbiol.* **49:**846–851.

Moosavi, A. N., Dawson, J. L., and King, R. A., 1986, The effect of sulphate-reducing bacteria on the corrosion of reinforced concrete, in: *Biologically Induced Corrosion* (S. C. Dexter, ed.), pp. 291–308, National Association of Corrosion Engineers, Houston, Tex.

Mountfort, D. O., and Bryant, M. P., 1982, Isolation and characterization of an anaerobic syntrophic benzoate-degrading bacterium from sewage sludge, *Arch. Microbiol.* **133:**249–256.

Mountford, D. O., Brulla, W. J., Krumholz, L. R., and Bryant, M. P., 1984, *Syntrophus buswellii* gen. nov., sp. nov.: A benzoate catabolizer from methanogenic ecosystems, *Int. J. Syst. Bacteriol.* **34**:216–217.

Nakai, Y., Kurahashi, H., Totsuka, N., and Wesugi, Y., 1982, Effect of corrosive environment on hydrogen induced cracking, CORROSION/82, Paper 132, National Association of Corrosion Engineers, Houston, Tex.

Nealson, K. H., 1983a, The microbial iron cycle, in: *Microbial Geochemistry* (W. E. Krumbein, ed.), pp. 159–190, Blackwell, Oxford,

Nealson, K. H., 1983b, The microbial manganese cycle, in: *Microbial Geochemistry* (W. E. Krumbein, ed.), pp. 191–221, Blackwell, Oxford.

Ng, T. K., and Kenealy, W. F., 1986, Industrial applications of thermostable enzymes, in: *Thermophiles: General, Molecular and Applied Microbiology* (T. D. Brock, ed.), pp. 197–215, John Wiley & Sons, Inc., New York.

Nichols, P. D., Guckert, J. B., and White, D. C., 1986, Determination of monounsaturated fatty acid double-bond position and geometry for microbial monocultures and complex consortia by capillary GC-MS of their dimethyl disulphide adducts, *J. Microbiol. Methods* **5**:49–55.

Obuekwe, C. O., and Westlake, D. W. S., 1982, Effect of reducible compounds (potential electron acceptors) on reduction of ferric iron by *Pseudomonas* species, *Microbiol. Lett.* **19**:57–62.

Obuekwe, C. O., Westlake, D. W. S., and Cook, F. D., 1981a, Effect of nitrate on reduction of ferric iron by a bacterium isolated from crude oil, *Can. J. Microbiol.* **27**:692–697.

Obuekwe, C. O., Westlake, D. W. S., Cook, F. D., and Costerton, J. W., 1981b, Surface changes in mild steel coupons from the action of corrosion-causing bacteria, *Appl. Environ. Microbiol.* **41**:766–774.

Obuekwe, C. O., Westlake, D. W. S., Plambeck, J. A., and Cook, F. D., 1981c, Corrosion of mild steel in cultures of ferric iron reducing bacterium isolated from crude oil. I. Polarization characteristics, *Corrosion* **37**:461–467.

Obuekwe, C. O., Westlake, D. W. S., Plambeck, J. A., and Cook, F. D., 1981d, Corrosion of mild steel in cultures of ferric iron reducing bacterium isolated from crude oil. II. Mechanism of anodic depolarization, *Corrosion* **37**:632–637.

Obuekwe, C. O., Westlake, D. W. S., and Cook, F. D., 1983, Corrosion of Pembina crude oil pipeline: The origin and mode of formation of hydrogen sulfide. *Eur. J. Appl. Microbiol. Biotechnol.* **17**:173–177.

Odom, J. M., and Peck, H. D., 1981, Hydrogen cycling as a general mechanism for energy coupling in the sulphate-reducing bacteria, *Desulfovibrio* sp., *FEMS Microbiol. Lett.* **12**:47–50.

Pankhania, I. P., 1988, Hydrogen metabolism in sulphate-reducing bacteria and its role in anaerobic corrosion, *Biofouling* **1**:27–47.

Pankhania, I. P., Moosavi, A. N., and Hamilton, W. A., 1986a, Utilization of cathodic hydrogen by *Desulfovibrio vulgaris* (Hidenborough), *J. Gen. Microbiol.* **132**:3357–3365.

Pankhania, I. P., Gow, L. A., and Hamilton, W. A., 1986b, The effect of hydrogen on the growth of *Desulfovibrio vulgaris* (Hidenborough) on lactate, *J. Gen. Microbiol.* **132**:3349–3356.

Parker, C. D., 1945a, The corrosion of concrete. 1. The isolation of a species of bacterium associated with the corrosion of concrete exposed to atmospheres containing hydrogen sulphide, *Aust. J. Exp. Biol. Med. Sci.* **23**:81–90.

Parker, C. D., 1945b, The corrosion of concrete. 2. The function of *Thiobacillus concretivorus* (nov. spec.) in the corrosion of concrete exposed to atmospheres containing hydrogen sulphide, *Aust. J. Exp. Biol. Med. Sci.* **23**:91–98.

Platt, R. M., Geesey, G. G., Davis, J. D., and White, D. C., 1985, Isolation and partial chemical analysis of firmly bound exopolysaccharide from adherent cells of a freshwater sediment bacterium, *Can. J. Microbiol.* **31**:675–680.

Pope, D. H., Zintel, T. P., Kuruvilla, A. K., Siebert, O. W., 1988, Organic acid corrosion of carbon steel: A mechanism of microbiologically influenced corrosion, CORROSION/88, Paper 79, National Association of Corrosion Engineers, Houston, Tex.

Postgate, J. R. (ed.), 1984, *The Sulphate-Reducing Bacteria,* 2nd ed., Cambridge University Press, Cambridge.

Ramaley, R. F., and Hixon, J., 1970, Isolation of a non-pigmented, thermophilic bacterium similar to *Thermus aquaticus, J. Bacteriol.* **103**:527–528.

Ridgway, H. F., and Olson, B. H., 1981, Scanning electron microscope evidence for bacterial colonization of a drinking-water distribution system, *Appl. Environ. Microbiol.* **41**:274–287.

Rigdon, J. H., and Beardsley, C. W., 1958, Corrosion of concrete by autotrophs, *Corrosion* **14**:206–208.

Rosanova, E. P., and Khudyakova, A. I., 1974, A new nonspore-forming thermophilic sulfate-reducing organism, *Desulfovibrio thermophilus* nov. sp., *Microbiologiya* **43**:1069–1075 (Engl. trans., pp. 908–912).

Rudd, T., Sterritt, R. M., and Lester, J. N., 1984, Formation and conditional stability constants of complexes formed between heavy metal and bacterial extracellular polymers, *Water Res.* **18**:379–384.

Salvarezza, R. C., de Mele, M. F. L., and Videla, H. A., 1983, Mechanisms of the microbial corrosion of aluminum alloys, *Corrosion* **39**:26–32.

Sand, W., and Bock, E., 1984, Concrete corrosion in the Hamburg sewer system system, *Environ. Technol. Lett.* **5**:517–528.

Savage, D. C., and Fletcher, M. (eds.), 1985, *Bacterial Adhesion,* Plenum Press, New York.

Schmitt, C. R., 1986, Anomalous microbiological tuberculation and aluminum pitting corrosion-case histories, in: *Biologically Induced Corrosion* (S. C. Dexter, ed.), pp. 69–75, National Association of Corrosion Engineers, Houston, Tex.

Semple, K. M., and Westlake, D. W. S., 1987, Characterization of iron-reducing *Alteromonas putrefaciens* strains from oil field fluids, *Can. J. Microbiol.* **3**:366–371.

Shelton, D. R., and Tiedje, J. M., 1984, Isolation and partial characterization of bacteria in an anaerobic consortium that mineralizes 3-chlorobenzoic acid, *Appl. Environ. Microbiol.* **48**:840–848.

Shilo, M. (ed.), 1979, *Strategies of Life in Extreme Environments,* Dahlem Konferenzen, Verlag Chemie, Weinheim.

Siporin, C., and Cooney, J. J., 1975, Extracellular lipids of *Cladosporium (Amorphotheca) resinae* grown on glucose or on n-alkanes, *Appl. Microbiol.* **29**:604–609.

Smith, J. S., and Miller, J. D. A., 1975, Nature of sulphides and their corrosive effects on ferrous metals: A review, *Br. Corros. J.* **10**:136–143.

Staehle, R. W., Hochmann, J., McCright, R. D., and Slater, J. E. (eds.), 1977, *Stress Corrosion Cracking and Hydrogen Embrittlement of Iron Base Alloys,* National Association of Corrosion Engineers, Houston, Tex.

Starkey, R. L., 1986, Anaerobic corrosion-perspectives about causes, in: *Biologically Induced Corrosion* (S. C. Dexter, ed.), pp. 3–7, National Association of Corrosion Engineers, Houston, Tex.

Stetter, K. O., 1986, Diversity of extremely thermophilic archaebacteria, in: *Thermophiles: General, Molecular and Applied Microbiology* (T. D. Brock, ed.), pp. 39–74, John Wiley & Sons, Inc., New York.

Stetter, K. O., Thomm, M., Winter, J., Wildgruber, G., Huber, H., Zillig, W., Janecovic, D., Konig, H., Palm, P., and Wunderl, S., 1981, *Methanothermus fervidus,* a novel extremely thermophilic methanogen isolated from an Icelandic hot spring, *Zbl. Bakt. Hyg. I Abt. Orig.* **2**:166–178.

Stetter, K. O., Konig, H., and Stackebrandt, E., 1983, *Pyrodictium* gen. nov., a new genus of submarine disc-shaped sulphur reducing archaebacteria growing optimally at 105°C, *Syst. Appl. Microbiol.* **4**:535–551.

Stieb, M., and Schink, B., 1985, Anaerobic oxidation of fatty acids by *Clostridium bryantii* sp. nov., a sporeforming obligately syntrophic bacterium, *Arch. Microbiol.* **140**:387–390.

Stoecker, J. G., and Pope, D. H., 1986, Study of biological corrosion in high temperature demineralized water, *Mater. Perform.* **25:**51–56.

Stoveland, S., and Lester, J. N., 1980, A study of the factors which influence metal removal in the activated sludge process, *Sci. Total Environ.* **16:**37–54.

Stranger-Johannessen, M., 1986, Fungal corrosion of the steel interior of a ship's holds, in: *Biodeterioration 6* (S. Barry and D. R. Houghton, eds.), C.A.B. International, Slough, United Kingdom.

Sutherland, I. W., 1972, Bacterial exopolysaccharides, *Adv. Microb. Physiol.* **8:**143–213.

Sutherland, I. W., 1984, Enzymes in the assay of microbial polysaccharides, *Proc. Biochem.* **18:**19–24.

Sutherland, I. W., 1985, Biosynthesis and composition of gram-negative bacterial extracellular and wall polysaccharides, *Annu. Rev. Microbiol.* **39:**243–270.

Tago, Y., and Aida, K., 1977, Exocellular mucopolysaccharide closely related to bacterial floc formation, *Appl. Environ. Microbiol.* **34:**308–314.

Tarasevich, M. R., 1979, Ways of using enzymes for acceleration of electrochemical reactions, *J. Electroanal. Chem.* **104:**587–597.

Tatnall, R. E., 1981a, Fundamentals of bacteria-induced corrosion, *Mater. Perform.* **20:**32–38.

Tatnall, R. E., 1981b, Case histories: Bacteria-induced corrosion, *Mater. Perform.* **20:**41–48.

Tiller, A. K., 1982, Aspects of microbial corrosion, in: *Corrosion Processes* (R. N. Parkins ed.), pp. 115–159, Applied Science Publishers, London, New York.

Tomei, F. A., and Mitchell, R., 1986, Development of an alternative method for studying the role of H_2-consuming bacteria in the anaerobic oxidation of iron, in: *Biologically Induced Corrosion* (S. C. Dexter, ed.), pp. 309–320, National Association of Corrosion Engineers, Houston, Tex.

Tomei, F. A., Maki, J. S., and Mitchell, R., 1985, Interactions in syntrophic associations of endospore-forming, butyrate-degrading bacteria and H_2-consuming bacteria, *Appl. Environ. Microbiol.* **50:**1244–1250.

Troiano, A. R., 1974, Introduction, in: *Hydrogen in Metals* (I. M. Bernstein and A. W. Thompson, eds.), pp. 3–15. American Society for Metals, Metals Park, Ohio.

Tuovinen, O. H., and Mair, D. M., 1985, Corrosion of cast iron pipes and associated water quality effects in distribution systems, in: *Biodeterioration 6* (S. Barry and D. R. Houghton, eds.), pp. 223–227, C.A.B. International, Slough, United Kingdom.

Tuovinen, O. H., Button, K. S., Vuorinen, A., Carlson, L., Mair, D. M., and Yut, L. A., 1980, Bacterial, chemical, and mineralogical characteristics of tubercles in distribution pipelines, *J. Am. Water Works Assoc.* **72:**626–635.

Tyler, P. A., and Marshall, K. C., 1967a, Hyphomicrobia—a significant factor in manganese problems, *J. Am. Water Works Assoc.* **59:**1043–1048.

Tyler, P. A., and Marshall, K. C., 1967b, Microbial oxidation of manganese in hydrooelectric pipelines, *Antonie van Leeuwenhoek J. Microbiol. Serol.* **33:**171–183.

Uhlinger, D. J., and White, D. C., 1983, Relationship between physiological status and formation of extracellular polysaccharide glycocalyx in *Pseudomonas atlantica*, *Appl. Environ. Microbiol.* **45:**64–70.

Videla, H. A., Guiamet, P. S., and DoValle, S., 1988, Effects of fungal and bacterial contaminants of kerosene fuels on the corrosion of storage and distribution systems, CORROSION/88, Paper 91, National Association of Corrosion Engineers, Houston, Tex.

von Wolzogen Kuhr, C. A. H., and van der Vlugt, L. S., 1934, Graphication of cast iron as an electrochemical process in anaerobic soils, *Water* **18:**147–165.

Wagner, C., and Traud, W., 1938, Uber die Deutung von Korrosionsvorgangen durch Uberlagerung von elektrochemischen Teilvorgangen und uber die Potentialbildung an Mischelektroden, *Z. Elektrochem.* **44:**391–454.

Wagner, P., and Little, B. J., 1986, Applications of a technique for the investigation of microbially induced corrosion, CORROSION/86, Paper 121, National Association of Corrosion Engineers, Houston, Tex.

Walch, M., 1986, The Microbial Ecology of Metal Surfaces, Ph.D. thesis, Harvard University, Cambridge, Mass.

Walch, M., and Mitchell, R., 1986, Microbial influence on hydrogen uptake by metals, in: *Biologically Induced Corrosion* (S. C. Dexter, ed.), pp. 201–208. National Association of Corrosion Engineers, Houston, Tex.

Weimer, P. J., 1986, Use of thermophiles for the production of fuels and chemicals, in: *Thermophiles: General, Molecular and Applied Microbiology* (T. D. Brock, ed.), pp. 217–255, John Wiley & Sons, Inc., New York.

Weimer, P. J., Van Kavelaar, M. J., Michel, C. B., and Ng, T. K., 1988, Effect of phosphate on the corrosion of carbon steel and on the composition of corrosion products on two-stage continuous cultures of *Desulfovibrio desulfuricans*, *Appl. Environ. Microbiol.* **54:**386–396.

Westlake, D. W. S., Semple, K. M., and Obuekwe, C. O., 1986, Corrosion by ferric iron-reducing bacteria isolated from oil production systems, in: *Biologically Induced Corrosion* (S. C. Dexter, ed.), pp. 193–200, National Association of Corrosion Engineers, Houston, Tex.

White, D. C., 1983, Analysis of microorganisms in terms of quantity and activity in natural environments, in: *Microbes in Their Natural Environments*, Society for General Microbiology Symposium 34 (J. H. Slater, R. Whittenbury, and J. W. T. Wimpenny, eds.), pp. 37–66, Society for General Microbiology, New York.

White, D. C., 1984, Chemical characterization of films, in: *Microbial Adhesion and Aggregation*, Dahlem Konferenzen Life Sciences Research Report 31 (K. C. Marshall, ed.), pp. 159–176, Springer-Verlag, Berlin.

Widdel, F., 1987, New types of acetate-oxidizing, sulphate-reducing *Desulfobacter species, D., hydrogenophilus* sp. nov., *D. latus* sp. nov., and *D. curvatus* sp. nov., *Arch. Microbiol.* **148:**286–291.

Widdel, F., and Pfennig, N., 1981, Studies on dissimilatory sulfate-reducing bacteria that decompose fatty acids. I. Isolation of a new sulfate-reducing bacteria enriched with acetate from saline environments. Description of *Desulfobacter postgatei* gen. nov., sp. nov., *Arch. Microbiol.* **129:**395–400.

Widdel, F., and Pfennig, N., 1982, Studies on dissimilatory sulfate-reducing bacteria that decompose fatty acids. II. Incomplete oxidation of propionate by *Desulfobulbus propionicus* gen. nov., sp. nov., *Arch. Microbiol.* **131:**360–365.

Widdel, F., Kohring, G. W., and Mayer, F., 1983, Studies on dissimilatory sulfate-reducing bacteria that decompose fatty acids. III. Characterization of the filamentous gliding *Desulfonema limicola* gen. nov., sp. nov., and *Desulfonema magnum* sp. nov., *Arch. Microbiol.* **134:**286–294.

Williams, A. G., and Wimpenny, J. W. T., 1977, Exopolysaccharide production by *Pseudomonas* NCIB11264 grown in batch culture, *J. Gen. Microbiol.* **102:**13–21.

Winter, J., Lerp, C., Zabel, H.-P., Wildenauer, F. X., Konig, H., and Schindler, F., 1984, *Methanobacterium wolfei*, sp. nov., a new tungsten-requiring, thermophilic, autotrophic methanogen, *Syst. Appl. Microbiol.* **5:**457–466.

Wrangstadh, M., Conway, P. L., and Kjelleberg, S., 1986, The production and release of an extracellular polysaccharide during starvation of a marine *Pseudomonas* sp. and the effect thereof on adhesion, *Arch. Microbiol.* **145:**220–227.

Zehnder, A. J. B., 1978, Ecology of methane formation, in: *Water Pollution Microbiology*, Vol. 2 (R. Mitchell, ed.), John Wiley & Sons, Inc., New York.

Zeikus, J. G., and Wolfe, R. S., 1972, *Methanobacterium thermoautotrophicum* sp. nov., an anaerobic, autotrophic extreme thermophile, *J. Bacteriol.* **109:**707–713.

Zeikus, J. G., Dawson, M. A., Thompson, T. E., Ingvorsen, K., and Hatchikian, E. C., 1983, Microbial ecology of volcanic sulphidogenesis: Isolation and characterization of *Thermodesulfobacterium commune* gen. nov. and sp. nov., *J. Gen. Microbiol.* **129:**1159–1169.

Zillig, W., Stetter, K. O., Schafer, W., Janekovic, D., Wunderl, S., Holz, I., and Palm, P., 1981, *Thermoproteales:* A novel type of extremely thermoacidophilic anaerobic archaebacteria isolated from Icelandic Solfataras, *Zbl. Bakt. Hyg. I Abt. Orig.* **2:**205–227.

Zillig, W., Stetter, K. O., Prangishvilli, D., Schafer, W., Wunderl, S., Janekovic, D., Holz, I., and
    Palm, P., 1982, *Desulfurococcaceae*, the second family of the extremely thermophilic, anaerobic,
    sulfur-respiring *Thermoproteales, Zbl. Bakt. Hyg. I Abt. Orig.* **3:**304–317.
Zinder, S. H., 1986, Thermophilic waste treatment systems, in: *Thermophiles: General, Molecular and
    Applied Microbiology* (T. D. Brock, ed.), pp. 257–277, John Wiley & Sons, Inc., New York.

Mathematical Modeling Of Nitrification Processes

J. I. PROSSER

1. Introduction

An understanding of the microbial cycling of nutrients in natural ecosystems requires knowledge of both the types and numbers of microorganisms involved and the nature of the processes they carry out. The study of microorganisms in natural environments is, however, notoriously difficult. In particular, the isolation, characterization, and enumeration of "typical," "dominant," or "significant" microbial populations is plagued by the lack of reliable *in situ* detection techniques, problems associated with nondestructive removal of cells, and choice of suitable media and cultural conditions for growth in the laboratory. These problems are exacerbated in studies of nitrifying bacteria. Autotrophic ammonia- and nitrite-oxidizing bacteria do not form visible colonies on solid medium, and although techniques that increase the apparent sizes of colonies of ammonia oxidizers are available (Soriano and Walker, 1973), the dilution plate technique is not convenient for routine use. Enumeration of viable cells therefore requires use of the most-probable-number (MPN) technique, which introduces an intrinsic statistical variability in addition to variability arising through cell extraction, experimental technique, and choice of suitable culture media. The length of incubation period is also critical; Matulewich *et al.* (1975) found that MPN counts of nitrite oxidizers had not reached a maximum after 100 days of incubation. Although the MPN technique has proved invaluable in many studies, the associated variability has led some workers to consider it qualitative rather than quantitative. Some of these problems have been overcome by use of fluorescent-antibody (FA) counting techniques (Belser, 1979), which currently pro-

J. I. PROSSER • Department of Genetics and Microbiology, Marischal College, University of Aberdeen, Aberdeen AB9 1AS, Scotland.

vide the most accurate means of assessing total nitrifier populations in natural environments but suffer from problems of specificity.

On the other hand, chemical techniques for the measurement of concentrations of substrates and products of microbial activity in natural environments are frequently more convenient, sensitive, and accurate. This applies particularly to the nitrogen cycle, where the different pools of nitrogen can be distinguished, and automated techniques are available for analysis of ammonia, nitrite, and nitrate. The nitrification process is therefore much more amenable to study in natural environments than the growth of nitrifying bacteria, and use of differential inhibitors and ^{15}N-labeling techniques permits study of interactions with other processes within the nitrogen cycle.

It is consequently much easier to describe quantitative aspects of the nitrogen cycle in terms of pool sizes and rates of transfer than in terms of changes in activities or concentrations of particular microorganisms. An understanding of ecosystems, as expressed in hypotheses or theories, depends on the extent to which predictions of those theories can be tested experimentally. Quantitative theories of microbial processes are therefore much better developed and more testable, although the individual organisms involved and their habitats must never be forgotten.

Quantitative descriptions and theories constitute mathematical models. In this chapter, discussion of the use of models in the study of nitrification will concentrate on studies at the process level, but introducing population dynamics where appropriate and relevant. A thorough understanding of the fundamental kinetics of nitrification is required for reliable measurement of rates of nitrification in natural environments and is most readily obtained by the study of pure cultures, which will be considered first. Such studies also provide insight into the quantitative effects of environmental factors such as pH, temperature, and inhibitors. The modeling of environmental and applied aspects of nitrification will then be discussed, with consideration of terrestrial and aquatic environments and sewage and wastewater treatment. Emphasis will be placed on the relationship between the kinetics of nitrification in pure culture and in natural environments and also on features common to the different approaches adopted by microbiologists, soil scientists, and chemical engineers.

2. Nitrification

Nitrification is the oxidation of reduced forms of nitrogen to nitrate. The major organisms responsible for this process are the autotrophic ammonia- and nitrite-oxidizing bacteria, which belong to five and four genera, respectively. All normally obtain energy from oxidation of ammonia or nitrite and obtain cell carbon from carbon dioxide. Ammonia oxidizers can also oxidize hydrocarbons, alcohols, and halocarbons (Wood, 1986), and nitrite oxidizers can grow heterotrophically (Bock *et al.*, 1986), but growth rates and yields under these conditions are negligible. A wide range of heterotrophic organisms oxidize organic reduced forms of nitrogen to nitrate. This process of heterotrophic nitrification occurs at a much lower rate in terms of amount of nitrogen converted per cell, but high concentrations of these organisms in environments unfavorable for autotrophic nitrification (e.g., low pH) may increase the significance of this process (Killham, 1986; Kuenen and Robertson, 1987). Heterotrophic nitrification has not been modeled and will not be considered further.

Nitrification is an aerobic process, optimal at mesophilic temperatures and slightly alkaline pH. In natural environments, it is important in linking the reduced and oxidized forms of nitrogen within the nitrogen cycle. Its significance in soil is related to plant nutrition, with both ammonia and nitrate acting as nitrogen sources but with nitrate usually preferred. Ammonia is adsorbed to negatively charged soil components, but nitrate is mobile and is readily leached from the soil. Nitrification can therefore lead to significant losses of ammonia-based fertilizers and subsequently to nitrate pollution of groundwaters, particularly in areas of intensive agriculture. Specific inhibitors of autotrophic nitrification are available to reduce such losses (e.g., nitrapyrin, allylthiourea, and acetylene) and are also used to distinguish autotrophic and heterotrophic processes (Oremland and Capone, 1988). In aquatic environments, nitrate is an important factor in eutrophication and also leads to production of nitrous and nitric oxides. Nitrification in sewage treatment prevents discharge of toxic levels of ammonium and, with denitrification, enables removal of nitrogen, reducing the risk of eutrophication.

3. Mathematical Modeling

A mathematical model is the representation of a system or process in mathematical form, most commonly as an equation or set of equations. The majority of nitrification models are deterministic; i.e., they consist of relationships between dependent variables (e.g., biomass concentration and ammonium concentration) and an independent variable, usually time. System behavior depends on the provision of constants found in the equations and on initial conditions. Stochastic models of nitrification, which incorporate statistical variation, are rare, but one example will be described.

The simplest nitrification models are algebraic equations describing, for example, changes in kinetics of substrate conversion as a function of time. Models may be empirical and descriptive, in which the equations are chosen to fit experimental data and serve only to describe those data quantitatively. These may be useful for comparing rates under different conditions (Robinson, 1985) and may give clues to mechanisms underlying growth and substrate conversion. Mechanistic models aim to predict system behavior from assumptions regarding controlling mechanisms and therefore represent hypotheses that may be tested by comparing predicted behavior with experimental data. Usually, many of these assumptions are necessary merely to simplify the theoretical description. For example, all cells may be considered identical and temperature and pH constant. Assumptions of this sort will never be completely obeyed experimentally but are commonly made, if only implicitly, in most laboratory studies. Experimental systems designed to test such models must obey these assumptions as closely as possible to allow isolation of assumptions regarding specific mechanisms. This modeling approach is therefore closely linked to the development of experimental models; rather than attempting to describe "reality," the aim is to study particular features of reality under controlled conditions.

Further classification of mathematical models is given by Roels and Kossen (1978), and general microbiological applications can be found in a number of textbooks [e.g., Bazin (1983)]. In practice, boundaries between the different types of model are frequently blurred, and mechanistic models may contain empirical components. This is particularly the case in large-system models such as those for the whole nitrogen cycle, in which detailed information on particular sections may be lacking.

In general, mechanistic models are more complex and consist of sets of ordinary or partial differential equations. Analytical solutions are rarely available, and generation of predictions (simulation) requires numerical approximation techniques and computer methods. These techniques, and their advantages and disadvantages are discussed by Prosser (1988). Unfortunately more advanced techniques such as stability analysis have rarely been applied to nitrification models, though they have provided important information in other microbiological applications.

4. Pure-Culture Studies

4.1. Basic Growth Kinetics

An understanding of the basic growth kinetics of nitrifying bacteria in pure culture is important for a number of reasons. From a technical and practical viewpoint, it is frequently necessary to compare rates of nitrification, for which reliable assessment of growth and activity is required. This involves measurement of changes in population size or, more frequently, in substrate or product concentrations. Interpretation of these changes requires a thorough understanding of growth kinetics, which may be obtained only in the absence of complicating factors resulting from varying environmental conditions and contaminating organisms. The kinetics observed in pure culture are applicable to nitrification in natural environments, often with similar rate constants, and deviation from "standard" kinetics highlights the nature of environmental effects.

The second, and important, reason for studying growth kinetics in pure culture is to gain information on the physiology of nitrifiers and on effects of environmental factors. The kinetics of inhibition provides clues to biochemical mechanisms of inhibitors, and variation in kinetics with pH may indicate ways in which pH affects growth and transport of metabolites. Here the rate equations represent mechanistic models of growth and may be too complex for routine measurements. They do, however, provide a basis for the use of simpler kinetic expressions and help define ranges over which their use may be justified.

4.2. Short-Term Measurements

Short-term batch experiments are frequently used to determine kinetic constants and quantitative effects of substrate concentration, inhibitors, etc., and typically last for several hours. The minimum doubling time reported for nitrifying bacteria is 8 hr (Skinner and Walker, 1961), and any growth in such experiments will consequently be negligible. Changes in substrate or product concentration therefore characterize activity rather than growth if incubation times are less than the lag period or the doubling time, and their use is appropriate for both intact cells and cell-free extracts.

The work of Boon and Laudelout (1962) provides an early example, in which oxygen uptake was used to assess activity of *Nitrobacter winogradskyi*. Nitrite oxidation followed Michaelis–Menten kinetics, and classical enzyme kinetics was used to determine saturation constants for nitrite and oxygen and to characterize effects of inhibitors, pH, and temperature (see Sections 4.6 and 4.7). Ammonia oxidation also follows Michaelis-Menten kinetics (Suzuki *et al.*, 1974), and again calculation of saturation constants for cell-free extracts and cell suspensions has yielded important information on effects of pH and inhibitors.

In such systems, product (nitrite or nitrate) concentration increases linearly. Variations in this linear increase with substrate concentration yield values for V_{max}, the maximum substrate-oxidizing activity, and K_m, the half-saturation constant for oxidation. This contrasts with growth studies in which product concentrations increase exponentially, and variation in specific growth rate with substrate concentration yields values for μ_m, the maximum specific growth rate, and K_s, the half-saturation constant for growth.

4.3. Measurement of Growth and Activity in Batch Culture

The basic growth kinetics of nitrifying bacteria in batch culture is no different from that of other bacteria. Growth is exponential, after a lag phase, and eventually enters deceleration and stationary phases. Cessation of nitrite oxidation usually results from complete utilization of nitrite; however, ammonia oxidizers produce nitrous acid, and reduction of pH limits growth unless media are buffered or the initial ammonia concentration is low. If initial substrate concentrations are high, growth may also be limited by substrate or product inhibition.

Techniques for measuring growth differ significantly from those used for heterotrophic organisms. Measurement of nitrifier biomass by absorbance techniques is difficult because of the low growth yields. The use of counting chambers to determine total cell concentration is tedious, while viable cell counts are inaccurate and require several months of incubation, as discussed above. The availability of automated techniques, however, facilitates measurement of substrate and product concentrations, and growth can also be measured by uptake of ^{14}C-labeled carbon dioxide or bicarbonate. Consequently, growth of nitrifying bacteria is usually measured by either substrate utilization or product formation.

Under conditions of balanced growth, e.g., during exponential growth in batch culture, product concentration increases exponentially. A semilogarithmic plot of nitrite or nitrate concentration versus time will be linear, with a slope equal to the maximum specific growth rate, as normally measured by turbidity. Slight deviations from this exponential relationship will occur during early growth if product has been carried over with the inoculum, but these deviations are rarely problematic. Specific growth rate calculated in this manner is similar to that calculated by cell counts and other techniques (Belser and Schmidt, 1980; Keen and Prosser, 1987a).

The situation can be represented by three differential equations,

$$\frac{dx}{dt} = \mu_m x \tag{1}$$

$$\frac{ds}{dt} = -\mu_m x/Y \tag{2}$$

$$\frac{dp}{dt} = kx \tag{3}$$

where x represents biomass or cell concentration, s and p represent substrate and product concentrations, k ($= \mu_m/Y$) is the activity per cell or per unit biomass, μ_m is the maximum specific growth rate, and Y represents biomass or cell yield. These equations assume that μ_m, Y, and activity per cell or per unit biomass are constant, as would be expected during balanced exponential growth, and they predict that proportional increases in biomass or cell concentration will equal those in product concentration. Belser and Schmidt (1980) demonstrated this for three strains of *Nitrosomonas europaea* (Fig. 1), *Nitrosospira,* and *Nitrosolobus,* with cell concentration determined by using counting chambers and the FA and MPN techniques. The FA technique was the most accurate, and differences between μ_m calculated from FA counts and changes in nitrite concentration were not significant for any of the five strains. Variability in μ_m values

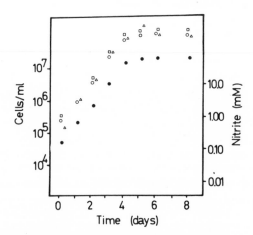

Figure 1. Changes in nitrite concentration (●) and cell concentration during growth of *Nitrosomonas europaea.* Cell concentration was determined by using counting chambers (□), FA labeling (○), and MPN techniques (△). [From Belser and Schmidt (1980), with permission.]

was assessed by using standard deviations calculated from successive pairs of data. Analysis of product concentrations gave lower standard deviations than did analysis of enumeration data and is therefore the technique of choice for measurement of μ_m in batch culture.

During balanced growth, specific growth rate is independent of cell activity. Cells with different activities, measured as amount of substrate converted per cell per unit time, may therefore have the same specific growth rate, although such differences will lead to differences in Y and the duration of growth. For example, Belser and Schmidt (1980) found two strains of *Nitrosomonas* with similar growth rates but cell activities of 0.009 and 0.023 pmol hr^{-1} cell^{-1}. The former strain was smaller, implying less efficient conversion of energy gained from ammonia oxidation into biomass.

4.4. Effect of Substrate Concentration

In many situations, activity per cell, specific growth rate, and yield coefficient are not constant. Cell activity, as discussed above, is governed by Michaelis–Menten kinetics, and the relationship between microbial specific growth rate and the concentration of a growth-limiting substrate (s) is generally described by the Monod equation:

$$\mu = \frac{\mu_m s}{K_s + s} \tag{4}$$

Laudelout *et al.* (1974, 1976) incorporated both growth and activity into a model for nitrification in batch culture based on the following equations:

$$\frac{dn_i}{dt} = \frac{\mu_m n_i s_i}{K_{s_i} + s_i} \tag{5}$$

$$\frac{ds}{dt} = \frac{-V_{max} s_i}{K_{mi} + s_i} \tag{6}$$

Equation (5) is equivalent to Eqs. (1) and (4), with biomass concentration replaced by cell concentration (n), V_{max} and K_m are the maximum rate and saturation constants of the Michaelis–Menten equation, and $i = 1$ for ammonium oxidation and 2 for nitrite oxidation.

V_{max} is related to cell number by the expression $kn = \mu/Y$, where k represents activity per cell. Modifications and predictions of this model are discussed in Sections 4.6 and 4.7. For now, the model indicates that both μ and k are not constant but vary with substrate concentration. Yoshioka *et al.* (1982) measured specific growth rate of ammonia and nitrite oxidizers in batch culture with initial substrate concentration ranging from 0.015 to 120 mM NH_4^+ and 20 to 100 mM NO_2^-. Growth was assessed by changes in product concentrations and total cell concentration by using the FA technique, and plots of μ versus s were used to estimate μ_m and K_s values. This approach is analogous to estimation of V_{max} for enzyme reactions from initial velocities determined at different initial substrate concentrations. However, the experiments of Yoshioka *et al.* (1982) involved several days of incubation during which substrate concentration, and therefore μ and k, would have been decreasing continuously. The inadvisability of calculating K_s from batch culture data is discussed by Nihtila and Virkkunen (1977); continuous-flow systems are always preferable. The situation is complicated further by variation in the efficiency with which energy from ammonia and nitrite oxidation is used to convert carbon dioxide into biomass. Glover (1985) found a decrease in the amount of carbon and nitrogen per cell during batch growth of both ammonia and nitrite oxidizers, suggesting a change in cell size. Chemostat studies (Section 4.5) suggested that this result was due to changes in substrate concentration, which emphasizes the fact that exponential increases in biomass concentration (true growth) and in nitrite concentration or even total cell concentration are equivalent only when growth is truly balanced.

Care must also be exercised in analyzing data from batch culture when specific growth rate is deliberately changed during exponential growth. Powell and Prosser (1986a) investigated inhibition of ammonia oxidation by nitrapyrin added to exponentially growing cells. In some cases, this led to exponential growth at a reduced rate. In calculating this reduced rate, the assumption of negligible initial nitrite concentration is no longer valid, and specific growth rate must be calculated by subtracting the concentration at the time of inhibitor addition from all subsequent concentrations. Unrestricted balanced but inhibited growth will then result in an exponential increase in product concentration.

In summary, the benefits of measuring specific growth rate of nitrifying bacteria by using changes in substrate and product concentrations are enormous. This applies particularly in complex environments, such as soil and biofilms, for which alternative measurements based on enumeration or determination of biomass are not possible. Extrapolation of this technique to such environments is discussed in Sections 4.8 and 5.

4.5. Growth in Continuous Culture

Skinner and Walker (1961) were the first to investigate nitrification in continuous culture and obtained steady-state cultures of *Nitrosomonas europaea* in an ammonia-limited chemostat. Further studies of nitrifiers in ammonia-limited (Glover, 1985; Keen and Prosser, 1987a) and nitrite-limited (Keen and Prosser, 1987a) chemostats have shown that nitrifying bacteria follow standard chemostat kinetics, represented by the equations

$$\frac{dx}{dt} = \frac{\mu_m s x}{K_s + s} - Dx \tag{7}$$

$$\frac{ds}{dt} = Ds_r - Ds - \frac{\mu_m s x}{Y(K_s + s)} - mx \tag{8}$$

where D is the dilution rate, s_r is the inflowing substrate concentration, and m represents the maintenance coefficient. In chemostat cultures of ammonia and nitrite oxidizers,

Glover (1985) found decreases in cellular carbon and nitrogen content, carbon dioxide fixation, enzyme activity, and organic carbon yield as dilution rate decreased. These effects were thought to result from changes in substrate concentration, and differences were also observed in yield in batch and continuous culture.

Belser (1984) used chemostat cultures of *Nitrosomonas europaea, Nitrosopira,* and *Nitrobacter* to determine variation in the ratio of bicarbonate uptake to carbon dioxide fixation. The ratio appeared to be independent of dilution rate and pH but was significantly affected by substrate concentration and some inhibitors. Both of these studies raise doubts regarding the validity of ^{14}C-labeled carbon dioxide or bicarbonate incorporation as a measure of nitrification rate in natural environments, where substrate concentrations will vary, and of use of specific inhibitors to differentiate carbon dioxide fixation by nitrifiers and by other autotrophs.

Keen and Prosser (1987a) analyzed growth of *Nitrosomonas europaea* and *Nitrobacter* in ammonia- and nitrite-limited chemostats, respectively, at a range of dilution rates. Growth was described well by Eqs. (7) and (8), and analysis of steady-state data yielded values for μ_m and K_s. Maximum specific growth rate calculated from batch and continuous culture differed significantly. For *Nitrosomonas,* the value obtained from continuous culture (0.039 hr^{-1}) was twice that from batch culture (0.017 hr^{-1}), while the situation was reversed for *Nitrobacter.* This phenomenon is commonly found in other organisms and reflects differences in metabolism during restricted (substrate limited) and unrestricted (substrate excess) growth. This should be borne in mind when μ/Y is used to assess maximum activity per cell in natural environments. Steady-state data also provide the values for true growth yield and maintenance coefficients for each organism. This is one of the few studies in which biomass, rather than cell, concentration was determined. Measurement of nitrifier biomass in natural environments is currently not feasible but should be attempted in liquid culture when appropriate. Expression of growth constants in terms of biomass rather than cell concentration is always preferable in eliminating variation due to differences in cell size and allowing reliable comparisons within and between populations and species.

Keen and Prosser (1987a) found no evidence for substrate inhibition in continuous culture at the concentrations studied and also demonstrated limitations of the Monod equation in describing transient growth. Step changes in dilution rate resulted in overshoots and undershoots in substrate and biomass concentrations rather than the predicted monotonic changes before establishment of subsequent steady states. This again is typical of other microorganisms, and accurate description of transient growth requires use of more complex, structured models. This work also provides the only determination of the saturation constant for growth in continuous culture. Other saturation constants relate to activity rather than growth; the relevance of this for natural populations is discussed below.

The work of Glover (1985) and others highlights an important point in studying nitrification kinetics. Short-term experiments, with incubation times less than the doubling time of the organism, measure activity, while long-term experiments measure growth. V_{max} and K_m and μ_m and K_s are related but different quantities. The former are properties of single enzyme reactions, while the latter are properties of a series of reactions following generation of energy from ammonia and nitrite oxidations. Environmental factors and inhibitors may significantly affect the linkage of ammonia and nitrite oxidation to growth. For example, Glover (1985) found variation in the biomass yield on carbon dioxide, and Belser and Schmidt (1980) found organisms with the same specific growth rate but different cell activities. Growth and activity measurements may each be appropriate in both pure-culture studies and natural environments, but the distinction is frequently not made. This leads to confusion in using constants obtained from pure-culture studies to predict nitrification rates in natural environments.

The distinction is equally important in physiological studies. For example, the accepted mechanism for inhibition of ammonia oxidation by nitrapyrin is chelation, by the inhibitor, of copper components of the cytochrome oxidase involved in ammonia oxidation (Campbell and Aleem, 1965). Evidence for this mechanism was the reversal of inhibition by copper in cell suspensions in which ammonium oxidation was measured by the rate of oxygen uptake in short-term experiments (50 min). Powell and Prosser (1986b) studied the effect of copper on inhibition of growth of $N.$ $europaea$ by nitrapyrin in batch culture experiments lasting up to 200 hr. Under these conditions, inhibition was increased in the presence of 0.046 μg of Cu^{2+} (as copper sulfate) ml^{-1}. These differences probably arise through differing effects on growth and activity, which emphasizes the need to distinguish these two types of study and dangers in extrapolating effects on activity to those on growth. Similar differences were also found between inhibition of activity and growth by chloropicolinic acid (Powell and Prosser, 1985).

4.6. Effect of Temperature on Nitrification

The effect of suboptimal temperatures on microbial specific growth rate is typically modeled by using the Arrhenius equation

$$\mu_m = Ae^{-E/RT} \qquad (9)$$

where A is a constant, E is the energy of activation, R is the gas constant, and T is the temperature in absolute degrees. This equation is based on a similar relationship describing the effect of temperature on the rate of chemical reactions. It has also been used to describe variation in K_s with temperature (Topiwala and Sinclair, 1971) and was incorporated into the model of Laudelout et al. (1974). Activation energies for both ammonia and nitrite oxidation were determined from experimental data, and simulations were carried out for nitrification in batch culture at several temperatures. These simulations predicted significant accumulation of nitrite at higher temperatures (20–30°C) due to differences in activation energies between the two processes and have implications for nitrification in natural environments. Similar temperature functions have been used in a number of computer models for sewage treatment (see Section 7), but there are few experimental data on the effect of temperature on growth of pure cultures of nitrifiers.

Recent studies on a wide range of heterotrophic organisms have shown the application of the Arrhenius equation to be invalid. These studies question the implicit assumption that the activation energy does not vary with temperature and propose instead the "square root" model (McMeekin et al., 1988). This empirical model provides a much better description of experimental data and suggests that the square root of the maximum specific growth rate is proportional to temperature up to the optimum, above which μ_m decreases exponentially with temperature. The relationship is represented by the equation

$$\mu_m = b(T - T_{\min}) (1 - e^{c(T - T_{\max})}) \tag{10}$$

where b and c are constants and $T_{\min}$ and $T_{\max}$ are intercepts on the x axis at temperatures below and above the optimum. $T_{\min}$ and $T_{\max}$ are notional minimum and maximum temperatures, and b and c are obtained from linear regression of experimental data. The model as yet has no mechanistic basis, but its application to the effect of temperature on nitrification should be considered in future studies.

4.7. Effect of pH and Inhibitors

The effect of pH on ammonia and nitrite oxidation has been quantified in short-term experiments involving both whole cells and cell-free extracts. Suzuki et al. (1974) analyzed such data for activity of N. europaea, assuming Michaelis–Menten kinetics (Fig. 2). They found that the K_m values for ammonia oxidation by both intact cells and cell-free extracts of N. europaea decreased with increasing pH if K_m was calculated in terms of total concentration of ammonia (i.e., both ionized and un-ionized forms). When

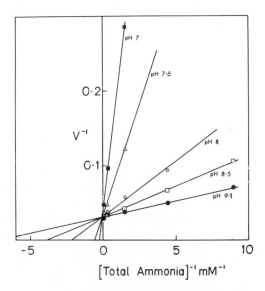

Figure 2. Effect of pH and ammonia concentration on the oxidation of ammonia by a cell suspension of *Nitrosomonas europaea*. The initial rate of oxidation (V) is expressed as nanomoles of oxygen consumed per minute. [From Suzuki *et al.* (1974), with permission.]

calculated in terms of un-ionized NH_3, K_m increased with pH for intact cells and was constant for cell extracts, indicating that NH_3, and not NH_4^+, acts as the substrate for ammonia oxidation. The increase in K_m for NH_3 with pH for intact cells was not thought by the authors to be significant. Ward (1987) observed a less marked decrease in the K_m for NH_4^+ for intact cells of *Nitrosococcus oceanus* but a significant increase in K_m for NH_3 with pH. This finding suggests that inhibition of ammonia oxidation by low pH is due to reduced NH_3 availability and additional effects on enzyme activity.

Boon and Laudelout (1962) used classical enzyme kinetics to demonstrate substrate inhibition of nitrite oxidation by intact cells and cell-free extracts of *Nitrobacter winogradskyi* and determined values for K_m and K_i. These yielded a nitrite concentration of 14.5 mM for optimal nitrite oxidation rate. Substrate inhibition was related to nitrous acid concentration and, in turn, to pH. This was confirmed by determination of the quantitative relationship between pH and K_i. Similar values for K_i in intact cells and cell-free extracts suggest that nitrite is freely permeable. Inhibition of nitrite oxidation at low and high pH is therefore due to noncompetitive inhibition by nitrous acid and competitive inhibition by hydroxyl ions, respectively.

Boon and Laudelout (1962) incorporated these features into a single equation and also derived an expression for the effect of pH on K_m and for noncompetitive product inhibition by nitrate. Laudelout *et al.* (1976) combined this with Eqs. (5) and (6) to describe the effects of pH on growth of nitrifying bacteria. H^+ was produced in direct relation to the oxidation of ammonia, and the authors also incorporated a term for oxygen limitation, assuming multiplicative double-substrate limitation kinetics. Experimental data fitted this model well and indicated that reduced oxygen supply will result in temporary accumulation of nitrite as a result of differences in saturation constants for oxygen. The significance for this in biofilms is discussed in Section 4.8.

Despite their potential commercial importance, the mechanisms and kinetics of inhibition by specific nitrification inhibitors have been little studied. Many are thought to act as chelating agents, and noncompetitive inhibition of nitrification might therefore be expected. Conflicting evidence for the mechanism of inhibition by nitrapyrin has been discussed above, and a detailed kinetic analysis of batch culture data has shown growing cells to be more sensitive than stationary-phase cells to inhibition by nitrapyrin (Powell and Prosser, 1986a).

In continuous culture, inhibition of *N. europaea* by potassium ethyl xanthate indicates the possibility of competitive inhibition rather than noncompetitive inhibition, which would be expected of chelating compounds (Underhill and Prosser, 1987). This again reflects a possible difference between growth and activity effects. At low concentrations of xanthate, stimulation of ammonia and nitrite oxidation occurred, with increases in cell concentration and yield coefficient but constant cell activity. Inhibition of *Nitrosomonas* was associated with increases in cell activity and yield coefficient, indicating increased efficiency of biomass formation. Inhibition of *Nitrobacter* increased cell activity and decreased the yield coefficient, and inhibition appeared to increase substrate utilization for maintenance.

Inhibition of ammonia oxidation by cell suspensions of *Nitrosococcus oceanus* has provided information on the biochemistry of nitrifiers. Glover (1982) observed competitive inhibition by methylamine, suggesting an active transport system for ammonium. Ward (1987) found inhibition of ammonium oxidation by oxygen at high ammonium concentrations but increased rates at low substrate concentrations. The rate of oxidation decreased with decreasing ammonia concentration, but the rate of decrease was less at saturating oxygen concentrations. Inhibition by methane was not fully described by kinetics for competitive inhibition, and the data suggested the presence of multiple active sites on the ammonia monooxygenase enzyme.

4.8. Surface Growth

In environments such as soil, sediments, and some sewage treatment processes, nitrifying bacteria grow on surfaces. It is therefore necessary to determine the effects of surface attachment on growth and activity. Nitrification again provides advantages for such studies in that kinetics of product formation may be used to determine activity and growth of attached microorganisms and growth rates may be determined without disturbing biofilms.

Keen and Prosser (1988) determined the kinetics of nitrite oxidation by *Nitrobacter* attached to glass microscope slides in batch culture and to ion exchange resins in continuous culture. Specific growth rates of adhered and free cell populations were similar, but cell activity was increased by the presence of glass surfaces in batch culture. Growth in continuous culture was studied in airlift column fermentors containing circulating ion exchange resin beads colonized by *Nitrobacter*. The system was operated at dilution rates greater than the critical dilution rate for free cells, and most cells were therefore attached, any free cells arising from detachment. After establishment of steady state, overall growth in such a system (i.e., biomass productivity) will be represented by the product of dilution rate and steady-state free cell biomass. Overall activity, or nitrate productivity, will equal the product of dilution rate and steady-state nitrate concentration. Such a steady state will be established only when the rate of colonization equals the rate of sloughing. In the system of Keen and Prosser (1988), this stage had probably not been reached even after 2700 hr, but nitrate productivity was used to estimate biomass concentration, assuming constant cell activity throughout the biofilm. Attached biomass calculated in this manner showed an increase typical of batch growth curves, with lag, exponential, and deceleration phases (Fig. 3). The specific colonization rate calculated from the exponential phase was 0.0011 hr^{-1}, significantly lower than the maximum specific growth rate in batch culture.

The activity of attached cells was less than that of free cells despite apparent stimulation of activity by surfaces in batch culture. This result may be due to increased cell damage, reduced viability caused by abrasion, or limitation of activity by diffusion of oxygen and other nutrients through the biofilm. Evidence also indicated uncoupling of growth and activity, with increases in nitrate productivity without associated increases in attached biomass, which is thought to be due to increased production of slime material involved in later stages of adhesion. This experimental system has also been used to determine the combined effects of attachment and pH on growth and activity of *Nitrobacter* (Keen and Prosser, 1987b).

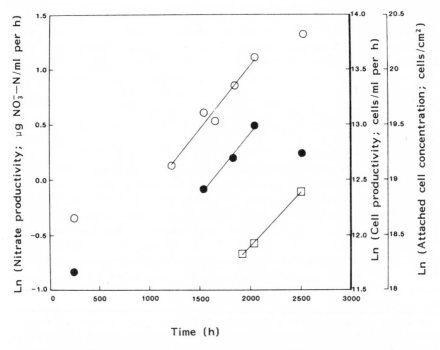

Time (h)

Figure 3. Changes in free cell productivity (●), nitrate productivity (○), and attached cell concentration (□) during growth of *Nitrobacter* in a continuous-flow airlift column fermentor containing ion exchange resin as a substratum for growth. [From Keen and Prosser (1988), with permission.]

In batch culture systems, the kinetics of nitrification in the presence of surfaces depends on the extent to which the surface is colonized. Product formation by *Nitrobacter* in the presence of uncolonized glass slides is exponential (Keen and Prosser, 1988), implying exponential surface growth. Similar kinetics are found for growth of *Nitrosomonas* on uncolonized clay minerals until reduction in pH limits growth (Armstrong and Prosser, 1988). Product formation in the presence of heavily colonized surfaces will be linear, since growth is likely to be limited by factors other than substrate concentration (e.g., space limitation and diffusion of nutrients through the biofilm).

More sophisticated models of surface growth of pure cultures of nitrifying bacteria have been provided by Bazin and Saunders (1973), Saunders and Bazin (1973), and Prosser and Gray (1977). These describe the situation in a packed column reactor in which particulate material (e.g., soil or glass beads) is fixed, providing a surface for microbial attachment. Medium flows continuously through the column and, for nitrifying bacteria, consists of inorganic medium containing ammonium. Effluent from the base of the column is analyzed for ammonia, nitrite, and nitrate, and ports allow sampling of soluble and particulate material at different depths. The system is therefore similar to a chemostat with washout of soluble components (e.g., substrates and products) but not of biomass and differs from the airlift column fermentor described above in that gradients of soluble components will be established. Initial modeling (Bazin and Saunders, 1973) consisted of modifications to Eqs. (7) and (8), omitting the term for maintenance and replacing the biomass washout term with the logistic equation

$$\mu = \mu_m(1 - x_{max}/x) \tag{11}$$

The logistic equation is based on the assumption that specific growth rate decreases as a linear function of population size, reaching zero at $x = x_{max}$, and is frequently used to describe limitation of microbial growth by space.

The model of Bazin and Saunders (1973) is valid only under conditions of nutrient excess. Substrate limitation results in variation in specific growth rate with both time (as the biofilm develops) and distance (as substrate concentration decreases during passage of material down the column). This decrease depends on biofilm development and therefore itself varies with time and must be described by partial differential equations as formulated by Saunders and Bazin (1973). Analytical solution of these equations is possible only under certain conditions, and simulation generally requires numerical approximation techniques. Prosser and Gray (1977) adopted such an approach to describe nitrite oxidation in a fixed glass bead column. The column was represented as a number of theoretical compartments, each considered homogeneous with respect to biomass and soluble components. Changes within compartments were described by chemostat-type equations, with output of soluble components from one compartment providing input to the compartment below but with no transfer of biomass. Fuller description of this system also requires consideration of dispersion and diffusion [see detailed discussion by Prosser and Bazin (1988)].

4.9. Summary

Theoretical and experimental studies of nitrification in pure culture provide information on the physiology of growth and activity of nitrifying bacteria and effects of environmental factors. They allow calculation of growth constants under ideal conditions and critical study of the different types of kinetics and situations in which they may operate. This information is necessary for study of more complex systems, providing a basis both for measurement of rates and activities in natural environments and for subcomponents of more complex mechanistic models.

5. Nitrification in Soil

5.1. Measurement of Rates of Nitrification

Nitrification rates in soil are traditionally measured by using incubation techniques. A quantity of soil or soil slurry is incubated at constant temperature, and soil is sampled at regular intervals for measurement of nitrite and nitrate concentrations. Ammonia may be added, or ammonia already present in the soil may act as the substrate. Pure-culture studies suggest three types of kinetics (Fig. 4).

First, zero-order kinetics, characterized by linear increases in nitrite plus nitrate concentration, will result from short-term (activity) measurements in the presence of nonlimiting ammonia concentrations. "Short term" may again be defined as an incubation period during which population size does not change significantly. Zero-order kinetics will also occur in enriched soils in which the maximum nitrifying population has been reached. Soil may then be considered equivalent to a standing crop of nitrifying bacteria or to a cell suspension, and the nitrification rate, obtained by linear regression of the increase in product concentration, is equivalent to V_{max}. In soil studies, this is termed the nitrification potential and represents the maximum rate of nitrification for a particular soil under the conditions of incubation.

Second, short-term measurements at limiting substrate concentrations will exhibit Michaelis–Menten kinetics, with first-order rate kinetics (with respect to substrate concentration) at ammonia or nitrate concentrations approaching the K_m value. As substrate concentration is increased, nitrification rate will approach V_{max} and kinetics will be zero order.

Third, in long-term incubations of nonenriched soil, growth will be unrestricted if the initial substrate concentration is significantly greater than the saturation constant for growth and if no other factors limit growth. This will result, as in liquid batch culture, in an exponential increase in nitrite plus nitrate concentration; i.e., product formation will be first order with respect to biomass and zero order with respect to substrate. A semilogarithmic plot of product concentration vs. time will therefore yield the maximum specific growth rate, μ_m. Unrestricted and exponential growth may not continue indefinitely. If complete ammonia utilization alone limits growth, exponential product formation will be followed by a sudden change to constant product levels. The buffering capacity of soil reduces growth limitation due to acid production, but other factors such as product inhibition and oxygen limitation will result in a prolonged deceleration phase and incomplete utilization of ammonia. Growth at submaximal rates will also occur if initial ammonia concentration is low with regard to the K_s value.

Examples of all three types of kinetics are found. The short-term nitrification activity technique of Schmidt (1982) for measurement of nitrification potential assumes zero-order kinetics, whereas first-order kinetics has been observed by Feigenbaum and Hadas (1980). Myrold and Tiedje (1986) evaluated the use of ^{15}N dilution techniques for simultaneous estimation of rate of nitrification, mineralization, immobilization, and denitrification. First-order kinetics were generally more appropriate but zero-order kinetics prevailed occasionally, particularly in short-term incubations, as expected from the analysis described above. Malhi and McGill (1982) determined initial rates of nitrification in soil incubated with a range of initial ammonium concentrations. Michaelis–Menten-type kinetics was observed, allowing calculation of V_{max} and K_m at various temperatures and soil moisture contents. These values were then used to simulate the time required for complete utilization of different levels of ammonium, and comparison with experimental data provided explanations for confusion regarding choice of zero- or first-order kinetics.

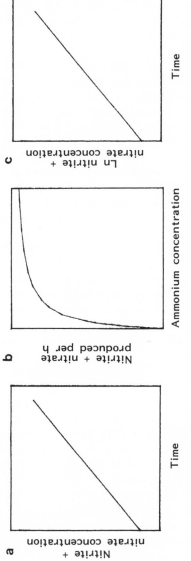

Figure 4. Three basic types of kinetics for nitrification in soil incubation studies. (a) Linear increases in product concentration during short-term incubations or in soil "saturated" with nitrifiers; (b) variation in short-term activity with ammonia concentration; (c) exponential product formation resulting from exponential growth.

The simple Michaelis–Menten model, however, was not appropriate under conditions of varying pH, soil moisture content, and other environmental factors. Powell and Prosser (1985) found exponential increases in nitrite concentration following inoculation of sterilized soil with a pure culture of *N. europaea*. The calculated specific growth rate was not significantly different from that of the same organism growing in liquid culture but was greater than rates estimated from changes in population sizes (Macdonald, 1979) and from reperfusion systems (Morrill and Dawson, 1962). These lower rates were obtained in unsterilized soil, where predation and competition for ammonia may reduce population growth.

The kinetics outlined above are idealized and discrepancies due to varying pH and moisture content have already been noted. Oxygen supply will also vary within the soil and with time, particularly during long-term studies, and reduced oxygen supply in soil slurries can affect even short-term nitrification activity. Other soil processes, such as mineralization, immobilization, and denitrification, will affect nitrate and ammonia concentrations (Addiscott, 1983), species variability may result in different kinetics, and initial ammonia concentration may be important in comparing soils of different cation exchange capacity. Yadvinder-Singh and Beauchamp (1985) found that more ammonia was required to saturate soils of higher cation exchange capacities and that this significantly affected nitrification kinetics.

These changes will produce complex kinetics which in some cases may be described empirically. For example, Hadas et al. (1986) described nitrate production and ammonia disappearance by using a modified version of the logistic equation. The rate of nitrification was assumed to decrease as a negative linear function of nitrate concentration, and nitrate concentration was therefore described as a function of time by integration of a modified form of Eq. (11) to yield

$$NO_3^- = \frac{a}{1 + (a/NO_{30}^- - 1) \exp[-ak(t - t_0)]} \tag{12}$$

where a and $NO_3^-{}_0$ are the initial and final nitrate concentrations, t_0 is the initial time, and k is a constant. A plot of nitrate concentration versus time generates a sigmoid curve, and Hadas et al. (1986) calculate K_{mx} as the maximal slope at the point of inflection, i.e., when the nitrate concentration equals $a/2$. Extrapolation of the line of slope K_{mx} to $NO_3^-{}_0$ allowed calculation of the delay period or lag period, t'. Values of a, k, and $NO_3^-{}_0$ were calculated by least squares fit of experimental nitrate and ammonia concentrations. Increases in nitrate concentration were equivalent to decreases in ammonia concentration, suggesting negligible mineralization or immobilization. The value of a will depend on initial ammonia concentration, but K_{mx} and t' may be used to compare nitrification in different soils and under different environmental conditions. Using this technique, Hadas et al. (1986) found that t' increased with soil depth, as might be expected if ammonia or oxygen concentration were greater nearer the soil surface.

Seifert (1980) proposed use of the Gompertz function for soil incubation studies

$$NO_3^- = NO_{3\ max}^- \exp\ [-\ m \exp\ (-\ kt)] \qquad (13)$$

where m and k are constants and $NO_{3\ max}^-$ is the final nitrate concentration. Such empirical models may serve descriptive and comparative functions and may fit experimental data better than simple linear or semilogarithmic plots. The requirement for such complex kinetic models, however, indicates deficiencies in experimental systems. These models are not based on a knowledge or understanding of the physiology of the organism or its interaction with the environment, and the constants within the equations have no biological meaning. Less complex kinetics, e.g., first-order and zero-order rate kinetics, incorporate environmental and biological assumptions that are effectively tested by incubation experiments. Deviations from such kinetics will actually provide important information regarding the system and deficiencies in it, and the fitting of arbitrary functions, such as the logistic and Gompertz functions, may mask useful information and prevent measurement of true nitrification rates. Another major problem is that the ability to distinguish between different types of kinetics depends on the accuracy of the experimental data. Even in pure-culture studies, variability may make such distinctions difficult, and data from soil samples are necessarily less accurate. This fact provides further arguments for use of simple kinetic models.

The models described above have been subjected to more fundamental criticisms. All assume that nitrifying bacteria are distributed homogeneously throughout the soil. Molina (1985) challenged this assumption, suggesting that nitrifiers exist in clusters which, when supplied with ammonium, produce nitrite and nitrate until self-inhibited by acid production. Nitrate production therefore reflects cumulative activity of such microbial clusters. A stochastic approach was then adopted which demonstrated that nitrate production followed the cumulative distribution of lag phases for soil microaggregates, each of which was estimated to contain 2.01 clusters of ammonia oxidizers. Further support for this approach was provided by Darrah *et al.* (1987). They considered microbial clusters to influence a surrounding sphere of soil, and kinetics of nitrate production varied with the relative size of cluster and surrounding soil spheres.

5.2. Effects of Environmental Factors

The effect of temperature on soil nitrification is most commonly described, as in liquid culture, by modifications of the Arrhenius equation. Addiscott (1983) calculated Arrhenius coefficients for a number of soils and compared them with coefficients for mineralization and with other published values. Seifert (1980) determined values of k in the Gompertz equation [Eq. 13 above] as a function of temperature. Experimental values were then fitted to the Arrhenius equation, and a technique was developed for predicting changes in nitrate concentration during periods of changing temperature. The Arrhenius equation may be sufficient if activity only is being measured, but if growth is occurring, the square root model described above should also be tested for applicability in soil.

Darrah *et al.* (1985) presented a model for growth of nitrifying bacteria, based on the Monod equation, that also considered diffusion of ammonia, ion exchange processes, and acid production resulting from nitrification. Theoretical predictions agreed well with experimental data obtained from a soil column system in which unamended soil was placed adjacent to soil amended with ammonium. Subsequent work (Darrah *et al.*, 1986a,b) considered effects of high ammonium chloride and sulfate concentrations, and an empirical model for the effect of pH on short-term nitrification activity was used to describe variation in specific growth rate with soil pH (Darrah *et al.*, 1986c).

The effects of other environmental factors have also been modeled (Bazin *et al.*, 1976; Belser, 1979), but surprisingly few attempts have been made to consider effects of nitrification inhibitors. Meikle (1979) developed an empirical model for predicting effects of nitrification inhibitors, in particular nitrapyrin, on nitrate accumulation in soil. Ammonia oxidation and nitrapyrin hydrolysis followed first-order kinetics, while nitrification increased as a simple linear function of both temperature (range, 2–35°C) and pH (range, 4.6–8.2). Proportionality constants were determined by multiple regression analysis of experimental data, and corrections were made for the effects of initial ammonia concentration. The rate of nitrapyrin degradation was related to temperature by an Arrhenius function and to organic matter content by a linear function. An empirical relationship was also determined for the effect of nitrapyrin on the rate of nitrification. The major environmental factor was considered to be temperature, whose variation throughout the year was described by a fourth-degree polynomial equation. The model was then used to predict the effect of nitrapyrin addition on loss of ammonia from two soils, with varying degrees of success. Gilmour (1984) modified this model by replacing the linear temperature function with an Arrhenius term, including effects of moisture content and defining the effect of pH on nitrification. This modified model described nitrification rate (NR) by the equation

$$NR = [\exp(A/TB)] \times CWC + D \times (E\mathrm{pH} - F)/0.95\, N_t \qquad (14)$$

where A through F are constants, WC is the gravimetric water content, and N_t is the initial ammonia concentration. An empirical relationship was also determined for the effect of organic matter content on inhibition,

$$I = J \times K^L \qquad (15)$$

where I represents percentage inhibition and K is organic carbon content. The author suggests that such models are of practical use in predicting effects of inhibitors in the field. This, however, will obviously depend on accuracy of both the model equations and the values of constants. The latter have no obvious biological meaning and must be calculated for each soil examined.

5.3. Other Nitrogen Transformations

The models discussed so far have considered nitrification in isolation from other soil nitrogen transformations, although the influence of ammonification and immobilization has been mentioned. Models of nitrification within the soil nitrogen cycle must

obviously consider the many other nitrogen transformations as well as soil physicochemical properties, and the above models serve as components of such large models. Conversely, larger-scale models allow prediction, and increase our understanding,
of the influence of other factors and processes on nitrification.

Such models may be broadly classified again as empirical or mechanistic. The
former attempt to factor out the rates and kinetics of different processes and effects of
environmental factors by statistical analysis of experimental data, usually by multiple
linear regression analysis. This produces simple relationships between pool sizes and
transfer rates, e.g., the models of Duffy *et al.* (1975) and Endelman *et al.* (1974),
discussed by Bazin *et al.* (1976). The mechanistic approach incorporates information
from processes studied in isolation into a composite model, which is then used to
generate predictions to be tested experimentally. Such models are more complex and
contain more constants, which must be obtained from experimental data. They are
therefore more difficult to test critically, but this disadvantage is counterbalanced by the
increased understanding of the soil nitrogen cycle that they provide.

Mechanistic models are usually based on zero- or first-order kinetics for nitrification and other processes. For example, Bhat *et al.* (1980) developed a comprehensive
model for transformation of nitrogen applied to farm wastes. Nitrification was described
by a zero-order rate constant that was related to temperature by an Arrhenius expression
and linearly related to moisture content and pH within certain ranges. It is therefore
similar to the model of Gilmour described above. This was incorporated into a large
model that considered ammonification, denitrification, immobilization, plant uptake,
and leaching of nitrogen. Inputs to the model were data on weather conditions (e.g.,
rainfall and temperature), soil properties (e.g. organic carbon and nitrogen content and
moisture characteristics), and the amount, composition, and timing of application of
farm waste. This was in addition to the rate constants and constants of proportionality
required by the individual equations. Model predictions were compared with experimental data from lysimeters with perennial rye-grass receiving four applications of pig slurry
per year. Changes in leachate volume throughout the year were predicted well by the
model, as were leachate nitrate concentrations (Fig. 5), though the latter were lower than
predicted. This was thought to be due to underprediction of mineralization of organic
nitrogen or overprediction of immobilization, plant uptake, or denitrification. Discrepancies between these and other aspects of predicted and experimental behavior suggested improvements in the model, particularly in terms of water mixing and leaching
through different soil types. Therefore, although the model was used to predict nitrification transformations, water movement, and plant yield, it also added significantly to an
understanding of the mechanisms controlling the system. The discrepancies could be
accounted for by specific failings in the model. With empirical models, such discrepancies arise solely through the inability to provide correct equations, which in themselves
have no biological meaning.

Other models favor the use of first-order rate kinetics of nitrification, and many are
based on the large-scale system models of Beek and Frissel (1973), van Veen and Frissel
(1981), and Mehran and Tanji (1974). Hsieh *et al.* (1981) used the latter as a basis for
predicting nitrogen transformations following applications of sewage sludge to soil.

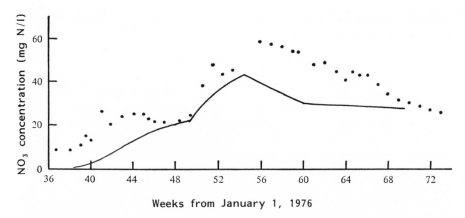

Weeks from January 1, 1976

Figure 5. Changes in leachate nitrate concentration observed experimentally (●) and predicted (———) by the model of Bhat *et al.* (1980). [From Bhat *et al.* (1980), with permission.]

Model predictions were good, but the authors suggested a requirement for inclusion of effects of temperature and moisture content. Cameron and Kowalenko (1976) extended Mehran and Tanji's model to account for exchange with clay-fixed ammonia and changes in microbial activity due to growth. The final model consisted of five equations describing changes in organic nitrogen, nitrate concentration, and soluble, free, and exchangeable ammonia. Experimental and predicted data on these components were then compared and the rate coefficients optimized to give the best fit. This fit was reasonable, but independent estimation of rate coefficients is preferable and sensitivity analysis helps indicate the validity of this approach. This involves determination of the effects on model predictions of similar proportional changes in each of the constants to determine which has the greatest effect. Such analysis is important in complex models, as the quality of fit increases with the number of constants. This proccdure was repeated at different temperatures and moisture contents. Regression analysis was then carried out to determine the effects of temperature and moisture content on rate coefficients for ammonification and nitrification, providing empirical equations for these effects.

A combination of empirical and mechanistic models was also used by Hirose and Tateno (1984) to investigate changes in soil nitrogen during successional changes accompanying colonization of *Polygonum cuspidatum* in a desert soil. Principle component analysis was used to determine relationships between bulk density, water content, pH, organic carbon, and nitrogen, ammonia, and nitrate concentrations during succession, allowing discrimination of two component factors controlling variation. The first was related to soil formation and was associated with a decrease in bulk density and increases in water content, organic carbon, organic nitrogen, and ammonium nitrogen. The second factor was related to nitrogen mineralization and nitrification, resulting in an increase in nitrate concentration. This second aspect was then investigated further, using a compartmental model involving plant uptake, nitrification, mineralization, and immobilization, and transfer reactions between compartments were described by first-order rate kinetics. Assuming equilibrium, which was considered valid in the short term, expressions were derived for the ratios of ammonium (N_2) to organic nitrogen and of nitrate (N_3) to ammonium as functions of rate constants, e.g.,

$$\frac{N_3}{N_2} = \frac{k_{23}(Q + R) + l_2R}{(k_{30} + k_{31})(Q + R) + l_3Q} \tag{16}$$

where k_{23}, k_{30}, and k_{31} are rate constants for transfer between compartments, l_2 and l_3 are leachate losses, and Q and R are constants. Ratios were predicted to decrease with succession due to increased microbial and plant activity, as was observed experimentally. In fact, the ratio was found to decrease linearly with increasing organic content, and the tested model therefore explains changes in compartment sizes with development of the ecosystem.

An excellent example of the application of a systems analysis approach to the soil nitrogen cycle is provided by Addiscott and Whitmore (1987). Their model incorporated several submodels for nitrate leaching, mineralization, crop growth and uptake, and nitrification, the latter being described by zero-order kinetics dependent on initial ammonium concentration, temperature, and moisture content. A good fit was obtained, using several criteria, between model predictions and experimental data on growth of winter wheat (Fig. 6). Sensitivity analysis was applied to determine the extent to which each model parameter affected different aspects of the system, thereby increasing understanding of the system and highlighting those rate constants and coefficients whose determination required greatest accuracy. The final model was then put to practical use for farmers by predicting adjustments to fertilizer applications required for optimization of nitrogen uptake and minimization of loss through leaching.

5.4. Soil Columns

The extensive use of reperfusion systems and continuous-flow soil columns has recently been reviewed by Prosser and Bazin (1988) and will not be discussed in detail here. Theoretical models of continuous-flow systems have been directed either toward physicochemical properties (e.g., diffusion, ion exchange, and hydrodynamic dispersion, with nitrification described by first-order kinetics) or toward microbiological aspects where microbial growth and activity are considered in detail. A combined theoretical approach is currently lacking.

These experimental systems have served two purposes. The first has been measurement of rate constants, diffusion coefficients, saturation constants, etc., associated with nitrification in the soil. The reliability and relevance of these values depend, respectively, on the validity of the assumptions on which they are based and the extent to which the experimental system represents the environmental situation. More correctly, soil columns have been used as experimental systems designed to test rigorously theoretical models, thereby leading to a greater understanding of soil nitrification and factors that

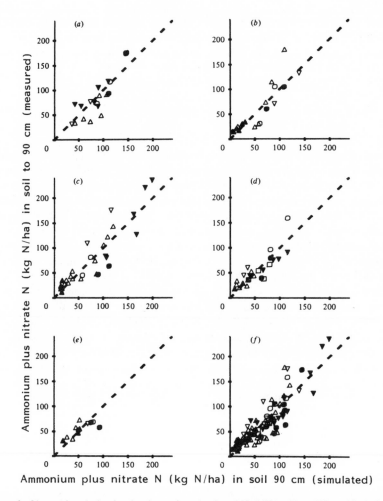

Figure 6. Observed and simulated values of total mineral-N ($NH_4^+ - N + NO_3^- - N$) during crop growth. (a)–(f) represent data from six sites (experimental and private farmland) in England. [From Addiscott and Whitmore (1987), with permission.]

affect it. This information may then be of applied use [see, for example, Tillotson and Wagenet (1982)].

Several workers have used cell conversion rates or cell activities obtained from soil column systems, and from pure cultures of nitrifying bacteria, in combination with rate measurements to assess numbers of nitrifiers in the soil and other environments (Belser, 1979). This approach should be adopted with extreme caution. Pure-culture studies have shown cell activities to vary widely among strains and under different cultural and environmental conditions. The extrapolation of activity values to natural environments is therefore dangerous and unlikely to produce results more reliable than those produced by existing enumeration techniques.

6. Nitrification in Aquatic Ecosystems

6.1. Measurement of Rates of Nitrification

Techniques for measuring nitrification in aquatic environments are similar in most respects to those already discussed, involving measurement of substrate disappearance or product formation, but with greater emphasis on use of ^{15}N-labeled ammonium and nitrate and incorporation of ^{14}C-labeled bicarbonate in the presence and absence of nitrification inhibitors. The techniques have been evaluated recently by Hall (1986) and Ward (1986) for freshwater and marine environments, respectively. Although some workers acknowledge the possibility of growth during rate measurements, addition of ammonium results in linear increases in product concentration over periods of time significantly greater than the expected doubling time, implying the absence of growth or "saturation" with nitrifying bacteria. The latter may occur in surface layers of sediment but is unlikely in water columns. In must therefore be assumed that other factors are limiting or preventing growth during rate measurements.

6.2. System Models

More detailed models of nitrification have been provided for river systems and for sediments. Cooper (1983a,b) estimated *in situ* nitrification activity (INA) and potential nitrification activity (PNA) in a stream receiving high levels of ammonia. Exponential increases in INA following removal of nitrifiers by storm events were used as a measure of *in situ* growth rate. This rate was found to be significantly less than that of pure cultures isolated from the same environments. PNA values were also found to be greater than INA values, and environmental stresses, particularly temperature, pH, and ammonia concentration, were considered important in reducing *in situ* growth rates. Comparison of PNA values and nitrifier counts in sediments and overlying waters indicated that only 2% of nitrification in streams resulted from benthic organisms.

Cooper (1984) quantified the effects of temperature, pH, and ammonia concentration by varying conditions for short-term nitrification assays. Changes in nitrification potential with temperature were described by an Arrhenius function, and activity decreased linearly with pH in the range 5.2–7.2. The effect of ammonia followed a Michaelis–Menten relationship, but with wide variations in K_m values at different sites. Values for the required constants were obtained at different depths, leading to a composite equation to predict total benthic nitrogenous oxygen consumption.

Use of simple kinetic models to predict effects of nitrification on dissolved oxygen concentrations in rivers is described by Cooper (1986). In deep rivers where planktonic nitrification may be significant, nitrification is described by a Monod-type function. Values for yield coefficient were obtained experimentally, and those for μ_m and K_s were obtained by computer fitting of data on ammonium removal. Prediction of dissolved oxygen concentration was achieved by considering both nitrification and reaeration. The applicability of the Monod equation depends on initial substrate concentration and on nitrifier biomass, as illustrated schematically in Fig. 7. Low ammonia concentrations will give rise to first-order or mixed-order kinetics which at low nitrifier population levels are related to growth. This analysis applies equally to nitrification in other environments.

Vanderborght and Billen (1975) modeled nitrification in aquatic sediments by using simple kinetic models for nitrification and denitrification based on experimental measurement of rates in sediments at different depths. In muddy sediments, redox potential

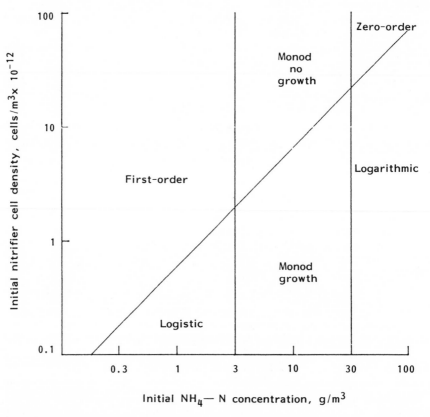

Figure 7. Applicability of six kinetic models of planktonic nitrification as a function of initial substrate concentration and cell density. [Redrawn from Cooper (1986).]

was too low for significant nitrification, but in sandy sediments the process occurred in the upper 1 cm. The rate of change of nitrate concentration (C) varied with both time (t) and distance (z) according to the equation

$$\frac{\delta C}{\delta t} = D \left(\frac{\delta^2 C}{\delta z^2} \right) - w \left(\frac{\delta C}{\delta z} \right) + \epsilon k_n - (1 - \epsilon) k_d C \qquad (17)$$

where D is the apparent diffusion coefficient, w is the sedimentation coefficient, k_n represents the zero-order rate coefficient for nitrification, and k_d is the first-order rate coefficient for denitrification. Nitrification and denitrification are considered mutually exclusive, the former occurring in the surface layers of the sediment ($\epsilon = 1$) and the latter occurring below this depth ($\epsilon = 0$). The model predicts accurately the nitrate profiles within sediments for values of k_n and k_d similar to those measured experimentally. It has been expanded by Vanderborght et al. (1977) and Jahnke et al. (1982) to incorporate ammonification, oxygen utilization, and sulfate reduction. Goloway and Bender (1982) and Christensen and Rowe (1984) adopted a similar approach but used a more complex function for nitrification rate, which was considered to decrease exponentially with depth. Their model was based on the decomposition of organic matter in sediments, releasing ammonium, and related nitrification rates to oxygen consumption rates. Detailed analysis of this model was used to determine the significance of nitrification in oxygen consumption in deep sediments.

7. Nitrification in Sewage and Waste Water Treatment Processes

The composition of organic matter in sewage treatment plants results in the release of ammonia, and high levels of ammonia may also be present in inflowing material, particularly from animal waste products. This leads to ammonia concentrations that would be toxic to fish if released into water courses. Process design is therefore directed to facilitate nitrification, and strategies are frequently adopted to encourage nitrogen removal by coupling nitrification and denitrification. A major problem in achieving nitrification is the slow growth rate of nitrifying bacteria in comparison to heterotrophs and in activated sludge systems the throughput time must be sufficiently low to prevent washout of nitrifiers. This problem is overcome in systems such as trickling filters, where nitrifiers form a component of a multispecies biofilm.

Nitrification in sewage treatment processes has recently been reviewed by Painter (1986). Emphasis here will be placed on kinetic and modeling approaches and their use in optimizing design of sewage treatment processes. (In this area the major input has been from chemical engineers.) In many cases, these approaches have increased our knowledge of the physiology of nitrifiers; in others, as in soil, they have highlighted biological problems of monitoring and assessing nitrifier growth and activity that frequently prevent complete and accurate quantitative descriptions of nitrification processes. The majority of work on nitrification in sewage relates most readily to activated sludge processes in which cells are freely suspended in an aerated "medium." These studies provide a link with, and are frequently based on, pure-culture studies; discussion of them will be followed by a description of systems in which nitrifying bacteria are present in biofilms.

7.1. Basic Kinetics

Most models of nitrification in sewage treatment are based on Eqs. (1) and (2), with specific growth rate described by the Monod equation. In most systems, ammonium concentration is significantly greater than the saturation constant for growth for ammonia and, as a simplification, ammonium conversion is usually considered zero order with respect to ammonium concentration and first order with respect to biomass. Maintenance of nitrifiers in flowthrough, suspended growth systems (equivalent to chemostats) requires a dilution rate less than μ_m. A more appropriate term for sewage treatment is sludge age, or sludge retention time, which is equivalent to $1/D$ and must be greater than the doubling time/ln 2. Both μ_m and μ will of course depend on factors such as temperature and pH, and a safety factor is therefore defined such that the ratio of maximum to required specific growth rate is greater than the ratio of peak nitrogen load to average load (Sharma and Ahlert, 1977). In systems in which sludge is recycled (equivalent to chemostat with feedback) sludge age is greater than $1/D$ and is defined as mixed liquor volatile suspended solids (MLVSS) in the reactor/MLVSS sludge yield or wastage per day.

The Monod equation describes growth rate limitation by a single substrate. Concentrations of other substrates, most importantly oxygen, may also be low in certain circumstances, and this has led to formulation of models for multiple substrate limitation. For example, Shah and Coulman (1978) modified the Monod equation, giving

$$\mu = \mu_m \left(\frac{s_1}{K_1 + s_1} \right) \left(\frac{s_2}{K_2 + s_2} \right) \tag{18}$$

where s_1 and s_2 are the concentrations of two substrates limiting growth, with saturation constants K_1 and K_2. This type of kinetics has been termed multiplicative because the $s/(K + s)$ terms are multiplied together. An alternative (threshold kinetics) in which the minimum $s/(K + s)$ value is chosen has been proposed for other systems, and its use seems to be supported by the scant experimental data available. Multiplicative kinetics is more commonly used, is more convenient mathematically, and has been used extensively in modeling nitrification. Argaman (1982) has extended the approach to limitation by three substrates, including carbon as a limiting component.

7.2. Measurement of Growth Constants

Values of three biological constants, μ_m, K_s, and Y, are required to describe basic kinetics, and their measurement raises significant practical and experimental problems because of the heterogeneous nature of sewage. A number of approaches have been adopted, the simplest being extrapolation of values obtained in pure-culture studies. While these may provide values in the correct range, *in situ* measurements are preferred.

Knowles *et al.* (1965) determined values of μ_m, K_s, and initial nitrifier cell concentration or biomass by comparison of experimental data on substrate utilization or product formation with those predicted by equations similar to Eqs. (1)–(4), using values of Y taken from the literature. Variation in Y was thought to affect the estimate of the initial cell concentration but not estimates of μ_m and K_s. This approach has subsequently been adopted by a number of other workers (Poduska and Andrews, 1975), although the disadvantages of calculating K_s from batch data have been discussed in Section 4.4. Measurement of μ_m using this technique is satisfactory; Beccari *et al.* (1979) also estimated μ_m by determining the critical dilution rate, D_{crit}, the dilution rate at which biomass washes out in a chemostat culture. This rate was determined in wastewater treatment plants by observing the sludge retention time at which nitrification ceased.

Respirometric methods have also been used to measure growth constants. These are short-term incubations in which V_{max} and K_m are estimated from differences between ammonium oxidation in the presence and absence of added ammonium (Charley *et al.*, 1980). Beccari *et al.* (1979) have also used a respirometric method to determine the yield coefficient and the sludge decay rate, the latter being important in recycle systems in which a proportion of the recycled biomass loses activity. Shieh and LaMotta (1979) measured nitrification rates in a two-stage chemostat system. Their aim was to use vigorous agitation to eliminate effects of reduced mass transport due to floc formation (termed internal diffusional resistance) and poor oxygen transfer in the bulk medium (termed external diffusional resistance). The authors claimed that this allowed measurement of the true intrinsic rate of nitrification, calculated from the rate of ammonium utilization in short-term (less than 5 hr) incubations, during which growth would be negligible. Both V_{max} and K_s were found to increase to a maximum as initial ammonium concentration increased, although both are normally considered constants. This variation is attributed to adaptation of the inoculum to different ammonium concentrations. The work may therefore be relevant to transient conditions but not to steady-state conditions. In addition, K_s was calculated from batch culture data and, as growth was negligible, represents K_m rather than K_s.

Novak (1979) found a decrease in the zero-order rate constant for nitrite oxidation by *Nitrobacter* as agitation and impeller power input increased. The most likely explanation for this result is increased cell damage at higher impeller speeds; however, the significance of this in sewage systems is unlikely to be great, as the experimental system consisted of a slurry containing glass beads, which had been shown to cause damage to nitrifiers during even gentle agitation conditions (Keen and Prosser, 1988).

Calculation of K_s in batch culture is dangerous, since cessation of growth is frequently due to factors other than substrate limitation. An alternative is use of a chemostat in series (Shieh and LaMotta, 1979) in which a chemostat is supplied with both ammonium and biomass from a second continuous-flow system. Variation in the inflowing substrate concentration and detention time for the second vessel resulted in Michaelis–Menten kinetics, but detention times were short, and growth constants calculated therefore referred to activity rather than growth. Further evidence of adaptation of the population to new conditions was found, but variation in K_m may have been due to the experimental system used. Although nitrifier growth may be significant in the second vessel, heterotrophic growth will increase in significance at longer detention times. Nitrifiers will therefore constitute a smaller proportion of MLVSS, leading to reduced values for V_{max}, calculated as activity per unit MLVSS.

Castens and Rozich (1986) used a similar experimental system to investigate substrate inhibition of ammonium oxidation. They measured ammonium conversion in a batch system for up to 100 hr but assumed that growth was never significant. It is likely, therefore, that although they used these data to calculate variation in μ with s, they were in fact measuring the effect of substrate concentration on cell activity.

Measurement of cell concentration or biomass concentration of nitrifying bacteria is difficult in sewage treatment, as it is in other natural environments. A crude estimate of nitrifier biomass may be obtained from the equations (Sharma and Ahlert, 1977)

$$f_1 = \frac{X_1}{X_1 + X_2 + X_4} \tag{19}$$

$$f_2 = \frac{X_2}{X_1 + X_2 + X_4} \tag{20}$$

where f_1 and f_2 are the fractions of ammonia and nitrite oxidizers in the total population and X_1, X_2, and X_4 are related by yield coefficient to the amount of ammonium oxidized, nitrate formed, and biological oxygen demand removed, respectively. An alternative method, with a basis similar to that discussed by Belser (1979) for estimation of nitrifier numbers or biomass in soil, involves calculation of the ratio of specific nitrifying activity [measured as milligrams of N oxidized per hour per milligram of cellular total Kjeldhal N (TKN)] of a sample (mixed culture) to that of a pure culture of nitrifying bacteria. TKN is used to estimate the total biomass concentration. Within the ranges used, Sharma and Ahlert (1977) found that nitrification was zero order for ammonia and first order for total biomass. Because of dangers inherent in relating pure-culture activities to those in mixed populations, this technique provides only a rough estimate of nitrifier biomass. Hall and Murphy (1980) modified this technique by inclusion of nitrapyrin during incubations to distinguish autotrophic and heterotrophic activity and by seeding the sample with a pure culture of *Nitrosomonas* or *Nitrobacter*, as appropriate. The basis of their technique is unclear because of a misquotation of the Monod equation and confusion between specific growth rate (μ) and specific substrate removal rate. Incubations were short term, such that growth would not be significant, and therefore specific growth rate cannot be calculated from their linear changes in substrate and product concentration. Their μ_m appears equivalent to specific activity defined by Srinath *et al.* (1976), and essentially they estimate nitrifier biomass as the activity of the pure culture divided by the difference in specific rate between seeded and unseeded samples. This method is therefore subject to the criticisms discussed in Section 5.4, and none of the techniques has apparently been tested by an alternative enumeration assay.

7.3. Effect of Biomass Concentration

In the simplest models, ammonium oxidation is considered zero order with respect to ammonia and first order with respect to biomass concentration. But there is evidence that the latter deviates at high biomass concentrations. Using suspended volatile solids (SVS) as an indicator of *Nitrosomonas* concentration, Wong-Chong and Loehr (1975) derived the empirical equation

$$k_2^* = \frac{k_{2\,\text{max}}^* k_2 S_{a_m}}{k_{2\,\text{max}}^* + k_2 S_{a_m}} \tag{21}$$

where k_2^* is the maximum ammonium oxidation rate, k_2 is the reaction rate constant, and S_a is equivalent to *Nitrosomonas* biomass concentration. The maximum rate, $k_{2\,\text{max}}^*$, was independent of temperature but varied with pH, while k_2 varied with both pH and temperature, the latter being described by the Arrhenius equation. It is not clear how k_2 is calculated, but it is defined as the oxidation rate per unit of sludge. Similar but less marked effects of increasing biomass concentration are reported by Wong-Chong and Loehr (1978) for nitrite oxidation by *Nitrobacter*. The authors suggest that their observations result from variation in yield coefficient rather than from reduction in oxygen or substrate concentrations or accumulation of inhibitory compounds at high sludge concentrations. It is difficult, however, to rule out such explanations in systems as complex as sewage. In addition, a major assumption is that SVS is a reliable measure of nitrifier population size, i.e., that nitrifiers occupy a constant proportion of SVS. Verification of this assumption requires significant improvements in enumeration techniques for nitrifying bacteria.

The models presented above consider biomass to be evenly distributed throughout the liquid phase. In activated sludge systems a significant proportion is in the form of flocs introducing substrate diffusion as a factor affecting the rate kinetics, dependent on floc size, diffusion coefficients, and flow patterns. Benefield and Molz (1984) attempted to model this situation, considering rates of conversion of organic carbon, nitrogen, phosphorous, and oxygen to be related by multiple substrate kinetic expressions similar to Eq. (18), with additional terms for utilization of oxygen for maintenance. Mass balance equations were derived that related rates of biological conversion of material within uniformly sized flocs and diffusion into flocs from surrounding medium. Both heterotrophs and nitrifiers were considered, and equations were incorporated into a one-dimensional flow system representing an activated sludge process. The final model consisted of five partial and four ordinary differential equations. The latter describe conditions within the floc and the former describe the material balances, using a dispersion model for flow pattern in the reactor. The two sets of equations were coupled by material balances between the liquid bulk phase and the solid floc phase.

Simulation of the model provides profiles of all nutrients within the flocs and also allows prediction of concentrations in the bulk phase during steady-state and transient growth conditions in systems with and without nitrification. These profiles suggest that nitrification can significantly affect the organic carbon content, the efficiency of organic conversion, and the substrate that is limiting growth. This model describes nitrification in these systems much more effectively but is too complex for routine use; its value lies in increasing basic understanding of the system under study.

7.4. Effects of Temperature, pH, and Inhibition

As in other environments, the effect of temperature on nitrification is typically described by the Arrhenius equation. Many examples are found in the literature of plots of ln μ_m versus $1/T$ (Fig. 8). As well as affecting the rate of nitrification, temperature can effect the nature of the process. For example, Randall et al. (1982) found accumulation of ammonia at below 10°C, of nitrite at between 10 and 16°C, and of nitrate at above 16°C due to differences in the relative activities of ammonia oxidizers, nitrite oxidizers, and ammonifiers.

Quinlan (1980) modeled the combined effects of substrate concentration and temperature on ammonia and nitrite oxidation, using previously published rate equations and constants. Isoconcentrate curves relating oxidation rate and temperature at a range of substrate concentrations were generated. The effects of substrate concentration on optimum temperature and maximum rate were quantified, and applications to control of industrial wastewater treatment systems were proposed. Benedict and Carlson (1974) considered the temperature dependence of both μ_m and K_s with four possible situations: neither or both independent of temperature, and either μ_m or K_s independent of temperature, with the other temperature dependent. These different situations resulted in qualitative variation in the temperature sensitivity coefficient with substrate concentration. This approach was illustrated for nitrification data in which μ_m and K_s had been determined at 15 and 20°C. An important prediction was that the increase in the rate of nitrification over this temperature range increases significantly with substrate concentration.

Nitrification at superoptimal temperature was investigated by Neufeld et al. (1986), who found a linear decrease in both maximum utilization rate and yield coefficient above the optimum and a logarithmic increase in K_m. Empirical equations derived from these plots were then used to construct a family of curves relating ammonium concentrations and sludge age at superoptimal temperatures.

Growth of both ammonia and nitrite oxidizers is inhibited by high concentrations of their respective substrates and products, in particular free ammonia and free nitrous acid. Dissociation of NH_4^+ and NO_2^- is pH dependent, and the optimum pH for nitrification will therefore depend on substrate and product concentrations and pH. Ammonia oxidizers are less sensitive to inhibition by free ammonia, and the situation is further complicated by acid production during ammonia oxidation. Anthonisen et al. (1976) represented schematically the combinations of pH and ammonia and nitrite concentrations permitting ammonia and nitrite oxidation and determined boundary concentrations by using experimental data from sewage treatment systems.

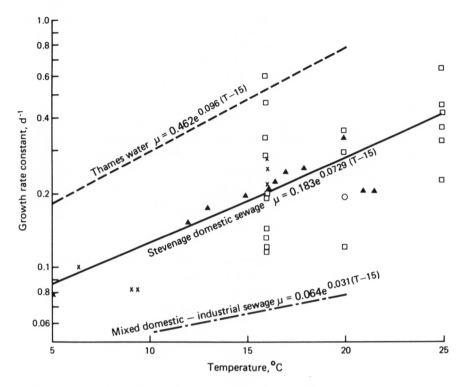

Figure 8. Growth rate constants of *Nitrosomonas* at various temperatures in Thames water and in sludge growth on sewage. [From Painter (1986), with permission.]

Quinlan (1984) assumed ammonium oxidation to be the rate-limiting step for nitrification and quantified the effect of pH by four equations,

$$V = V_m f_1 \text{ (H)} \tag{22}$$

$$K = K_m f_1 \text{ (H)} f_2 \text{(H)} \tag{23}$$

$$f_1 = 1/[1 + (H/K_1) + (K_2/H)] \tag{24}$$

$$f_2 = 1 + (H/K_3) \tag{25}$$

where V_m and K are the velocity and saturation constants, with maxima V_m and K_m, and are functions of both H^+ concentration (H) and temperature. K_1, K_2, and K_3 are rate constants. The function f_1 describes the effect of ampholytic ionization of the enzyme–substrate complex, and f_2 describes ionization of substrate. These equations provide a good fit to experimental data from a number of sources for both whole cells and cell-free extracts. Further analysis predicts that optimum pH decreases linearly with $\log N/(K_m^* + N)$, where K_m^* equals $K_1 K_m/K_3$. Thus, from the practical point of view, nitrification will not be optimized by maintaining pH at a constant value but must be adjusted in relation to ammonia concentration to achieve optimal pH.

Substrate inhibition of nitrifying bacteria has been investigated by Rozich and Castens (1986), using a two-stage chemostat system. The first chemostat was used to determine specific growth rate at relatively low substrate concentrations and to supply a second chemostat with biomass. The second vessel was also supplied with high ammonium concentrations, and data were fitted to the Haldane equation for substrate inhibition of enzyme reactions. A value of 20 mg of N ml^{-1} was obtained for K_i, but data for specific growth rate and high substrate concentrations were quite variable, and the fits of these data to the Monod and Haldane equations were not substantially different.

Inhibitors of any sort can have important effects on nitrification in continuous-flow sewage treatment plants, which are frequently operated at retention times close to those that would result in washout of slow-growing nitrifying bacteria. Neufeld *et al.* (1986) quantified empirically the effects of phenolic compounds and free and complexed cyanides by the equation

$$\log V_m = k_1 - k_2 \log C \tag{26}$$

where k_1 and k_2 are constants and C is the concentration of inhibitor. An exception was free cyanide, where $\log V_m$ was directly proportional to inhibitor concentration. For many of the compounds investigated, K_m varied in a similar manner. Quantification allowed ranking of these compounds in order of toxicity and prediction of operating conditions, in particular sludge age, required for nitrification at any particular toxic concentration. Problems arising from inhibitors are reduced in feedback systems and in those in which nitrifiers are present in a fixed biofilm, which reduces the possibility of washout.

7.5. Surface Growth

Nitrification in suspended-growth, wastewater treatment systems requires operation at sludge retention times sufficiently long to avoid washout of the slow-growing nitrifying bacteria. An alternative solution that allows faster throughput of material is the use of biofilm reactors in which nitrifiers are attached to solid surfaces and are less susceptible to washout. Such systems include fluidized beds, trickling filters, and rotating biological contractors, and their description requires two major modifications to basic nitrification models. First, attachment, growth, and detachment of cells must be considered; second, diffusion of nutrients, particularly ammonium and oxygen into the biofilm, must be modeled. Diffusion has already been considered in floc formation, but the greater nitrifying biomass and activity found in fixed film systems increases oxygen demand and introduces oxygen limitation as a significant factor.

Tanaka *et al.* (1981) modeled nitrification in a fluidized bed reactor by using a mass balance approach. For example, the mass balance for nitrogen was described by the equation

$$V\frac{dC_{N,2}}{dt} = F(C_{N,1} - C_{N,2}) + r_N V_R \tag{27}$$

where V represents total system volume, F is flow rate, $C_{N,1}$ and $C_{N,2}$ are inlet and outlet nitrogen concentrations, and V_R is reactor volume. Similar equations were derived for the oxygen balance and were coupled to those for nitrogen by stoichiometric relationships between nitrogen oxidation and oxygen utilization. The rate of nitrogen oxidation, r_N, was described by double-substrate limitation. In batch culture, rearrangement of the basic model equations predicted a linear relationship between $\ln (C_{N,0}/C_N)/(C_{N,0})$ versus time, with an intercept of $1/K_N$ and a slope of V/K_N, allowing calculation of both parameters. ($C_{N,0}$ represents initial concentration.) To avoid problems associated with measuring ammonia and nitrite concentrations in the K_s region, initial rates of oxygen utilization were measured for different starting concentrations of ammonia and nitrite. Parameters therefore reflect activity and not growth. The saturation constant for ammonium was determined in this way for attached organisms and for those detached by shearing. A lower saturation constant for oxygen was found for attached cells, presumably due to reduced diffusion and transport of oxygen within the biofilm. In batch culture supplied with air, nitrite accumulated, but this was prevented in nonoxygen-limited systems supplied with pure oxygen. In continuous-flow systems there was no oxygen limitation, and the effect of flow rate on conversion efficiency was determined. Plug flow was considered to be a better description of the fluidized bed system, modeled by using a "tank-in-series" approach equivalent to the finite difference approach of Prosser and Gray (1977), and predictions fitted well with experimental data from the batch reactor.

Tanaka and Dunn (1982) considered the nature of the biofilm in more detail and adopted a diffusion reaction approach analogous to a chemical reaction taking place within a porous catalyst. In particular, they used a modeling approach to predict whether oxygen or ammonium would limit nitrification within the film. Thus, the ammonium and oxygen concentrations within the film would depend on their concentrations in the bulk phase, the diffusion coefficients within the film (D_s), and their rates of utilization, r_s, i.e.,

$$\frac{ds}{dt} = D_s \frac{d^2s}{dz^2} + r_s \tag{28}$$

where z represents depth. Dimensional analysis of this equation demonstrated that when diffusion coefficients for ammonium and oxygen were equal, the penetration of ammonium relative to that of oxygen was determined by the oxygen/ammonium concentration ratio in the bulkphase. Which of the two compounds limited growth depended on the stoichiometric oxygen requirement for ammonium (or nitrite) oxidation. If the oxygen/ammonium ratio was less than this requirement, ammonia penetrated to a greater depth and the overall process was predicted to be oxygen limited. This relationship was independent of the precise reaction rate kinetics, and an expression was also derived for the case of different diffusion coefficients. Experimental measurements of concentration profiles within the biofilm were not attempted, but predictions of bulk phase ammonium, nitrite, and oxygen concentrations were compared with experimental data from batch and continuous-flow systems. Nitrite accumulation occurred in both systems at lower oxygen concentrations but not under oxygen excess, due to either greater K_{O_2} for nitrite oxidation than for ammonium oxidation or to a greater proportion of nitrite oxidizers at greater biofilms depths. Provision of organic carbon in the feed caused growth of heterotrophs within the film, reducing oxygen supply and ammonium and nitrite oxidation rates. Again, nitrite oxidation was affected to a greater extent than ammonium oxidation.

Simulation of the model demonstrated that if the saturation constant for oxygen was greater for ammonia than for nitrite oxidation, as measured experimentally by Tanaka *et al.* (1981), nitrite accumulation could occur only if the proportion of nitrite to ammonium oxidizers increased with depth of biofilm. Nitrite accumulation could, however, occur with uniform distribution of both types of organisms if the difference in saturation constants was reversed. The authors highlight the need for more accurate measurement of saturation constants to distinguish between these explanations.

Gujer and Boller (1986) proposed a similar model for trickling filter systems. Mass balance equations were derived based on diffusion of oxygen and nitrogen into the biofilm, and expressions were obtained for substrate penetration and for determination of whether oxygen or nitrogen was rate limiting. Predictions were generated for the ammonium flux into the biofilm and for ammonium and oxygen concentrations in the bulk phase as functions of temperature. These agreed well with the limited experimental data available. The model was also used to consider reactor design in the light of other factors, such as pH, nitrite concentration, recirculation, and reactors connected in series.

Surface growth is also important when one considers nitrification in rotating biological contactors. Models for these systems will not be described in detail here, but they adopt approaches similar to those discussed above. Examples include the work of Poon *et al.* (1981), Watanabe *et al.* (1984), and Gonenc and Harremoes (1985).

7.6. Summary

The discussion presented above provide a review of the basic kinetics and mathematical models used to describe nitrification in sewage treatment plants. It is beyond the scope of this chapter to discuss the application of these models to specific problems and specific treatment systems. The general aim of such applications is, however, to determine values for constants required for simulation of models and to enable prediction of operating conditions that permit active nitrification and also to predict the degree of nitrification. Examples of such modifications are inclusion of decay constants for feedback systems, incorporation into more general models involving denitrification and degradation of organic carbon and nitrification of industrial and domestic wastes, and comparisons of one- and two-stage sludge systems and of systems involving plug flow and recycling. Some of the larger models may be very complex. For example, Gromiec *et al.* (1982) modeled a six-phase nitrification–denitrification system with intermittent aeration and simultaneous precipitation of phosphorus with ferrous sulfate. This model consisted of 84 differential equations and is therefore unlikely to be of direct applied use, while its predictive use is limited by the large number of growth constants and other model parameters, realistic values of which are required for simulation.

8. Concluding Remarks

Mathematical models of nitrification can be seen to range from simple kinetic descriptions of substrate utilization and product formation to complex sets of differential equations describing interactions between nitrifiers and their physicochemical and biological environments. The simplicity of the former is essential for measurements of rates of nitrification, particularly in natural environments. Modifications to describe experimental data more completely can conceal environmental effects of significance and deficiencies in experimental systems. The more complex mechanistic and systems models, on the other hand, serve to consider such complicating factors and must be tested in carefully designed and controlled experimental systems. Between these two extremes lies a range of approaches that reflect differences in the functions of models and different degrees of compromise. For example, in designing sewage treatment processes for nitrification, environmental factors of less significance must be ignored if useful predictions are to be obtained.

The different applied aspects of nitrification have also led to a number of different approaches. Nitrification in sewage treatment is typically modeled by chemical engineers, soil nitrification by soil chemists or soil microbiologists, and nitrification in pure-culture laboratory systems by microbial physiologists and biochemists. Each must learn from the methodological and conceptual approaches adopted by the different disciplines for full appreciation of the quantitative effects of both physicochemical and biological features of the environment on nitrification. A problem common to all is the difficulty in accurately enumerating nitrifying bacteria. This problem is not unique to nitrifiers but is highlighted by the ease and accuracy with which the process of nitrification can be measured. Equivalent techniques for enumeration in the laboratory and in natural environments would present nitrification as a model system for detailed ecological studies with great potential.

References

Addiscott, T. M., 1983, Kinetics and temperature relationships of mineralization and nitrification in Rothamsted soils with differing histories, *J. Soil Sci.* **34**:343.

Addiscott, T. M., and Whitmore, A. P., 1987, Computer simulation of changes in soil mineral nitrogen and crop nitrogen during autumn, winter and spring, *J. Agric. Sci.* **109**:141.

Anthonisen, A., Loehr, R. C., Prakasam, T. B. S., and Srinath, E. G., 1976, Inhibition of nitrification by ammonia and nitrous acid, *J. Water Pollut. Control Fed.* **48**:835.

Argaman, Y., 1982, Single sludge nitrogen removal from industrial wastewater, *Water Sci. Technol.* **14**:7.

Armstrong, E. F., and Prosser, J. I., 1988, Growth of *Nitrosomonas europaea* on ammonia-treated vermiculite, *Soil Biol. Biochem.* **20**:409.

Bazin, M. J., 1983, *Mathematics in Microbiology*, Academic Press, London.

Bazin, M. J., and Saunders, P. T., 1973, Dynamics of nitrification in a continuous flow system, *Soil Biol. Biochem.* **5**:531.

Bazin, M. K., Saunders, P. T., and Prosser, J. I., 1976, Models of microbial interactions in the soil, *CRC Crit. Rev. Microbiol.* **5**:464.

Beccari, M., Marani, D., and Ramadori, R. 1979, A critical analysis of nitrification alternatives, *Water Res.* **13**:185.

Beek, J., and Frissel, M. J., 1973, *Simulation of Nitrogen Behaviour in Soils*, Pudoc, Wageningen.

Belser, L. W., 1979, Population ecology of nitrifying bacteria, *Annu. Rev. Microbiol.* **33**:309.

Belser, L. W., 1984, Bicarbonate uptake by nitrifiers: Effects of growth rate, pH, substrate concentration and metabolic inhibitors, *Appl. Environ. Microbiol.* **48**:1100.

Belser, L. W., and Schmidt, E. L., 1980, Growth and oxidation kinetics of three genera of ammonia oxidisers. *FEMS Microbiol. Lett.* **7**:213.

Benedict, A. H., and Carlson, D. A., 1974, Rational assessment of the Streeter-Phelps temperature coefficient, *J. Water Pollut. Control Fed.* **46**:1792.

Benefield, L., and Molz, F., 1984, A model for the activated sludge process which considers wastewater characteristics, floc behavior, and microbial populations, *Biotechnol. Bioeng.* **26**:352.

Bhat, K. K. S., Flowers, T. H., and O'Callaghan, J. R., 1980, A model for the simulation of the fate of nitrogen in farm wastes on land application, *J. Agric. Sci.* **94**:183.

Bock, E., Koops, H.-P., and Harms, S. H., 1986, Cell biology of nitrifying bacteria, in *Nitrification*, (J. I. Prosser, ed.) pp. 17–38, IRL Press, Oxford.

Boon, B., and Laudelout, H., 1962, Kinetics of nitrite oxidation by *Nitrobacter winogradskyi*, *Biochem. J.* **85**:440.

Cameron, D. R., and Kowalenko, C. G., 1976, Modelling nitrogen processes in soil: Mathematical development and relationships, *Can. J. Soil Sci.* **56**:71.

Campbell, N. E., and Aleem, M. I. H., 1965, The effect of two-chloro, 6-(trichloromethyl) pyridine on the chemoautotrophic metabolism of nitrifying bacteria, *Antonie van Leeuwenhoek J. Microbiol. Serol.* **31**:124.

Castens, D. J., and Rozich, A. F., 1986, Analysis of batch nitrification using substrate inhibition kinetics, *Biotechnol. Bioeng.* **28**:461.

Charley, R. C., Hooper, D. G., and McLee, A. G., 1980, Nitrification kinetics in activated sludge at various temperatures and dissolved oxygen concentrations, *Water Res.* **14**:1387.

Christensen, J. P., and Rowe, G. T., 1984, Nitrification and oxygen consumption in north west Atlantic deep-sea sediments, *J. Mar. Res.* **42**:1099.

Cooper, A. B., 1983a, Effect of storm events on benthic nitrifying activity, *Appl. Environ. Microbiol.* **46**:957.

Cooper, A. B., 1983b, Population ecology of nitrifiers in a stream receiving geothermal inputs of ammonium, *Appl. Environ. Microbiol.* **45**:1170.

Cooper, A. B., 1984, Activities of benthic nitrifiers in streams and their role in oxygen consumption, *Microb. Ecol.* **10**:317.

Cooper, A. B., 1986, Developing management guidelines for river nitrogenous oxygen demand, *J. Water Pollut. Control Fed.* **58**:845.

Darrah, P. R., White, R. A., and Nye, P. H., 1985, Simultaneous nitrification and diffusion in soil. I. The effects of addition of low levels of ammonium chloride, *J. Soil Sci.* **36**:281.

Darrah, P. R., White, R. A., and Nye, P. H., 1986a, Simultaneous nitrification and diffusion in soil. II. The effects at levels of ammonium chloride which inhibit nitrification. *J. Soil Sci.* **37**:41.

Darrah, P. R., White, R. A., and Nye, P. H., 1986b, Simultaneous nitrification and diffusion in soil. III. The effects of the addition of ammonium sulphate, *J. Soil Sci.* **37**:53.

Darrah, P. R., Nye, P. H., and White, R. E., 1986c, Simultaneous nitrification and diffusion in soil. V. The effects of pH change, following the addition of ammonium sulphate, on the activity of nitrifiers, *J. Soil Sci.* **37**:479.

Darrah, P. R., White, R. E., and Nye, P. H., 1987, A theoretical consideration of the implications of cell clustering for the prediction of nitrification in soil, *Plant Soil* **99**:387.

Duffy, J., Chung, C., Boast, C., and Franklin, M., 1975, A simulation model of biophysicochemical transformations of nitrogen in tile-drained corn belt soil, *J. Environ. Qual.* **4**:477.

Endelman, F. J., Box, G. E. P., Boyle, J. R., Hughes, R. R., Kenney, D. R., Northup, M. L., and Saffigna, P. G., 1974, The Mathematical Modelling of Soil-Water-Nitrogen Phenomena, I. B. P. Report EDFB-IBP-74-7, Oak Ridge National Laboratory, Oak Ridge, Tenn.

Feigenbaum, S., and Hadas, A., 1980, Utilisation of fertiliser nitrogen-nitrogen-15 by field-grown alfafa, *Soil Sci. Soc. Am. J.* **44**:1006.

Gilmour, J. T., 1984, The effects of soil properties on nitrification and nitrification inhibition, *Soil Sci. Soc. Am. J.* **48**:1262.

Glover, H. E., 1982, Methylamine, an inhibitor of ammonium oxidation and chemoautotrophic growth in the marine nitrifying bacterium *Nitrosococcus oceanus*, *Arch. Microbiol.* **132**:37.

Glover, H. E., 1985, The relationship between inorganic and organic carbon production in batch and chemostat cultures of marine nitrifying bacteria, *Arch. Microbiol.* **142**:45.

Goloway, F., and Bender, M., 1982, Diagenetic models of interstitial nitrate profiles in deep sea suboxic sediments, *Limnol. Oceanogr.* **27**:624.

Gonenc, E. I., and Harremoes, P., 1985, Nitrification in rotating disc systems. I. Criteria for transition from oxygen to ammonia rate limitation, *Water Res.* **19**:1119–1127.

Gromiec, M., Valve, M., and Liponkoskie, M., 1982, Nutrient Removal from Waste waters by Single Sludge Systems, Tech. Res. Cent. Finl. Research Report, pp. 1–126. Department of Water Management, Institute of Meteorology and Water Management, Warsaw, Poland.

Gujer, W., and Boller, M., 1986, Design of a nitrifying tertiary trickling filter based on theoretical concepts, *Water Res.* **20**:1353.

Hadas, A., Feigenbaum, N., Feigin, A., and Portnoy, R., 1986, Nitrification rates in profiles of differently managed soil types, *Soil Sci. Soc. Am. J.* **50**:633.

Hall, E. R., and Murphy, K. L., 1980, Estimation of nitrifying biomass and kinetics in wastewater, *Water Res.* **14**:297.

Hall, G. H., 1986, Nitrification in lakes, in: *Nitrification* (J. I. Prosser, ed.), pp. 127–156, IRL Press, Oxford.

Hirose, E., and Tateno, M., 1984, Soil nitrogen patterns induced by colonisation of *Polygonum cuspidatum* on Mt. Fuji, *Oecologia* (Berlin) **61:**128.

Hsieh, Y. P., Douglas, L. A., and Motto, H. L., 1981, Modelling sewage sludge decomposition in soil: II. Nitrogen transformations, *J. Environ. Qual.* **10:**59.

Jahnke, R. A., Emerson, S. R., and Murray, J. W., 1982, A model of oxygen reduction, denitrification, and organic matter mineralisation in marine sediments, *Limnol. Oceanogr.* **27:**610.

Keen, G. A., and Prosser, J. I., 1987a, Steady state and transient growth of autotrophic nitrifying bacteria, *Arch. Microbiol.* **147:**73.

Keen, G. A., and Prosser, J. I., 1987b, Interrelationship between pH and surface growth of *Nitrobacter, Soil Biol. Biochem.* **19:**665.

Keen, G. A., and Prosser, J. I., 1988, The surface growth and activity of *Nitrobacter. Microb. Ecol.* **15:**21.

Killham, K., 1986, Heterotrophic nitrification, in: *Nitrification* (J. I. Prosser, ed.), pp. 117–126, IRL Press, Oxford.

Knowles, G., Downing, A. L., and Barrett, M. J., 1965, Determination of kinetics constants for nitrifying bacteria in mixed culture, with the aid of an electronic computer, *J. Gen. Microbiol.* **38:**263.

Kuenen, J. G., and Robertson, L. A., 1987, Ecology of nitrification and denitrification, *Symp. Soc. Gen. Microbiol.* **42:**162.

Laudelout, H., Lambert, R., Fripiat, J. L., and Pham, M. L., 1974, Effet de la temperature sur la vitesse d'oxydation de l'ammonium en nitrate par des culture mixtes de nitrifiants, *Ann. Microbiol. Inst. Pasteur* **125B:**75.

Laudelout, H., Lambert, R., and Pham, M. L., 1976, Influence du pH et de la pression partielle d'oxygene sur la nitrification, *Ann. Microbiol. Inst. Pasteur* **127A:**367.

Macdonald, R. M., 1979, Populations dynamics of the nitrifying bacterium, *Nitrosolobus* in soil, *J. Appl. Ecol.* **16:**529.

Malhi, S. S., and McGill, W. B., 1982, Nitrification in three Alberta soils: Effect of temperature, moisture and substrate concentration, *Soil Biol. Biochem.* **14:**393.

Matulewich, V. A., Strom, P. F., and Finstein, M. S., 1975, Length of incubation for enumerating nitrifying bacteria present in various environments, *Appl. Microbiol.* **29:**265.

McMeekin, T. A., Olley, J., and Ratkowsky, D. A., 1988, Temperature effects on bacterial growth rates, in: *Physiological Models in Microbiology* (M. J. Bazin and J. I. Prosser, eds.), pp. 75–89, CRC Press, Boca Raton, Fla.

Mehran, M., and Tanji, K. K., 1974, Computer modelling of nitrogen transformation in soils, *J. Environ. Qual.* **3:**391.

Meikle, R. W., 1979, Prediction of ammonium nitrogen fertiliser disappearance from soils in the presence and absence of N-serve nitrogen stabilisers, *Soil Sci.* **127:**292.

Molina, J. A. E., 1985, Components of rates of ammonium oxidation in soil, *Soil Sci. Soc. Am. J.* **49:**603.

Morrill, L. G., and Dawson, J. E., 1962, Growth rates of nitrifying chemoautotrophs in soil, *J. Bacteriol.* **83:**205.

Myrold, D. D., and Tiedje, J. M., 1986, Simultaneous estimation of several nitrogen cycle rates using 15N: Theory and application. *Soil Biol. Biochem.* **18:**559.

Neufeld, R., Greenfield, J., and Rieder, B., 1986, Temperature, cyanide and phenolic nitrification inhibition, *Water Res.* **20:**633.

Nihtila, M., and Virkkunen, J., 1977, Practical identifiability of growth and substrate consumption models, *Biotechnol. Bioeng.* **19:**1831.

Novak, L. T., 1979, Role of agitation conditions in nitrification, *Biotechnol. Bioeng.* **21:**1457.

Oremland, R. S., and Capone, D. G., 1988, Use of "specific" inhibitors in biogeochemistry and microbial ecology, in: *Advances in Microbial Ecology*, Vol. 10 (K. C. Marshall, ed.), pp. 285–383, Plenum Press, New York.

Painter, H. A., 1986, Nitrification in the treatment of sewage and waste-waters, in: *Nitrification* (J. I. Prosser, ed.), pp. 185–211, IRL Press, Oxford.

Poduska, R. A., and Andrews, J. F., 1975, Dynamics of nitrification in the activated sludge process, *J. Water Pollut. Control Fed.* **47**:2599.

Poon, C. P. C., Chin, H. K., Smith, E. D., and Mikaucki, W. J., 1981, Upgrading with rotating biological contactors for ammonia nitrogen removal, *J. Water Pollut. Control Fed.* **53**:1158–1165.

Powell, S. J., and Prosser, J. I., 1985, The effect of nitrapyrin and chloropicolinic acid on ammonium oxdiation by *Nitrosomonas europaea*, *FEMS Microbiol. Lett.* **28**:51.

Powell, S. J., and Prosser, J. I., 1986a, Inhibition of ammonium oxidation by nitrapyrin in soil and liquid culture, *Appl. Environ. Microbiol.* **52**:782.

Powell, S. J., and Prosser, J. I., 1986b, The effect of copper on inhibition by nitrapyrin of growth of *Nitrosomonas europaea*, *Curr. Microbiol.* **14**:177.

Prosser, J. I., 1988, Mathematical modelling and computer simulation, *in Computers in Microbiology—A Practical Approach* (T. Bryant and J. W. T. Wimpenny, eds.), pp. 125–159, IRL Press, Oxford.

Prosser, J. I., and Bazin, M. J., 1988, The use of packed column reactors to study microbial nitrogen transformations in the soil, in: *Handbook of Laboratory Systems for Microbiol Ecosystem Research* (J. W. T. Wimpenny, ed.), pp. 31–49, CRC Press Inc., Boca Raton, Fla.

Prosser, J. I., and Gray, T. R. G., 1977, Use of finite difference method to study a model system of nitrification at low substrate concentrations, *J. Gen. Microbiol.* **102**:119.

Quinlan, A. V., 1980, The thermal sensitivity of nitrification as a function of the concentration of nitrogen substrate, *Water Res.* **14**:1501.

Quinlan, A. V., 1984, Prediction of the optimum pH for ammonia-N oxidation by *Nitrosomonas europaea* in well-aerated natural and domestic-waste waters, *Water Res.* **18**:561.

Randall, C. W., Benefield, L. D., and Buth, D., 1982, The effects of temperature on the biochemical reaction rates of the activated sludge process, *Water Sci. Technol.* **14**:413.

Robinson, J. A., 1985, Determining microbial kinetic parameters using nonlinear regression analysis. Advantages and limitations in microbial ecology, in: *Advances in Microbial Ecology*, Vol. 8 (K. C. Marshall, ed.), pp. 61–114, Plenum Press, New York.

Roels, J. A., and Kossen, N. W. F., 1978, On the modelling of microbial metabolism, *Prog. Ind. Microbiol.* **14**:95.

Rozich, A. F., and Castens, D. J., 1986, Inhibition kinetics of nitrification in continuous-flow reactors, *J. Water Pollut. Control Fed.* **58**:220.

Saunders, P. T., and Bazin, M. J., 1973, Non steady state studies of nitrification in soil: Theoretical considerations, *Soil Biol. Biochem.* **5**:545.

Schmidt, E. L., 1982, Nitrification in soil, *Agronomy* **22**:253.

Seifert, J., 1980, The effect of temperature on nitrification intensity in soil, *Folia Microbiol.* **25**:144.

Shah, D. B., and Coulman, G. A., 1978, Kinetics of nitrification and denitrification reactions, *Biotechnol. Bioeng.* **20**:43.

Sharma, B., and Ahlert, R. C., 1977, Nitrification and nitrogen removal, *Water Res.* **11**:897.

Shieh, W. K., and LaMotta, E. J., 1979, The intrinsic kinetics of nitrification in a continuous flow suspended growth reactor, *Water Res.* **13**:1273.

Skinner, F. A., and Walker, N., 1961, Growth of *Nitrosomonas europaea* in batch and continuous culture, *Arch. Mikrobiol.* **38**:339.

Soriano, S., and Walker, F., 1973, The nitrifying bacteria in soils from Rothamsted classical fields and elsewhere, *J. Appl. Bacteriol.* **36**:523.

Srinath, E. G., Loehr, R. C., and Prakasam, T. B. S., 1976, Nitrifying organism concentration and activity, *J. Environ. Eng. Div.* **102**:449–463.

304   J. I. Prosser

Suzuki, I., Dular, U., and Kwok, S. C., 1974, Ammonia or ammonium ion as substrate for oxidation by *Nitrosomonas europaea* cells and extracts, *J. Bacteriol.* **120**:556.

Tanaka, H., and Dunn, I. J., 1982, Kinetics of biofilm nitrification, *Biotechnol. Bioeng.* **24**:669.

Tanaka, H., Uzman, S., and Dunn, I. J., 1981, Kinetics of nitrification using a fluidised sand bed reactor with attached growth, *Biotechnol. Bioeng.* **23**:1683.

Tillotson, W. R., and Wagenet, R. J., 1982, Simulation of fertiliser nitrogen under cropped situations, *Soil Sci.* **133**:133.

Topiwala, H. H., and Sinclair, C. G., 1971, Temperature relationships in continuous culture, *Biotechnol. Bioeng.* **13**:795.

Underhill, S. E., and Prosser, J. I., 1987, Inhibition and stimulation of nitrification by potassium ethyl xanthate, *J. Gen. Microbiol.* **133**:3237.

Vanderborght, J.-P., and Billen, G., 1975, Vertical distribution of nitrate concentration in interstitial water of marine sediments with nitrification and denitrification, *Limnol. Oceanogr.* **20**:953.

Vanderborght, J.-P., Wollast, R., and Billen, G., 1977, Kinetic models of diagenesis in disturbed sediments. Part 2. Nitrogen diagenesis, *Limnol. Oceanogr.* **22**:794.

van Veen, J. A., and Frissel, M. J., 1981, Simulation model of the behaviour of N in soil, in: *Simulation of Nitrogen Behaviour of Soil-Plants Systems* (M. J. Frissel and J. A. van Veen, eds.), pp. 126–144, Pudoc, Wageningen.

Ward, B. B., 1986, Nitrification in marine environments, in: *Nitrification* (J. I. Prosser, ed.), pp. 157–184, IRL Press, Oxford.

Ward, B. B., 1987, Kinetic studies on ammonia and methane oxidation by *Nitrosococcus oceanus*, *Arch. Microbiol.* **147**:126.

Watanabe, Y., Masuda, S., Nishidome, K., and Wantawin, C., 1984, Mathematical model of simultaneous organic oxidation, nitrification and denitrification in rotating biomass biological contactors, *Water Sci. Technol.* **17**:385–397.

Wong-Chong, G. M., and Loehr, R. C., 1975, The kinetics of microbial nitrification, *Water Res.* **9**:1099.

Wong-Chong, G. M., and Loehr, R. C., 1978, Kinetics of microbial nitrification: Nitrite-nitrogen oxidation, *Water Res.* **12**:605.

Wood, P., 1986, Nitrification as a bacterial energy source, in: *Nitrification* (J. I. Prosser, ed.), pp. 39–62, IRL Press, Oxford.

Yadvinder-Singh, and Beauchamp, E. G., 1985, Alternate method for characterising nitrifier activity in soil, *Soil Sci. Soc. Am. J.* **49**:1432.

Yoshioka, T., Terai, H., and Saiyo, Y., 1982, Growth kinetic studies of nitrifying bacteria by the immunofluorescent counting method. *J. Gen. Microbiol.* **28**:169.

Physiological Ecology and Regulation of N₂ Fixation in Natural Waters

HANS W. PAERL

1. Historical and Current Perspectives

Biological nitrogen fixation, the enzyme (nitrogenase)-catalyzed process by which certain procaryotes reduce atmospheric dinitrogen (N_2) to ammonia (NH_3), is of fundamental importance in mediating the availability of utilizable nitrogen in the biosphere (Delwiche, 1970; Carpenter and Capone, 1983; Howarth *et al.*, 1988). This process is of particular relevance in ecosystems exhibiting deficiencies in nitrogen availability; in this regard, it is well established that geographically and trophically diverse freshwater lakes, rivers, and reservoirs as well as estuarine, coastal, and oceanic habitats exhibit chronic nitrogen deficiencies (Ryther and Dunstan, 1971; Eppley *et al.*, 1973; Parsons *et al.*, 1977; Mann, 1982; Goldman and Horne, 1983; Wetzel, 1983). Among these waters, newly formed combined nitrogen inputs attributable to N_2 fixation may regulate productivity and fertility (Horne and Fogg, 1970; Horne and Viner, 1971; Horne and Goldman, 1972; Brezonik, 1973; Mague and Holm-Hansen, 1975; Wiebe *et al.*, 1975; Lean *et al.*, 1978; Paerl *et al.*, 1981; Martinez *et al.*, 1983).

Despite its obvious importance in aquatic nitrogen transformations and its crucial role in potentially alleviating nitrogen limitation, the dynamics of N_2 fixation has only recently been the focus of intense and diverse geochemical and biological studies in freshwater and marine ecosystems (Dugdale and Goering, 1967; Carpenter and Capone, 1983). The relatively late emphasis on N_2 fixation studies in aquatic ecology and chemistry is somewhat surprising, since the biogeochemical importance of this process in terrestrial environments has been duly recognized and intensively studied for well over a century. Early in the 1800s, biologists and chemists recognized that not all

HANS W. PAERL • Institute of Marine Sciences, University of North Carolina, Chapel Hill, Morehead City, North Carolina 28557.

nitrogen requirements for plant growth could be met by combined nitrogen supplied in soils or groundwater (see Saussure, 1804). Although Liebig concluded in 1842 that all carbon derived for plant growth originated in the atmosphere, several contemporaries suspected that a portion of plant nitrogen requirements might also be derived from the vast reservoir of atmospheric N_2 demonstrated in 1771 by Priestley. Lawes *et al.* (1861) and Schultz-Lupitz (1887) formally postulated that certain plants possessed the ability to directly utilize atmospheric N_2 for growth, calling these organisms Stickstofffressers (N_2 eaters). In 1888, Hellriegel and Wilfarth convincingly demonstrated the need for leguminous plants to satisfy their nitrogen requirements through N_2 fixation; the fixed nitrogen being derived from symbiotic heterotrophic bacteria located in the nodulated roots of the host plants. In the same year, Beyerinck (1888) isolated *Rhizobium* strains catalyzing this reaction. Soon afterwards, the microbiology and ecology of terrestrial N_2-fixing microorganisms flourished and diversified.

In part because of the economic importance of leguminous plants as protein sources, the ubiquity of symbiotic (*Rhizobium*) and asymbiotic (*Azotobacter, Azospirillum*) N_2 fixers, and the ease with which N_2 fixation could be quantified in biomass-rich soils, the biogeochemical importance of this process was initially recognized in terrestrial ecosystems. Although naturalists reported on the rather puzzling and paradoxical means by which massive blooms of cyanobacteria could periodically invade and dominate seemingly pristine and nutrient-poor (presumably including nitrogen deficiency) surface waters [Leeuwenhoek, cited in Fogg (1969) and Darwin (1858)], recognition and confirmation of such microorganisms as N_2 fixers had to await the pioneering studies of Burris and Wilson (1946), Kamen and Gest (1949), Fogg (1942, 1944), and others (see Fogg, 1974). In concert, these studies revealed that numerous photoautotrophs (including heterocystous cyanobacteria and photosynthetic bacteria) were capable of utilizing atmospheric N_2 as their sole nitrogen source. In the early 1950s, aided by analytical techniques (micro-Kjeldahl equipment, mass spectrometry of ^{15}N as a tracer, and low-level colorimetric inorganic nitrogen analyses) as well as improved culturing capabilities, microbiologists and aquatic biologists began to recognize the diversity, ubiquity, and potential biogeochemical importance of aquatic N_2 fixation. In particular, the bloom-forming freshwater cyanobacterial genera *Anabaena* and *Nostoc* were the subjects of early studies (Stewart, 1973; Fogg, 1969, 1974).

However, because of the depauperate nutritive and resultant oligotrophic nature of many aquatic ecosystems, additional discoveries of novel asymbiotic and symbiotic N_2-fixing genera relied on development of more sensitive techniques for detecting and ultimately quantifying N_2-fixing activities. More routine utilization of the stable isotope ^{15}N in laboratory N_2 assimilation studies (Wilson and Burris, 1947) and the development of the highly sensitive and easily executed acetylene reduction technique (Stewart *et al.*, 1967) for field (*in situ*) and laboratory assays of N_2 fixation (nitrogenase activity) greatly facilitated the search for novel N_2-fixing genera and allowed ecosystem-level assessments of nitrogen inputs via this process (Dugdale *et al.*, 1961; Dugdale and Goering, 1967; Horne and Fogg, 1970; Bunt *et al.*, 1970; Horne and Goldman, 1972). The short-lived radioactive isotope ^{13}N has, over the past decade, received additional application as a method of measuring internal pathways of nitrogen metabolism and resolving symbiotic relationships involving nitrogen-fixing microorganisms (Meeks *et al.*, 1985). Most recently, the application of immunochemical (nitrogenase-specific antibodies) and molecular [DNA–nitrogen-fixing (*nif*) gene probe hybridizations; RNA transcription–translation] techniques (Rice *et al.*, 1982; Hazelkorn, 1986; Bergman *et al.*, 1986) has allowed investigators to characterize and screen as well as quantify a wide range of naturally occurring and isolated microorganisms for their N$_2$-fixing potentials. Complementary applications of these techniques will be instrumental in discriminating aquatic microorganisms in which the nitrogenase enzyme complex has been synthesized (and expressed) from those solely having the genetic potential (presence of structural DNA *nif* genes, but not translated into protein synthesis). Finally, once deployed in the field, molecular and immunochemical techniques will help clarify and delineate population and community dynamics among naturally occurring N$_2$-fixing assemblages and consortia. Future applications of these techniques will also further our general knowledge and interpretations regarding environmental regulation of aquatic N$_2$ fixation.

2. Aquatic N$_2$-Fixing Microorganisms: Their Diversity and Habitats

The past 30 years has seen much progress in identification and characterization of diverse aquatic N$_2$-fixing microbiota. Eubacterial and cyanobacterial taxa known to be capable of N$_2$ fixation and their respective habitats are listed in Tables I and II. Large informational gaps exist in ecological, biogeochemical, and physiological evaluations of these taxa. The list of confirmed N$_2$ fixers is steadily expanding. Accordingly, these tables will no doubt require updating shortly after publication of this article.

It will become evident in the ensuing discussion of the biochemistry and physiology of N$_2$ fixation that this process is an energy-demanding one. Energy requirements can be met by photolithoautotrophic, chemolithoautotrophic, and/or heterotrophic means. Nitrogen fixers using photolithoautotrophy as a chief energy source include obligate anaerobes such as the purple and green sulfur bacteria (H$_2$S oxidizers) and certain oxygenic cyanobacterial genera (H$_2$O oxidizers). Several facultative photolithoautotrophs, also capable of heterotrophic and photoheterotrophic growth, fix N$_2$ (Kamen and Gest, 1949). These include *Rhodospirillum* and *Rhodobacter* spp. and some endosymbiotic cyanobacteria. Among filamentous, potentially heterotrophic cyanobacteria, *Anabaena azollae* (recently designated a *Nostoc* species) has been shown to be capable of utilizing several sugars as energy sources supporting N$_2$ fixation (Rozen *et al.*, 1988). From both energetic and ecological perspectives, it would be advantageous for various endosymbiotic cyanobacteria (including *Richelia* spp., present in the diatom *Rhizoselenia;* *Nostoc* spp., present in the cycad *Gunnera; Calothrix* and *Oscillatoria,* located in the siphonous green seaweed *Codium;* numerous filamentous and coccoid taxa comprising complex intertidal and benthic microbial mat communities; and deep-living planktonic communities) to contemporaneously or alternatively utilize photolithoautotrophy and heterotrophy as energy sources for N$_2$ fixation.

By far, the best-documented aquatic N$_2$-fixing cyanobacteria are the filamentous heterocystous genera, including the freshwater genera *Anabaena, Aphanizomenon,*

Table I. Known N_2-Fixing Eubacterial Genera Found in Aquatic Habitats (Including Terrestrial Genera Flushed into Receiving Waters)

Functional group	Genus	Habitat
Photosynthetic	*Chlorobium* *Chromatium* *Ectothiorhodospira* *Rhodobacter* (formerly *Rhodopseudomonas*) *Rhodospirillum*	Freshwater: anoxic hypolimnetic "plates" within the euphotic zone; benthic mats and epilithic/epiphytic communities; geo- thermal (hot spring) mats Marine: benthic epilithic/epiphytic communi- ties; intertidal/subtidal microbial mats; exposed mangrove roots; reefs; littoral zones of marshes, mudflats, and tide pools
Chemolithoautotrophic	*Thiobacillus*	Freshwater/marine: benthic mats and epili- thic/epiphytic communities; possible geo- thermal (vent) mats; mudflats and marshes
Heterotrophic Aerobic	*Arthrobacter* +O *Azomonas* + *Azorhizobium* X *Azospirillium* +O *Azotobacter* +O *Azotococcus* +OX *Beijerinckia* +OX *Bradyrhizobium* X *Corynebacterium* +O *Derxia* +O *Methylobacter* +O *Methylococcus* +O *Methylocystis* +O *Methylosinus* +O *Spirillum* +OX	Freshwater/marine: +, planktonic in oxic waters; O, benthic/epilithic/epiphytic in oxic waters; X, associated (asymbiotic) with higher plants (leaves and roots)
Facultative anaerobes	*Bacillus* *Campylobacter* *Enterobacter* *Escherichia* *Klebsiella* *Vibrio*	Freshwater/marine: associated with epili- thic/epiphytic and root surfaces, benthic mats, biofilms, surfaces of invertebrates and vertebrates, digestive tracts of inver- tebrates and vertebrates, decomposing or- ganics (benthic/planktonic detritus)
Anaerobic	*Beggiatoa* *Clostridium* *Desulfovibrio*	Freshwater/marine: decomposing organic matter; sediments; benthic mats; epilithic surfaces; biofilms

Nostoc, Gloeotrichia, Nodularia, and *Cylindrospermum,* the marine genera *Calothrix, Scytonema,* and *Tolypothrix,* and the hot spring genera *Fisherella* and *Mastigocladus* (Geitler, 1932; Fogg *et al.,* 1973; Rippka *et al.,* 1979; Carr and Whitton, 1982). It has been shown that the bulk of N_2 fixation occurs in the thick-walled, O_2-devoid hetero-cysts (Wolk, 1982), which occur at fairly regular intervals either along or at ends of filaments at frequencies ranging from 0.5 to 15% of total cell numbers, depending on

Table II. Known N₂-Fixing Cyanobacterial Genera Found in Aquatic Habitats

Major group	Genus	Habitats
Heterocystous filamentous	*Anabaena*+ ●	+ Freshwater: planktonic
	Aphanizomenon +	○ Freshwater: benthic/epilithic
	Calothrix ○□⊕⊠	⊕ Freshwater: epiphytic
	Cylindrospermum ⊕⊠	● Freshwater: endosymbiotic
	Fischerella ○⊕	X Marine[a]: planktonic
	Gloeotrichia +○⊕	□ Marine: benthic/epilithic
	Hapalosiphon ○⊕	⊠ Marine: epiphytic
	Mastigocladus ○□	■ Marine: endosymbiotic
	Microchaete ⊕	
	Nodularia +X	
	Nostoc +X○□⊕●	
	Richellia ■	
	Rivularia ○⊕□	
	Scytonema +○⊕□	
	Stigonema ○⊕	
	Tolypothrix +○⊕□	
	Westiellopsis ○	
Nonheterocystous filamentous	*Lyngbya* +○⊕□⊠	
	Microcoleus □⊠	
	Oscillatoria ○⊕□⊠	
	Phormidium ○⊕□⊠	
	Plectonema +○⊕□⊠	
	Raphidiopsis +○	
	Trichodesmium X	
Unicellular	*Gloeocapsa* (Gloeothece)+○⊕	
	Synechococcus X□	

[a]Marine includes brackish waters.

degrees of ambient nitrogen depletion and individual taxa. During heterocyst differentiation, photosystem II, containing the O_2-evolving machinery of photosynthesis associated with vegetative cells, is "lost" (Donze *et al.*, 1972). However, the light-capturing and electron transport components of photosystem I are retained during heterocyst differentiation (Wolk and Simon, 1969; Wolk, 1982). Other notable features of mature heterocysts include the absence of phycobillin accessory pigments, enhanced respiration, and a thickened cell wall (Wolk, 1982). These characteristics function jointly to exclude O_2 while maintaining the ability to capture and transfer radiant energy and reductant (fixed carbon compounds) provided by neighboring vegetative cells (Wolk, 1968). In return, fixed nitrogen (in the form of glutamine) is transported from heterocysts to vegetative cells (Wolk, 1982). Accordingly, heterocysts appear to be a highly successful physiological and structural adaptation satisfying the need for contemporaneous N_2 fixation and oxygenic photosynthesis in oxygen-rich environments.

Nonheterocystous N_2-fixing cyanobacteria are also present, and at times abundant, in natural waters. The most frequently cited are members of the genus *Oscillatoria*. In particular, the filamentous colonial planktonic marine *Oscillatoria*, *Trichodesmium* (three commonly found species), often occurs in geographically diverse subtropical and tropical nitrogen-deficient marine waters. It is generally agreed that, owing to its widespread distribution, *Trichodesmium* can be a significant contributor of fixed nitrogen in the marine environment (Carpenter and Capone, 1983). Consequently, the physiological mechanisms promoting N_2 fixation in this oxygenic photoautotroph have received considerable attention over the past decade (Carpenter and McCarthy, 1975; Carpenter and Price, 1976; Mague *et al.*, 1977). On the basis of extensive field observations, Carpenter, McCarthy, and colleagues (see Carpenter and Capone, 1983) concluded the following concerning the occurrence and regulation of N_2 fixation in *Trichodesmium* spp.: (1) N_2 fixation is strongly light mediated, (2) rates of N_2 fixation are directly related to the ability of filaments (trichomes) to aggregate into either fusiform flakes or spheres, and (3) maximum cellular N_2 fixation rates occur during calm (nonturbulent) periods, when buoyant trichomes readily aggregate near the sea surface. When aggregates are physically disrupted (by shaking or mild sonication), cellular N_2 fixation rates (under either illuminated or dark conditions) decrease abruptly, while vigorous shaking suppresses N_2 fixation. Microautoradiographs revealed that photosynthetic $^{14}CO_2$ incorporation was largely confined to terminal portions of trichomes. On the basis of these findings, Carpenter and Price (1976) speculated that *Trichodesmium* was able to spatially separate N_2 fixation from oxygenic photosynthesis, confining the former to internal regions of aggregates where O_2 minima might occur. Physical separation of aggregates would break down internal O_2 gradients, exposing the entire length of each trichome to potentially inhibitory ambient O_2 tensions. Subsequent studies using localized tetrazolium salt reduction (Bryceson and Fay, 1981; Paerl and Bland, 1982) confirmed the presence of highly reduced conditions in internal regions of aggregates. More recently, microelectrodes (having resolution on the order of 10 to 50 μm) were deployed to directly measure internal versus external O_2 gradients in a variety of *Trichodesmium* aggregates during an N_2-fixing bloom in coastal Atlantic Ocean waters off eastern North Carolina (Paerl and Bebout, 1988). This study demonstrated that internal O_2 depletion was sufficient for N_2 fixation to take place, even during illuminated periods. Also, it was shown that large aggregates contemporaneously exhibited lowest internal O_2 conditions and maximum cellular rates of N_2 fixation (acetylene reduction).

   Aggregation is a phenomenon common to diverse nonheterocystous cyanobacteria suspected (but in some cases not proven) of fixing N_2. It is tempting to speculate that, with *Trichodesmium*, aggregation reflects a need to localize nitrogenase in internal O_2-depleted microzones, while photosynthetic reductant is supplied in peripheral regions of aggregates. Among other filamentous nonheterocystous genera, nitrogenase activity (NA) has been associated with axenic marine *Microcoleus* and *Lyngbya* (Haystead *et al.*, 1970; Pearson *et al.*, 1981). While it can be shown that NA is often strongly light mediated in benthic and planktonic mats as well as epilithic–epiphytic fouling communities dominated by these genera, it is not certain that NA is exclusively or even partly confined to the cyanobacteria themselves. Certainly, close, highly specific associations between heterotrophic bacteria and photosynthetically active cyanobacteria are commonplace in natural waters (Fig. 1). Mucilagenous sheaths surrounding individual or commonly aggregated trichomes or colonies are often heavily epiphytized by bacteria. Such epiphytization does not necessarily reflect senescence of host cyanobacteria. On the contrary, hosts generally exhibit maximum photosynthetic rates and growth while epiphytized (Fig. 2). Among such associations, NA could reflect indirect stimulation of epiphytic bacterial N$_2$ fixation. Cyanobacterial photosynthate excretion could support bacterial heterotrophy, resultant respiration (localized O$_2$ consumption), and hence NA. Furthermore, diel N$_2$–CO$_2$ fixation studies conducted on intertidal North Carolina (coastal Atlantic Ocean waters) mats dominated by *Microcoleus chthonoplastes* bundles and/or *Lyngbya aestuarii* tufts reveal periods of enhanced dark-mediated NA, particularly during evening and nighttime hours (Bebout *et al.*, 1987; Paerl *et al.*, 1988). Similar results have been obtained among epiphytic fouling communities in marine waters (Paerl and Carlton, 1989). These results support the hypothesis that heterotrophic N$_2$ fixation may at times dominate such communities. In some cases, dominant cyanobacterial genera have been grown axenically under N-free conditions with partial success (Pearson *et al.*, 1981; Stal and Krumbein, 1985), supporting the contention that certain strains among these genera are capable of N$_2$ fixation. Immunochemical (immunofluorescence–immunogold) studies, using nitrogenase-specific polyclonal and monoclonal antibodies, will provide a more definitive and direct answer to the potential "division of labor" among such N$_2$-fixing consortia (Bergman *et al.*, 1986). Regardless of the relative roles played by cyanobacteria and eubacteria, studies thus far completed implicate O$_2$ tension as an important (and at times overriding) regulator of community N$_2$ fixation rates within nitrogen-deficient systems.

The last notable group of potential N$_2$-fixing cyanobacteria are unicellular nonheterocystous genera. Among these, *Gloeocapsa* and *Synechococcus* have received recent attention in freshwater and marine studies, respectively (Gallon and Chaplin, 1988; Mitsui *et al.*, 1987). Gallon and Chaplin (1988) unequivocally demonstrated NA in axenic clones of *Gloeocapsa alpicola,* with NA being temporally separated from oxygenic photosynthesis. Similarly, Leon *et al.* (1986) have shown a cyclic inverse relationship between N$_2$ fixation and photosynthesis in some marine *Synechococcus* species. In both cases, NA rapidly (within minutes) commenced after cessation of photosynthesis (in darkness). Apparently, members of both genera are capable of reactivation of previously synthesized but O$_2$-inactivated (during illumination) nitrogenase, since the sudden appearance of NA would appear too fast to be exclusively attributable to *de novo* synthesis (doubling times of both genera are on the order of several hours under optimal growth conditions). Recently, Mitsui and co-workers have shown certain marine *Synechococcus* strains to be capable of temporally alternating NA and photosynthetic O$_2$ evolution even under continuous illumination (Mitsui *et al.*, 1987), a feat truly remarkable for a unicell in which both processes are in very close proximity. The physiological mechanisms controlling such "rhythmic avoidance" of O$_2$ inhibition of NA are currently under investigation; speculation focuses on intracellular buildups of carbohydrate-rich photosynthate which, as "macromolecular end products," may allosterically control photosynthetic and N$_2$-fixing activities.

From an ecological perspective, recent findings of NA in *Gloeocapsa* and *Syne-*

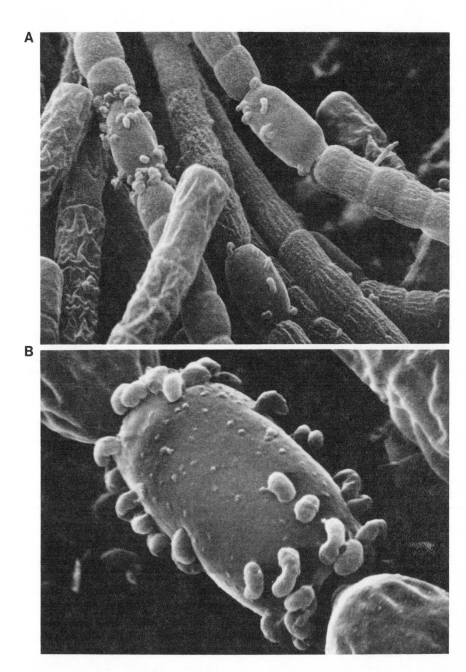

Figure 1. (A) Unidentified bacteria specifically associated with the heterocysts of the freshwater cyanobacterium *Aphanizomenon flos-aquae*. Sample obtained by A. J. Horne from Clear Lake, California, during an N_2-fixing bloom in June 1974. [Photograph from Paerl and Kellar (1978).] (B) Populations of the heterotrophic bacterium *Pseudomonas aeruginosa* specifically associated with a heterocyst of the freshwater cyanobacterium *Anabaena oscillarioides*. Sample obtained from Lake Rotongaio, New Zea-

C

D

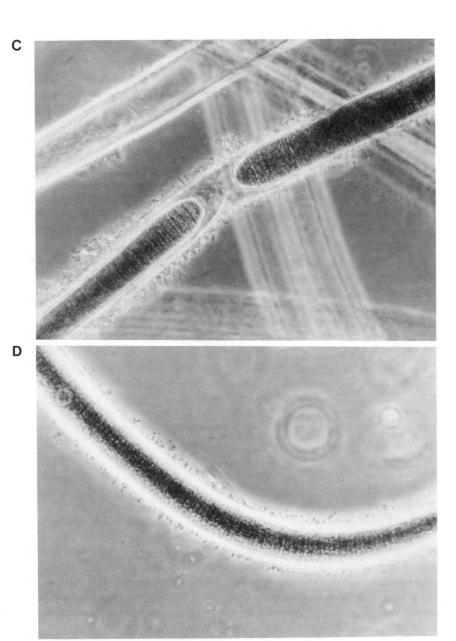

land, during an actively growing, N_2-fixing bloom. It was later shown that *P. aeruginosa* is chemotactically attracted to the heterocysts (Paerl and Gallucci, 1985). Terminal regions of heterocysts are preferred attachment sites; these are also sites of cellular fixed carbon and nitrogen "leakage" from intact *A. oscillarioides* filaments. (C and D) Attachment of unidentified bacteria to mucilaginous sheaths surrounding filaments of the nonheterocystous marine cyanobacterium *Lyngbya aestuarii*. Both N_2 fixation (acetylene reduction) and photosynthetic growth rates are high during bacterial epiphytization, indicating that this association involves metabolically active rather than senescent N_2 fixers.

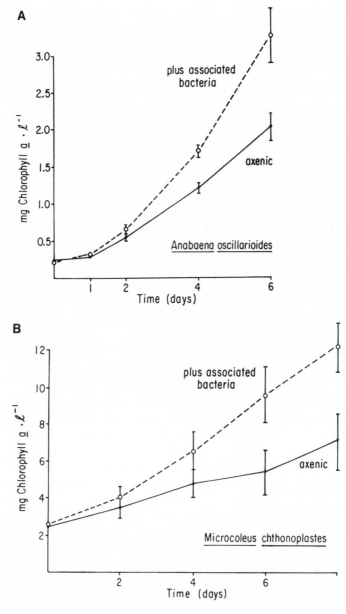

Figure 2. (A) Chlorophyll *a* (as an indicator of cyanobacterial biomass)-specific growth of axenic versus bacteria-associated populations of the heterocystous cyanobacterium *Anabaena oscillarioides* over a 6-day period under nitrogen-free (N_2-fixing) conditions. Triplicate treatments (variability among replicates is shown) were incubated under a 12-hr light/12-hr dark cycle. (B) Chlorophyll *a*-specific growth of axenic versus bacteria-associated populations of the marine nonheterocystous filamentous cyanobacterium *Microcoleus chthonoplastes* under nitrogen-free conditions. Error bars reflect variability among triplicates. Growth was under a 12-hr light/12-hr dark cycle.

chococcus are of importance, since both genera have been shown to be significant contributors to microalgal biomass and primary production in freshwater and marine planktonic and benthic environments, respectively (Waterbury *et al.*, 1979; Takahashi and Hori, 1984; Stockner and Antia, 1986; Stockner, 1988). Various other unicellular genera, including *Synechocystis, Gloeothece, Aphanocapsa,* and *Chroococcus,* are currently under investigation as potential N₂ fixers.

Chemolithoautotrophic microorganisms capable of N₂ fixation without heterocysts occupy a wide range of aquatic habitats. Virtually all chemolithoautotrophs must reside in anoxic or microaerophilic habitats that offer refuge from potentially inhibitory (to nitrogenase) levels of O₂ while providing abundant supplies of reduced inorganic compounds. Oxidation of reduced forms of sulfur (H_2S, S^{2-}), iron (Fe^{2+}), manganese (Mn^{2+}), carbon (CH_4), and hydrogen (H_2) constitute major sources of reductant for the conversion of N₂ to NH_3. Habitats abundant in chemolithotrophic N₂ fixers include anoxic marine and freshwater sediments, biofilms, mats, mangroves, swamps and wetlands, geothermal springs, vents and abyssal trench communities, decomposing organic matter (detritus), and faunal gut communities (Stewart, 1975; Yates, 1977; Jorgensen, 1980).

A largely unrecognized but potentially important (from nitrogen flux and budget perspectives) N₂-fixing community is the vast array of heterotrophic eubacterial genera inhabiting aquatic habitats of varying O₂ tension. Some familiar genera occupying both planktonic and benthic habitats fall into this category. In addition to their known genetic potentials for fixing N₂, these genera play fundamentally important roles in nutrient (C, N, P. S) cycling. Relevant genera include *Pseudomonas, Bacillus, Vibrio, Azotobacter, Spirillum, Klebsiella,* and *Clostridium* (Postgate, 1978). Both laboratory isolates and field populations of these genera have revealed N₂-fixing capabilities, either as active (detectable acetylene reduction and/or ¹⁵N₂ fixation rates) or as potentially active (structural genes present and/or nitrogenase synthesized but inactive) populations. Frequently, N₂-fixing heterotrophs can be isolated from planktonic habitats, decomposing plant litter, sediments, animal remains, fecal material, and suspended detritus (including dead phytoplankton and zooplankton as well as amorphous marine and freshwater aggregates of organic and inorganic nature, such as "marine snow") (Kawai and Sugahara, 1971; McClung and Patriquin, 1980; Guerinot and Patriquin, 1981; Guerinot and Colwell, 1985; Urdaci *et al.*, 1988). Little is known about the quantitative contributions of microheterotrophs to aquatic nitrogen budgets. To a large extent, this knowledge gap exists because (1) rates of N₂ fixation in microhabitats or "microzones" supporting heterotrophs may fall below analytical detection limits, (2) microheterotrophs and their substrates are patchily distributed and hence only infrequently encountered, (3) attachment substrates supporting microheterotrophs may be perturbed or altered by sampling and manipulation for N₂ fixation assays, hence potentially destroying community structure and O₂-depleted microzones, and (4) microheterotrophic N₂ fixation may be closely linked to photosynthetic organic matter and oxygen production, thereby giving rise to spatial–temporal patterns hitherto unrecognized and inadequately considered in field assessments.

In terms of geographic distribution, taxonomically diverse heterotrophic N₂ fixers have been isolated from disparate benthic and planktonic habitats, including pelagic Pacific Ocean waters (Maruyama *et al.*, 1970), the Sea of Japan (Kawai and Sugahara, 1971), the Caribbean Sea (Guerinot and Colwell, 1985), coastal Atlantic Ocean waters (Wynn-Williams and Rhodes, 1974), Australian, New Zealand, European, and North American estuaries (Bohlool and Wiebe, 1978; Carpenter and Capone, 1983), and numerous large and small lakes (Stewart, 1975; Postgate, 1978). Surely, more water bodies and new taxa are likely to be identified. In part, our inability to thus far adequately assess aquatic heterotrophic N_2 fixation potentials and characteristics is linked to difficulties applying conventional plating and liquid enrichment culturing techniques to isolation and enumeration of relevant microorganisms. Survival and growth of hetero-

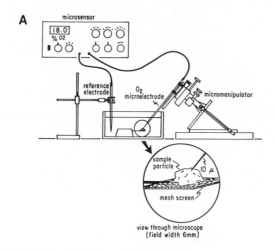

Figure 3. (A) Deployment of microelectrode (10-μm tip size) for determining O_2 gradients associated with detrital particles and submersed surfaces. Note that the microelectrode is mounted on a micromanipulator to allow for small-scale (10–100 μm) resolution of gradients. [From Paerl and Prufert (1987).] (B) Microelectrode determination of O_2 gradients associated with N_2-fixing marine microbial aggregates (maltose-enriched bacterial aggregate). As an abiotic control, some aggregates were poisoned with 4% (wt/vol) formalin (to arrest microbial O_2 consumption) before microelectrode determinations. In concert, these determinations reveal distinct internal (as well as external boundary layer) O_2 depletion in biologically active aggregates. Furthermore, the presence of localized O_2 depletion was directly related to N fixation (acetylene reduction) potentials in these aggregates. (C) Nitrogenase activity associated with various particle types examined in parallel with O_2 microelectrodes. Note that among all particles (initially added as sterile suspensions and subsequently colonized by naturally occurring microbiota) maximum activity corresponded to strongest internal O_2 gradients. Nitrogenase activities were detected on suspensions having approximately equal concentrations of particles. (D) Photomicrograph (phase contrast at ×400) of the edge of a maltose-enriched marine aggregate. Note ameboid protozoan grazers associated with the surface of this aggregate. (E) High-magnification (phase contrast at ×1000) view of the edge of a maltose-enriched marine aggregate. Note bacteria lodged in the mucilaginous matrix comprising much of this aggregate.

trophs under nitrogen-deficient conditions require specific environmental conditions. These include highly individualistic ambient O_2 requirements, ranging from obligate anaerobiosis to aerobiosis (*Azotobacter*, for instance), and adequate energy supplies in the form of specific organic substrates. It has been our experience, using microaerophilic stab inoculations in soft, nitrogen-free marine agar (purified and supplemented with

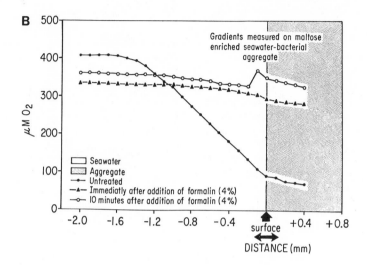

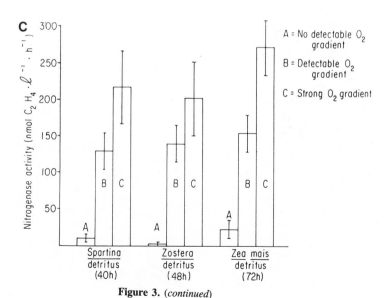

Figure 3. (*continued*)

D

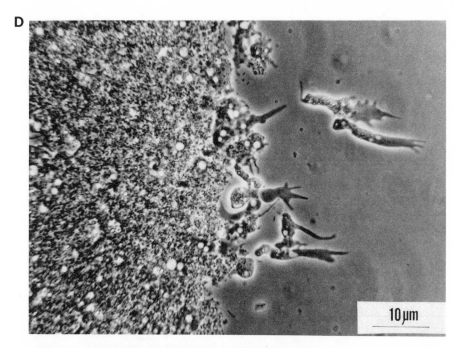

10 μm

E

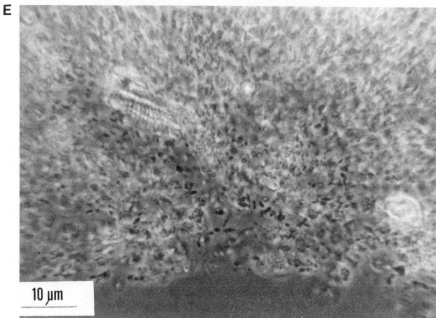

10 μm

Figure 3. (*continued*)

0.01 M mannitol as the sole organic carbon source), that active N$_2$-fixing microheterotrophs can be isolated from virtually any water sample collected along North
Carolina's nitrogen-depleted Atlantic coast and adjacent estuaries. The establishment,
proliferation, and maintenance of heterotrophic N$_2$-fixing conditions are tightly controlled by well-defined microscale oxygen regimes (as measured with microelectrodes)
concomitant with specific organic matter enrichment (Paerl and Prufert, 1987; Paerl et
al., 1987; Bebout et al., 1987; Paerl and Bebout, 1988) (Fig. 3). Such requirements can
seldom be maintained for more than a few days in the laboratory. Finally, we have noted
that to elicit N$_2$ fixation under these conditions, interacting microbial consortia (usually
at least three to four morphologically distinct species) must first be established. Separation of these consortia by further subculturing of individual species commonly leads to
cessation of N$_2$ fixation.

It is therefore concluded that because of complex and highly individualistic environmental requirements, standard microbiological isolation and culturing techniques
have thus far yielded gross underestimates of marine and freshwater heterotrophic N$_2$-
fixing species. Coupling these factors with sampling and analytical limitations, we can
further conclude that N$_2$ fixation rates associated with these organisms likely represent
underestimates. These methodological constraints underscore the need for alternative
and complementary techniques, including highly specific and sensitive immunochemical and molecular assessments of N$_2$ fixation activities and potentials.

3. The Physiological Ecology of Aquatic N$_2$ Fixation

Nitrogen fixation is mediated by the highly conserved (molecularly and phylogenetically) enzyme complex nitrogenase in procaryotes (Fig. 4). This complex is
composed of two subunits, an iron-containing protein dinitrogenase reductase and a
molybdenum–iron-containing protein dinitrogenase (Burgess, 1984).

Both proteins are composed of subunits, dinitrogenase reductase being a dimer
(two identical subunits of ca. 30 kilodaltons each) and dinitrogenase being a tetramer
(each 61-kilodalton subunit containing a 4Fe–4S cluster and Fe–Mo cofactor). The
manner in which Fe and Mo are organized in dinitrogenase remains unclear. Until the
early 1980s, the above-described complex was thought to be present in all N$_2$-fixing
procaryotes (Bothe, 1982). Recently, however, Bishop et al. (1980, 1982), Eady et al.
(1987), and Dilworth et al. (1987) have described "alternative" nitrogenases in which
molybdenum (in dinitrogenase) has been replaced by vanadium or in which molybdenum and/or vanadium are absent, being replaced by iron (the so-called Fe–Fe protein)
(Bishop et al., 1988). These findings, besides being of obvious biochemical, molecular,
and evolutionary interest, are also of ecological significance, since microorganisms
having these alternative nitrogenases would also exhibit distinctly different metal requirements; implications of these findings will be discussed more thoroughly in Section
4. Furthermore, microorganisms possessing alternative nitrogenases do not reduce acety-

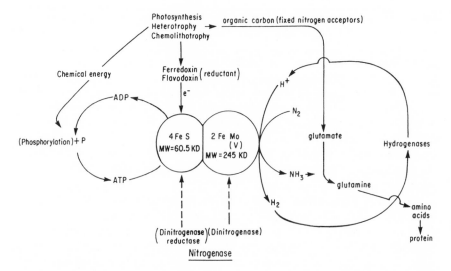

Figure 4. Schematic diagram of the N_2 fixation process as catalyzed by the nitrogenase complex. The energy, reductant, and carbon sources supporting this process as well as associated H_2 cycling are shown. MW, Molecular weight; KD, kilodaltons.

lene (see Section 4) exclusively to ethylene, as do Mo-containing strains. Among cells with Fe–Fe-containing nitrogenases, acetylene reduction appears less efficient (per unit reducing equivalents), with some acetylene reduced to ethane instead of ethylene (Dilworth *et al.*, 1987). Hence, detectability and estimation of NA are quite individualistic among such microorganisms.

A unique feature of all nitrogenases is their capability to cleave the highly stable triple bond of the dinitrogen molecule (Stiefel, 1977). To accomplish this with concomitant reduction of N_2, reduced conditions are required. It follows that anoxic conditions are an absolute requirement for N_2 fixation to proceed. The mere presence of O_2 in cell-free extracts irreversibly inactivates nitrogenase (Yates, 1977; Stiefel, 1977). Among intact microorganisms, O_2 inactivation appears to at least be partly reversible (Stewart, 1975; Postgate, 1978; Paerl, 1978); this reversibility may be linked to intracellular O_2 protection mechanisms, including a hydrogenase-coupled Knallgas (O_2 respiration) reaction (Bothe *et al.*, 1978), photorespiration (Burris, 1977), superoxide dismutase, catalase, and peroxidase O_2 and free-radical scavenging enzyme systems (Di Guiseppi and Fridovich, 1984). It is well known that in addition to catalyzing the reduction of N_2 to NH_3, nitrogenase shunts some fraction of its reducing power (electrons and protons) to H_2 formation (Stiefel, 1977; Yates, 1977) (Fig. 4). While this represents a waste of reducing power in terms of N_2-fixing potential, such H_2 production may be coupled to specific uptake hydrogenases involved in O_2 respiration (Knallgas reaction, for example), thereby using H_2 recycling as a means of locally protecting nitrogenase from O_2 inactivation (Bothe *et al.*, 1978).

In addition to its high sensitivity to O_2, nitrogenase activity and synthesis are readily inhibited in the presence of accumulated ammonia or ammonium ions. Under these conditions, glutamine readily accumulates, and this early product of N_2 fixation is thought to be the "end product inhibitor" of this process (Stewart, 1975; Yates, 1977). Ammonia-ammonium also inhibits differentiation of vegetative cells into heterocysts. Ammonia-ammonium suppression (through glutamine accumulation) of both NA and cell differentiation appears to be readily reversible among intact microorganisms once nitrogen-deficient conditions reappear.

Under appropriate conditions, nitrogenase reduces triple-bonded compounds other than N_2. For example, cyanide ($C\equiv N$) is readily reduced under N_2-free conditions (Stiefel, 1977). This characteristic has proven particularly fortuitous in the case of acetylene reduction. Acetylene (C_2H_2) is readily reduced to ethylene (C_2H_4), and because C_2H_2 is at least 60 times more soluble than N_2 in water, the former substrate (when added at saturating concentrations) readily outcompetes N_2 for binding and reducing sites on nitrogenase (Stewart *et al.*, 1967; Stewart, 1975). The theoretical molar ratio of C_2H_2 to N_2 reduction is $3:1$, since three bonds must be cleaved per molecule of N_2, while C_2H_2-to-C_2H_4 reduction involves the cleavage of only one of three bonds (Stiefel, 1977). In practice, this ratio is seldom achieved, in large part because of variable amounts of H_2 production (which would increase the ratio) (Bothe *et al.*, 1978). Despite an inexact C_2H_2/N_2 reduction ratio, C_2H_2 reduction is often preferred over other techniques of estimating N_2 fixation because both C_2H_2 and C_2H_4 are readily detected in small amounts by gas chromatography and the C_2H_2 reduction technique is easily used in the field. The C_2H_2 reduction technique is considered the standard method for estimating aquatic NA.

Nitrogen fixation is an appreciable energy-requiring process. Energy is obtained from the photosynthetic, chemolithoautotrophic, or heterotrophic (catabolic) conversion of radiant and reduced chemical energy into the phosphorylated nucleotide ATP (Fig. 4). From 12 to 15 ATPs are required for the reduction of one N_2 molecule (Bothe, 1982). Accordingly, adequate cellular phosphorus (phosphate) and reductant supplies are essential. Furthermore, demands exist for specific metals serving as cofactors in the enzymatic synthesis and functioning of nitrogenase as well as electron transport systems (ETS) supporting NA. Relevant metals include iron, a structural cofactor of nitrogenase, a cofactor in oxidative phosphorylation, and a constituent of ferredoxin (Hill reaction of photosynthesis); cobalt, present in ETS and vitamins; molybdenum, a structural cofactor of nitrogenase; vanadium, an (alternative) cofactor of nitrogenase; and zinc in ETS, nitrogenase enzyme synthesis and vitamins. In Section 4, I will specifically address potential roles that restricted supplies of these nutrients may play in the regulation of aquatic N_2 fixation.

Both its conservative molecular nature and its universal sensitivity to O_2 are compelling reasons for believing nitrogenase to be an evolutionarily ancient enzyme system. Evidence for its early inclusion in the evolution and diversification of procaryotic life exists in its omnipresence among a wide variety of anaerobic photolithoautotrophs and photoheterotrophs (*Chromatiaceae, Chlorobiaceae, Rhodospirillaceae*), methanogens, sulfate reducers, hydrogen bacteria, and obligate anaerobic heterotrophs, all of which played key roles in production and mineralization processes in the Precambrian (2.5–3.5 billion years ago) seas (Schopf, 1968, 1975; Cloud, 1976; Schopf and Walter, 1982). Well-preserved Precambrian stromatolitic rock formations in Canada (Gunflint iron formation), South Africa (Onverwacht group), and Australia (Warrawcona and George Creek groups) contain abundant fossil evidence pointing to the abundance of heterocystous cyanobacteria, providing further direct evidence for diversification of N_2-fixing procaryotes at this early stage in biospheric evolution (Schopf and Walter, 1982). The N_2-fixing cyanobacteria represent "breakthrough" photolithoautotrophs, having ushered in and dominated the early transition from anoxic to contemporary O_2-rich biospheric conditions (Margulis, 1982). This transition called for a host of molecular, physiological, and structural adaptations facilitating and optimizing N_2 fixation in the presence of oxygenic photosynthesis in an increasingly oxygenated atmosphere. For this reason, diverse cyanobacteria as well as O_2-tolerant eubacteria (*Azotobacter*) are the focus of studies aimed at elucidating cellular O_2 protection mechanisms (Stewart, 1975; Postgate, 1978).

Symbiotic associations in part reflect intraorganismal strategies aimed at cooptimizing N_2 fixation and biomass production processes in nitrogen-depleted waters. In many such associations, the host plant or animal provides respiratory protection from O_2 and adequate nutrition (phosphorus, metals, organics) for endosymbionts, thereby circumventing environmental constraints on N_2 fixation. In return, the host thrives, despite apparent ambient nitrogen limitation. Perhaps the best examples of high-productivity symbioses residing in nitrogen-depleted waters are extensive mats of *Rhizoselenia* (diatom)–*Richelia* (N_2-fixing endosymbiotic cyanobacterium) in ultraoligotrophic seas, endosymbiotic N_2-fixing eubacteria and cyanobacteria in and on corals (Coelenterata), sponges (Porifera), didemnid ascidians (tunicates-Protochordata), shipworms and shrimps (Arthropoda) protozoans, ferns (*Azolla*), cycads (*Cycas*), and macroalgae (Chlorophyta, Phaeophyta) Margulis, 1981; Stewart *et al.*, 1980; Ahmadjian and Paracer, 1986). Without endosymbiotic N_2 fixers, both the geographic ranges and standing stocks of these hosts would be confined to more nitrogen-enriched waters. In addition, a potentially important suite of microbe–microbe interactions, involving specific attachment to and orientation of N_2-fixing procaryotes with a variety of heterotrophs as well as primary producers, can be found in both marine and freshwater habitats. Examples of such exosymbioses include heterotrophic bacteria specifically oriented toward and around heterocysts of N_2-fixing cyanobacteria (Fig. 1). This association leads to optimal cell-specific N_2 fixation rates in host cyanobacteria (Fig. 5), bacterial epiphytes located within sheaths and mucilagenous matrices in cyanobacteria–mat and aggregate systems, which through complex interactions among laminated populations of photoautotrophs, chemolithotrophs, and heterotrophs form energy-rich, O_2-depleted microzones harboring N_2-fixing procaryotes (Paerl, 1982; Paerl and Prufert, 1987; Bebout *et al.*, 1987; Paerl and Carlton, 1988). In summary, there exists ample evidence that closely coordinated intracellular physiological mechanisms and intercellular consortia operate contemporaneously and contiguously to optimize aquatic N_2 fixation in the presence of ambient physical–chemical constraints.

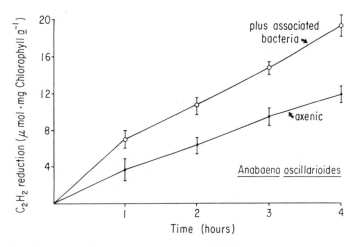

Figure 5. Cell (chlorophyll *a*)-specific rates of N₂ fixation (acetylene reduction) in axenic versus bacteria-associated populations of the heterocystous cyanobacterium *Anabaena oscillarioides*. Triplicate treatments (variability among replicates is shown) were incubated for 3 days under a 12-hr light/12-hr dark illumination cycle before acetylene reduction determinations (conducted under illuminated conditions). Subsamples were withdrawn every 4 hr.

4. Environmental Constraints and Limitations on Aquatic N₂ Fixation

In Section 3, I dealt with energetic, nutritive, and biotic requirements for microbial N₂ fixation. I will now turn to the question of environmental supplies versus biotic demands. More specifically, is it possible to point to a set of common variables applicable to the regulation of both freshwater and marine N₂ fixation? Judging from the diverse communities of procaryotes known to be capable of N₂ fixation, it is certain that among natural assemblages, the presence of structural *nif* genes is widespread. Therefore, *a priori*, it seems unlikely that the apparent absence or paucity of N₂ fixation in vast numbers of aquatic ecosystems is due to an absence of genetic potential. Hence, we are left to consider a suite of environmental physical–chemical limitations. Among nutritional factors, phosphorus limitation is of prime importance in regulating freshwater N₂ fixation potentials (Stewart and Alexander, 1971; Healy, 1982). Direct relationships between magnitudes of phosphorus-loading and N₂ fixation rates are nowhere as evident as in eutrophied freshwater lakes, reservoirs, and rivers periodically plagued by planktonic blooms of the filamentous, heterocystous cyanobacterial genera *Anabaena, Aphanizomenon, Gloeotrichia, Nodularia,* and *Nostoc* (Fogg, 1969; Granhall and Lundgren, 1971; Vollenweider, 1976; Schindler *et al.,* 1978). Long-term coupling of excessive phosphorus loading to bloom potentials has been documented for numerous lakes exhibiting water quality problems associated with such blooms: Lakes Washington (near Seattle) (Edmondson, 1970), Erie (Beeton, 1965; Burns and Ross, 1972), and Mendota (Wisconsin) (Birge and Juday, 1922; Fallon and Brock, 1980), Green Bay, Wisconsin (Vanderhoef et al., 1974), and Clear Lake, California (Horne and Goldman, 1972), in North America; Lakes Trummen (Granhall and Lundgren, 1971; Cronberg et al., 1975), Lough Neagh (Gibson et al., 1971), and Esthwaite Water (Reynolds and Walsby, 1975) in Europe; Lake Kasumigaura in Japan (Seki et al.,1980); and numerous rural (agricultural regions) lakes, ponds, and rivers in Africa, Australia, New Zealand, Asia, and South America. Agricultural (fertilizer), municipal (sewage, either treated or untreated), and industrial discharges have repeatedly been implicated as quantitatively significant freshwater phosphorous sources (Fogg, 1969; Vollenweider, 1976; Likens, 1972). Fortunately, timely recognition of the importance of controlling phosphorus inputs has led to a reversal in eutrophication and associated N_2-fixing cyanobacterial blooms in freshwater habitats, with lakes Washington, Erie, and Trummen being prime examples.

The temporal and spatial relationships between phosphorus loading and cyanobacterial N_2 fixation potentials have been well documented in Canada's Experimental Lakes Area (ELA), where Schindler and colleagues (1978) followed phytoplankton community composition, productivity, and N_2 fixation responses to an array of phosphorus (P), nitrogen (N), as well as P-plus-N additions. Manipulative whole-lake nutrient experiments clearly showed that dominance of and proliferation by N_2-fixing genera were heavily dependent on both P supply rates and ratios of P versus N inputs (Schindler et al., 1978). Flett et al. (1980) have elaborated on the predictability of and dominance by N_2-fixing cyanobacteria over a range of P versus N loading scenarios at ELA. Although some degree of individuality among species composition and magnitudes of N_2 fixation was evident, upper and lower limits of P loading as well as P versus N input rates could be utilized to predict N_2 fixation potentials among these lakes. Smith (1983) has elaborated further by generating models predicting the likelihood of cyanobacterial (grouping N_2-fixing with non-N_2-fixing cyanobacteria) dominance based on relative proportions of N versus P inputs among geographically diverse lakes. Such modeling efforts have at times proven useful in formulating specific phosphorus (and nitrogen) input constraints aimed at controlling nuisance blooms. The absolute amounts of P and N loading are of additional value in considering management strategies. In the face of excessive nitrogen (N-sufficient condition) loading, decreases in P loading alone may not effectively decrease cyanobacterial dominance, since nuisance non-N_2 -fixing genera (including scum-forming Microcystis, Lyngbya, and some Oscillatoria species) often dominate under such conditions.

Conversely, excessive P loading (relative to N) does not necessarily signal the onset or proliferation of N_2-fixing cyanobacterial bloom species. Constraints in iron availability have been implicated in limiting primary productivity among diverse phytoplankton assemblages. Iron demands for both synthesis and function of nitrogenase are appreciable, being second only to phosphorus (Hutner, 1972). However, iron limitation of N_2 fixation has been reported in only a few instances (Wurtzbaugh and Horne, 1983); infrequent reports of iron limitations may reflect a lack of experimental evidence rather than a lack of ubiquity. In light of its chemical behavior in aquatic habitats, iron deficiencies should receive more serious consideration as a potential modulator of N$_2$ fixation, especially in aerobic surface waters having adequate phosphorus inputs. Under oxidized conditions, iron is commonly bound as ferric hydroxides and oxides, forms which are not readily available to phytoplankton (Stumm and Morgan, 1981). Perhaps because of constraints in iron availability, numerous N$_2$-fixing cyanobacterial (and eubacterial) genera excrete potent siderophore chelators, capable of sequestering iron in waters exhibiting trace concentrations and competing (for iron binding) against natural chelators (humic and fulvic acids) (Neilands, 1967; Murphy et al., 1976).

Constraints in iron availability may be most severe in large lake and oceanic environments where allochthonous inputs of new iron are limited or nonexistent (Martin and Fitzwater, 1988), while autochthonous precipitated and particle-bound iron would settle out of the euphotic zone, remaining entrained in nonmixed hypolimnetic waters. It has been suggested that under these conditions, atmospheric inputs of iron (as dust from distant places, air pollutants, or volcanic ash) constitute an increasingly important source of new iron (Rueter, 1982; J. G. Rueter, personal communication).

Deficiencies among other required trace metals have also been considered in the regulation of aquatic N$_2$ fixation potentials. Foremost in this group is molybdenum, a structural component of nitrogenase. In contrast to the common insoluble forms in which iron exists, molybdenum is readily available in the highly soluble anionic form molybdate (Stumm and Morgan, 1981). The first report of molybdenum deficiency controlling phytoplankton primary production was that of Goldman (1964) in a small meso-oligotrophic cirque (glacial) lake, Castle Lake, California. Molybdenum impacts on N$_2$ fixation in this nitrogen-limited lake were not examined, although molybdenum enrichment was shown to significantly stimulate NA in the root nodules of alder (*Alnus tenufolia*) trees surrounding the lake (J. C. Priscu, unpublished data). However, Axler et al., (1980) subsequently showed molybdenum additions (as MoO$_4^{2-}$) stimulated nitrate reductase activity in this lake. It is therefore likely that stimulation of primary productivity was the result of enhanced NO$_3^-$ assimilation (NO$_3^-$ is readily available during much of the year) and utilization in this lake. Recently, Wurtzbaugh (1988) found that molybdenum additions stimulated both photosynthesis and NA in several Utah reservoirs known to contain extremely low (undetectable by atomic absorption spectrometry) levels of molybdenum. A host of other, largely unpublished studies (including several by this author) have thus far failed to demonstrate molybdenum limitation of NA in a variety of lakes and rivers periodically hosting N$_2$-fixing cyanobacterial taxa (Paerl et al., 1987).

Among marine studies, Howarth and Cole (1985) proposed that molybdenum deficiencies might help explain the apparent paucity of N$_2$ fixation in oceanic habitats. They suggested and subsequently showed (Cole et al., 1986) that a structural analog of molybdate, sulfate, could effectively compete for uptake sites on cell membranes, consequently potentially blocking molybdenum uptake in sulfate-rich (~28 mM) seawater. However, the fact that sulfate may potentially outcompete molybdate at uptake sites is not equivocal evidence for molybdenum limitation of N$_2$ fixation. As long as enough molybdenum is taken up to satisfy requirements for nitrogenase synthesis and function, limitation will be avoided. In addition, recent studies (Collier, 1985; P. Hansen, unpublished data) have shown molybdate concentrations to be reasonably high (relative to cellular demands) and remarkably constant (60–100 nM) in diverse oceanic regions (including sulfate-rich areas), arguing against large-scale depletion of this micronutrient. Finally, a large number of nutrient addition bioassays conducted over the past 7 years in nitrogen-deficient North Carolina Atlantic coastal waters, as well as in Caribbean oceanic waters, have consistently failed to show molybdenum limitation of either the establishment or proliferation (of existing N_2-fixing assemblages) of N_2 fixation. Therefore, if present, molybdenum limitation does not appear to be a widespread feature of marine or fresh waters.

Investigations of other trace metal (Cu, Zn, B, Mn, Co) deficiencies among aquatic N_2-fixing assemblages are rare. Limited results thus far have been uniformly negative. In the 7-year North Carolina study cited above, we failed to verify either limitation or regulation of N_2 fixation by any of these trace metals (Paerl et al., 1987). Despite the negative results thus far obtained, additional insights as to potential trace metal impacts on freshwater and marine N_2 fixation potentials are needed.

Physical conditions such as irradiance, temperature, and turbulence (vertical and horizontal mixing) must be considered as parallel limiting factors that, in concert with nutritional requirements, dictate aquatic N_2 fixation potentials. Most N_2-fixing cyanobacterial genera (especially bloom genera) prefer warm, vertically stratified, high-irradiance conditions (Reynolds and Walsby, 1975; Paerl, 1988). A deficiency in any or all of these conditions could override nutritional regulation of N_2-fixing assemblages. In particular, persistent highly mixed conditions can effectively negate dominance among N_2-fixing (and non-N_2-fixing) cyanobacteria. Using gas vacuolation as a buoyancy-adjusting mechanism, most cyanobacterial bloom species can readily and rapidly (up to 20 m·hr^{-1}) adjust their vertical orientation in the water column (Walsby, 1972; Reynolds and Walsby, 1975), the relative position in the water column ultimately reflecting mutually satisfying nutritional and energy demands for growth. In vertically stratified, gently mixed or intermittently stratified waters, buoyancy alteration ensures a competitive advantage for cyanobacteria. In contrast, well-mixed conditions can override vertical buoyancy adjustments, the results being suboptimal growth and N_2-fixing conditions and a competitive advantage for non-N_2-fixing genera. Enhanced vertical mixing can redistribute previously entrained (in hypolimnetic waters) nitrogenous (and other) nutrients into the euphotic zone, thereby providing previously limiting nitrogen to surface-dwelling non-N_2-fixing taxa. The introduction of extensive hypolimnetic nitrogen reserves, particularly ammonia, can also lead to a rapid cessation of N_2 fixation through end product suppression of nitrogenase and the differentiation of vegetative cells to heterocysts. According to this scenario, conditions favoring N_2-fixing cyanobacterial dominance can be quickly terminated through physical forcing (mixing) accompanied by altered nutrient concentrations and balances (Fig. 6).

Exposed coastal and pelagic marine environments are often characterized by high magnitudes and lengthy periods of wind- and tide-induced turbulence; such conditions readily override the ability of the filamentous nonheterocystous N_2-fixing cyanobacterium *Trichodesmium* spp. to form large aggregates (necessary for N_2 fixation to proceed) accumulating near the sea surface (through buoyancy) as blooms. The result is a suppression of N_2-fixing potential despite the fact that ambient N/P ratios might otherwise favor N_2 fixation. This scenario appears to be a feature of vast regions of the

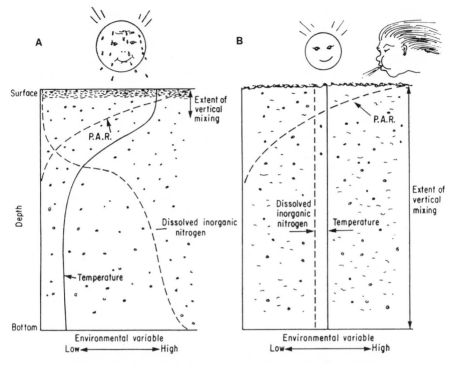

Figure 6. Schematic representation of the orientation of buoyant N$_2$-fixing cyanobacterial bloom species (*Anabaena, Aphanizomenon,* and *Trichodesmium*) during thermally stratified (A) versus vertically mixed (B) conditions in the water column. Generalized vertical profiles of photosynthetically active radiation (PAR), temperature, and inorganic nitrogen concentrations accompany the relative distribution of N$_2$-fixing cyanobacteria (small filaments) versus non N$_2$-fixing eucaryotic species (●).

world's oceans. The following question then arises: in light of the fragile nature of *Trichodesmium* aggregates and the necessity to remain aggregated for N$_2$ fixation to take place, why are not more O$_2$-resistant heterocystous species prevalent and dominant in these waters? At present, no satisfactory answer exists. It is apparent that adequate phosphorus supplies exist for *Trichodesmium* to form massive N$_2$-fixing blooms in such waters, physical conditions permitting. Why then cannot a more turbulent and hence O$_2$-tolerant analog N$_2$-fixing species occupy this apparently empty niche? Consideration of additional nutritional requirements may shed some light on this paradox.

5. Roles of Organic Matter and Microzone Formation

In virtually all waters where potential inorganic nutrient (P, Fe, Mo) limitations of N$_2$ fixation are alleviated (either through anthropogenic or natural inputs), N$_2$ fixation still supplies only a portion of these ecosystems' nitrogen demands (Horne and Fogg, 1970; Horne and Goldman, 1972; Flett *et al.*, 1980; Howarth *et al.*, 1988). This is true in both marine and freshwater habitats, although marine habitats exhibit the most chronic disparities between N_2 fixation rates (supplies) and biological nitrogen demands (Carpenter and Capone, 1983; Howarth *et al.*, 1988). Collectively, these observations imply that factors other than inorganic nutrient limitation may at times regulate the establishment, development, and proliferation of aquatic N_2 fixers. The parallel importance of favorable physical conditions has previously been discussed as crucial for the optimization of N_2 fixation in any aquatic habitat. The efficacy of physical conditioning of N_2 fixation revolves around two factors: (1) provision of adequate energy supplies and (2) promotion of localized O_2 depletion either on surfaces of particles or biofilms or through the creation of O_2 gradients either in water column or benthic regions. Specifically, physical factors such as ambient temperature (Fogg, 1969), irradiance (Van Liere and Walsby, 1982), and water column stability (Ganf and Horne, 1975) have been shown to regulate magnitudes of N_2 fixation. In addition, early observations by Pearsall (1932) and others (see Fogg, 1969) revealed a possible linkage between organic matter enrichment and the establishment of known N_2-fixing cyanobacterial genera. Although these studies hinted that organic matter (either allochthonous or autochthonous in origin) may play a role in enhancing aquatic N_2 fixation, contemporary research efforts have only recently focused on mechanisms underlying potential relationships. Both marine and freshwater studies conducted by this laboratory have shown that sterile organic matter enrichment (1) leads to development of N_2 fixation in waters previously void of this process and (2) stimulates N_2 fixation in both planktonic and benthic habitats known to support this process. Below, I will describe the role of organic matter in regulating N_2 fixation in full-salinity, nitrogen-depleted Atlantic coastal waters bordering North Carolina. These waters represent a paradoxical situation in that despite chronic and widespread nitrogen limitation, biological N_2 fixation is either absent or present at ecologically insignificant rates.

Bioassay evidence consistently and convincingly indicated that the absence of planktonic and benthic microbial N_2 fixation in these waters could not be attributable to deficiencies (lack of availability) of phosphorus, iron, or molybdenum (Paerl, 1985; Paerl *et al.*, 1987) (Fig. 7). Furthermore, the ranges of salinities commonly encountered in these waters (20–35 ppt) were by themselves not inhibitory to either growth or NA of cyanobacterial or eubacterial N_2 fixers that we were able to find in or isolate from these waters (Paerl *et al.*, 1987; H. W. Paerl and B. M. Bebout, manuscript in preparation). Also, commonly encountered pH regimes (7.8–8.5) proved to have no bearing on either the absence or presence of N_2 fixation. Finally, adequate supplies of photosynthetically active radiation necessary for light-mediated NA in cyanobacteria were ensured among the near-surface and intertidal habitats examined.

We did, however, identify and specifically examine local habitats where N_2 fixation readily and consistently occurred. Included were subtidal regimes of estuaries, sounds, intertidal regions of barrier islands, sand spits that supported microbial (cyanobacteria-dominated) mats, seagrass (*Zostera*) beds, and *Spartina*-dominated marshes (N_2 fixation occurred both in benthic regions and on planktonic detrital particles). In all cases, N_2

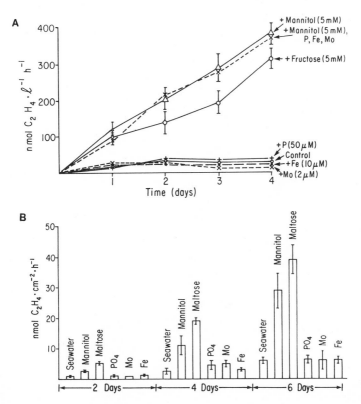

Figure 7. (A) Relative impacts of soluble inorganic nutrients, phosphorus as (NaH₂PO₄), iron (as EDTA-chelated FeCl₃), molybdenum (as Na₂MoO₄), the organics mannitol and fructose, and a combination of mannitol and inorganic nutrients on N₂ fixation (acetylene reduction) potentials of coastal Atlantic Ocean (North Carolina) water containing natural concentrations of *Spartina* detritus. Triplicate samples (variability among replicates is shown) were incubated under illuminated conditions (at 1-m depth) in an outdoor pond. Subsamples were taken daily after initial nutrient enrichment. (B) Effects of nutrient enrichment on N₂ fixation (acetylene reduction) potentials in an intertidal benthic cyanobacterial mat community located on Shackleford Banks (coastal Atlantic Ocean), North Carolina. "Seawater" denotes control conditions in which no nutrient enrichments were administered. All samples were incubated (in triplicate for each treatment) under natural illumination and temperature conditions. Subsamples were withdrawn every 2 days after initial nutrient enrichment; these were assayed (under illuminated conditions) for 3 hr for acetylene reduction activity. Concentrations of added nutrients were identical to those for panel A. Note the consistent stimulatory effect of organic enrichment in both planktonic (A) and benthic (B) habitats, while the addition of inorganic nutrients appears to have had no bearing on N₂ fixation potentials in either habitat.

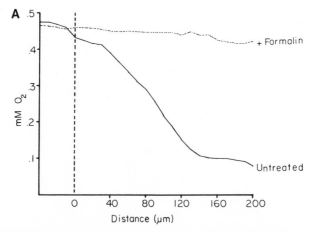

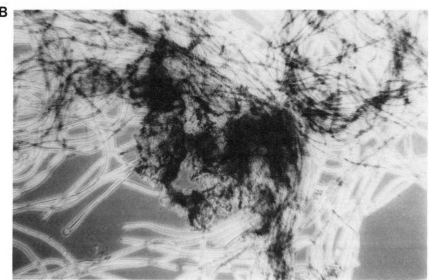

Figure 8. (A) Localized O_2 depletion within an illuminated planktonic marine cyanobacterial (*Oscillatoria* spp.)-bacteria aggregate as determined by a microelectrode. A parallel formalin (2%, wt/vol)-killed control was run to account for abiotic O_2 depletion (none appears evident) in aggregates. The dimension and shape of the internal O_2 gradients are shown inside the aggregate (right side of vertical dashed line); O_2 tension in bulk-phase seawater is shown to the left of the vertical dashed line. (B) Internal O_2-depleted microzones in an illuminated cyanobacterial (*Phormidium* spp.)-bacteria marine aggregate as shown by localized reduction of the tetrazolium salt TTC. Intra- and intercellular reduced TTC crystals appear as black patches dispersed throughout aggregates. (C) Inverse relationship of O_2 saturation (due to photosynthesis as measured by a microelectrode) and nitrogenase activity in a marine biofilm colonized by a mixed cyanobacterial (*Oscillatoria* spp., *Lyngbya chthonoplastes*)-bacteria community. The microelectrode was placed in a constant position in the biofilm while illumination was switched on (left of vertical line) or off (right of vertical line). Nitrogenase activity (acetylene reduction) was determined on biofilm subsamples at 20-min intervals during either light or dark cycles.

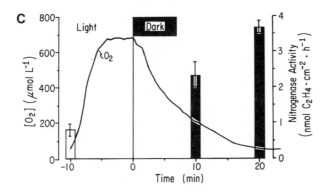

Figure 8. (*continued*)

fixation appeared confined to surficial microenvironments (microzones) exhibiting both organic matter (living and nonliving) enrichment as well as strong dissolved oxygen gradients (including localized regions of oxygen depletion) (Paerl, 1985; Paerl et al., 1987; Paerl and Carlton, 1988) (Fig. 3). From these findings, we conclude that marine N_2 fixation in nitrogen-depleted environments is closely mediated by microzones harboring trophic states and dissolved O_2 concentrations distinct from the ambient environment. These trophic states are a reflection of the production and utilization of organic matter.

We have deployed microelectrode technology and localized tetrazolium salt [specifically, 2,3,5-triphenyl-3-tetrazolium chloride (TTC)] reduction experiments to examine the development of O_2-depleted (reduced) microzones, both in parallel with bioassay experiments and among naturally occurring N_2-fixing sites (detritus, aggregates, microbial mats). Microelectrode O_2 profiles in suspended aggregates composed of *Spartina–Zostera* detritus as well as among cyanobacterial (*Oscillatoria* spp.)–bacteria aggregates and biofilms revealed that such communities had lower internal O_2 tensions than surrounding waters (Paerl and Prufert, 1987; Paerl and Carlton, 1988) (Fig. 8). TTC reduction demonstrated that reduced zones existed both within microbes and extracellularly in patchy matrices within microbial aggregates (Paerl and Bland, 1982) (Fig. 8). Both techniques showed that microbial (bacterial and cyanobacterial) N_2 fixation (acetylene reduction) proceeded when O_2-depleted (reduced) microzones were detected. When such microzones were destroyed, either by ambient O_2 supersaturation or excessive shaking (physically disrupting microzones), N_2 fixation ceased. In concert, these findings suggest that N_2 fixation is mediated by the availability and appropriate types of reduced microzones.

Among benthic N_2-fixing habitats, cyanobacteria-dominated mats contribute substantially to primary productivity in geographically diverse N-depleted, shallow-water temperate and tropical marine environments (Whitton and Potts, 1982; Cohen et al.,

1984; Bebout *et al.*, 1987). We have measured significant rates of N_2 fixation (acetylene reduction) in an intertidal mat located on Shackleford Banks, North Carolina (Atlantic Ocean) (Bautista and Paerl, 1985; Bebout *et al.*, 1987). Diel (24-hr) studies showed a temporal separation of photosynthesis (O_2 evolution) and N_2 fixation, with photosynthesis being maximized during early to mid-morning hours, while N_2 fixation (acetylene reduction) was maximized during evening (dark) hours (Fig. 9). From a quantitative (budgetary) standpoint, these results showed that nighttime N_2 fixation rates may exceed daytime rates by as much as two- to threefold. Hence, accurate and realistic estimates of N inputs via N_2 fixation in these systems must take into account highly significant (if not dominant) nighttime contributions.

Oxygen supersaturation during periods of high photosynthetic activity inhibits N_2 fixation at the surface of the mat (Bebout *et al.*, 1987). TTC reduction and O_2 vertical microelectrode profiles have shown the existence of highly reduced microzones within the mat. Addition of 3(3,4-dichlorophenyl)-1,1-dimethylurea (DCMU), an inhibitor of photosystem II, led to a rapid decrease in mat O_2 concentrations (Fig. 10). Simultaneously, N_2 fixation revealed dramatic increases, a demonstration of the *in situ* suppressing effect of O_2 on nitrogenase activity (Fig. 10). Nitrogen fixation was usually higher under dark than under light conditions. As with suspended aggregates, the addition of simple sugars resulted in an increase in N_2 fixation, whereas after the addition of inorganic nutrients (P, Fe, Mo), no detectable impact on N_2 fixation was observed (Fig. 7). These results suggest either that N_2 fixation in the mat is primarily eubacterial and strongly mediated by localized O_2 tension or that locally high rates of bacterial activity (including O_2 consumption) stimulate cyanobacterial N_2 fixation. In

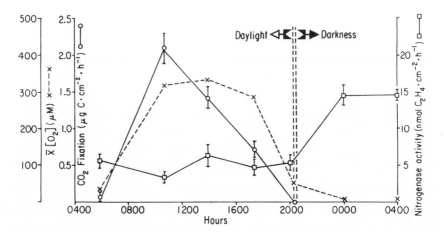

Figure 9. Diel interactions of photosynthetic CO_2 fixation (○), resultant O_2 saturation (x), and N_2 fixation (acetylene reduction) (□) in the upper 5 mm of a marine benthic cyanobacterial mat community located on Shackleford Banks, North Carolina. Samples were incubated in triplicate under naturally occurring illuminated/dark periods. Data points are midway during 3-hr incubations.

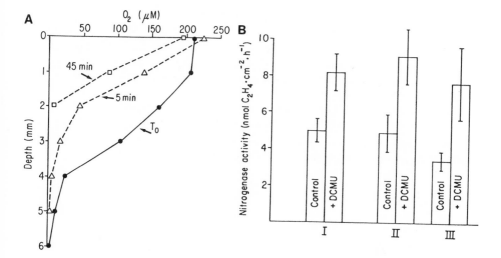

Figure 10. Short-term effects of the photosystem II (O_2 evolution) inhibitor DCMU (2×10^{-5} M) on O_2 saturation (A) and N_2-fixing (acetylene reduction) activities (B) of the cyanobacterial mat community located on Shackleford Banks. DCMU was added 15 min before acetylene reduction assays, which were conducted in three separate locations under illumined (photosynthetic) conditions. Controls received no DCMU. Error bars represent variability among triplicate 2-hr assays at each location.

each case, relatively high rates of N_2 fixation are made possible through the presence and maintenance of highly reduced microzones.

6. Evolutionary and Ecological Considerations

Much useful information and research direction can be gained by jointly considering environmental and evolutionary constraints on microbial N_2 fixation under present-day biospheric conditions. It is well known that anaerobic conditions prevailed during the development and early evolution of microbial life on Earth (Cloud, 1976). Because N_2 fixation is an obligate anaerobic process, it is not surprising that this process is confined to procaryotes, which first appeared, proliferated, and evolved during the O_2-devoid Precambrian periods (Stanier, 1977; Knoll, 1977; Schopf and Walter, 1982). Ancient oceans and freshwater habitats are thought to have been enriched with geochemically formed organic matter (primordial soup) far exceeding present-day concentrations (Cloud, 1976; Knoll, 1977; Holland, 1978; Walker et al., 1982). The disparity between Precambrian and current concentrations of organic matter is particularly evident among pelagic oceanic and large oligotrophic lake environments.

Two major environmental alterations have thus taken place since the development of N_2 fixation on earth: (1) biospheric O_2 levels have increased dramatically, largely owning to the advent of oxygenic photosynthetic plants (initially cyanobacteria), and (2) organic matter has been steadily reduced by heterotrophic microorganisms as well as by higher-ranked heterotrophs. These events have led to an incompatibility between the N_2-fixing process (nitrogenase) and modern-day (O_2-rich) biospheric conditions. As a result, N_2-fixing microorganisms have had to confine their activities to relatively reduced (O_2-poor) microzones in aerobic waters or develop specialized cells that exclude oxygen. Such environments include surfaces, interstitial regions, and endosymbioses. Localized respiration of organic matter in these microzones aids in maintaining subsaturated O_2 conditions. In this regard, it is interesting to note that aerobic eucaryotic algae and higher plants and animals are incapable of N_2 fixation unless they harbor microzones for N_2-fixing microorganisms.

7. Ecosystem-Level Regulation of N_2 Fixation: Are There Fundamental Differences between Freshwater and Marine Habitats?

From the previous discussions, it is evident that N_2 fixation is established, promoted, and sustained when several environmental factors and constraints are concomitantly optimized or alleviated. An example of such synergism can be found in temperate and tropical eutrophic, phosphorus-enriched lakes revealing periodic vertical stratification of the water column. Frequently, such lakes host massive near-surface and surface blooms of N_2-fixing heterocystous cyanobacterial genera. While contemporaneous and contiguous nitrogen depletion and phosphorus sufficiency favor the establishment of such blooms, the development and duration of blooms are frequently controlled by intensity and duration of irradiance as well as turbulence induced by wind and riverine advection. Appropriately, these physical factors dictate thermal stratification characteristics of the epilimnion. As mentioned previously, vertical stratification is a strong determinant in cyanobacterial dominance, particularly among gas-vacuolated (buoyant) genera (Fogg, 1969; Reynolds and Walsby, 1975). Generally, the more well-defined and permanent the stratification, the greater the advantage to these genera, which can readily seek optimal depths for photosynthetic production and N_2 fixation purposes through diel buoyancy adjustments. The ability to freely migrate through the water column also facilitates replenishment of essential nutrients. For example, periodic downward migration near dark (aphotic) but phosphorus- and trace metal-rich hypolimnetic waters allows for assimilation and storage of large nutrient quotas. Subsequent migration into euphotic epilimnia ensures adequate photosynthetic production and N_2 fixation, making use of previously sequestered inorganic nutrients for growth and replication.

Under strongly mixed (destratified) conditions, N_2-fixing cyanobacteria often lose numerical dominance because (1) enrichment of epilimnetic nitrogen from deeper waters may suppress nitrogenase activity, hence negating a previously held competitive advantage, and (2) previously buoyant surface-dwelling N_2 fixers heavily dependent on high radiant energy flux may forcibly be mixed into dimly lit or aphotic waters (Fig. 6).

While small nutrient-enriched eutrophic lakes, particularly those sheltered from wind mixing, offer the best refuge for such N$_2$ fixers, they are by no means the only habitats exemplifying positive synergistic interactions between nutrient adequacy and favorable physical conditions. Periodically (hourly to daily), phosphorus-sufficient estuarine, large lake, and oceanic surface waters can usher in and sustain sizable blooms of planktonic N$_2$-fixing cyanobacteria. During the 1950s and 1960s, under the influence of enhanced phosphorus loading, pelagic sections of Lake Erie (one of the U.S. Great Lakes) were plagued with massive and persistent *Anabaena* blooms during periodically stratified summer months (Beeton, 1965; Burns and Ross, 1972). As a brackish water analog, the heterocystous genus *Nodularia* forms extensive blooms in phosphorus-sufficient regions of the Baltic Sea during late summer months (Niemi, 1979; Howarth *et al.*, 1988). *Nodularia* and *Oscillatoria* blooms can frequent portions of estuaries periodically exhibiting salinity and/or thermal stratification (Huber, 1986). In N-depleted tropical and subtropical oceans, N$_2$-fixing *Oscillatoria* spp. (*Trichodesmium* spp.) form surface blooms extending for hundreds of square kilometers, key prerequisites again being phosphorus sufficiency and calm sea state. Calm weather conditions lasting from less than a day to weeks can trigger *Trichodesmium* blooms in the tropical Caribbean Sea, the subtropical Sargasso Sea, and neighboring Atlantic Ocean waters off the southeastern U.S. coast, off the eastern coasts of China and Japan in diverse regions of the South China Sea (near Taiwan), and in waters bordering the Australian Great Barrier Reef or the Central American Barrier Reef (Carpenter, 1983). Despite the thus far unexplained fact that heterocystous taxa are replaced by nonheterocystous taxa as bloom species in oceanic and near-shore marine waters, in each case N$_2$-fixing blooms are tightly controlled by a similar set of chemical and physical conditions.

Epilithic and epiphytic N$_2$-fixing microbial communities are cosmopolitan in diverse freshwater and marine habitats. A common feature is their ability to thrive and persist in waters either too turbulent or too nutrient deficient to support active planktonic N$_2$-fixing assemblages. North American lacustrine examples include Castle Lake, Lake Tahoe (California–Nevada), and Crater Lake (Oregon); these N-depleted lakes exhibit an absence of planktonic N$_2$ fixation, while diverse and periodically rich epilithic microbial communities, including heterocystous N$_2$-fixing cyanobacteria (and possibly eubacteria), are a common feature of both lakes (Reuter *et al.*, 1986; J. E. Reuter, unpublished data). Oceanic analogs include epilithic hard- and soft-bottom N$_2$-fixing communities (dominated by the cyanobacterial genera *Calothrix, Rivularia, Gloeotrichia, Tolypothrix, Nostoc, Microcoleus,* and *Oscillatoria*) in subtidal coastal and reef habitats as well as intertidal lagoons (mats) (Whitton and Potts, 1982; Cohen *et al.*, 1984); overlying planktonic communities consistently fail to exhibit N$_2$-fixing activity (Carpenter and Capone, 1983).

Similar comparisons can be drawn among epiphytic communities. Marine macrophytes, including the seagrasses *Zostera* (eelgrass) and *Thallasia* (turtle grass), harbor epiphytic communities rich in N$_2$-fixing cyanobacterial (both heterocystous and nonheterocystous) and eubacterial taxa (Gotto and Taylor, 1976; Capone and Taylor, 1977, 1980; Penhale and Capone, 1981). Such epiphytism is both geographically and temporally (seasonally) widespread. Freshwater macrophytes, including notorious "weed" species such as *Elodea* and *Myriophyllum,* are at times (especially mid to late summer) heavily epithytized by heterocystous and nonheterocystous N_2-fixing cyanobacteria (Finke and Seeley, 1978; Paerl and Kellar, 1979). Studies in respective ecosystems have commonly concluded that N_2 fixation rates associated with epiphytic communities are relatively high, constituting a significant nitrogen source for host macrophytes and associated micro- and macrobiota (Capone and Taylor, 1980; Carpenter and Capone, 1983).

Micro- and macroalgae in N-depleted waters are also frequently colonized by both symbiotic and saprophytic N_2 fixers. Like macrophytic epiphytes, microalgal epiphytes include both cyanobacteria and eubacteria (Head and Carpenter, 1975; Rosenberg and Paerl, 1980; Paerl, 1982). Macrophytes and micro- or macroalgae are potential hosts for N_2-fixing epiphytes and saprophytes since they excrete (either actively or passively) significant fractions of photosynthate as extracellular mucous slimes and soluble organic compounds (Hellebust, 1965; Fogg, 1971; Fogg and Burton, 1977; Nalewajko, 1978). Such compounds characteristically have high C/N ratios and are readily biodegraded (exceptions being tannic, humic, and some phenolic exudates). Moreover, following macrophyte or algal death, a substantial amount of organic matter is available as energy and carbon sources for associated heterotrophic N_2 fixers (eubacteria). Accordingly, during various stages of decomposition, macrophyte, macroalgal, and microalgal detritus are frequently sites of locally high rates of N_2 fixation (Paerl, 1984; Paerl *et al.,* 1987).

Taking previously discussed microzone dynamics into consideration, it is not surprising that epiphytic, epilithic, and epipsamnic (mat) communities are often sites of intense N_2-fixing activities. These communities have key commonalities, all of which tend to support and promote microzone-oriented N_2 fixation: (1) physical and structural stability in the form of surficial biofilms, (2) localized nutrient enrichment, particularly of phosphorus, trace metals, and organic compounds, and (3) localized O_2 depletion resulting from biochemical oxidation (respiration) of organic matter. In a synergistic fashion, organic matter enrichment, enhanced respiration, and enhancement of N_2 fixation and resultant alleviation of N-limited conditions work together in these microzones. It has been argued that phosphorus enrichment alone could sustain such tropic stimulation in aquatic habitats, since resultant stimulation of N_2 fixation in N-limited waters should lead to enhanced production of organic matter, which in turn facilitates microzone formation (Smith, 1984). Experimental verification of these events has pointed to the additional and fundamental importance of specific surficial substrates serving as nuclei or loci for the establishment and proliferation of N_2-fixing microorganisms (Paerl, 1985; Paerl and Prufert, 1987; Paerl and Carlton, 1988). It is noteworthy that structurally and functionally, both endo- and exosymbiotic associations reflect many of the prerequisites for surficial, microzone N_2 fixation.

8. Conclusions

Despite substantial, and at times enigmatic, differences among N_2-fixing taxa, striking ecological similarities exist in the range of habitats amenable (or prohibitory) to the establishment and proliferation of N_2-fixing microorganisms among diverse fresh-water and marine habitats. Physical, chemical, and biotic constraints on N_2 fixation potentials appear similar across salinity gradients. Often physical and chemical constraints are long lasting in oceanic (pelagic) environments. Both the proximity and relative contributions of allochthonous (terrigenous runoff, anthropogenic inputs) phosphorus and/or organic matter loading are key determinants in realizing such discrepancies. Clearly, small, eutrophic impoundments in agricultural and urban catchments receive large inputs of essential nutrients, while pelagic regions of large lakes and oceanic gyre systems are deficient in such inputs. Accordingly, relative contributions by N_2 fixation to nitrogen demands are far greater in the former than latter systems. From a biogeochemical perspective, the degree to which N_2 fixation can supplant nitrogen requirements plays a key role in determining the trophic state and fertility in individual aquatic environments, regardless of salinity. Future applications of immunological and molecular methods should help elucidate further the roles of individual microbes in aquatic ecosystems.

ACKNOWLEDGMENTS. Funding for research activities was provided by the National Science Foundation (BSR 86-14951, OCE 85-00740, OCE8820036), a North Carolina sea grant (RMER-5), the North Carolina Water Resources Research Institute (Project 70025), and the North Carolina Biotechnology Center. The assistance of J. Garner, B. Bright, V. Page and H. Page in preparation of the manuscript is greatly appreciated. J. Priscu and E. Kuenzler critically reviewed drafts of the manuscript.

References

Ahmadjian, V., and Paracer, S., 1986, *Symbiosis: An Introduction to Biological Associations*, University Press of New England, Hanover.

Axler, R., Gersberg, R. M., and Goldman, C. R. G., 1980, Stimulation of nitrate uptake and photosynthesis by molybdenum in Castle Lake, California, *Can. J. Fish. Aquat. Sci.* **37**:707–712.

Bautista, M. F., and Paerl, H. W., 1985, Diel N_2 fixation in an intertidal marine cyanobacterial mat community, *Mar. Chem.* **16**:369–377.

Bebout, B., Paerl, H. W., Crocker, K. M., and Prufert, L. E., 1987, Diel interactions of oxygenic photosynthesis and N_2 fixation (acetylene reduction) in a marine microbial mat community, *Appl. Environ. Microbiol.* **53**:2353–2362.

Beeton, A. M., 1965, Eutrophication of the St. Lawrence Great Lakes, *Limnol. Oceanogr.* **10**:240–254.

Bergman, B., Lindblad, P., and Rai, A., 1986, Nitrogenase in free-living and symbiotic cyanobacteria: Immunoelectron microscope localization, *FEMS Microbiol. Lett.* **35**:75–78.

Beyerinck, M. W., 1888, *Bot. Ztg.* **46**:725–735, 741–750, 757–790, 797–804.

Birge, E. A., and Juday, C., 1922, The inland waters of Wisconsin. The plankton. 1. Its quality and chemical composition, *Wis. Geol. Nat. Hist. Surv. Bull.* **64**:1–222.

Bishop, P. E., Jarlenski, D. M. L., and Hetherington, D. R., 1980, Evidence for an alternative nitrogen fixation system in *Azotobacter vinelandii*, *Proc. Nat. Acad. Sci. USA* **77**:7342–7345.

Bishop, P. E., Jarlenski, D. M. L., and Hetherington, D. R., 1982, Expression of an alternative nitrogen-fixing system in *Azotobacter vinelandii*, *J. Bacteriol.* **150**:1244–1246.

Bishop, P. E., Premakumar, R., Joerger, R. D., Jacobson, M. R., Dalton, D. A., Chisnell, J. R., and

338 Hans W. Paerl

Wolfinger, E. D., 1988, Alternative nitrogen fixing systems in *Azotobacter vinelandii*, in: *Nitrogen Fixation: Hundred Years After* (H. Bothe, F. J. de Bruyn, and W. E. Newton, eds.), pp. 71–80, Gustav Fischer, Stuttgart.

Bohlool, B. B., and Wiebe, W. J., 1978, Nitrogen fixing communities in an intertidal ecosystem, *Can. J. Microbiol.* **24**:932–938.

Bothe, H., 1982, Nitrogen fixation, in: *The Biology of Cyanobacteria* (N. G. Carr and B. A. Whitton, eds.), pp. 87–104, Blackwell Scientific Publications, Oxford.

Bothe, H., Distler, E., and Eisbrenner, G., 1978, Hydrogen metabolism in blue-green algae, *Biochemie* **60**:277–289.

Brezonik, P. L., 1973, Nitrogen Sources and Cycling in Natural Waters, U.S. Environmental Protection Agency Report 660/3-73-002.

Bryceson, I., and Fay, P., 1981, Nitrogen fixation in *Oscillatoria (Trichodesmium) erythraea* in relation to bundle formation and trichome differentiation, *Mar. Biol.* **61**:159–166.

Bunt, J. S., Cooksey, K. E., Keeb, M. A., Lee, C. C., and Taylor, B. F., 1970, Assay of algal nitrogen fixation in the marine subtropics by acetylene reduction, *Nature* (London) **227**:1163–1164.

Burgess, B. K., 1984, Structure and reactivity of nitrogenase: An overview, in: *Advances in Nitrogen Fixation Research* (W. E. Newton and C. Veeger, eds.), pp. 103–130, Nyhoff/Junk, The Hague.

Burns, N. M., and Ross, C., 1972, Project Hypo: An Intensive Study of the Lake Erie Central Basin Hypolimnion and Related Surface Water Phenomena, Can. Centre Inland Waters, Paper 6, U.S. Environmental Protection Agency Technical Report TS-05-71-208-24.

Burris, J. E., 1977, Photosynthesis, photorespiration, and dark respiration in eight species of algae, *Mar. Biol.* **39**:371–379.

Burris, R. H., and Wilson, P. W., 1946, Characteristics of the nitrogen-fixing enzyme in *Nostoc muscorum*, *Bot. Gaz.* **108**:254–262.

Capone, D. G., and Taylor, B. F., 1977, Nitrogen fixation (acetylene reduction) in the phyllosphere of *Thalassia testudinum*, *Mar. Biol.* (Berlin) **40**:12–28.

Capone, D. G., and Taylor, B. F., 1980, Microbial nitrogen cycling in a seagrass community, in: *Estuarine Perspectives* (V. S. Kennedy, ed.), pp. 153–161, Academic Press, New York.

Carpenter, E. J., 1983, Nitrogen fixation by marine *Oscillatoria (Trichodesmium)* in the world's oceans, in: *Nitrogen in the Marine Environment* (E. J. Carpenter and D. G. Capone, eds.), pp. 65–104, Academic Press, New York.

Carpenter, E. J., and Capone, D. G. (eds.), 1983, *Nitrogen in the Marine Environment*, Academic Press, New York.

Carpenter, E. J., and McCarthy, J. J., 1975, Nitrogen fixation and uptake of combined nitrogenous nutrients by *Oscillatoria (Trichodesmium) thiebautii* in the western Sargasso Sea, *Limnol. Oceanogr.* **20**:389–401.

Carpenter, E. J., and Price, C. C., IV, 1976, Marine *Oscillatoria (Trichodesmium):* An explanation for aerobic nitrogen fixation without heterocysts, *Science* **191**:1278–1280.

Carr, N. G., and Whitton, B. A. (eds.), 1982, *The Biology of Cyanobacteria*, Blackwell Scientific Publications, Oxford.

Cloud, P., 1976, Beginnings of biospheric evaluation and their biogeochemical consequences, *Paleobiology* **2**:351–387.

Cohen, Y., Castenholz, R. W., and Halvorson, H. O., 1984, *Microbial Mats: Stromatolites*, A. R. Liss, New York.

Cole, J. J., Howarth, R. W., Nolan, S. S., and Marino, R., 1986, Sulfate inhibition of molybdate assimilation by planktonic algae and bacteria: Some implications for the aquatic nitrogen cycle, *Biogeochemistry* **2**:179–196.

Collier, R. W., 1985, Molybdenum availability in the Northeast Pacific Ocean, *Limnol. Oceanogr.* **30**:1351–1354.

Cronberg, G., Grelin, C., and Larsson, K., 1975, The Late Trummen restoration project. II. Bacteria, phytoplankton and phytoplankton productivity, *Verh. Int. Ver. Limnol.* **19**:1088–1096.

Darwin, C., 1858, *Narrative of the Surveying Voyages of H.M.S. Adventure and Beagle (1840–1843)*, Oxford University Press, Oxford.

Delwiche, C. C., 1970, The nitrogen cycle, *Sci. Am.* **223:**136–146.

Di Guiseppi, J., and Fridovich, I., 1984, Toxicology of molecular oxygen, *Crit. Rev. Toxicol.* ·**12:**315–342.

Dilworth, M. J., Eady, R. R., Robson, R. L., and Miller, R. W., 1987, Ethane formation from acetylene as a potential test for vanadium nitrogenase *in vivo*, *Nature* (London) **327:**167–169.

Donze, M., Haveman, J., and Schiereck, P., 1972, Absence of photosystem 2 in heterocysts of the blue-green alga *Anabaena*, *Biochim. Biophys. Acta* **256:**157–161.

Dugdale, R. C., and Goering, J. J., 1967, Uptake of new and regenerated forms of nitrogen in primary productivity, *Limnol. Oceanogr.* **12:**196–206.

Dugdale, R. C., Menzel, D. W., and Ryther, J. H., 1961, Nitrogen fixation in the Sargasso Sea, *Deep-Sea Res.* **7:**298–300.

Eady, R. R., Robson, R. L., Richardson, T. H., Miller, R. W., and Hawkins, M., 1987, The vanadium nitrogenase of *Azotobacter chroococcum:* Purification and properties of the VFe protein, *Biochem. J.* **244:**197–199.

Edmondson, W. T., 1970, Phosphorus, nitrogen, and algae in Lake Washington after diversion of sewage, *Science* **169:**690–691.

Eppley, R. W., Renger, E. H., Venrick, E. L., and Mullin, M. M., 1973, A study of plankton dynamics and nutrient cycling in the Central Gyre of the North Pacific Ocean, *Limnol. Oceanogr.* **18:**534–551.

Fallon, R. D., and Brock, T. D., 1980, Planktonic blue-green algae: Production, sedimentation, and decomposition in Lake Mendota, Wisconsin, *Limnol. Oceanogr.* **25:**72–88.

Finke, L. R., and Seeley, H. W., Jr., 1978, NItrogen fixation (acetylene reduction) by epiphytes of freshwater macrophytes, *Appl. Environ. Microbiol.* **36:**129–138.

Flett, R. J., Schindler, D. W., Hamilton, R. D., and Campbell, E. R., 1980, Nitrogen fixation in Canadian Precambrian Shield Lakes, *Can. J. Fish. Aquat. Sci.* **37:**494–505.

Fogg, G. E., 1942, Studies on nitrogen fixation by blue-green algae. I. Nitrogen fixation by *Anabaena cylindrica* Lemn., *J. Exp. Biol.* **19:**78–87.

Fogg, G. E., 1944, Growth and heterocyst production in *Anabaena cylindrica* Lemn., *New Phytol.* **43:**164–175.

Fogg, G. E., 1969, The physiology of an algal nuisance, *Proc. R. Soc. Lond. B* **173:**175–189.

Fogg, G. E., 1971, Extracellular products of algae in freshwater, *Arch. Hydrobiol. Beih. Ergebn. Limnol.* **5:**1–25.

Fogg, G. E., 1974, Nitrogen fixation, in: *Algal Physiology and Biochemistry* (W. D. P. Stewart, ed.), pp. 650–582, Blackwell Scientific Publications, Oxford.

Fogg, G. E., and Burton, N. F., 1977, Utilization of glycolate by bacteria epiphytic on seaweeds, *J. Phycol. Suppl.* **13:**22.

Fogg, G. E., Stewart, W. D. P., Fay, P., and Walsby, A. E., 1973, *The Blue-Green Algae*, Academic Press, London.

Gallon, J. R., and Chaplin, A. E., 1988, Recent studies on N₂ fixation by nonheterocystous cyanobacteria, in: *Nitrogen Fixation: Hundred Years After* (H. Bothe, F. J. de Bruyn, and W. E. Newton, eds.), pp. 183–188, Gustav Fischer, Stuttgart.

Ganf, G. G., and Horne, A. J., 1975, Diural stratification, photosynthesis and nitrogen fixation in a shallow, equatorial lake (Lake George, Uganda), *Freshwater Biol.* **5:**13–39.

Geitler, L., 1932, Cyanophycaea, in: *Rabenherst's Kryptogamenflora von Deutschland, Osterreich und der Schweiz* (R. Kolkwitz, ed.), Vol. 14, Akademische Verlagsgesell-Schaft, Leipzig.

Gibson, C. E., Wood, R. B., Dickson, E. L., and Jenson, D. H., 1971, The succession of phytoplankton in Lough Neagh, 1968–1970, *Mitt. Int. Ver. Theor. Angew. Limnol.* **19:**140–160.

Goldman, C. R., 1964, Primary productivity and micronutrient limiting factors in some North American and New Zealand lakes, *Int. Ver. Theor. Angew. Limnol. Verh.* **15:**365–374.

Goldman, C. R., and Horne, A. J., 1983, *Limnology*, McGraw Hill, New York.

Gotto, J. W., and Taylor, B. F., 1976, N_2 fixation associated with decaying leaves of the red mangrove (*Rhizophora mangle*), *Appl. Environ. Microbiol.* **31**:781–783.

Granhall, V., and Lundgren, A., 1971, Nitrogen fixation in Lake Erken, *Limnol. Oceanogr.* **16**:711–719.

Guerinot, M. L., and Colwell, R. R., 1985, Enumeration, isolation, and characterization of N_2-fixing bacteria from seawater, *Appl. Environ. Microbiol.* **50**:350–355.

Guerinot, M. L., and Patriquin, D. G., 1981, N_2 fixing vibrios isolated from the gastrointestinal tract of sea urchins, *Can. J. Microbiol.* **27**:311–347.

Haystead, A., Robinson, R., and Stewart, W. D. P., 1970, Nitrogenase activity in extracts of heterocystous and non-heterocystous blue-green algae, *Arch. Mikrobiol.* **74**:235–243.

Hazelkorn, R., 1986, Organization of the genes for nitrogen fixation in photosynthetic bacteria and cyanobacteria, *Annu. Rev. Microbiol.* **40**:525–547.

Head, W. D., and Carpenter, E. J., 1975, Nitrogen fixation associated with the marine macroalgae *Codium fragile*, *Limnol. Oceanogr.* **20**:815–823.

Healy, F. P., 1982, Phosphate, in: *The Biology of Cyanobacteria* (N. G. Carr and B. A. Whitton, eds.), pp. 105–124, Blackwell Scientific Publications, Oxford.

Hellebust, J. A., 1965, Excretion of some organic compounds by marine phytoplankton, *Limnol. Oceanogr.* **10**:192–206.

Hellriegel, H., and Wilfarth, H., 1888, Beilageheft Z. Ver. Rubenzuckerind, D. Reiches.

Holland, H. D., 1978, *The Chemistry of the Atsmophere and Oceans*, John Wiley & Sons, Inc., New York.

Horne, A. J., and Fogg, G. E., 1970, Nitrogen fixation in some English Lakes, *Proc. R. Soc. Lond. B* **175**:351–366.

Horne, A. J., and Goldman, C. R., 1972, Nitrogen fixation in Clear Lake, California; 1. Seasonal variation and the role of heterocysts, *Limnol. Oceanogr.* **17**:678–692.

Horne, A. J., and Viner, A. B., 1971, Nitrogen fixation and its significance in tropical Lake George, Uganda, *Nature* (London) **232**:417–418.

Howarth, R. W., and Cole, J. J., 1985, Molybdenum availability, nitrogen limitation, and phytoplankton growth in natural waters, *Science* **229**:653–655.

Howarth, R. W., Marino, R., Lane, J., and Cole, J., 1988, Nitrogen fixation in freshwater, estuarine and marine ecosystems. I. Rates and importance, *Limnol. Oceanogr.* **33**:688–701.

Huber, A. L., 1986, Nitrogen fixation by *Nodularia spumigena* Mertens (Cyanobacteriaceae). I. Field studies on the contribution of blooms to the nitrogen budget of the Peel-Harvey Estuary, Western Australia, *Hydrobiologia* **131**:193–203.

Hutner, S. H., 1972, Inorganic nutrition, *Annu. Rev. Microbiol.* **26**:313–346.

Jorgensen, B. B., 1980, Mineralization and the bacterial cycling of carbon, nitrogen and sulfur in marine sediments, in *Contemporary Microbial Ecology* (D. C. Ellwood, J. N. Hedger, M. J. Latham, J. M. Lynch, and J. H. Slater, eds.), pp. 239–252, Academic Press, London.

Kamen, M. D., and Gest, H., 1949, Evidence for a nitrogenase system in the photosynthetic bacterium *Rhodospirillum rubrum*, *Science* **109**:560.

Kawai, A., and Sugahara, I., 1971, Microbiological studies on nitrogen fixation in aquatic environments. II. On the nitrogen fixing bacteria in offshore regions, *Bull. Jpn. Soc. Sci. Fish.* **37**:981–985.

Knoll, A. H., 1977, Paleomicrobiology, in: *Handbook of Microbiology*, 2nd ed., pp. 8–29, CRC Press, Boca Raton, Fla.

Lawes, J. B., Gilbert, J. H., and Pugh, E., 1861, *Phil. Trans. R. Soc. Lond.* **151**:431–577.

Lean, D. R. S., Liao, C. F., Murphy, T. P., and Painter, D. S., 1978, The importance of nitrogen fixation in lakes, *Ecol. Bull.* **26**:41–51.

Leon, C., Kumazawa, S., and Mitsui, A., 1986, Cyclic appearance of aerobic nitrogenase activity during synchronous growth of unicellular cyanobacteria, *Curr. Microbiol.* **13**:149–153.

Liebig, J. V., 1842, Die organische Chemie in ihrer Anwendung auf Agrikulture und Physiologie, Vieweg Braunschweig.

Likens, G. E. (ed.), 1972, Nutrients and eutrophication, *Am. Soc. Limnol. Oceanogr. Spec. Symp. 1.*

Mague, T. H., and Holm-Hansen, O., 1975, Nitrogen fixation on a coral reef, *Phycologia* **14:**87–92.

Mague, T. H., Mague, F. C., and Holm-Hansen, O., 1977, Physiology and chemical composition of nitrogen fixing phytoplankton in the central North Pacific Ocean, *Mar. Biol.* **41:**213–227.

Mann, K. H., 1982, *Ecology of Coastal Waters,* University of California Press, Berkeley.

Margulis, L., 1981, *Symbiosis in Cell Evolution,* W. H. Freeman Co., San Francisco.

Margulis, L., 1982, *Early Life,* Science Books International, Boston.

Martin, J. H., and Fitzwater, S. E., 1988, Iron deficiency limits phytoplankton growth in the north-east Pacific subarctic, *Nature* (London) **331:**341–343.

Martinez, L. A., Silver, M. W., King, J. M., and Alldredge, A. L., 1983, Nitrogen fixation by floating diatom mats: A source of new nitrogen to oligotrophic ocean waters, *Science* **221:**152–154.

Maruyama, Y., Toga, N., and Matsuda, O., 1970, Distribution of nitrogen fixing bacteria in the central Pacific Ocean, *J. Oceanogr. Soc. Jpn.* **26:**360–366.

McClung, C. R., and Patriquin, D. G., 1980, Isolation of a nitrogen fixing Campylobacter species from the roots of *Spartina alterniflora* Loisel, *Can. J. Microbiol.* **26:**881–886.

Meeks, J. C., Enderlin, C. S., Joseph, C. M., Steinberg, N., and Weeden, Y. M., 1985, Use of ^{13}N to study N₂ fixation and assimilation by cyanobacterial-lower plant associations, in: *Nitrogen Fixation Research Progress* (H. J. Evans, P. J. Bottomley and W. E. Newton, eds.), pp. 301–307, Martinus Nyhoff, Dordrecht.

Mitsui, A., Kumazana, S., Takahashi, A., Ikemoto, H., Cao, S., and Arai, T., 1987, Strategy by which nitrogen fixing unicellular cyanobacteria grow photoautotrophically, *Nature* (London) **323:**720–722.

Murphy, T. O., Lean, D. R. S., and Nalewajko, C., 1976, Blue-green algae: Their excretion of iron-selective chelators enables them to dominate other algae, *Science* **192:**900–902.

Nalewajko, C., 1978, Release of organic substances, in: *Handbook of Phycological Methods* (J. A. Helebust and J. S. Cragie, eds.), pp. 389–398, Cambridge University Press, Cambridge.

Neilands, J. B., 1967, Hydroxamic acids in nature, *Science* **156:**1443–1447.

Niemi, A., 1979, Blue-green algal blooms and N : P ratio in the Baltic Sea, *Acta Bot. Fenn.* **110:**57–61.

Paerl, H. W., 1978, Light-mediated recovery of N₂ fixation in the blue-green algae *Anabaena* spp. in O₂-supersaturated waters, *Oecologia* (Berlin) **32:**135–139.

Paerl, H. W., 1982, Interactions with bacteria, in: *The Biology of Cyanobacteria* (N. G. Carr and B. A. Whitton, eds.), pp. 441–462, Blackwell Scientific Publications, Oxford.

Paerl, H. W., 1984, Alteration of microbial metabolic activities in association with detritus, *Bull. Mar. Sci.* **35:**393–408.

Paerl, H. W., 1985, Mcrozone formation: Its role in the enhancement of aquatic N₂ fixation, *Limnol. Oceanogr.* **30:**1246–1252.

Paerl, H. W., 1988, Nuisance phytoplankton blooms in coastal, estuarine and inland waters, *Limnol. Oceanogr.* **33:**823–847.

Paerl, H. W., and Bebout, B. M., 1988, Direct measurements of O₂-depleted microzones in marine *Oscillatoria (Trichodesmium):* Relation to N₂-fixation, *Science* **241:**442–445.

Paerl, H. W., and Bland, P. T., 1982, Localized tetrazolium reduction in relation to N₂ fixation, CO₂ fixation, and H₂ uptake in aquatic filamentous cyanobacteria, *Appl. Environ. Microbiol.* **43:**218–226.

Paerl, H. W., and Carlton, R. G., 1988, Control of N₂ fixation by oxygen depletion in surface-associated microzones, *Nature* (London) **332:**260–262.

Paerl, H. W., and Gallucci, K. K., 1985, Role of chemotaxis in establishing a specific nitrogen-fixing cyanobacterial-bacterial association. *Science* **227:**647–649.

Paerl, H. W, and Kellar, P. E., 1978, Significance of bacterial (Cyanophyceae) *Anabaena* associations with respect to N₂ fixation in freshwater. *J. Phycol.* **14:**254–260.

Paerl, H. W., and Kellar, P. E., 1979, Study of the Importance of Nitrogen Fixation to the Growth of *Myriophyllum spicatum*, Report to National Water Research Institute, Burlington, Ontario, Canada.

Paerl, H. W., and Prufert, L. E., 1987, Oxygen-poor microzones as potential sites of microbial N_2 fixation in nitrogen-depleted aerobic marine waters, *Appl. Environ. Microbiol.* **53**:1078–1087.

Paerl, H. W., Webb, K. L., Baker, J., and Wiebe, W. J., 1981, Nitrogen fixation in waters, in *Nitrogen Fixation*, Vol. 1, *Ecology* (W. J. Broughton, ed.), Clarendon, New York.

Paerl, H. W., Crocker, K. M., and Prufert, L. E., 1987, Limitation of N_2 fixation in coastal marine waters: Relative importance of molybdenum, iron, phosphorus and organic matter availability, *Limnol. Oceanogr.* **32**:525–536.

Paerl, H. W., Bebout, B. M., and Prufert, L. E., 1989, Naturally occurring patterns of oxygenic photosynthesis and N_2 fixation in a marine microbial mat: Physiological and ecological ramifications, in: *Microbial Mats: Physiological Ecology of Benthic Microbial Communities* (Y. Cohen and E. Rosenberg, eds.), pp. 326–341, American Society for Microbiology, Washington, D.C.

Parsons, T. R., Takahashi, M., and Hargrave, B. T., 1977, *Biological Oceanographic Processes*, 2nd ed., Pergamon Press, Oxford.

Pearsall, W. H., 1932, Phytoplankton in the English Lakes. 2. The composition of the phytoplankton in relation to dissolved substances, *J. Ecol.* **20**:241–262.

Pearson, H. W., Malin, G., and Howsley, R., 1981, Physiological studies on *in vitro* nitrogenase activity by axenic cultures of *Microcoleus chthonoplastes*, *Br. Phycol. J.* **16**:139–143.

Penhale, P., and Capone, D. G., 1981, Primary productivity and nitrogen fixation in two macroalgae-cyanobacterial associations, *Bull. Mar. Sci.* **31**:164–169.

Postgate, J. R., 1978, Nitrogen fixation, *Studies in Biology No. 92*, E. Arnold, London.

Reuter, J. E., Loeb, S. L., and Goldman, C. R., 1986, Inorganic nitrogen uptake by epilithic periphyton in an N-deficient lake, *Limnol. Oceanogr.* **31**:149–160.

Reynolds, C. S., and Walsby, A. E., 1975, Water blooms, *Biol. Rev.* **50**:437–481.

Rice, D., Mazur, B. J., and Hazelkorn, R., 1982, Isolation and physical mapping of nitrogen fixing genes from the cyanobacterium *Anabaena* 7120, *J. Biol. Chem.* **257**:13157–13163.

Rippka, R., Deruelles, J., Waterbury, J. B., Herdman, M., and Stanier, R. Y., 1979, Generic assignments, strain histories and properties of pure cultures of cyanobacteria, *J. Gen. Microbiol.* **111**:1–61.

Rosenberg, G., and Paerl, H. W., 1980, Nitrogen fixation by blue-green algae associated with the siphonous green seaweed *Codium decorticatum*: Effects on ammonium uptake, *Mar. Biol.* **61**:151–158.

Rozen, A., Schonfeld, M., and Tel-or, E., 1988, Fructose-enhanced development and growth of the N_2-fixing cyanobiont *Anabaena azollae*, *Z. Naturforsch.* **43**:43–46.

Rueter, J. G., 1982, Theoretical Fe limitations of microbial N_2 fixation in the oceans, *Eos* **63**:495.

Ryther, J. H., and Dunstan, W. M., 1971, Nitrogen, phosphorus and eutrophication in the coastal marine environment, *Science* **171**:1008–1012.

Saussure, T. de, 1804, *Recherches Chimiques sur la Vegetation*, Paris.

Schindler, D. W., Fee, E. J., and Ruszczynski, T., 1978, Phosphorus input and its consequences for phytoplankton standing crop and production in the Experimental Lakes Area and in similar lakes, *J. Fish. Res. Bd. Can.* **35**:190–196.

Schopf, J. W., 1968, Microflora of the Bitter Springs Formation, Late Precambrian, central Australia, *J. Paleontol.* **42**:651–688.

Schopf, J. W., 1975, Precambrian paleobiology: Problems and perspectives, *Ann. Rev. Earth Planet. Sci.* **3**:213–249.

Schopf, J. W., and Walter, M. R., 1982, Origin and early evolution of cyanobacteria: The geological evidence, in: *The Biology of Cyanobacteria* (N. G. Carr and B. A. Whitton, eds.), pp. 543–564, Blackwell Scientific Publications, Oxford.

Schultz-Lupitz, A., 1887, *Landw. Jahrb.* **10**:777–848.

Seki, H., Takahashi, M., Hasa, Y., and Ichimura, S., 1980, Dynamics of dissolved oxygen during algal bloom in Lake Kasumigaura, Japan, *Water Res.* **14:**179–183.

Smith, S. V., 1984, Phosphorus versus nitrogen limitation in the marine environment, *Limnol. Oceanogr.* **29:**1149–1160.

Smith, V. H., 1983, Low nitrogen to phosphorus ratios favor dominance by blue-green algae in lake phytoplankton, *Science* **221:**669–671.

Stal, L. J., and Krumbein, W. E., 1985, Nitrogenase activity in the non-heterocystous cyanobacterium *Oscillatoria* sp. grown under alternating light-dark cycles, *Arch. Microbiol.* **143:**67–71.

Stanier, R. Y., 1977, The position of cyanobacteria in the world of phototrophs, *Carlsberg Res. Commun.* **42:**77–98.

Stewart, W. D. P., 1973, Nitrogen fixation by photosynthetic microorganisms, *Annu. Rev. Microbiol.* **27:**283–316.

Stewart, W. D. P. (ed.), 1975, *Nitrogen Fixation by Free-Living Microorganisms,* Cambridge University Press, London.

Stewart, W. D. P., and Alexander, G., 1971, Phosphorus availability and nitrogenase activity in aquatic blue-green algae, *Freshwater Biol.* **1:**389–401.

Stewart, W. D. P., Fitzgerald, G. P., and Burris, R. H., 1967, In situ studies on N_2 fixation using the acetylene reduction technique, *Proc. Natl. Acad. Sci. USA* **58:**2071–2078.

Stewart, W. D. P., Rowell, P., and Rai, A. N., 1980, Symbiotic nitrogen-fixing cyanobacteria, in: *Proceedings of an International Symposium on Nitrogen Fixation* (W. D. P. Stewart and J. R. Gallon, eds.), pp. 239–277, Oxford University Press, Oxford.

Stiefel, E. I., 1977, The mechanism of nitrogen fixation, in: *Recent Developments in Nitrogen Fixation* (W. Newton, J. R. Postgate, and C. Rodriguez-Barrueco, eds.), pp. 69–108, Academic Press, London.

Stockner, J. G., 1988, Phototrophic picoplankton: An overview from marine and freshwater ecosystems. *Limnol. Oceanogr.* **33**(4, part 2):765–775.

Stockner, J. G., and Antia, N. J., 1986, Algal picoplankton from marine and freshwater ecosystems: A multidisciplinary approach, *Can. J. Fish. Aquat. Sci.* **43:**2472–2503.

Stumm, W., and Morgan, J. J., 1981, *Aquatic Chemistry,* 2nd ed., John Wiley & Sons, Inc., New York.

Takahashi, M., and Hori, T., 1984, Abundance of picoplankton in the subsurface chlorophyll maximum layer in subtropical and tropical waters, *Mar. Biol.* **79:**177–186.

Urdaci, M. C., Stal, L. J., and Marchand, M., 1988, Occurrence of nitrogen fixation among *Vibrio* spp., *Arch. Microbiol.* **150:**224–229.

Vanderhoef, L. N., Huang, C., and Musil, R., 1974, Nitrogen fixation (acetylene reduction) by phytoplankton in Green Bay, Lake Michigan, in relation to nutrient concentrations, *Limnol. Oceanogr.* **19:**119–125.

Van Liere, L., and Walsby, A. E., 1982, Interactions of cyanobacteria with light, in: *The Biology of Cyanobacteria* (N. G. Carr and B. A. Whitton, eds.), pp. 9–46, Blackwell Scientific Publications, Oxford.

Vollenweider, R. A., 1976, Advances in defining crucial loading levels for phosphorus in lake eutrophication, *Mem. Ist. Ital. Idrobiol.* **33:**53–83.

Walker, J. C. G., Klein, C., Schidlowski, M., Schopf, J. W., Stevenson, D. J., and Walter, M. R., 1982, Environmental evolution of the Archean-Early Proterozoic Earth, in: *Origin and Evolution of Earth's Earliest Biosphere* (J. W. Schopf, ed.), Princeton University Press, Princeton, N.J.

Walsby, A. E., 1972, Structure and function of gas vacuoles, *Bacteriol. Rev.* **36:**1–32.

Waterbury, J. B., Watson, S. W. Guillard, R. R. L., and Brand, L. E., 1979, Widespread occurrence of a unicellular, marine, planktonic cyanobacterium, *Nature* (London) **277:**293–294.

Wetzel, R. G., 1983, *Limnology,* 2nd ed., W. B. Saunders, Philadelphia.

Whitton, B. A., and Potts, M., 1982, Marine littoral, in: *The Biology of Cyanobacteria* (N. G. Carr and B. A. Whitton, eds.), pp. 515–542, Blackwell Scientific Publications, Oxford.

Wiebe, W. J., Johannes, R. E., and Webb, K. L., 1975, Nitrogen fixation in a coral reef community, *Science* **188**:257–259.

Wilson, P. W., and Burris, R. H., 1947, The mechanism of biological nitrogen fixation, *Bacteriol. Rev.* **11**:41–73.

Wolk, C. P., 1968, Movement of carbon from vegetative cells to heterocysts in *Anabaena cylindrica, J. Bacteriol.* **96**:2138–2143.

Wolk, C. P., 1982, Heterocysts, in: *The Biology of Cyanobacteria* (N. G. Carr and B. A. Whitton, eds.), pp. 359–386, Blackwell Scientific Publications, Oxford.

Wolk, C. P., and Simon, R. D., 1969, Pigment and lipids of heterocysts, *Planta* (Berlin) **86**:92–97.

Wurtzbaugh, W. A., 1988, Iron, molybdenum and phosphorus limitation of N_2 fixation maintains nitrogen deficiency of plankton in the Great Salt Lake drainage (Utah, U.S.A.), *Verh. Int. Verein. Limnol.* **23**:121–130.

Wurtzbaugh, W. A., and Horne, A. J., 1983, Iron in eutrophic Clear Lake, California: Its importance for algal nitrogen fixation and growth, *Can. J. Fish. Aquat. Sci.* **40**:1419–1429.

Wynn-Williams, D. D., and Rhodes, M. E., 1974, Nitrogen fixation in seawater, *J. Appl. Bacteriol.* **37**:203–216.

Yates, M. G., 1977, Physiological aspects of nitrogen fixation, in: *Recent Developments in Nitrogen Fixation* (W. Newton, J. R. Postgate, and C. Rodrigues-Barrueco, eds.), pp. 219–270, Academic Press, London.

9

Organic Sulfur Compounds in the Environment
Biogeochemistry, Microbiology, and Ecological Aspects

DON P. KELLY and NEIL A. SMITH

1. Introduction

1.1. Background

More than a decade has elapsed since the review in *Advances* by Bremner and Steele (1978) of the role of microorganisms in the atmospheric sulfur cycle. In the intervening decade or so, the dawning realization in the 1970s that volatile organic sulfur compounds are major components of the global sulfur cycle has developed from informed speculation to the status of established fact, supported by ever-accumulating data.

It is our purpose in this review to present current views and information on the biogenic sources, transformations and sinks, and the microbiological degradation of organic sulfur compounds and to illustrate the ramifications and importance of these processes to the biosphere as a whole. In reviewing the sources of natural sulfur compounds that may become substrates for microbial growth, we are naturally including information on sources other than microbial ones; such information tends to be widely scattered in the biological and chemical literature. We hope that bringing at least some of this together in one place will enable a broader view to be taken of the metabolic challenges posed to microorganisms by sulfur in the natural environment, as well as broadening the horizons of microbial ecology with respect not only to the diversity of the

DON P. KELLY and NEIL A. SMITH • Department of Biological Sciences, University of Warwick, Coventry CV4 7AL, England.

potential substrates available but also to potential novel habitats in which sulfur compounds occur.

1.2. The Atmospheric Component of the Sulfur Cycle

Any model for the global biogeochemical cycle of sulfur must contain volatile or gaseous sulfur compounds to enable the steady-state flow of sulfur between the terrestrial and marine environments. This cardinal point has been recognized for many years (Conway, 1942; Junge, 1960; Eriksson, 1963; Kellogg *et al.*, 1972; Kelly, 1980, 1988), and the oceans have long been regarded as a major source of such compounds, the favored candidate for the atmospheric link in the sulfur cycle being hydrogen sulfide (Rodhe and Isaksen, 1980). Hydrogen sulfide arises primarily by dissimilatory sulfate reduction probably in any organic-rich anaerobic environment on earth, partly from heterotrophic organic sulfur compound catabolism and assimilatory sulfur metabolism, and partly from the chemical reduction of seawater sulfate by ferrous iron in the basalts over which the water passes in hydrothermal systems of the submarine oceanic ridges (von Damm, 1983). Until the last decade or so, hydrogen sulfide has continued to be regarded as the major biogenic sulfur gas, with an estimated annual global production into the atmosphere from all sources of about 142 million tonnes (142 Tg) S (Robinson and Robbins, 1970). This figure is currently probably a maximum estimate of the contribution by H_2S to atmospheric sulfur, and its importance as an atmospheric sulfur gas has been questioned (Rasmussen, 1974). There is no doubt that the cycling of sulfate-sulfur through sulfide is immensely important in anaerobic environments and at oxic–anoxic aquatic and soil interfaces, but the amount of sulfide turned over at such interfaces and by photosynthetic bacteria is probably vastly in excess of that escaping to the atmosphere, especially if diffusion through a column of oxygenated water must take place. Definitive measurements of ocean surface concentrations and the flux of H_2S to and from marine, freshwater, and terrestrial sources have in fact yet to be undertaken. Hydrogen sulfide is thus an important component of the closed sulfur cycle of terrestrial and aquatic sulfureta, although its contribution to the atmosphere is now thought to lie in the range 5–58 Tg S/year of a total biogenic input of 65/125 Tg/year (Table I; Andreae, 1985). Hydrogen sulfide also falls outside this discussion of organic sulfur compounds in global ecology and will not be mentioned further. The microbiology of inorganic sulfur has been reviewed elsewhere (Kelly, 1982, 1988; Postgate and Kelly, 1982).

As is discussed in the appropriate sections, it is now believed that the major, and probably predominant, sulfur gases, filling the role required for a perfect biogeochemical cycle, are the methylated sulfides, carbonyl sulfide and carbon disulfide. A generalized global estimate of the biogenic contribution of the known sulfur gases (i.e., excluding sulfur dioxide) to the flux of sulfur to the atmosphere would rank these as 75% dimethyl sulfide, 15% hydrogen sulfide, and 10% carbon disulfide + carbonyl sulfide, with minor contributions from other gases such as methanethiol, dimethyl disulfide, dimethyl sulfoxide, and higher molecular weight alkyl sulfides. These estimates are derived from a survey of many studies, and the degree of uncertainty in estimating either gross amounts, or relative contributions, is still very great.

Table I. Ranges of Estimated Rates of Emission of Volatile Sulfur Compounds to the Atmosphere from Natural Sources[a]

Source	Sulfur compound released (Tg S/yr)						
	SO_2	H_2S	DMS	DMDS (and others)	CS_2	COS	Total
Oceanic		0–15	38–40	0–1	0.3	0.4	38.7–56.7
Salt marsh		0.8–0.9	0.58	0.13	0.07	0.12	1.7–1.8
Inland swamps		11.7	0.84	0.2	2.8	1.85	17.4
Soil and plants		3–41	0.2–4.0[b]	1[b]	0.6–1.5	0.2–1.0	5.0–48.5
Burning of biomass	7	0–1		0–1		0.11	7.1–9.1
Volcanoes and fumaroles	8	1		0–0.02	0.01	0.01	9.0
Total	15	16.5–70.6	39.6–45.4	1.3–3.4	3.8–4.7	2.7–3.5	78.9–142.6

[a] Based on Andreae (1985), Steudler and Peterson (1984), Suylen (1988) and studies cited in the text.
[b] As more data become available, these figures could be revised upward to a significant degree (see Section 2.1).

2. Organic Sulfur Compounds in the Natural Environment

Sulfur occurs in an amazing diversity of forms in the environment. Many of its compounds, such as cysteine, cystine, glutathione, methionine, taurine, and coenzyme A, are very familiar to all biologists. Three heterocyclic sulfur-containing rings are of central importance in all living organisms: the thiazole ring of thiamine, the substituted tetrahydrothiophene ring of biotin, and the oxidized disulfide ring of lipoic acid (Fig. 1). Even these compounds serve to illustrate the variety of chemical combinations in which sulfur, with its several possible oxidation states, can occur. In addition, there are probably thousands of other sulfur compounds of biological origin whose existence is recognized by very few microbiologists. Some of these are of great global consequence as mobile components of the biogeochemical cycle of sulfur. Others undergo synthesis and degradation by processes that are as yet only vaguely understood. That there is a paucity of definitive biochemical data on the mechanisms of synthesis and degradation of all but a few of the less glamorous organic sulfur compounds may be a consequence in part of the antisocial odors associated with many of them but is more attributable to the facts that many have been accepted as being of major ecological importance only in recent

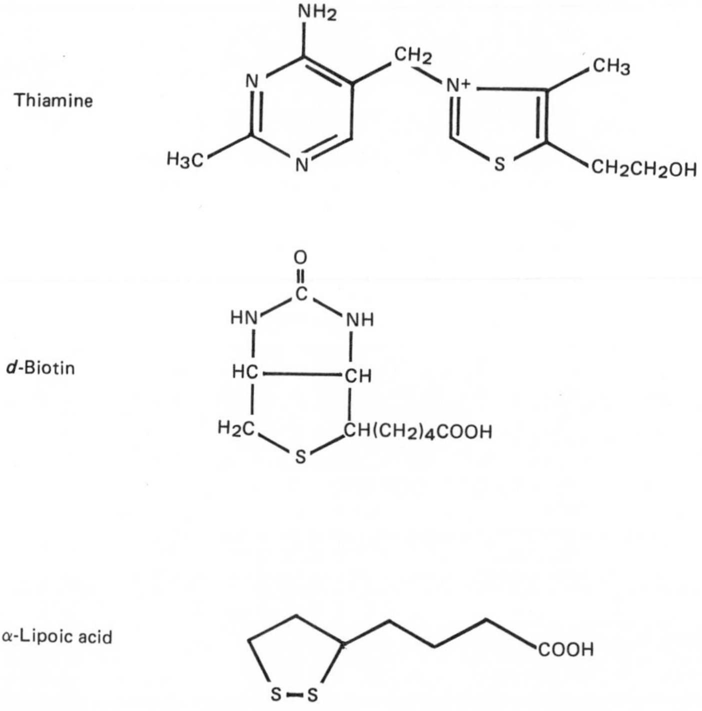

Figure 1. Structures of some metabolically important complex sulfur compounds.

years and little work has yet been done. When one considers that the final stages of the biosynthetic assembly of such a central sulfur metabolite as thiamine, namely, the insertion of the sulfur atom and mechanism of formation of the thiazole ring, are still not completely understood (Tazuya *et al.*, 1987), it is not surprising that the microbial ecology and biochemistry of degradation of such compounds are far from well established.

The variety of naturally occurring sulfur compounds is described in part in reviews by Challenger (1959), Young and Maw (1974), Kadota and Ishida (1972), Kjaer (1977), Bremner and Steele (1978), and Krouse and McCready (1979).

2.1. Methylated Sulfides, Carbon Sulfide, and Carbonyl Sulfide

> There is, therefore, another matter of the volatile kind which waters called sulphureous commonly contain: viz. a subtile vapour, which it is impossible to obtain in a solid form and imparts to these waters, when present in large quantities, the smell of hepar sulphurs, putrid eggs or the scouring of a gun. (Joshua Walker, 1784, in *Walker on Waters*).

Of such compounds, the longest known as products in the natural environment are dimethyl sulfide (DMS), dimethyl disulfide (DMDS), methanethiol (MT), carbon disulfide (CS_2), and carbonyl sulfide (COS). These are all of low boiling point, all but CS_2 being gases at normal temperatures, highly toxic (Ljunggren and Norberg, 1943; Moubasher *et al.*, 1974; Ashworth *et al.*, 1977; Rodgers *et al.*, 1980; Sweetnam *et al.*, 1987), and malodorous (Selyuzhitskii, 1972; Sivelä, 1980). In low concentrations, DMS is an important odor and flavor component of many foodstuffs, including tea, cocoa, beers, milk, and many cooked vegetables (Anness, 1980; Anness *et al.*, 1979), but at higher concentrations it contributes to spoilage (Keenan and Lindsay, 1968; Bills and Keenan, 1968). For both flavor production and spoilage, DMS formation is, of course, in many cases the result of microbial activity. The odor threshold concentration (parts per billion) for DMS is in the range of 0.6–40, compared with 0.1–3.6 for DMDS, 0.9–8.5 for MT, and 8.5–1000 for H_2S (Selyuzhitskii, 1972; Windholz, 1983).

The five compounds named above make up the greater part of the volatile organic sulfur compound fraction of the biosphere. Tentative global estimates for the release rates and mean environmental concentrations of these compounds are given in Tables I and II. The oceans are a major source of volatile sulfur, but some estimates put terrestrial production of volatile organic sulfur compounds at an equivalent level. Total land emissions of DMS, DMDS, MT, CS_2, and COS were estimated at 0.43 g $S/m^2/year$ (Adams *et al.*, 1981). Assuming that half of the earth's nonoceanic surface area of about 148 million km^2 produces these compounds at this rate, an extrapolation of about 32 Tg S/year from the terrestrial environment can be made.

The sulfur content of the atmosphere is small compared with that of the hydrosphere and pedosphere and has been estimated at about 1.8 million tonnes (1.8 Tg S; Krouse and McCready, 1979). Atmospheric sulfur arises from both abiotic (anthropogenic and natural) and biotic sources, the former being considerably easier to measure. Biological sulfur fluxes are particularly difficult to estimate since they are so

Table II. Ranges of Concentration of Some Volatile Organic Sulfur Compounds Reported in the Environment[a]

Compound	Freshwater (μg liter^{-1})	Seawater (ng liter^{-1})	Salt marsh (ng liter^{-1})	Air (pg liter^{-1})
DMS	62–70	3–310 (97,000)[b]	60–3800	6–200
DMDS	—[c]	14–19	94–282	—
MT	—	14–19	48–144	—
CS_2	—	0.5–19	76–228	20–1200
COS	—	14–19	60–180	510

[a]Data are from Andreae and Barnard (1984), Andreae and Raemdonck (1983), Barnard et al. (1982), Bechard and Rayburn (1979), Deprez et al. (1986), Graedel et al. (1981), Kim and Andreae (1987), Lovelock (1974), Lovelock et al. (1972), Maroulis and Bandy (1976), Sandalls and Penkett (1977), Toon et al. (1987), and Wakeham et al. (1984).
[b]Wakeham et al., 1984.
[c]—, Insufficient data available but undoubtedly present in the environment at low concentrations.

variable with location and season. As discussed below, the major sources of methylated sulfides are biotic, while CS_2 and COS are also produced in significant quantity from volcanoes and, to a small extent, from the chemical industry (Steudler and Peterson, 1984; Krouse and McCready, 1979; Babich and Stotzky, 1978). The natural abundance of methylated sulfides, CS_2, and COS is very low, making accurate estimates of atmospheric concentrations on a global basis difficult; in aquatic and soil environments and in local atmospheric environments, considerable diversity can be found. The most reliable data are for DMS (see Section 2.1.1), but, as Tables I and II show, these estimates contain considerable margins for doubt: taking the upper estimates for DMS, COS, and CS_2 in the atmosphere (Table II), COS is the most abundant trace organic sulfur gas, being about 2.5 × the concentration of DMS and CS_2. The combined mass of these gases (maximally 900 pg/liter air) compares with the estimate of 6000 pg H_2S/liter air (Hitchcock, 1976). Locally very high anthropogenic levels of methylated sulfur gases can occur: effluent gases from a sulfate–cellulose mill were, for example, reported to contain MT, DMS, and DMDS at 94, 16.6, and 21.7 ppm, respectively (Sivelä and Sundman, 1975; Sivelä, 1980). Low natural atmospheric concentrations of the volatile organic sulfides indicate either that they have little or no role in global sulfur turnover, the low concentrations reflecting low rates of production, or that their turnover rates are high and what is measured is a relatively small but very active pool. From the data of Tables I and II, the relationship of DMS production and its atmospheric pool size can be calculated: taking the total mass of the atmosphere as 5 × 10^9 Tg at a density of 1.293 g/liter, the relative volume of the atmosphere can be taken as 3.87 × 10^{21} liters, indicating (Table II) a maximum of about 0.77 Tg DMS in the global atmosphere. Given a production rate of about 45 Tg/year (Table I), a rapid turnover of DMS is indicated: given current estimates of a DMS residence time of only hours to days (see Section 2.1.3), the steady-state pool of DMS is a very dynamic one. It thus reflects a prodigious flow of sulfur in this form through the biosphere.

2.1.1. Dimethyl Sulfide in the Atmosphere

Currently, the total biogenic sulfur emissions to the atmosphere are estimated at about 70 Tg S/year (Andreae and Raemdonck, 1983; Andreae, 1985). DMS was first detected in ocean surface waters by Lovelock *et al.* (1972), who proposed that it could be the dominant gaseous component of the sulfur cycle. This confirmed the much earlier observation that DMS was produced by many living organisms, especially marine algae (Challenger, 1959). Since then, measurements of DMS in the marine atmosphere and in seawater, together with estimates of flux rates into the air (Liss and Slater, 1974; Nguyen *et al.*, 1978; Barnard *et al.*, 1982; Andreae and Raemdonck, 1983; Cline and Bates, 1983; Andreae, 1980a, 1985, 1986; Dacey and Wakeham, 1986; Ferek *et al.*, 1986), demonstrate that oceanic export of DMS to the atmosphere accounts globally for about 40 Tg S/year, or around half the total natural input of sulfur to the air. Taking this value, and the presumed predominance of DMS as the volatile sulfur product of the oceans, DMS accounts for about 90% of the volatile sulfur output of the oceans (excluding particulate sulfur in the form of sulfate spray). Adding to it the estimated DMS flux from continental sources (Adams *et al.*, 1981), a total global DMS flux of 52 Tg S/year is indicated against a global biogenic sulfur flux of 103 Tg S/year, which compares with an anthropogenic sulfur dioxide flux of about 104 Tg S/year (Andreae and Raemdonck, 1983). At this stage of our knowledge, it should not be forgotten that the range of estimates in the literature for the global production of sulfur gases from the oceans is 34–190 Tg S/year (Andreae and Raemdonck, 1983), indicating that the quantitative importance of DMS, though not here disputed, may be reduced by further research. The important point is that there is an immense flux of DMS into the atmosphere. Where does it come from, what happens to it, and where does microbial ecology stand in relation to it?

2.1.2. Sources of Dimethyl Sulfide

The relatively high concentrations of DMS in seawater compared with those in the atmosphere (Table II) drive a significant ocean–atmosphere flux of DMS (Dacey *et al.*, 1987). DMS concentration in surface waters is proportional to the marine primary productivity of those waters and rapidly declines at depth, indicating surface rather than sedimentary origins for the DMS. A meromictic, hypersaline Antarctic lake contained DMS up to a remarkable maximum of 97 μg/liter, this being observed just above the oxic–anoxic interface (Franzmann *et al.*, 1987). Consistent with the observed predominance of DMS in most environments, those waters also contained MT, DMDS, COS, and CS_2, but at concentrations of 14–19 ng/liter (Deprez *et al.*, 1986)

Land masses and some areas with dense vegetation also contribute greatly to the sulfur flux, but data on local concentrations are sparse. Emission rates of DMS-sulfur from soils have been estimated at 11–45 pg/g dry weight/hr, corresponding to extrapolated global outputs of 1.5–4.9 Tg/year (Hitchcock, 1975). Emission rates from forests ranged from 2 to 43 pg/g dry weight/hr (Hitchcock, 1975; Lovelock *et al.*, 1972). A recent study of DMS emission from the wetlands of Ontario, Canada (Nriagu *et al.*,

1987), indicated emission rates of 147 mg DMS/m^2 and a total Canadian emission of up to 0.18 Tg DMS/year.

Globally, salt marshes release up to 1.7 Tg S/year, about half in the form of DMS, while mud flats produce more COS and H_2S than DMS, with average fluxes of 0.2 g S/m^2/year (Aneja et al., 1979a). While salt marshes exhibit production rates of volatile organic sulfur compounds that are up to two orders of magnitude greater than those from the oceans or inland soils, their area means that they contribute only 1.6–5.4% of total global biogenic emissions (Steudler and Peterson, 1984; Wakeham et al., 1984; Dacey et al., 1987; Aneja et al., 1979a).

2.1.2.a. The Role of Dimethylsulfonium Propionate. The major source of DMS in the marine environment appears to be the enzymatic cleavage of dimethylsulfonium propionate (DMSP; dimethyl-β-propiothetin) into DMS and acrylate. DMSP occurs, apparently as an osmoregulator, in marine algae (Challenger, 1959; Dickson et al., 1980, 1982; White, 1982; Reed, 1983; Vairavamurthy et al., 1985; Ackman et al., 1966; Tocher and Ackman, 1966; Tocher et al., 1966) and is also found in various plants, including *Zostera marina* and some *Spartina* species (Larher et al., 1977; White, 1982; Dacey et al., 1987), in which it can reach 30–280 μmol/g dry weight. It has been established that the major sulfur compound in most algae is DMSP (Table III). The concentrations in the large algae range widely from 0.06 to 108 μmol/g dry weight (White, 1982). Its cleavage to DMS was first demonstrated by Challenger and Simpson (1948) and subsequently demonstrated by numerous workers (Cantoni and Anderson, 1956; Wagner and Stadtman, 1962; Kadota and Ishida, 1972; Kiene and Visscher, 1987). One of the earliest reports of the liberation of DMS from seaweed was that of Haas (1935), who demonstrated its production from *Polysiphonia fastigiata* and *P. nigrescens*. His observation was subsequently well substantiated and extended (Kadota and Ishida, 1972; Bremner and Steele, 1978; Bechard and Rayburn, 1979). At least part of this liberation is a result of the continuous slow cleavage of DMSP within the algal

Table III. Cyanobacteria and Algae That Have Been Shown to Contain DMSP or to Produce DMS[a]

Cyanobacteria	*Anacystis, Microcoleus, Oscillatoria, Phormidium, Plectonema, Synechococcus*
Bacillariophyta	*Skeletonema, Thalassiosira*
Chlorophyta	*Carteria, Chlorococcum, Cladophora, Codium, Enteromorpha, Micromonas, Monostroma, Oedogonium, Prasinocladus, Spongomorpha, Tetraselmis, Ulothrix, Ulva*
Dinophyta	*Amphidinium, Gymnodinium, Gyrodinium, Prorocentrum*
Haptophyta	*Emiliania, Hymenomonas, Phaeocystis, Syrachosphaera*
Phaeophyta	*Dictyopteris, Egregia, Endatachne, Halidrys, Laminaria, Macrocystis, Pelvetia*
Prasinophyta	*Platymonas*
Rhodophyta	*Ceramium, Corallina, Gelidium, Gigartina, Gracilaria, Piocamium, Polysiphonia, Soliera*

[a] After Suylen (1988) and Kiene and Taylor (1988a,b).

tissues, producing methylated sulfides that accumulate in the algae (Bechard and Rayburn, 1979).

Recently (Keller et al., 1989), a comprehensive review of the occurrence of DMSP and DMS in algae showed that the major producers of DMS are the Dinophyceae (dinoflagellates), especially *Prorocentrum,* and the Prymnesiophyceae (coccolithophores), especially *Phaeocystis* and *Hymenomonas.* Most Chlorophyta (containing chlorophylls a and b), the Cyanobacteria, and the Cryptomonads (with the exception of a *Cryptomonas* sp.) are minor producers of DMS, whereas the Bacillariophyceae and Chrysophyceae (which contain chlorophylls a and c), especially *Ochromonas* and *Chrysamoeba,* contain and release significant amounts of DMS. The paramount importance of some groups has potentially great ecological significance: for example, the abundance of coccolithophores in the ocean drives much of the oceanic DMS output, but lack of knowledge of the factors affecting coccolithophore numbers, or the role of DMSP in their ecophysiology, means that their response to climatic change or pollutants cannot be predicted. This means that the factors regulating a major biogenic source of volatile sulfur in the sulfur biogeochemical cycle are unknown.

The rate of DMS production by phytoplankton has been shown to increase about 24-fold during grazing by zooplankton; this could be attributable to release of DMS from ruptured algal cells, but may also result from DMSP breakdown in the zooplankton gut or from bacterial degradation of zooplankton feces (Dacey and Wakeham, 1986). Stresses such as increased salinity or nitrogen limitation can also result in greater DMS output (Vairavamurthy et al., 1985; Turner and Liss, 1987). A strong correlation has been found between the vertical profiles of DMSP and DMS in the water column of a seasonally stratified salt pond, and there is correspondence of phytoplankton abundance and seawater DMS concentrations (Cline and Bates, 1983; Barnard et al., 1984; Wakeham et al., 1984). There is also strong evidence to suggest much DMS release comes from the bacterial breakdown of senescent algae (Zinder et al., 1977; Bechard and Rayburn, 1979; Bremner and Steele, 1978). There is direct evidence for DMSP conversion to DMS by both bacteria and phytoplankton; for example, a marine mud *Clostridium* could grow on DMSP, releasing the stoichiometric amount of DMS (1 : 1), and a cell-free extract of the dinoflagellate *Gyrodinium* produced equimolar DMS and acrylate from DMSP (Kadota and Ishida, 1972).

An elegant study by Dacey et al. (1987) demonstrated that DMS production by salt marshes could not be attributed exclusively to its production by the surface sediments because this would require turnover rates of up to 300 times per day at pore water concentrations 50 nM. Their data demonstrated that *Spartina* leaves could account for the observed DMS output if only 0.4% of the leaf DMSP turned over per day. Thus, direct input of DMS from plants and algae, without passage through any anaerobic metabolism, is a major source of DMS at least in some environments.

2.1.2.b. Dimethyl Sulfoxide. Dimethyl sulfoxide (DMSO) occurs in natural waters at concentrations of 1–200 nM, whereas DMS is generally present at much lower concentrations in the same waters (Andreae, 1980a). DMSO has some anthropogenic sources, including paper manufacture, but arises in part from atmospheric oxidation of DMS (Cox and Sandalls, 1974; see Section 2.1.3). It is also formed from the anaerobic phototrophic oxidation of DMS, a process that may lead to significant production of DMSO in anaerobic ponds and marshes (Zeyer et al., 1987). A significant source of DMS is believed to be the bacterial reduction of DMSO to DMS, with DMSO being used in low-oxygen or anaerobic environments as a respiratory electron acceptor (Ando et al., 1957; Yen and Marrs, 1977; Zinder and Brock, 1978a,b; McEwan et al., 1983; Bilous and Weiner, 1985; King et al., 1987; Kiene and Capone, 1988). The occurrence of DMS in beer appears to be due to reduction of DMSO by yeasts (Anness, 1980; Anness and Bamforth, 1982).

2.1.2.c. *Methionine and Other Sulfonium Compounds.* Bremner and Steele (1978) thoroughly reviewed the extensive work conducted with soils, both in their natural state and supplemented with sulfur-containing amino acids. DMS production was stimulated by homocysteine, S-methylcysteine, methionine, and methionine sulfone and sulfoxide, although, except with homocysteine, DMDS production exceeded that of DMS. DMS was produced, often as the major sulfur volatile, in many soils supplemented with sulfate or complex sulfur sources such as plant material and sewage sludges. MT (see Section 2.1.4) is known to be a major product of methionine degradation and probably gives rise to DMS secondarily by microbial methylation. Other sulfonium compounds believed to give rise to DMS in some soils (and other environments) include S-methylmethionine, which is present in higher plants and some marine algae (Bills and Keenan, 1968; Hattula and Granroth, 1974; White, 1982), phosphatidyl sulfocholine, which occurs in some diatoms (Anderson et al., 1976), S-adenosylmethionine, and trimethylsulfonium chloride (TMS). A *Pseudomonas* was shown to degrade TMS to DMS and methylenetetrahydrofolate by means of a TMS : tetrahydrofolate methyltransferase (Kadota and Ishida, 1972). DMS production by rumen microorganisms has been reported (Salsbury and Merricks, 1975), and DMS in the breath of normal human beings (Cooper, 1983) is presumably also of microbial origin. In rat liver (Weisiger et al., 1980), DMS appears to be produced by a thiol S-methyltransferase from MT and S-adenosylmethionine. The capacity to produce DMS thus seems to be ubiquitous among living organisms, and all habitats and organisms are likely to be exposed to low levels of DMS at some time.

2.1.3. *Atmospheric Oxidation of Dimethyl Sulfide*

DMS undergoes photochemically driven destruction in the atmosphere; the principal reaction is of DMS with photochemically produced OH radicals, producing sulfur dioxide (20–40%) and methane sulfonic acid (MSA; up to 50%), along with other minor products such as DMSO, dimethyl sulfone, and sulfuric acid (Grosjean and Lewis, 1982; Panter and Penzhorn, 1980; Hatakeyama et al., 1982, 1985; Harvey and Lang, 1986; Ferek et al., 1986). In the marine atmosphere, destruction by NO_x is negligible (Ferek et al., 1986), but the reaction with NO_3 radical [producing sulfur dioxide and other unknown products (Ferek et al., 1986)] could be significant. The destruction of DMS by NO_3 radical is likely to be especially important in urban areas with high levels of NO_x air pollution. It seems likely to be as significant as a nonphotochemical nighttime atmospheric sink as is reaction with OH radical in daytime (Winer et al., 1984). Its effect on DMS in the marine atmosphere is probably important only when there is input of NO_x from continental air masses (Winer *et al.*, 1984). The lifetime of DMS is hours or days in the trophosphere (Andreae and Barnard, 1984), meaning that little DMS enters the stratosphere and hence this compound does not contribute to stratospheric sulfate (Ferek *et al.*, 1986). The lifetimes of the other minor oxidation products are less certain, but further transformations could occur in the atmosphere. MSA is a stable but reactive compound, not being hydrolyzed by boiling water or hot aqueous alkali, and has a vast literature in relation to synthetic, catalytic, and industrial chemistry. Its stability means that it is deposited in precipitation along with sulfur dioxide (Klockow *et al.*, 1978). Measurements of MSA in Antarctic ice cores dating from recent times to about 30,000 years ago showed MSA to occur in all samples (Saigne and Legrand, 1987). The MSA concentrations were two to five times greater during the last ice age, which is consistent with its biological origin from the greatly enhanced marine biogenic activity at that time, leading to substantial marine production of DMS. These findings show that MSA must be regarded as a major biogenic organic sulfur compound whose behavior in environments less inert than Antarctic ice needs study (see Section 3.3).

2.1.4. Methanethiol and Dimethyl Disulfide

Both MT and DMDS are generated in salt marsh sediments, soils, freshwater algae, decomposing algal mats, industrial effluents, activated sludge, and sewage systems (Rasmussen, 1974; Bechard and Rayburn, 1979; Zinder *et al.*, 1977; Bremner and Steele, 1978; Koenig *et al.*, 1980; Tomita *et al.*, 1987; Headley, 1987). A major source of MT is the microbial degradation of methionine, on which there is a substantial literature (e.g., Kadota and Ishida, 1972; Ferchichi *et al.*, 1985, 1986). MT appears to be produced ubiquitously by the majority of heterotrophic bacteria by means of enzymatic sulfide-dependent thiol methyltransferase activity, using, for example, *S*-adenosylmethionine as the methyl donor for sulfide methylation (Drotar *et al.*, 1987a,b). MT is also reported to be produced anaerobically from DMS (and 3-methiolpropionate) in salt marshes (Kiene and Taylor, 1988a,b). DMDS is frequently reported as the major product from, for example, soils supplemented with methionine and its sulfoxide and sulfone or *S*-methylcysteine. This is likely to result at least in part from the oxidation and consequent dimerization of MT to DMDS, and a large number of MT- and DMDS-forming bacteria have been isolated from activated sludge (Tomita *et al.*, 1987). DMDS is produced as an attractant pheromone in the female hamster (Singer *et al.*, 1976).

2.1.5. Organic Sulfur Transformations in Salt Marsh and Marine Environments: Interrelations of Methylated Sulfides with Dimethylsulfoniopropionate, 3-Mercaptopropionate, Hydrogen Sulfide, Dimethyl Sulfoxide, and Methane

Dimethylsulfoniopropionate (DMSP) as a primary marine precursor of DMS has different origins in seawater and salt marsh environments: as an algal metabolite in marine systems but arising mainly from *Spartina alterniflora* in DMS-producing salt marshes (Dacey *et al.*, 1987). In anaerobic muds and sediments, there is also usually active sulfate reduction to H_2S, producing concentrations of H_2S that can be 10–100 mg/liter of pore water, the fate of which is usually regarded as a combination of immobilization in the sediments as metal sulfides (especially iron sulfides, eventually stabilizing as pyrite) and chemical and biological oxidation at the oxic/anoxic interface in the water column or by phototrophic bacteria in illuminated anoxic waters or mud surfaces. Organic sulfur has long been known to be an important component of the reduced sulfur pool of oceanic and other sediments, accounting for up to 10% of the total reduced sulfur, but the nature and origin of this were obscure (Volkov and Rozanov, 1983). While some of this fraction is long-lived complex compounds (such as thiophene derivatives; see Section 2.4.2), recent work has shown that there are interactions between H_2S and organic compounds in sediments, leading to the formation of organic sulfides. Most important of these to be identified to date is the reaction of sulfide with acrylate (generated from DMSP cleavage to DMS) to form 3-mercaptopropionate, which has been shown to be widely distributed in marine coastal sediments (Vairavamurthy and Mopper, 1987). This is potentially an important route in sedimentary diagenesis for the incorporation of sulfur into an organic form, but it is also likely to have a more dynamic importance for the continuous turnover of low-molecular-weight organic sulfur in modern sediments, muds, and anoxic soil environments. Mercaptopropionate has, for example, been shown to be a sulfur source for growth of *Chlorella fusca* (Krauss and Schmidt, 1987).

There has recently been intensive study of these systems, and the microbial and chemicoecological relationships of the organic sulfur compounds in these important environments are becoming clearer. The work by Kiene, Oremland, and colleagues (Kiene, 1988; Kiene and Capone, 1988; Kiene and Taylor, 1988a,b; Kiene *et al.*, 1986) has demonstrated a second source for 3-mercaptopropionate in coastal sediments and has identified complex microbiological interrelationships between the producers and consumers of the methylated sulfides. DMSP cannot only be converted to DMS and acrylate, but can also be demethylated to 3-methiolpropionate, which is further demethylated to give 3-mercaptopropionate. It remains to be established which process for 3-mercaptopropionate formation predominates in sediments. In the same sediments, DMS (derived from DMSP) was shown by substrate feeding experiments to be demethylated to produce MT; this appeared to be the major route for MT formation in these sediments. 3-Mercaptopropionate was metabolized by sediment organisms, but its further processing, products derived from it, and the organisms involved are as yet apparently completely unknown. In salt marsh sediments, DMSO and DMDS were reduced to DMS and MT, respectively (Kiene and Capone, 1988), and there was evidence of interconversion of DMS and MT by methylation and demethylation. In these anaerobic muds, both methanogens and sulfate-reducing bacteria appear to be involved in the transformations of methylated sulfur compounds. Low concentrations appeared to be competed for by these groups (Kiene, 1988; Kiene *et al.*, 1986): in slurry incubation experiments, about 80% of the DMS consumption was attributed to the activity of sulfate-reducing bacteria and 20% to methanogenic metabolism. The importance of DMS to the two physiological types is, however, masked by these figures; DMS in fact contributed about 28% to the total methane production rate from the sediment, while only about 1% of the sulfate reduction was DMS consuming. A methylotrophic methanogen has been isolated and shown to use DMS as a growth substrate (Kiene *et al.*, 1986; Oremland and Zehr, 1986), producing methane and carbon dioxide at ratios close to 3:1. The methanogens must therefore be added to the complex of organisms transforming methylated sulfur anaerobically. The scheme in Fig. 2 summarizes the metabolic interrelationships of the methylated sulfur compounds found in sediments, including the aerobic reactions that also occur at the oxic interfaces or when such sediments are artificially oxygenated. At present, we have little microbiological or biochemical detail with which to explain the activity of these complex microbial ecosystems. It is also likely that at present we are aware only of the major, or most easily traced, processes and that numerous other reactions occur. It is known, for example, that at least 20 thiols occur in sediment pore waters (Dacey *et al.*, 1987) and that thiomalic and other thio-acids may arise from the

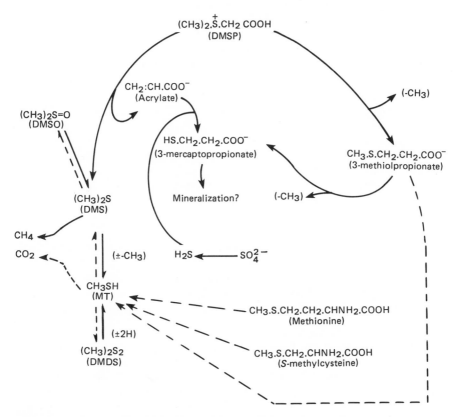

Figure 2. Generalized scheme of the interrelationships of the methylated sulfides and their precursors and products in salt marsh and sediment environments.

reaction of sulfide with a range of unsaturated aliphatic acids present in sediments (Kiene and Taylor, 1988a,b).

2.1.6. Availability and Microbiological Turnover of Methylated Sulfides in the Environment: Organic Sulfur "Minicycles"

The low concentrations of methylated sulfides encountered in most environments could be used to support the argument that they are of little metabolic significance in microbial ecology. This argument is not tenable; local concentrations of DMS in the surface ocean can reach 60 nM and can be 100 nM at sediment–water interfaces or 25 nM in swamp waters, and they reach extremely high levels in some situations, such as 640 nM in Organic Lake, Antarctica, and 200–1000 nM during the rotting of algal blooms or their grazing by zooplankton (Andreae, 1980a,b, 1985; Andreae and Barnard, 1984; Bechard and Rayburn, 1979; Rasmussen, 1974; Howes et al., 1985; Dacey and Wakeham, 1986; Nriagu et al., 1987). In some cases, extraordinarily high concentrations occur, such as the 0.1–0.2 mM DMSP reported by Kiene (1988) in the surface of salt marsh tidal flats. Concentrations of these orders are sufficient to enable DMS (and other sulfides at nanomolar concentrations) to be regarded as substrates. For example, seawater and sediment water concentrations of known metabolites, such as amino acids and keto-acids (1–4000 nM), methane (2–1000 nM), and formate (40 μM), also vary widely but contribute to microbial (and animal) nutrition in the open ocean as well as in more concentrated sources (Scranton and Brewer, 1977; Dando et al., 1986; Winfrey et al., 1981; Kieber and Mopper, 1983; Fuhrman and Ferguson, 1986). The finding of low concentrations of individual metabolites may well be indicative of very rapid consumption of those metabolites by the biota (Fuhrman and Ferguson, 1986), indicating very high affinities for the substrates (or large standing populations of organisms capable of rapid metabolism). The affinity constants for MT, DMS, and DMDS oxidation by some bacteria (see Section 3.1) are in the micromolar range, and steady-state chemostat cultures of thiobacilli growing on DMS or DMDS do not contain detectable substrates, indicating great avidity in their consumption by the bacteria.

As with many closed environments, sediment pore waters, stratified waters, soils, and oxic–anoxic interfaces may well show localized very high concentrations of some unusual substrates, and future in situ work should endeavor to establish whether there are microbial ecosystems that depend exclusively on the cyclical transformations of compounds such as DMS, DMSO, and their precursors and products (see Section 2.1.5).

2.2. Carbon Disulfide and Carbonyl Sulfide

The global concentrations and production rates for CS_2 and COS are shown in Table I and II. Total annual biogenic plus anthropogenic outputs of these compounds to the atmosphere can be estimated at values of 4 to 5 million tons of CS_2 and around 3 million tons of COS, with about one-fifth to one-quarter of the CS_2 being anthropogenic (Table I; Khalil and Rasmussen, 1984; Steudler and Peterson, 1984; Turco et al., 1980).

2.2.1. Natural Sources and Transfer in the Environment

The biogenic origins of these compounds are less well known than those of the methylated sulfides, but are probably no less diverse. Soils supplemented with cysteine, cystine, homocystine, lanthionine, or djenkolic acid, or plant materials, manures, or sludges produced one or both of these compounds (Banwart and Bremner, 1976; Bremner and Steele, 1978).

CS_2 was reported by Lovelock (1974) in diverse marine samples, with strong evidence that it originated in anaerobic marine muds (where about 30 ng/l were detected) and was stable for at least 10 days in oxic seawater, indicating it could transfer sulfur from the sea bed to surface waters and the atmosphere. Production of CS_2 from anaerobic sediments has also recently been proposed by Wakeham et al., (1984). CS_2 has been found in Antarctic lake waters along with COS (Deprez et al., 1986). The greatest production rates of CS_2 and COS so far reported are for salt marshes (Tables I and II; Steudler and Peterson, 1984), and much of this output seems to be attributable to Spartina alterniflora (Aneja et al., 1979b). COS appears to be fairly uniform in concentration (at 10–100 pM) in surface seawaters, where some of it is hydrolyzed to produce hydrogen sulfide (Elliott et al., 1989). The oceans are estimated to be a global source of COS, liberating around 0.6 Tg/year (Rasmussen et al., 1982a). Recently (Guenther et al., 1989), COS has been reported as a dominating sulfur gas among the terrestrial biogenic sulfur emissions: proportions found were COS (38%), DMS (35%, rising to 41% in summer), H_2S (21%), and CS_2 + DMDS (6%).

Carbon disulfide has been shown to be emitted by a number of higher plants, including Brassica, Quercus, Medicago, and Zea mays (see Haines et al., 1989). Recently it has been shown that the major sulfur gas emitted by Stryphnodendron excelsum trees, in a central American rainforest, is carbon disulfide (Haines et al., 1989). The principal source of the CS_2 is the tree roots, and the amount emitted increases if roots are disturbed or if the local concentration of CS_2 is lowered artificially. The amounts potentially released through the soil surface are so great (up to 5 μg S/m^2 per min) that the trees can be detected by their odor! Carbon disulfide has also been reported to be present in the roots of Acacia pulchella (Whitfield et al., 1981). The immediate precursors for its production are not known in the Acacia (F. B. Whitfield, personal communication) or the Stryphnodendron (B. Haines, personal communication). It seems likely that CS_2 emission by roots of these plants must influence the bacterial and fungal species composition of the biological communities in the rhizosphere: the biocidal effects of CS_2 are known to extend to the nitrifying bacteria and some pathogenic fungi, as well as to root-infesting nematodes. The general bactericidal effect of CS_2 probably means that many more soil species are directly regulated by the plant system. Soils subjected to a constant input of CS_2 must also provide an enrichment environment for the selection of CS_2-degrading bacteria, but this has not yet been investigated.

Carbon disulfide and COS are released from volcanic eruptions, giving concentrations of CS_2 in the plume as high as 19 ng/liter (Rasmussen et al., 1982b; Stoiber et al., 1971).

Atmospheric concentrations of CS_2 are lower over remote, nonvolcanic, regions

(20–200 pg/liter) than over heavily populated and industrialized ones (200–1200 pg/liter). A mean value for the lower 2 km of the trophosphere is 100 pg/liter (Bandy et al., 1981), and at 12 km the CS_2 concentration is 18 pg/liter in both the Northern and Southern hemispheres (Tucker et al., 1985). In the atmosphere, CS_2 reacts with OH radicals to produce COS, but the rate of reaction of COS with OH is much slower, resulting in its presence at higher global concentrations in the trophosphere and stratosphere. The atmospheric lifetimes of CS_2 and COS are estimated at 7–12 days and 1–2 years, respectively, with COS being the most abundant sulfur gas in the atmosphere (Stotzky and Schenk, 1976; Khalil and Rasmussen, 1984; Steudler and Peterson, 1984).

COS shows supersaturation in the surface waters of the Pacific Ocean, and its production in marine surface waters is a photochemical process involving COS formation from organic sulfur compounds, including DMSP, methionine, cysteine, and glutathione (Ferek and Andreae, 1983, 1984). Salt marshes also produce COS, and studies of coastal sediments and sediment cores showed that COS production, although occurring in darkness, was greatly stimulated by light (Aneja et al., 1979a,b; Steudler and Peterson, 1984; Jørgensen and Okholm-Hansen, 1986). Biogenic production of COS also occurs: it has been shown that cysteine can be converted, via mercaptopyruvate, to thiocyanate (Cooper, 1983), and thiocyanate (and isothiocyanate) can give rise to COS (Munnecke et al., 1962; Sivelä, 1980).

2.2.2. Anthropogenic Carbon Disulfide

Worldwide annual commercial production of CS_2 ran at between 1 and 1.7 million tonnes between 1977 and 1987 (Timmerman, 1978; Thomel, 1987). Of this, up to 65% was used in the manufacture of viscose silk, 15–25% for making tetrachloromethane, 10–18% for cellophane production, and smaller amounts for fungicides, fumigants, vulcanization accelerators, and solvent applications (Lay et al., 1986; Timmerman, 1978; Thomel, 1987). Viscose rayon manufacture requires 290–320 kg CS_2 per tonne produced (Lay et al., 1986), and 100–300 kg of this is subsequently lost to the atmosphere (Hoeven et al., 1986; Katalyse, 1987). These data allow an estimate that maximally 1 million tonnes of CS_2 could enter the atmosphere annually from this industry worldwide. This figure is an overestimate, in part because up to 50% of waste CS_2 may be scrubbed from the exhaust air before venting to the atmosphere. CS_2 also escapes into the air from other processes (including its own manufacture) and from coal blast furnaces and oil refining (Ackermann et al., 1980).

2.3. Sulfoxides and Sulfonates

A primary atmospheric source of DMSO is from the photochemical oxidation of DMS, but the quantitative importance of this process is unknown. Given a global production of DMS of up to 45 Tg S/year, a DMSO production rate of 1–5 Tg S/year can be postulated. The very extensive use of DMSO in the chemical industry means that there is an anthropogenic input to the environment, but this input has not been precisely quantified. Other sulfoxides, sulfones, etc., occur naturally (e.g., of methionine), both of anthropogenic and of natural sources (Kjaer, 1977). The earliest of these to be discovered, in 1951, was the garlic amino acid alliin, but a huge variety are now known, the genesis of which is attributed to sulfide oxidations (Kjaer, 1977).

Input of sulfonates into the environment is considerable. MSA (see Section 2.1.3) arises from the atmospheric oxidation of DMS, MT, and DMDS (Hatakeyama et al., 1982) and could have a production rate of 20–25 Tg S/year. More complex sulfonates are common; the plant sulfolipid component, 6-sulfo-D-quinovose, has been called a major component of the sulfur cycle and is synthesized by plants and algae to the extent of at least 3.6×10^4 Tg/year (Harwood and Nicholls, 1979; Shibuya et al., 1963). Cysteic acid and various sulfo-aliphatic acids are known, and taurine (2-aminoethanesulfonate) and related compounds are excreted by humans in amounts estimated to be at least 40,000 tonnes/year (Plennart and Heine, 1973). Linear alkylbenzenesulfonates and naphthalenesulfonates occur in wastewaters, from origins such as detergents, with annual release quantities of thousands of tonnes (Biedlingmaier and Schmidt, 1983; Soeder et al., 1987; Zuerrer et al., 1987).

2.4. Higher Alkyl and Aromatic Sulfides and Polysulfur Compounds

A great variety of sulfides appear to be generated by living organisms and are widely distributed, especially among plants. Aromatic sulfides such as thiophenol and mercaptobenzoic acid are among the sulfur compounds in coal and oil, and aromatic thiols occur in Bachu leaf oil (Kjaer, 1977). Diallyl- and dipropyl sulfide and di- and trisulfides occur in garlic and onion (Sparnins et al., 1988), and dimethyl tri- and tetrasulfides have been found in sewage wastewaters (Koenig et al., 1980). Numerous nonsymmetric disulfides and trisulfides [R^1-S-(S)-S-R^2] occur in plants, such as 2-butylpropenyldisulfide and other disulfides in asafetida oil and allylmethyl and propylmethyl di- and trisulfides in garlic and onions (Kjaer, 1977; Sparnins et al., 1988). Seeds of the neem tree (Azadirachta indica) contain 22 volatile sulfur compounds, including dimethylthiophenes, and various di-, tri-, and tetrasulfides (Balandrin et al., 1988). The major volatile constituent of crushed seeds (76% of the headspace volatiles) was dipropyl disulfide, which was shown to be larvicidal to the yellow fever mosquito, tobacco budworm, and corn earworm. The production of such chemicals was proposed to be a defense mechanism against invasive microorganisms and herbivores. Products of the neem tree have been used since antiquity in Hindu medicine, and their efficacy may well reside in part in the known antibacterial effects of neem oil (see Balandrin et al., 1988). The curative and insect-repellent properties of the plant may thus be due to its sulfur compounds. Plants producing such compounds could therefore have a very positive selective effect on the kind of microflora associated with their own surfaces and soils containing exudates from them. Numerous ring systems containing disulfur links are known, for example in the mangroves. Among the most unusual is lenthionine (from the mushroom Lentinus), which consists of a ring with S—S and S—S—S components linked by single carbon atoms (Kjaer, 1977). It seems much more common to find disulfides than free thiols in nature, presumably because of the ease with which thiols oxidize to dithiols (see also Section 2.1.4). Ethanethiol apparently occurs in the urine of rabbits that have eaten cabbage (cited in Windholz, 1983).

2.5. Naturally Occurring Aromatic and Heterocyclic Sulfur Compounds

The structures of some naturally occurring heterocyclic sulfur compounds are given in Fig. 1.

2.5.1. Thiamine and Lipoic Acid

Few naturally occurring sulfur compounds are aromatic, and thiophenol and aromatic thiophenes are mentioned elsewhere. Possibly the most important and most abundant aromatic sulfur-containing metabolite is thiamine (Leder, 1975; Fig. 1); this compound contains a thiazole ring, in which the sulfur is derived from cysteine, which does not, however, contribute carbon to the molecule (Tazuya *et al.*, 1987). α-Lipoic acid, unusually, has a ring containing an S—S bond that opens to the reduced dithiol form in the metabolic reactions in which lipoate participates.

2.5.2. Thiophene and Its Derivatives

Although relatively few thiophene compounds are known in nature, some are very abundant, and the ring system is ubiquitous in living organisms. For example, *d*-biotin (Eisenberg, 1975; Fig. 1) is a very unusual molecule containing a fusion of an imidazole ring with a tetrahydrothiophene ring (with a valeric acid side chain). It is particularly interesting that the sulfur-oxidizing extreme thermoacidophile *Sulfolobus*, which is reportedly able to degrade dibenzothiophene (see Section 3.4), contains a unique caldariellaquinone that is based on the benzothiophene structure (Kjaer, 1977).

Many thiophene derivatives occur in coffee aroma (Gianturco *et al.*, 1968; Stoll *et al.*, 1967) and in plant and fungal material (Fig. 3). It should be noted that many of these contain thiophene rings that are substituted or linked at the 2- (and 5-) carbon positions. A diversity of thiophenes, benzothiophenes, and dibenzothiophenes occur in crude oil, coal, and peat (along with sulfur, thiols, sulfides, and disulfides) (Fig. 3). Crude oils may contain as much as 14% sulfur, with some fractions comprising 70% sulfur-containing compounds (Thompson *et al.*, 1965; Cripps, 1971). Condensed thiophenes commonly comprise at least half the organic sulfur in coal and oil, and compounds like dibenzothiophene, released from oil into the natural environment, have been found to be accumulated by algae, mosquitoes, and snails (Ensley, 1975).

While the resistance of the thiophene ring to degradation in the natural environment is well illustrated by the presence of relict thiophene structures in oil and coal and in marine sediments of tertiary and quaternary age (Brassell *et al.*, 1986), there are reports of the microbial degradation of both simple and complex thiophenes. Thiophene itself does not seem to have been the subject of successful cultivation of any organism using it as a growth substrate, but it is destroyed when subjected to activated sludge treatment (H. A. Painter, personal communication). Degradation of thiophene compounds is discussed in Section 3.5.

2.6. Mammalian Odors: Pheromones, Attractants, and Repellants

Mammalian odors is a field that has been neglected by the microbial ecologist; the diversity of characteristic, and in many cases exotic, compounds produced in scent glands or sacs, vaginal secretions, and the urine by mammals is very great. Specialized organs such as the anal sacs of some mammals constitute very special potential habitats for microbial colonization. This is particularly true for the organic sulfur compounds that are among the attractant (or repellant) odor components in some animals.

Organosulfur compounds are sometimes excreted in the urine and thereby represent a sulfur input to the environment. These include DMDS, dimethyl sulfone, propyl sulfide, allyl isothiocyanate, thiophene, and possibly thiolan-2-one in human urine and (giving it its characteristic odor) isopentenyl methyl sulfide, phenylethyl methyl sulfide, and 3-methylbutyl methyl sulfide in red fox urine (Albone, 1984). Urine from a female beagle was found to contain methyl propyl sulfide and methyl butyl sulfide as major volatile constituents, with minor constituent sulfur compounds being methyl "pentenyl" sulfide and methylpropyl, methylbutyl, dipropyl, and propylbutyl disulfides (Schultz *et al.*, 1985). Mouse urine contains sex-specific sulfur compounds not yet found in other species: 2-isopropyl- and 2-*sec*-butyl-4,5-dihydrothiazoles.

One of the best-known examples of organosulfur production in mammals is the presence in the vaginal secretion of the hamster (*Mesocricetus auratus*) of DMDS, which acts as an attractant pheromone (Singer *et al.*, 1976; Johnson, 1983; Albone, 1984).

The anal sac secretion of the striped hyena (*Hyaena hyaena*) contains about 1% of 5-thiomethylpentane-2,3-dione (Wheeler *et al.*, 1975; Albone, 1984). Recent work by A. J. Buglass, F. M. C. Darling and J. S. Waterhouse (F. Darling, personal communication) has shown that this compound is only a relatively minor component of the secretion of the brown hyena (*Hyaena brunnea*), and that it is absent from that of the spotted hyena (*Crocuta crocuta*). The secretion of the striped hyena also contained dimethy trisulfide (F. Darling, personal communication), a compound whose only other reported "natural" sources are wastewaters and sewage gases (Koenig *et al.*, 1980).

Possibly the most interesting anal sac secretions from the point of view of the microbial ecology of organic sulfur are those of the Mephitinae and Mustelidae: the striped skunk (*Mephitis mephitis*) secretion contains around 150 thiols, sulfides, and disulfides, particularly of C_4 and C_5 alkyl and alkenyl groups, of which 3-methylbutane-1-thiol, *trans*-2-butene-1-thiol and *trans*-2-butylenyl methyl disulfide predominate in approximately equal amounts (Andersen and Bernstein, 1975; Andersen *et al.*, 1982): the finding of the nonsymmetric crotonyl methyl disulfide illustrates that such compounds are not exclusive to the plant kingdom (see Section 2.3). Butanethiol, once

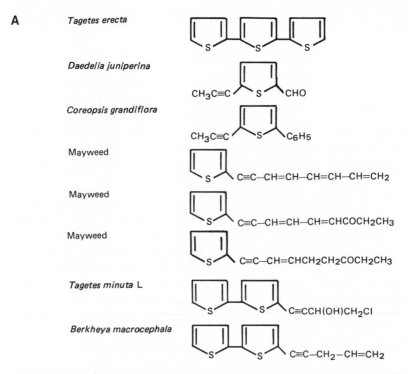

Figure 3. Representative thiophene structures found in plants and fungi (A), coffee aroma (B), and fossil fuels (crude oil and coal) (C). R, Alkyl group. [After Amphlett (1968), Cripps (1971), Challenger (1959), Gianturco *et al.* (1968), Sørensen (1961), and Stoll *et al.* (1967).]

regarded as the main odor component of the skunk secretion, is now known to be only a minor component.

Extensive studies on the mink (*Mustela vison*), and of the stoat and ferret (summarized by Albone, 1984), showed that their anal sac secretions contain the thietane (SC$_3$) and 1,2-dithiolane (S$_2$C$_3$) ring systems. The mink produces principally 2,2-dimethylthietane, with lesser amounts of isomeric 2,3-dimethylthietanes and 3,3-dimethyl-1,2-dithiolane. The stoat (*Mustela erminea*) produced mainly 2-propylthietane, with some 2-pentenylthietane and 3-propyl-1,2-dithiolane, while the ferret (*Mustela putorius furo*) produced several dimethylthietanes, dimethyl-dithiolanes, and 3-propyl-1,2-dithiolane, with the predominant component being *trans*-2,3-dimethylthietane (Crump, 1980a,b). Mink secretions also contain dimethyl- and dibutyl-disulfides and some mixed disulfides. Interestingly, traces of the fox compound, isopentenyl methyl sulfide, were also present, indicating a possible biosynthetic relationship between this compound and the thietanes (which to date are unique to the mustelids), between which there could be easy interconversion (Albone, 1984). Recently, the anal sac secretion of the weasel (*Mustela*

B

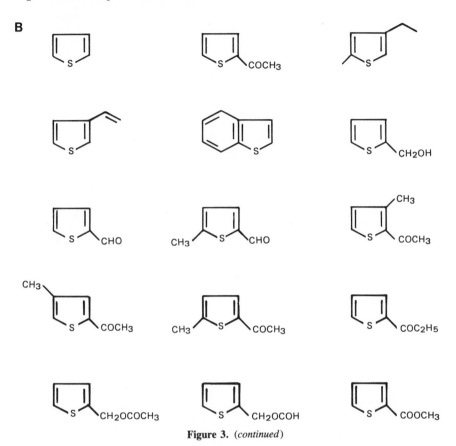

Figure 3. (*continued*)

nivalis) has also been shown to contain a wide range of sulfur compounds (F. Darling, personal communication). These include 2,4-dimethylthietane, *cis* and *trans* 2,3-dimethyl-thietane, 2-ethylthietane, *cis* and *trans* 2-ethyl-3-methylthietane, 3,3-dimethyl-1,2-dithiolane, *trans* (major) and *cis* (minor) 3,4-dimethyl-1,2-dithiolane, 3-ethyl-4-methyl-1,2-dithiolane, and dimethyl disulfide.

Providing this detailed information has two justifications: first, the biochemistry of the processes by which these attractant and repellant compounds are produced is totally unknown. There are indications that some stages of the biosynthesis of these substances may be effected by microorganisms carried, effectively as symbionts, by the animals in their anal sacs (Albone, 1984; Albone *et al.*, 1977). Certainly a variety of types of bacteria have been isolated from, for example, the fox sac (Albone *et al.*, 1978), and these organs can be regarded as microaerophilic fermentation vessels, in a loose way analogous to the rumen, termite gut, or other animal organs in which key roles are played by microbes. Clearly the microbial ecologist and physiologist need to study these

C

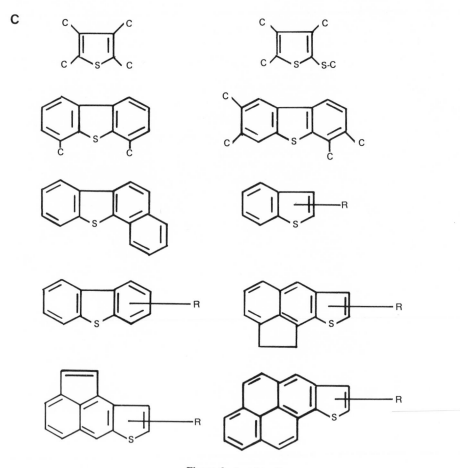

Figure 3. (*continued*)

systems. Secondly, these organs represent a concentrated environmental source of unusual sulfur compounds and as such would be expected to be selectively colonized by bacteria able to use compounds such as thietanes, complex thiols, and disulfides. An ecophysiological study of these habitats not only could provide information on the animal–microbe interactions but could provide model data on the mechanisms of sulfur compound degradation, whether by single organisms or consortia, and establish whether such compounds are used as unique substrates or require cometabolism.

2.6.1. Halitosis and Normal Production of Methylated Sulfides in Humans

The gas respired by all normal healthy human beings contains hydrogen sulfide, MT, DMS, and (in some individuals) DMDS (Chen *et al.*, 1970; Conkle *et al.*, 1975;

Manolis, 1983; Tonzetich, 1977). Between 80 and 90% of these gases arise in the oral cavity from the microbial degradation of free sulfur-amino acids (especially methionine) and from thiol and disulfide groups released from proteins (Tonzetich, 1977). Virtually all of the volatile sulfur compounds released from saliva come from the particulate proteinaceous matter in it: sloughed epithelial cells, leukocytes, and dead bacterial cells; little if any volatile sulfur arises from the inorganic sulfur in the nonparticulate, soluble fraction of saliva (Tonzetich, 1977, 1978).

Halitosis is defined as unpleasant breath arising from physiological and pathological causes from oral and systemic sources (Tonzetich, 1977). In 85–90% of cases, halitosis can be attributed to hydrogen sulfide and MT arising actually in the oral cavity and not elsewhere in the body (Tonzetich, 1977; Rosenberg, 1989; Rosenberg and Gabbay, 1987). A strong correlation between the strength of oral malodor and the concentrations of volatile sulfur present has been shown in numerous studies; MT is regarded as the principal malodorous component, followed by hydrogen sulfide, with DMS usually contributing little to halitosis (Kaizu, 1976a,b; Kaizu *et al.*, 1978; Rosenberg and Gabbay, 1987; Schmidt *et al.*, 1978; Tonzetich, 1977, 1978). MT (and hydrogen sulfide) are generated rapidly when saliva samples are incubated, and addition of methionine to such incubations greatly enhances MT production (Ishikawa *et al.*, 1984; Tonzetich, 1977).

Clinically healthy people with oral malodor showed a high degree of coating of the tongue; removal of this coating by tongue scraping markedly decreased MT production (Kaizu, 1976b; Kaizu *et al.*, 1978). In oral disease, a major source of mouth odor is the putrefaction of proteinaceous material in periodontal pockets (Kaizu, 1976a,b; Kaizu *et al.*, 1978; Rosenberg and Gabbay, 1987).

Organic sulfides in the breath can have other origins in some pathological conditions and in at least some normal individuals. Increased DMS, MT, and ethanethiol were found in patients with liver cirrhosis (Kaji *et al.*, 1978; Manolis, 1983). Breath odor in such patients, due to DMS, was increased on feeding with methionine, which was metabolized by intestinal bacteria to produce DMS that could not be metabolized by the liver (Chen *et al.*, 1970; Manolis, 1983). There is, however, also evidence for normal production of DMS by human metabolism (Blom *et al.*, 1988); work with human (and rat) hepatocytes demonstrated that methionine could be metabolized by a transamination pathway leading via 4-methylthio-2-oxobutyrate to 3-methylthiopropionate (MTP). This compound could either be split to yield MT, which, by methylation, could produce DMS, or the MTP could itself be methylated to form DMSP, which can give rise directly to DMS (see Sections 2.1.2a and 2.1.5).

Other sulfur gases can arise in human beings. Ethanethiol, higher alkanethiols, and alkysulfides occur in human breath (Manolis, 1983), and CS_2 has been shown to be respired as a breakdown product of the alcohol-aversion drug disulfiram (Wells and Koves, 1974; Manolis, 1983):

$$CH_3CH_2NCSSSSCNCH_2CH_3 \rightarrow 2CH_3CH_2NCSSH \rightarrow 2CH_3CH_2NH + 2CS_2$$

These details serve to illustrate that the major environmental sulfur gases are also produced in significant amounts by metabolism of normal human organs (and by the saprophytic bacteria of the organism) and also presumably by all other mammals. Calculating the mean of the data for DMS output by normal males (Conkle et al., 1975) gives a value of 20 ($\pm$16) μg DMS/hr. This converts to an output of 175 mg DMS/ year/person, or a speculative figure of 875 tonnes DMS/year for a world population of 5 billion. As DMS accounts for less than 10% of the total volatile sulfur compounds in expired air (Tonzetich, 1977), the bulk being MT and hydrogen sulfide (in comparable amounts), global human MT production will be greater. Taking a figure [estimated from Tonzetich (1977)] of 5×10^{-7} g MT/liter of expired air and a daily expiration of 20,000 liters (Anonymous, 1955), MT output could be 3.65 g/year/person or 1825 tonnes/year from the world population. Output from other animal sources would probably greatly exceed these figures. The animal kingdom must thus be regarded as a major source of volatile environmental organic sulfur compounds, in addition to providing diverse niches in which bacteria using such compounds as substrates are very likely to reside. This is a virtually unexplored field of microbial ecology.

3. Microbiological Degradation of Organic Sulfur Compounds

In many cases, degradation of organic sulfur compounds by pure or mixed cultures or samples such as soil or sewage sludge has been shown, but the mechanisms of degradation have not been established. The following sections are devoted primarily to those systems for which at least some mechanistic information is available.

3.1. Methylated Sulfides

The first demonstration of the oxidation of a methylated sulfide (DMS) by a bacterial pure culture was by Sivelä (Sivelä and Sundman, 1975; Sivelä, 1980), who isolated a *Thiobacillus* that was reported to obtain energy from oxidizing DMS and to assimilate carbon both by ribulose bisphosphate-dependent carbon dioxide fixation and by a variant of the serine pathway for more direct assimilation of the one-carbon units derived from the methyl groups of DMS. This organisms was obligately chemolithotrophic and could not grow methylotrophically on one-carbon compounds such as methylamine, methanol, and formaldehyde (Sivelä, 1980). Unfortunately, this organism is no longer available (Suylen and Kuenen, 1986; Suylen, 1988), and others with the same mix of metabolic capabilities and limitations have not yet been isolated.

However, several thiobacilli that use DMS, DMDS, and MT as energy substrates for growth have been isolated (Kanagawa and Kelly, 1986; Smith and Kelly, 1988a,b). These organisms oxidize the substrates completely to sulfate and carbon dioxide and assimilate carbon exclusively by carbon dioxide fixation through the Calvin cycle. They do, however, obtain energy from the oxidation of the methyl groups as well as from the sulfide of these compounds.

The first methylated sulfur compound whose degradative pathway was established was DMSO. This compound was found to be a substrate for two isolates of *Hyphomicrobium* (De Bont *et al.*, 1981; Suylen and Kuenen, 1986; Suylen *et al.*, 1986, 1987; Suylen, 1988). DMSO is reduced to DMS by an NADH-dependent reductase, and then DMS is cleaved to methanethiol and formaldehyde by a monooxygenase. Formaldehyde then follows a conventional oxidation pathway via formaldehyde dehydrogenase to formate and formate dehydrogenase to carbon dioxide, while MT is converted to sulfide by an oxidase. Sulfide is oxidized to sulfate by as yet uncharacterized reactions in the hyphomicrobia. Carbon assimilation by the hyphomicrobia is by the serine pathway, and the species studied so far are methylotrophs that are incapable of autotrophic carbon dioxide fixation and chemolithotrophic growth on inorganic sulfur compounds. They do, however, derive energy from the oxidation of the sulfide moiety of DMSO and DMS and can obtain energy simultaneously from methylamine and thiosulfate when grown on these as dual substrates in the chemostat (Suylen *et al.*, 1986; Suylen, 1988).

The central metabolic pathway for DMS, DMDS, and MT oxidation is apparently identical in *Thiobacillus thioparus* strains E6 and TK-m to that established in the hyphomicrobia (Kanagawa and Kelly, 1986; Smith and Kelly, 1988a,b; Kelly, 1988). DMDS is reductively cleaved into two molecules of MT that follow the pathway as shown in Fig. 4.

3.2. Other Sulfides

There is little information on the biodegradation mechanisms of other organic sulfides. Recently, a number of strains of Gram-negative and Gram-positive (mainly coryneform) heterotrophs have been isolated from diverse habitats, including New Zealand hot spring soils, volcanic soils (crater rim and steam vent on Mt. Erebus, Antarctica), White Island volcano in New Zealand, and the oxycline of Lake Fryxell, Antarctica (C. G. Harfoot and D. P. Kelly, unpublished data). These were enriched on substrates including DMS, DMDS, and diethylsulfide (DES) and will oxidize a number of longer-chain alkyl sulfides. One culture grown on DES oxidized DMS, DES, dipropylsulfide, DMDS, and diethyldisulfide (DEDS), ethane- and butanethiols, and CS_2. With the exception of dipropylsulfide (27%) and butanethiol (66%), these compounds were oxidized completely, oxygen uptake being consistent with conversion to sulfate and carbon dioxide. This DES-grown culture also completely oxidized formaldehyde and acetaldehyde. K_m values for the oxidation of DES, DEDS, and ethanethiol by cell suspensions were in the range of 0.5–2 μM, showing that these organisms would be able to act on these substrates if they occurred at concentrations in the 100–1000 nM concentration range in their habitats. As yet, the biochemistry of these organisms and their mode of growth on these substrates are unknown.

Using an uncharacterized pseudomonad isolated from oil-contaminated soil, Koehler *et al.* (1978) demonstrated degradation of dibenzylsulfide, which was degraded to produce an unidentified acid- and water-soluble sulfur compound(s). Mercap-

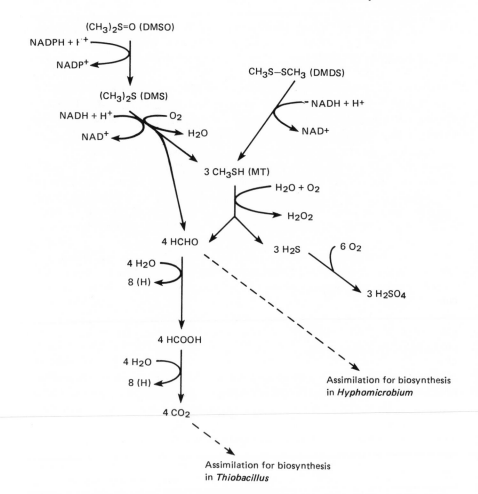

Figure 4. Generalized scheme for the bacterial oxidation of methylated sulfides, as deduced from studies with hyphomicrobia and thiobacilli.

toethanol, mercaptoethanesulfonate, and mercaptopyruvate were among the sulfur compounds used as sulfur sources for growth by *Chlorella fusca* (Krauss and Schmidt, 1987), indicating ability to derive sulfur for biosynthesis from these compounds, but others such as ethanethiol and thioacetic acid were not used.

3.3. Carbon Disulfide and Carbonyl Sulfide

Until recently, there was no information on the microbial metabolism and degradation of either CS_2 or COS other than that CS_2 was known to be an inhibitor of soil nitrification (Ashworth *et al.*, 1977) and that some methanogens could use it as a sulfur source (Rajagopal and Daniels, 1986). CS_2 is now known to be oxidized by some heterotrophs (see Section 3.2) and *Thiobacillus thioparus* strain TK-m (Kelly, 1988; Smith and Kelly, 1988c). This strain grows autotrophically on thiocyanate (S=C=N), CS_2 (S=C=S), and COS (O=C=S), the metabolism of each of which requires the scission of a S=C bond. The organism produced COS and H_2S when incubated anaerobically with CS_2, leading to the hypothesis that the first step of normal aerobic CS_2 metabolism was its hydrolytic cleavage to COS and hydrogen sulfide:

$$S=C=S + H_2O = O=C=S + H_2S$$

The COS appeared to undergo similar hydrolysis to carbon dioxide and hydrogen sulfide:

$$O=C=S + H_2O = CO_2 + H_2S$$

As yet, enzymes for these proposed steps have not been sought. This scheme is in contrast to that proposed for the oxidative degradation of CS_2 by rat liver, which involves an NADPH-dependent monooxygenase containing cytochrome P-450 (Chengelis and Neal, 1987). Clearly, a monooxygenase was not responsible for the anaerobic production of COS and H_2S by the *Thiobacillus*. It seems likely that different mechanisms for CS_2 metabolism will be found, some possibly involving an initial oxidation by molecular oxygen. Anaerobes (in addition to the methanogens) degrading CS_2 are likely to exist and will prove biochemically interesting.

The only source of energy available from the dissimilation of CS_2 is the oxidation to sulfate of the hydrogen sulfide produced from it. CS_2 and COS must therefore be regarded exclusively as chemolithotrophic substrates, in contrast to the methylated sulfides, which can be substrates for both chemolithotrophic and methylotrophic growth. This was demonstrated by comparison of the growth yields of the *Thiobacillus* on thiosulfate, thiocyanate, and COS, which were 5.6–6.1 g cell-carbon mol^{-1}, indicating sulfur oxidation to be the only source of metabolic energy in each case.

It is of interest that while this *Thiobacillus* can grow autotrophically on COS as an energy substrate, COS is known to be a specific inhibitor of autotrophic carbon dioxide assimilation. Both inorganic carbon transport and ribulose bisphosphate carboxylase have been shown to be inhibited by COS, which also acts as an alternative substrate for the latter enzyme (Laing and Christeller, 1980; Lorimer and Pierce, 1990; Ogawa and Togasaki, 1988).

Cometabolism of CS_2 by heterotrophs has not been shown, but it seems feasible to propose that it could be assimilated under sulfur-limiting conditions in the natural environment.

Plants are reported to be a major biogenic sink for COS (Brown and Bell, 1986), potentially consuming 2–5 Tg/year, compared with COS destruction by soil at a maximum of only 0.04 Tg/year. Some of this is likely to be by conversion of COS to H_2S

(by processes akin to those by which plants convert excess sulfate, sulfur dioxide, or sulfur amino acids to sulfide), but a role for the phylloplane microflora in COS oxidation is also worthy of investigation.

3.4. Sulfonates

The importance of MSA in the natural cycling of sulfur has already been stressed (see Section 2.1.3), but virtually nothing is known of its biological fate. Undoubtedly, it enters into chemical combination with soil and water constituents when deposited in rain or snow, but the methanesulfonate group must still be available for metabolism. MSA was used as a sulfur source for growth by *Chlorella fusca*, but it was the least well used of all the C_1- to C_8-n-alkyl-1-sulfonates (Biedlingmaier and Schmidt, 1983; Krauss and Schmidt, 1987). MSA was also used a sulfur source for growth by some bacteria (Cook and Huetter, 1982). There appears to be no significant literature on MSA metabolism, presumably desulfonation being the only reaction catalyzed by the organisms using it for sulfur. The search for methylotrophs and sulfur bacteria able to derive energy from its degradation has recently revealed such organisms (S. Baker and D. P. Kelly, unpublished observations).

A wide range of sulfonates, sulfinates, and sulfones was used by *Chlorella*, leading to the suggestion that the green algae could be important in the biodegradation of such compounds (especially xenobiotics) in nature (Krauss and Schmidt, 1987). There is a large body of literature on the biodegradation of pollutant sulfonates; these are particularly the naphthalenesulfonic acids, which are present in many wastewaters and have been cited as 10% of the pollution burden of the Rhine (Zuerrer et al., 1987; Soeder et al., 1987). Naphthalene and benzene sulfonic acids are desulfonated (and used as sulfur sources for growth) by several bacteria and algae, but these compounds are persistent in the environment, and it seems likely that organisms showing unique metabolic use of them for energy metabolism, for example, are uncommon. The sulfanilic acid 4-aminobenzenesulfonate (4-ABS) supported the growth of a binary mixed culture (consisting of a Gram-negative aerobic rod and *Agrobacterium radiobacter*), for which it provided all carbon, nitrogen, sulfur, and energy requirements (Feigel and Knackmuss, 1988). Catechol-4-sulfonic acid was identified as the initial metabolite of 4-ABS by the two-species consortium, the aerobic rod hydroxylating the 4-ABS in the 2,3-position and releasing ammonia by a dioxygenase reaction. Following ring cleavage, the sulfonate group was released, probably as sulfite, which was recovered as sulfate (equimolar to the initial 4-ABS concentration) at the end of growth.

3.5. Thiophenes and Aromatic Sulfur Compounds

The bulk of the literature on thiophene degradation is concerned with the complex thiophenes, benzothiophene and dibenzothiophene (DBT), that occur in oil and coal (Monticello and Finnerty, 1985; Kargi and Robinson, 1984; Fedorak et al., 1988;

Kanagawa and Kelly, 1987), the interest being both in the commercial desulfurization of such fuels and the determination of the environmental fate of pollutants from oil spills. There are very few reports of complete degradation of the benzothiophenes. *Sulfolobus* was reported to convert DBT to sulfate, meaning that the thiophene ring has to be degraded; A. T. Knecht [Ph.D. thesis, Louisiana State University, 1961, cited by Cripps (1971)] obtained a mixed culture of an *Arthrobacter* and a pseudomonad that produced reducible inorganic sulfur from DBT. The pathway proposed involved phenylpropionate, benzoate, catechol, and salicylate as intermediates, with sulfate being the sulfur end product. In general, the condensed thiophenes are not degraded in this way, and the products of their bacterial metabolism are water-soluble materials in which the thiophene ring is still intact. In virtually all cases, the organisms used have been grown on complex media and DBT was provided as a cometabolic substrate. A series of oxidations of the aromatic ring(s) leads to hydroxylated intermediates, including *cis*-DBT-dihydrodiol, DBT-diol, DBT-5-oxide, 3-hydroxy-2-formylbenzothiophene-5-oxide, 4-[2-(3-hydroxy)-thionaphthenyl]-2-oxo-3-butenoic acid, and others (Ensley, 1984; Laborde and Gibson, 1977; Monticello *et al.*, 1985). The ability in some *Pseudomonas* species to degrade DBT to water-soluble complex products is carried on a 55-megadalton plasmid (Monticello *et al.*, 1985), indicating that this ability could become widely distributed among the microbiota in response to DBT contamination from oil spillage, for example.

Mormile and Atlas (1988) used a medium containing *trans*-4-[2-(3-hydroxy)-benzothiophene]-2-oxo-butenoic acid and 3-hydroxy-2-formylbenzothiophene (generated by the action of *Pseudomonas putida* on DBT) for the enrichment of mixed cultures that actively degraded the latter compound. Carbon dioxide was also released from the enrichment cultures, but no sulfate was produced. Supplementing the enrichment cultures with DBT-5-oxide or DBT-sulfone also resulted in production of carbon dioxide (after a 12-day lag), but no sulfate formation. This work demonstrated that some mineralization of carbon in the DBT ring structure occurred but suggests that the thiophene ring structure was not degraded in these aerobic cultures.

The bacterial and fungal degradation of n-alkylated tetrahydrothiophenes has also been reported, but the biochemical details have yet to be established (Fedorak *et al.*, 1988).

Under anaerobic conditions, DBT, dibenzylsulfide, thianaphthene, and butyl and octyl sulfides were desulfurized to various degrees by a hydrogen-oxidizing *Desulfovibrio* (Koehler *et al.*, 1984). Thiophene was reported to be desulfurized by cell-free extracts of the organism. It is clear that the process of converting DBT and similar compounds to simpler thiophenes, ultimately producing complete mineralization, is still not understood and clearly involves cometabolism and the sequential activities of various microbial types in aerobic and anaerobic habitats.

There has been more success in isolating organisms degrading the simple thiophene-2-carboxylic acid (T-2-C). Three bacteria (a *Flavobacterium*, a pseudomonad, and a *Rhodococcus*) have been isolated that can use T-2-C as the sole substrate for growth and effect complete oxidation of the molecule to carbon dioxide and sulfate (Cripps, 1971, 1973; Amphlett, 1968; Kanagawa and Kelly, 1987; Kelly, 1988). Few other thiophene derivatives could be metabolized by these bacteria; the *Rhodococcus* could also use the 5-methyl-2-carboxylate and thiophene-2-acetic acid but not thiophene itself or a wide range of 2- and 3-substituted thiophenes (Kanagawa and Kelly, 1987). The biodegradation pathway for T-2-C proposed by Cripps (1973) involved the initial formation of the coenzyme A-ester, followed by hydroxylation, oxidation, and ring opening (S—C_5 bond), leading to the formation of 2-oxoglutarate, which enters the Krebs cycle. The sulfur atom of the ring was released as sulfate by the organisms used by Cripps (1973) and Kanagawa and Kelly (1987) but as sulfide by the *Flavobacterium* (Amphlett, 1968).

It is clear that too little is known about the fate of the thiophene ring in natural environments, given its occurrence in many forms in many habitats.

The aromatic sulfur compounds thianthrene and thioxanthene (each of which contains two aromatic rings linked, respectively, by two —S— bridges and by —S— and —C—) were reported to be degraded, with the release of sulfate, by *Sulfolobus* (Kargi, 1987). Thianthrene was oxidized (in the presence of hydrogen peroxide) to thianthrene monosulfoxide by the liginase from *Phanerochaete chrysosporium* (Schreiner *et al.*, 1988), indicating that solubilization of sulfur from such heterocycles in coal could be catalyzed naturally or commercially by fungal enzymes.

3.6. Miscellaneous Sulfur Compounds

There are numerous isolated examples of the degradation of unusual sulfur compounds that serve to illustrate the metabolic potential in the environment for the microbiological dissimilation of the simpler organic sulfur compounds. Among these are the degradation of primary alkyl sulfates, including the release of sulfate from methyl sulfate under aerobic or denitrifying conditions (White *et al.*, 1987), the degradation of dimethylphosphorodithioate (DMPT), diethylphosphorodithioate, dimethylphosphorothioate, and dimethyl- and diethylphosphates by activated sludge and the degradation of DMPT by a consortium of a *Pseudomonas* and a *Thiobacillus* (Kanagawa *et al.*, 1980, 1982); use of taurine, cysteic acid, and mercaptosuccinic acid as sole substrates for the growth of soil organisms (Stapley and Starkey, 1970; Hall and Berk, 1968); and the decomposition of thiourea by *Penicillium* (Lashen and Starkey, 1970).

4. Some Effects of Organic Sulfur Compounds on Global Ecology

The role of organosulfur compounds in the biogeochemical turnover of sulfur and local closed sulfur cycles is dealt with throughout this review, but there is a recently established link also to climatic and regional effects. Globally, a correlation has been established between atmospheric aerosol particle populations (condensation nuclei) and atmospheric DMS (Bates *et al.*, 1987). It was proposed that DMS could have a role in the control of climate through a cloud albedo feedback mechanism; variation in DMS output could thus be climatically significant. The DMS produced from phytoplankton (and converted photochemically to methane sulfonic acid) was also suggested to be responsible for 30–50% of the sulfur acids falling as acid rain on Scandinavia during the bloom period of May to August (Cocks and Kallend, 1988; Fletcher, 1989, Pearce, 1988). This background input of biogenic sulfur is obviously of relatively greater importance in regions remote from anthropogenic sulfur inputs to the atmosphere. It has been observed that more than 60% of the sulfur in rain (other than that from sea salt) falling in remote regions of Scotland cannot be attributed to individual anthropogenic emissions and presumably contains much sulfur of biogenic origin (Cocks and Kallend, 1988). Similarly, up to 30% of the acidifying sulfur burden of the atmosphere in remote areas of Canada has been attributed to biogenic DMS (Nriagu et al., 1987). As discussed earlier (see Section 2.1.3), the major products of the reaction of DMS with OH and NO_3 radicals in the atmosphere are sulfur dioxide and MSA. Both of these might give rise to sulfate in rain, but MSA itself is also deposited in rain and snow (see Section 2.1.3) and could perhaps be used as an indicator of biogenic sulfur in precipitation. The presence of MSA in the atmosphere of the Southern Hemisphere indicates the biological input of sulfur to air relatively unpolluted by human release of sulfur oxides (Cocks and Kelland, 1988). Increased algal blooms, releasing large amounts of DMS, could therefore have several direct effects on the global environment, and these comments serve to emphasize the need for greater understanding of the ecological balance and control of DMS emission from the natural environment and the fate of compounds such as MSA both in the atmosphere and after deposition.

ACKNOWLEDGMENTS. We are grateful to Dr. M. Amphlett and Dr. Trudy Suylen for letting us see copies of their theses on organic sulfur biochemistry and to Ann Wood, Simon Baker, Mel Rosenberg, and George Dodd for assistance with the literature survey. D. P. K. wishes to thank Prof. Geoffrey Shellam and Mrs. Peta Locke for their hospitality and facilities in the Department of Microbiology, University of Western Australia, where part of this review was prepared.

References

Ackermann, D. G., Haro, M. T., Richard, G., Takata, A. M., Weller, P. J., Bean, D. J., Cornaby, B. W., Mihlan, G. J., and Rogers, S. E., 1980, Health Impacts, Emissions, and Emission Factors for Noncriteria Pollutants Subject to De Minimus Guidelines and Emitted from Stationary Conventional Combustion Processes, U.S. Environmental Protection Agency, Report EPA-450/2-80-074.

Ackman, R. G., Tocher, C. S., and McLachlan, J., 1966, Occurrence of dimethyl-β-propiothetin in marine phytoplankton, J. Fish. Res. Board Can. 23:357.

Adams, D. F., Farwell, S. O., Robinson, E., Pack, M. R., and Bamesberger, W. L., 1981, Biogenic sulfur source strengths, Environ. Sci. Technol. 15:1493.

Albone, E. S., 1984, Mammalian Semiochemistry, Wiley, Chichester.

Albone, E. S., Gosden, P. E., and Ware, G. C., 1977, Bacteria as a source of chemical signals in mammals, in: Chemical Signals in Vertebrates (D. Mueller-Schwarze and M. M. Mozell, eds.), pp. 35–43, Plenum Press, New York.

Albone, E. S., Gosden, P. E., Ware, G. C., Macdonald, D. W., and Hough, N. G., 1978, Bacterial action and chemical signalling in the red fox (*Vulpes vulpes*) and other mammals, in: *Flavor Chemistry of Animal Foods* (R. W. Bullard, ed.), pp. 78–91, ACS Symposium Series 67, American Chemical Society, Washington, D.C.

Amphlett, M. J., 1968, The Microbiological Transformation of Sulphur-Containing Aromatic Compounds, Ph.D. thesis, University of Wales.

Andersen, K. K., and Bernstein, D. T., 1975, Some chemical constituents of the scent of the striped skunk *Mephitis mephitis*, *J. Chem. Ecol.* **1**:493.

Andersen, K. K., Bernstein, D. T., Caret, R. L., and Romanczyk, L. J., 1982, Chemical constituents of the defensive secretion of the striped skunk *Mephitis mephitis*, *Tetrahedron* **38**:1965.

Anderson, R., Kates, M., and Volcani, B. E., 1976, Sulphonium analogues of lecithin in diatoms, *Nature (London)* **263**:51.

Ando, H., Kumagai, M., Karashimada, T., and Iida, H., 1957, Diagnostic use of dimethyl sulfoxide reduction test within Enterobacteriaceae, *Jpn. J. Microbiol.* **1**:335.

Andreae, M. O., 1980a, Dimethyl sulfoxide in marine and fresh waters, *Limnol. Oceanogr.* **25**:1054.

Andreae, M. O., 1980b, The production of methylated sulfur compounds by marine phytoplankton, in: *Biogeochemistry of Ancient and Modern Environments* (P. A. Trudinger and M. R. Walter, eds.), pp. 253–259, Springer Verlag, Berlin.

Andreae, M. O., 1985, The emission of sulfur to the remote atmosphere: A background paper, in: *The Biogeochemical Cycling of Sulfur and Nitrogen in the Remote Atmosphere* (J. N. Galloway, ed.), pp. 5–25, Reidel, New York.

Andreae, M. O., 1986, The ocean as a source of atmospheric sulfur compounds, in: *The Role of Air-Sea Exchange in Geochemical Cycling* (P. Buat-Menard, ed.), pp. 331–362, Reidel, New York.

Andreae, M. O., and Barnard, W. R., 1984, The marine chemistry of dimethyl sulfide, *Mar. Chem.* **14**:267.

Andreae, M. O., and Raemdonck, H., 1983, Dimethyl sulfide in the surface ocean and the marine atmosphere: A global view, *Science* **221**:744.

Aneja, V. P., Overton, J. H., Cupitt, L. T., Durham, J. L., and Wilson, W. E., 1979a, Direct measurements of emission rates of some atmospheric biogenic sulfur compounds, *Tellus* **31**:174.

Aneja, V. P., Overton, J. H., Cupitt, L. T., Durham, J. L., and Wilson, W. E., 1979b, Carbon disulphide and carbonyl sulphide from biogenic sources and their contributions to the global sulphur cycle, *Nature (London)* **282**:493.

Anness, B. J., 1980, The reduction of dimethylsulphoxide to dimethyl sulphide during fermentation, *J. Inst. Brewing* **86**:134.

Anness, B. J., and Bamforth, C. W., 1982, Dimethyl sulphide—a review, *J. Inst. Brewing* **88**:244.

Anness, B. J., Bamforth, C. W., and Wainwright, T., 1979, The measurement of dimethyl sulfoxide in barley and malt and its reduction to dimethyl sulfide by yeast, *J. Inst. Brewing* **85**:346.

Anonymous, 1955, Recommendations of the International Commission on Radiological Protection, *Br. J. Radiol.*, Suppl. 6.

Ashworth, J., Briggs, G. G., Evans, A. A., and Matula, J., 1977, Inhibition of nitrification by nitrapyrin, carbon disulphide and trithiocarbonate, *J. Sci. Food Agric.* **28**:673.

Babich, H., and Stotzky, G., 1978, Atmospheric sulfur compounds and microbes, *Environ. Res.* **15**:513.

Balandrin, M. F., Lee, S. M., and Klocke, J. A., 1988, Biologically active volatile organosulfur compounds from seeds of the neem tree, *Azadirachta indica* (Meliaceae), *J. Agric. Food Chem.* **36**:1048.

Bandy, A. R., Maroulis, P. J., Shalaby, L., and Wilner, L. A., 1981, Evidence for a short trophospheric residence time for carbon disulfide, *Geophys. Res. Lett.* **8**:1180.

Banwart, W. L., and Bremner, J. M., 1976, Evolution of volatile sulfur compounds from soils treated with sulfur-containing organic materials, *Soil Biol. Biochem.* **8**:439.

Barnard, W. R., Andreae, M. O., Watkins, W. E., Bingemer, H., and Georgii, H. W., 1982, The flux of dimethyl sulfide from the oceans to the atmosphere, *J. Geophys. Res.* **87**:8787.

Barnard, W. R., Andreae, M. O., and Iverson, R. L., 1984, Dimethylsulfide and *Phaeocystis poucheti* in the south-eastern Bering Sea, *Continental Shelf Res.* **3**:103.

Bates, T. S., Charlson, R. J., and Gammon, R. H., 1987, Evidence for the climatic role of marine biogenic sulphur, *Nature* (London) **329**:319.

Bechard, M. J., and Rayburn, W. R., 1979, Volatile organic sulfides from freshwater algae, *J. Phycol.* **15**:379.

Biedlingmaier, S., and Schmidt, A., 1983, Alkylsulfonic acids and some S-containing detergents as sulfur sources for growth of *Chlorella fusca, Arch. Microbiol.* **136**:124.

Bills, D. D., and Keenan, T. W., 1968, Dimethyl sulfide and its precursor in sweetcorn, *J. Agric. Food Chem.* **16**:643.

Bilous, P. T., and Weiner, J. H., 1985, Dimethyl sulfoxide reductase activity by anaerobically grown *Escherichia coli* HB101, *J. Bacteriol.* **162**:1151.

Blom, H. J., van den Elzen, J. P. A. M., Yap, S. H., and Tangerman, A., 1988, Methanethiol and dimethylsulfide formation from 3-methylthiopropionate in human and rat hepatocytes, *Biochim. Biophys. Acta* **972**:131.

Brassell, S. C., Lewis, C. A., de Leeuw, J. W., de Lange, L., and Sinninghe Damste, J. J., 1986, Isoprenoid thiophenes: Novel products of sediment diagenesis, *Nature (London)* **320**:160.

Bremner, J. M., and Steele, C. G., 1978, Role of microorganisms in the atmospheric sulfur cycle, *Adv. Microb. Ecol.* **2**:155.

Brown, K. A., and Bell, J. N. B., 1986, Vegetation—the missing sink in the global cycle of carbonyl sulphide, *Atmos. Environ.* **20**:537.

Cantoni, G. L., and Anderson, D. G., 1956, Enzymatic cleavage of dimethylpropiothetin by *Polysiphonia lanosa, J. Biol. Chem.* **222**:171.

Challenger, F., 1959, *Aspects of the Organic Chemistry of Sulphur,* Butterworths, London.

Challenger, F., and Simpson, M. I., 1948, Studies on biological methylation, *Arch. Biochem.* **69**:514.

Chen, S., Zieve, L., and Mahadevan, V., 1970, Mercaptans and dimethyl sulfide in the breath of patients with cirrhosis of the liver. Effect of feeding methionine, *J. Lab. Clin. Med.* **75**:628.

Chengelis, C. P., and Neal, R. A., 1987, Oxidative metabolism of carbon disulfide by isolated rat liver hepatocytes and microsomes, *Biochem. Pharmacol.* **36**:363.

Cline, J. D., and Bates, T. S., 1983, Dimethylsulfide in the equatorial Pacific Ocean: A natural source of sulfur to the atmosphere, *Geophys. Res. Lett.* **10**:949.

Cocks, A., and Kallend, T., 1988, The chemistry of atmospheric pollution, *Chem. Br.* **24**:884.

Conkle, J. P., Camp, B. J., and Welch, B. E., 1975, Trace composition of human respiratory gas, *Arch. Environ. Health* **30**:290.

Conway, E. J., 1942, Mean geochemical data in relation to oceanic evolution, *Proc. R. Ir. Acad. Sect. A* **48**:119.

Cook, A. M., and Huetter, R., 1982, Ametyne and prometyne as sulfur sources for bacteria, *Appl. Environ. Microbiol.* **43**:781.

Cooper, A. J. L., 1983, Biochemistry of sulfur-containing amino acids, *Annu. Rev. Biochem.* **52**:187.

Cox, R. A., and Sandalls, F. J., 1974, The photooxidation of hydrogen sulfide in air, *Atmos. Environ.* **8**:1269.

Cripps, R. E., 1971, Microbial Metabolism of Aromatic Compounds Containing Sulphur, Ph.D. thesis, University of Warwick, Coventry, United Kingdom.

Cripps, R. E., 1973, The microbial metabolism of thiophen-2-carboxylate, *Biochem. J.* **134**:353.

Crump, D. R., 1980a, Thietanes and dithiolanes from the anal gland of the stoat, *Mustela erminea, J. Chem. Ecol.* **6**:759.

Crump, D. R., 1980b, Anal gland secretion of the ferret (*Mustela putorius* forma *furo*), *J. Chem. Ecol.* **6**:837.

Dacey, J. W. H., and Wakeham, S. G., 1986, Oceanic dimethylsulfide: Production during zooplankton grazing on phytoplankton, *Science* **233**:1314.

Dacey, J. W. H., King, G. M., and Wakeham, S. G., 1987, Factors controlling emission of di-
methylsulphide from salt marshes, *Nature (London)* **330**:643.

Dando, P. R., Southward, A. J., Southward, E. C., and Barrett, R. L., 1986, Possible energy sources for
chemosynthetic prokaryotes symbiotic with invertebrates from Norwegian fjord, *Ophelia* **26**:135.

De Bont, J. A. M., van Dijken, J. P., and Harder, W., 1981, Dimethyl sulphoxide and dimethyl sulphide
as a carbon, sulphur and energy source for growth of *Hyphomicrobium* S, *J. Gen. Microbiol.*
127:315.

Deprez, P. P., Franzmann, P. D., and Burton, H. R., 1986, Determination of reduced sulfur gases in
Antarctic lakes and seawater by gas chromatography after solid absorbent preconcentration, *J.
Chromatogr.* **362**:9.

Dickson, D. M., Wyn Jones, R. G., and Davenport, J., 1980, Steady state osmotic adaptation in *Ulva
lactuca*, *Planta* **150**:158.

Dickson, D. M., Wyn Jones, R. G., and Davenport, J., 1982, Osmotic adaptation in *Ulva lactuca* under
fluctuating salinity regimes, *Planta* **155**:409.

Drotar, A., Burton, G. A., Tavernier, J. E., and Fall, R., 1987a, Widespread occurrence of bacterial thiol
methyltransferases and the biogenic emission of methylated sulfur gases, *Appl. Environ. Microbiol.*
53:1626.

Drotar, A., Fall, L. R., Mishalanie, E. A., Tavernier, J. E., and Fall, R., 1987b, Enzymatic methylation
of sulfide, selenide, and organic thiols by *Tetrahymena thermophila*, *Appl. Environ. Microbiol.*
53:2111.

Eisenberg, M. A., 1975, Biotin, in: *Metabolic Pathways*, Vol. 7, The *Metabolism of Sulfur Compounds*
(D. M. Greenberg, ed.), pp. 27–56, Academic Press, New York.

Elliott, S., Lu, E., and Sherwood-Rowland, F., 1989, Hydrogen sulfide in oxic seawater, in: *Biogenic
Sulfur in the Environment* (E. S. Saltzman and W. J. Cooper, eds.), pp. 314–326, American
Chemical Society (Symposium Series 393), Washington, D.C.

Ensley, B. D., 1975, Microbial metabolism of condensed thiophenes, in: *Metabolic Pathways*, Vol. 7,
The *Metabolism of Sulfur Compounds* (D. M. Greenberg, ed.), pp. 309–317, Academic Press, New
York.

Ensley, B. D., 1984, Microbial metabolism of condensed thiophenes, in: *Microbial Degradation of
Organic Compounds* (T. D. Gibson, ed.), pp. 309–317, Marcel Dekker, Inc., New York.

Eriksson, E., 1963, The yearly circulation of sulfur in nature, *J. Geophys. Res.* **68**:4001.

Fedorak, P. M., Payzant, J. D., Montgomery, D. S., and Westlake, D. W. S., 1988, Microbial degradation of N-alkyl tetrahydrothiophenes found in petroleum, *Appl. Environ. Microbiol.* **54**:1243.

Feigel, B. J., and Knackmuss, H.-J., 1988, Bacterial catabolism of sulfanilic acid via catechol-4-sulfonic
acid, *FEMS Microbiol. Lett.* **55**:113.

Ferchichi, M., Hemme, D., Nardi, M., and Pamboukian, N., 1985, Production of methanethiol from
methionine by *Brevibacterium linens* CNRZ 918, *J. Gen. Microbiol.* **131**:715.

Ferchichi, M., Hemme, D., and Nardi, M., 1986, Induction of methanethiol production by *Brevibacterium linens* CNRZ 918, *J. Gen. Microbiol.* **132**:3075.

Ferek, R. J., and Andreae, M. O., 1983, The supersaturation of carbonyl sulfide in surface waters of the
Pacific Ocean, *Geophys. Res. Lett.* **10**:393.

Ferek, R. J., and Andreae, M. O., 1984, Photochemical production of carbonyl sulphide in marine
surface waters, *Nature (London)* **307**:148.

Ferek, R. J., Chatfield, R. B., and Andreae, M. O., 1986, Vertical distribution of dimethyl sulphide in
the marine atmosphere, *Nature (London)* **320**:514.

Fletcher, I., 1989, North Sea DMS emissions as a source of background sulfate over Scandinavia: a
model, in: *Biogenic Sulfur in the Environment* (E. S. Saltzman and W. J. Cooper, eds.), pp. 489–
501, American Chemical Society (Symposium Series 393), Washington, D.C.

Franzmann, P. D., Deprez, P. P., Burton, H. R., and van den Hoff, J., 1987, Limnology of Organic

Lake, Antarctica, a meromictic lake that contains high concentrations of dimethyl sulfide, *Aust. J. Freshwater Res.* **38**:409.

Fuhrman, J. A., and Ferguson, R. L., 1986, Nanomolar concentrations and rapid turnover of dissolved free amino acids in seawater: Agreement between chemical and microbiological measurements, *Mar. Ecol. Prog. Ser.* **33**:237.

Gianturco, M. A., Giammarino, A. S., and Friedel, P., 1968, Volatile constituents of coffee V, *Nature (London)* **210**:1358.

Graedel, T. E., Kammlott, G. W., and Franey, J. P., 1981, Carbonyl sulfide: Potential agent of atmospheric sulfur corrosion, *Science* **212**:663.

Grosjean, D., and Lewis, R., 1982, Atmospheric photooxidation of methyl sulfide, *Geophys. Res. Lett.* **9**:1203.

Guenther, A., Lamb, B., and Westberg, H., 1989, U.S. National biogenic sulfur emissions inventory, in: *Biogenic Sulfur in the Environment* (E. S. Saltzman and W. J. Cooper, eds.), pp. 14–30, American Chemical Society, (Symposium Series 393), Washington, D.C.

Haas, P., 1935, CLVII, The liberation of methyl sulfide by seaweed, *Biochem. J.* **29**:1297.

Haines, B., Black, M., and Bayer, C., 1989, Sulfur emissions from roots of the rain forest tree *Stryphnodendron excelsum*, in: *Biogenic Sulfur in the Environment* (E. S. Saltzman and W. J. Cooper, eds.), pp. 58–69, American Chemical Society (Symposium Series 393), Washington, D.C.

Hall, M. R., and Berk, R. S., 1968, Microbial growth on mercaptosuccinic acid, *Can. J. Microbiol.* **14**:515.

Harvey, G. R., and Lang, R. F., 1986, Dimethylsulfoxide and dimethylsulfone in the marine atmosphere, *Geophys. Res. Lett.* **13**:49.

Harwood, J. J., and Nicholls, R. G., 1979, The plant sulpholipid—a major component of the sulphur cycle, *Biochem. Soc. Trans.* **7**:440.

Hatakeyama, S., Okuda, M., and Akimoto, H., 1982, Formation of sulfur dioxide and methanesulfonic acid in the photooxidation of dimethyl sulfide in the air, *Geophys. Res. Lett.* **9**:583.

Hatakeyama, S., Izumi, K., and Akimoto, H., 1985, Yield of SO_2 and formation of aerosol in the photooxidation of DMS under atmospheric conditions, *Atmos. Environ.* **19**:135.

Hattula, T., and Granroth, B., 1974, Formation of dimethyl sulfide from S-methylmethionine in onion seedlings (*Allium cepa*), *J. Sci. Food Agric.* **25**:1517.

Headley, J. V., 1987, GC/MS identification of organosulfur compounds in environmental samples, *Biomed. Environ. Mass Spectrom.* **14**:275.

Hitchcock, D. R., 1975, Dimethyl sulfide emissions to the global atmosphere, Chemosphere No. 3, pp. 137–138, Pergamon Press, New York.

Hitchcock, D. R., 1976, Atmospheric sulfates from biological sources, *J. Air Pollut. Control Assoc.* **26**:210.

Hoeven, J. C. M. van der, Mak, J. K., Flohr, P. M., and Knippenberg, J. A. J. van, 1986, *Review of Literature on Carbon Disulfide*, Bericht der NOTOX, 's-Hertogenbosch, und D.H.V., Raumplanung und Umwelt, Bereich Umweltschutz.

Howes, B. L., Dacey, J. W. H., and Wakeham, S. G., 1985, Effects of sampling technique on measurements of pore water constituents in salt marsh sediments, *Limnol. Oceanogr.* **30**:221.

Ishikawa, M., Shibuya, K., Tokita, F., and Koshimizu, M., 1984, A study of bad breath. (2) The evaluation of bad breath by methyl mercaptan production from methionine, *Koku Eisei Gakkai Zasshi* **34**:124.

Johnson, R. E., 1983, Chemical signals and reproductive behavior, in: *Pheromones and Reproduction in Mammals* (J. G. Vandenbergh, ed.), pp. 3–37, Academic Press, New York.

Jørgensen, B. B., and Okholm-Hansen, B., 1986, Emissions of biogenic sulfur gases from a Danish estuary, *Atmos. Environ.* **19**:1737.

Junge, C. E., 1960, Sulfur in the atmosphere, *J. Geophys. Res.* **65**:227.

Kadota, H., and Ishida, Y., 1972, Production of volatile sulfur compounds by microorganisms, *Annu. Rev. Microbiol.* **26**:127.

Kaizu, T., 1976a, Source of foul breath and its control, *Nippon Shika Ishikai Zasshi* **29**:228.

Kaizu, T., 1976b, Analysis of volatile sulfur compounds in exhaled air by gas chromatography, *Nippon Shishubyo Gakkai Kaishi* **18**:1.

Kaizu, T., Tsunoda, M., Aoki, H., and Kimura, K., 1978, Analysis of volatile sulfur compounds in mouth air by gas chromatography, *Bull. Tokyo Dent. Coll.* **19**:43.

Kaji, H., Hisamura, M., Saito, N., and Murao, M., 1978, Evaluation of volatile sulfur compounds in the expired alveolar gas in patients with liver cirrhosis, *Clin. Chim. Acta* **85**:279.

Kanagawa, T., and Kelly, D. P., 1986, Breakdown of dimethyl sulphide by mixed cultures and by *Thiobacillus thioparus*, *FEMS Microbiol. Lett.* **34**:13.

Kanagawa, T., and Kelly, D. P., 1987, Degradation of substituted thiophenes by bacteria isolated from activated sludge, *Microb. Ecol.* **13**:47.

Kanagawa, T., Dazai, M., and Takahara, Y., 1980, Degradation of O,O-dimethyl phosphorodithioate by activated sludge, *Agric. Biol. Chem.* **44**:2631.

Kanagawa, T., Dazai, M., and Fukuoka, S., 1982, Degradation of O,O-dimethyl phosphorodithioate by *Thiobacillus thioparus* TK-1 and *Pseudomonas* AK-2, *Agric. Biol. Chem.* **46**:2571.

Kargi, F., 1987, Biological oxidation of thianthrene, thioxanthene and dibenzothiophene by the thermophilic organism *Sulfolobus acidocaldarius*, *Biotechnol. Lett.* **9**:478.

Kargi, F., and Robinson, J. M., 1984, Microbial oxidation of dibenzothiophene by the thermophilic organism *Sulfolobus acidocaldarius*, *Biotechnol. Bioeng.* **26**:687.

Katalyse, 1987, BUND, Oko-Institut, ULF: Chemie am Arbeitsplatz. Gefahrliche Arbeitsstoffe, Berufskrankheiten und Auswege, Rowohlt, Reinbeck.

Keenan, T. W., and Lindsay, R. C., 1968, Evidence for a dimethyl sulfide precursor in milk, *J. Dairy Sci.* **51**:112.

Keller, M. D., Bellows, W. K., and Guillard, R. R. L., 1989, Dimethyl sulfide production in marine phytoplankton, in: *Biogenic Sulfur in the Environment* (E. S. Saltzman and W. J. Cooper, eds.), pp. 167–182, American Chemical Society (Symposium Series 393), Washington, D.C.

Kellogg, W. W., Cadle, R. D., Allen, E. R., Lazrus, A. L., and Martell, E. A., 1972, The sulfur cycle, *Science* **175**:587.

Kelly, D. P., 1980, The sulphur cycle: Definitions, mechanisms and dynamics, in: *Sulphur in Biology,* Ciba Foundation Symposium 72 (new series), pp. 3–18, Excerpta Medica, Amsterdam.

Kelly, D. P., 1982, Biochemistry of the chemolithotropic oxidation of inorganic sulphur, *Phil. Trans. R. Soc. London Sect. B* **298**:499.

Kelly, D. P., 1988, Oxidation of sulphur compounds, *Soc. Gen. Microbiol. Symp.* **42**:65.

Khalil, M. A., and Rasmussen, R. A., 1984, Global sources, lifetimes and mass balances of carbonyl sulfide (COS) and carbon disulfide in the earth's atmosphere, *Atmos. Environ.* **18**:1805.

Kieber, D. J., and Mopper, K., 1983, Reversed phase high performance liquid chromatographic analysis of alpha-keto acid quinoxalinole derivatives: Optimization of technique and application to natural samples, *J. Chromatogr.* **281**:135.

Kiene, R. P., 1988, Dimethylsulfide metabolism in salt marsh sediments, *FEMS Microbiol. Ecol.* **53**:71.

Kiene, R. P., and Capone, D. G., 1988, Microbial transformations of methylated sulfur compounds in anoxic salt marsh sediments, *Microb. Ecol.* **15**:275.

Kiene, R. P., and Taylor, B. F., 1988a, Biotransformations of organosulfur compounds in sediments via 3-mercaptopropionate, *Nature (London)* **332**:148.

Kiene, R. P., and Taylor, B. F., 1988b, Demethylation of dimethylsulfoniopropionate and production of thiols in anoxic marine sediments, *Appl. Environ. Microbiol.* **54**:2208.

Kiene, R. P., and Visscher, P. T., 1987, Production and fate of methylated sulfur compounds from methionine and dimethylsulfoniopropionate in anoxic salt marsh sediments, *Appl. Environ. Microbiol.* **53**:2426.

Kiene, R. P., Oremland, R. S., Catena, A., Miller, L. G., and Capone, D. G., 1986, Metabolism of reduced methylated sulfur compounds in anaerobic sediments and by a pure culture of an estuarine methanogen, *Appl. Environ. Microbiol.* **52:**1037.

Kim, K.-H., and Andreae, M. O., 1987, Determination of carbon disulfide in natural waters by adsorbent preconcentration and gas chromatography with flame photometric detection, *Anal. Chem.* **59:**2670.

King, G. F., Richardson, D. J., Jackson, J. B., and Ferguson, S. J., 1987, Dimethylsulphoxide and trimethylamine-N-oxide as bacterial electron transport acceptors: Use of nuclear magnetic resonance to assay and characterise the reductase system in *Rhodobacter capsulatus, Arch. Microbiol.* **149:**47.

Kjaer, A., 1977, Low molecular weight sulphur-containing compounds in nature: A survey, *Pure Appl. Chem.* **49:**137.

Klockow, D., Bayer, W., and Faigle, W., 1978, Gas chromatographic determination of traces of low molecular weight carboxylic and sulfonic acids in aqueous solutions, *Fresenius Z. Anal. Chem.* **292:**385.

Koehler, M., Genz, I., Babenzien, H.-D., Eckardt, V., Hieke, W., 1978, Mikrobielle Abbau organischer Schwefelverbindungen, *Z. Allg. Mikrobiol.* **18:**67.

Koehler, M., Genz, I.-L., Schicht, B., and Eckart, V., 1984, Mikrobielle Entschwefelung von Erdoel und schweren Erdoelfraktionen, *Zbl. Bakterial.* **139:**239.

Koenig, W. A., Ludwig, K., Sievers, S., Rinken, M., Stotting, K. H., and Guenther, W., 1980, Identification of volatile organic sulfur compounds in municipal sewage systems by GC/MS, *J. High Res. Chromatogr. Chromatogr. Commun.* **3:**415.

Krauss, F., and Schmidt, A., 1987, Sulphur sources for growth of *Chlorella fusca* and their influence on key enzmymes of sulphur metabolism, *J. Gen. Microbiol.* **133:**1209.

Krouse, H. R., and McCready, R. G. L., 1979, Biogeochemical cycling of sulfur, in: *Biogeochemical Cycling of the Mineral-Forming Elements* (P. A. Trudinger, ed.), pp. 401–430, Elsevier, Amsterdam.

Laborde, A. L., and Gibson, D. T., 1977, Metabolism of dibenzothiophene by a *Beijerinckia* species, *Appl. Environ. Microbiol.* **34:**783.

Laing, W. A., and Christeller, J. T., 1980, A steady-state kinetic study on the catalytic mechanism of ribulose bisphosphate carboxylase from soybean, *Arch. Biochem. Biophys.* **202:**592.

Larher, F., Hamelin, J., and Stewart, G. R., 1977, L'acide dimethyl sulfonium-3-propanoique de *Spartina anglica, Phytochemistry* **18:**1396.

Lashen, E. S., and Starkey, R. L., 1970, Decomposition of thiourea by a *Penicillium* species and soil and sewage-sludge microflora, *J. Gen. Microbiol.* **64:**139.

Lay, M. D. S., Sauerhoff, M. W., and Saunders, D. R., 1986, Carbon Disulfide, in: *Ullmann's Encyclopaedia of Industrial Chemistry,* 5th ed., Vol. A 5, pp. 185–195, Verlag Chemie, Weinheim.

Leder, I. G., 1975, Thiamine, biosynthesis and function, in: *Metabolic Pathways,* Vol. 7, *The Metabolism of Sulfur Compounds* (D. M. Greenberg, ed.), pp. 57–73, Academic Press, New York.

Liss, P. S., and Slater, P. G., 1974, Flux of gases across the air-sea interface, *Nature (London)* **247:**181.

Ljunggren, G., and Norberg, B. O., 1943, On the effect of toxicity of dimethyl sulfide, dimethyl disulfide and methyl mercaptan, *Acta Physiol. Scand.* **5:**248.

Lorimer, G. H., and Pierce, J., 1990, Carbonyl sulfide: An alternative substrate for but not an activator of ribulose-1,5-bisphosphate carboxylase, *Biochemistry* [cited as in press by Ogawa and Togasaki (1988)].

Lovelock, J. E., 1974, CS_2 and the natural sulphur cycle, *Nature (London)* **248:**625.

Lovelock, J. E., Maggs, R. J., and Rasmussen, R. A., 1972, Atmospheric dimethyl sulphide and the natural sulphur cycle, *Nature (London)* **237:**452.

Manolis, A., 1983, The diagnostic potential of breath analysis, *Clin. Chem.* **29:**5.

Maroulis, P. J., and Bandy, A. R., 1976, Estimate of the contribution of biologically produced dimethyl sulfide to the global sulfur cycle, *Science* **196**:247.

Maw, G. A., 1981, The biochemistry of sulphonium salts, in: *The Chemistry of the Sulphonium Compounds* C. J. M. Stirling, ed.), Part 2, pp. 703–771, Wiley, New York.

McEwan, A. G., Ferguson, S. J., and Jackson, J. B., 1983, Electron flow to dimethylsulphoxide or trimethylamine-N-oxide generates a membrane potential in *Rhodopseudomonas capsulata, Arch. Microbiol.* **136**:300.

Monticello, D. J., and Finnerty, W. R., 1985, Microbial desulfurization of fossil fuels, *Annu. Rev. Microbiol.* **39**:371.

Monticello, D. J., Bakker, D., and Finnerty, W. R., 1985, Plasmid-mediated degradation of dibenzothiophene by *Pseudomonas* species, *Appl. Environ. Microbiol.* **49**:756.

Mormile, M. R., and Atlas, R. M., 1988, Mineralization of the dibenzothiophene biodegradation products 3-hydroxy-2-formyl benzothiophene and dibenzothiophene sulfone, *Appl. Environ. Microbiol.* **54**:3183.

Moubasher, A. H., Elnaghy, M. A., and Abdel-Hafez, S. I., 1974, Effect of fumigation of three grains with formalin and carbon disulfide on the grain-borne fungi, *Bull. Fac. Sci. Assiut Univ.* **3**:13.

Munnecke, D. E., Domsch, J. H., and Eckert, J. W., 1962, Fungicidal activity of air passed through columns of soil treated with fungicides, *Phytopathology* **52**:1298.

Nguyen, B. C., Gaudry, A., Bonsang, B., and Lambert, G., 1978, Reevaluation of the role of dimethyl sulphide in the sulphur budget, *Nature (London)* **275**:637.

Nriagu, J. O., Holdway, D. A., and Coker, R. D., 1987, Biogenic sulfur and the acidity of rainfall in remote areas of Canada, *Science* **237**:1189.

Ogawa, T., and Togasaki, R. K., 1988, Carbonyl sulfide: An inhibitor of inorganic carbon transport in cyanobacteria, *Plant Physiol.* **88**:800.

Oremland, R. S., and Zehr, J. P., 1986, Formation of methane and carbon dioxide from dimethylselenide in anoxic sediments and by a methanogenic bacterium, *Appl. Environ. Microbiol.* **52**:1031.

Panter, R., and Penzhorn, R. D., 1980, Alkyl sulfonic acids in the atmosphere, *Atmos. Environ.* **14**:149.

Pearce, F., 1988, Phytoplankton shares the blame for sulphur pollution, *New Sci.* 11 Feb. 1988, p. 25.

Plennart, W., and Heine, W., 1973, *Normalwerte*, 4. Aufl., VEB Verlag, Berlin.

Postgate, J. R., and Kelly, D. P., 1982, *Sulphur Bacteria,* The Royal Society, London.

Rajagopal, B. S., and Daniels, L., 1986, Investigation of mercaptans, organic sulfides, and inorganic sulfur compounds as sulfur sources for the growth of methanogenic bacteria, *Curr. Microbiol.* **14**:137.

Rasmussen, R. A., 1974, Emission of biogenic hydrogen sulfide, *Tellus* **26**:254.

Rasmussen, R. A., Khalil, M. A. K., and Hoyt, S. D., 1982a, The oceanic source of carbonyl sulfide (OCS), *Atmos. Environ.* **16**:1591.

Rasmussen, R. A., Khalil, M. A. K., Dalluge, R. W., Penkett, S. A., and Jones, B., 1982b, Carbonyl sulfide and carbon disulfide from the eruptions of Mount St. Helens, *Science* **215**:665.

Reed, R. H., 1983, Measurement and osmotic significance of β-dimethylsulfoniopropionate in marine macroalgae, *Mar. Biol. Lett.* **34**:173.

Robinson, E., and Robbins, R. C., 1970, Gaseous sulfur pollutants from urban and natural sources, *J. Air Pollut. Control Assoc.* **20**:233.

Rodgers, G. A., Ashworth, J., and Walker, N., 1980, Recovery of nitrifier populations from inhibition by nitrapyrin or CS_2, *Zbl. Bakteriol. II Abt.* **135**:477.

Rohde, H., and Isaksen, I., 1980, Global distribution of sulfur compounds in the trophosphere estimated in a height/latitude transport model, *J. Geophys. Res.* **85**:7401.

Rosenberg, M., 1989, Microbial films in the mouth: Some ecologically relevant observations, in: *Microbial Mats: Ecological Physiology of Benthic Microbial Communities* (Y. Cohen and E. Rosenberg, eds.), pp. 245–250, American Society for Microbiology, Washington, D.C.

Rosenberg, M., and Gabbay, J., 1987, Halitosis—a call for affirmative action, *Dent. Med.* **5**:13.

Saigne, C., and Legrand, M., 1987, Measurements of methanesulphonic acid in Antarctic ice, *Nature (London)* **330:**240.

Salsbury, R. L., and Merricks, D. L., 1975, Production of methane thiol and dimethyl sulfide by rumen microorganisms, *Plant Soil* **43:**191.

Sandalls, F. J., and Penkett, S. A., 1977, Measurements of carbonyl sulfide and carbon disulfide in the atmosphere, *Atmos. Environ.* **11:**197.

Schmidt, N. F., Missan, S. R., Tarbet, W. J., and Cooper, A. D., 1978, The correlation between organoleptic mouth odor ratings and levels of volatile sulfur compounds, *Oral Surg. Oral Med. Oral Pathol.* **45:**560.

Schreiner, R. P., Stevens, S. E., and Tien, M., 1988, Oxidation of thianthrene by the ligninase of *Phanerochaete chrysosporium*, *Appl. Environ. Microbiol.* **54:**1858.

Schultz, T. H., McKenna Kruse, S. M., and Flath, R. A., 1985, Some volatile constituents of female dog urine, *J. Chem. Ecol.* **11:**169.

Scranton, M. I., and Brewer, P. G., 1977, Occurrence of methane in the near-surface waters of the western subtropical North Atlantic, *Deep Sea Res.* **24:**127.

Selyuzhiyskii, G. B., 1972, Experimental data used to determine the maximum permissible concentration of methyl mercaptan, dimethyl sulphide and dimethyl disulphide in the air of the production area of paper and pulp plants, *Gig. Tr. Prof. Zabol.* **16:**46.

Shibuya, I., Yagi, T., and Benson, A. A., 1963, Sulfonic acids in algae, in: *Microalgae and Photosynthetic Bacteria*, pp. 627–636, University of Tokyo Press, Tokyo.

Singer, A. G., Agosta, W. C., O'Connell, R. J., Pfaffmann, C., Bowen, D. V., and Field, F. H., 1976, Dimethyl disulphide; an attractant pheromone in hamster vaginal secretion, *Science* **191:**948.

Sivelä, S., 1980, Dimethyl sulphide as a growth substrate for an obligately chemolithotrophic *Thiobacillus*, in: *Commentationes Physico-Mathematicae, Dissert. No. 1* (L. Simons, ed.), pp. 1–69, Societas Scientareum Fennica, Helsinki.

Sivelä, S., and Sundman, V., 1975, Demonstration of *Thiobacillus*-type bacteria, which utilize methyl sulphides, *Arch. Microbiol.* **103:**303.

Smith, N. A., and Kelly, D. P., 1988a, Isolation and physiological characterization of autotrophic sulphur bacteria oxidizing dimethyl disulphide as sole source of energy, *J. Gen. Microbiol.* **134:**1407.

Smith, N. A., and Kelly, D. P., 1988b, Mechanism of oxidation of dimethyl disulphide by *Thiobacillus thioparus* strain E6, *J. Gen. Microbiol.* **134:**3031.

Smith, N. A., and Kelly, D. P., 1988c, Oxidation of carbon disulphide as the sole source of energy for the autotrophic growth of *Thiobacillus thioparus* strain TK-m, *J. Gen. Microbiol.* **134:**3041.

Soeder, C. J., Hegewald, E., and Kneifel, H., 1987, Green algae can use naphthalenesulfonic acids as sources of sulfur, *Arch. Microbiol.* **148:**260.

Sørensen, N. A., 1961, Structural patterns of polyacetylenic compounds from the plant family Compositae, *Pure Appl. Chem.* **2:**569.

Sparnins, V. L., Baraby, G. and Wattenberg, L. W., 1988, Effect of organosulfur compounds from garlic and onions on benzo[a]pyrene-induced neoplasia and glutathione S-transferase activity in the mouse, *Carcinogenesis* **9:**131.

Stapley, E. O., and Starkey, R. L., 1970, Decomposition of cysteic acid and taurine by soil microorganisms, *J. Gen. Microbiol.* **64:**77.

Steudler, P. A., and Peterson, B. I., 1984, Contribution of gaseous sulphur from salt marshes to the global sulphur cycle, *Nature (London)* **311:**455.

Stoiber, R. E., Leggett, D. C., Jenkins, T. F., Murrmann, R. P., and Rose, W. I., 1971, Organic compounds in volcanic gas from Santiaguito volcano, Guatemala, *Bull. Geol. Soc. Am.* **82:**2299.

Stoll, M., Winter, M., Gaukschi, F., Flament, I., and Willhalm, B., 1967, Recherches sur les aromes. Sur l'arome de café I, *Helv. Chim. Acta* **50:**628.

Stotzky, G., and Schenk, S., 1976, Volatile organic compounds and microorganisms, *Crit. Rev. Microbiol.* **4**:353.

Suylen, G. M. H., 1988, Microbial metabolism of dimethyl sulphide and related compounds, Proefschrift, Technical University of Delft, Delft, The Netherlands.

Suylen, G. M. H., and Kuenen, J. G., 1986, Chemostat enrichment and isolation of *Hyphomicrobium* EG, *Antonie von Leeuwenhoek J. Microbiol. Serol.* **52**:281.

Suylen, G. M. H., Stefess, G. C., and Kuenen, J. G., 1986, Chemolithotrophic potential of a *Hyphomicrobium* species, capable of growth on methylated sulphur compounds, *Arch. Microbiol.* **146**;192.

Suylen, G. M. H., Large, P. J., van Dijken, J. P., and Kuenen, J. G., 1987, Methyl mercaptan oxidase, a key enzyme in the metabolism of methylated sulphur compounds by *Hyphomicrobium* EG, *J. Gen. Microbiol.* **133**:2989.

Sweetnam, P. M., Taylor, S. W., and Elwood, P. C., 1987, Exposure to carbon disulphide and ischaemic heart disease in a viscose rayon factory, *Br. J. Ind. Med.* **44**:220.

Tazuya, K., Yamada, K., Nakamura, K., and Kumaoka, H., 1987, The origin of the sulfur atom of thiamin, *Biochim. Biophys. Acta* **924**:210.

Thomel, F., 1987, Synthesen mit Schwefelkohlenstoff, *Chem. Z.* **111**:285.

Thompson, C. J., Coleman, H. J., Hopkins, R. L., and Rall, H. T., 1965, Hydrocarbon analysis, p. 329, ASTM STP 389, *American Society for Testing and Materials.*

Timmerman, R. W., 1978, Carbon disulfide, in: *Encyclopaedia of Chemical Technology* (R. E. Kirk and D. F. Othmer, eds.), pp. 742–757, Wiley, New York.

Tocher, C. S., and Ackman, R. G., 1966, The identification of dimethyl-β-propiothetin in the algae *Syracosphaera carterae* and *Ulva carterae* in relation to sulfur source and salinity variations, *Limnol. Oceanogr.* **30**:59.

Tocher, C. S., Ackman, R. G., and McLachlan, J., 1966, The identification of dimethyl-β-propiothetin in the algae *Syracosphaera carterae* and *Ulva lactuca, Can. J. Biochem.* **44**:519.

Tomita, B., Inoue, H., Chaya, K., Nakamura, A., Hamamura, N., Ueno, K., Watanabe, K., and Ose, Y., 1987, Identification of dimethyl disulfide-forming bacteria isolated from activated sludge, *Appl. Environ. Microbiol.* **53**:1541.

Tonzetich, J., 1977, Production and origin of oral malodor: A review of mechanisms and methods of analysis, *J. Periodontol.* **48**:13.

Tonzetich, J., 1978, Oral malodor: An indicator of health status, *Int. Dent. J.* **28**:309.

Toon, O. B., Kasting, J. F., Turco, R. P., and Liu, M. S., 1987, The sulfur cycle in the marine atmosphere, *J. Geophys. Res.* **D92**:943.

Tucker, B. J., Maroulis, P. J., and Bandy, A. R., 1985, Free trophospheric measurements of carbon disulfide over a 45°N to 45°S latitude range, *Geophys. Res. Lett.* **12**:9.

Turco, R. P., Whitten, R. C., Toon, O. B., Pollack, J. B., and Hamill, P., 1980, Stratospheric aerosols and climate, *Nature (London)* **283**:283.

Turner, S. M., and Liss, P. S., 1987, Dimethyl sulphide and dimethyl sulphoniopropionate studies in European coastal waters, American Chemical Society, Division of Environmental Chemistry, 194th National Meeting (New Orleans), Vol. 27, no. 2, pp. 1–4.

Vairavamurthy, A., and Mopper, K., 1987, Geochemical formation of organosulphur compounds (thiols) by addition of H_2S to sedimentary organic matter, *Nature (London)* **329**:623.

Vairavamurthy, A., Andreae, M. O., Iversen, R. L., 1985, Biosynthesis of dimethyl sulfide and dimethyl propiothetin by *Hymenomonas carterae* in relation to sulfur source and salinity variations, *Limnol. Oceanog.* **30**:59.

Volkov, I. I., and Rozanov, A. G., 1983, The sulphur cycle in oceans. I. Reservoirs and fluxes, in: *The Global Biogeochemical Sulphur Cycle* (M. V. Ivanov and J. R. Freney, eds.), pp. 353–448, Wiley, Chichester.

von Damm, K. L. 1983, Chemistry of Submarine Hydrothermal Solutions at 21° North, East Pacific Rise and Guayamas Basin, Gulf of California, Ph.D. thesis, Massachusetts Institute of Technology, Cambridge, Mass.

Wagner, C., and Stadtman, E. R. 1962, Bacterial fermentation of dimethyl-β-propiothetin, *Arch. Biochem. Biophys.* **98**:331.

Wakeham, S. G., Howes, B. L., Dacey, J. W. H., Schwarzenbach, R. P., and Zeyer, J., 1984, Biogeochemistry of dimethylsulfide in a seasonally stratified coastal salt pond, *Geochim. Cosmochim, Acta* **51**:1675.

Weisiger, R. A., Pinkus, L. M., and Jakoby, W. B., 1980, Thiol S-methyltransferase: Suggested role in detoxication of intestinal hydrogen sulfide, *Biochem. Pharmacol.* **29**:2885.

Wells, J., and Koves, E., 1974, Detection of carbon disulphide (a disulfiram metabolite) in expired air by gas chromatography, *J. Chromatogr.* **92**:442.

Wheeler, J. W., von Endt, D. W., and Wenmer, C., 1975, 5-thiomethylpentane-2,3-dione. A unique natural compound from the striped hyaena, *J. Amer. Chem. Soc.* **97**:441.

White, G. F., Dodgson, K. S., Davies, I., Matts, P. J., Shapleigh, J. P., and Payne, W. J., 1987, Bacterial utilisation of short-chain primary alkyl sulphate esters, *FEMS Microbiol. Lett.* **4**:173.

White, R. H., 1982, Analysis of dimethyl sulfonium compounds in marine algae, *J. Marine Res.* **40**:529.

Whitfield, F. B., Shea, S. R., Gillen, K. J., and Shaw, K. J., 1981, Volatile components from the roots of *Acacia pulchella* R. Br. and their effect on *Phytophthora cinnamomi* Rands. *Aust J. Bot.* **29**:195.

Windholz, M. (ed.), *The Merck Index*, 10th ed., Merck & Co., Inc., Rahway, N.J.

Winer, A. M., Atkinson, R., and Pitts, J. N., 1984, Gaseous nitrate radical: Possible nighttime atmospheric sink for biogenic organic compounds, *Science* **224**:156.

Winfrey, M. R., Marty, D. G., Bianchi, A. J. M., and Ward, D. M., 1981, Vertical distribution of sulphate reduction, methane production and bacteria in marine sediments, *Geomicrobiol. J.* **2**:341.

Yen, H. C., and Marrs, B., 1977, Growth of *Rhodopseudomonas capsulatus* under anaerobic dark conditions with dimethyl sulfoxide, *Arch. Biochem. Biophys.* **181**:411.

Young, L., and Maw, G. A., 1974, *The Metabolism of Sulphur Compounds*, Methuen, London.

Zeyer, J., Eicher, P., Wakeham, S. G., and Schwarzenbach, R. P., 1987, Oxidation of dimethyl sulfide to dimethyl sulfoxide by phototrophic purple bacteria, *Appl. Environ. Microbiol.* **53**:2026.

Zinder, S. H., and Brock, T. D., 1978a, Dimethyl sulfoxide as an electron acceptor for anaerobic growth, *Arch. Microbiol.* **116**:35.

Zinder, S. H., and Brock, T. D., 1978b, Dimethyl sulfoxide reduction by microorganisms, *J. Gen. Microbiol.* **105**:335.

Zinder, S. H., Doemel, W. N., and Brock, T. D., 1977, Production of volatile sulfur compounds during the decomposition of algal mats, *Appl. Environ. Microbiol.* **34**:859.

Zuerrer, D., Cook, A. M., and Leisinger, T., 1987, Microbial desulfonation of substituted naphthalenesulfonic acids and benzenesulfonic acids, *Appl. Environ. Microbiol.* **53**:1459.

10

The Ecology of Chitin Degradation

GRAHAM W. GOODAY

1. Chitin and Its Occurrence

1.1. Chitin Structure

Chitin is the (1→4)-β-linked homopolymer of *N*-acetyl-D-glucosamine (Fig. 1). The individual polymer chains can be thought of as helices, as each sugar unit is inverted with respect to its neighbors. This leads to their stabilization as rigid ribbons by 03—H . . . 05 and 06—H . . . 07 hydrogen bonds. The commonest form of chitin is α-chitin. Its unit cell is of two *N*,*N*′-diacetylchitobiose units of two chains in an antiparallel arrangement. Thus, adjacent polymer chains run in opposite directions, held together by 06—H . . . 06 hydrogen bonds, and the chains are held in sheets by 07 . . . H—N hydrogen bonds (Minke and Blackwell, 1978). This gives a statistical mixture of CH_2OH orientations, equivalent to half oxygens on each residue, each forming inter- and intramolecular hydrogen bonds. This results in two types of amide groups; all are involved in the interchain C=O . . . H—N bonds, while half of the groups also serve as acceptors for 06—H . . . O=C intramolecular bonds. This extensive intermolecular hydrogen bonding leads to a very stable structure, the individual polymer chains eventually giving rise to microfibrils if allowed to crystallize (Gooday, 1983). A less common form of chitin is β-chitin, in which the unit cell is of one *N*,*N*′-diacetylchitobiose unit, giving a polymer stabilized as a rigid ribbon, as for α-chitin, by 03—H . . . 05 intramolecular bonds (Gardner and Blackwell, 1975). Chains are then held together in sheets by C=O . . . H—N hydrogen bonding of the amide groups and by the CH_2OH side chains, forming intersheet hydrogen bonds to the carbonyl oxygens on the next chains (06—H . . . 07). This gives a structure of parallel poly-*N*-acetylglucosamine chains with no intersheet hydrogen bonds. The parallel arrangement of poly-

GRAHAM W. GOODAY • Department of Genetics and Microbiology, Marischal College, University of Aberdeen, Aberdeen AB9 1AS, Scotland.

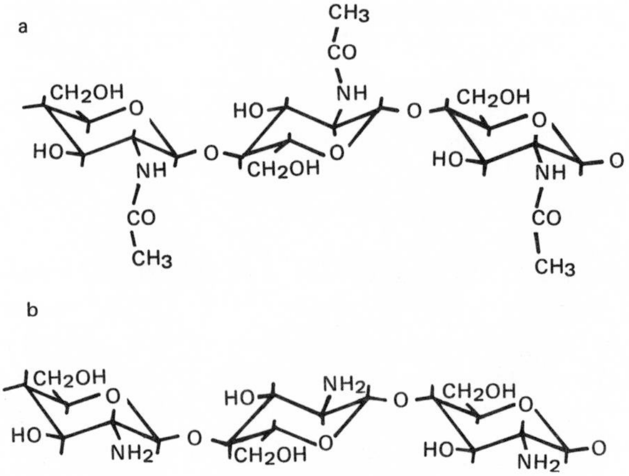

Figure 1. Structures of chitin (a) and chitosan (b).

mer chains in β-chitin allows for more flexibility than the antiparallel arrangement of α-chitin, but the resultant polymer still has immense strength (Lindsay and Gooday, 1985a). A third form is γ-chitin, with mixed parallel and antiparallel orientations (Rudall and Kenchington, 1973).

With only one exception, the β-chitin of diatoms, chitin is always found cross-linked to other structural components. In fungal walls, it is found covalently bonded to glucans, either directly, as in *Candida albicans* (Surarit *et al.*, 1988), or via peptide bridges (Sietsma *et al.*, 1986). In insects and other invertebrates, the chitin is always associated with specific proteins, with both covalent and noncovalent bonding, to produce the observed ordered structures (Rudall and Kenchington, 1973; Giraud-Gaille and Bouligand, 1986; Kramer and Koga, 1986; Blackwell and Weih, 1984). There are often also varying degrees of mineralization, in particular calcification, and sclerotization, involving interactions with phenolic and lipid molecules (Poulicek *et al.*, 1986a; Peter *et al.*, 1986). In both fungi and invertebrates, there are varying degrees of deacetylation, giving a continuum of structure between chitin (fully acetylated) and chitosan (fully deacetylated) (Fig. 1) (Datema *et al.*, 1977; Davis and Bartnicki-Garcia, 1984; Aruchami *et al.*, 1986; Gowri *et al.*, 1986).

1.2. Occurrence of Chitin

Chitin is utilized as a structural material by most species alive today. Its distribution phylogenetically is clearly defined. Save for one possible exception as a component of *Streptomyces* spore walls (Smucker and Pfister, 1978; Smucker, 1984), chitin is unreported from procaryotes.

Chitin occurs sporadically among the Protista: in cyst walls of some ciliates, flagellates, and amebae (Bussers and Jeuniaux, 1974; Ward *et al.*, 1985; Arroyo-

Begovich and Carabez-Trejo, 1982); in the lorica walls of some ciliates and chrysophyte algae (Herth *et al.*, 1986); in the flotation spines of some centrix diatoms (Herth *et al.*, 1986; Lindsay and Gooday, 1985a; Smucker and Dawson, 1986); and in the walls of some filamentous chlorophyte algae and oomycete fungi (Pearlmutter and Lembi, 1978; Aronson and Lin, 1978; Campos-Takai *et al.*, 1982).

Save for a very few exceptions, chitin is ubiquitous in the Fungi: in the hyphochytridiomycetes, chytridiomycetes, zygomycetes, ascomycetes, basidiomycetes, and deuteromycetes (Bartnicki-Garcia and Lippman, 1982). In the fungal walls, it characteristically occurs as α-chitin microfibrils in the innermost layer (Gooday and Trinci, 1980). The microfibrils of particular walls have characteristic structures and arrangements. Thus, in dimorphic fungi such as *Candida albicans*, in the yeast and mycelial walls they occur as short stubby fibrils arranged randomly (Gow and Gooday, 1983); in vegetative hyphal walls of the filamentous fungus *Neurospora crassa*, they are long and arranged randomly except at apices of narrow elongating germ hyphae, where they tend to be arranged axially (Burnett, 1979); and in *Coprinus cinereus*, they are long and randomly oriented in vegetative hyphal walls but long and wound in shallow helices around the cells of the stipe of the mushroom (Gooday, 1983). Chitins purified from these different sources, however, have indistinguishable chemical properties (Gow *et al.*, 1987). The chitin in the fungal wall also varies in its crystallinity (i.e., degree of internal hydrogen bonding), its degree of covalent bonding to other wall components such as glucans, and, especially in the zygomycetes, in its degree of deacetylation. Thus, Vermeulen and Wessels (1984) describe the newly synthesized chitin in hyphal apices of *Schizophyllum commune* as being very susceptible to degradation by chitinase and solubilization by dilute mineral acid. They ascribe this to its nascent state, before extensive hydrogen bonding and covalent bonding have occurred.

Chitin has a very wide distribution as a tough skeletal material among invertebrates (Jeuniaux, 1963, 1982). It is found as α-chitin in the calyces of hydrozoa, the egg shells of nematodes and rotifers, the radulae of molluscs, and the cuticles of arthropods and as β-chitin in the shells of brachiopods and molluscs, in cuttlefish "bone" and squid "pen," in the peritrophic membranes of annelids and some arthropods, and in tubes of pogonophora. Chitin is found in some insect cocoons and peritrophic membranes and in stomach cuticle linings of squid (Rudall and Kenchington, 1973). The importance of physical form to biological function is indicated by squid, *Loligo,* having α-chitin in its beak, β-chitin in its pen, and γ-chitin in its stomach lining, suggesting toughness, rigidity, and flexibility as the major properties in each case.

1.3. Fossil Chitin

It will be stressed several times in this review that native chitinous materials are relatively resistant to degradation in nature. Coupled with the widespread occurrence of such materials in the biosphere, this fact leads to the idea that chitin should survive in fossils (e.g., Wyckoff, 1972). Although there are reports of fossil chitin, e.g., in pogonophora (Carlisle, 1964) and insect wings from amber (Abderhalden and Heyns, 1933), chitin seems to survive fossilization much less well than do the proteins associated with it (Florkin, 1965). Fossils of crustacea were found to contain only traces of chitin on chemical analysis and no chitin-like microfibrils on electron microscopy (Brumioul and Voss-Foucart, 1977). Direct experimentation by Voss-Foucart et al. (1974) showed that the chitin of Nautilus shell survived "simulated fossilization" by heating at 160°C at 442 atm for 53 days but could not be detected with certainty from fossil Jurassic nautiloids, although proteinaceous residues were detected readily. In view of the chitinase activities detected in soils and sediments referred to later, the apparent widespread loss of chitin during fossilization could be the result of long-term enzymatic lysis as well as possible chemical diagenesis.

1.4. Annual Production of Chitin

The annual production of chitin is enormous, but just how enormous is difficult to say. Most estimates have been based on the commercial fisheries of Antarctic krill, especially *Euphausia superba*. Everson (1977) estimates annual production of this one species at 940–1350 million tons year^{-1}. Using conversion figures given by Allan et al. (1978) gives a production of 5.8–8.4 million tons of chitin year^{-1}. These values are equivalent to 355–514 mg chitin m^{-2} year^{-1}. In a separate study, Ikeda and Dixon (1982) estimate moults of E. *superba* to contribute 450 mg (dry weight) cm^{-2} year^{-1} to the oceanic detritus, equivalent to about 100 mg chitin cm^{-2} year^{-1}. Jeuniaux et al. (1989) present calculated values for a range of aquatic systems, using their own results and those from other authors, expressed as milligrams of chitin per square meter per year. For seawater, there are values of 4.5 for krill in the Atlantic Ocean and North Sea and 1001 for zooplankton in Calvi Bay, Mediterranean Sea. For marine benthic communities, there are values of 230–1000 in Calvi Bay and 1500 for lobsters on the Natal Coast, South Africa. For freshwaters, values range from 130–3200 for zooplankton in eutrophic lakes in Holland and Denmark to 21,000 for bryozoans in a Belgian Pond to 51,000 for arthropods in a Japanese water chestnut lake. The input of chitin into the estuarine sediment of the River Ythan by one crustacean alone, *Corophium volutator,* has been estimated as 3700 mg m^{-2} year^{-1} (K. Hillman, G. W. Gooday, and J. I. Prosser, unpublished observations).

The enormous production of chitin revealed in these estimates has lead to detailed considerations of its commercial exploitation. Evaluations of invertebrate and microbial sources are presented by Allan et al. (1978), Berkeley (1979), and Muzzarelli (1985). Currently only a tiny fraction of this potential is utilized (chiefly as chitosan, made by chemically deacetylating chitin), mainly for water purification. If chitin and chitosan do become more widely used by industry, then worries about their fate in the environment will open up a new area of chitin ecology.

1.5. Amount of Chitin in the Biosphere

Consideration of this enormous production leads to the question: how much is recycled and how much is accumulating? Unfortunately at present there is insufficient information to answer this question, although since there are no obvious massive accumulations of chitin in nature and it does not appear much in fossils (Section 1.3), it seems likely that most chitin is relatively rapidly recycled in most environments.

There are few estimates of chitin contents in nature. Some obtained from particular localized habitats are referred to later where appropriate. In the rocky benthos in Calvi Bay, Jeuniaux *et al.* (1986) report a total biomass benthic cover in seaweed communities (excluding decapod crustaceans) of 0.71 g m^{-2}, with decapod contribution varying from 0 to 1.8 g m^{-2}, and in sciaphilous communities in semidark caves of 0.27 g m^{-2}. Poulicek and Jeuniaux (1989) report analyses of 100 samples of marine sediments from a range of origins. Their chitin contents varied from 2 to 2800 μg g^{-1} decalcified sediment. The richest sediments were bryozoan and shelly sands and gravels. Cross (1985) reports mean annual contents of particulate chitin as 30.8, 44.1, and 46.5 μg liter^{-1} in river water, estuarine water (River Don), and inshore seawater (Hareness Inlet, Aberdeen). Cross points out that this seawater figure, if extended over the volume of the oceans, yields a total of 6.65 × 10^{13} metric tons. Clearly, it is not valid to extend this neritic value to the oceans. But even if the value for the oceans is one or two orders of magnitude lower, when considered together with amounts of larger marine chitin particles and chitinous animals and of freshwater and terrestrial fungi and invertebrates, this estimate confirms that the amount of chitin in the world is immense. A working estimate, meantime, for both annual production and steady-state amount of chitin is of the order of 10^{10}–10^{11} tons.

2. Pathways of Chitin Degradation

With such large quantities of chitin being produced annually, recycling is important to prevent a sink in the global carbon and nitrogen. Mineralization of chitin is certainly primarily a microbial process, and chitin can provide sole carbon and nitrogen sources for many microbes. Possible pathways of degradation are discussed by Davis and Eveleigh (1984).

When the pathway for the degradation is not known, the process is best termed chitinoclastic (chitin breaking). When the pathway involves the initial hydrolysis of the (1→4)-β-glycosidic bond, the process is termed chitinolytic. Hydrolysis of this bond is accomplished by chitinase (Fig. 2). *exo*-Chitinase cleaves diacetylchitobiose units from the nonreducing end of the chitin chain. *endo*-Chitinase cleaves glycosidic linkages randomly along the chitin chain, eventually giving diacetylchitobiose as the major product, together with some triacetylchitotriose, which may be slowly degraded to disaccharide and monosaccharide. There may not always be a clear distinction between these two activities (see Davis and Eveleigh, 1984), and the site of hydrolysis depends very much on the nature of the substrate. For example, the pure crystalline β-chitin of diatom spines is degraded only from the end of the spine by *Streptomyces* chitinase, with the release of only diacetylchitobiose, whereas colloidal chitin is degraded to oligomers and disaccharide (Lindsay and Gooday, 1985a; Smucker and Dawson, 1986). Lysozyme has a low activity toward chitin. Diacetylchitobiose (often called chitobiose, but beware confusion with the product of chitosanase) is hydrolyzed to *N*-acetylglucosamine by

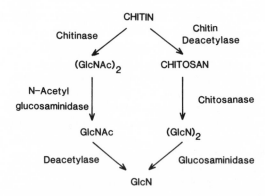

Figure 2. Probable major pathways of chitin degradation.

β-*N*-acetylglucosaminidase. Diacetylchitobiose and *N*-acetylglucosamine may be transported into cells. Further metabolism of *N*-acetylglucosamine may be via phosphorylation to *N*-acetylglucosamine-6-phosphate, or via deacetylation to glucosamine, and then deamination.

An alternative pathway involves the deacetylation of chitin to chitosan. The importance of this alternative pathway in some environments is shown by the finding of considerable amounts of chitosan in estuarine sediments (Hillman *et al.*, 1989). Chitosanase can then hydrolyze (1→4)-β-glycosidic bonds to give chitobiose (glucosaminyl-glucosaminide) (Fig. 2; Davis and Eveleigh, 1984). This, in turn, can be hydrolyzed to glucosamine by glucosaminidase. Additional evidence for this alternative pathway also comes from the detection of chitin deacetylase activity in estuarine sediments (Hillman *et al.*, unpublished observations), but the major sources of this activity are yet to be identified.

Pure cultures of bacteria with chitin as the nutrient source indicate a variety of patterns of chitin catabolism. *Clostridium* strain 9.1 showed a transient appearance of diacetylchitobiose and an accumulation of *N*-acetylglucosamine during chitin fermentation. Pel and Gottschal (1986a) suggest that the nutrient substrate for the bacterium is the disaccharide, which results from chitinase action and is transported into the cell, while the monosaccharide cannot be utilized and accumulates as a minor product of chitinolytic action on trisaccharide. *Chromobacterium violaceum* showed a transient appearance of *N*-acetylglucosamine and acetate and an accumulation of ammonia during aerobic growth on chitin (Streichsbier, 1983). Accumulation of ammonia in pure cultures of marine chitinoclastic bacteria, sometimes with no visible disintegration of the chitin, was observed by ZoBell and Rittenberg (1938). *Cytophaga johnsonae* strain C31 showed accumulations of glucosamine and *N*-acetylglucosamine, with peaks at 24 and 48 hr, respectively (Sundarraj and Bhat, 1972). Growth on *N*-acetylglucosamine gave rise to glucosamine and glucosaminic acid in the culture filtrates, suggesting a pathway of chitin to diacetylchitobiose to *N*-acetylglucosamine to glucosamine to glucosaminic acid to gluconic acid. Veldkamp (1955) also reported accumulations of *N*-acetyl-glucosamine and glucosamine in culture media of *"Pseudomonas chitinovorans"* (=*Aeromonas caviae* NCIMB 8500) and *Cytophaga johnsonae* (=*Lysobacter* sp. NCIMB 8501) that had been grown with chitin.

3. Chitin Digestion by Microbes

As is evident throughout this review, the use of chitin as a nutrient is quite wide-spread among microbes. Most notably among the procaryotes are the gliding bacteria, pseudomonads, vibrios, *Photobacterium,* enteric bacteria, actinomycetes, bacilli, and clostridia (Gooday, 1979). Surveys of species are provided by Berkeley (1978), Clarke and Tracey (1956), and Monreal and Reese (1969). Among the eucaryotic microbes digesting chitin, we find chitin utilization widespread among myxomycetes, chytrids, zygomycetes, deuteromycetes, ascomycetes, and basidiomycetes (Gooday, 1979; Monreal and Reese, 1969). One notable fungus is the freshwater chytrid *Karlingia asterocysta,* an obligate chitophile (Murray and Lovett, 1966). Tracey (1955) reports chitinolysis by soil amebae, and probably many other protozoa have this ability.

4. Chitin Degradation in the Sea

As we have seen, the sea produces vast amounts of chitin, chiefly as carapaces of zooplankton. Studies of the fate of this chitin have borne out the statement of ZoBell and Rittenberg (1938) that "bacteria are probably responsible for the disintegration of much chitin in the sea." Most of the chitin is produced near to the surface, and studies have shown that recycling of chitinous material raining to the bottom occurs both in the water column and in the sediments. The rate of degradation will be enhanced by phenomena of adherence of chitinolytic microflora (Section 8) and passage through animal guts (Section 9). The importance of these processes is highlighted by the repeated finding of chitinolytic bacteria associated with zooplankton and particulate matter (Lear, 1963; Seki, 1965a; Seki and Taga, 1963a, 1965a; Jones, 1958; Hood and Meyers, 1977a).

4.1. Degradation in the Water Column

Estimates of chitinoclastic bacteria in seawater vary considerably, reflecting in part the methodological problems inherent in such work such as the propensity of the bacteria to adhere to particulate matter. Seki (1966) and Seki and Taga (1963a) report between 10^2 and 10^4 heterotrophic bacteria per ml in surface waters of Aburatsubo Inlet, with higher numbers in summer than winter, of which less than 0.4% were chitinoclastic. In a study in Sagami Bay, Seki and Taga (1965b) found that chitinoclastic bacteria formed the major population, often comprising 100% of the heterotrophic bacteria isolated, but total counts were much lower, of the order of 1 cell per ml. Their numbers increased in summer by as much as 10-fold. There was a marked diurnal fluctuation in their depth profile, which Seki and Taga attributed to vertical diurnal migrations of zooplankton. Okutani (1975) and Okutani and Kitada (1971) reported that in seawater of Maizura Inlet, between 1 and 2% of bacteria, i.e., about 10^2 cells per ml, were chitin decomposers. Lear (1963) emphasized the variability and unpredictability of counts of marine pelagic bacteria. He presented depth profiles of bacteria in pelagic waters off southern California as counts of numbers of colonies on chitin agar, but only 0.3% of colonies were chitinolytic, i.e., showed clearing zones on the plates. In contrast, 19% of isolates from surface water at La Jolla were chitinolytic. Counts based on halos of clearing may, however, be misleading, as there may be clearing hidden below the colony, for example by *Cytophaga* strains that require contact with chitin for its degradation. Brisou *et al.* (1964) presented results from the Atlantic Ocean, with the percentages of chitinolytic bacteria being 14 and 9.6 in the surface waters off Africa and Europe and 2.3 in deep waters. Herwig *et al.* (1988) reported chitinolytic bacteria, predominantly psychrophilic *Vibrio* species, at low concentration (approximately 1 cell per ml) in water columns off the Antarctic Peninsula. Heterotrophic counts ranged from 10 to 630 cells per ml and were generally highest in surface waters, declining with depth. In surface seawater at Hareness Inlet, Aberdeen, Cross (1985) found 295 chitinolytic isolates per ml, comprising 22% of the total count of heterotrophs. Monthly monitoring over 18 months showed higher counts of both chitinolytic cells and total heterotrophs in summer and autumn.

Poulicek (1985) implicates chitin and its microbial degradation as important features of the biogeochemical cycling of heavy metals in the ocean by virtue of its cation-binding properties. He suggests that chitinous particles from the surface plankton, such as their moults and fecal pellets, will adsorb cations such as iron and manganese in the surface waters and then sink. The cations will be released in deeper waters as the chitin is degraded, and some will reach the sediment to be degraded there. The net effect is a downward movement of heavy-metal cations from surface to sediment.

4.2. Degradation in Sediments

Consistently higher counts of chitinolytic bacteria, expressed both as total counts and as percentage of total heterotrophs, have been reported from marine sediments than from seawater. ZoBell and Rittenberg (1938) reported 10^3 cells per g in surface sediments off California; Hock (1940) reported 125 cells per g at the top of mud sediments 878 m deep off Woods Hole, dropping to 5 cells per g at 5 and 11 cm below the surface. A littoral sand sample had 6×10^4 per g at the surface, dropping to 6×10^3 and 6×10^2 per g at 5 and 11 cm, respectively, below the surface. Herwig *et al.* (1988) found much higher levels of chitinolytic bacteria in their Antarctic sediment samples than in seawater, 10^4–10^5 per g as opposed to 1 cell per ml. Rittenberg *et al.* (1937) reported high recoveries of anaerobic chitinoclastic bacteria from marine sediments. Seki (1965b) isolated aerobic and anaerobic chitinoclasts from the marine mud of Aburatsubo Inlet. Neither count showed appreciable seasonal variation, and chitinoclastic bacteria were from 10^5 to 10^7 per g (wet weight) throughout the year, a much higher value relative to the water above. The low counts of anaerobic chitinoclasts, at 10^1 to 10^2 per g, led Seki

(1965b) to conclude that aerobic bacteria in the top centimeters were largely responsible for the mineralization of chitin in marine mud. Chitinolytic isolates of *Clostridium* species have been reported by Billy (1969) and Timmis *et al.* (1974). A group that must not be forgotten that may play a role in chitin mineralization in aerobic sediments are the Actinomycetes. Weyland (1981) described the occurrence of these in seven marine sediments. He used chitin agar as one of his enumeration media, so presumably at least some of his isolates were chitinolytic. He found predominantly streptomyces, nocardioforms, and micromonosporae in sediment at 0–200, 200–2000, and over 2000 m, respectively. Other marine chitinolytic bacteria have been described by Kihara and Morooka (1962) and Campbell and Williams (1951). Reichardt (1988) studied characteristics of chitinolytic bacteria isolated at 0°C from Antarctic shelf sediments and classified them as psychrotrophs and psychrophiles. The distribution of the bacteria was patchy, being especially high next to tubes and burrows inhabited by animals; sediment at Station I yielded 33 colony-forming units cm^{-3}, whereas the wall of a burrow gave a count of 8000. The proportion of psychrophiles was also higher from samples of sediments next to tubes and burrows, being 93% of isolates, compared with 35% of isolates from uncolonized sediment. Brisou *et al.* (1964) report that 23 and 12% of total bacteria isolated were chitinolytic in sediments from Port d'Alger, Mediterranean Sea, and the Atlantic Ocean off Portugal, respectively. Okutani (1975) and Okutani and Kitada (1971) reported that about 1% of isolates of bacteria from the surface of sediment in Maizura Inlet were chitinolytic, at a count of about 10^3 cells per g wet mud. Boland (1987) reports chitinolytic bacteria in marine sediments from the Atlantic and Mediterranean as being 0–15% of the total count. Poulicek and Jaspar-Versali (1981), Poulicek and Jeuniaux (1982), and Poulicek *et al.* (1986b) described the colonization and degradation of mollusc shells in sediment in the Bay of Calvi, Corsica, by a range of microbes. Poulicek and Jaspar-Versali (1984) followed the process in more detail, proposing a sequence of initial colonization by chitinolytic and proteolytic bacteria, allowing penetration of decalcifying cyanobacterial filaments into the shell, followed by the penetration of chitinolytic and proteolytic fungal hyphae. Poulicek and Jeuniaux (1989) reported a linear correlation between ln number of chitinolytic bacteria and chitin content of decalcified sediment for 31 diverse sediments. Liston *et al.* (1965) presented results of one experiment studying the degradation of chitin in a "model sea-bed" in the laboratory. The number of chitinoclastic bacteria peaked at about 10^6 ml^{-1} after 21 days but then rapidly fell to about 10^3 ml^{-1} by day 25. The authors attributed this fall to the action of protozoa and of bacteriophages.

4.3. Degradation in the Deep Sea

Most of the productivity in the deep sea is dependent on detritus raining down. Much of this, including dead centric diatoms and zooplankton, molted invertebrate exoskeletons, and fecal pellets, will contain chitin. Deming (1985, 1986) suggests that a considerable portion of this material is recycled while still suspended. She found that on enrichment with 3% chitin, bacterial activity was stimulated to a much greater extent in samples from suspended sediment traps than from sediment cores. Poulicek and

Jeuniaux (1982) and Poulicek *et al.* (1986b) present estimates of chitin contents of deep-sea sediments showing no marked accumulations, values ranging from 60–160 μg (g decalcified sediment)$^{-1}$ being similar to those of shallow-sea sediments. Thus, it is to be expected that degradation processes will be similar irrespective of depth, and this has been borne out by direct experimental results.

Seki and Taga (1963b) incubated pieces of chitin with chitinolytic bacteria (probably *Vibrio* species) for 4 days at 25°C at 1, 200, 400, and 600 atm. The amounts of chitin decomposed were 84, 35, 28, and 17 mg, respectively. Helmke and Weyland (1986) presented results of chitinase formation and activity from 14 strains of psychrophilic and psychrotrophic marine Antarctic bacteria under simulated deep-sea conditions. The bacteria were *Vibrio* species isolated from the upper layer of deep and shallow sediments and from animal guts. All isolates were able to produce chitinase at a hydrostatic pressure of 400 bars at 3°C. The formation of chitinase was diminished in cultures at 400 bars compared with cultures at 1 bar, but this inhibition was much less for the deep-sea psychrophilic isolates than for shallow-sea isolates. In all cases, whether from isolates from deep or shallow water, the chitinases were barotolerant, showing unchanged or enhanced activity on increasing pressure from 1 to 1000 bars at pH 6.8. At pH 5.0, however the activities of some chitinases were inhibited with increasing pressure, but this response was not correlated with depth of isolation of the bacteria. Kim and ZoBell (1972) reported that the activity of chitinase from a barophobic actinomycete was not affected appreciably by pressure up to 1000 bars. Similarly, Koch and Disteche (1986) reported a small increase, by a factor of about 1.4 in values of K_m and V_{max}, in activity of chitinase at 22°C from the mesophilic bacterium *Serratia marcescens* on increasing the hydrostatic pressure from 1 to 1000 bars. Helmke and Weyland (1986) conclude that indigenous bacteria are capable of decomposing chitin particles in the depth of the Antarctic Ocean, as are chitinases produced in surface waters and transported down by sinking particles.

Wirsen and Jannasch (1976) deposited packages of chitinous material (cleaned sections of lobster tails and crab claws) at a depth of 5300 m, temperature of 2°C, for 7 months and at a depth of 38 m, temperature of 1–12°C, for 5 months. On retrieval, they found almost complete degradation of both deep-water and shallow-water samples. The samples were enclosed in bottles with perforations of 13-mm diameter, so the contribution of animal grazing to the degradation is uncertain.

Poulicek (1983) and Poulicek *et al.* (1986c) described weathering of calcareous mollusc shells in deep-sea sediments and found processes similar to those described for shells in shallow-sea sediments by Poulicek and Jeuniaux (1982), Poulicek (1982), and Poulicek *et al.* (1986b). Scanning electron micrographs showed fungi and bacteria, including actinomycetes, boring into the shells, a negative linear correlation between remaining percentage of organic matter and density of microborers (up to 300,000 borings cm^{-2}), and a positive linear correlation between relative surface-to-volume ratio and number of microborings (i.e., the surface area accessible to dissolution was increased by the microborers). Poulicek *et al.* (1986c) also observed accretions of metal crystals on scanning electron micrographs of deep-sea mollusc shells. They suggested that these may have been nucleating around chitinous material, since chitin and chitosan show strong binding of heavy-metal cations (Muzzarelli, 1977). They further suggested that such aggregates and their associated microbial flora may play a role in the genesis of the deep-sea manganese nodules.

As with other marine habitats, the guts of deep-sea animals contain chitinolytic bacteria (Schwarz *et al.*, 1976) of genera that are characteristically chitinolytic, such as *Photobacterium* and *Vibrio* (Ohwada *et al.*, 1980), and so it is to be expected that comminution and exposure to the chitinases of the animals themselves and of their gut microflora will be major factors in the recycling of chitin the in the deep sea.

Morita (1979) described the isolation of chitin-digesting bacteria from necrotic lesions from carapaces of the deep-sea crab *Chionoecetes tanneria*, which lives at depths of 500 m to greater than 4000 m. The majority of female crabs had more than 100 necrotic lesions, from which strains of *Vibrio, Pseudomonas,* and *Moraxella* were isolated. Thus, we can expect the pathogenesis of invertebrates by chitin-degrading bacteria to be similar in the deep sea to that in shallow waters.

The remarkable ecosystems of the deep-sea thermal vents should be rich areas for the study of chitin degradation; as by analogy with their shallow-water relatives, the clam shells, crab carapaces, and, in particular, the pogonophoran tubes should be rich in chitin.

4.4. Rates of Degradation

Attempts have been made to quantify rates of chitin degradation in the sea. Cross (1985) suspended pieces of squid pen chitin–protein (40% chitin : 60% protein) in litterbags of 20-μm nylon mesh in the water column at Hareness Inlet, Aberdeen, for various periods over a year with a mean temperature of 7.2°C. The mean degradation rate at 7.4 mg g^{-1} day^{-1} was unexpectedly low in comparison, for example, with rates observed for nearby estuarine or riverine samples, and Cross suggested that this finding reflected the artificial nature of this experiment, with meio- and macrofauna being excluded. Jeuniaux (1981) and Poulicek and Jeuniaux (1982) monitored the rates of degradation of samples of mollusc shell on a shelly gravel sediment 37 m deep in the Bay of Calvi, in the Mediterranean Sea. Three types of shell material behaved differently. The chitin in mother-of-pearl from *Mytilus edulis* initially disappeared rapidly, 90% after the first 6 months, but then very slowly; that from the prismatic layer of *Pinna nobilis* disappeared much more slowly, only 60% after 2 years; and that from the crossed-lamellar layer of *Tridacna gigas* initially disappeared slowly, 10% after 6 months, then rapidly, 90% over 12 months, and then very slowly. These differing patterns are presumably related to differences in the molecular architectures of the shells. Seki and Taga (1963b) and Seki (1965a) estimated the times for mineralization of chitin in the sea by incubating chitin strips in the laboratory at various temperatures with cultures of bacteria (probably *Vibrio* species). They showed that rates were higher for smaller strips, i.e., for higher unit surface area. Choosing results using their smallest strips, less than 330 μm, they calculated a degradation rate of 27 mg day^{-1} g chitin $^{-1}$ at 25°C. Herwig *et al.* (1988) estimated chitin mineralization rates by incubating samples with ^{14}C-labeled chitin and measuring rates of release of radioactive carbon dioxide. Rates were extremely low in the seawater, 0.00085–0.0019% in 48 hr at 0°C, and higher in marine sediments, 0.0039–0.010%. These very low rates, bearing in mind the very high input of chitin from krill, led the authors to suggest that chitin may be accumulating in Antarctic marine sediments; but, as with the experiment of Cross (1985), these values do not take into account the probable major synergistic effect of meio- and macrofauna in enhancing chitin degradation rates.

5. Chitin Degradation in Estuaries

Estuaries are important sites for chitin recycling, receiving input from terrestrial, freshwater, and marine sources as well as from their own productivity. Chitin is therefore likely to be a major substrate for microbial productivity in aerobic and anaerobic environments.

5.1. Estuarine Chitinoclastic Microbes

Reichardt *et al.* (1983) isolated 103 strains of chitinoclastic bacteria from the estuarine upper Chesapeake Bay. Of these, 44 were yellow-orange pigmented *Cytophaga*-like bacteria, with a range of salt requirement. The largest phenon was of facultatively anaerobic flexirubin pigment-producing isolates and included reference strains of *Cytophaga johnsonae* and *Cytophaga aquatilis*. Other bacteria isolated included vibrios, pseudomonads, and *Chromobacterium* strains. Sixteen percent of the isolates, including 10 of 44 *Cytophaga* isolates and 4 of 8 fermentative chitin degraders, showed the phenomenon of salinity-induced shift-up of growth temperature range; i.e., growth on seawater agar was noted at a higher temperature than when seawater-free agar was used. The authors comment that this phenomenon has rarely been reported before (never for *Cytophaga* isolates) and may be a major factor in the dispersal and survival of bacteria in the fluctuating chemical and physical environment in an estuary. Chan (1970) presented results of a study of chitinolytic bacteria from Puget Sound. Counts were about 10^2–10^4 ml^{-1} in the water and 10^4–10^5 g^{-1} in the sediment. Genera identified, in decreasing order of abundance, were *Vibrio, Pseudomonas, Aeromonas, Cytophaga, Streptomyces,* and *Photobacterium; Bacillus* and *Chromobacterium* were also found. The dominant chitinolytic bacteria isolated by Cross (1985) from degrading squid pen samples in the Don Estuary, Aberdeen, at a mean temperature of 8.5°C were *Pseudomonas* species. Hillman, Gooday, and Prosser (unpublished observations) isolated chitinolytic *Streptomyces* strains from the nearby Ythan estuary. Lakshmanaperumalsamy (1983) identified 162 isolates of chitinoclastic bacteria from the Vellar Estuary, India, as follows (in decreasing order of abundance): *Vibrio, Aeromonas, Pseudomonas,* micrococci, *Flavobacterium/Cytophaga, Moraxella, Achromobacter, Bacillus,* and *Cornybacterium.*

Boyer (1986) described the isolation of both facultative and obligate anaerobic chitin-digesting bacteria from Massachussetts salt marsh sediments. He showed that in anaerobic culture on chitin, they produced CO_2, H_2, and various types of volatile fatty acids. Facultatively anaerobic isolate CD3 produced (millimoles per 100 mmol of chitin) acetate (48), CO_2 (12), and H_2 (11) and could be grown in coculture with sulfate-reducing and/or methanogenic bacteria. Pel and Gottschal (1986a–c) have studied chitin degradation by anaerobic bacteria from the Em-Dollard Estuary, Holland. In the tidal flats of this estuary, they estimated that anaerobic mineralization of carbon amounted to 30% of the total mineralization and surmised that chitin must be an important substrate for a considerable part of the microbial sediment population. In anaerobic mixed cultures from the upper layer of the mud, they observed rapid fermentation of chitin, with the formation of acetate. The rate of chitin fermentation was unchanged in the presence or absence of sulfate in the growth medium, but in its presence the acetate was utilized by sulfate-reducing bacteria. They isolated eight strains of similar chitinolytic bacteria and chose one, designated *Clostridium* sp. strain 9.1, for further study. Its major fermentation products (millimoles per 100 mmol of N-acetylglucosamine equivalents of chitin) were H_2 (106), CO_2 (150), acetate (117), ethanol (80), formate (10), and NH_4^+ (75), with a carbon recovery of 10%. Supplementation of the medium with dithionate (0.2 mM) lowered the E_h of the culture fluid from -160 to -380 mV and greatly stimulated the rate of chitin fermentation but not the total amount of chitin fermented (Fig. 3a). Diacetylchitobiose accumulated initially and then soon disappeared as N-acetylglucosamine accumulated (Fig. 3b). Glucosamine and N-acetylglucosamine only supported very slow growth of *Clostridium* 9.1, but diacetylchitobiose and a mixture of oligomers (tri-, tetra-, and pentasaccharides) supported growth to an amount similar to that supported by chitin. The authors suggest that these bacteria are specialized utilizers of the disaccharide, possibly via a phosphorylating enzyme, and that the accumulation of N-acetylglucosamine (Fig. 3b) represents accumulation of nonutilizable monomers appearing during the random hydrolysis of chitin oligomers. Further experiments led Pel and Gottschal (1987) to suggest that the stimulation of chitin degradation by dithionate was via the maintenance of protein sulfhydryl groups essential for the chitinolytic system. Addition of p-chloromercuribenzoic acid (pCMB), a membrane-permeable sulfhydryl agent, completely blocked the fermentation and partially inhibited chitin hydrolysis. Subsequent addition of excess sodium sulfide relieved inhibition of chitin hydrolysis but not of fermentation; as a result, diacetylchitobiose accumulated in the culture. In contrast, the inhibition of both chitin hydrolysis and fermentation by the addition of p-chloromercuriphenylsulfonic acid, a membrane-permeable analog of pCMB, was relieved by the subsequent addition of sodium sulfide. Treatment with a thiol-oxidizing agent, ferricyanide or 5% oxygen, strongly inhibited fermentation but not diacetylchitobiose accumulation, presumably by inhibiting the uptake system of the disaccharide. The authors suggest that this phenomenon may have ecological significance for this bacterium in the upper layer of the mud: if exposed to oxygen, the *Clostridium* species would still hydrolyze the chitin to disaccharide, which could provide substrate for the growth of nearby facultative aerobic bacteria, which would then consume oxygen to render the microenvironment anaerobic again. Pel and Gottschal (1989) develop the idea of synergistic interspecies interactions enhancing chitin degradation by showing that fermentation by *Clostridium* sp. strain 9.1 proceeded up to eight times faster in mixed culture than in pure culture. Stimulation also resulted from addition of spent

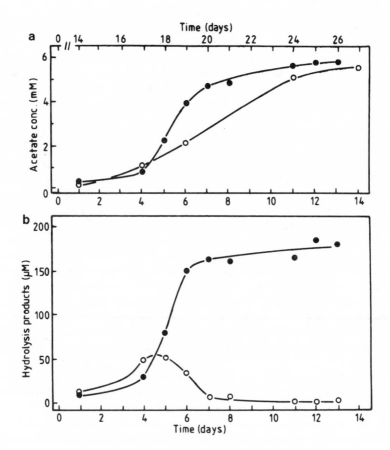

Figure 3. (a) Rate of acetate formation during chitin fermentation by *Clostridium* sp. strain 9.1 (see Fig. 7) in the absence (○; upper time axis) and presence (●; lower time axis) of 0.2 mM dithionite. Note the difference in scale on the upper and lower time axes. (b) Accumulation of *N*-acetylglucosamine (●) and diacetylchitobiose (○) in the culture fluid of strain 9.1 during chitin fermentation in the presence of dithionite. [From Pel and Gottschal (1986b).]

media from such mixed cultures to pure cultures of strain 9.1. They attributed this phenomenon to the production of a thioredoxin-type compound that maintained the reduced state of the postulated essential sulfhydryl groups in the chitinolytic system.

5.2. Rates of Degradation

Cross (1985) presents results of estimations of the rates of degradation of squid pen chitin–protein (40 : 60) in litterbags of 20-μm nylon mesh at two sites in the estuary of the River Don, Aberdeen, for various periods over a year with a mean temperature of

8.5°C. Mean degradation rates were 44.8 and 9.9 mg g^{-1} day^{-1} at the sediment–water interfaces at the upstream and downstream sites, respectively, 25.7 and 28.3 mg g^{-1} day^{-1} in the two anaerobic sediment sites (5 cm deep), and 81.0 mg g^{-1} day^{-1} in the flowing water column of the downstream site. In these studies and in the parallel riverine and marine studies, Cross found that the chitin and protein components of the squid pen were degraded at the same rate, suggesting synergism between chitinolytic and proteolytic systems in attacking this chitin–protein structure.

Rates of chitin degradation *in vitro* and *in situ* in Borataria Bay estuary, Louisiana, have been estimated by Hood and Meyers (1973, 1977b, 1978). For studies *in vitro*, purified chitin flakes were incubated in closed flasks at 22°C with sediment samples or with estuarine water. Degradation rates were 47 and 22 mg g^{-1} month^{-1}, respectively. Very much higher degradation rates, however, were observed *in situ*, when pieces of native chitin from white shrimp (i.e., calcified chitin–protein complex) were suspended in acetate bags (pore size, 60 × 10 μm). Rates were dependent on temperature, the maximum observed being 118 mg g^{-1} day^{-1} at 23°C the minimum being 35 mg g^{-1} day^{-1} at 7°C. Rates were similar for samples at the water–sediment interface and in the water column. The rate was higher for smaller particles; for example, that for particles of 0.25 cm^2 was about double that for particles of 1.0 cm^2. Native chitin was degraded faster than purified chitin. The native chitin was also more heavily and rapidly colonized than purified chitin. In one experiment *in situ*, total bacterial biomass after 4 days was 2.7 × 10^9 cells g^{-1} native chitin, of which 8.9% were chitinoclastic, rising to 4.3 × 10^9 cells g^{-1} after 8 days, of which 15.6% were chitinoclastic. The chitin was almost totally degraded after 23 days.

With this work as a basis, Portier and Meyers (1981, 1984) developed continuous-flow microcosms based on microbial chitin utilization to study the effects and fate of pesticides and other xenobiotic compounds. They showed that application of azinphosmethyl to their system resulted in increased populations of total heterotrophs and chitinoclasts and that methyl parathion–chitin and phenol–chitin combinations could result in increased degradation rates for both substrates, depending on salinity. They conclude that "chitin provides an advantageous surface for microbial colonization and toxicant adsorption" and that this characteristic might encourage cometabolic degradation of the xenobiotic.

Boyer and Kator (1985) describe a method for measuring the rate of degradation of native chitin *in vitro*, by incubating estuarine water with ^{14}C-labeled crabshell from crabs treated with radioactive *N*-acetylglucosamine as a specific precursor of chitin. Particle size of this native chitin (i.e., calcified and containing protein) was 0.5–1 mm^2. Mineralization rate was estimated by measuring radioactivity evolved as carbon dioxide, and degradation rate was estimated by measuring combustion of remaining solid matter to carbon dioxide and determining its radioactivity. There was a good relationship between the two values, with 96% of the added label being recovered as the sum of both phases. The maximum rate of mineralization was 207 mg day g^{-1} at 20°C. The highest count of chitinoclastic bacteria was observed at the time of maximum degradation and mineralization, between 2 and 4 days.

Radioactive chitin has been used by Smucker (1982) to estimate the chitinase hydrolytic potential of estuarine sediments. Sediment was suspended in an equal volume of buffer, filtered, and incubated at 30°C for 4 hours with tritiated chitin, made by reacetylating chitosan with tritiated acetic anhydrides; soluble products were then counted. Values for two sediments were 6 and 14 nmol disaccharide equivalents ml^{-1}.

The fate of chitin in sediments in the Ythan Estuary, Aberdeen, has been studied by Hillman, Gooday, and Prosser (unpublished observations). The major input of chitin into this estuary is the productivity of the major invertebrate species, the crustacean *Corophium volutator*, adults of which contain 8.4% chitin. The annual input from this source alone was estimated as 3.7 g chitin m^{-2}. Chitin and chitosan were detected in the sediments, both being considerably higher in winter than summer. Values for chitin and chitosan in the top 2 cm of sediment were 4.3 and 55.8 mg (g dry sediment)$^{-1}$, respectively, in summer, and 36.2 and 156.7 mg g^{-1} in winter. Chitin contents declined rapidly with depth, being very low in the anaerobic sediment below 5 cm. Chitosan levels showed little variation with depth in the upper sediment but were lower in the anaerobic sediment. It was suggested that much of the chitosan represented degradation of chitin produced in the sediment, whereas much of the chitin was from surface deposition. Chitinase and chitin deacetylase activities were detected in the sediment, both being higher in summer than winter. Values for chitinase were 93 and 31 mg chitin hydrolyzed hr^{-1} (g dry sediment)$^{-1}$, respectively, for summer and winter; for chitin deacetylase, values were 33 and 5 mg chitin deacetylated h^{-1} g^{-1} for summer and winter. All assays were at 15°C. Chitinase activities showed little variation with depth, whereas chitin deacetylase activities were highest in the upper sediment. The finding of considerable levels of chitosan and chitin deacetylase suggests that in these sediments a considerable amount of chitin degradation is via deacetylation as well as by hydrolysis to chito-oligosaccharides, but the major agents of this deacetylation have yet to be identified. Litterbag experiments (nylon, 20-μm mesh), using purified chitin, chitosan, and cellulose for comparison, gave rates of degradation for 2 months in the summer as 10.3, 7.0, and 4.4 mg (g polysaccharide)$^{-1}$ day^{-1}, respectively. Hillman *et al.* (1989) reported the operation of a model system designed to imitate the natural estuarine environment and have used it to study chitin degradation. Results obtained were remarkably similar to those for the natural system; for example, relative rates for degradation of chitin, chitosan, and cellulase for 1 month at 15°C were 2.4:2.0:1, compared with 2.3:1.6:1 in the Ythan Estuary.

6. Chitin Degradation in Freshwaters

In compared with marine and estuarine systems, the fate of chitin in freshwaters has received less attention, probably chiefly because here the recycling of chitin is less important than that of the green plant polysaccharides cellulose and hemicelluloses. Nevertheless, the chitin in freshwaters, in the form of invertebrates and fungi, is a carbon and nitrogen source that should not be ignored.

6.1. Freshwater Chitinoclastic Microbes

Warnes and Rux (1982) sampled a man-made lake in Indiana for chitinolytic bacteria. During spring, streptomycetes and myxobacteria were the major types; during summer, *Serratia* species were important, especially *Serratia marcescens*, which was the most commonly isolated Gram-negative bacterium. Pseudomonads were also isolated during warmer periods, together with strains of *Moraxella, Flavobacterium*, and *Chromobacterium*. A very few isolates of *Aeromonas hydrophila* were seen throughout the year. Chitinolytic genera encountered by Dondereski (1984) in Polish lakes were *Achromobacter, Bacillus, Nocardia, Pseudomonas, Flavobacterium, Cytophaga*, and enterobacteria. Streichsbier (1982) has isolated the following chitinolytic bacteria from a creek near Vienna: *Pseudomonas* species, *Chromobacterium violaceum, Janthinobacterium lividum, Serratia marcescens, Bacillus thuringiensis*, and *Bacillus cereus*. *Pseudomonas* species have also been reported by Smucker *et al.* (1985) and Ruschke (1967), and *Chromobacter violaceum* has been reported by Cross (1985), who found it to be more abundant in winter in the River Don, Aberdeen. Reichardt and Morita (1982) describe a marked effect of temperature on both aerobic and anaerobic enrichments of chitinoclastic *Cytophaga johnsonae* strains from sediments from Lake Tahoe, Oregon. Using colloidal chitin (200 mg liter^{-1}) and 10 mM N-acetylglucosamine, they observed good enrichment at 5 and 25°C, but with colloidal chitin alone, enrichment was only obtained at 5°C. The authors suggest that this finding reflects the physiological nature of psychrotrophic acclimatization by this organism. When colloidal chitin selective media have been used, actinomycetes have been isolated readily from freshwaters, particularly in sediments (Johnstone and Cross, 1976). In an extensive survey in the lakes of the English Lake District, these authors found that the numbers isolated from the muds correlated with the lakes' productivity status. High numbers of *Micromonospora, Streptomyces* and nocardioforms were isolated from all of the lakes.

6.2. Rates of Degradation

Estimates of rates of degradation of chitin have been made by Cross (1985) by suspending pieces of squid pen chitin–protein in nylon litterbags (20-μm mesh) in the River Don, Aberdeen, for extended periods. The mean annual rate was 90.4 mg g^{-1} day^{-1} at a mean temperature of 7.6°C. The squid pen was colonized rapidly. The mean count of chitinolytic colonizers was 2.1×10^8 cells (g squid pen)$^{-1}$, and these were 54.2% of the total count of heterotrophic bacteria; in comparison, the chitinolytic bacteria in the surrounding water column were 5.6% of the corresponding heterotrophic count. Aumen (1980) studied the degradation of the chitinous exoskeleton of the crayfish *Procambarus versutus* in an acidic woodland stream in Florida. The counts of total sediment bacteria ranged from 5×10^4 to 15×10^4 (g wet weight sediment)$^{-1}$, of which 3 to 11% were chitinoclastic. In contrast, in the later stages of decomposition, the proportion of chitinolytic bacteria reached 88% of the total heterotrophic count. *Streptomyces* species developed rapidly after 11 days, completely covering the surface in scanning electron micrographs. Filamentous fungi were observed by light microscopy after 25 days.

Donderski *et al.* (1984) isolated chitinolytic bacteria from six Polish lakes and found wide differences in total counts and in proportion of chitinolytic isolates to total isolates. The lowest proportions were 3.3 and 9.6% of total heterotrophic count in the water column and sediment, respectively, for one lake and 42.8 and 34.2%, respectively, in another. Okutani and Kitada (1971) and Okutani (1975) reported less than 1% of heterotrophic bacteria as chitinolytic in both water and surface sediment of Lake Biwar, Japan, but in the water column the proportion of chitinolytic isolates did increase with increasing depth.

Chitin degradation has been studied in Lake Erie by Warnes and Randles (1977, 1980), who followed the microbial succession on chitin-coated slides in the lake water or sediment samples. Bacterial activity was greatest in the water column from 2 to 7 days, whereas sediment populations showed greatest numbers at 7 days in the aerobic zone and 13 days in the anaerobic zone. Warnes and Rux (1982) studied the degradation of chitin particles in nylon mesh litterbags in the water column of an Indiana lake. Degradation rates were similar for bags with mesh sizes of 10, 25, and 100 μm. Of the two particles sizes used, 0.7–1.0 and 1–1.3 mm, the smaller particles were degraded faster, with a rate of 43.6 mg g^{-1} day^{-1} at 28°C for the first 20–30 days. Rates were much higher in summer than in spring or autumn. Yamamoto and Seki (1979) have estimated the rate of chitin degradation in a water chestnut ecosystem in Lake Kasumigaura, Japan, by measuring input of chitin as biomass of planktonic arthropods and particulate chitin and output of chitin as assessed by density of chitinoclastic bacteria and their chitinoclastic rate. The latter was estimated as 6.0×10^{-8} mg carbon (mgC) chitin bacteria count^{-1} day^{-1} at 30°C and 2.5×10^{-9} mgC chitin bacteria count^{-1} day^{-1} at 20°C, the latter figure being similar to that in coastal waters. The annual peak of planktonic crustacea, at the beginning of August, was soon followed by peaks of total particulate chitin in the water and population density of chitinoclastic bacteria. Chitin content of the surface sediment showed little annual fluctuation. It was estimated that about one-third of the chitin produced was decomposed in the water column and that the rest was sedimented to the lake bottom for further degradation. The annual budget for chitin in the ecosystem was estimated (expressed as gC per square meter) as inputs of 26 (endogenous) and 10 (exogenous, e.g., insect corpses) and outputs of 23 (mineralization in the water column) and 24 (in the sediment surface).

Sturz and Robinson (1986) have studied the anaerobic degradation of chitin in sediments of Blelham Tarn, England. Using nylon litterbags containing strips of chitin purified from scampi carapace, they determined degradation rates of 10.7 and 9.3 mg g^{-1} day^{-1} for samples in the 0- to 6-cm and 42- to 48-cm horizon respectively, at 6–8°C, representing aerobic and anaerobic decomposition. The numbers of anaerobic chitinolytic bacteria isolated were highest in the surface 1 cm, at 10^6 (g dry weight sediment)$^{-1}$, quickly dropping to 10^5 g^{-1} from 1 to 11 cm. In contrast, numbers of aerobic isolates were about 10^6 g^{-1} throughout the top 11 cm. Sturz and Robinson also estimated rates of anaerobic mineralization *in vitro* by incubating sediment samples with ^{14}C-labeled chitin prepared from *Drosophila* larvae fed with radioactive glucose and measuring total ^{14}C in the gas space after combustion to carbon dioxide. This experi-

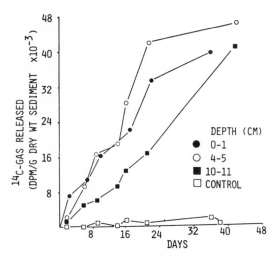

Figure 4. Release of ^{14}C-labeled gas (chiefly CO_2 and CH_4) from ^{14}C-labeled chitin incubated in sediment from various depths below the sediment–lakewater interface of Blelham Tarn. [From Sturz and Robinson (1986).]

ment showed appreciable rates of degradation of chitin, particularly in the upper sediment samples (Fig. 4).

7. Chitin Degradation in Soil

The soil contains many chitinous invertebrates, protozoa, and fungi as its normal living components. It also receives input of chitin of dead organisms from the air and from overlying vegetation. But, as stated by Parsons (1980), "We know little about the amino sugar-containing polymers in soils, but the information available suggests they are closely associated with neutral sugars and amino acids. There is no evidence either for the presence or absence of chitin but intuitively it is difficult to believe it is not present."

7.1. Microbial Degradation

Chitin-degrading microbes can be isolated readily from soils. The numbers and types reported vary greatly with the different soils and conditions, but it is difficult to judge the extent to which these variations are caused by variations in methods used by the investigators. Skinner and Dravis (1937), estimating the range of chitin-degrading microbes in a variety of soils by dilutions in liquid medium containing chitin strips, found that nonfilamentous bacteria were the most abundant, with fewer molds and only small numbers of actinomycetes. When they isolated at random from agar plates, they found actinomycetes to be the most abundant degraders of chitin. They favored the validity of their liquid culture method, going as far as saying about chitin agar that although such media are useful for isolation or organisms, "they are useless for quantitative plating." Veldkamp (1955), while agreeing that very different results are obtained with different techniques, favored the use of chitin agar. He concluded that although actinomycetes are only a small percentage of the soil microbes, they nevertheless form the major part of the chitinivorous microbes, with other bacteria being less important at 1–3% and fungi forming less than 1% of the population of chitinivorous microbes. He describes differences in different soils: there were more chitinivorous microbes and more "total microbes" in a sandy soil treated with farmyard manure than in the same type of soil treated with mineral fertilizers. The smallest number of both types of microbe was found in a sandy soil of acid pH, and the largest numbers were found in a clay soil treated with farmyard manure.

Mihaly (1960) reported that about 20% of bacterial isolates, typically rods, from rhizospheres of crop plants were chitinivorous and that they retained this property for a longer period in culture. Okafor (1966a,b) used microscopy to study the colonization and degradation of chitinous materials in soils. When he used 0.5-cm squares of deproteinized chitin from cuttlefish shell buried for up to 100 days at 10°C, nonfilamentous bacteria and fungi were the major colonizers, while at 29°C there were also actinomycetes, nematodes, and protozoa (Okafor, 1966a). At both temperatures, fungal hyphae observed in his temperate soil were chiefly aseptate, while in his tropical soil they were chiefly septate. At both temperatures, the rate of decomposition of the chitin was faster in his temperate soil than in his tropical soil. When he used pieces of locust wing, either untreated, dewaxed, or dewaxed and deproteinized, buried for up to 300 days at 25°C, there were clear differences in their colonization by fungi, nonfilamentous bacteria, actinomycetes, and protozoa (Okafor, 1966b). Nonfilamentous bacteria were most abundant on all three substrates, but there was more fungal colonization on untreated wings than on dewaxed and deproteinized wings. The rate of degradation, as assessed by recovery of material, was faster for treated than for untreated wing pieces. Thus, all pieces of untreated wing but no pieces of deproteinized wing were recovered from the soil after 300 days. The chitin in the wing is presumably protected from rapid degradation by the waxy surface of the wing. The major organisms implicated in chitin degradation in this system were a fungus, *Mortierella,* a *Pseudomonas* species, and two *Streptomyces* species.

Okafor (1967) characterized the fungi and bacteria growing at 10 and 29°C on chitin in his temperate and tropical soils in more detail. The major genera represented in the isolates that grew on chitin pieces and cleared chitin agar (i.e., produced sufficient extracellular chitinolytic activity to give a halo around the cultures) for the fungi from the temperate soil were *Mortierella, Trichoderma, Penicillium, Verticillium, Humicola,* and *Chaetomium;* those from the tropical soil were *Emericella, Aspergillus, Malustella,* and *Thielavia;* those that cleared chitin agar for the bacteria from the temperate soil were *Beneckea (Vibrio), Pseudomonas,* and *Streptomyces;* those from the tropical soil were *Beneckea (Vibrio), Bacillus,* and *Streptomyces.* Some other fungi and bacteria grew on the chitin pieces but did not clear the chitin agar. In some cases, however, clearing of chitin agar was observed in cocultures of two bacterial isolates streaked across each other, for example, *Arthrobacter* plus *Cytophaga* or *Arthrobacter* plus *Achromobacter.*

Veldkamp (1955) records the following genera as the major chitinivorous microbes isolated from his soils: *Achromobacter, Flavobacterium, Chromobacterium, Bacillus, Cytophaga, Pseudomonas, Streptomyces, Micromonospora,* and *Nocardia* among the bacteria and *Mortierella* and *Aspergillus* among the fungi. His two major bacterial isolates have been reidentified, *Pseudomonas chitinovorans* as *Aeromonas caviae* NCIMB 8500 and *Cytophaga johnsonae* as *Lysobacter* sp. NCIMB 8501. Gray and Bell (1963) and Gray and Baxby (1968) suggest that fungi are the most important chitin degraders in forest soils, in part perhaps because their potential for rapid filamentous growth would enable them to colonize particulate substrates. Gray and Baxby (1968) give extensive lists of microbes, particularly fungi, implicated in the degradation of naturally occurring and artificially buried chitin in pine forest soils. Among these, *Verticillium*, *Mortierella*, and *Trichoderma* were the most important in the acid horizons, and *Mortierella*, *Paecilomyces*, *Gliomastix*, *Verticillium*, *Pseudomonas*, *Bacillus*, and *Streptomyces* were most important in the alkaline horizon. The same three bacterial genera were found as the major chitin degraders in litter from the silver fir, *Abies albus*, by Faure-Raynaud (1981) and, together with *Arthrobacter*, *Flavobacterium*, *Trichoderma*, *Penicillium*, and *Humicola* from traps baited with fungi, insect cuticle, or chitin by Fargues *et al.* (1977).

Of the organisms isolated in all of these studies, the ability to degrade chitin is one of the most characteristic features of soil streptomycetes (Williams and Robinson, 1981). For example, chitin media are used for selective isolation and enumeration of streptomycetes (Lingappa and Lockwood, 1962; Baxby and Gray, 1968; Hsu and Lockwood, 1975). Williams and Robinson (1981) studied the chitinolytic streptomycetes of an acidic pine forest soil and classified them as acidophilic (growth at pH 4.5 but not at pH 7.0), acidoduric (growth at both pH values), or neutrophilic (growth at pH 7.0 but not at pH 4.5). Laboratory experiments compared two systems: the F_2-H horizon, about 100% organic matter, 30% humic acid, 1% nitrogen, 0.4% hexosamine, and strong buffering capacity; and the A_1 horizon, about 5% organic matter, 0.6% humic acid, 0.02% nitrogen, 0.04% hexosamine, and weak buffering capacity. In both systems, amendment with either chitin or ground dried fungal tissue at 2.5% (wt/wt) resulted in significant increases in counts of streptomycetes in all cases. The numbers of chitinolytic streptomycetes increased more or less in parallel with the total population, the proportion of chitinolytic isolates usually starting above 70% in unamended soils and not increasing in amended systems. The major differences between the two soils are clearly related to the greater buffering capacity of the F_2-H horizon. Following the populations with time showed that degradation of the chitin was accompanied by an alkalinization and a total replacement of acidophilic streptomycetes by neutrophilic streptomycetes (Fig. 5).

Bull (1970) studying the relative recalcitrance of some fungal materials in soil, showed that chitinases are inhibited by melanin. He correlated this with observations that melanized cell walls were notably more recalcitrant than nonmelanized walls. This observation also could have relevance to consideration of the fate of invertebrate remains in soil, as they are often pigmented.

As well as chitin-degrading bacteria, chitosan-degrading bacteria can readily be isolated from soils. Davis and Eveleigh (1984) screened soils from barnyard, forest, and salt marsh, and 5.9, 1.5, and 7.4%, respectively, of total heterotrophic isolates were chitosan degraders, compared with 1.7, 1.2, and 7.4% of chitin degraders. Monaghan *et al.* (1973) and Monaghan (1975) showed that chitosanase activity is widespread among

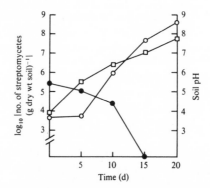

Figure 5. Changes in numbers of neutrophilic (○) and acidophilic (●) streptomyces and in soil pH (□) in soil from the A_1 horizon of pine woodland, amended with 2.5 % (wt/wt) sterile chitin flakes. [From Williams and Robinson (1981).]

soil microbes, occurring in gliding bacteria such as *Sporocytophaga* sp. and myxobacteria, in *Bacillus* species, and in *Arthrobacter* and *Streptomyces* species among the bacteria and in a range of fungi, including zygomycetes, ascomycetes, basidiomycetes, and deuteromycetes. Fenton *et al.* (1978) tested chitosan-degrading organisms from the soil for the ability also to degrade chitin and cellulose and found that some could utilize only chitosan, some chitosan and chitin, some chitosan and cellulose, and some all three. Davis and Eveleigh (1984) have studied the chitosanase of a soil isolate of *Bacillus circulans* and shown that it is specifically inducible by chitosan, and not by chitin or carboxymethylcellulose, and specifically active on chitosan. Hedges and Wolf (1974), however, reported an enzyme from a soil isolate of *Myxobacter* species that was active against both chitosan and carboxymethylcellulose. Some chitosan production in the soil will be as a major cell wall constituent of zygomycetes such as *Mucor, Rhizopus,* and *Mortierella* species, but some may be via deacetylation of chitin (Section 2). If the deacetylation of chitin is an important pathway for degradation, then, as for estuarine sediments (Section 5.2), the major agents of this process are yet to be identified.

Gould *et al.* (1981) used a model soil system to study the effect of both pure cultures and mixed cultures on the decomposition of chitin. They used carbon dioxide evolution from soil amended by addition of chitin (3 mg g^{-1} chitin-C) as a measure of chitin decomposition, as suggested by Okafar (1966c). They detected an increase in carbon dioxide evolution with the addition of chitin with an actinomycete isolate and with *Fusarium oxysporum* but not with any of three bacterial isolates, even though all five microbes were isolated from the native soil as chitin degraders. All five, however, gave a net increase in NH_4^+ in the amended soil.

7.2. Soil Chitinase Activity

Chitinase activity has been detected in soil by Rodriguez-Kabana *et al.* (1983) by incubating toluene-treated soil with a colloidal chitin suspension and assaying release of

N-acetylglucosamine. Chitinase activities diminished with soil depth, halving in activity by about 30–40 cm. Activities were higher in soils growing corn or soyabean than those growing cotton. Activities were higher in fertilized than in unfertilized soils. Addition of chitin at a rate of 0.8% or higher resulted in sharp increases in activity. Activities increased with time, with maximal activities at 6–10 weeks of the 10-week experiment. These enzyme activities presumably reflect the activities of the indigenous biomass. The microbial flora of these soils with chitin amendments was assessed through the 10-week experiment in comparison with the chitinase activities. Three groups were studied: bacteria, actinomycetes, and fungal propagules. Of these, only the count of fungal propagules showed a significant positive correlation with chitinase activity for the whole range of amendments of 0–4% (wt/wt) chitin. The counts of actinomycetes showed an increase in soils treated with 0–1% chitin, but counts in soils with 4% were no higher than those with no addition of chitin. Chitinase activities also have been measured in soils of tussocks of arctic sedge by Linkins and Neal (1982). They found highest activities from the tussocks that had the most growth of mycorrhizal shrubs. These soils also had the highest respiration rates, which the authors took as a measure of microbial activity.

7.3. Effects of Addition of Chitin to Soil

Several of the studies already referred to, and several of those on biological control discussed below, describe microbial and chemical consequences of adding chitin to soil. Veldkamp (1955) studied this in detail. In experiments with water-saturated soils in soil perfusion columns, the numbers of chitinovorous microbes rose specifically and dramatically in the soils amended with 1.25% (wt/wt) chitin but declined to very low levels after 3 weeks in the control soils. Veldkamp comments on "a remarkable feature of these experiments" in that chitin-decomposing actinomycetes were nearly absent in fluid reservoirs and in the soil column, amounting to less than 1% of the total organisms recovered, the remainder being unicellular bacteria; in the native soils, in contrast, actinomycetes formed well over 90% of the recoverable chitinovorous microbes. Another unexpected observation was that most of the chitin-decomposing bacteria, 12 of 15 different strains, that were isolated from the chitin–soil columns irreversibly lost the ability to degrade chitin after repeated subculture in the absence of chitin. In experiments with relatively dry soils, different results were obtained. In all cases, addition of chitin resulted in a considerable increase in microbial population, compared with relatively small changes in population in soils without added chitin. In all cases, the changes resulted in a predominance of actinomycetes over other bacteria. In a clay soil and a sandy soil, both of which had been treated previously with farmyard manure, actinomycetes formed the major part of chitinovorous as well as "total" microbes soon after addition of chitin. Veldkamp monitored nitrification as a result of chitin decomposition in both of these experimental systems, i.e., in water-saturated soils and in relatively dry soils (10% water content of sandy soils and 17% of clay soil). In all cases, there was considerable recovery of the added chitin-N as nitrate-N, amounting up to 60% after 48 days. Polyanskaya et al. (1985) described the dynamics of growth and survival of *Rhizobium leguminosarum* when added to soil together with chitin and the chitinolytic *Streptomyces alivocinereus*. In model experiments, they reported that the nodule bacteria and the actinomycete grew well in coculture in liquid medium with chitin as the sole carbon source. In experiments in nonsterile soil, growth of the nodule bacteria was initially stimulated, but by 51 days they had been eliminated.

7.4. Addition of Chitin to Soil for Biological Control of Disease

Soil treatment with chitin or chitinous material has been suggested as a means of controlling both soil-borne fungal and nematode plant pathogens. The rationale is to favor an increase in a microbial flora capable of attacking the chitinous hyphal walls of pathogenic fungi or the chitinous shells of nematode eggs. Mitchell and Alexander (1962) and Mitchell (1963) describe the effect of adding chitin to soil infested with *Fusarium solani* f. *phaseoli* on the development of bean root rot. Significant reduction in disease was achieved with both chitin and ground lobster shell, each at a level equivalent to 500 lb/acre. Diffusates from the treated soil, probably lytic enzymes, both inhibited the growth of the *F. solani* and caused lysis of its mycelium. This was attributed to the fact that addition of the chitin to the soil stimulated the development of a mycolytic microbial flora. Similarly, Buxton *et al.* (1965), Khalifa (1965), and Van Eck (1978) and Henis *et al.* (1967) and Sneh and Henis (1970) reported some control of *Fusarium* and *Rhizoctonia* pathogens, respectively, by chitinous additions to soil but suggested that this may have been due to stimulation of production of inhibitory substances rather than the development of a mycolytic flora. Okafor (1970), however, reported no change in abundance of *Fusarium* in chitin-amended soils sown with wheat seeds but an increase in abundance of *Mortierella* and a decrease in root length of the wheat seedlings.

Mian *et al.* (1982), Godoy *et al.* (1983), and Culbreath *et al.* (1986) describe the effect of adding chitin to soils infested with the nematode *Meloidogyne arenaria* on the development of galls on the roots of summer crookneck squash, *Cucurbita pepo*. In their initial experiments, chitin was added to the soil at 0–4% (wt/wt) and allowed to decompose for 3 weeks, and then the soil was planted with the squash. After 6 weeks of growth, plants from soils with 1% or more of chitin had no galls, while control plants had many galls. Soils with chitin amendments had increased activities of chitinase, urease, and aryl phosphatase. They also developed a specific microbial flora, in particular the fungi *Fusarium solani, Fusarium udum, Thielaria basicola, Humicola fuscoatra*, and *Pseudeurotium ovale*, all of which parasitized the eggs of the nematode *in vitro* (Mian *et al.*, 1982). A later experiment allowed the chitin to decompose for 10 weeks before planting (Godoy *et al.*, 1983). In this case, chitin treatments at 0.4% and above reduced galling but those at 0.8% and above were phytotoxic. In line with the findings of Veldkamp (1955), chitin amendments resulted in increases in pH, conductivity, and ammonia and nitrate contents as well as in chitinase activity. The predominant fungus was *Malbranchea aurantiaca*, which parasitized the nematode eggs *in vitro*. Culbreath *et al.* (1986) then assessed the effect of adding a fungus, *Paecilomyces lilacinus,* together with the chitin at 0–1% (wt/wt). The chitin was allowed to decompose for 2 weeks, the soil was planted with squash, which were harvested after 6 weeks, and the soil was replanted with tomato seedlings, which were harvested after a further 6 weeks. In this experiment, chitin amendment resulted in increasing galling of the squash roots, *P. lilacinus* by itself had no effect, but infection of both squash and tomato was reduced by the combined treatment of fungus and chitin.

Spiegel *et al.* (1986) studied the effects of chitin addition to soils planted with beans and tomatoes and inoculated with the root knot nematode *Meloidogyne javanica*. In the first month there was some phytotoxicity and a small reduction in galling, but on replanting with fresh plants there was an increase in fresh weight and almost total inhibition of galling. Spiegel *et al.* (1987) repeated a similar experiment, comparing results in irradiated soil and nonirradiated soil. In the first cycle of plant growth, in irradiated soil there was a similar rate of galling with or without chitin treatment but a reduction of galling with chitin in nonirradiated soil. In the second cycle, chitin treatment reduced galling in both soils but to a much greater extent in nonirradiated soil. In the nonirradiated soil there was an increase in chitinolytic microbes, especially actinomycetes, with addition of chitin. Brown *et al.* (1982) reported control of the root knot nematode *Meloidogyne incognita* on tomato plants by soil amendments with chitinous shrimp waste, particularly when it had been pretreated by alkali extraction or had the addition of a slurry of actinomycetes pregrown on chitin. They found that additions to the soil of either chitin or shrimp waste resulted in a large increase in the total microbial population, especially of actinomycetes.

It should be clear from this account that amendment of soils with chitinous material does offer promise for the control of fungal and nematode plant pathogens. However, when conflicting observations are reported, for example *Fusarium* species diminishing, increasing, or unchanging in abundance, it is also clear that protocols would have to be tailored individually to the type of soil and crop regime under study.

Hadwiger *et al.* (1984) have developed a different approach to the control of soil-borne fungal pathogens, by using chitosan as a seed treatment. This approach is based on their observations that chitosan is a strong eliciter of plant defense mechanisms and also that it is directly toxic to a range of plant-pathogenic fungi. McCormick and Anderson (1984) presented preliminary results testing the idea that pesticides could be applied to soil in a form giving controlled release by being covalently bonded to chitin. The rate of release presumably would be dependent to a large extent on the activity of soil chitinolytic microbes.

8. Adhesion of Microbes to Chitin

Chitin and chitosan have very strong adsorptive properties, both for a range of organic molecules and for metal cations (Muzzarelli, 1977). They are also positively charged, increasingly so with increasing deacetylation. Therefore, it is no surprise that chitin-rich material is rapidly colonized by microbes in nature. Cross (1985) suspended small squares of squid pen ([chitin–protein, 40 : 60 (wt/wt)] in nylon bags (20-μm mesh) in a range of aquatic environments (riverine, estuarine, and marine) and found that they were covered by biofilms within a few days (Fig. 6). These biofilms were not

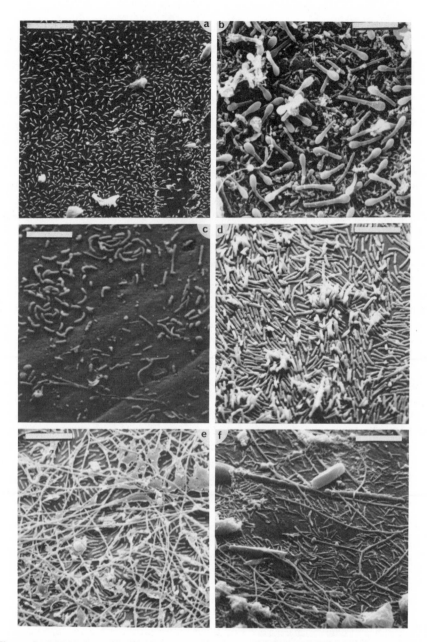

Figure 6. Colonization of squid pen chitin to protein in 200-μm nylon mesh litterbags, shown by scanning electron microscopy (samples fixed with 5% glutaraldehyde, dehydrated, and gold coated). (a) River Don water column, 7 days of immersion; (b) River Don sediment, 21 days of burial; (c) Don Estuary sediment, 7 days of burial (note apparent etching of surface in bacterium-shaped patches); (d) Don Estuary sediment, 21 days of burial; (e, f) Hareness Inlet seawater column, 42 and 112 days of immersion, with a variety of bacterial rods and filaments as well as diatoms. (M. G. Cross and G. W. Gooday, unpublished observations). Scale bars represent 20 μm for panel a and 10 μm for panels b to f.

necessarily chitinolytic/proteolytic, since in the marine sites, for example, the squid pen remained relatively intact for many months. However, as chitin is a particulate substrate, it is to be expected that chitin-degrading microbes will attach to it to obtain their nutrients. The importance of the physical nature of the substrate is shown in experiment by Ou and Alexander (1974). They found that admixture with glass microbeads enhanced the rate of mineralization of chitin particles by pure cultures of a *Pseudomonas* species and by soil inocula, whereas separation of chitin and bacteria by layers of microbeads greatly reduced the rate.

Particular attention has been paid to the adsorption of the pathogenic but also chitinolytic *Vibrio* species. Kaneko and Colwell (1975) described a strong adsorption of *Vibrio parahaemolyticus,* but not *Escherichia coli* or *Pseudomonas fluorescens,* onto chitin particles and the chitinous surface of copepods. Kaneko and Colwell suggested that this phenomenon has ecological significance for this estuarine organism, as the adsorption was reduced by increasing salinity and pH from estuarine values to seawater values, which would favor retention of the bacteria within the estuary. Kaneko and Colwell (1978) correlate the annual cycle of *V. parahaemolyticus* in Chesapeake Bay with that of the zooplankton. The maximum numbers occur in the summer months on the zooplankton, in the water column, and in the sediment. Kaneko and Colwell suggest that the *V. parahaemolyticus* associated with the zooplankton are involved in their subsequent mineralization. They define two main clusters of isolates: Group I-A are chitinolytic and would play a significant ecological role in the initial colonization of copepods during spring and summer months; Group II-A do not possess a chitinase but can utilize *N*-acetylglucosamine and so may be important in the complete decomposition or mineralization of copepods.

Similarly, the adsorption of *Vibrio cholerae* onto chitin and chitinous exoskeletons of invertebrates has been described in detail by several authors. Huq *et al.* (1983, 1984) reported that cells of *V. cholerae,* both O1 and non-O1 serovars, attached to live copepods, especially the oral region and egg sacs, but not to killed animals. Survival of the bacteria was extended in the presence of live copepods. Nalin *et al.* (1979) described the adherence of *V. cholerae* to chitin, suggesting that this might decrease the dose needed to establish infection, and Dietrich *et al.* (1984) extended this observation to record adherence of the bacteria to shell of blue crab, *Callinectes sapidus.* Adherence was affected by temperature, salt concentration, and pH, being maximum at 35°C, 2%, and pH 7. MacDonell *et al.* (1984), however, reported that in their experiments, cells of *V. cholerae* strain WF110 did not adhere to chitin from nutrient-deprived culture but only to chitin from nutrient-enriched culture, and then only about half the cells adhered. Amako *et al.* (1987) described the protective effect of chitin against the killing of *V. cholerae* strains at low temperature. In the presence of 0.5% chitin, the values for percent survival of cells in saline at 0°C were 73, 42, 23, and 30 at days 1,2,3, and 4, in comparison with 0% survival at day 1 in the control or in experiments with a range of other additions, including *N*-acetylglucosamine, starch, and peptidoglycan. The minimum concentration of particulate chitin giving this protective effect was 0.04% (wt/vol).

Belas and Colwell (1982) present a detailed study of the importance of flagellation to adsorption of cells of *V. cholerae, V. parahaemolyticus,* and a range of other bacteria to chitin. Both species have a polar flagellum, but under some environmental conditions

V. parahaemolyticus cells produce numerous lateral flagella. Adsorption of laterally flagellated *V. parahaemolyticus* to chitin particles at 15, 25, 37, and 45°C followed the Langmuir adsorption isotherm; i.e., bacteria saturated the chitin surface so that, as the number of bacteria added to the assay was increased, the number of bacteria bound to the chitin reached a plateau and did not increase further. The adsorption was strongest at 25°C. At 4°C, however, the bacteria were only polarly flagellated, and proportional adsorption kinetics were observed; i.e., the number of bacteria bound to chitin was directly proportional to the number of organisms added to the system. The same proportional adsorption kinetics were observed with four of six strains of polarly flagellated *V. cholerae*. The authors present further results from scanning electron microscopy and from competition experiments that suggest that laterally flagellated cells will saturate the binding sites on the chitin surface, whereas polarly flagellated cells will continue adsorbing, eventually clumping to cells already bound. The authors further suggest that the formation of lateral flagella, after initial reversible adsorption of a polarly flagellated cell, could increase forces holding the bacterium to the surface and be an intermediate stage to the irreversible adsorption involving the formation of a polysaccharide glycocalyx.

Other chitin-digesting microbes also bind to their substrate. ZoBell and Rittenberg (1938) describe "obligate periphytism" among their isolates of marine chitinoclastic bacteria:

> For example, one culture covered the chitin strip with a heavy orange growth while the surrounding menstruum remained quite colorless. When a loopful of the liquid was transferred to another tube of chitin medium no growth developed. However, when a bit of the orange growth scraped from the original chitin strip was used as the inoculum, bacteria began to develop at once on the new chitin strip.

Cytophaga johnsonae, a ubiquitous soil organism (Stanier, 1947; Christensen, 1977), characteristically binds to chitin as it degrades it, but Sundarraj and Bhat (1972) divided strains into two groups: those only hydrolyzing chitin in close contact with it, and those that release cell-free chitinases. An isolate of the first type, C35, grew more slowly than an isolate of the second type, C31, in an unshaken liquid medium with chitin as the carbon source, and the cells adhered to the chitin particles. Shaking the cultures slowed the growth of C31 and prevented growth of C35. Wolkin and Pate (1985) describe a class of nonmotile mutants of *C. johnsonae* with an interesting pleiotropy. In addition to being nonmotile and failing to move latex beads over their surface, they are all unable to digest and utilize chitin (used by the parental strain), and all are resistant to phages that infect the parental strain and have relatively nonadherent and nonhydrophobic surfaces compared with wild-type strains. The authors conclude that all characteristics associated with the pleiotropy require moving cell surfaces, and thus chitin digestion requires some feature of this, presumably involving enzymatic contact between bacterium and substrate.

Pel and Gottschal (1986a) illustrate the direct contact between the anaerobic chitin-digesting bacterium *Clostridium* sp. strain 9.1 and chitin microfibrils (Fig. 7). As in the case of cellulolytic *Clostridium* species, this may involve specific enzymatic structures on the cell surface.

Filamentous chitinolytic microbes can be observed adhering to chitinous material and burrowing into it. Thus, baiting of soil and fresh water with chitin will yield attached chitinolytic chytrids such as the *Chytriomyces* species described by Reisert and Fuller (1962), which will penetrate the material by their rhizoids. Penetration of chitinous mollusc shells by a range of filamentous organisms, including fungi, is illustrated by Poulicek *et al.* (1986c; Fig. 8).

Figure 7. *Clostridium* sp. strain 9.1 growing on chitin fibers in an anaerobic medium containing 0.1% (wt/vol) chitin. (a) Scanning electron micrograph showing close association between cells and chitin, 5 days of growth. (b) Electron micrograph of a cell negatively stained with uranyl acetate. Arrows point to chitin fibers attached to the cell surface. Scale bars represent 0.5 μm. [From Pel and Goltschal (1986a).]

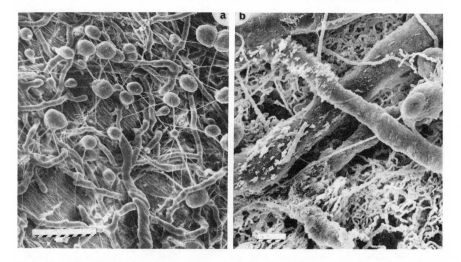

Figure 8. Two scanning electron micrographs showing diversity of microbes colonizing and boring into mollusc shells after 4 years of immersion at 37 m deep in the Bay of Calvi, Mediterranean Sea. Scale bars represent 10 μm. [Micrographs from M. Poulicek (see Poulicek *et al.*, 1986c).]

9. Chitin Digestion in Animals

Chitin in fungi and invertebrates composes a considerable part of the diet of many herbivorous and carnivorous animals. There can be three sources of chitinolytic enzymes in the animal's digestive system: from the animal itself, from the endogenous commensal gut microflora, or from the ingested food.

Most work has been done with fish. The typical marine fish gut microflora is dominated by bacteria that are characteristically chitinolytic, notably *Vibrio* and *Photobacterium* species and *Enterobacteriaceae* (Hamid *et al.*, 1979; Lindsay and Gooday, 1985b; MacDonald *et al.*, 1986; Mowlah *et al.*, 1979; Okutani, 1966). This finding led to the suggestion that these bacteria played a symbiotic role in the fish gut analogous to that of cellulolytic microbes in the rumen. Thus, Goodrich and Morita (1977a,b) ascribe the gut chitinolytic activities in the marine fish studied by them as being the products of their microbial flora. This view, however, is not borne out by extensive experiments with a wide range of other fish. With cod, *Gadus morhua*, Danulat and Kausch (1984), Danulat (1986), and Lindsay and Gooday (1985b) found no correlation between chitinase activities of stomach or intestine and the bacterial flora. Further, Lindsay and Gooday (1985b) and Lindsay (1984) found no correlations between feeding and either enzyme activities or bacterial floras. Thus, they concluded that in cod the major chitinolytic activity is from constitutive chitinases produced by fish stomach and fish intestine. Similar conclusions were reached by Kono *et al.* (1987a,b) for red sea bream *Pagrus major*, Japanese eel *Anguilla japonica*, and yellowtail *Seriola quinqueradiata*.

Both Kono *et al.* (1987b) with bream and Lindsay *et al.* (1984) with rainbow trout *Salmo gairdneri* showed that treatment with antibacterial antibiotics greatly reduced the numbers of gut bacteria but had no effect on chitinase activities nor, in the trout, did feeding with chitin in the diet or with chitin plus live chitinolytic bacteria (*Vibrio alginolyticus*). Furthermore, in the bream and eel, feeding with chitin had no effect on the numbers of total bacteria or chitinolytic bacteria in the digestive systems. In Japanese sea bass *Lateolabrox japonicus* and in Dover sole *Solea solea*, Okutani (1966) and Clark *et al.* (1988), respectively, describe a stomach chitinase activity from the fish and one intestinal activity that may in part be contributed by bacteria.

It is now clear from direct sampling of fish tissue that a wide range of fish produce constitutive chitinase activities in their digestive tracts (Yoshida and Sera, 1970; Micha *et al.*, 1973; Fange *et al.*, 1979; Jeuniaux *et al.*, 1982; Lindsay, 1984; Rehbein *et al.*, 1986; Benmouna *et al.*, 1986). These activities, however, show no correlation with the diet of the fish; rather, they correlate with the mode of feeding; fish that ingest prey whole (gulpers) have high activities, fish that mechanically disrupt prey (grinders) have low activities (Lindsay, 1984). A further feature is that despite constitutive stomach and intestine chitinases and (variable) chitinolytic gut microflora, Lindsay *et al.* (1984) showed in rainbow trout that chitin is poorly digested and is a very poor nutrient. The conclusion is that the fish chitinases primarily are food-processing enzymes; i.e., they are for shelling shrimps.

Among other vertebrates, there are reports of chitinases in gastrointestinal tracts. Studies of activities from gastric mucosal extracts and thus presumably of animal origin include those from the primate *Perodicticus potto,* dog, fox, pig, mouse, mole, hedgehog, bat, insectivorous birds, reptiles, and amphibians (Jeuniaux, 1963, 1971; Dandrifosse *et al.*, 1965; Dandrifosse, 1975; Cornelius *et al.*, 1975, 1976). Studies that suggest the involvement of chitinolytic bacteria in chitin digestion include those on baleen whales, rat, and chicken. Seki and Taga (1965c) reported that all of the bacteria isolated from samples of stomachs of fin whale and sperm whale that had been stored frozen for several months were chitinolytic. Herwig *et al.* (1984) present evidence for a rumen-type fermentation in baleen whales and suggest that chitin could be a substrate. Kuhl *et al.* (1978) found elevated cecal weights in chitin-fed rats and this finding, together with examination of feces, suggested the participation of intestinal bacteria in chitin degradation. Austin *et al.* (1981) fed chickens with diets supplemented with chitin and whey and reported that bacterial lysis of the chitin produced growth factors for *Bifidobacterium bifidus,*which then utilized the lactose in the whey, resulting in enhanced growth of the chickens. Patton and Chandler (1975) and Patton *et al.* (1975) describe digestion of chitin by calves and steers, implying a chitinolytic flora in the bovine rumen.

Among invertebrates, chitin digestion is probably widespread with or without the participation of a microbial chitinolytic flora (Jeuniaux, 1963, 1971; Elyakova, 1972; Smucker, 1982; Smucker and Wright, 1984). In chitin-producing organisms, it will of course be important to distinguish between chitinase activities of digestive significance and those involved in morphogenesis. Chitinases involved in digestion include those of the crystalline styles of mussels and oysters (Birkbeck and McHenery, 1984; Smucker and Wright, 1984) and of the digestive fluids of a spider (Mommsen, 1980) and a snail (Lunblad *et al.*, 1976). In a variety of invertebrates, there is evidence that commensal chitinolytic gut bacteria play an important role in chitin digestion: white shrimp *Panaeus setiforus* (Hood and Meyers, 1977a), octopus *Polypus vulgaris* and squid *Loligo edulis* (Seki and Taga, 1963b), a deep-sea amphipod (Schwarz *et al.*, 1976), and copepods (Lear, 1963; Sochard *et al.*, 1979).

Whether or not chitinous structures are completely degraded by the animal enzymes and/or the gut microflora during passage through the gut, this passage in itself will be of importance in enhancing the recycling of the chitin, for example by comminution, by decalcification, and by incubation with microbes. Thus, any chitin in the feces will certainly be more biodegradable than that in the original food.

10. Involvement of Chitin Degradation in Pathogenesis and Symbiosis

Pathogens of chitin-producing organisms are usually found to produce chitinases. These presumably aid the invasion of the pathogen, as well as providing nutrients directly in the form of amino sugar and indirectly by exposing other host material to enzymatic degradation. Examples include the oomycete *Aphanomyces astaci,* a pathogen of crayfish (Unestam, 1966, 1968; Soderhall and Unestam, 1975); the fungus *Paecilomyces lilacius,* a pathogen of nematode eggs (Culbreath *et al.,* 1986); the entomopathogenic fungi *Beauveria bassiana, Metarrhizium anisophiae,* and *Verticillium lecanii* (Smith and Grula, 1983; St Leger *et al.,* 1986a,b); mycophilic fungi, *Cladobotryum* species (G. W. Gooday, unpublished observations; Zhloba *et al.,* 1980); the bacterial *Serratia* species, insect pathogens (Ewing *et al.,* 1973; Lysenko, 1976; Grimont *et al.,* 1988); and a *Photobacterium* species causing exoskeleton lesions of the tanner crab (Baross *et al.,* 1978). Iverson *et al.* (1984) describe the symbiotic association of chitinolytic bacteria *Serratia lignefaciens* and *Serratia marcescens* with the sugar beet root maggot. They show the penetration of the inner puparial surface by the bacteria and suggest that chitin digestion by the bacteria aids in imago emergence. Among the protozoa, the ciliate *Ascophyrs rodor* is an ectoparasite of the red shrimp, attaching itself to the exoskeleton, into which it sinks by digesting the area underneath itself and which forms its sole diet (Bradbury *et al.,* 1987).

The fungal gardens of insects, in the nests of macrotermitinids and attine ants and in tunnels in wood excavated by ambrosia beetles, involve the digestion of the fungal chitin by the insects. In the case of attine ants, this involves the predigestion of some of the chitin via deposition of feces rich in chitinase on the fungal mycelium (Martin *et al.,* 1973, 1981) as well as digestion in the insect's infrabuccal pocket by chitinases from the labial glands (Febvay *et al.,* 1984).

A rapid response of plants to infection, wounding, or treatment with ethylene is the induction of antifungal chitinase activity (Metraux and Boller, 1986; Schlumbaum *et al.,* 1986; Mauch *et al.,* 1988). This can be seen as a first-line defense against fungal attack. Likewise, seeds of maize, wheat, and barley have antifungal chitinase activities (Roberts and Selitrennikoff, 1988). The insectivorous plants characteristically produce chitinases as digestive enzymes (e.g., Amagase *et al.*, 1972). It may be that the insectivorous habit in plants has arisen by adaptation of their chitinase defense system. Some animals have chitinase activities in their blood and other tissues, e.g., bovine serum (Lunblad *et al.*, 1979) and turbot plasma (Manson *et al.*, 1989), and it may be that these activities play a defense role against potential pathogens.

11. Summary

This review has shown the degradation of chitin as a process of major importance in the biosphere. The chitin is produced in all natural environments, globally in enormous quantities, and its mineralization is primarily a microbial process. The final generalization may be made that the rate of degradation of chitin is higher (1) with increased environmental temperature, (2) with smaller particle size, (3) with involvement of meio- and macrofauna, both for commination and enzymatic action, and (4) in its native state, i.e., with associated proteins, than in a purified state. On the other hand, the degradation rate is lower if the chitin is protected by melanin or a water-repellent coating. It is clear, however, that in all environments nearly all of the chitin is recycled in the carbon and nitrogen cycles despite its relatively refractory nature.

The recycling of the chitin carbon is epitomized by Primo Levi (1984), who traces the history of one atom via many adventures into chitin of an insect which

> lays its eggs, and dies: the small cadaver lies in the undergrowth of the woods, it is emptied of its fluids, but the chitin carapace resists for a long time almost indestructible. The snow and sun return above it without injuring it: it is buried by the dead leaves and the loam, it has become a slough, a "thing," but the death of atoms, unlike ours, is never irrevocable. Here are at work the omnipresent, untiring, and invisible gravediggers of the undergrowth, the microorganisms of the humus. The carapace, with its eyes by now blind, has slowly disintegrated, and the ex-drinker, ex-cedar, ex-wood worm has once again taken wing.

References

Abderhalden, E., and Heyns, K., 1933, Nachweis von Chitin in Flügelresten von Coleopteren des oberen Mitteleocäns. *Biochem. Z.* **259:**320–321.

Allan, G. G., Fox, J. R., and Kong, N., 1978, A critical evaluation of the potential sources of chitin and chitosan, in: *Proceedings of the First International Conference on Chitin/Chitosan* (R. A. A. Muzzarelli and E. R. Pariser, eds.), pp. 64–78, MIT Sea Grant Report 78-7.

Amagase, S., Mori, M., Nakayama, S., 1972, Digestive enzymes in insectivorous plants. IV. Enzymatic digestion of insects by *Nepenthes* secretion and *Drosera peltata* extract: Proteolytic and chitinolytic activities, *J. Biochem.* **72:**765–767.

Amako, K., Shimodori, S., Imoto, T., Miake, S., and Umeda, A., 1987, Effects of chitin and its soluble derivatives on survival of *Vibrio cholerae* O1 at low temperature, *Appl. Environ. Microbiol.* **53:**603–605.

Aronson, J. M., and Lin, C. C., 1978, Hyphal wall chemistry of *Leptomitus lacteus, Mycologia* **70:**363–369.

Arroyo-Begovich, A., and Carabez-Trejo, A., 1982, Location of chitin in the cyst wall of *Entamoeba invadens* with colloidal gold tracers, *J. Parasitol.* **68:**253–258.

Aruchami, M., Sundara-Rajulu, G., and Gowri, N., 1986, Distribution of deacetylase in arthropods, in: *Chitin in Nature and Technology* (R. A. A. Muzzarelli, C. Jeuniaux, and G. W. Gooday, eds.), pp. 263–265, Plenum Press, New York.

Aumen, N. G., 1980, Microbial succession on a chitinous substrate in a woodland stream, *Microbial Ecol.* **6:**317–327.

Austin, P. R., Brine, C. J., Castle, J. E., and Zikakis, J. P., 1981, Chitin—new facets of research, *Science* **212:**749–753.

Baross, J. A., Tester, P. A., and Morita, R. Y., 1978, Incidence, microscopy and etiology of exoskeleton lesions in the tanner crab *Chionectes tanner, J. Fish. Res. Board Can.* **35:**1141–1149.

Bartnicki-Garcia, S., and Lippman, E., 1982, Fungal cell wall composition, in: *CRC Handbook of Microbiology*, 2nd ed., Vol. IV, *Microbial Composition: Carbohydrates, Lipids and Minerals* (A. J. Laskin and H. A. Lechevalier, eds.), pp. 229–252, CRC Press, Boca Raton.

Baxby, P., and Gray, T. R. G., 1968, Chitin decomposition in soil. I. Media for isolation of chitinoclastic micro-organisms from soil. *Trans. Br. Mycol. Soc.* **51:**287–242.

Belas, M. R., and Colwell, R. R. 1982, Adsorption kinetics of laterally and polarly flagellated Vibrio, *J. Bacteriol.* **151:**1568–1580.

Benmouna, M., Jaspar-Versali, M. F., Toussaint, C., and Jeuniaux, C., 1986, A comparative study of chitinase activity in digestive tract of *Serranus cabrilla* and *Serranus scriba, Biochem. Syst. Ecol.* **14:**435–437.

Berkeley, R. C. W., 1978, Chitinolytic and chitosanolytic microorganisms and the potential biodeterioration problem in the commercial application of chitin and its derivatives, in: *Proceedings of the First International Conference on Chitin/Chitosan* (R. A. A. Muzzarelli and E. R. Pariser, eds.), pp. 570–577, MIT Sea Grant Report 78-7.

Berkeley, R. C. W., 1979, Chitin, chitosan and their degradative enzymes, in: *Microbial Polysaccharides and Polysaccharases* (R. C. W. Berkeley, G. W. Gooday, and D. C. Ellwood, eds.), pp. 205–236, Academic Press, London.

Billy, C., 1969, Etude d'une bacterie chitinolytique anaerobie *Clostridium chitinophilum* n.sp., *Ann. Inst. Pasteur* **6:**75–82.

Birkbeck, T. H., and McHenery, J. G., 1984, Chitinase in the mussel, *Mytilus edulis* (L.), *Comp. Biochem. Physiol.* **77B:**861–865.

Blackwell, J., and Weih, M. A., 1984, The structure of chitin-protein complexes, in: *Chitin, Chitosan and Related Enzymes* (J. P. Zikakis, ed.), pp. 257–272, Academic Press, New York.

Boland, A., 1987, Chitinolyse dans les sediments marins, *Ann. Soc. R. Zool. Belg.* **117:**111.

Boyer, J. N., 1986, End products of anaerobic chitin degradation by salt marsh bacteria as substrates for dissimilatory sulfate reduction and methanogenesis, *Appl. Environ. Microbiol.* **52:**1415–1418.

Boyer, J. N., and Kator, H. I., 1985, Method for measuring microbial degradation and mineralisation of ^{14}C-labelled chitin obtained from the blue crab, *Callinectes sapidus, Microb. Ecol.* **11:**185–192.

Bradbury, P., Deroux, G., and Campillo, A., 1987, The feeding apparatus of a chitinivorous ciliate, *Tissue Cell* **19:**351–363.

Brisou, J., Tysset, C., de Rautlin, de la Roy, Y., Curcier, R., and Moreau, R., 1964, Etude sur la chitinolyse a milieu marin, *Ann. Inst. Pasteur* **106:**469–478.

Brown, L. R., Brown-Skrobot, S., Teichart, C., Blasingame, D. J., and Ladner, C. M., 1982, The use of chitinous seafood wastes for the control of plant parasitic nematodes, in: *Chitin and Chitosan* (S. Hirano and S. Tokura, eds.), pp. 227–232, The Japanese Society of Chitin and Chitosan, Tottori.

Brumioul, D., and Voss-Foucart, M. F., 1977, Substances organiques dans kles carapaces de crustaces fossiles. *Comp. Biochem. Physiol.* **57B:**171–175.

Bull, A. T., 1970, Inhibition of polysaccharidases by melanin: Enzyme inhibition in relation to mycolysis, *Arch. Biochem. Biophys.* **137:**345–356.

Burnett, J. H., 1979, Aspects of the structure and growth of hyphal walls, in: *Fungal Walls and Hyphal Growth* (J. H. Burnett and A. P. J. Trinci, eds.) pp. 1–25, Cambridge University Press, Cambridge.

Bussers, J. C., and Jeuniaux, C., 1974, Recherche de la chitine dans les productions métaplasmatiques de quelques ciliés, *Protistologica* 10:43–46.

Buxton, E. W., Khalifa, O., and Ward, V., 1965, Effect of soil amendment with chitin on pea wilt caused by *Fusarium oxysporum* f. *pisi*, *Ann. Appl. Biol.* 55:83–88.

Campbell, L. L., and Williams, O. B., 1951, A study of chitin-decomposing microorganisms of marine origin, *J. Gen. Microbiol.* 5:894–905.

Campos-Takai, G. M., Dietrich, S. M. C., and Mascaranhas, Y., 1982, Isolation and characterisation of chitin from the cell walls of *Achlya radiosa*, *J. Gen. Microbiol.* 128:207–209.

Carlisle, D. B., 1964, Chitin in the Cambrian fossil *Myolithellus*, *Biochem. J.* 90:1c.

Chan, J. G., 1970. The Occurrence, Taxonomy and Activity of Chitinolytic Bacteria from Sediment, Water and Fauna of Puget Sound, Ph.D. thesis, University of Washington, Seattle.

Christensen, P. J., 1977, The history, biology and taxonomy of the Cytophaga group, *Can. J. Microbiol.* 23:1599–1653.

Clark, J., Quayle, K. A., Macdonald, N. L,, and Stark, J. R., 1988, Metabolism in marine flatfish—V. Chitinolytic activities in dover sole *Solea solea* (L.), *Comp. Biochem. Physiol.* 90B:379–384.

Clarke, P. H., and Tracey, M. V., 1956, The occurrence of chitinase in some bacteria, *J. Gen. Microbiol.* 14:188–196.

Cornelius, C., Dandrifosse, C., and Jeuniaux, C., 1975, Biosynthesis of chitinase by mammals of the order Carnivora, *Biochem Syst. Ecol.* 3:121–122.

Cornelius, C., Dandrifosse, C., and Jeuniaux, C., 1976, Chitinolytic enzymes of the gastric mucosa of *Perodictus potto* (primate prosimian): Purification and enzyme activity, *Eur. J. Biochem.* 7:445–448.

Cross, M., 1985, Microbial Colonisation and Degradation of Chitin in Aquatic Environments, Ph.D. thesis, University of Aberdeen, Aberdeen, Scotland.

Culbreath, A. K., Rodriguez-Kabana, R., and Morgan-Jones, G., 1986, Chitin and *Paecilomyces lilacinus* for control of *Meloidogyne arenaria*, *Nematropica* 16:153–166.

Dandrifosse, G., 1975, Purification of chitinases contained in pancreas or gastric mucosa of frog, *Biochemie* 57:829–831.

Dandrifosse, G., Schoffeniels, E., and Jeuniaux, C., 1965, Secretion de chitinase par la muqueuse gastrique isolee, *Biochim. Biophys. Acta* 94:153–164.

Danulat, E., 1986, Role of bacteria with regard to chitin degradation in the digestive tract of the cod *Gadus morhua*, *Mar. Biol.* 90:335–343.

Danulat, E., and Kausch, H., 1984, Chitinase activity in the digestive tract of the cod *Gadus morhua* (L.), *J. Fish Biol.* 24:125–133.

Datema, R., Ende, van den M., and Wessels, J. G. H., 1977, The hyphal wall of *Mucor mucedo*. 2. Hexosamine-containing polymers, *Eur. J. Biochem.* 80:621–626.

Davis, B., and Eveleigh, D. E., 1984, Chitosanases: Occurrence, production and immobilization, in: *Chitin, Chitosan and Related Enzymes* (J. P. Zikakis, ed.), pp. 161–179, Academic Press, Orlando.

Davis, L. L., and Bartnicki-Garcia, S., 1984, The co-ordination of chitosan and chitin synthesis in *Mucor rouxii*, *J. Gen. Microbiol.* 130:2095–2102.

Deming, J. W., 1985, Bacterial growth in deep-sea sediment trap and boxcore samples, *Mar. Ecol. Prog. Ser.* 25:305–312.

Deming, J. W., 1986, Ecological strategies of barophilic bacteria in the deep ocean, *Microbiol. Sci.* 3:205–211.

Dietrich, M. A., Hackney, C. R., and Grodner, R. M., 1984, Factors affecting the adherence of *Vibrio cholerae* to blue crab (*Callinectes sapidus*) shell, in: *Vibrios in the Environment* (R. R. Colwell, ed.), pp. 601–611, John Wiley & Sons, New York.

Donderski, W., 1984, Chitinolytic bacteria in water and bottom sediments of two lakes of different trophy, *Acta Microbiol. Pol.* **33**:163–170.

Donderski, W., Bylinska, T., Czajkowski, H., Gardocka, B., Kowalkowska, D., Myzyk, G., and Poziemska, E., 1984, Initial studies on heterotrophic bacteria capable of decomposition of some macromolecular compounds in waters and bottom sediments of six lakes in Ilawa Lake District, *Acta Univ. Nicolai Copernici* **57**:75–82.

Elyakova, L. A., 1972, Distribution of cellulases and chitinases in marine invertebrates, *Comp. Biochem. Physiol.* **43B**:67–70.

Everson, I., 1977, The Living Resources of the Southern Ocean, FAO Southern Ocean Fisheries Survey Programme, GLO/50/77/1.

Ewing, W. H., Davis, B. R., Fife, M. A., and Lessel, E. F., 1973, Biochemical characterization of *Serratia liquefaciens* (Grimes and Hennerty) Bascomb et al. (formerly *Enterobacter liquefaciens*) and *Serratia rubidaca* (Stapp) comb. nov. and designation of type and neotype strains, *Int. J. Syst. Bacteriol.* **23**:217–225.

Fange, R., Lunblad, G., Lind, J., and Slettengren, K., 1979, Chitinolytic enzymes in the digestive system of marine fishes, *Mar. Biol.* **53**:317–321.

Fargues, J., Kilbertus, G., Reisinger, O., and Olah, G. M., 1977, Chitinolytic activity in soils, in: *Soil Biology and Conservation of the Biosphere* (J. Szegi, ed.), pp. 257–260, Akademiai Kiado, Budapest.

Faure-Raynaud, M., 1981, Determination de l'activite chitinolytique de microorganismes, bacteria et levures, de la litiere du sapin *Abies alba* Mill, *Ann. Microbiol. (Inst. Pasteur)* **132B**:267–279.

Febvay, G., Decharme, M., and Kermarrec, A., 1984, Digestion of chitin by the labial glands of *Acromyrmex octospinosus* Reich (Hymenoptera: Formicidae), *Can. J. Zool.* **62**:229–234.

Fenton, D., Davis, B., Rotgers, C., and Eveleigh, D. E., 1978, Enzymatic hydrolysis of chitosan, in: *Proceedings of the First International Conference on Chitin/Chitosan* (R. A. A. Muzzarelli and E. R. Pariser, eds.), pp. 525–541, MIT Sea Grant Report 78-7.

Florkin, M., 1965, Paléoproteines, *Acad. R. Belg. Bull. Cl. Sci.* **51**:156–169.

Gardner, K. H., and Blackwell, J., 1975, Refinement of the structure of β-chitin, *Biopolymers* **14**:1581–1595.

Giraud-Gaille, M. M., and Bouligand, Y., 1986, Chitin-protein molecular organization in arthropod, in: *Chitin in Nature and Technology* (R. A. A. Muzzarelli, C. Jeuniaux, and G. W. Gooday, eds.), pp. 29–35, Plenum Press, New York.

Godoy, G., Rodriguez-Kabana, R., Shelby, R. A., and Morgan-Jones, G., 1983, Chitin amendments for control of *Meloidogyne arenaria* in infested soils. 2. Effects of microbial population, *Nematropica* **13**:63–74.

Gooday, G. W., 1979, A survey of polysaccharase production: A search for phylogenetic implications, in: *Microbial Polysaccharides and Polysaccharases* (R. C. W. Berkeley, G. W. Gooday, and D. C. Ellwood, eds.), pp. 437–460, Academic Press, London.

Gooday, G. W., 1983, The microbial synthesis of cellulose, chitin and chitosan, *Prog. Ind. Microbiol.* **18**:85–127.

Gooday, G. W., and Trinci, A. P. J., 1980, Wall structures and biosynthesis in fungi, in: *The Eukaryotic Microbial Cell* (G. W. Gooday, D. Lloyd, and A. P. J. Trinci, eds.), 30th Symposium of Society for General Microbiology, Cambridge University Press, Cambridge.

Goodrich, T. D., and Morita, R. Y., 1977a, Incidence and estimation of chitinase activity associated with marine fish and other estuarine samples, *Mar. Biol.* **41**:349–353.

Goodrich, T. D., and Morita, R. Y., 1977b, Bacterial chitinase in the stomachs of marine fishes from Yaquina Bay, Oregon, USA, *Mar. Biol.* **41**:335–360.

Gould, W. D., Bryant, R. J., Trofymov, J. A., Anderson, R. V., Elliott, E. T., and Coleman, D. C., 1981, Chitin decomposition in a model soil system. *Soil Biol. Biochem.* **13**:487–492.

Gow, N. A. R., and Gooday, G. W., 1983, Ultrastructure of chitin in hyphae of *Candida albicans* and other dimorphic and mycelial fungi, *Protoplasma* **115**:52–58.

Gow, N. A. R., Gooday, G. W., Rosser, J. D., and Wilson, M. J., 1987, Infrared and X-ray diffraction data on chitins of variable structure, *Carbohydr. Res.* **165**:105–110.

Gowri, N., Aruchami, M., and Sundara-Rajulu, G., 1986, Natural deacetylation of the cuticle in *Sacculina rotundata*, in: *Chitin in Nature and Technology* (R. A. A. Muzzarelli, C. Jeuniaux, and G. W. Gooday, eds.), pp. 266–268, Plenum Press, New York.

Gray, T. R. G., and Baxby, P., 1968, Chitin decomposition in soil. The ecology of chitoclastic micro-organisms in forest soil, *Trans. Br. Mycol. Soc.* **51**:293–309.

Gray, T. R. G., and Bell, T. F., 1963, The decomposition of chitin in an acid soil, in *Soil Organisms* (J. Doeksen and J. van der Drift, eds.), pp. 222–230, North Holland Publishing Co., Amsterdam.

Grimont, P. A. D., Jackson, T. A., Ageron, E., and Noonan, M. J., 1988, *Serratia entomophila* sp. nov. associated with amber disease in the New Zealand grass grub *Costerytra zealandica*, *Int. J. Syst. Bacteriol.* **38**:1–6.

Hadwiger, L. A., Fristensky, B., and Riggelman, R. C., 1984, Chitosan, a natural regulator in plant-fungal pathogen interactions, increases crop yields, in: *Chitin, Chitosan and Related Enzymes* (J. P. Zikakis, ed.), pp. 291–302, Academic Press, Orlando.

Hamid, A., Sakuda, T., and Kakimoto, D., 1979, Microflora in the alimentary tract of gray mullet—4. Estimation of enzymic activity of the intestinal bacteria. *Bull. Jpn. Soc. Sci. Fish.* **45**:99–106.

Hedges, A., and Wolfe, R. S., 1974, Extracellular enzyme from myxobacter AL-1 that exhibits both β-1,4-glucanase and chitosanase activities. *J. Bacteriol.* **120**:844–853.

Helmke, E., and Weyland, H., 1986, Effect of hydrostatic pressure and temperature on the activity and synthesis of chitinases of Antarctic Ocean bacteria. *Mar. Biol.* **91**:1–7.

Henis, Y., Sneh, B., and Katan, J., 1967. Effect of organic amendments on *Rhizoctonia* and accompanying microflora in soil, *Can. J. Microbiol.* **13**:643–649.

Herth, W., Mulisch, M., and Zugenmaier, P., 1986, Comparison of chitin fibril structure and assembly in three unicellular organisms, in: *Chitin in Nature and Technology* (R. A. A. Muzzarelli, C. Jeuniaux, and G. W. Gooday, eds.), pp. 107–120, Plenum Press, New York.

Herwig, R. P., Staley, J. T., Nerini, M. K., and Braham, H. W., 1984, Baleen whales: Preliminary evidence for forestomach microbial fermentation, *Appl. Environ. Microbiol.* **47**:421–423.

Herwig, R. P., Pellerin, N. B., Irgens, R. L., Maik, J. S., and Staley, J. T., 1988, Chitinolytic bacteria and chitin mineralization in the marine waters and sediments along the Antarctic Peninsula, *FEMS Microbiol. Ecol.* **53**:101–112.

Hillman, K., Gooday, G. W., and Prosser, J. I., 1989, A simple model system for small scale *in vitro* study of estuarine sediment ecosystems, *Lett. Appl. Microbiol.* **8**:41–44.

Hock, C. W., 1940, Decomposition of chitin by marine bacteria, *Biol. Bull.* **79**:199–206.

Hood, M. A., and Meyers, S. P., 1973, The biology of aquatic chitinolytic bacteria and their chitinolytic activities, *Mer* **11**:213–219.

Hood, M. A., and Meyers, S. P., 1977a, Microbiological and chitinoclastic activities associated with *Panaeus setiferus*, *J. Oceanogr. Soc. Jpn.* **33**:235–241.

Hood, M. A., and Meyers, S. P., 1977b, Rates of chitin degradation in an estuarine environment, *J. Oceanogr. Soc. Jpn.* **33**:328–334.

Hood, M. A., and Meyers, S. P., 1978, Chitin degradation in estuarine environments and implications in crustacean biology, in: *Proceedings of the First International Conference on Chitin/Chitosan* (R. A. A. Muzzarelli and E. R. Pariser, eds.), pp. 563–569, MIT Sea Grant Report 78-7.

Hsu, S. C., and Lockwood, J. L., 1975, Powdered chitin as a selective medium for enumeration of actinomycetes in water and soil, *Appl. Microbiol.* **29**:422–426.

Huq, A., Small, E. B., West, P. A., Huq, M. I., Rahman, R., and Colwell, R. R., 1983, Ecological relationships between *Vibrio cholerae* and planktonic crustacean copepods, *Appl. Environ. Microbiol.* **45**:275–283.

Huq, A., West, P. A., Small, E. B., Huq, M. I., and Colwell, R. R., 1984, Influence of water temperature, salinity and pH on survival and growth of toxigenic *Vibrio cholerae* serovar O1 associated with live copepods in laboratory microcosms, *Appl. Environ. Microbiol.* **48**:420–424.

Ikeda, T., and Dixon, P., 1982, Observations on moulting in Antarctic krill (*Euphausia superba* Dana), *Aust. J. Mar. Freshwater Res.* **33**:71–76.

Iverson, K. L., Bromel, M. C., Anderson, A. W., and Freeman, T. P., 1984, Bacterial symbionts in the sugar beet root maggot *Tetanops myopaeformis* (von Röder), *Appl. Environ. Microbiol.* **47**:22–27.

Jeuniaux, C., 1963, *Chitine et Chitinolyse*, Masson et Cie, Paris.

Jeuniaux, C., 1971, On some biochemical aspects of regressive evolution in animals, in: *Biochemical Evolution and the Origin of Life* (E. Schoffeniels, ed.), pp. 304–313, North Holland Publishing Co., Amsterdam.

Jeuniaux, C., 1981, Faunistique et ecologie chimique des peuplements benthique sur substrats dur et des sediments de la Baie de Calvi (Corse), *Bull. Soc. R. Sci. Liege* **11–12**:446–452.

Jeuniaux, C., 1982, La chitine dans le régne animal, *Bull. Soc. Zool. France* **107**:363–386.

Jeuniaux, C., Dandrifosse, G., and Micha, J. C., 1982, Caracteres et evolution de enzymes chitinolytiques chez les vertebrates inferieurs. *Biochem. Syst. Ecol.* **10**:365–372.

Jeuniaux, C., Bussers, J. C., Voss-Foucart, M. F., and Poulicek, M., 1986, Chitin production by animals and natural communities in marine environment, in *Chitin in Nature and Technology* (R. A. A. Muzzarelli, C. Jeuniaux, and G. W. Gooday, eds.), pp. 515–522, Plenum Press, New York.

Jeuniaux, C., Voss-Foucart, M., Poulicek, M., and Bussers, J., 1989, Sources of chitin, estimated from new data on chitin biomass and production, in: *Proceedings of the 4th International Conference on Chitin/Chitosan* (G. Skjak-Braek, T. Anthonsen, and P. A. Sandford, eds.), pp. 3–12, Elsevier, Amsterdam.

Johnstone, D. W., and Cross, T., 1976, The occurrence and distribution of actinomycetes in lakes of the English Lake District, *Freshwater Biol.* **6**:457–463.

Jones, G. E., 1958, Attachment of marine bacteria to zooplankton, *U.S. Fish. Wildlife Serv. Spec. Sci. Rep.* **279**:77–78.

Kaneko, T., and Colwell, R. R., 1975, Adsorption of *Vibrio parahaemolyticus* onto chitin and copepods, *Appl. Microbiol.* **29**:269–274.

Kaneko, T., and Colwell, R. R., 1978, The annual cycle of *Vibrio parahaemolyticus* in Chesapeake Bay. *Microb. Ecol.* **4**:135–155.

Khalifa, O., 1965, Biological control of *Fusarium* wilt of peas by organic soil amendments, *Ann. Appl. Biol.* **56**:129–137.

Kihara, K., and Morooka, N., 1962, Studies on marine chitin-decomposing bacteria. (I) Classification and description of species, *J. Oceanogr. Soc. Jpn.* **18**:41–46.

Kim, J., and ZoBell, C. E., 1972, Agarase, amylase, cellulase and chitinase activity at deep sea pressures, *J. Oceanogr. Soc. Jpn.* **28**:131–137.

Koch, B., and Disteche, A., 1986, The influence of pressure and temperature on the hydrolysis of prawn chitin in seawater by chitinase from *Serratia marcescens, Oceanol. Acta* **9**:515–517.

Kono, M., Matsui, T., and Shimizur, C., 1987a, Chitin-decomposing bacteria in digestive tracts of cultured red sea bream and Japanese eel, *Nippon Suisan Gakkaishi* **53**:305–310.

Kono, M., Matsui, T., and Shimizur, C., 1987b, Effect of chitin, chitosan and cellulose as diet supplements on the growth of cultured fish, *Nippon Suisan Gakkaishi* **53**:125–129.

Kramer, K. J., and Koga, D., 1986, Insect chitin. Physical state, synthesis, degradation and metabolic regulation, *Insect Biochem.* **16**:851–877.

Kuhl, J., Nittinger, and Siebert, G., 1978, Vervetang von Krillschalen in Futterungsversuchen an de Ratte, *Arch. Fischereiwiss.* **29**:99–103.

Lakshmanaperumalsamy, P., 1983, Preliminary studies on chitinoclastic bacteria in Vellar Estuary, *Mahasagar Bull. Nat. Inst. Oceanogr.* **16**:293–298.

Lear, D. W., 1963, Occurrence and significance of chitinolastic bacteria in pelagic waters and zooplankton, in: *Symposium on Marine Microbiology* (C. H. Oppenheimer, ed.), pp. 594–610, Charles C. Thomas, Springfield, Ill.

Levi, P., 1984, *The Periodic Table,* (R. Rosenthal, trans.), Shocken Books, Pantheon Books of Random House, New York.

Lindsay, G. J. H., 1984, Distribution and function of digestive tract chitinolytic enzymes in fish, *J. Fish. Biol.* **24:**529–536.

Lindsay, G. J. H., and Gooday, G. W., 1985a, Action of chitinase in spines of the diatom *Thalassiosira fluviatilis, Carbohydr. Polymers* **5:**131–140.

Lindsay, G. J. H., and Gooday, G. W., 1985b, Chitinolytic enzymes and the bacterial microflora in the digestive tract of cod, *Gadus morhua, J. Fish. Biol.* **26:**255–265.

Lindsay, G. J. H., Walton, M. J., Adron, J. W., Fletcher, T. C., Cho, C. Y., and Coway, C. B., 1984, The growth of rainbow trout (*Salmo gairdneri*) given diets containing chitin and its relationships to chitinolytic enzymes and chitin digestibility, *Aquaculture* **37:**315–334.

Lingappa, Y., and Lockwood, J. L., 1962, Chitin media for selective isolation and culture of actinomycetes, *Phytopathology* **52:**317–323.

Linkins, A. E., and Neal, J. L., 1982, Soil cellulase, chitinase, and protease activity in *Eriophorum vaginatum* tussock tundra at Eagle Summit, Alaska, *Holarctic Ecol.* **5:**135–138.

Liston, J., Wiebe, W. J., and Lighthart, B., 1965, Activities of marine benthic bacteria, *Res. Fish. Coll. Fish. Contrib. Univ. Wash.* **1965:**39–41.

Lunblad, G., Elander, M., and Lind, J., 1976, Chitinase and β-*N*-acetylglucosaminidase in the digestive juice of *Helix pomatia, Acta Chem. Scand.* **30:**889–894.

Lunblad, G., Elander, M., Lind, J., and Slettengren, 1979, Bovine serum chitinase, *Eur. J. Biochem.* **100:**455–460.

Lysenko, O., 1976, Chitinase of *Serratia marcescens* and its toxicity to insects, *J. Invertebr. Pathol.* **27:**385–386.

MacDonald, N. L., Stark, J. R., and Austin, B., 1986, Bacterial microflora in the gastro-intestinal tract of Dover sole (*Solea solea L.*), with emphasis on the possible role of bacteria in the nutrition of the host, *FEMS Microbiol. Lett.* **35:**107–111.

MacDonell, M. T., Baker, R. M., Singleton, F. L., and Hood, M. A., 1984, Effects of surface association and osmolarity on seawater populations of an environmental isolate of *Vibrio cholerae,* in: *Vibrios in the Environment* (R. R. Colwell, ed.), pp. 535–549, John Wiley & Sons, New York.

Manson, F. D. C., Gooday, G. W., and Fletcher, T. C., 1989, The development of gastric and blood chitinase activity in the turbot *Scophthalmus maximus* (L.), in *Proceedings of the 4th International Conference on Chitin/Chitosan* (G. Skag-Braek, T. Anthonsen, and P. A. Sandford, eds.), pp. 243–253. Elsevier, Amsterdam.

Martin, M. M., Gieselmann, M. J., and Martin, J. S., 1973, Rectal enzymes of attine ants: α-amylase and chitinase, *J. Insect. Physiol.* **19:**1409–1416.

Martin, M. M., Kukor, J. J., Martin, J. S., O'Toole, T. E., and Johnson, M. W., 1981, Digestive enzymes of fungus-feeding beetles, *Physiol. Zool.* **54:**137–145.

Mauch, F., Hadwiger, L. A., and Boller, T., 1988. Antifungal hydrolases in pea tissue, *Plant Physiol.* **87:**325–333.

McCormick, C. L., and Anderson, R. W., 1984, Synthesis and characterisation of chitin pendently substituted with the herbicide metribuzen, in: *Chitin, Chitosan and Related Enzymes* (J. P. Zikakis, ed.), pp. 41–53, Academic Press, Orlando.

Metraux, J. P., and Boller, T., 1986, Local and systemic induction of chitinase in cucumber *Cucumis sativus* cultivar Wisconsin plants in response to viral, bacterial and fungal infections, *Physiol. Mol. Plant Pathol.* **28:**161–170.

Mian, J. H., Godoy, G., Shelby, R. A., Rodriguez-Kabana, R., and Morgan-Jones, G., 1982, Chitin amendments for control of *Meloidogyne arenaria* in infested soil, *Nematropica* **12:**71–84.

Micha, J. C., Dandrifosse, G., and Jeuniaux, C., 1973, Distribution et localisation tissulaire de la synthese des chitinases chez les vertebres inferieurs, *Arch. Int. Physiol. Biochim.* **81:**439–451.

Mihaly, K., 1960, Chitin breakdown of rhizosphere bacteria, *Nature (London)* **188:**251.

Minke, R., and Blackwell, J., 1978, The structure of α-chitin, *J. Mol. Biol.* **120:**167–181.

Mitchell, R., 1963, Addition of fungal cell-wall components to soil for biological disease control, *Phytopathology* **53:**1068–1071.

Mitchell, R., and Alexander, M., 1962, Microbiological processes associated with the use of chitin for biological control, *Soil Sci. Soc. Am. Proc.* **26:**556–558.

Mommsen, T. P., 1980, Chitinase and β-*N*-acetylglycosaminidase from the digestive fluid of the spider, *Biochim. Biophys. Acta* **612:**361–372.

Monaghan, R. L., 1975, The Discovery, Distribution and Utilisation of Chitosanase, Ph.D. Thesis, Rutgers University, New Brunswick, N.J.

Monaghan, R. L., Eveleigh, D. E., Tewari, R. P., and Reese, E. T., 1973, Chitosanase, a novel enzyme, *Nature (London) New Biol.* **245:**78–80.

Monreal, J., and Reese, E. T., 1969, The chitinase of *Serratia marcescens, Can. J. Microbiol.* **15:**689–696.

Morita, R. Y., 1979, Current status of the microbiology of the deep sea, *Ambio Spec. Rep.* **6:**33–36.

Mowlah, A. H., Sakata, T., and Kakimoto, D. M., 1979, Microflora in the alimentary tract of gray mullet—V. Studies on the chitinolytic enzymes of *Enterobacter* and *Vibrio, Bull. Jpn. Soc. Sci. Fish.* **45:**1313–1317.

Murray, C. L., and Lovett, J. S., 1966, Nutritional requirements of the chytrid, *Karlingia asterocysta*, an obligate chitinophile, *Am. J. Bot.* **53:**469–476.

Muzzarelli, R. A. A., 1977, *Chitin*, Pergamon Press, Oxford.

Muzzarelli, R. A. A., 1985, Chitin, in: *The Polysaccharides* (G. O. Aspinall, ed.), pp. 417–450, Academic Press, Orlando.

Nalin, D. R., Reid, V. D., Levine, M. M., and Cisneros, L., 1979, Adsorption and growth of *Vibrio cholerae* on chitin, *Infect. Immun.* **25:**768–770.

Ohwada, K., Tabor, P. S., and Colwell, R. R., 1980, Species composition and barotolerance of gut microflora of deep sea benthic macrofauna collected at various depths in the Atlantic Ocean, *Appl. Environ. Microbiol.* **40:**746–755.

Okafor, N., 1966a, Ecology of micro-organisms in chitin buried in soil, *J. Gen. Microbiol.* **44:**311–327.

Okafor, N., 1966b, The ecology of micro-organisms on, and the decomposition of insect wings in the soil, *Plant Soil* **25:**211–237.

Okafor, N., 1966c, Estimation of the decomposition of chitin in soil by the method of carbon dioxide release, *Soil Sci.* **102:**140–142.

Okafor, N., 1967, Decomposition of chitin by micro-organisms isolated from a temperate soil and a tropical soil, *Nova Hedwigia* **13:**209–226.

Okafor, N., 1970, Influence of chitin on mycoflora and length of roots of wheat seedlings, *Trans. Br. Mycol. Soc.* **55:**483–485.

Okutani, K., 1966, Studies of chitinolytic systems in the digestive tracts of *Lateolabrux japonicus, Bull. Misaki Mar. Biol. Inst. Kyoto Univ.* **10:**1–47.

Okutani, K., 1975, Microorganisms related to mineralization of chitin in aquatic environments, in: *Nitrogen Fixation and Nitrogen Cycle* (H. Takahashi, ed.), pp. 147–154, University of Tokyo Press, Tokyo.

Okutani, K., and Kitada, H., 1971, The distribution of chitin and *N*-acetylglucosamine-decomposing bacteria in aquatic environments, *Tech. Bull. Fac. Agric. Kagawa Univ.* **23:**127–136.

Ou, L., and Alexander, M., 1974, Effect of glass microbeads on the microbial degradation of chitin, *Soil Sci.* **118:**164–167.

Parsons, J. W., 1980, Chemistry and distribution of amino sugars in soils and soil organisms, in: *Soil Biochemistry,* Vol. 5 (E. A. Paul and J. N. Ladd, eds.), pp. 197–227, Marcel Dekker, Inc., New York.

Patton, R. S., and Chandler, P. T., 1975, *In vivo* digestibility evaluation of chitinous material, *J. Dairy Sci.* **58:**397–403.

Patton, R. S., Chandler, P. T., and Gonzalez, O. G., 1975, Nutritive value of crab meal for young ruminating calves, *J. Dairy Sci.* **58**:404–409.

Pearlmutter, N. L., and Lembi, C. A., 1978, Localization of chitin in algal and fungal cell walls by light and electron microscopy, *J. Histochem. Cytochem.* **26**:782–791.

Pel, R., and Gottschal, J. C., 1986a, Mesophilic chitin-degrading anaerobes isolated from an estuarine environment, *FEMS Microbiol. Ecol* **38**:39–49.

Pel, R., and Gottschal, J. C., 1986b, Chitinolytic communities from an anaerobic estuarine environments, in: *Chitin in Nature and Technology* (R. A. A. Muzzarelli, C. Jeuniaux, and G. W. Gooday, eds.), pp. 539–546, Plenum Press, New York.

Pel, R., and Gottschal, J. C., 1986c, Stimulation of anaerobic chitin degradation in mixed cultures, *Antonie van Leeuwenhoek J. Microbiol. Serol.* **52**:359–360.

Pel, R., and Gottschal, J. C., 1987, The effect of oxygen and sulfydryl reagents on the hydrolysis and the fermentation of chitin by *Clostridium* 9.1, *FEMS Microbiol. Lett.* **44**:59–62.

Pel, R., and Gottschal, J. C., 1989, Interspecies interaction based on transfer of a thioredoxin-like compound in anaerobic chitin-degrading mixed cultures, *FEMS Microbiol. Ecol.* **62**:349–358.

Peter, M. G., Kegel, G., and Keller, R., 1986, Structural studies on sclerotized insect cuticle, in: *Chitin in Nature and Technology* (R. A. A. Muzzarelli, C. Jeuniaux, and G. W. Gooday, eds.), pp. 21–28, Plenum Press, New York.

Polyanskaya, L. M., Kozhevin, P. A., and Zvyagintsev, D. G., 1985, Stimulation and elimination of nodule bacteria in soil after the introduction of an actinomycete and chitin, *Microbiology* **53**:830–833.

Portier, R. J., and Meyers, S. P., 1981, Chitin transformation and pesticide interactions in a simulated aquatic microenvironmental system, *Dev. Ind. Microbiol.* **22**:543–55.

Portier, R. J., and Meyers, S. P., 1984, Coupling of in situ and laboratory microcosm protocols for ascertaining fate and effect of xenobiotics, in: *Toxicity Screening Procedures using Bacterial Systems* (D. Liu and B. J. Outka, eds.), pp. 345–379, Marcel Dekker, Inc., New York.

Poulicek, M., 1982, Coquilles et Autre Structures Squelettiques des Mollusques: Composition, Chimique, Biomasse et Biodegradation en Milieu Marin, Ph.D. thesis, University of Liege, Liege, Belgium.

Poulicek, M., 1983, Patterns of mollusk shell biodegradation in bathyal and abyssal sediments, *J. Mollusc Stud. Suppl.* **12A**:136–141.

Poulicek, M., 1985, Importance of chitin in the biogeochemical cycling of heavy metals in oceanic environment, in: *Proceedings, Metal Cycling in the Environment*, Brussels, pp. 153–165., SCOPE, Brussels, Belgium.

Poulicek, M., and Jaspar-Versali, M. F., 1981, Etude experimentalle de la degradation des coquilles de mollusques au niveau des sediments marins, *Bull. Soc. R. Sci. Liege* **50**:513–518.

Poulicek, M., and Jaspar-Versali, M. F., 1984, Biodegradation de la trame organique des coquilles de mollusques en mileau marin: action des microorganismes endolithes, *Bull. Soc. R. Sci. Liege* **53**:114–126.

Poulicek, M., and Jeuniaux, C., 1982, Biomass and biodegradation of mollusk shell chitin in some marine sediment, in: *Chitin and Chitosan* (S. Hirano and S. Tokura, eds.), pp. 196–199, The Japanese Society of Chitin and Chitosan, Tottori.

Poulicek, M., and Jeuniaux, C., 1989, Chitin biomass in marine sediments, in: *Proceedings of the 4th International Conference on Chitin/Chitosan* (G. Skjak-Braek, T. Anthronsen, and P. A. Sandford, eds.), Elsevier, Amsterdam.

Poulicek, M., Voss-Foucart, M. F., and Jeuniaux, C., 1986a, Chitinoproteic complexes and mineralization in mollusk skeletal structures, in: *Chitin in Nature and Technology* (R. A. A. Muzzarelli, C. Jeuniaux, and G. W. Gooday, eds.), pp. 7–12, Plenum Press, New York.

Poulicek, M., Goffinet, G., Voss-Foucart, M. F., Bussers, J. C., Jaspar-Versali, M. F., and Toussaint, C., 1986b, Chitin degradation in natural environment (mollusk shells and crab carapaces), in: *Chitin in Nature and Technology* (R. A. A. Muzzarelli, C. Jeuniaux, and G. W. Gooday, eds.), pp. 547–550, Plenum Press, New York.

Poulicek, M., Machiroux, R., and Toussaint, C., 1986c, Chitin diagenesis in deep-water sediments, in: *Chitin in Nature and Technology* (R. A. A. Muzzarelli, C. Jeuniaux, and G. W. Gooday, eds.), pp. 523–530, Plenum Press, New York.

Rehbein, H., Danulat, E., and Leineman, M., 1986, Activities of chitinase and protease and concentration of fluoride in the digestive tract of Antarctic fishes feeding on krill *Euphausia superba* (Dana), *Comp. Biochem. Physiol.* **85A:**545–551.

Reichardt, W., 1988, Impact of the antarctic benthic fauna on the enrichment of biopolymer degrading psychrophilic bacteria, *Microb. Ecol.* **15:**311–321.

Reichardt, W., and Morita, R. Y., 1982, Influence of temperature adaptation on glucose metabolism in a psychrotrophic strain of *Cytophaga johnsonae, Appl. Environ. Microbiol.* **44:**1282–1288.

Reichardt, W., Gunn, B., and Colwell, R. R., 1983, Ecology and taxonomy of chitinoclastic *Cytophaga* and related chitin-degrading bacteria isolated from an estuary, *Microb. Ecol.* **9:**273–294.

Reisert, P. S., and Fuller, M. S., 1962, Decomposition of chitin by *Chytridiomyces* species, *Mycologia* **54:**647–657.

Rittenberg, S. C., Anderson, D. Q., and ZoBell, C. E., 1937, Studies on the enumeration of marine anaerobic bacteria, *Proc. Soc. Exp. Biol. Med.* **35:**652–653.

Roberts, W. K., and Selitrennikoff, C. P., 1988, Plant and bacterial chitinases differ in antifungal activity, *J. Gen. Microbiol.* **134:**169–176.

Rodriguez-Kabana, R., Godoy, G., Morgan-Jones, G., and Shelby, R. A., 1983, The determination of soil chitinase activity: Conditions for assay and ecological studies, *Plant Soil* **75:**95–106.

Rudall, K. M., and Kenchington, W., 1973, The chitin system, *Biol. Rev.* **48:**597–636.

Ruschke, R., 1967, Uber weitere chitinolytische Bakterien aus dem Feldsee, *Arch. Hydrobiol. Suppl.* **23:**115–120.

Schlumbaum, A., Mauch, F., Vogeli, U., and Boller, T., 1986, Plant chitinases are potent inhibitors of fungal growth, *Nature (London)* **324:**365–367.

Schwarz, J. R., Yayanos, A. A., and Colwell, R. R., 1976, Metabolic activities of the intestinal microflora of a deep sea invertebrate, *Appl. Environ. Microbiol.* **31:**46–48.

Seki, H., 1965a, Microbiological studies on the decomposition of chitin marine environment—XI. Rough estimation of chitin decomposition in the ocean, *J. Oceanogr. Soc. Jpn.* **21:**17–24.

Seki, H., 1965b, Microbiological studies on the decomposition of chitin marine environment—X. Decomposition of chitin in marine sediments, *J. Oceanogr. Soc. Jpn.* **21:**25–32.

Seki, H., 1966, Seasonal fluctuation of heterotrophic bacteria in the sea of Aburatsubo Inlet, *J. Oceanogr. Soc. Jpn.* **22:**15–26.

Seki, H., and Taga, N., 1963a, Microbial studies on the decomposition of chitin in marine environment—I. Occurrence of chitinoclastic bacteria in the neritic region. *J. Oceanogr. Soc. Jpn.* **19:**101–108.

Seki, H., and Taga, N., 1963b, Microbial studies on the decomposition of chitin in marine environment—III. Aerobic decomposition of chitin by isolated chitinolytic bacteria, *J. Oceanogr. Soc. Jpn.* **19:**143–157.

Seki, H., and Taga, N., 1965a, Microbial studies on the decomposition of chitin in marine environment—V. Chitinoclastic bacteria as symbionts. *J. Oceanogr. Soc. Jpn.* **19:**158–161.

Seki, H., and Taga, N., 1965b, Microbial studies on the decomposition of chitin in marine environment—VIII. Distribution of chitinolytic bacteria in the pelagic and neritic waters, *J. Oceanogr. Soc. Jpn.* **21:**174–187.

Seki, H., and Taga, N., 1965c, Microbial studies on the decomposition of chitin in marine environment—VI. Chitinoclastic bacteria in the digestive tract of whales from the Antarctic Ocean. *J. Oceanogr. Soc. Jpn.* **20:**272–277.

Sietsma, J. H., Vermuelen, C. A., and Wessels, J. G. H., 1986, The role of chitin in hyphal morphogenesis, in: *Chitin in Nature and Technology* (R. A. A. Muzzarelli, C. Jeuniaux, and G. W. Gooday, eds.), pp. 63–69, Plenum Press, New York.

Skinner, C. E., and Dravis, F., 1937, A quantitative determination of chitin destroying microorganisms in soil, *Ecology* **18**:391–397.

Smith, R. J., and Grula, E. A., 1983, Chitinase is an inducible enzyme in *Beauvaria bassiana*, *J. Invertebr. Pathol.* **42**:319–326.

Smucker, R. A., 1982, Determination of chitin hydrolytic potential in an estuary, in: *Chitin and Chitosan* (S. Hirano and S. Tokura, eds.), pp. 135–139, The Japanese Society of Chitin and Chitosan, Tottori.

Smucker, R. A., 1984, Biochemistry of the *Streptomyces* spore sheath, in: *Biological, Biochemical and Biomedical Aspects of Actinomycetes* (L. Ortiz-Ortiz, L. F. Bojalib, and V. Yakoleff, eds.), pp. 171–177, Academic Press, Orlando.

Smucker, R. A., and Dawson, R., 1986, Products of photosynthesis by marine phytoplankton: Chitin in TCA 'protein' precipitates, *J. Exp. Mar. Biol. Ecol.* **104**:143–152.

Smucker, R. A., and Pfister, R. M., 1978, Characteristics of *Streptomyces coelicolor* A3(2) aerial spore rodlet mosaic, *Can. J. Microbiol.* **24**:397–408.

Smucker, R. A., and Wright, D. A., 1984, Chitinase activity in the crystalline style of the American oyster *Crassostrea virginica*, *Comp. Biochem. Physiol.* **77A**:239–241.

Smucker, R. A., Warnes, C. E., and Haviland, C. J., 1985, Chitinase production by a freshwater Pseudomonad, in: *Biodeterioration 6* (S. Barry, D. R. Houghton, G. C. Llewellyn, and C. E. O'Rear, eds.), pp. 549–553, C.A.B. International Institute, London.

Sneh, B., and Henis, Y., 1970, Production of antifungal substances against *Rhizoctonia solani* in chitin amended-soil, *Phytopathology* **62**:595–600.

Sochard, M. R., Wilson, D. F., Austin, B., and Colwell, R. R., 1979, Bacteria associated with the surface and gut of marine copepods. *Appl. Environ. Microbiol.* **37**:750–759.

Soderhall, K., and Unestam, T., 1975, Properties of extracellular enzymes from *Aphanomyces astaci* and their relevance in the penetration process of crayfish cuticle. *Physiol. Plant.* **35**:140–146.

Spiegel, Y., Cohn, E., and Chet, I., 1986, Use of chitin for controlling plant parasitic nematodes. I. Direct effects on nematode reproduction and plant performance, *Plant Soil* **95**:87–96.

Spiegel, Y., Chet, I., and Cohn, E., 1987, Use of chitin for controlling plant parasitic nematodes. II. Mode of action, *Plant Soil* **98**:337–346.

Stanier, R. Y., 1947, Studies on non-fruiting myxobacteria. I. *Cytophaga johnsonae* n.sp., a chitin-decomposing myxobacterium, *J. Bacteriol.* **53**:207–315.

St Leger, R. J., Charnley, A. K., and Cooper, R. M., 1986a, Cuticle degrading enzymes of entomopathogenic fungi: Mechanisms of interaction between pathogen enzymes and insect cuticle, *J. Invertebr. Pathol.* **47**:295–302.

St Leger, R. J., Cooper, R. M., and Charnley, A. K., 1986b, Cuticle-degrading enzymes of entomopathogenic fungi: Regulation of production of chitinolytic enzymes, *J. Gen. Microbiol.* **132**:1509–1517.

Streichsbier, F., 1982, Okologische Untersuchungen zum mikrobiellen Chitinabbau in stadtnahen Fliessgewässern, *Wasser Abwasser* **25**:53–70.

Streichsbier, F., 1983, Utilization of chitin as sole carbon and nitrogen source by *Chromobacterium violaceum*, *FEMS Microbiol. Lett.* **19**:129–132.

Sturz, H., and Robinson, J., 1986, Anaerobic decomposition of chitin in freshwater sediments, in *Chitin in Nature and Technology* (R. A. A. Muzzarelli, C. Jeuniaux, and G. W. Gooday, eds.), pp. 531–538, Plenum Press, New York.

Sundarraj, N., and Bhat, J. V., 1972, Breakdown of chitin by *Cytophaga johnsonae*, *Arch. Mikrobiol.* **85**:159–167.

Surarit, R., Gopal, P. K., and Shepherd, M. G., 1988, Evidence for a glycosidic linkage between chitin and glucan in the cell wall of *Candida albicans*, *J. Gen. Microbiol.* **134**:1723–1730.

Timmis, K., Hobbs, G., and Berkeley, R. C. W., 1974, Chitinolytic clostrida isolated from marine mud, *Can. J. Microbiol.* **20:**1284–1285.

Tracey, M. V., 1955, Cellulase and chitinase in soil amoebae, *Nature (London)* **175:**815.

Unestam, T., 1966, Chitinolytic, cellulolytic and pectinolytic activity in vitro of some parasitic and saprophytic oomycetes, *Physiol. Plant.* **19:**15–30.

Unestam, T., 1968, Some properties of unpurified chitinase from crayfish plague fungus, *Aphanomyces astaci, Physiol. Plant.* **21:**137–147.

Van Eck, W. H., 1978, Autolysis of chlamydospores of *Fusarium solani* f. sp. *cucurbitae* in chitin and laminarin amended soils, *Soil Biol. Biochem.* **10:**89–92.

Veldkamp, H., 1955, A study of the aerobic decomposition of chitin by microorganisms, *Meded. Landbouwhogesch. Wageningen* **55:**127–174.

Vermeulen, C. A., and Wessels, J. G. H., 1984, Ultrastructural differences between wall apices of growing and non-growing hyphae of *Schizophyllum commune, Protoplasma* **120:**123–130.

Voss-Foucart, M. F., Jeuniaux, C., and Greyoire, C., 1974, Resistance de la chitine de la nacre du nautile (Mollusque Cephalopode) a l'action de certains facteurs intervenant au cours de la fossilisation, *Comp. Biochem. Physiol.* **48B:**447–451.

Ward, H. D., Alroy, J., Lev, B. I., Keusch, G. T., and Pereira, M. E. A., 1985, Identification of chitin as a structural component of *Giardia* cysts, *Infect. Immun.* **49:**629–634.

Warnes, C. E., and Randles, C. I., 1977, Preliminary studies on chitin decomposition in Lake Erie sediments, *Ohio J. Sci.* **77:**224–230.

Warnes, C. E., and Randles, C. I., 1980, Succession in a microbial community associated with chitin in Lake Erie sediment and water, *Ohio J. Sci.* **80:**250–255.

Warnes, C. E., and Rux, T. P., 1982, Chitin mineralisation in a freshwater habitat, in: *Chitin and Chitosan* (S. Hirano and S. Tokura, eds.), pp. 191–195, The Japanese Society of Chitin and Chitosan, Tottori.

Weyland, H., 1981, Distribution of actinomycetes on the sea floor, *Zentralbl. Bakteriol.* **11:**185–193.

Williams, S. T., and Robinson, C. S., 1981, The role of streptomycetes in decomposition of chitin in acidic soils, *J. Gen. Microbiol.* **127:**55–63.

Wirsen, C. O., and Jannasch, H. W., 1976, Decomposition of solid organic materials in the deep sea, *Environ. Sci. Technol.* **10:**880–886.

Wolkin, R. H., and Pate, J. L., 1985, Selection for nonadherent or nonhydrophobic mutants co-selects for non-spreading mutants of *Cytophaga johnsonae* and other gliding bacteria, *J. Gen. Microbiol.* **131:**737–750.

Wyckoff, R. W. G., 1972, *The Biochemistry of Animal Fossils,* Scientechnica Publ., Bristol.

Yamamoto, H., and Seki, H., 1979, Impact of nutrient enrichment in a waterchestnut ecosystem at Takahama-Iri bay of Lake Kasumigaura, Japan, *Water Air Soil Pollut.* **12:**519–527.

Yoshida, Y., and Sera, H., 1970, On chitinolytic activities in the digestive tracts of several species of fishes and mastication and digestion of foods by them. *Bull. Jpn. Soc. Sci. Fish.* **36:**751–754.

Zhloba, N. M., Tiunova, N. A., and Sidorova, I. I., 1980, Extracellular hydrolytic enzymes of mycophilic fungi, *Mikol. Fitopatol.* **14:**496–499.

ZoBell, C. E., and Rittenberg, S. C., 1938, The occurrence and characteristics of chitinoclastic bacteria in the sea, *J. Bacteriol.* **35:**275–287.

Microbial Plasticity
The Relevance to Microbial Ecology

E. TERZAGHI and M. O'HARA

Natural selection . . . works like a tinkerer—a tinkerer who does not know exactly what he is going to produce but uses whatever he finds around him whether it be pieces of string, fragments of wood, or old cardboard; in short it works like a tinkerer who uses everything at his disposal to produce some kind of workable object. For the engineer, the realization of his task depends on his having the raw materials and the tools that fit exactly his project. The tinkerer, in contrast, always manages with odds and ends. What he ultimately produces is generally to no special project, and it results from a series of contingent events, of all the opportunities he had to enrich his stock. . . . Evolution does not produce novelties from scratch. It works on what already exists, either transforming a system to give it new functions or combining several systems to produce a more elaborate one. . . . It is at the molecular level that the tinkering aspect of natural selection is perhaps the most apparent.

Jacob (1977)

1. Introduction

Classical microbial taxonomy rests solidly on assumptions of phenotypic stability and constancy of bacterial species. This phenotypic stability, in turn, not only has been presumed to reflect a corresponding physiological and genotypic stability but has also carried the tacit assumption that there is a manifest correspondence between genotype and phenotype. The discoveries first of plasmids and then of various classes of both

E. TERZAGHI and M. O'HARA • Department of Microbiology and Genetics, Massey University, Palmerston North, New Zealand.

autonomous and nonautonomous mobile genetic elements presented a serious challenge to this comfortable view of microbial stability. And now, as the techniques of contemporary microbiology and molecular genetics are being focused on an increasingly wide variety of microbial groups, mounting evidence is suggesting with increasing insistence that the microbial genome, and hence phenotype, is even more plastic and adaptable than initially imagined.

With interest increasing in the engineering of microorganisms for specified tasks in both industrial and field situations, it is important to understand the inherent potential for change of specific microbial groups. On the one hand, utilizing this inherent potential for change should permit a judicious selection of organism and experimental strategy for production of a strain to carry out a specified industrial process. On the other hand, a bacterial strain designed and constructed in the laboratory to do a specific task may change, even dramatically, in response to biological or physical factors in the new industrial or field environment (Lechevallier et al., 1987; Walter et al., 1987). In the assessment of environmental impact of field release, there will be increasing pressure to understand how specific strains will respond to environmental challenge (Alexander, 1986; Levin et al., 1987; National Academy of Sciences, 1987; Office of Technology Assessment, 1988).

There are several dimensions of the problem of microbial variability. Intracellular rearrangement of genetic information and intercellular exchange of genetic information can result in a great deal of short-term microbial variability. However, the distinction should be made between this short-term, perhaps adaptive variation and that long-term variation which becomes incorporated into the evolutionary history of a group of microorganisms. This distinction is of critical importance in view of the consistency and coherence of the results of contemporary taxonomy based upon nucleic acid sequence homologies (total genomic and ribosomal RNA) as well as numerical methods (Schleifer and Stackebrandt, 1983). Furthermore, maintenance of gross genomic structure across fairly broad taxonomic groups (Riley, 1985) places very distinct limits on the extent of variability at this level of organization.

Gene and genome evolution has been considered by a number of workers and reviewed by Riley (1985). Many of the events that we suggest are manifestations of genomic plasticity are also associated with evolutionary divergence of groups of microorganisms. However, aside from a few very general observations presented in Section 5, we do not want to venture onto the treacherous ground of speculation about the relationship between contemporary variation and long-term evolutionary divergence.

The primary aim of this review is to assemble the published information relating to phenotypic and genomic variability of microorganisms that is not simply a consequence of conventional gene regulation or intercellular transfer of genetic information and that does not necessarily relate to events of evolutionary significance. Though we wish to focus on the inherent plasticity manifested by at least some groups of bacteria, for reasons of completeness we will very briefly discuss at the outset the broad implications of intercellular transfer and refer the reader to recent reviews of the topic. We will end with a brief discussion of the practical implications of microbial plasticity.

2. Interorganismal Exchange

Reanney *et al.* (1982, 1983) have imaginatively and forcefully presented a picture of microbial ecosystems as forming genetically open communities with potentially extensive promiscuous exchange mediated by transformation, transduction, and conjugation. Documentation of extensive flow of genetic information between groups of microorganisms in the field situation is difficult because of the paucity of readily scorable markers and the inherent intractibility of natural ecosystems. Nevertheless, Trevors *et al.* (1987) have recently reviewed the current evidence supporting the notion of widespread genetic exchange in natural microbial ecosystems. A not dissimilar situation has been proposed for viruses. Campbell and Botstein (1983) and Botstein (1980), following an early suggestion by Szybalski and Szybalski (1974), have drawn a picture of lambdoid phage evolution and contemporary genetic relationships in which novelty may be generated by the exchange of blocks or modules of genetic information between possibly dissimilar phage or even by the rescue of isolated modules residing in the host chromosome. Recent work with selected phage of the lactic streptococci suggests similar interesting relationships between different phage isolates (Coveney *et al.*, 1987; Jarvis and Meyer, 1986). In the very different flu virus system, a similar mechanism has been proposed by Gething *et al.* (1980) and discussed more explicitly by Van Rompuy *et al.* (1983) to account for the antigenic shifts that have led to flu pandemics.

Rearrangements of at least small segments of the genome of some bacteria have been recognized for a long time as a source of phenotypic variation. Currently, many diverse examples of genomic rearrangements are documented and in some instances well characterized and understood. Borst and Greaves (1987), in a thorough review of selected procaryotic and eucaryotic examples, made the useful distinction between programmed and unprogrammed rearrangements, which format we will follow.

Programmed rearrangements are inversion, transposition, deletion, and insertion events that occur at precisely specified end points and that are mediated by site-specific recombinases. These events may be regulated by specific environmental signals or may occur simply with stochastic regularity; the phenotypic consequences of the rearrangement, while known for the majority of the better-studied cases, remain unknown for others. Unprogrammed rearrangements in general appear to lack the topological definition of the programmed class of events and are random in respect to time. It may be safely assumed that a number of rearrangements that are currently recognized and described below but are not yet well characterized will shift toward the programmed class as they become better understood. In only one case (*nifHDK* of *Anabaena; nif* is the name given to genes whose products are directly involved in nitrogen fixation) has it been demonstrated that the rearrangement is regulated in the sense of being a response to a specific environmental signal (Haselkorn *et al.*, 1986). The phenotypic consequences and significance of the unprogrammed events, while often recognized as mutations, usually remain unclear; these are discussed in Sections 6 and 7. Regardless of the details of mechanism, we wish to view genotypic and phenotypic plasticity from an adaptive perspective and would like to raise the possibility that at least some classes of rearrangements may be under a global control that is sensitive to some kind of yet to be identified environmental signal, in a fashion analogous to the SOS response to DNA damage.

3. Programmed Rearrangement

3.1. Topology of Rearrangement

Irrespective of the agent of breakage and reunion of DNA molecules, there are a few basic topological principles of inter- and intramolecular recombination (Fig. 1). Rearrangement events frequently involve the nucleotide sequence- and polarity-specific interaction of DNA segments and lead to inversions, deletions or insertions, sequence substitutions, and amplification or deamplification. Rearrangements have also been shown to occur, albeit rarely, in the absence of demonstrable sequence-specific interactions. We will now look at several specific examples of rearrangements that have been relatively well characterized.

3.2. Inversions

The best-characterized family of rearrangements is a group of invertible elements that control the switching between the expression of two alternative genes. The salient properties of the *hin, gin,* and *cin* systems found, respectively, in *Salmonella typhimurium,* phage Mu, and phage P1, have been reviewed by Simon and Silverman (1983). In all of these cases, the inversion of a region (H, G, and C, respectively) bounded by IR (inverted repeat) sequences is catalyzed by the site specific recombinase (products of the *hin, gin,* and *cin* genes; Fig. 2a and b). There is extensive sequence homology between the three pairs of IR sequences, the three recombinases are able to complement one another and, finally, the genes encoding the recombinases lie adjacent to one of the boundary IR sequences in which may be found an element of the respective promoter sequences for those genes. In each case, inversion leads to production of different structural proteins, thus changing the antigenic properties of the organism and, in the case of the prophages, their host range. In the case of *S. typhimurium,* the antigenic alteration is effected by movement of a promoter; in the case of Mu and P1, the promoter is stationary outside one of the IR sequences, and it is the alternative genes encoding the protein that are brought into the correct juxtaposition. More recent work has identified *cis*-acting elements that modulate the frequency of *cin*-mediated inversion (P1; Huber *et al.,* 1985) and of *hin*-mediated inversion (*S. typhimurium;* Johnson *et al.,* 1987). In addition, Iida and Hiestand-Nauer (1986) have demonstrated that the *cin* function can mediate inversion at several other sites, albeit at a lower frequency, and in the process, in a gene conversion-like event, can alter the target recombinational overlap sequence to one that conforms with the normal target sequence.

Two further invertible regions have been found in the *Escherichia coli* genome. The P region, and its associated *pin* gene encoding the appropriate site-specific recombinase, resides within a defective prophage (Plasterk and van de Putte, 1985) and is homologous

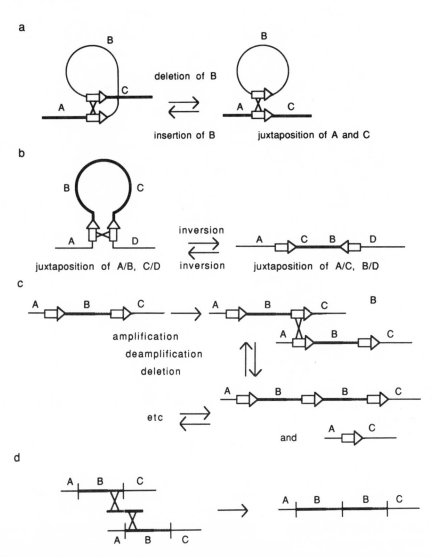

Figure 1. Topological consequences of recombination between repeated sequences. (a) Recombination between direct repeats, leading to deletion of the included material, or reversal of the recombinational event, leading to insertion of a sequence. In both cases, there is an alteration of sequence neighborhoods. (b) Recombination between inverted repeat sequences, leading to inversion of the included sequences and alteration of sequence neighborhoods. (c) Recombination between direct repeats, on separate molecules, immediately following duplication. Sequential repetitions of this event can result in rapid amplification or deamplification of the included sequence or even deletion of that sequence. (d) Recombination between a small fragment bearing a novel junction site and two strands bearing the appropriate unjoined sequences can regenerate the rearrangement, in this case a duplication.

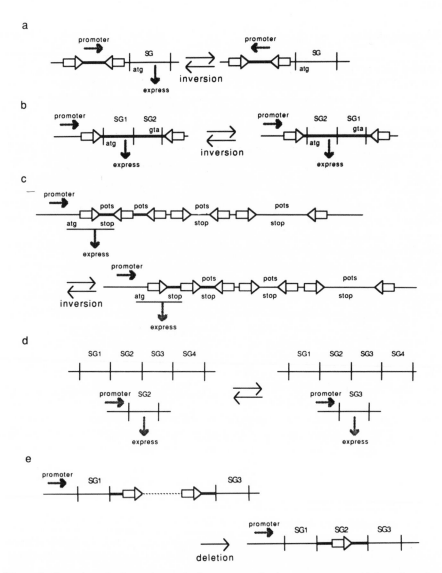

Figure 2. General mechanisms of control of gene expression by rearrangement. (a) Inversion of a promoter element, bracketed by inverted repeat elements, activates or deactivates an adjacent structural gene, as seen in the control of *fim* in *E. coli* or, in a modified format, with *hin* in *S. typhimurium*. (b) Inversion of a structural gene(s) bracketed by inverted repeat elements, downstream of a promoter element, oriented such that either a single gene is activated or deactivated or alternative genes are expressed. This organization of the invertible block is seen in the *gin* and *cin* systems of coliphages Mu and P1, respectively. (c) Demonstration that inversion of blocks bracketed by any pair of inverted repeat element can generate as many as seven different polypeptides from the single promoter element, as found in the "shufflon" on R64 of *E. coli*. (d) Transposition of coding sequences from a battery of such sequences into an expression site provided with a promoter sequence. This organization is responsible for the rapid variation of surface antigen structure of *N. gonorrhoeae* and *B. hermsii*. (e) Deletion of a large block of DNA inserted in the middle of a structural gene allows normal expression, as seen in *nifD* of *Anabaena*.

to the *hin-*, *gin-*, and *cin*-mediated systems. The sequence bracketed by the IR sequences contains an ORF (open reading frame) to which no function has yet been attributed. The synthesis of type I fimbriae is regulated by an invertible element containing a promoter-like sequence that is found just upstream of an ORF (*fimA*), the sequence of which corresponds with the known primary structure of the fimbriae (Abraham *et al.*, 1985). Inversion is mediated by a site-specific recombinase encoded by a pair of nearby genes.

A more complex set of overlapping inversions has been described by Komano *et al.* (1987a). A region of the broad-host-range plasmid R64 was identified that contains an upstream translational start sequence and seven downstream repeated sequences in both orientations, as well as a number of appropriately oriented stop signals (Fig. 2c). By a series of site-specific recombinational events, as many as seven different polypeptides could be synthesized from this region, all having in common a short N-terminal sequence. Rearrangement is catalyzed by a presumed site-specific recombinase encoded by the *rci* gene located nearby. No function has yet been assigned to this region, which has been designated a "shufflon" by the authors. Analysis of other group IncI plasmids has shown this element to be widely distributed (Komano *et al.*, 1987b). Kubo *et al.* (1988) have just demonstrated that the *rci* gene, which is responsible for shufflon rearrangement, has sequence homology with the integrase family of genes.

An interesting point to be made about the programmed inversions is that at least two apparently different site-specific recombinases have been recruited to play a role in the regulation of expression of very different alternative structural proteins. Although in each case the key regulatory event is the inversion by recombination between pairs of IR sequences, the detailed regulatory consequences of inversion vary (Fig. 2). The *hin* gene (and hence the *gin, cin,* and *pin* genes) has been shown to share some sequence homology with the transposon Tn*3* and TN*1000*($\gamma\delta$) resolvases, site-specific recombinases implicated in transposon transposition. Thus, the *hin* family of site-specific recombinases is linked with a totally different class of agent of DNA rearrangement (Johnson and Simon, 1987). The type I fimbrial site-specific recombinase has structural and functional similarity to the λ *int* (integrase) gene (Dorman and Higgins, 1987) and hence the extended and distantly related family of lambdoid site-specific recombinases (Argos *et al.*, 1986). Lewin (1987) pointed out that the targets of the λ *int* function and the Tn*3 res* (resolvase) functions show extensive homology, suggesting an otherwise unsuspected link between these two representative site-specific recombination systems.

3.3. Transposition and Deletion

A second mode of programmed genotypic alteration that leads to defined phenotypic modifications entails either the deletion or the transposition of blocks of coding sequence.

A number of pathogenic bacteria are well known for their ability to elude the vertebrate immune system by means of alteration of their surface antigens. Two particularly well studied instances of this variation are *Borrelia hermsii* and *Neisseria gonorrhoeae*. In the former case, it has been demonstrated (Plasterk *et al.*, 1985) that linear plasmids, shown by Barbour and Garon (1987) to have covalently closed ends, carry an expression site into which can be replicatively inserted any one of an array of surface antigen gene cassettes carried on other linear plasmids (Fig. 2d). To date, only the switch between display of two specific antigens (designated 7 and 21) has been studied in detail.

In the case of *Neisseria*, all pilus antigens examined share a common amino-terminal sequence but differ in the rest of the molecule, which consists of variable and hypervariable regions. As with *Borrelia*, it has been demonstrated that antigenic variation is effected by replicative transposition of modules encoding the variable and hypervariable portions of the pilus monomer units into expression sites that include the coding sequences for the N-terminal constant portion of the pilin monomer (Haas and Meyer, 1986). These modules reside in clusters of silent sequences between which conversion-like events can increase sequence diversity. One (phase) variant has been identified as totally lacking pili because of a module carrying an assembly missense mutation (Swanson *et al.*, 1986). In neither *Borrelia* or *Neisseria* has the molecular mechanism of replicative transposition or gene conversion been elucidated, although it has been suggested that in the latter organism, genetic change is mediated by transformation (Seifert and So, 1988). A second group of phase and antigen variation functions, the *opacity* proteins, have been identified in *Neisseria*, for which a novel transcriptional control, in addition to transposition or gene conversion, was proposed (Stern *et al.*, 1986; Meyer, 1987).

Haselkorn *et al.* (1986) have elucidated the mechanism by which the *nifHDK* operon of *Anabaena* is activated. Under conditions of nitrogen sufficiency, the *D* gene is split by an 11.6-kb sequence that is bounded by DR (direct repeat) sequences and includes a gene (*xisA*) encoding a site-specific recombinase recognizing these bracketing DR sequences. Under conditions of nitrogen deprivation, the *xisA* gene is induced, the 11.6-kb block is removed and the key *nifHDK* operon is expressed. They have more recently shown (Haselkorn *et al.*, 1988) that another deletion (55 kb bracketed by a different pair of repeated sequences) leads to the presumably functional juxtaposition of several more genes related to nitrogen fixation.

In none of these three cases has the molecular mechanism of rearrangement been elucidated, and it will be interesting to see whether they are mechanisms shared with other functionally unrelated systems. We will turn now to a consideration of events that may lead to expression of new functions but for which there seem to be no specific structural predisposing factors.

4. Unprogrammed Rearrangement

4.1. Agents of Change

Nucleotide sequence alteration, by base pair transition, transversion, insertion, or deletion, is relatively well understood and well documented (Schaaper *et al.*, 1986). Shapiro (1985) has thoroughly and concisely reviewed the variety of known molecular mechanisms of DNA rearrangement in *E. coli*. The stated objective of that review was to foster a more dynamic view of genome structure than had hitherto prevailed. And although it is safe to assume that the basic cellular mechanisms of nucleic acid management are universal, it is probably equally safe to assume that as our understanding of a broader spectrum of bacteria begins to match that of *E. coli*, there will come to light a variety of strategies and mechanisms for conferring the degree of plasticity that is appropriate for given groups of microbes.

General homologous recombination (often defined as RecA mediated) operating upon a pair of separated homologous sequences can in principle yield a variety of genetic rearrangements, discussed briefly above and outlined in Fig. 1. A point that has not been systematically investigated is the relationship between (1) length and degree of homology and (2) frequency of recombination. Qualitative experimental work has indicated that a length of between 30 and 150 bp of at least 97% sequence homology is required for normal RecA-mediated recombination (Bianchi and Radding, 1983; Gonda and Radding, 1983). However, it is probably safe to assume that to a first approximation the probability of a RecA-mediated event will be a direct function of the length and homology of the region undergoing exchange. A poorly understood "general non-homologous," presumably RecA-independent recombination (Shapiro, 1985) can occur at very low frequencies and requires a short (5–20 bp) region of sequence homology. In a study of tandem chromosomal duplications, Anderson and Roth (1978) showed that RecA-dependent events manifested definite breakpoint site preference, whereas the RecA-independent events, occurring at approximately half the frequency, showed no such site preference. Finally, in a systematic study of the actual nucleotide sequences involved in various classes of spontaneous rearrangement events, Schaaper *et al.* (1986) found that deletion and primary duplication events can occur in the absence of bracketing repeated sequences but do seem to be associated with sequences capable of generating novel local secondary structure.

Since plasmids are common agents of intercellular transfer of genetic material, it would be of interest to know whether recombination between plasmids or between plasmids and chromosomal DNA occurs with equal facility. Niaudet *et al.* (1984) have reported that plasmid sequences in *Bacillus subtilis* participate in recombination at a much higher frequency than do similar chromosomal sequences. In *Haemophilus influenzae* there is a much higher level of plasmid recombination (both RecA dependent and independent) after transformation than after introduction by conjugation (Balganesh and Setlow, 1986). Neither finding has been subjected to further systematic examination.

All of the rearrangements provide mechanisms for the novel juxtaposition of regulatory sequences and coding sequences, or indeed for the fusion of coding sequences to form chimeric genes encoding proteins of novel function. In addition, the amplification or deamplification of a coding sequence provides a mechanism for increasing or decreasing the level of expression of that sequence by virtue of gene copy number.

Finally, it was pointed out by Campbell (1965) that inherent to novel junctions generated by rearrangements is the capacity of a short sequence, including that junction, to induce the formation of that rearrangement in a RecA-mediated event, as illustrated for a duplication in Fig. 1d. This point has important implications for the storage or transfer of determinants of specific rearrangements that may be of functional or adaptive significance (Mori *et al.,* 1986).

In early studies of repetitive DNA sequences and genome structure (Britten and Kohne, 1968), it was generally believed that bacterial DNA consists entirely of single-copy sequence. However, since then it has become increasingly apparent that many, if not all, microbial groups have at least a modest amount of repeated sequence. While any pair of even very short repeated sequences is potentially recombinogenic, of particular interest in this context are the transposable elements IS (insertion sequences) and Tn (transposons).

The various properties of IS and Tn elements have been most thoroughly elucidated in *E. coli*, and comprehensive reviews have been published (Bennett, 1985; Cullum, 1985; Shapiro, 1985). Each transposable element has its own characteristic features in terms of structural details and biological activities. Typically, IS elements are composed of IR sequences bracketing a presumed transposase gene, the product of which is responsible for catalyzing transposition events. Tn elements of class I consist of a pair of IS-like elements bracketing other genes encoding generally nonessential functions such as drug resistance. Class II Tn elements consist of a pair of simple IR sequences bracketing the transposition function(s) and other function(s), as in the case of class I. It should be pointed out that the most common marker found associated with Tn elements is drug resistance for the simple reason that this sort of marker is very easy to identify and to select. Indeed, any function can be bracketed by a pair of transposable elements, thereby becoming endowed with the mobility of that pair of elements and creating a new transposon.

Transposition is usually a duplicative event, with a copy of the element occupying both the original and a new position, though precise excision of transposable elements can occur at a much lower frequency. Different elements differ enormously in respect to specificity of target sequences, with some appearing to be able to insert anywhere and others having a very restricted range of potential sites of insertion. It is evident that there exist at least two mechanisms of transposition (perhaps corresponding to the two structural classes of transposons), for which molecular models have been proposed (Shapiro, 1985). In all cases described so far, a short (<10 bp) host sequence, corresponding to the insertion site, is duplicated, with one copy found at each end of the transposable element.

IS and Tn elements may be associated in their immediate vicinities with an elevated frequency of sequence inversion and deletion. The latter event may be a precise excision of a block of genetic material from one or the other side of the element, or it may be a deletion of the element including genetic material on one or both sides of the element. Many of these events are interpretable as replicative transposition to a nearby site, in either inverted or direct orientation, followed by homologous recombination (as outlined in Fig. 1). However, there are instances that are not readily interpretable in this fashion and may be more direct, perhaps accidental consequences of transposition.

Early work on transposable elements suggested that the usual effect of insertion was the interruption of the normal function at the target site. However, it appeared that some transposable elements carry, in at least one orientation, a good promoter sequence that can serve to switch on transcription, in an unregulated fashion, of sequences downstream of the site of insertion (Saedler *et al.*, 1974) and hence act as a mobile promoter (Charlier *et al.*, 1982). However, it has been suggested that in addition, transposable elements can, when placed in the appropriate sequence context, provide a partial promoter sequence (Ciampi *et al.*, 1982; Hinton and Musso, 1982; Jaurin and Normark, 1983) or even an "enhancer-like" sequence (Schnetz *et al.*, 1987) which can greatly stimulate a nearby, otherwise very weak, promoter.

In Table I are listed the variety of microorganisms in which IS or Tn elements have been positively identified. It should be noted that some, and perhaps many, of these elements are able to function in a broad range of hosts, being spread by natural plasmid vectors. The list may be expected to expand rapidly as more microbial groups are subjected to the appropriate molecular analysis. Indeed, one can predict with confidence that most groups of bacteria will be found to carry some form of mobile genetic element. Campbell (1983) has explored the possible evolutionary significance of transposons.

We can see that there is a wide variety of mechanisms by which changes in the primary structure of the DNA can be wrought. These changes can range from single nucleotide replacements to wholesale rearrangement of blocks of sequences and can lead to substantial and unexpected phenotypic consequences, to be examined below.

Table I. Documented Occurrence of Microbial Transposable Elements

Mobile element	Organism	Reference
Tn*10*	*Escherichia coli*	Kleckner (1983)
Tn*3*	*E. coli*	Bennett (1985)
	Salmonella paratyphi	Bennett (1985)
	S. typhimurium	Bennett (1985)
	Pseudomonas aeruginosa	Bennett (1985)
	P. vulgaris	Bennett (1985)
	Providencia	Bennett (1985)
TnPc*1*	*Pseudomonas cepacia*	Scordilis *et al.* (1987)
Tn*3951*	*Staphylococcus aureus*	Townsend *et al.* (1984)
Tn*4551*	*Bacteroides fragilis*	Smith and Spiegel (1987)
Tn*4351*	*B. fragilis*	Rasmussen *et al.* (1987)
Tn*4556*	*Streptomyces fradiae*	Chung (1987)
Tn*918*	*Streptococcus faecalis*	Clewell *et al.* (1985)
Tn*919*	*S. sanguis*	Fitzgerald and Clewell (1985)
Tn*4451*, Tn*4452*	*Clostridium perfringens*	Abraham and Rood (1987)
IS*1*, IS*2*, IS*3*, IS*4*, IS*5*	*E. coli*	Cullum (1985)
IS*402*, IS*403*, IS*404*, IS*405*	*Pseudomonas cepacia*	Scordilis *et al.* (1987)
ISH*26*	*Halobacterium halobium*	Ebert *et al.* (1987)
ISH*51*	*H. volcanii*	Hofman *et al.* (1986)
ISRm*1*	*Rhizobium meliloti*	Ruvkun *et al.* (1982)
IS*476*	*Xanthomonas campestris*	Kearney *et al.* (1988)
Unnamed	*Corynebacterium diphtheriae*	Rappuoli *et al.* (1987)

4.2. Evolution of Function

Random alterations in nucleotide sequences or insertions of IS or Tn elements generally lead to loss of function. However, there is a steadily growing literature reporting the gain of function in a variety of organisms as a consequence of one or a combination of these agents or by sequence amplification. In Table II are listed a representative sampling of documented cases of genuine acquisition of a function that is normally not found or expressed in a particular organism.

In a pioneering series of studies with *Aerobacter* (now *Klebsiella*) *aerogenes*, Lerner *et al.* (1964) and Wu *et al.* (1968) described a series of mutational steps by which an organism could acquire the capacity to metabolize and utilize for growth a novel substrate, in this case xylitol. The one precondition for this and a number of similar studies is that the organism possess a catabolic enzyme with at least a vestige of activity for the novel substrate. The first mutational alteration observed was the constitutive synthesis of the normally inducible ribitol dehydrogenase, allowing weak growth on xylitol. Improved growth was found after mutations that increased the specific activity of the ribitol dehydrogenase for xylitol and allowed constitutive expression of the D-arabitol operon, whose permease transports the xylitol into the cell. In parallel work, Hartley *et al.* (1976) found that ribitol dehydrogenase activity could be increased by gene duplication or promoter-up mutations. Thus, strains of *K. aerogenes* that normally do not utilize xylitol can be altered to grow well on this sugar alcohol by means of a combination of regulatory and structural gene mutations as well as by opportunistic utilization of elements of two different operons. Mortlock (1981, 1984) has reviewed the extensive literature relating to the mutational alteration of the range of pentoses and pentitols metabolizable by the *Enterobacteriaceae*.

1,2-Propanediol is normally produced by *E. coli* during anaerobic growth on L-fucose as a result of reduction of lactaldehyde by NADH, but it cannot be used as a substrate for anaerobic growth. Selection for growth on this novel substrate yielded strains that were no longer inducible for the fucose-utilizing enzymes and had become constitutive for the key propanediol oxido-reductase and lactaldehyde dehydrogenase enzymes. Thus, merely by reversing the normal direction of flow of metabolites through a pathway, as a consequence of regulatory changes, growth on a novel carbon and energy source is permitted (Hacking *et al.*, 1978). Similarly, Clark and Rod (1987) report the selection and characterization of a strain of *E. coli* that grows on butanol. They have shown that this novel metabolic capacity is the consequence of a series of regulatory mutations in the pathway leading to ethanol.

Short-chain aliphatic amides are good carbon and nitrogen sources for growth of *Pseudomonas aeruginosa* and serve as inducers for an aliphatic amidase. Longer-chain or aromatic amides do not induce the appropriate enzymes, nor can they be deamidated. Furthermore, the system is subject to general catabolite repression and specific repression by some of the nonmetabolizable amides. Betz *et al.* (1974) described an extensive series of selection experiments in which the range of amides metabolizable for growth was increased to include longer-chain and aromatic molecules. Analysis of the mutations allowing the expanded range of substrates showed that, as observed in other systems, the

Table II. Representative Examples of Mechanisms for Generating Metabolic Novelty

Acquired capacity	Organism	Alteration	Reference
Xylitol utilization	*Klebsiella aerogenes*	Constitutive *rbt* operon	Lerner *et al.* (1964)
Improved xylitol utilization	*K. aerogenes*	Mutation of *RDH* gene	Wu *et al.* (1968)
	K. aerogenes	Duplication of *RDH* gene	Hartley *et al.* (1976)
	K. aerogenes	Promoter-up of *RDH* gene	Hartley *et al.* (1976)
	K. aerogenes	Constitutive D-ara operon	Wu *et al.* (1968)
L-Galactose and D-arabinose utilization	*Escherichia coli*	Mutation in L-fucose pathway inducer	Zhu and Lin (1986)
Propanediol utilization	*E. coli*	Changes of fucose pathway regulation	Hacking and Lin (1977)
Citrate utilization	*E. coli*	*citA* and *citB* mutations	Hall (1982)
Butanol utilization	*E. coli*	Successive regulatory mutations	Clark and Rod (1987)
β-Lactamase resistance	*Pseudomonas cepacia*	IS activation of introduced gene	Scordilis *et al.* (1987)
Expanded range of amide utilization	*P. aeruginosa*	Constitutive expression	Betz *et al.* (1974)
	P. aeruginosa	Mutation of amidase gene	Betz *et al.* (1974)
	P. aeruginosa	Catabolite irrepressible	Smyth and Clarke (1975)
	P. aeruginosa	Butyramide inducible	Turberville and Clarke (1981)
β-Glucoside utilization	*E. coli*	*bglR* or *bglY* mutations	Reynolds *et al.* (1985)
	E. coli	*gyrA* or *gyrB* mutations	Reynolds *et al.* (1985)
	E. coli	Cap site mutations	Reynolds *et al.* (1985)
	E. coli	Constitutive *cel*	Kricker and Hall (1984)
	E. coli	Deletion of insert	Parker *et al.* (1988)
Suppression of ΔleuD	*E. coli*	Mutation in *newD*	Kemper (1984)
Suppression of Δlac	*E. coli*	Mutations in *ebg* operon	Hall (1983)
Lactobionate utilization	*E. coli*	Recombination of *ebgA* mutants	Hall (1983)

first step was the constitutive synthesis of the amidase, which has weak activity on the normally nonmetabolizable substrate. Depending on the exact conditions of selection, among the constitutive variants found, there were some that were no longer catabolite repressible (Smyth and Clarke, 1975) and/or repressible by specific amides (Brown and Clarke, 1975). Further mutations yielded single amino acid replacements in the amidase, which extended the range of substrate specificity of the enzyme well beyond that shown by the original wild-type enzyme. Of particular interest in this series of studies was the discovery of a repressor gene mutation with altered inducer specificity (Turberville and Clarke, 1981). Following on from the early observations that even single amino acid replacements were sufficient to alter enzyme substrate specificity, it is now routine to modify almost at will various parameters of enzyme catalysis by means of both *in vivo* and *in vitro* (site-directed) mutagenesis [reviewed by Inouye and Sarma, (1986)].

An alternative experimental approach to the acquisition of new function is to eliminate a known well-characterized function by means of a genetic deletion and then ask if a suitable replacement function can be retrieved or evolved from the remaining genetic material. Hall and his colleagues have conducted an extended study of β-galactoside utilization in strains of *E. coli* deleted for the *lacZ* gene (Hall, 1983). They found that such strains could be readily isolated and were the consequence of a mutation to constitutive expression of a hitherto unrecognized *ebg* operon that encodes an enzyme with low lactose-hydrolyzing activity. Selection on a series of β-galactosides yielded mutants of varying substrate specificities, and, in at least one instance, recombination between two such mutants yielded a double mutant with a substrate specificity not seen in either of the single mutants. In the same vein, Kemper (1984) found that prototrophy could be readily recovered in a strain of *E. coli* deleted for the *leuD* gene by means of inactivation of *supQ*, which in turn liberated the product of *newD* to functionally substitute for *leuD*. Presumably, in the wild-type cell the products of *supQ* and *newD* strongly interact, so that the latter product is not free to participate in the leucine biosynthetic pathway. The normal function of *supQ/newD* is unknown.

4.3. Cryptic Functions

Yet another category of acquisition of new functions includes those that appear fully operational as the result of single events that activate hitherto nonexpressible (cryptic) genes (Hall *et al.*, 1983). Although the *ebg* operon (see above) encodes a function that is cryptic in that it is normally not expressed, it did not add to the generally recognized metabolic repertoire of *E. coli*. On the other hand, single alterations in the *bgl* operon conferred on the mutant strain of *E. coli* the novel capacity to metabolize certain β-glucosides. Of particular interest is that manifestation of the *bgl* operon can be mediated by a variety of events: point mutation, alteration of local superhelicity and insertion of IS*1* or IS*5* (Reynolds *et al.*, 1985), and deletion (Parker *et al.*, 1988). Of particular significance in the latter report is the observation that the frequency of the deletion event is higher on old plates, suggesting that the event may be a measured response to environmental challenge. Following the initial discovery of the *bgl* operon in *E. coli*, several other hitherto unrecognized genes for β-glucoside utilization have been identified, including most notably the *cel* operon (Kricker and Hall, 1987). When expressed, this operon confers the ability to utilize the β-glucoside cellobiose (Kricker and Hall, 1984).

Hall and his group (Hall *et al.*, 1983; Hall and Betts, 1987) have developed a quantitative hypothesis to explain the retention of cryptic genes in a microbial population over seemingly long periods of time. The key points are that the cryptic function is required periodically and that when it is not required, those organisms that have cryptified the unneeded function are at a selective advantage. Although Li (1984) argued that the later stipulation is unnecessary, Kricker and Hall (1984) present experimental evidence supporting the contention that cryptification of unneeded functions is selectively advantageous. Hall and co-workers view the ensemble of cryptic genes as an internal reservoir of periodically useful genes that can be activated by single mutational or insertional events. It is the last characteristic which operationally identifies cryptic genes as a subclass of the widely recognized family of sequences known as pseudogenes.

That cryptic genes may be of wide significance is suggested by observations from a number of systems. Morishita *et al.* (1981) have reported that several species of *Lactobacillus* are normally auxotrophic in respect to a number of amino acids and vitamins. However, after an extensive program of mutagenesis and screening, prototrophs for most of the amino acids and vitamins were isolated. In addition, some isolates were found that, although not prototrophic, were able to utilize biosynthetic intermediates not used by the parental strains. Juni and Heym (1980) observed that natural isolates of *Neisseria gonorrhoeae* differ substantially from one another in respect to auxotrophy for a number of amino acids. While the direct experiments have not been done, the implication is that any given isolate of *N. gonorrhoeae* is potentially prototrophic for most amino acids, though many of these functions may in fact be cryptic at the time of isolation.

As part of an effort to understand the well-known wide catabolic potential of *Pseudomonas cepacia,* Lessie and Gaffney (1986) described a number of different transposable elements in that organism. These elements were identified by virtue of mediation of the expression of otherwise nonfunctional foreign genes (*lac* and *bla*) introduced on plasmids (Scordilis *et al.*, 1987). Although there is as yet no evidence bearing on this point, these transposable elements could equally well be turned to activating resident cryptic genes or metabolically useful plasmid-borne genes introduced from exogenous sources. An alternative mechanism for activation of a cryptic function is seen with the appearance of the capacity to metabolize *meta*-chlorobenzoate by *Pseudomonas* sp. A 4.3-kb genomic fragment encoding the appropriate enzymes was shown to have been reversibly amplified in the competent derivatives (Rangnekar, 1988). A similar mechanism was reported by Potekhin and Danilenko (1985) for the conversion of a kanamycin-sensitive strain of *Streptomyces rimosus* into a resistant strain.

Arico and Rappuoli (1987) report that *Bordetella parapertussis* and *B. bronchiseptica* carry the pertussis toxin genes, which are responsible for *B. pertussis* pathogenesis. However, in the two former species, these genes are silent for lack of a promoter sequence but appear otherwise functional and hence should be regarded as cryptic functions. Similarly, *Neisseria meningitidis* C114 has been shown to carry unexpressed truncated pilin genes that are homologous to those that can be expressed in *N. gonorrhoeae* (Perry *et al.*, 1988).

An interesting variation on the theme of acquisition of new function mediated by an IS element has been reported by Kearney *et al.* (1988). They observed that a strain of *Xanthomonas campestris* could become pathogenic on a hitherto resistant host by virtue of insertional inactivation (IS476) of the *avr*Bs$_1$ gene, which is responsible for eliciting the host defense mechanism.

Schmid (1988) suggests that there may be several levels of organization of the microbial genome, access to any one of which may be environmentally controlled. As an example of a hitherto unrecognized level of organization, Downs and Roth (1987) have reported that after perturbation of purine metabolism, strains of *Salmonella typhimurium* can liberate phage indistinguishable from P22. The strains are not lysogenic for P22 by the usual criteria, and no evidence for P22 nucleotide sequences can be found in the genome by conventional Southern blotting procedures. Furthermore, the phages are not induced by the normal range of agents known to trigger the SOS response. The authors suggest that the phage sequences may be sequestered in the bacterial genome in a form that is lost or is undetectable in the course of the usual DNA extraction and detection procedures and have termed such DNA "archival." An intriguing extension of these observations is that there may be whole classes of cryptic functions in the archival state and hence undetectable by conventional techniques but that may be activated by environmental or nutritional signals yet to be recognized. Cairns *et al.* (1988) and Hall (1988) have very recently published data that raise the possibility of directed mutation in the sense of preferential generation of adaptive genetic alterations. While a molecular mechanism has not been proposed, the implication is that the environmental challenge somehow elicits a specific genetic response.

It is clear that a variety of unanticipated stable phenotypes can be generated by a variety of mechanisms. Some of these, like point mutation, have been well understood for a long time. Others, like recovery of archival sequences, invoke mechanisms for which no precedents are yet known.

5. Evidence for Genomic Change

5.1. Chromosomal

Contemporary techniques of examining total genomic DNA quite independently of phenotypically scorable markers now permit a direct assessment of sequence variability. Sapienza and Doolittle (1982) observed that the genome of the prototypical archaebacterium *Halobacterium* carries an exceptionally high percentage of repeat sequences, and it was found that these sequences are associated with a very high frequency of major genomic rearrangement (Sapienza *et al.*, 1982). Indeed, at every division at least 20% of the progeny differed from the parent by at least one rearrangement event. Since the initial observation, it has been demonstrated by Hofman *et al.* (1986) that at least some of these repeated sequences appear to be IS elements. The functional or adaptive significance of the genomic rearrangements remain unknown.

By the criteria of genomic restriction fragment patterns, *Mycoplasma ovipneumoniae* has been shown by Mew *et al.* (1985) to manifest a very high level of diversity of genome structure among isolates from within a limited geographic range. This genomic diversity was not demonstrably associated with variation of Polyacrylamide gel electrophoresis protein profile or serological characteristics. Under laboratory conditions of cultivation (outside of the living host animal), the genome appears to be stable through many passages. Similar, though perhaps less extreme, variability has been observed in *M. gallosepticum* (Santha *et al.*, 1988). The molecular basis of the observed variability has not yet been fully established. A contrasting situation has been observed with *Leptospira,* for which natural isolates of specific serovars collected across broad geographic ranges showed identical or near identical genomic restriction fragment patterns (Marshall *et al.*, 1981; Thiermann *et al.*, 1985). Similar stability of genomic restriction fragment patterns has been observed among *Rickettsiae* by Regnery and Spruill (1984) and among *Brucella* species by O'Hara *et al.* (1985).

Nisen and Shapiro (1980) have reported that in *Caulobacter crescentus* regions bounded by IR sequences undergo developmentally related rearrangement events. More recently, Gilson *et al.* (1987) have demonstrated that as much as 1% of the *E. coli* genome may consist of highly repetitive palindromic units of 20–40 nucleotides each. Although the function of these units remain unknown, it has been demonstrated that members of this class of sequences mark the borders of chromosomal blocks that have undergone rearrangement. In *Campylobacter coli,* reversible antigenic variation has been shown to be correlated with genomic rearrangements (Guerry *et al.*, 1988) that the authors suggest may be of the programmed variety. From rearrangements seen in total genomic DNA, we turn now to rearrangements involving plasmid DNA.

5.2. Plasmid

Plasmid structure is a fertile area for assessing genomic rearrangements because the relatively small size and supercoiled state allow ready size measurement, isolation, and restriction map determination. Many workers have reported genetic rearrangements involving plasmid DNA, often identified or selected because of an associated alteration in a clear phenotypic characteristic.

The genus *Pseudomonas* has attracted a great deal of attention because of the great diversity of substrates found to allow growth of members of this group of organisms. Of particular interest has been the capacity to metabolize unsaturated ring systems and their chlorinated derivatives, the genes for which have generally been found on plasmids. An extensive series of rearrangements involving both plasmid and chromosomal sequences that accompany alterations in the range of metabolizable substrates have been described in *P. putida* (Chatterjee and Chakrabarty, 1982; Pickup and Williams, 1982; Williams and Jeenes, 1981). A similarly extensive series of cryptic plasmid and chromosomal rearrangements were described in *P. cepacia* and *P. syringae,* respectively (Gaffney and Lessie, 1987; Szabo and Mills, 1984), many of these involving either repeated sequences or IS elements.

Rhizobium (now *Bradyrhizobium) japonicum* and *R. phaseoli* can undergo rearrangements involving resident plasmids (Berry and Atherley, 1984; Soberon-Chavez *et* al., 1986). In the former case, introduction of RP1 led to transfer of approximately 55 kb from the chromosome to one of the normally resident plasmid and integration of RP1 into the chromosome. In the latter case, a sequence of alterations involving both the symbiotic (pSym) plasmids and chromosomal DNA generated first the loss of the Nod+ phenotype and then its subsequent recovery. It was suggested that the multiple copies of *nifH* were involved in these rearrangements. A number of *Rhizobium* and *Agrobacterium* species have small families of repeated sequences scattered throughout the genome (Flores *et al.*, 1987). Further work showed that at least in some clonal phenotypic variants, there were genomic rearrangements involving repeated sequences (Flores *et al.*, 1988).

In *Bradyrhizobium japonicum* the region carrying the principle determinants for symbiotic nitrogen fixation asymmetrically accumulate reiterated sequences, and these sequences are involved in at least one half of the *nod* (nodulation)/*nif* deletion events observed (Hahn and Hennecke, 1987). By comparison of a series of *B. japonicum* strains, Haugland and Verma (1981) concluded that in the course of the divergence of these strains, extensive exchange between plasmid and chromosomal DNA had taken place.

Murphy and Novick (1981) have described a series of both Rec-mediated and apparent site-specific recombinational events between several closely related penicillinase plasmids of *Staphylococcus aureus* that led to duplications, inversions, and deletions. Some of these rearrangements resulted in the loss of the penicillinase phenotype. Plasmid rearrangements were also associated with the loss of photosynthetic ability in *Rhodopseudomonas sphaeroides* (Nano and Kaplan, 1984). In *Streptomyces coelicolor,* an interesting module can exist either free as a member of a series of plasmids or, after site-specific recombination, in the integrated state (Omer and Cohen, 1986). No function has yet been assigned to this rearrangement. Valla *et al.* (1987) have reported ready rearrangement of plasmid and chromosomal DNA in *Acetobacter xylinum,* some of which rearrangement was associated with the loss of cellulose synthetic capacity. Recovery of this capacity was not reported.

In view of the great potential for genomic change, it is reasonable to ask about the evolutionary significance of this variability. To make really meaningful generalizations in this area would require a vastly greater data base than is currently available. However, Riley (1985) reviewed the information relating to genomic divergence among *Enterobacteriaceae* and has commented on the genomic divergence seen in other groups of organisms. Comparisons of *E. coli* and *S. typhimurium* show largely congruent genetic maps in spite of an estimated divergence time of 50 million years. *P. aeruginosa* and *P. putida* were cited as examples of two relatively closely related species with substantially divergent genetic maps. Holloway and Morgan (1986) have pointed out that the maps of the latter two organisms show much less functional organization than is found among *Enterobacteriaceae* and have suggested that the genus *Pseudomonas* may have evolved by accretion of pieces of genetic information, perhaps borne by plasmids, rather than by a more orderly evolutionary process. No systematic evolutionary work has been done on those groups of demonstrable high genomic variability such as *Halobacterium* and *Mycoplasma.*

Similar comparisons of the structure of plasmids from geographically and temporally diverse origins yield parallel mixed conclusions. Chatterjee and Chakrabarty (1983) report that chlorobenzoate-degradative plasmids isolated from geographically and temporally dispersed *Pseudomonas* spp. show virtually identical structures. Smith and Thomas (1987) and Villarroel *et al.* (1983) compared the structure of a whole series of P group plasmids and concluded that they all share a common backbone but differ in the spectrum and location of insertion elements they carry. Shalita *et al.* (1980) studied an extended series of plasmids from the Gram-positive bacterium *Staphylococcus aureus* isolated from geographically diverse strains. Half of the isolates could be shown to be evolutionarily related through a sequence of insertion events, and the rest fell into a number of apparently unrelated groups.

With such a limited data base, the only conclusion to be drawn from the broadly evolutionary perspective is that for a given degree of divergence (as measured by nucleotide sequence data), there are big differences in extent of gross chromosomal rearrangements between different lines of descent. Plasmids manifest a similar spectrum of diversity in respect to sequence conservation.

Thus, by direct examination of genomic DNA, substantial sequence rearrangement has been detected in a wide variety of organisms. Not too surprisingly, plasmids, a common agent of intercellular exchange of DNA, are often implicated in rearrangement events. In some instances, rearrangement events have been correlated with detectable phenotypic changes, whereas often enough, under appropriate experimental conditions, rearrangements have occurred without detected phenotypic alterations. It must be noted that the absence of detectable phenotypic changes may well be a reflection on the observer rather than on the organism concerned.

We turn now to a report of microbial plasticity on an unprecedented scale. By a combination of UV irradiation and nutritional stress, *Rhizobium* spp. could be induced to undergo a sequence of transformations that yielded in turn five classes of organisms distinctly different from one another at the generic level of discrimination (Heumann *et al.*, 1984). The differences included gross physiologic changes, changes in the restriction/modification systems, alterations in plasmid profile, variations in GC content from 57.5% to 72.1%, and interclass DNA homology in the neighborhood of 50%. Alterations of this magnitude would require selective amplification and deamplification and/or transfer of DNA in and out of an archival state on an unprecedented scale. General acceptance of these results, which are at total variance with conventional wisdom, will require careful confirmation by other groups. Such confirmation has yet to appear in the literature.

6. Phenotypic Changes of Unknown Genetic Basis

In addition to genomic changes either in association with discernible phenotypic changes or not, there have been reported a wide variety of phenotypic variations, for some of which the underlying genetic basis has not yet been determined. Extensive studies of microbial physiology in a few selected organisms have revealed a variety of coordinated responses to specific identified signals. Examples of such patterned response are the SOS response and catabolite repression [reviewed by Gottesman (1984)] and regulation by key energy intermediates [reviewed by Hellingswerf and Konings (1985)]. In none of these cases has there been a need to invoke mechanisms beyond conventional gene regulation, chemical kinetics, and thermodynamics. Whereas some additional cases, upon further examination, may well turn out to be manifestations of already understood events, investigation of others may lead to new insights into genomic plasticity.

Among pathogens, a general and widely recognized phenomenon is attenuation, whereby an isolate gradually loses pathogenicity during laboratory propagation. It is not known whether this change simply reflects a mutational loss of function or whether it is part of an orderly and regulated genotypic alteration in response to altered circumstances. Another generally recognized characteristic of pathogens is enormous variety and variability of antigenic properties, presumably evolved as a means of evasion of the host immune systems. Whereas very few cases have been examined from the molecular genetic perspective (discussed above), the vast majority of examples cited remain as intriguing instances of microbial variability of an unknown genetic basis (Birkbeck and Penn, 1986; Clegg and Gerlach, 1987; Harris et al., 1987).

Mucoid alginate production by P. aeruginosa is associated with pathogenesis and has been shown to be a cryptic trait in the sense that, under appropriate in vivo or in vitro selective conditions, normally nonmucoid strains can generate mucoid derivatives (Flynn and Ohman, 1988). Ready loss of this character is also observed, suggesting a binary switching mechanism of the sort discussed earlier. P. atlantica also manifests variable synthesis of extracellular polysaccharide, with expression being correlated with the reversible insertion of a 1.2-kb sequence into a key gene of the pathway (Bartlett et al., 1988). Serratia marcescens manifests well-known reversible variations in colony pigmentation, which has been shown recently to be correlated with alterations in the structure of a flagellar antigen (Paruchuri and Harshey, 1987). It was suggested that perhaps a programmed rearrangement is involved.

A class of phenotypic change of totally unknown mechanism has been described by Shapiro (1985, 1986, 1987). He has developed techniques for examining systematic alterations in gene expression during the course of growth (development) of a bacterial colony, with particular attention to events that occur in very mature colonies on old agar plates. Some details of the patterns of cell growth and gene expression suggest that regulation is a simple function of developmental age of the colony. Other observations are most readily explained by an age- or nutrition-dependent alteration in genotypic state, which may involve genomic rearrangements. There are as yet no direct observations published that bear on the latter suggestion.

In a somewhat similar vein, Wanner (1985) and Wanner et al. (1988) have described clonal variation of expression of alkaline phosphatase in E. coli phoR (phosphatase) mutants. On the basis of combined genetic and physiological data, the authors suggest that the observed clonal variation may be a response to environmental signals, in this case phosphate deprivation, reflecting a higher order of regulation than the conventional tight functional coupling between signal and response. Using Mudlac-directed operon fusions in *S. typhimurium*, Spector *et al.* (1988) have identified a number of functions that are expressed only under extreme starvation conditions. As yet, there is no evidence to suggest that either of these examples represent manifestations of rearrangement.

To test the general thesis of this review, that microorganisms may have intrinsic hidden reserves of plasticity, defined strains must be subjected to natural environmental stress in the absence of other organisms with which they could interact. Few such experiments have been reported, although Trevors (1988) has reviewed this general area of endeavor. Using *P. putida*, Jain *et al.* (1987), by the technique of colony hybridization, were able to reach the very limited conclusion that identifiable blocks of DNA (ranging in size from large plasmids to chromosomes) were recoverable after inoculation into microcosm ecosystems. Palmer *et al.* (1984) described dramatic genotypic and phenotypic changes in *E. coli* upon introduction into a sterilized microcosm aquatic ecosystem to which had been added trace amounts of various toxic compounds. Upon growth of these altered organisms on normal nutrient agar, the original characteristics were recovered. Munro *et al.* (1987) reported similar marked physiological alterations in *E. coli* upon prolonged incubation in sterilized and supplemented seawater microcosms. There has been no further work reported on the molecular events underlying the observed alterations. The hope and expectation with such experiments is to elucidate the mechanisms by which microorganisms cope with the ensemble of stresses presented by natural ecosystems. The key question is whether well-understood physiological mechanisms suffice or whether there exist additional, more drastic, and perhaps less coordinated responses.

Roszak and Colwell (1987) have very thoroughly reviewed the literature relating to viability, survivability, morphological variation, and survival strategies of bacteria in natural ecosystems. Two important cautionary points were made: our techniques for accurately assessing microbial populations in natural ecosystems are inadequate; and the characteristics of a given organism determined under laboratory conditions may differ dramatically from characteristics of the same organism under field conditions.

7. Significance of Microbial Plasticity to Microbial Ecology

From the wide range of microbial literature that we have discussed, there are several interrelated observations and conclusions that we wish to make.

First, the enormous success of classical microbiology and microbial genetics has to a considerable degree depended on selection and utilization of strains of organisms that behave in a reliable and consistent manner on laboratory media that usually bore little or no resemblance to the natural environment. Furthermore, the whole objective of microbial research was to discern order in a seemingly chaotic Nature; thus, elements of disorder were, by and large, perforce ignored. These two aspects of microbiological research perhaps conspired to impose a regularity and predictability upon Nature which may be misleading if we wish to understand how microorganisms behave outside the confines of the laboratory test tube or petri plate.

Second, techniques for manipulating and examining DNA have allowed us to see individual and populational genotypic variability independent of overt phenotypic variability.

Third, beyond the primary metabolic pathways and a few global regulatory circuits, generally both determined of necessity in a relatively limited set of domesticated laboratory organisms, our ignorance is profound about what makes a bacterium successful in a specific, and perhaps varying, niche. Indeed, for most bacteria, we do not even know what the microniche looks like.

We also wish to suggest that beyond the capacity for intercellular exchange of genetic information, many (or perhaps most) groups of bacteria possess hidden reserves of genetic information that may be manifested either under appropriate environmental conditions or merely by chance. Some mechanisms of activation involving inversion, transposition, mutation, and transposable elements have been described in detail in a very few organisms, and substantial genomic and phenotypic variation is beginning to be seen in numerous others. Are there major mechanisms of response to environmental challenge, leading to global regulation or genomic rearrangement, that we have yet to recognize? Or can all of the manifestations of genotypic and phenotypic plasticity seen and yet to be seen be accommodated within the conceptual framework that currently exists? We believe that there is much to be learned in this area of fundamental interest to microbial ecology.

The implications of these considerations for biotechnology and assessment of environmental impact of microbial release are substantial. On the one hand, the task of constructing strains for performing specific industrial tasks could be made much easier if one were to make an accurate assessment of the cryptic capacities of the organism and full advantage taken of this aspect of microbial plasticity rather than think only in terms of introducing foreign genes. On the other hand, it is important to understand how organisms may change, either in the industrial or in the natural environment, in response to challenges unrealizable at the laboratory bench.

We have been struck by the lack of cross-referencing, even between closely related work in the general area of this review. In addition, many of the significant observations have been relegated to specialist journals and hence have, on occasion, been difficult to track down. We fear that as a consequence of these two factors, we have overlooked many interesting examples of microbial plasticity. If this is the case, our apologies to the neglected parties.

Note added in proof: Hybridization analysis of *Agrobacterium tumefaciens* and species of *Rhizobium* has suggested DNA reiteration is a common feature in these genera and high-frequency rearrangements have been detected involving some of the reiterated DNA families [Palacios, R., Flores, M., Brom, S., Martinez, E., Gonzalez, V., Frenk, S., Quinto, C., Cevallos, M. A., Segovia, L., Romero, D., Garciarrubio, A., Pinero, D., and Davila, G., 1987, Organization of the *Rhizobium phaseoli* genome, in: *Molecular Biology of the Plant-Microbe Interaction* (A. Puhler, ed.), pp. 151–156, Martinus Nijhoff, The Hague]. Subsequent characterization of a strain of *R. phaseoli* has suggested that a hyper-recombinant mutant, in which frequencies of plasmid rearrangement are significantly higher than the wild type, is phenotypically similar to certain *E. coli* mutants affected in DNA replication or repair (Soberon-Chavez, G., and Najera, R., 1989, Symbiotic plasmid rearrangement in a hyper-recombinant mutant of *Rhizobium leguminosarum* biovar *phaseoli*, *J. Gen. Microbiol.* **135**:47).

ACKNOWLEDGMENTS. We wish to thank the Massey University Agricultural Research Foundation for the generous support of Dr. A. N. MacGregor for the work in rhizobial plasticity (manuscript in preparation), which led to the assembly of the material presented in this review. We wish also to thank the Stanford University Biological Sciences Library for the use of their facilities by E. T. and Dr. B. E. Terzaghi for her encouragement in this endeavor and her critical reading of the manuscript. Comments on the manuscript by colleagues in the Department of Microbiology and Genetics (Massey University) were very helpful and are gratefully acknowledged.

References

Abraham, J. M., Freitag, C. S., Clements, J. R., and Eisenstein, B. I., 1985, An invertible element of DNA controls phase variation of type I fimbriae of *Escherichia coli, Proc. Natl. Acad. Sci. USA* **82**:5724.

Abraham, L. J., and Rood, J. I., 1987, Identification of Tn*4451* and Tn*4452*, chloramphenicol resistance transposons from *Clostridium perfringens, J. Bacteriol.* **169**:1579.

Alexander, M., 1986, Ecological concerns relative to genetically engineered microorganisms, in: *Microbial Communities in the Soil* (V. Jensen, A. Kjoller, and L. H. Sorensen, eds.), pp. 347–354, Elsevier Applied Science Publications, London.

Anderson, R. P., and Roth, J., 1978, Tandem chromosomal duplications in *Salmonella typhimurium*: fusion of histidine genes to novel promoters, *J. Mol. Biol.* **119**:147.

Argos, P., Laudy, A., Abremski, K., Egan, J. B., Haggard-Ljungquist, E., Hoess, R. H., Kahn, M. L., Kalionis, B., Narayana, S. V. L., Pierson III, L. S., Sternberg, N., and Leong, J. M., 1986, The integrase family of site-specific recombinases: regional similarities and global diversity, *EMBO J.* **5**:433.

Arico, B., and Rappuoli, R., 1987, *Bordetella parapertussis* and *Bordetella bronchiseptica* contain transcriptionally silent pertussis toxin genes, *J. Bacteriol.* **169**:2847.

Balganesh, M., and Setlow, J. K., 1986, Plasmid-to-plasmid recombination in *Haemophilus influenzae, J. Bacteriol.* **165**:308.

Barbour, A. G., and Garon, C. F., 1987, Linear plasmids of the bacterium *Borrelia burgdorferii* have covalently closed ends, *Science* **237**:409.

Bartlett, D. H., Wright, M. E., and Silverman, M., 1988, Variable expression of extracellular polysaccharide in the marine bacterium *Pseudomonas atlantica* is controlled by genome rearrangement, *Proc. Natl. Acad. Sci. USA* **85**:3923.

Bennett, P., 1985, Bacterial transposons, in: *Genetics of Bacteria* (J. Scaife, D. Leach, and A. Galazzi, eds.), pp. 97–115, Academic Press, New York.

Berry, J. O., and Atherley, A. G., 1984, Induced plasmid-genome rearrangements in *Rhizobium japonicum, J. Bacteriol.* **157**:218.

Betz, J. L., Brown, P. R., Smyth, M. J., and Clarke, P. H., 1974, Evolution in action, *Nature (London)* **247**:261.

Bianchi, M. E., and Radding, C. M., 1983, Insertions, deletions and mismatches in heteroduplex DNA made by RecA protein, *Cell* **35**:511.

Birkbeck, T. H., and Penn, C. W. (eds.), 1986, *Antigenic Variation in Infectious Diseases,* Society for General Microbiology, Special Publication 19, IRL Press, Washington, D.C.

Borst, P., and Greaves, D. R., 1987, Programmed gene rearrangements altering gene expression, *Science* **235**:658.

Botstein, D., 1980, A theory of modular evolution for bacteriophages, *Ann. N.Y. Acad. Sci.* **354**:4841.

Britten, R. J., and Kohne, D. E., 1968, Repeated sequences in DNA, *Science* **161**:529.

Brown, J. E., and Clarke, P. H., 1975, Mutations in a regulator gene allowing *Pseudomonas aeruginosa* 8602 to grow on butyramide, *J. Gen Microbiol.* **64**:329.

Cairns, J., Overbaugh, J., and Miller, S., 1988, The origin of mutants, *Nature (London)* **335**:142.

Campbell, A., 1965, The steric effect in lysogenization by bacteriophage lambda. 1. Lysogenization of a partially diploid strain of *Escherichia coli* K12, *Virology* **27**:329.

Campbell, A., 1983, Transposons and their evolutionary significance, in: *Evolution of Genes and Proteins* (M. Nei and R. K. Koehn, eds.), pp. 258–279, Sinauer & Associates, Sunderland, Mass.

Campbell, A., and Botstein, D., 1983, Evolution of the lambdoid phages, in: *Lambda II* (R. W. Hendrix, J. W. Roberts, F. W. Stahl, and R. A. Weisberg, eds.), pp. 365–380, Cold Spring Harbor Laboratory, Cold Spring Harbor, N.Y.

Charlier, D., Piette, J., and Glansdorf, N., 1982, IS3 can function as a mobile promoter in *Escherichia coli, Nucleic Acids Res.* **10**:5935.

✓ Chatterjee, D. K., and Chakrabarty, A. M., 1982, Genetic rearrangements in plasmids specifying total degradation of chlorinated benzoic acids, *Mol. Gen. Genet.* **188**:279.

✓ Chatterjee, D. K., and Chakrabarty, A. M., 1983, Genetic homology between independently isolated chlorobenzoate-degradative plasmids, *J. Bacteriol.* **153**:532.

Chung, S.-T. 1987, TN*4556,* a 6.8-kilobase-pair transposable element of *Streptomyces fradiae, J. Bacteriol.* **169**:4436.

Ciampi, M. S., Schmid, M. B., and Roth, J. R., 1982, Transposon Tn10 provides a promoter for transcription of adjacent sequences, *Proc. Natl. Acad. Sci. USA* **79**:5016.

Clark, D. P., and Rod, M. L., 1987, Regulatory mutations that allow the growth of *Escherichia coli* on butanol as a carbon source, *J. Mol. Evol.* **25**:151.

Clegg, S., and Gerlach, G. F., 1987, Enterobacterial fimbriae, *J. Bacteriol.* **169**:934.

Clewell, D. B., An, F.-Y., White, B. A., and Gawron-Burke, C., 1985, *Streptococcus faecalis* sex pheromone (CAM373) also produced by *Staphylococcus aureus* and identification of a conjugative transposon Tn*918, J. Bacteriol.* **162**:1212.

Coveney, J. A., Fitzgerald, G. F., and Daly, C., 1987, Detailed characterization and comparison of four lactic streptococcal bacteriophages based on morphology, restriction mapping, DNA homology and structural protein analysis, *Appl. Environ. Microbiol.* **53**:1439.

Cullum, J., 1985, Insertion sequences, in: *Genetics of Bacteria* (J. Scaife, D. Leach, and A. Galizzi, eds.), pp. 85–96, Academic Press, New York.

Dorman, C. J., and Higgins, C. F., 1987, Fimbrial phase variation in *Escherichia coli:* dependence on integration host factor and homologies with other site-specific recombinases, *J. Bacteriol.* **169**:3840.

Downs, D. M., and Roth, J. R., 1987, A novel P22 prophage in *Salmonella typhimurium, Genetics* **117**:367.

Ebert, K., Hanke, C., Delius, H., Goebel, W., and Pfeifer, F., 1987, A new insertion element ISH26 from *Halobacterium halobium. Mol. Gen. Genet.* **206**:81.

Fitzgerald, G. F., and Clewell, D. B., 1985, A conjugative transposon (Tn*919*) in *Streptococcus sanguis, Infect. Immun.* **47**:415.

Flores, M., Gonzalez, V., Brom, S., Martinez, E., Pinero, D., Romero, D., Davila, G., and Palacios, R., 1987, Reiterated sequences in *Rhizobium* and *Agrobacterium* spp., *J. Bacteriol.* **169**:5782.

Flores, M., Gonzalez, V., Pardo, M. A., Leija, A., Martinez, E., Romero, D., Pinero, D., Davila, G., and Palacios, R., 1988, Genomic instability in *Rhizobium phaseoli, J. Bacteriol.* **170**:1191.

Flynn, J. L., and Ohman, D. E., 1988, Cloning of genes from mucoid *Pseudomonas aeruginosa* which control spontaneous conversion to the alginate production phenotype, *J. Bacteriol.* **170**: 1452.

Gaffney, T. D., and Lessie, T. G., 1987, Insertion sequence-dependent rearrangements of *Pseudomonas cepacia* plasmid pTGL1, *J. Bacteriol.* **169**:224.

Gething, M. J., Bye, J., Skehel, J., and Waterfield, M., 1980, Cloning and DNA sequence of double-stranded copies of haemagglutinin genes from H2-strains and H3-strains elucidates antigenic shift and drift in human influenza virus, *Nature (London)* **287**:301.

Gilson, E., Clement, J. M., Perrin, D., and Hoffnung, M., 1987, Palindromic units: A case of highly repetitive DNA sequences in bacteria, *Trends Genet.* **3**:226.

Gonda, D. K., and Radding, C. M., 1983, By searching processively RecA protein pairs DNA molecules that share a limited stretch of homology, *Cell* **34**:647.

Gottesman, S., 1984, Bacterial regulation: Global regulatory networks, *Annu. Rev. Genet.* **18**:415.

Guerry, P., Logan, S. M., and Trust, T. J., 1988, Genomic rearrangements associated with antigenic variation in *Campylobacter coli, J. Bacteriol.* **170**:316.

Haas, R., and Meyer, T. F., 1986, The repertoire of silent pilus genes in *Neisseria gonorrhoeae: evidence for gene conversion, Cell* **44**:107.

Hacking, A. J., and Lin, E. C. C., 1977, Regulatory changes in the fucose system associated with the evolution of a catabolic pathway for propanediol in *Escherichia coli, J. Bacteriol.* **130**:832.

Hacking, A. J., Aquilar, J., and Lin, E. C. C., 1978, Evolution of propanediol utilization in *Escherichia coli:* mutant with improved substrate-scavenging power, *J. Bacteriol.* **136**:522.

Hahn, M., and Hennecke, H., 1987, Mapping of a *Bradyrhizobium japonicum* DNA region carrying genes for symbiosis and an asymmetric accumulation of reiterated sequences, *Appl. Environ. Microbiol.* **53**:2247.

Hall, B. G., 1982, Chromosomal mutation for citrate utilization by *Escherichia coli* K-12, *J. Bacteriol.* **151**:269.

Hall, B. G., 1983, Evolution of new metabolic functions in laboratory organisms, in: *Evolution of Genes and Proteins* (M. Nei and R. K. Koehn, ed.), pp. 234–257, Sinauer & Associates, Sunderland, Mass.

Hall, B. G., 1988, Adaptive evolution that requires multiple spontaneous mutations. I. Mutations involving an insertion sequence. *Genetics* **120**:887.

Hall, B. G., and Betts, P. W., 1987, Cryptic genes for cellobiose utilization in natural isolates of *Escherichia coli, Genetics* **115**:431.

Hall, B. G., Yokoyama, S., and Calhoun, D. H., 1983, Role of cryptic genes in microbial evolution, *Mol. Biol. Evol.* **1**:109.

Harris, L. A., Logan, S. M., Guerry, P., and Trust, T. J., 1987, Antigenic variation of *Campylobacter* flagella, *J. Bacteriol.* **169**:5066.

Hartley, B. S., Altosaar, I., Dothie, J. M., and Neuberger, M. S., 1976, Experimental evolution of a xylitol dehydrogenase, in: *Proceedings of the Third John Innes Symposium* (R. Markham and R. W. Horne, eds.), pp. 191–200, North-Holland, Amsterdam.

Haselkorn, R., Golden, J. W., Lammers, P. J., and Mulligan, M. E., 1986, Developmental rearrangement of cyanobacterial nitrogen-fixation genes, *Trends Genet.* **2**:255.

Haselkorn, R., Bukema, W. J., Golden, J. W., Lammers, P. J., and Mulligan, M. E., 1988, Rearrangement of nitrogen fixation genes during heterocyst differentiation in the cyanobacterium *Anabaena* 7120 abstr. L010, UCLA Symp. Mol. Basis Plant Dev., p. 131.

Haugland, R., and Verma, D. P. S., 1981, Interspecific plasmid and genomic DNA sequence homologies and localization of *nif* genes in effective and ineffective strains of *Rhizobium japonicum, J. Mol. Appl. Genet.* **1**:205.

Hellingswerf, K. J., and Konings, W. N., 1985, The energy flow in bacteria: The main free energy intermediates and their regulatory role, *Adv. Microb. Physiol.* **26**:125.

Heumann, W., Rosch, A., Springer, R., Wagner, E., and Winkler, K. P., 1984, In *Rhizobiaceae* five different species are produced by rearrangements of one genome, induced by DNA-damaging agents, *Mol. Gen. Genet.* **197**:425.

Hinton, D. M., and Musso, R. E., 1982, Transcription initiation sites within an IS2 insertion in a Gal-constitutive mutant of *Escherichia coli, Nucleic Acids Res.* **10**:5015.

Hofman, J. D., Schalkwyk, L. C., and Doolittle, W. F., 1986, ISH51: A large, degenerate family of insertion sequence-like elements in the genome of the archaebacterium *Halobacterium volcanii, Nucleic Acids Res.* **14**:6983.

Holloway, B. W., and Morgan, A. P., 1986, Genome organization in *Pseudomonas, Annu. Rev. Microbiol.* **40**:79.

Huber, H. E., Iida, S., Arber, W., and Bickle, T. A., 1985, Site-specific DNA inversion is enhanced by a DNA sequence element in cis, *Proc. Natl. Acad. Sci. USA* **82**:3776.

Iida, S., and Hiestand-Nauer, R., 1986, Localized conversion at the crossover sequences in the site-specific DNA inversion system of bacteriophage P1, *Cell* **45**:71.

Inouye, M., and Sarma, R. (eds.), 1986, *Protein Engineering: Applications in Science, Medicine and Industry,* Academic Press, New York.

Jacob, F., 1977, Evolution and tinkering, *Science* **196**:1161.

Jain, R. K., Sayler, G. S., Wilson, J. T., Houston, L., and Pacia, D., 1987, Maintenance and stability of introduced genotypes in groundwater aquifer material, *Appl. Environ. Microbiol.* **53**:996.

Jarvis, A. W., and Meyer, J., 1986, Electron microscopic heteroduplex study and restriction endo-nuclease cleavage analysis of the DNA genomes of three lactic streptococcal bacteriophages, *Appl. Environ. Microbiol.* **51**:566.

Jaurin, B., and Normark, S., 1983, Insertion of IS2 creates a novel *amp* C promoter in *Escherichia coli, Cell* **32**:809.

Johnson, R. C., and Simon, M. I., 1987, Enhancers of site-specific recombination in bacteria, *Trends Genet.* **3**:262.

Johnson, R. C., Glasgow, A. C., and Simon, M. I., 1987, Spatial relationship of the Fis binding sites for Hin recombinational enhancer activity, *Nature (London)* **329**:462.

Juni, E., and Heym, G. A., 1980, Studies of some naturally occurring auxotrophs of *Neiserria gonor-rhoeae, J. Gen. Microbiol.* **121**:85.

Kearney, B., Ronald, P. C., Dahlbeck, D., and Staskawicz, B. J., 1988, Molecular basis for evasion of plant host defence in bacterial spot disease of pepper, *Nature (London)* **332**:541.

Kemper, J., 1984, Gene recruitment for a subunit of isopropylmalate isomerase, in: *Microorganisms as Model Systems for Studying Evolution* (R. P. Mortlock, ed.), pp. 255–284, Plenum Press, New York.

Kleckner, N., 1983, Transposon Tn10, in: *Mobile Genetic Elements* (J. Shapiro, ed.), pp. 261–299, Academic Press, New York.

Komano, T., Kubo, A., and Nisioka, T., 1987a, Shufflon: Multi-inversion of four contiguous DNA segments of plasmid R64 creates seven different open reading frames, *Nucleic Acids Res.* **15**:1165.

Komano, T., Kim, S. R., and Nisoka, T., 1987b, Distribution of shufflon among Incl plasmids, *J. Bacteriol.* **169**:5317.

Kricker, M., and Hall, B. G., 1984, Directed evolution of cellobiose utilization in *Escherichia coli* K12, *Mol. Biol. Evol.* **1**:171.

Kricker, M., and Hall, B. G., 1987, Biochemical genetics of the cryptic gene system for cellobiose utilization in *Escherichia coli* K12, *Genetics* **115**:419.

Kubo, A., Kusukawa, A., and Komano, T., 1988, Nucleotide sequence of the *rci* gene encoding shufflon-specific DNA recombinase in the Incl 1 plasmid R64: Homology to the site-specific recombinases of the integrase family. *Mol. Gen. Genet.* **213**:30.

Lechevallier, M. W., Camper, A. K., Broadaway, S. C., Henson, J. M., and McFeters, G. A., 1987,

Sensitivity of genetically engineered organisms to selective media, *Appl. Environ. Microbiol.* **53**:606.

Lerner, S. A., Wu, T. T., and Lin, E. C. C., 1964, Evolution of a catabolic pathway in bacteria, *Science* **146**:1313–1315.

Lessie, T. G., and Gaffney, T., 1986, Catabolic potential of *Pseudomonas cepacia*, in: *The Bacteria*, Vol. ✓ X (J. R. Sokatch, ed.), pp. 439–482, Academic Press, New York.

Levin, M. A., Seidler, R., Borguin, A. W., Fowle III, J. R., and Barkay, T., 1987, Developing methods to assess environmental release, *BioTechnology* **5**:38.

Lewin, B., 1987, *Genes III*, p. 603, John Wiley & Sons, New York.

Li, W.-H., 1984, Retention of cryptic genes in microbial populations, *Mol. Biol. Evol.* **1**:213.

Marshall, R. B., Wilton, B. E., and Robinson, A. J., 1981, Identification of *Leptospira* serovars by restriction endonuclease analysis, *J. Med. Microbiol.* **14**:163.

Mew, A. J., Ionas, G., Clarke, J. K., Robinson, A. J., and Marshall, R. B., 1985, Comparison of *Mycoplasma ovipneumoniae* isolates using bacterial restriction endonuclease DNA analysis and SDS-PAGE, *Vet. Microbiol.* **10**:541.

Meyer, T. F., 1987, Molecular basis of surface antigen variation in *Neisseria*, *Trends Genet.* **3**: 319.

Mori, M., Tanimoto, A., Yoda, K., Harada, S., Koyama, N., Hashiguchi, K., Obinata, M., Yamasaki, M., and Tamura, G., 1986, Essential structure in the cloned transforming DNA that induces gene amplification of the *Bacillus subtilis amyE-tmrB* region, *J. Bacteriol.* **166**:787.

Morishita, T., Deguchi, Y., Yajima, M., Sakurai, T., and Yura, T., 1981, Multiple nutritional requirements of lactobacilli: Genetic lesions affecting amino acid biosynthetic pathways, *J. Bacteriol.* **148**:64.

Mortlock, R. P., 1981, Catabolism of five carbon sugars, in: *Microbiology—1981* (D. Schlessinger, ed.), pp. 151–155, American Society for Microbiology, Washington, D.C.

Mortlock, R. P., 1984, The utilization of pentitols in studies of the evolution of enzyme pathways, in: *Microorganisms as Model Systems for Studying Evolution* (R. P. Mortlock, ed.), pp. 1–22, Plenum Press, New York.

Munro, P. M., Gauthier, M. J., and Laumond, F. M., 1987, Changes in *Escherichia coli* cells starved in seawater or grown in seawater-wastewater mixtures, *Appl. Environ. Microbiol.* **53**:1476.

Murphy, E., and Novick, R. P., 1980, Site-specific recombination between plasmids of *Staphylococcus aureus*, *J. Bacteriol.* **141**:316.

Nano, F. E., and Kaplan, S., 1984, Plasmid rearrangements in the photosynthetic bacterium *Rhodopseudomonas sphaeroides*, *J. Bacteriol.* **158**:1094.

National Academy of Science Committee on the Introduction of Genetically Engineered Organisms into the Environment, 1987, *Introduction of Recombinant DNA-Engineered Organisms into the Environment: Key Issues*, National Academy Press, Washington, D.C.

Niaudet, B., Janniere, L., and Ehrlich, S. D., 1984, Recombination between repeated DNA sequences occurs more often in plasmids than in the chromosome of *Bacillus subtilis*, *Mol. Gen. Genet.* **197**:46.

Nisen, P., and Shapiro, L., 1980, Inverted-repeat nucleotide sequences in *Escherichia coli* and *Caulobacter crescentus*, *Cold Spring Harbor Symp. Quant. Biol.* **45**:81.

Office of Technology Assessment, 1988, *New Developments in Biotechnology—Field-Testing Engineered Organisms: Genetic and Ecological Issues*, U.S. Government Printing Office, Washington, D.C.

O'Hara, M., Collins, D. M., and De Lisle, G. W., 1985, Restriction endonuclease analysis of *Brucella ovis* and other *Brucella* species, *Vet. Microbiol.* **10**:425.

Omer, C. A., and Cohen, S. N., 1986, Structural analysis of plasmid and chromosomal loci involved in site-specific excision and integration of the SLP1 element of *Streptomyces coelicolor*, *J. Bacteriol.* **166**:999.

Palmer, L. M., Baya, A. M., Grimes, D. J., and Colwell, R. R., 1984, Molecular genetic and phenotypic alteration of *Escherichia coli* in natural water microcosms containing toxic chemicals, *FEMS Microbiol. Lett.* **21**:169.

Parker, L. L., Betts, P. W., and Hall, B. G., 1988, Activation of a cryptic gene by excision of a DNA fragment, *J. Bacteriol.* **170**:218.

Paruchuri, D. K., and Harshey, R. M., 1987, Flagellar variation in *Serratia marcescens* is associated with color variation, *J. Bacteriol.* **169**:61.

Perry, A. C. F., Nicholson, I. J., and Saunders, J. R., 1988, *Neisseria meningitidis* C114 contains silent truncated pilin genes that are homologous to *Neisseria gonorrhoeae pil* sequences, *J. Bacteriol.* **170**:1691.

Pickup, R. W., and Williams, P. A., 1982, Spontaneous deletions in the TOL plasmid pWW20 which give rise to the B3 regulatory mutants of *Pseudomonas putida* MT20, *J. Gen Microbiol.* **128**:1385.

Plasterk, R. H. A., and van de Putte, P., 1985, The invertible P-DNA segment in the chromosome of *Escherichia coli*, *EMBO J.* **4**:237.

Plasterk, R. H. A., Simon, M. I., and Barbour, A. G., 1985, Transposition of structural genes to an expression sequence on a linear plasmid causes antigenic variation in the bacterium *Borrelia hermsii*, *Nature (London)* **318**:257.

Potekhin, V. A., and Danilenko, V. N., 1985, The determinant of kanamycin resistance of *Strepto myces rimosus:* amplification in the chromosome and reversed genetic instability, *Mol. Biol.* **19**:672.

Rangnekar, V. M., 1988, Variations in the ability of *Pseudomonas* sp. strain B13 cultures to utilize *meta*-chlorobenzoate is associated with tandem amplification and deamplification of DNA, *J. Bacteriol.* **170**:1907.

Rappuoli, R., Perugini, M., and Ratti, G., 1987, DNA element of *Corynebacterium diphtheriae* with properties of an insertion sequence and usefulness for epidemiological studies, *J. Bacteriol.* **169**:308.

Rasmussen, J. L., Odelson, D. A., and Macrina, F. L., 1987, Complete nucleotide sequence of insertion element IS*4351* from *Bacteroides fragilis, J. Bacteriol.* **169**:3573.

Reanney, D. C., Roberts, W. P., and Kelly, W. J., 1982, Genetic interactions among microbial communities, in: *Microbial Interactions and Communities* (A. T. Ball and J. H. Slater, eds.), pp. 287–323, Academic Press, London.

Reanney, D. C., Gowland, P. C., and Slater, J. H., 1983, Genetic interactions among communities, in: *Microbes in the Natural Environment* (R. Whittenbury and J. W. T. Wimpenny, eds.), Cambridge University Press, Cambridge.

Regnery, R. L., and Spruill, C. L., 1984, Extent of genetic heterogeneity among human isolates of *Rickettsiae prowazekii* as determined by restriction endonuclease analysis of rickettsial DNA, in: Microbiology—1984 (D. Schlessinger and L. Leive, eds.), pp. 297–304, American Society for Microbiology, Washington, D.C.

Reynolds, A. E., Mahadevan, S., Felton, J., and Wright, A., 1985, Activation of the cryptic *bgl* operon: Insertion sequences, point mutations and changes in superhelicity affect promoter strength, in: *Genome Rearrangements* (M. Simon and I. Herskowitz, eds.), A. R. Liss, New York.

Riley, M., 1985, Discontinuous processes in the evolution of the bacterial genome, in: *Evolutionary Biology*, Vol. 19 (M. K. Hecht, B. Wallace, and G. T. Praner, eds.), pp. 1–36, Plenum Press, New York.

Roszak, D. B., and Colwell, R. R., 1987, Survival strategies of bacteria in the natural environment, *Microbiol. Rev.* **51**:365.

Ruvkun, G. B., Long, S. R., Meade, H. M., van den Bos, R. C., and Ausubel, F. M., 1982, ISRml: a *Rhizobium meliloti* insertion sequence that transposes preferentially into nitrogen fixation genes, *J. Mol. Appl. Genet.* **1**:405.

Saedler, H., Reif, H. J., Hu, S., and Davidson, N., 1974, IS2, a genetic element for turn-off and turn-on of gene activity in *Escherichia coli, Mol. Gen. Genet.* **132**:265.

Santha, M., Lukacs, K., Burg, K., Bernath, S., Rasko, L., and Stopkovits, L., 1988, Intraspecies genotypic heterogeneity among *Mycoplasma gallisepticum* strains, *Appl. Environ. Microbiol.* **54**:607.

Sapienza, C., and Doolittle, W. F., 1982, Unusual physical organization of the *Halobacterium* genome, *Nature (London)* **295**:384.

Sapienza, C., Rose, M. R., and Doolittle, W. F., 1982, High-frequency genomic rearrangements involving archaebacterial repeat sequence elements, *Nature (London)* **299**:182.

Schaaper, R. M., Danforth, B. N., and Glickman, B. W., 1986, Mechanisms of spontaneous mutagenesis: An analysis of the spectrum of spontaneous mutation in the *Escherichia coli lac* I gene, *J. Mol. Biol.* **189**:273.

Schleifer, K. H., and Stackebrandt, E., 1983, Molecular systematics of prokaryotes, *Annu. Rev. Microbiol.* **37**:143.

Schmid, M. B., 1988, Structure and function of the bacterial chromosome, *Trends Biochem. Sci.* **13**:131.

Schnetz, K., Toloczyki, C., and Rak, B., 1987, β-Glucoside (*bgl*) operon of *Escherichia coli* K-12: Nucelotide sequence, genetic organization, and possible evolutionary relationship to regulatory components of two *Bacillus subtilis* genes, *J. Bacteriol.* **169**:2579.

Scordilis, G. E., Ree, H., and Lessie, T. G., 1987, Identification of transposable elements which activate gene expression in *Pseudomonas cepacia*, *J. Bacteriol.* **169**:8.

Seifert, H. S., and So, M., 1988, Genetic mechanisms of bacterial antigenic variation, *Microbiol. Rev.* **52**:327.

Shalita, Z., Murphy, E., and Novick, R. P., 1980, Penicillinase plasmids of *Staphylococcus aureus:* structural and evolutionary relationships, *Plasmid* **3**:291.

Shapiro, J. A., 1985, Mechanisms of DNA reorganisation in bacteria, *Int. Rev. Cytol.* **93**:25.

Shapiro, J. A., 1986, Control of *Pseudomonas putida* growth on agar surfaces, in: *The Bacteria*, Vol. X, (J. R. Sokatch, ed.), pp. 27–69, Academic Press, New York.

Shapiro, J. A., 1987, Organization of developing *Escherichia coli* colonies viewed by scanning electron microscopy, *J. Bacteriol.* **169**:142.

Simon, M. I., and Silverman, M., 1983, Recombinational regulation of gene expression in bacteria, in: *Gene Function in Procaryotes* (J. Beckwith, J. Davies, and J. A. Gallant, eds.), pp. 211–227, Cold Spring Harbor Laboratory, Cold Spring Harbor, N.Y.

Smith, C. A., and Thomas, C. M., 1987, Comparison of the organization of the genome of phenotypically diverse plasmids of incompatibility group P: Members of the IncP β subgroup are closely related, *Mol. Gen. Genet.* **206**:419.

Smith, C. J., and Spiegel, H., 1987, Transposition of Tn4551 in *Bacteroides fragilis:* Identification and properties of a new transposon from *Bacteroides* spp., *J. Bacteriol.* **169**:3450.

Smyth, P. F., and Clarke, P. H., 1975, Catabolite repression of *Pseudomonas aeroginosa* amidase: Isolation of promoter mutants, *J. Gen. Microbiol.* **90**:91.

Soberon-Chavez, G., Najera, R., Olivera, H., and Segovia, L., 1986, Genetic rearrangements of a *Rhizobium phaseoli* symbiotic plasmid, *J. Bacteriol.* **167**:487.

Spector, M. P., Park, Y. K., Tirgari, S., Gonzalez, T., and Foster, J. W., 1988, Identification and characterization of starvation-regulated genetic loci in *Salmonella typhimurium* by using Mu d-directed *lacZ* operon fusions, *J. Bacteriol.* **170**:345.

Stern, A., Brown, M., Nickel, P., and Meyer, T. F., 1986, Opacity genes in *Neisseria gonorrhoeae:* control of phase and antigenic variation, *Cell* **47**:61.

Swanson, J., Bergstrom, K., Robbins, S., Barrera, O., Corwin, D., and Koomey, J. M., 1986, Gene conversion involving the pilin structural gene correlates with pilin+ pilin− changes in *Neisseria gonorhoeae, Cell* **47**:267.

Szabo, L. J., and Mills, D., 1984, Integration and excision of pMC7105 in *Pseudomonas syringae* pv. *phaseolicola:* Involvement of repetitive sequences, *J. Bacteriol.* **157**:821.

Szybalski, W., and Szybalski, E. H., 1974, Visualization of the evolution of viral genomes, in: *Viruses, Evolution and Cancer* (E. Kurstak and K. Maramorosch, eds.), Academic Press, New York.

Thiermann, A. B., Hansaker, A. L., Mosely, S. L., and Kingscote, B., 1985, New method for classification of leptospiral isolates belonging to serogroup Pomona by restriction endonuclease analysis: serovar kennewicki, *J. Clin. Microbiol.* **21**:585.

Townsend, D. E., Ashdown, N., Greed, L. C., and Grubb, W. B., 1984, Transposition of gentamicin resistance to staphylococcal plasmids encoding resistance to cationic agents, *J. Antimicrob. Chemother.* **14**:115.

Trevors, J. T., 1988, Use of microcosms to study genetic interactions between microorganisms, *Microbiol. Sci.* **5**:132.

Trevors, J. T., Barkay, T., and Bourquin, A. W., 1987, Gene transfer among bacteria in soil and aquatic environments: A review, *Can. J. Microbiol.* **33**:191.

Turberville, C., and Clarke, P. H., 1981, A mutant of *Pseudomonas aeruginosa* PAC with an altered amidase inducible by the novel substrate, *FEMS Microbiol. Lett.* **10**:87.

Valla, S., Coucheron, D. H., and Kjosbakken, J., 1987, The plasmids of *Acetobacter xylinum* and their interaction with the host chromosome, *Mol. Gen. Genet.* **208**:76.

Van Rompuy, L., Min Jou, W., Verhoeyen, M., Huylebroeck, D., and Fiers, W., 1983, Molecular variation of influenza surface antigens, *Trends Biochem. Sci.* **8**:414.

Villarroel, R., Hedges, R. W., Maenhaut, R., Leemans, J., Engler, G., Van Montagu, M., and Schell, J., 1983, Heteroduplex analysis of P plasmid evolution: The role of insertion and deletion of transposable elements, *Mol. Gen. Genet.* **189**:390.

Walter, M. V., Porteous, A., and Seidler, R. J., 1987, Measuring genetic stability in bacteria of potential use in genetic engineering, *Appl. Environ. Microbiol.* **53**:105.

Wanner, B. L., 1985, Phase mutants: Evidence of a physiologically regulated "change-in-state" gene system in *Escherichia coli,* in: *Genome Rearrangement* (M. Simon and I. Herskowitz, eds.), pp. 103–122, A. R. Liss, New York.

Wanner, B. L., Wilmes, M. R., and Hunter, E., 1988, Molecular cloning of the wild-type *phoM* operon in *Escherichia coli* K-12, *J. Bacteriol.* **170**:279.

Williams, P. A., and Jeenes, D. J., 1981, Origin of catabolic plasmids, in: *Microbiology—1981* (D. Schlessinger, ed.), pp. 144–147, American Society for Microbiology, Washington, D.C.

Wu, T. T., Lin, E. C. C., and Tanaka, S., 1968, Mutants of *Aerobacter aerogenes* capable of utilizing xylitol as a novel carbon source, *J. Bacteriol.* **96**:447.

Zhu, Y., and Lin, E. C. C., 1986, An evolvant of *Escherichia coli* that employs the L-fucose pathway also for growth on L-galactose and D-arabinose, *J. Mol. Evol.* **23**:259.

12

Microbial Mats in Australian Coastal Environments

GRAHAM W. SKYRING and JOHN BAULD

1. Introduction

Microbial mats are intrinsically important to microbial ecology because many of the characteristics describing quantitative and qualitative relationships between constructing microorganisms and their aquatic environments occur over vertical sections of a few millimeters or, at most, a centimeter. Their fossilized counterparts, stromatolites, are found in many parts of the world (Schopf, 1983). However, there are spectacular examples of large and very ancient stromatolite structures and accompanying well-preserved fossil microorganisms in Australia (Schopf, 1968; Oehler, 1976, 1978; Oehler *et al.*, 1979; Walter *et al.*, 1980; Walter, 1983; Awramik *et al.*, 1983; Schopf *et al.*, 1987). Contemporary interest in the biology of microbial mats is indicated by the presentation of 75 contributions at two recent international conferences that were concerned specifically with microbial mats (Cohen *et al.*, 1984; Cohen and Rosenberg, 1989). Various kinds of microbial mats occur in Australian coastal environments; however, cyanobacterial and diatomaceous mats are the most extensive and most thoroughly investigated. In this review, we have taken coastal environments to encompass embayments, estuaries, and coastal lakes. We have also included coral reefs, since these habitats are significant and important features of Australian coastal environments.

Microbial mats are more than an ecological curiosity. Micropaleontological and biogeochemical evidence has indicated that microbes morphologically analogous to cyanobacteria were significant colonizers of aquatic environments in the Archean [3.8–2.5 billion years before present (BP)], Proterozoic (2.5–0.58 billion years BP), and

GRAHAM W. SKYRING • CSIRO Division of Water Resources, Canberra, ACT 2601, Australia.   JOHN BAULD • Division of Continental Geology, Bureau of Mineral Resources, Canberra, ACT 2601, Australia.

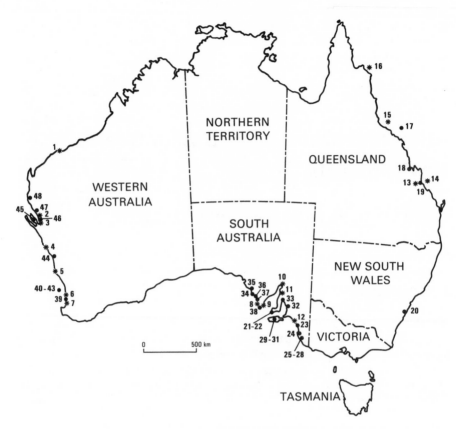

Figure 1. Locations of microbial mats in Australian coastal environments. Sites 1–16 (∗) have been or continue to be subject to detailed taxonomic and/or ecological studies. The remainder (17–48) are localities where the presence of microbial mats has been recorded but detailed investigations have yet to be carried out. Following is the name of each numbered locale, together with a reference in parentheses (upper-case letter), the key to which follows the locality list. Key to locales: (1) Dampier Archipelago (T); (2) Gladstone Embayment, Shark Bay (M); (3) Hamelin Pool, Shark Bay (Q); (4) Hutt Lagoon (B); (5) Lake Thetis (O); (6) Lake Clifton (S); (7) Lake Hayward (H); (8) Lake Damascus, Eyre Peninsula (E); (9) Pillie Lake, Eyre Peninsula (N); (10) Mambray Creek, Spencer Gulf (F); (11) Fisherman Bay, Spencer Gulf (D); (12) Hydromagnesite Lakes, Coorong (U); (13) Bajool salt works (K); (14) One Tree Island, Great Barrier Reef (G); (15) Central Great Barrier Reef (V); (16) Lizard Island, Great Barrier Reef (J); (17) outer reef flat, Great Barrier Reef (R); (18) Broad Sound (L); (19) Port Alma salt works (P); (20) Lake Illawarra (C); (21) Deep Lake (C); (22) Lake Inneston (C); (23) Pellet Lake (C); (24) Lake Fellmongery (C); (25) Lake St. Clair (E); (26) Lake Robe (E); (27) Lake Eliza (I); (28) Salt Dip (E); (29) Lake Grainger (E); (30) Lake KI-27 (E); (31) Lake KI-30 (E); (32) Dry Creek salt works (P); (33) Port Clinton, St. Vincent's Gulf (A); (34) Lake Hamilton (E); (35) Lake Newlands South (E); (36) Lake Tungketta (E); (37) Lake Malata (E); (38) Sleaford Mere (E); (39) Martins Tank (E); (40) Salmon Swamp (E); (41) Serpentine Lake (E); (42) Government House Lake (E); (43) Herschell Lake (E); (44) Leeman Lagoon (A); (45) Lharidon Bight, Shark Bay (Q); (46) Hutchinson Embayment, Shark Bay (Q); (47) Wooramel Delta, Shark Bay (M); (48) Lake McLeod (C). Key to references: (A) Bauld (unpublished

early Phanerozoic, (0.58–0.28 billion years BP) and that they were responsible for the formation of stromatolite reefs that were similar in size and form to the sponge, algal, and coral reefs of the Phanerozoic (Grotzinger, 1989). Even today, microbial mats play an important role in nutrient cycling on barrier and atoll coral reefs. Fossil stromatolite reefs, formed during the Proterozoic, have been found near large mineral sulfide deposits (Mendelsohn, 1976), suggesting some causal relationship. Oils of Proterozoic age from Australia (Jackson et al., 1986; Summons et al., 1988), North America (Imbus et al., 1988), the USSR (Fowler and Douglas, 1987), and Oman (Gorin et al., 1982; Grantham et al., 1987) are associated with stromatolitic facies and are thought to be of microbial origin. Because these geological observations were important for mineral exploration research, the Baas Becking Geobiological Laboratory (BBGL) commenced investigations in 1977 into the biogeochemistry of intertidal cyanobacterial mats on the northeastern coast of Spencer Gulf and on Shark Bay (Fig. 1 and Table I) to examine the following for potential mineral sulfide and hydrocarbon formation in the mats and associated sediments: (1) the rates of mat photosynthesis and the contribution of photosynthesis to the energy required for sulfate reduction; (2) rates of sulfate reduction and metal sulfide formation; (3) environmental factors controlling or affecting both photosynthesis and the degradation of organic matter; (4) the geochemically important products of microbial processes occurring in the mats; and (5) the stable sulfur isotope distribution in the sulfates and sulfides of mat environments.

We here document and discuss investigations of the structure, biology, and biogeochemistry of microbial mats occurring in Australian coastal environments. Many of the more recent investigations were an integral part of research programs of the BBGL (The BBGL closed June 30, 1987. Most of the work reported here was done while the authors were members of the BBGL.) The core of this review presents a comprehensive and integrated assemblage of contributions, some of which are being prepared for publication, that members of the BGL have made to the ecology and biogeochemistry of microbial mats. Other investigations of microbial mats are included where appropriate.

2. Occurrence and Characteristics of Australian Microbial Mats

2.1. Location and Habitat

Microbial mats of various types have been reported from about 50 Australian coastal localities and generally appear to be common components of (shallow) perma-

---

data); (B) Bauld (in preparation); (C) Bauld (1981a, Table 8.3); (D) Bauld (1984a); (E) Bauld (1986a, Table I); (F) Bauld et al. (1980); (G) Borowitzka et al. (1978); (H) C. M. Burke (personal communication, in progress); (I) Burne and Ferguson (1983); (J) Burris (1976); (K) M. Coleman, personal communication, in progress); (L) Cook and Mayo (1977); (M) Davies (1970); (N) De Deckker et al. (1982); (O) Grey et al. (1990); (P) Jones et al. (1981); (Q) Logan et al. (1974); (R) Monty (1979); (S) Moore (1987); (T) Paling (1986); (U) Walter et al. (1973); (V) Wilkinson et al. (1984).

Table I. Representative Selection of Microbial Mats from Australian Coastal Environments and Information about Their Habitats and Microbial Populations

Mat designation (synonym)	Location (Fig. 1 ref. no.)	Habitat characteristics (approx. salinity range)[a]	Major agent of construction and characteristics[b]	Other significant microbial components	Mat characteristics and/or peculiarities[c]	Reference
Tufted (reticulate)	Fisherman Bay, Spencer Gulf (11)	M, InEx Es, Ds, Ii, D-W	*Lyngbya aestuarii* Cb, Fil, Sh	*Microcoleus chthonoplastes*	CS, LD, PSB SP	Bauld (unpublished)
Smooth (flat, stratiform)	Mambray Creek, Spencer Gulf (10)	M, InEx Es, Ds (Dm), Fi (Fr), H-D (ca. 45–100‰)	*Microcoleus chthonoplastes* Cb, Fil, Sh	*Phormidium* spp.	CS, Ld (LD)	Bauld *et al.* (1980), Skyring *et al.* (1983), Bauld (1984a)
	Gladstone Embayment, Shark Bay (2)	M, InEx (P?) Es, Dr, Ii, W+ (ca. 55–120+%)	*Microcoleus chthonoplastes* Cb, Fil, Sh	*Phormidium* spp.; seasonal surface development of coccoid Cb	CS, LD, PSB cf. *Chloroflexus* sp.; accumulation to ca. 30 cm	Bauld and D'Amelio (in preparation)
Pustular (mamillate,	Hamelin Pool, Shark Bay (3)	M, InEx Es, Ds (Dm), Ii	*Entophysalis major* Cb, Co, Sh		CM, LX SP	Bauld *et al.* (1979), Bauld

cinder)		(Fr), H-D (ca. 65–100+‰)					(1984a), Bauld (in preparation)
Flocculent	Lake Thetis, Swan Coastal Plain (5)	L, P	Em, Do (Dr) (ca. 39–53‰)	Diatoms	*Oscillatoria* spp. Cb, Fil, ±Sh	CF, LX (LD), PSB No FeS precipitates; accumulation to ca. 50 cm	Grey *et al.* (1990)
Colloform (diatomaceous)	Hamelin Pool, Shark Bay (3)	M, P (δInEx)	Em, Do (Dr) (ca. 60/65–70‰)	*Mastogloia* spp. *Amphora* spp. and other diatoms	*Mastogloia* cf. *ha-lophila Brachysira apo-nina* Di, Sta, ±Sh	CF (CM), Ld (LX) Subsurface carbonate cementation occurs	Bauld *et al.* (1979), Bauld (1984a), John and Bauld (in preparation)

[a] M, Marine; L, lacustrine; P, permanently submerged; InEx, alternating periods of inundation and exposure; Es, substantial evaporative increase in salinity; Em, moderate evaporative increase in salinity; Ds, desiccation is frequent or for long periods; Dm, desiccation, though frequent, is rarely for long periods; Dr, desiccation occurs infrequently; Do, desiccation never occurs. For those mats subject to InEx, inundation regime can be characterized by frequency, regularity, and duration: F, inundation is a frequent event (time scale ca. hours to days); I, inundation is infrequent (time scale ca. weeks to months); r, inundation is a regular event; i, inundation is irregular; H, duration of each inundation event is hours; D, duration is days; W, duration is weeks (months?). δ, low probability of reaching designated status; +, significant probability of greater magnitude; ?, considered possible but not recorded or known; ±, frequent but not universal. Parenthetic abbreviations indicate less commonly recorded status.

[b] Cb, cyanobacterium; Di, diatom; Fil, filamentous; Co, coccoid; Sta, stalked; Sh, ensheathed.

[c] Cohesion: CS, strongly cohesive (leathery); CM, moderately cohesive; CF, fragile. Lamination: LD, well laminated; Ld, poorly laminated; LX, not laminated. PSB, macroscopically visible layer(s) of purple sulfur bacteria; SP, major agent of construction possesses pigmented sheath. Parenthetic abbreviations indicate less commonly recorded status.

Figure 2. Some representative examples of Australian mats, together with constructing microbes and habitats. (a) Low-level, oblique aerial view of intertidal mats in Fisherman Bay, Spencer Gulf (Fig. 1, no. 11). This area is subjected to low-frequency flooding via tidal channels during "king" tides; in the lower areas near the tidal channels, mats may be immersed for periods of days to weeks. (b) Smooth mat from Mambray Creek, Spencer Gulf, constructed primarily by *Microcoleus chthonoplastes,* a filamen-

nently submerged to intermittently exposed benthic habitats in relatively protected littoral marine areas (Figs. 1 and 2). Coral reefs, mangroves and their tidal channels, samphire flats, saline coastal lakes, and even the concentrator ponds of evaporative salt works host microbial mats (see Bauld, 1981a). The geographic distribution (Fig. 1) of these cohesive benthic microbial communities may well prove to reflect the exigencies of contemporary scientific effort rather than a genuine ecological range. This caveat aside, it should be noted that most occurrences, and the great preponderance of detailed studies, are in habitats hypersaline with respect to "normal" seawater (ca. 35‰) and that all of the remainder are in saline waters (>3–5‰; Williams, 1981).

The majority of Australian mat habitats are, in sedimentological terms, carbonate environments or provinces; i.e., much of the sediment deposited or precipitated is carbonate in its various mineralic forms (Logan *et al.*, 1970; Walter *et al.*, 1973; Cook and Mayo, 1977; Marshall and Davies, 1978; Bauld *et al.*, 1980; Burne and Colwell, 1982; De Deckker *et al.*, 1982; Moore, 1987). Microbial mat localities are also generally characterized by high insolation, which promotes high evaporation rates and, in intermittently exposed habitats, significant diel temperature variation within the surface layers, where midday summer temperatures may reach 50–55°C. Given that marine salts provide the bulk of the major anions and cations present in these habitats, it is not surprising that aquatic pHs fall in the neutral–alkaline range.

The overwhelming impression, therefore, is that of an ecological niche having circum-neutral pH and under only moderate diel thermal stress, primarily controlled by salinity and/or desiccation (i.e., the osmotic and matric aspects of water activity; see Section 3.4). The consequences of these strictures will become apparent as we examine these microbial communities, their dominant components, the environmental optima and ranges of their metabolic activities, and the physiological strategies enabling their survival and continued mat construction.

---

tous cyanobacterium (see panel c). The mat is laminated (inset) and, as shown by this water-saturated example, very cohesive. The vertical section is 2.4 cm deep. (c) Phase-contrast photomicrograph of *M. chthonoplastes* from smooth mat (shown in panel b). The trichomes, which are mobile by gliding, are shown within the hyaline sheath; the latter has a sticky surface to which sediment particles readily adhere. (d) Pustular mat constructed by *Entophysalis major* (see panel e), which covers large areas of the intertidal zone of Hamelin Pool (Fig 1, no. 3). The irregular topography of the mat surface and the lack of lamination (not shown here) are a consequence of the morphology and growth habitat of *E. major*. (The checkerboard scale is 10 cm long.) (e) Phase-contrast photomicrograph of *E. major* from Hamelin Pool pustular mat (panel d). The coccoid cells of this cyanobacterium are encased within multiple sheaths. Sediment particles can adhere to the surface of the sheaths, but *E. major*, lacking both gliding motility and filamentous morphology, is relatively inefficient (compare with panels b and c) at sediment trapping and binding. Note that the sheaths are stained with a brown pigment. (f) Oblique view of permanently submerged stromatolites from Shark Bay (these examples are at a depth of ca. 2–3 m, Hamelin Pool; Fig. 1, no. 3); the tallest is ca. 0.5 m. The lithified structures are covered by colloform mat constructed by a mixed cyanobacterial–microalgal community dominated by diatoms. The lower sides are also colonized by a beard of *Acetabularia* sp. The inset shows an example of a stalked diatom (length ca. 45μm) from such a community. Panels b–e have been previously published by J. Bauld and are reproduced with permission; panel f (excluding the inset) is by courtesy of P. E. Playford.

2.2. Community Structure and Microbial Components

Benthic microbial communities range from those which are noncohesive and often macroscopically invisible to those which are demonstrably, and frequently strikingly, cohesive (microbial mats) to those in which carbonate cementation occurs, resulting in lithification and concomitant structural preservation. Organic preservation of microbial components is the exception rather than the rule in the absence of silicification. The efficacy of structural preservation is apparent in the fossil record, which contains examples (referred to variously as stromatolites, thrombolites, and microbialites) ranging in age from several hundred years BP [e.g., Hamelin Pool (Logan, 1961; Playford and Cockbain, 1976)] to several billion years BP [e.g., Archean Warrawoona Group, Pilbara Block, Western Australia (Walter, 1983)].

Microbial mats are visibly distinguishable from other benthic microbial communities by their cohesion (Fig. 2b) and frequently, but not invariably, by their laminated structure. The latter is often made macroscopically visible by microbial pigments and/or trapped or precipitated sediments, which reflect the underlying vertical stratification of both physicochemical parameters and microbial populations (e.g., Bauld, 1986a; Pierson et al., 1987). Even where a microbial mat is not obviously laminated, vertical stratification of biotic and abiotic properties is still present and measurable at a resolution of tens to hundreds of micrometers (e.g., Revsbech and Jørgensen, 1986). The presence of even a very thin microbial mat or film significantly modifies the sediment–water interface (Marshall, 1976). Both permeability and porosity are decreased, and the exchange of water and gases across the interface is thereby impeded (Bubela, 1980).

Properties that characterize and distinguish one mat from another include morphology, fabric, lamination, the degree of cohesion, the extent of lithification, and the nature of the microbial community. Although representatives of many physiological groups can be isolated from mat communities, the character of any given mat is largely determined by those particular microbes constructing the mat; i.e., those species responsible for cohesion (Table I), in contradistinction to species that are merely residential but may (occasionally) contribute the greatest biomass to the community. The suites of characteristic constructing microbes are regulated primarily by environmental constraints such as salinity and/or inundation regime for marine intertidal habitats. The latter is, in turn, usually determined by elevation but can be modified by sediment microtopography and even by mat morphology (J. Bauld and R. V. Burne, unpublished data). Several representative examples of microbial mats from well-studied localities (Fig. 1) are listed in Table I. Some microbial mats (e.g., smooth, tufted) are of cosmopolitan occurrence and have been studied in both Australian and other, Northern Hemisphere localities (Cohen et al., 1980, 1984). Within Australia, Shark Bay is notable for both the variety and areal extent of mat types (Logan et al., 1974; Bauld, 1984a).

Mats are created by organisms possessing physical or behavioral attributes that promote cohesion, i.e., that can trap and/or bind sediment particles and other mat inhabitants (Fig. 2). The most cohesive mats are those constructed in intermittently exposed and inundated (shoreline) habitats by filamentous cyanobacteria capable of gliding motility and possessing enveloping sheaths (Figs. 2b and c). The sheaths are polysaccharide-rich (W. F. Dudman and J. Bauld, unpublished data) and have a "sticky" surface to which sediment particles bind; the filamentous morphology enables the development of a living "meshwork" that promotes sediment trapping. Even empty abandoned sheaths buried below the surface cyanobacterial layer continue to provide significant cohesive strength in the absence of viable trichomes (Bauld, 1981b; Grey et al., 1990). Fattom and Shilo (1984) have demonstrated the hydrophobic nature of the surface of benthic cyanobacteria, a property that inhibits dispersal of all but their hydrophilic hormogonia. The capacity for gliding permits movement and optimal repositioning on a diurnal schedule through tactic responses to environmental stimuli such as light, oxygen, and sulfide. In situations where sediment input is relatively constant, the imprint of diurnal variation in light intensity is recorded by the cyclic repetition (i.e., lamination) in the orientation and density of cyanobacterial filaments of the accreting mat [e.g., domal mats from the outer reef flats (Monty, 1979)]. Accretion rates are generally controlled by sediment input and for intertidal mats constructed by filamentous cyanobacteria are commonly in the range of 1–10 mm per year (Bauld et al., in press).

In the low intertidal and subtidal zones (or their lacustrine equivalents), the dominance of cyanobacteria is considerably less than in the mid to high intertidal zones, and populations of eucaryotic microalgae dominated by diatoms commonly assume prominence (Table I and Fig. 2f). Cohesion in these benthic microbial communities is frequently derived from the properties and activities of certain species of diatoms, particularly those species in which the siliceous frustules are surrounded by mucilaginous capsules and are borne on polysaccharide-containing stalks, thus providing a trapping and binding function analogous to that of cyanobacterial filaments. Many other species of diatoms inhabit, but do not construct, these mats (J. John and J. Bauld, in preparation), which appear to be widespread in Australian coastal saline lakes and in Hamelin Pool (Table I, colloform), where they are estimated to cover about 140 km^2 of the sublittoral platform. Compared with cyanobacterial mats, the diatom-constructed mats are much more fragile and prone to erosion and less obviously laminated, though their propensity to carbonate cementation (see below) has preserved a record of their past activity in Hamelin Pool (e.g., Logan, 1961; Playford and Cockbain, 1976).

For every generalization there is at least one interesting exception, and in this context of Australian coastal microbial mats the exception is provided by *Entophysalis major,* a cyanobacterium that constructs mats dominating the areal cover of the Hamelin Pool intertidal zone (Figs. 2d and e; Logan et al., 1974; J. Bauld, in preparation). In contrast to other mat types (Table I), where nonfilamentous organisms are usually residential rather than structural members of the benthic microbial community, pustular mat is both constructed and overwhelmingly dominated by the coccoid cyanobacterium *E. major.* Each cell is encased within a series of multilamellate sheaths (Fig. 2e) that build up macroscopic "pustules" or colonies. The presence of other organisms was detected by phase-contrast and transmission electron microscopy (TEM) (J. Bauld and E. D'Amelio, in preparation) and by the isolation of aerobic chemoheterotrophs and anoxygenic phototrophic bacteria (J. Bauld, unpublished data). The utility of TEM as a means of detecting microbes distinguishable by their fine structure is illustrated by the discovery of a (presumptive) *Chloroflexus* sp. (Table I) and other, as yet unidentified filamentous phototrophic procaryotes (J. Bauld and E. D'Amelio, unpublished data).

Our understanding of interactions between components of microbial mat communities is characterized by observations and measurements from which it is possible to construct testable models; some progress has been made in this area (Fig. 3). Moriarty (1983) estimated that smooth mat, tufted mat, and colloform mat from Hamelin Pool had bacterial populations of 3.9, 8.0, and 1.0×10^6 mm^{-3}, respectively, but the bacterial numbers in pustular mat were significantly less (0.2×10^6 mm^{-3}); he indicated, however, that counting was difficult because filamentous bacteria were particularly abundant. Moriarty (1983) also determined bacterial productivities and concluded that most heterotrophic bacterial activity occurs in the upper 10 cm of the mat–sediment ecosystem. From bacterial productivities and primary productivities (Bauld *et al.*, 1979), Moriarty (1983) calculated that 20–30% of the primary production is used by the heterotrophic bacteria [these estimates may not include the sulfate-reducing bacteria (SRB); D. J. W. Moriarty and G. W. Skyring, (unpublished data)]. Chambers (1982; see also Skyring *et al.*, 1983) calculated from sulfur isotope fractionation data (Chambers *et al.*, 1975) an SRB population of around 10^9 cm^{-3} of interstitial water in intertidal cyanobacterial sediments from Mambray Creek. This would constitute 20–30% of the bacterial population.

Benthic microbial communities associated with Australian coral reefs, variously referred to as mats, crusts/pavements, and algal turfs, are reported from permanently submerged and intertidal, periodically emergent habitats. They preferentially colonize lithified substrates such as dead coral heads and carbonate plates in both subtidal and

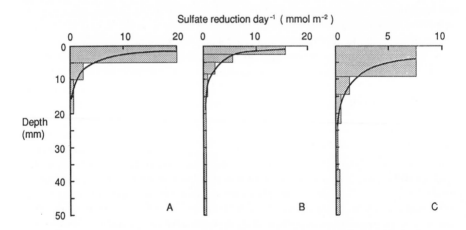

Figure 3. Sulfate reduction rates from experiments with smooth mat from Spencer Gulf in which [^{35}S]sulfate was omitted from successive layers below the mat. Rates were calculated by measuring differences between these and control samples in which [^{35}S]sulfate was added to all layers (Skyring *et al.*, 1983). Depth profiles of sulfate reduction rates for 50-mm smooth mat and associated sediment show that (A) ca. 90% of the sulfate reduction occurs in the mat and the upper 5 mm of associated sediment and (B) ca. 75% occurs within the cyanobacterial mat (ca. 2 mm thick). (C) Sulfate reduction rates in smooth mat from Gladstone Embayment (Shark Bay) showing that ca. 86% of sulfate reduction occurs in the upper 10 mm of mat and associated sediment; the mat in this area was ca. 5 mm thick.

intertidal areas (Burris, 1976; Monty, 1979; Wilkinson *et al.*, 1984). Some, particularly those present in frequently emergent intertidal areas, are dominated by filamentous cyanobacteria such as *Scytonema* sp. (Burris, 1976). Monty (1979) has reported the occurrence of gelatinous, laminated domes constructed by *Phormidium hendersonii* in the high-energy outer reef flats; but although cyanobacteria can be abundant in rarely emergent and subtidal communities, other eucaryotic phototrophs such as red algae assume greater importance and are often dominant (Borowitzka *et al.*, 1978; Wilkinson *et al.*, 1984).

In addition to the various microscopic and culturing techniques normally used to establish the presence of physiologic or taxonomic groups, their presence can be confirmed or inferred by the detection of diagnostic biomarker signatures. For example, the differences in microbial components between cyanobacterial and diatomaceous mats (Table I) determined by microscopic taxonomy are confirmed by analysis of community carotenoid pigments. Mats, such as smooth, tufted, and pustular, are dominated by the carotenoids echinenone and myxoxanthophyll, which are diagnostic for cyanobacteria, whereas colloform mats contain fucoxanthin, diadinoxanthin, and diatoxanthin, which are diagnostic for diatoms (Palmisano *et al.*, 1989). Hydrocarbon analysis confirms the presence of cyanobacteria (C_{15}–C_{19} normal alkanes and alkenes together with abundant monomethyl alkanes) and eucaryotic microalgae (sterols) in the appropriate mats; it also demonstrates the unexpected presence of abundant and unusual highly branched C_{20} and C_{25} alkene and alkane hydrocarbons in colloform mats from Hamelin Pool and in other diatomaceous mats from Australian coastal saline lakes (R. E. Summons, personal communication). The microbial source of these unique hydrocarbon biomarkers is presently unknown.

The role of microbes in the formation of carbonate cements and hence lithification of mat communities remains a vexing question. Despite the impressive number of instances in which carbonate precipitation is reported to be associated with microbes resident within mat communities (Pentecost and Riding, 1986; Burne and Moore, 1987; Lazar *et al.*, 1989), it is yet to be unequivocally demonstrated by experiments under carefully controlled conditions that vital biological processes cause the carbonate to precipitate (Pentecost and Riding, 1986; Pentecost and Bauld, 1988). For example, under laboratory conditions, it is possible to nucleate calcite crystals both on the sheaths of cyanobacteria that are associated with calcification in nature and on the sheaths of those that are not. Furthermore, nucleation is initiated equally well on the sheaths of intact, photosynthetically active trichomes and on empty sheaths of the same cyanobacterium (Pentecost and Bauld, 1988).

3. Phototrophic Activity and Environmental Constraints

3.1. Light

Light is the environmental factor that assumes overriding, central importance for all mat communities except those of deep-sea hydrothermal vents (e.g., Jannasch and

Wirsen, 1979) and other aphotic habitats. Light is the obligate energy source for the phototrophic microbes constructing these benthic mats, and their species composition, metabolic activity, vertical stratification, and, indeed, their survival depend on the quality and quantity of the light arriving at the mat surface and the effectiveness of the strategies with which various microbial components cope under any particular regime. The light-trapping and/or shielding pigments in each successive microbial layer decrease the light intensity passing to the next layer. More important, selective absorption constrains both the pigments that can be effectively used to capture utilizable light energy and the classes of microbes able to make a living within each successively depleted light regime (Jørgensen *et al.*, 1987; Pierson *et al.*, 1987).

Microbes occupying the surface layers of mat communities are exposed to a conflict between the necessity for light energy and the associated hazards of exposure to dangerously high light intensities, particularly during summer, and concomitant potentially damaging UV-B and UV-C irradiation (see Knoll and Bauld, 1989). Several strategies appear to have evolved to deal with this problem. Certain cyanobacteria exposed at the surface of intertidal mats in Spencer Gulf and Shark Bay produce pigments that are localized in their sheaths (Fig. 2e; Figs. 5 and 2h in Bauld, 1981b and 1986a, respectively). These sheath pigments (e.g., scytonemine) appear to function as highly efficient light screens that preferentially absorb in the near-UV portion of the spectrum in organisms such as *Entophysalis major* and *Lyngbya aestuarii* (Bauld, in preparation; R. W. Castenholz, personal communication). Incomplete separation of *E. major* mat samples into surface and subsurface layers shows that photosynthetic $^{14}CO_2$ fixation in the deeply pigmented surface fraction can sustain near-maximal rates [ca. 85–90% of maximum photosynthetic rate (MPR)] at full incident light intensities (ca. 2300 μE sec^{-1} m^{-2}), whereas the less-pigmented subsurface fraction is inhibited (ca. 30% MPR) at these intensities (Bauld, in preparation). Both layers, however, appear to be light saturated at relatively low intensities. Even the heavily pigmented surface fraction is saturated at ca. 700 μE sec^{-1} m^{-2}, whereas the subsurface fraction is saturated at ca. 100 μE sec^{-1} m^{-2}, a value that is consistent with measurements for many cyanobacteria cultured in the laboratory (van Liere and Walsby, 1982).

In diatom-dominated subtidal colloform mats, intracellular accessory photosynthetic pigments, the carotenoids, appear to provide the necessary screening; Palmisano *et al.* (1989) found that these mats exhibit very high carotenoid/chlorophyll ratios during the summer. The "umbrella" strategy is particularly appropriate for mats in which the primary agents of construction are not motile. On the other hand, observations suggest that evasion via phototactic gliding motility is the strategy used by *Microcoleus chthonoplastes*. This cyanobacterium, which does not possess any sheath pigments (e.g., Fig. 2c in Bauld, 1981b), exhibits low carotenoid/chlorophyll ratios (Palmisano *et al.*, 1989) and has been observed to exhibit negative phototaxis under high incident light intensities. *M. chthonoplastes* is normally found slightly below the surface, where it forms a bright blue-green layer. The lack of photoinhibition at intensities approaching full incident sunlight in correctly oriented intact mat cores is achieved by the *Microcoleus* population gliding down (or up, under low light intensities), away from the increasing light. The organism thus positions itself as close as possible to depths where optimal light intensities prevail, and this behavior explains the noninhibitory response at incident intensities exceeding saturation (Bauld, in preparation).

3.2. Photoautotrophic Activity and Primary Productivity

As with other ecosystems, microbial mats compromise both producers and consumers of organic matter (Fig. 4). In the habitats under consideration here, primary production occurs via photoautotrophic CO_2 fixation, and the range of phototrophic microbes that can be detected in Shark Bay mats, either by direct microscopic observation or by culture techniques (i.e., cyanobacteria, diatoms, and other eucaryotic microalgae as well as purple and green bacteria) is consistent with the occurrence of both anoxygenic and oxygenic photosynthesis. CO_2 fixation is presumably by way of the Rubisco pathway (Calvin cycle), although the presence of presumptive *Chloroflexus* and *Chlorobium* (D'Amelio and Bauld, unpublished data) suggests that other mechanisms (Fuchs *et al.*, 1980; Holo and Sirevag, 1986) may also be active in certain mat communities. Mats dominated by cyanobacteria and/or diatoms do indeed carry out oxygenic photosynthesis, as determined by their sensitivity to the specific inhibitor DCMU [3-(3,4-dichlorophenyl)-1,1-dimethylurea]. Deeper layers, visibly dominated by purple sulfur bacteria, are DCMU insensitive, predictably consistent with their anoxygenic photosynthesis (Bauld, 1984a and in preparation).

As might be expected, the situation in nature appears to be somewhat more complicated than these DCMU data would suggest. Mat communities that one might intuitively expect to be oxygenic, and therefore DCMU sensitive, are observed to be incompletely sensitive. An interesting example of this is provided by mats constructed by *E. major* (Table I), in which significant (ca. 10–15%) activity remains after DCMU treatment

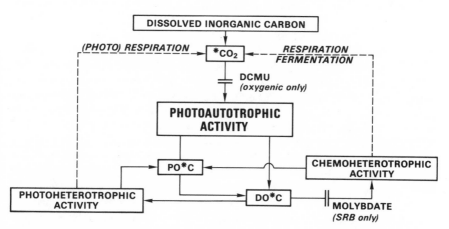

Figure 4. Simplified carbon flow in microbial mat communities with emphasis on phototrophic organisms.

(Bauld, in preparation). Although it is possible that a portion of the applied DCMU is inactivated, for example, by binding to sediment particles, it is probable that the residual is real and represents either the expression of facultative anoxygenic photosynthesis by *Entophysalis* or continued obligate anoxygenic photosynthesis by less obvious mat residents. There is presently no evidence to either support or refute the possibility of facultative anoxygenic photosynthesis. Consistent with obligate anoxygenesis is the isolation of purple sulfur anoxygenic phototrophs not only from reducing sediments nearby (Bauld *et al.*, 1987) but also from *Entophysalis* mats (presumptive *Chromatium* spp.; Bauld, unpublished data). Small procaryotes, possessing photosynthetic lamellae, within the interstices of the tightly packed *Entophysalis* colonies have been observed by TEM (D'Amelio and Bauld, in preparation).

Mats constructed by the filamentous cyanobacterium *M. chthonoplastes* also exhibit similar DCMU-insensitive $^{14}CO_2$ fixation (Bauld, 1984a). The possibilities discussed above also pertain, but with a difference: *M. chthonoplastes* is reported to be capable of facultative anoxygenic photosynthesis (Cohen *et al.*, 1986). However, earlier experiments with natural populations of *M. chthonoplastes* mats failed to demonstrate the induction of facultative anoxygenic photosynthesis (Bauld, 1984a). This was attributed to the use of either inadequate induction periods or significantly nonoptimal sulfide concentrations. A culture of *M. chthonoplastes* from Spencer Gulf was later found by Y. Cohen (Hebrew University of Jerusalem, Eilat, Israel) to be capable of facultative anoxygenic photosynthesis (Cohen *et al.*, 1986).

Primary production has been measured in several of these mats; in Spencer Gulf mats, such measurements have been made for several mat types at regular intervals over time spans in excess of 1 year (see Skyring *et al.*, 1983). The range of rates reported for primary productivity in Spencer Gulf and Shark Bay mats (ca. 0.1–3.0 g C m^{-2} day^{-1}; Bauld *et al.*, 1979; Skyring *et al.*, 1983; Bauld, 1984a) is similar to those found in comparable mats and habitats (Table II). It also falls well within the range reported for all mats from lacustrine, geothermal, and marine habitats—a range that spans three orders of magnitude without any discernible pattern (Castenholz *et al.*, in press) and of which at least some of the variation is attributable to the variety of methodologies used and their inherent biases (Bauld, 1984a). Productivity by microbial mats is comparable to published data for phytoplankton productivity on an areal basis (Boynton *et al.*, 1983). For Spencer Gulf smooth mat (*M. chthonoplastes*), primary production rates were highest during winter and spring (see Fig. 5 of Skyring *et al.*, 1983), a phenomenon attributed to increased seasonal immersion and the associated diminution of desiccation stress (Bauld *et al.*, 1980; Skyring *et al.*, 1983; see also Section 3.4).

Evidence from field and laboratory experiments with both intact mats and homogenates of mat microbes demonstrates that a measurable and significant proportion of the [^{14}C]carbonate fixed by various communities is subsequently released as extracellular dissolved organic carbon (DO^{14}C). The proportion of fixed carbon subsequently excreted was commonly in the range of 1–6% but sometimes higher (Bauld and Chambers, 1983; Bauld, 1984a). For Shark Bay tufted mat, constructed primarily by *Lyngbya aestuarii*, the proportion excreted was usually in the range of 2–3% over common environmental salinities and greatest at very high or very low salinities (Bauld, in

Table II. Primary Production in Representative Microbial Mats from Various Aquatic Habitats

Location (Fig. 1 ref. No.)	Habitat	Major components of community	Rate (g C m^{-2} day^{-1})	Reference
Spencer Gulf (10)	Marine Intertidal	Filamentous cyanobacteria	0.2–3.1	Skyring et al. (1983), Bauld (1984a)
Hamelin Pool (3)	Marine Subtidal	Stalked and pennate diatoms	0.5	Bauld et al. (1979)
Solar Lake	Lacustrine (saline) Littoral	Filamentous cyanobacteria	8.0–12.0	Krumbein et al. (1977)
Tague Bay, St. Croix	Coral reef Subtidal	Filamentous cyanobacteria and chlorophytes	2.1–3.1	Carpenter (1985)
Algal Lake, Ross Island, Antarctica	Lacustrine (freshwater) Littoral	Filamentous cyanobacteria	1.9–3.6	Goldman et al. (1972)
Tecopa Bore, Mohave Desert	Geothermal stream	Filamentous cyanobacteria	1.8–4.5	Naiman (1976)

preparation). Microscopic examination of mat samples reveals the close physical proximity of identifiably phototrophic and presumptive degradative bacteria. Furthermore, the isolation of aerobic chemoheterotrophs (Bauld, unpublished data), anaerobic photoheterotrophs (Bauld *et al.*, 1987; Bauld, unpublished data), and acetate- and lactate-utilizing SRB (G. W. Skyring and Li Yaquin, unpublished data), together with estimates of heterotrophic bacterial productivity (Moriarty, 1983), suggests that DOC released during photosynthesis is likely to be rapidly utilized by consumers adapted to such substrates (Fig. 4). Consequently, as reported earlier for geothermal mats (Bauld and Brock, 1974), measurements in the absence of appropriate inhibitors are likely to be gross underestimates of the actual rate of excretion because of simultaneous heterotrophic consumption. Our data are consistent with this interpretation. In situations where heterotrophic consumption is constrained (e.g., sulfate reduction by sulfate deprivation or by molybdate inhibition), there is a significant increase in the proportion of fixed carbon released as DO^{14}C (Bauld and Chambers, 1983; see also Sections 3.5–3.9). The increase in percent DOC, which occurs as mat suspensions are diluted (Bauld, unpublished data) and as the diffusive distance between producers and consumers increases, is also consistent with this view and with data obtained from geothermal mats (Bauld and Brock, 1974).

3.3. Photoheterotrophic Activity

Photoheterotrophy has been little studied in natural microbial communities and rarely in benthic habitats, though representatives of mat phototrophs (e.g., the purple sulfur bacteria) show strong photoheterotrophic capabilities (see Bauld, 1984b; Bauld *et al.*, 1987). Photoheterotrophic activity has been estimated in Australian coastal microbial mat communities by measuring its prerequisite, photoassimilation (i.e., light-dependent uptake), with suitable ^{14}C-labeled organic compounds. Substrates photoassimilated by Spencer Gulf and Shark Bay mats under field conditions included acetate, glucose, malate, and glycolate (Bauld, 1987, in preparation). Photoheterotrophic activity in these benthic communities was in the range of 30–50% of total (chemotrophic plus phototrophic) heterotrophic uptake, comparable to that found during analogous experiments with lacustrine planktonic microbial communities (Ellis and Stanford, 1982; McKinley and Wetzel, 1979).

Subsequently, moie detailed uptake studies of the central metabolic intermediate, acetate, were undertaken in various geothermal and hypersaline lacustrine and marine habitats (Bauld, 1987, in preparation). Photoassimilation of [*carboxyl*-^{14}C]acetate, as a proportion of total uptake, was usually in the range of 30–50% significantly greater than the 0–30% observed for [*methyl*-^{14}C]acetate. However, the interpretation of data derived from experiments using substrates with radiolabeled carboxyl groups should be undertaken with some care. DCMU partially inhibits [1-^{14}C]acetate photoassimilation in some oxygenic mat communities, a phenomenon which suggests that the DCMU-sensitive portion of measured "photoassimilation" results from (heterotrophic) liberation of $^{14}CO_2$ and its subsequent fixation by DCMU-sensitive oxygenic photosynthesis (Bauld, 1987, in preparation). This process accounted for 56% of 1-^{14}C photoassimilation in

Shark Bay *Entophysalis* mats. On the other hand, in mat communities dominated by *M. chthonoplastes*, [1-^{14}C]acetate photoassimilation was totally DCMU insensitive. Laboratory cultures of this filamentous cyanobacterium were also DCMU insensitive (Bauld, 1987, in preparation), indicating that the acetate carboxyl group was directly incorporated into cyanobacterial biomass. In contrast, 93% of [1-^{14}C]acetate photoassimilation in natural populations of *Oscillatoria boryana* from Rotorua geothermal mats was DCMU sensitive (Bauld, 1987; Castenholz *et al.*, 1988), suggesting that carboxyl release and oxygenic photosynthetic fixation preceded incorporation. The quantitative significance of both true and pseudo photoheterotrophy is presently unknown for benthic microbial communities. However, the demonstrable capacity for photoassimilation in natural populations and laboratory cultures suggests that photomixotrophic growth may be important at environmental concentrations of DOC and dissolved inorganic carbon (DIC), with the emphasis shifting to photoheterotrophy as available light energy diminishes (Van Baalen *et al.*, 1971).

3.4. Environmental Controls on Phototrophic Processes

With the exception of light (see Section 3.1 above), water activity (a_w) in its various manifestations probably exerts the most significant environmental control over phototrophic microbial communities in benthic coastal habitats. Even in relation to a_w, solar flux plays an important indirect role by raising the salinity of overlying waters through evaporation and by promoting desiccation in mats exposed to the atmosphere. Fluctuation in a_w is obviously of less consequence to permanently submerged mats, where the strategy is more likely to be that of coping with high but constant or only slowly fluctuating salinities. In the presence of free liquid water, the osmotic component of a_w proscribes the physiological strategies that can be used, whereas the matric component assumes overriding importance as water is removed (Brown, 1976; Griffin, 1981; Potts and Bowman, 1985). The metabolic response of natural mat communities to salinity (osmotic) and desiccation (matric) has been investigated in Australian coastal habitats by measuring photosynthetic $^{14}CO_2$ fixation under ambient conditions of light and temperature.

The intertidal habitats of Spencer Gulf and Shark Bay are characterized by irregular and commonly infrequent inundation, thus subjecting the colonizing mats to wide fluctuations in salinity and to intermittent desiccation (Skyring *et al.*, 1983). The duration and severity of these events depends on such factors as elevation relative to Australian height datum (AHD) and local microtopography and is exacerbated by high local (sea)water salinities and high summer insolation. Commensurate with their intertidal habitat, smooth mat (*Microcoleus*) from Spencer Gulf and tufted mat (*Lyngbya*) from Shark Bay showed broad salinity response curves—rates greater than 75% of MPR occurring over salinity ranges of ca. 35–140‰ (Bauld, 1984a) and ca. 12–115‰ (Bauld, 1983), respectively. On the other hand, diatomaceous subtidal mats from Hamelin Pool, Shark Bay, evinced a much narrower salinity response, exhibiting a distinct peak at the environmental salinity (60‰) together with a 75% or greater MPR over a relatively narrow salinity range of 35–70‰, consistent with the limited seasonal fluctuation in this subtidal locality (Bauld, 1983). In some cases, the relationship between environmental salinities and organismal tolerances are less obvious. *Chromatium vinosum* strain HPC is an anoxygenic phototroph (purple sulfur bacterium) isolated from the intertidal zone of Hamelin Pool, and although the organism is an obligate halophile able to grow at salinities of between 5 and 85‰, its optimum salinity is actually 35‰ (Bauld *et al.*, 1987), well below the salinity of the local Hamelin Pool seawater (65–70‰) periodically flooding the area. Subsequent field measurements explained this apparent discrepancy; low-salinity (16–18‰) groundwaters seep from the base of beach dunes adjacent to this locality and, mixing with the local seawater, produce highly variable salinities within the range of 16–70‰.

As tides recede, or as the water table falls below the mat surface, desiccation commences and continues, aided by high insolation, which commonly raises temperatures in emergent mat surfaces to 48–50°C or higher (Bauld, 1984a). Under such conditions water loss is rapid; if the loss is sustained, mats develop characteristic polygonal desiccation cracks and may attain significant mechanical strength (Figs. 5 and 7 in Bauld, 1984a), though this is frequently associated with considerable deformation. Desiccation (and rewetting) of intact subsamples of smooth mat from the Fisherman Bay area of Spencer Gulf was followed under controlled laboratory conditions by monitoring changes in moisture content and photosynthetic $^{14}CO_2$ fixation during drying at temperatures of 35–45°C, which closely mimicked summer mat conditions during daytime. Using excised mat cores of standard dimensions (0.282 cm^2), it was observed that moisture loss was rapid and complete within about 40 min, whereas photosynthetic activity declined more slowly, becoming undetectable only after 240 min (Bauld, 1983, in preparation). Larger mat samples, more representative of intact field mats, lost moisture much more slowly, attaining constant weight only after 24–48 hr, at which time their moisture content (Karl Fischer determination) was ca. 2–3%. The retention of these small amounts of moisture suggests that tightly held water, possibly residing in the thick sheaths of intertidal mat-constructing cyanobacteria, may play a role in desiccation survival (Bauld, 1983, 1984a, in preparation). Immersion of desiccated mat is rapidly followed by water uptake and renewed photosynthetic activity, which commences instantaneously or at most within minutes of immersion (e.g., Fig. 6 of Bauld, 1984a). These phenomena are consistent with a survival and growth strategy of rapid response to fluctuations of unpredictable frequency and duration in both osmotic and matric components of a_w.

3.5. Interactions with Other Biota

We are still largely ignorant of the factors that determine whether microbial communities will successfully colonize any particular sedimentary environment. Significant factors include sedimentary regime, the presence of grazing and/or burrowing fauna, and the array of potential microbial colonizers. In the absence of grazing pressure, the successful colonizers will be those possessing the appropriate strategies to survive and reproduce under local abiological environmental constraints. As we might infer from Section 3.4, the most successful colonizers and constructors of mats in these conditions are usually the filamentous, ensheathed cyanobacteria (however, see Section 2.2). In marginal habitats where erratic immersion–exposure cycles are the rule, the current consensus, developed largely from observation rather than experimentation, is that such conditions are sufficiently severe with respect to salinity and/or desiccation to exclude benthic grazing fauna (e.g., Garrett, 1970) and thus to permit the development of stable microbial mats. For example, cerithid gastropods, which are known to limit mat development (Garrett, 1970; Javor and Castenholz, 1984), suffer significant mortality at salinities of >70‰ (Javor and Castenholz, 1984).

The argument that mat and/or stromatolite development is constrained by way of salinity-limited fauna is difficult to sustain for those habitats that are permanently submerged, or only infrequently exposed, under waters that do not exceed normal or hypersaline marine salinities. Such salinities do not preclude grazing and burrowing fauna, but microbial mats, often less physically robust than their intertidal counterparts, have been recorded with increasing frequency from habitats of moderate salinity (Dravis, 1983; Moore et al., 1983; Dill et al., 1986; see also summary in Ward et al., in press). The argument also appears less convincing than it might even for habitats such as Hamelin Pool, indisputably hypersaline relative to normal seawater. Large expanses of the sublittoral platform are covered by mats (Playford and Cockbain, 1976; Bauld, 1984a) apparently forming stromatolites, yet far from being a biological desert there is a variety of grazing and burrowing fauna (Walter and Bauld, 1986).

Those factors determining the outcome of competition between mat-forming benthic microbial communities and other photosynthetic communities in permanently submerged or only infrequently exposed habitats remains a significant gap in our understanding (Bauld, 1986a). There are four major competitors, corals, charophytes, aquatic macrophytes, and phytoplankton, with the first two constrained by high salinities. Charophytes occupy lacustrine habitats, but even the most halophilic species (*Lamprothamnium papulosum*), which appears to be healthy at salinities up to ca. 70‰, cannot germinate zygotes above ca. 50‰ (Burne et al., 1980). The upper limit for growth (and reproduction?) of aquatic macrophytes appears to lie somewhere in the range of ca. 60–160‰ (Bauld, 1986a and references therein). Potential competition from phytoplankton, on the other hand, is not constrained by salinity. The outcome of competition at low salinities where all these phototrophs can coexist may well be influenced by light (both quantity and quality), itself controlled by turbidity, self-shading, nutrient status, meiofauna, and other grazers.

The C/P ratios of mat-associated organic matter from Hamelin Pool are in the range of 4000–9000 (Atkinson, 1987; M. J. Atkinson, personal communication), indicating that these communities rely on efficient recycling of phosphate for growth and maintenance (Skyring, 1984; Atkinson, 1987). The very low phosphate concentrations in Hamelin Pool water (<0.02 μM) and sediments (<1 μmol g^{-1}) are a consequence of phosphate removal by the sea grass meadows on the Faure Sill as water flows in from the Indian Ocean (Atkinson, 1987). This situation is exacerbated by the long residence time of at least 1 year for Hamelin Pool water (Smith and Atkinson, 1983), severely reducing the total quantity of phosphate imported into the pool. Nevertheless, in this phosphate-depleted environment, the available phosphate is retained within the mat structure (Skyring, 1984). Sediment phosphate concentrations in the vicinity of the cyanobacterial mats are <1 μmol P g^{-1} (Atkinson, 1987), whereas the phosphate concentration in the cyanobacterial mats is 13 μmol pg^{-1} (Skyring, 1984); this also indicates that the mats concentrate phosphate against a steep gradient. It is possible that this colonizing behavior of forming cohesive mats and the ability to concentrate phosphate enables these, and perhaps other types of cyanobacterial mats, to be maintained in environments where the productivity of phytoplankton is severely restricted (Kimmerer et al., 1985). Lyons et al. (1984) showed that phosphate was very rapidly cycled in cyanobacterial mats from Solar Lake (Sinai) and Bonaire (the Netherlands Antilles) and suggested that this was important in sustaining continued growth of microbial mats.

3.6. Nitrogen Fixation

Atkinson (1987) and Bauld et al. (1979) record inorganic nitrogen concentrations in Hamelin Pool water of 0.6 (maximum) and 20 μM, respectively. The latter is high compared with the average of many analyses (0.4 μM) by Atkinson (1987). The high value obtained by Bauld et al. (1979) was probably the result of the disturbance and suspension of sediment detritus during heavy weather caused by Cyclone Alby, which passed through the area just 2–3 days before sampling. Cyclonic disturbances of this magnitude are infrequent at Shark Bay, and although strong winds cause seiching and wave action during February and March (I. A. Johns, personal communication), it is probable that the waters of Hamelin Pool usually contain inorganic nitrogen concentrations of around 0.6 μM.

Calculations from primary productivity and average C/N ratios of cyanobacterial mats from Hamelin Pool (Atkinson, 1987 and personal communication) indicate that the daily nitrogen requirements of 1.0 m^2 of the intertidal photosynthesizing mats would be around 6 mmol. Even if the mats were 100% efficient at scavenging nitrogenous nutrients, it seems unlikely that tidal waters in Hamelin Pool could transport enough inorganic nitrogen to sustain these photosynthetic activities. Although there may be some transport of nutrients to the intertidal zones by groundwater, Smith and Atkinson (1983) concluded that this would not contribute significantly to nutrient budgets. It seems likely, therefore, that the nitrogen requirements of the intertidal mat and possibly the subtidal mats and columnar stromatolites of Hamelin Pool are supplied via N_2 fixation.

Skyring et al. (1988, 1989) showed that smooth, tufted, and pustular mats from Hamelin Pool (Table I) were capable of reducing acetylene (indicating nitrogenase activity) and that the reduction was light dependent. However, Lyngbya isolated from tufted mat could not reduce acetylene, and it appeared that Microcoleus had the active nitrogenase. Acetylene reduction was completely stopped by photosystem I inhibitors, confirming that the nitrogenase activity of intact mat was light dependent (Skyring et al., 1989). DCMU (an inhibitor of photosystem II) inhibited acetylene reduction in tufted mat but caused a stimulation in both smooth and pustular mats, and light-dependent nitrogenase activity was lower in air than in argon for all mats. Collectively, the results indicated that the nitrogenase in the mats was O_2 sensitive and that the activity was within photosynthetic organisms, presumably the dominant cyanobacteria. Qualitatively, these results are compatible with those for cyanobacterial mats in other parts of the world (Stal et al., 1984; Bebout et al., 1987). Extrapolation of the quantitative data to mats in situ at Hamelin Pool and Spencer Gulf was not appropriate because of the perturbations caused during experimentation. However, demonstration of the widespread capacity for Hamelin Pool cyanobacterial mats to fix N_2 in a nutrient-poor habitat is consistent with the notion that N_2 fixation is vital for their maintenance and growth.

Wiebe (1985) reviewed nitrogen dynamics on coral reefs and concluded that N_2 fixation is an important process on coral reefs. Investigations into N_2 fixation on Australian coral reefs (Burris, 1976; Wiebe et al., 1975; Wilkinson et al., 1984) indicate that cyanobacterial mats (or turfs) are the most important ecosystems in which N_2 fixation occurs. The highest rates of acetylene reduction occurred on the outer shelf of the central Great Barrier Reef ($2.5–4.5$ nmol cm^{-2} hr^{-1}), whereas algal biomass was greatest for the inner shelf (Wilkinson et al., 1984). The reasons for the different acetylene reduction rates in inner- and outer-shelf mats was not elucidated, but Wilkinson et al. (1984) suggest differences in species or nitrogenase activity. Making "many assumptions," they estimated N_2 fixation rates for inner-, middle-, and outer-shelf reefs at 4.4, 14.8, and 20.4 kg N ha^{-1} yr^{-1}, respectively, and concluded that cyanobacterial turf (mats) on reef rubble and reef flat is the major source of nitrogen to a coral reef (see also Johannes et al., 1972; Webb et al., 1975). Cyanobacteria within sponge tissue also fixed N_2 and probably contributes significantly to the nitrogen dynamics of a coral reef (Wilkinson, 1979).

4. Degradative Processes and Environmental Constraints

4.1. Aerobic Processes

Much of the biogeochemical data for smooth and tufted mats from Spencer Gulf and Shark Bay indicates that anaerobic processes are dominant in the degradation of organic matter. However, it is probable that aerobic processes are dominant in the intertidal pustular mats of Hamelin Pool. This interpretation is partially supported by the absence of detectable sulfate reduction, the presence of nitrate (ca. 6.0 μM), the absence of ammonia in surface pore waters, and the presence of appreciable quantities of oxygen (3.0 ppm at 20 cm) in the subsurface pore waters associated with these mats (Skyring, unpublished data; Fig. 5). F. S. Lupton (personal communication) also showed that anaerobic hydrogenase activity in pustular mat was low. These data are in contrast to those for the tufted, smooth, and colloform mats of the lower intertidal zones in Hamelin Pool, where nitrate is undetectable, ammonia is often present in the associated pore waters, and the oxygen depth profile is very steep (Fig. 5). There was also some evidence for the presence of aerobic bacteria from measurements of aerobic hydrogenase activity in smooth mat from Gladstone Embayment (Table I); however, this was around four times less than the anaerobic hydrogenase activity (F. S. Lupton, unpublished data).

Aerobic thiobacilli, which differ from the majority of neutrophilic thiobacilli in

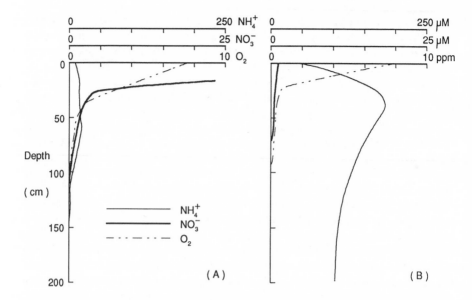

Figure 5. Oxygen, nitrate, and ammonia concentrations in the pore water of sediments associated with (A) pustular mat and (B) smooth mat at Nilemah Embayment (Shark Bay). The pustular mats were infrequently inundated with tidal waters; therefore, we have given surface values for O_2 and ammonia that were determined on another occasion. Phosphate concentrations were <1 μM in all pore water samples.

being unable to produce or oxidize polythionates, have been isolated from cyanobacterial mats from the intertidal and subtidal zones of Hamelin Pool. It is possible that these bacteria are actively involved in sulfide turnover in cyanobacterial mats at Shark Bay (P. A. Trudinger and L.-E. Hogie, personal communication).

4.2. Fermentation and Sulfate Reduction

Fermentation or respiration of the photosynthetic products of cyanobacterial mats during dark periods may be expected, since this is usual for all photosynthetic ecosystems. However, there is now substantial evidence from investigations into microbial processes occurring within cyanobacterial mats from Australian and other locations, that fermentation occurs also during photosynthetic periods (Jørgensen and Cohen, 1977; Nedwell and Abram, 1978; Bauld et al., 1979, 1980; Cohen et al., 1980; Bauld and Chambers, 1983; Lyons et al., 1983; Skyring et al., 1983, 1988, 1989; Skyring, 1984, 1989; Chambers, 1985; Anderson et al., 1987; King, 1988). At least for some cyanobacterial mats, it appears that the anaerobic diagenesis of cyanobacterial cell substance accounts for a significant part, if not all, of the processes by which organic components are remineralized. Fermentation rates in natural habitats are difficult to measure, however, minimum fermentation rates may be estimated from sulfate reduction rates because the SRB derive their energy for growth from the metabolic end-products of fermentation (see Skyring, 1987). However, it may not always be assumed that the rate at which sulfate reduction occurs is dependent upon the rate at which fermentation supplies energy substrates for the SRB. Skyring and Lupton (1984) showed that the fermenters, which produced volatile fatty acids from added glucose in sediments from Shark Bay, were less sensitive to increases in salinity than were the indigenous SRB. This phenomenon may explain the initial burst of sulfate reduction which occurred just prior to the main peak that was stimulated by tidal flooding in Mambray Creek mats (Skyring *et al.*, 1983). Between flooding cycles of around 16 days these mats slowly desiccated and this resulted in an increase in salinity of the surrounding porewaters (e.g., from 110 to 150‰).

There are several methods for measuring sulfate reduction rates in a variety of coastal environments (see Skyring, 1987), however, we found that they were not particularly suitable for our investigations of the laminated mats from Spencer Gulf. A novel technique in which solid $Na_2{}^{35}SO_4$, contained in a sodium silicate film dried onto a glass rod, implanted to transverse the mat laminations (Skyring *et al.*, 1979, 1983; Skyring and Chambers, 1980; Skyring, 1984, 1987), enabled the determination of sulfate reduction rates over millimeter distances. This subsequently led to the development of a sensitive silver wire method for quantifying sulfate reduction in Solar Lake (Sinai) cyanobacterial mats over a micrometer scale (Y. Cohen, personal communication). Examples of sulfate reduction rates, determined for several cyanobacterial mat communities at several locations in Spencer Gulf and Shark Bay, are given in Table III (see Skyring, 1987); there is a wide range of rates for the various mats, with the highest rates occurring in smooth and tufted mats. Sulfate reduction also occurred at high rates in the colloform flat mat communities that colonized the near-shore subtidal sediments and also the columnar stromatolites in Hamelin Pool (Skyring, 1985a). Bauld *et al.* (1987) isolated photosynthetic sulfur bacteria and detected bacteriochlorophyll from these mats,

Table III. Sulfate Reduction Rates in Some Australian Cyanobacterial Mat Communities

Place	Mat type	Sulfate reduction (mmol m^{-2})	Rate per day (nmol/g)	Reference
Mambray Creek	Smooth	2–104		Skyring *et al.* (1983)
	Tufted	2–21		
Hamelin Pool	Smooth	0.4–5		Bauld *et al.* (1979), Skyring (1987)
	Tufted	0.3–2		
	Pustular	0		
	Colloform (intertidal)	29		Skyring (unpublished)
	Colloform (stromatolite)		19	Skyring (1985a)

inferring the occurrence of sulfide (from sulfate reduction) and strictly anaerobic micro-niches. On the other hand, the fact that sulfate reduction was not detected in the intertidal/supratidal pustular mat that covered extensive areas of the upper- and mid-intertidal zones of Hamelin Pool was interpreted by Skyring (1984) as being due to the well-aerated coarse sediments that the *Entophysalis* of the pustular mat preferred to colonize.

The expression of sulfate reduction rates on an areal basis is generally adequate for comparisons of rates between various mat ecosystems. Compared with the depth profiles for sulfate reduction in other marine environments, however, the profiles for cyanobac-terial mats with few exceptions, exhibit a geometric or power function ($y = ax^b$) (Fig. 3). This indicates that all of the microbial processes, from the photosynthesis of cyanobacterial cell mass to the remineralization of that organic matter to carbon dioxide by respiring cyanobacteria and the heterotrophic bacteria, including the SRB, occur either within or in very close proximity to the oxygenic component in the mat (Fig. 4; Skyring, 1988; Skyring and Johns, 1980; Skyring et al., 1983, 1988, 1989). This is now a well-recognized phenomenon for many mat types (Cohen et al., 1984; Skyring, 1987; Cohen and Rosenberg, 1989).

The sulfate reduction rates in the smooth mats of the intertidal sediments at Mambray Creek indicated that there were seasonal trends with, unexpectedly, the high-est sulfate reduction rates occurring during the winter and spring months (1977–78). It was shown by multivariate analysis that the low summer activity was due to the effects of high salt concentrations on both the photosynthetic and sulfate reduction rates during the dry and highly evaporative summer months (Skyring et al., 1983) The temperature coefficient for the SRB at Mambray Creek was similar (2–2.4) to that for the SRB in other environments (Skyring et al., 1983) and this, in conjunction with the seasonal distribution of [32]S in the acid volatile sulfides (Chambers, 1982; Skyring et al., 1983), indicated that the average sulfate reduction rate for SRB was slower in the winter than in the summer and autumn months and that the effect of high salt concentrations was to reduce the number (or biomass) of SRB during the summer. Similar salt effects on sulfate reduction in cyanobacterial mats in coastal lagoons were observed by Cohen (1989).

Skyring et al. (1983) concluded from a multivariate analysis of the Mambray Creek data that 80–100% of the organic carbon synthesized by the cyanobacteria in the mat was eventually oxidized by the SRB; sulfate reduction rates correlated positively with the primary productivity of the mat when the productivity data were transformed by a time series equation that accounted for the rate of decomposition of photosynthate. Most of the photosynthate was oxidized within 1–2 months of formation. They also estimated that the molar ratio between photosynthetically fixed carbon and reduced sulfate was $2:1$ ($\pm 20\%$), which is consistent with the following stoichiometry for the oxidation of organic carbon by the SRB (Gieskes, 1973):

$$(CH_2O)_x\,(NH_3)_y\,(H_2PO_4)_z + \tfrac{x}{2}SO_4^{2-} \rightarrow x\,CO_2 + \tfrac{x}{2}\,S^{2-} + y\,NH_3 + z\,H_3PO_4 + x\,H_2O$$

A polynomial equation describing the quantitative relationships between the sulfate reduction rate, available organic carbon, temperature, pore water content, and salinity was derived, and the model was independently consistent for the Mambray Creek cyanobacterial ecosystem. Also, the model may have general applicability for those marine ecosystems in which productivity is high and the SRB populations are numerically similar (see Skyring, 1984, 1987).

Quantitative relationships between primary productivity and sulfate reduction were first calculated for Solar Lake cyanobacterial mats by Jørgensen and Cohen (1977), who found that 99% of the daily primary productivity was oxidized by the SRB. These and other investigators have also shown that most of the sulfate reduction in cyanobacterial mat ecosystems generally occurs at very high rates in close proximity to or even within the mat (Howarth and Marino, 1984; Lyons et al., 1983; Nedwell and Abram, 1978). The results of Bauld et al. (1979), Skyring (1984), and Skyring et al. (1983, 1988, 1989) were consistent with this remarkable phenomenon.

Although the SRB oxidize a range of organic substrates, current evidence suggests that the acetate-utilizing SRB are an important component of the microbial population in cyanobacterial mats. For example, enrichment cultures for SRB in cyanobacterial mats from Shark Bay using acetate as the energy source were the quickest to show positive growth, inferring that the population of acetate-utilizing SRB was present in large numbers (Trudinger, 1982; Trudinger, personal communication).

The excretion of DOC and its utilization by chemoheterotrophs, including the anaerobic SRB, has been established (see Sections 3.2 and 3.8). The heterotrophic uptake of acetate, a central metabolite, and other substrates such as glucose is carried out by several classes of phototrophic mat components (see Sections 3.2 and 3.3). Also, Bauld and Chambers (1983) and Chambers (1985) showed that acetate was a major component of the fatty acid content of the pore waters in Shark Bay cyanobacterial mats, and they showed that by molybdate inhibition experiments that acetate was rapidly used by the SRB present in the mat. They observed much lower concentrations of propionate and butyrate. Furthermore, Bauld (1984a) showed that the same cyanobacterial mats produced 1–6% DOC during photosynthesis, and subsequently Chambers (1985) showed by molybdate inhibition studies that this DOC was also rapidly oxidized by the SRB.

Phase-contrast microscopy (Fig. 2), TEM (Bauld, unpublished data; D'Amelio and Bauld, in preparation), and experimental data (Skyring et al., 1989) are consistent with the idea that DOC and DIC are transferred rapidly between adjacent "complementary" physiological groups and that both carbon and sulfur participate in syntrophic interactions. This organizationally satisfying producer–consumer flow may well be complicated by inter- or intraniche competition. For example, under low-light conditions, organisms logically syntrophic for sulfur, e.g., the phototrophic purple sulfur bacteria and the heterotrophic SRB (Bauld, 1986b), may well compete simultaneously with each other for organic carbon (Bauld, 1984b), since it is probable that photoheterotrophy is favored over photoautotrophy at very low light intensities (see Section 3.3).

Gaseous hydrogen may play an important role as an intermediary metabolite during fermentation in marine and other aquatic ecosystems (see Ward, 1984; Skyring, 1984, 1987; Anderson et al., 1987). Large quantities of H_2 can be produced during an anaerobic degradation of algal organic matter (Bubela et al., 1975). Significant H_2 production also occurred in smooth mats from Spencer Gulf and smooth and tufted mat from Shark Bay, and oxidation of the H_2 by the SRB may account for around 20% of the sulfate reduced during dark periods (Skyring et al., 1988, 1989). In these mat communities, the H_2 was derived by fermentation from organic carbon synthesized by the cyanobacterial community. It also appears that cyanobacterial mats have a relatively high concentration of hydrocarbons and that during anaerobic diagenesis, the patterns of hydrocarbon alter with time (Trudinger, personal communication).

Sulfide, the end product of bacterial sulfate reduction, may be precipitated initially as FeS in natural environments. Seasonal changes in ^{34}S-isotope distribution in acid-volatile sulfide from cyanobacterial mats and associated sediment, collected at monthly intervals from Mambray Creek, suggested a short residence time for sulfide in the sediments. The sulfur isotope analyses also showed that the intertidal cyanobacterial ecosystem was essentially open with respect to sulfate (Chambers, 1982). The range of isotope values ($\delta^{34}S$) for the sulfides (-9 to $-27‰$) was similar to the range noted for the sulfides of similar mats in Shark Bay (Bauld et al., 1979). The sulfur isotope data from detailed studies of Australian cyanobacterial mats provide new information for interpreting and modeling sedimentary sulfide ore formation (see Chambers, 1982).

4.3. Methanogenesis

Because methanogens are poor competitors with SRB for substrate (see Henrichs and Reeburgh, 1987), methanogenesis in most marine environments containing sulfate is barely detectable. This appears to be the situation for cyanobacterial mats in a marine setting, since J. Hansen (see Skyring, 1984) showed that methanogenesis occurred only at very low rates in smooth mat from Mambray Creek (0.3 mmol m^{-2} day^{-1}), and F. S. Lupton (personal communication) also found that methanogens could be enriched from cyanobacterial mats from Hamelin Pool, but methanogenesis was not an active process. These observations contrast with the high rates of methanogenesis in a *Spirulina* mat occurring in a hypersaline, sulfate-rich pond at St. Croix, U.S.A., where interactions between salinity and the presence of methylamines may lead to this unusual situation (King, 1988).

5. Conclusions

The driving forces behind our ecological studies of microbial mats were economic. The objective of BBGL research was to construct and/or test paleoenvironmental models of biogenic sulfide ore genesis (primarily those associated with stromatolites and carbonates) and, more recently, to apply the same approach to the formation of some types of petroleum source rocks, to be used in Australian exploration for these commodities. Before the BBGL research program was initiated, there were a number of ore genesis models that invoked microbial mats (Renfro, 1974; Garlick, 1981), but these lacked essential information about, and consequently a clear understanding of, the production and early diagenesis of organic matter in these benthic microbial communities. Our aim was to supply that information and understanding.

5.1. Ecology

The information we have presented shows that there is a wide variety of microbial mats in several types of Australian coastal environments, each having ecological significance. For example, the cyanobacterial mats of Spencer Gulf play an important role in binding and consolidating sandy intertidal sediments and those of the Great Barrier Reef are responsible for maintaining nitrogen requirements for the entire reef community. However, it is the enormous expanse and wide variety of microbial mats of Shark Bay that must make this area one of the most significant assemblages of phototrophic microbial ecosystems in the world. The range of mats and their lithified variants (stromatolites) of Shark Bay are living fossils and are probably very similar to the primitive microbial ecosystems that heralded the beginning of life on earth (Schopf, 1983). Although the mat communities within a particular mat type may vary (Table I), there are commonalities among similar mat communities in widely separated areas. For example, most photosynthetically active smooth mats from Spencer Gulf and Shark Bay are only 1–2 mm thick, but they are tough and cohesive and composed essentially of intertwined sheaths of *Microcoleus;* they are generally associated with fine-grained sediments that are invariably anoxic, supporting high sulfate reduction rates. This fits the general description for smooth mats in other parts of the world, although some smooth mats cover the surfaces of quite thick sequences of mat remnants, particularly in low-energy environments (e.g., Solar Lake, Gladstone Embayment). Tufted mats are composed of *Microcoleus* and *Lyngbya,* they are tough and cohesive, and they are generally associated with anoxic sediments where sulfate reduction occurs. On the other hand, pustular mats from Shark Bay are delicate friable mats composed of organically cemented *Entophysalis* colonies; the coarse-grained sediments that this community colonizes are oxic, and sulfate reduction is generally not a characteristic feature.

5.2. Minerals, Petroleum, and Modeling

Research at the BBGL resulted in an assemblage of information and knowledge that enabled various models for mineral sulfide formation to be tested for their realistic environmental settings and quantitative components. For example, Skyring *et al.,* (1983) and Skyring (1981, 1985b) calculated that from primary productivity and sulfate reduction rates for smooth mat from Spencer Gulf, sufficient sulfide could be produced for ore-grade mineralization (see Trudinger *et al.,* 1972), but the preservation of iron sulfide, often a precursor to heavy-metal sulfide formation, was poor. Ferguson *et al.* (1988), using the descriptive and quantitative ecological information from Shark Bay and Spencer Gulf, subsequently constructed four feasible models for biogenic sulfide mineral formation in coastal settings in which the environments were dominated by a variety of cyanobacterial mats.

Studies of the production and degradation of microbial organic matter in Spencer Gulf and Shark Bay also prompted closer examination of coastal salt lakes for the occurrence of microbial mats and their lithified counterparts (Bauld, 1981a). This led to consideration of juxtaposed microbial mat–stromatolite systems in lacustrine or quasi-lacustrine paleoenvironments as progenitors of petroleum source rock–reservoir rock associations (Burne and Bauld, 1985; Grey et al., 1990; R. V. Burne and J. Bauld, in preparation). Microbial mats have been found in hypersaline Lake Eliza (South Australia; Fig. 1 and Table I); the organic-rich, laminated sediments of the upper 30 cm are described (Burne and Ferguson, 1983) as cryptalgal (i.e., stromatolitic and probably cyanobacterial), and it is possible that cyanobacteria are (or were) the main contributors of organic matter to those sediments. A seasonal study of organic carbon fluxes showed that high inferred primary productivity combined with continuous but low sulfate reduction rates (to maintain a permanent anoxic state) were important biological factors controlling the rate of organic matter degradation and preservation; high salinities also played a major role in retarding the rates of organic matter degradation (Skyring and Lupton, 1986).

5.3. Serendipity, Controversy, and Speculation

From our various investigations have come a variety of spinoffs. As yet uncompleted studies of cyanobacterial sheath composition (Dudman and Bauld, unpublished data) and consideration of possible reasons for their apparent resistance to degradation (Bauld, 1981b) led to experiments which indicated that carbonate precipitation can be initiated on sheaths in the presence or absence of cyanobacterial trichomes (Pentecost and Bauld, 1988). Our understanding of lacustrine microbial mats also led to a reconciliation of the apparently anomalous association of stromatolitic carbonates with Proterozoic glacial deposits (Walter and Bauld, 1983), a long-standing source of controversy within the geological community. Application of our expanding knowledge to shallow, ephemeral (playa) lakes, of which Australia has an abundance and some of which host microbial mats, has found potential utility in the field of solar system exploration, specifically in the case of Mars, where the emphasis has shifted from the previous attempts to detect extant life forms (the Viking missions) to the future search for evidence of now-extinct benthic microbial communities (see Bauld, 1988).

Several lines of evidence suggested that the cyanobacterial mats of Shark Bay and Spencer Gulf became established in environments that are nutrient poor. Since most of the mats fix N_2, their nitrogenous requirements are not constrained. Phosphate is limited by the amount in inundating waters, however and the physiological response of the mats is to concentrate it 1000-fold and retain it against a steep gradient. Stromatolitic phosphorites are known from the Georgina Basin (Australia, Cambrian, ca. 540 million years BP; Southgate, 1980), and Soudry and Southgate (1989) have that shown these exhibit primary enrichment of phosphate. Cyanobacterial reefs and intertidal stromatolites are prominent in the geological record until the early Phanerozoic (Grotzinger, 1989). It is possible that the seas and oceans during this period were low in phosphate, thus facilitating the colonization of marine substrata within the photic zone by microbial mat communities, especially the cyanobacteria. Cook and Shergold (1986) review phosphate deposits of the world and discuss evidence indicating that the first appearance of metazoans corresponds with the onset of a period of extensive phosphogenesis during the Late Proterozoic–Early Cambrian. Could it be that the decreasing incidence of stromatolites from this period was due more to a loss of colonizing advantages of microbial mat communities caused by increasing seawater phosphate concentrations than to the evolution of grazing metazoans?

Microbial mats find commercial application in solar evaporative salt ponds (see Bauld, 1981a). Development of cyanobacterial mats has been shown to effectively seal the bottom sediments and hence the (expensive) seepage loss of water in salinas in the Bahamas (Davis, 1978) and Australia (Jones *et al.*, 1981), among others. Algal mats are used very successfully to poise the chemical and physical properties of the water of the Great Barrier Reef Aquarium operated in Townsville (Queensland) by the Great Barrier Reef Marine Park Authority (Goertemiller, 1988; Morrissey and Jones, 1988). It seems to us that there is significant potential for other kinds of commercial applications. For example, we now know that microbial mats are able to concentrate many compounds against high concentration gradients, and Tobschall and Dissanayake (1986) have shown that the cyanobacterial mats on the tidal flats of Mannar (Sri Lanka) are highly enriched in gold and platinum, suggesting a potential for commercial exploitation.

5.4. Some Unsolved Problems and Unanswered Questions

5.4.1. Factors Enabling the Establishment of Microbial Mats

In modern habitats, metazoans graze the microbial mats of coral reefs and, to a lesser extent, those of Shark Bay, but in neither situation are they grazed beyond sustainable proportions. The high salinity of Hamelin Pool may limit but does not exclude metazoan development, and the colonization of some low-salinity environments by cyanobacterial mats (Moore *et al.*, 1983; Dravis, 1983; Dill *et al.*, 1986) are clearly not "salt" protected. Our research, together with that of others (Atkinson, 1987; Smith and Atkinson, 1983), suggests that because cyanobacterial mat communities have a capacity to fix N_2 and extract and conserve nutrients, particularly phosphate, against very low concentrations gradients, they are able to become established in coastal environments that are depleted in essential nutrients and therefore compete effectively against other potential inhabitants (e.g., phytoplankton and aquatic macrophytes).

5.4.2. Photoheterotrophy and Carbon Flux in Microbial Mats

Photoheterotrophic growth has been shown to occur in laboratory cultures of various cyanobacteria, and photoheterotrophic uptake of suitable substrates has been observed in natural populations of both phytoplankton and benthic microbial communities (see Bauld, 1986a). It may be surmised that, for energetic reasons, photoheterotrophy might assume greater significance under low-light conditions (Van Baalen *et al.*, 1971), but the quantitative contribution of this activity to phototrophic carbon flux in microbial mats remains undetermined.

5.4.3. Constraints on the Degradation of Autochthonous Organic Matter

We and others have shown that there are several interacting factors which constrain the rates of organic matter degradation in natural environments: refractory parts of the biomass, high salinity, low water activity, low temperature and redox conditions, and sulfide production. However, it was also evident that in some environments, one or other factor was dominant (e.g., high salinities in Shark Bay and Spencer Gulf and sulfide in Lake Eliza; Skyring and Lupton, 1986). It would be useful, especially for modeling hydrocarbon production and preservation, to construct several scenarios in which realistic ecological settings and the most important biological and physical constraints on diagenetic reactions were detailed.

5.4.4. Paradoxes

The measurement of *in situ* processes in microbial mats is fraught with many of the difficulties encountered in soils and sediments, with the added complication that both phototrophic and chemotrophic activities may occur in very close physical proximity. For example, we and others have evidence for the presence and activity of strictly anaerobic bacteria in photosynthesizing oxygenic cyanobacterial mats. The existence of anaerobic microorganisms in apparently aerobic environments is usually attributed to the presence of anaerobic microniches. The high photosynthetic rates of these mats and the resultant local high concentrations of O_2 stretch the imagination to invoke an anaerobic microniche explanation. Our experience over the last decade has demonstrated the need for more knowledge about the location of processes and physical dimensions of flows and fluxes within the mat. Major achievements have been the application of microelectrodes to the measurement of some microbial processes in mats, films, and sediments (Revsbech and Jørgensen, 1986) and a refinement of the silicate/rod method (Skyring *et al.*, 1983) for measuring sulfate reduction rates over micrometer distances in microbial mats (Y. Cohen, personal communication). One major objective should now be the *in situ* simultaneous measurement of interacting processes, probably requiring the use of as yet undeveloped specific inhibitors, and continuous nondestructive, microprocessor-controlled monitoring that could be assessed from remote locations. Such an approach would enable real-time, short-interval monitoring of multiple fluxes and pool sizes over successive diurnal cycles for sustained periods, which would be of importance in, for example, mat communities subjected to alternating inundation and exposure. The resultant data base would then allow detailed predictive mathematical modeling of interactive C-S-N-P-O-H fluxes and pools for both synthetic and degradative processes. Reliable and inexpensive remote monitoring communications systems were developed over the last decade in the BBGL to monitor meteorological and hydrological parameters

(Johns and Jacobson, 1988), and the adaption of such systems to monitoring mat processes could make a major contribution.

5.4.5. Reflections

There are many Australian environments that offer potential for research into a wide range of microbial mats. However, sustained and integrated research into the microbial mats of Shark Bay, one of the most significant ecosystems of all modern environments, has temporarily ceased. The area is a heritage of global significance, esthetically and scientifically, and worthy of a sustained national research program.

ACKNOWLEDGMENTS. We wish to thank our colleagues from the Baas Becking Geobiological Laboratory who permitted us personal communications. Much of the work we report here was made possible only by the dedication of many of our technical colleagues; we would like to thank in particular Ian Johns, Svetlana Dibb, Jackie Luck, Andrea Blanks, Lea-Ellen Hogie, Vanessa Grey, Valmai Hicks, Craig Manning, Garth Trengove, Dan Ho, and the late Michael Reed.

The late Dr. Arthur Gaskin, chief of the former CSIRO Division of Mineralogy, always supported and encouraged our research; we wish to record, posthumously, our sincere thanks. Dr. P. A. Trudinger and Dr. M. R. Walter, former directors of the Baas Becking Geobiological Laboratory, were keenly appreciative and supportive of the research strategies used to execute this kind of microbial ecology. We also thank Phil Trudinger and Malcolm Walter for reading the manuscript and making suggestions that resulted in considerable improvements. John Bauld publishes with the permission of the Director, Bureau of Mineral Resources. A significant proportion of the work reported here was done at the Baas Becking Geobiological Laboratory, which was supported by the BMR, CSIRO, and the Australian Mineral Industries Research Association Ltd. Finally, we would like to dedicate this chapter to our wives, Bonnie Bauld and Fay Skyring, who were deprived of our scintillating (^{35}S, ^{14}C, ^{3}H!) company for many months during the last decade.

References

Anderson, K. L., Tayne, T. A., and Ward, D. M., 1987, Formation and fate of fermentation products in hot spring cyanobacterial mats, *Appl. Environ. Microbiol.* **53:**2343–2352.

Atkinson, M. J., 1987, Low phosphorus sediments in a hypersaline marine environment, *Estuarine Coastal Shelf Sci.* **24:**335–347.

Awramik, S. M., Schopf, J. W., and Walter, M. R., 1983, Filamentous fossil bacteria from the Archean of Western Australia, *Precambrian Res.* **20:**357–374.

Bauld, J., 1981a, Occurrence of benthic microbial mats in saline lakes, *Hydrobiologia* **81:**87–111.

Bauld, J., 1981b, Geobiological role of cyanobacterial mats in sedimentary environments: Production and preservation of organic matter, *BMR J. Aust. Geol. Geophys.* **6:**307–317.

Bauld, J., 1983, Response of microbial mats to salinity and desiccation, *Int. Symp. Microb. Ecol. Abstr.* **3:**42.

Bauld, J., 1984a, Microbial mats in marginal marine environments: Shark Bay, Western Australia, and Spencer Gulf, South Australia, in: *Microbial Mats: Stromatolites* (Y. Cohen, R. W. Castenholz, and H. O. Halvorson, eds.), pp. 39–58, Alan R. Liss, Inc., New York.

Bauld, J., 1984b, Role of Photoheterotrophic Bacteria in Carbon and Sulfur Cycling in Benthic Microbial Communities, discussion paper, SCOPE-UNEP Workshop on Global Sulfur Cycle, Tallinn, Estonian SSR.

Bauld, J., 1986a, Benthic microbial communities of Australian salt lakes, in: *Limnology in Australia* (P. De Deckker and W. D. Williams, eds.), pp. 95–111, CSIRO, Melbourne.

Bauld, J., 1986b, Transformation of sulfur species by phototrophic and chemotrophic microbes, in: *The Importance of Chemical "Speciation" in Environmental Processes* (M. Bernhard, F. E. Brinckman, and P. J. Sadler, eds.), pp. 255–273, Dahlem Konferenzen LS 33, Springer-Verlag, Berlin.

Bauld, J., 1987, (Photo)heterotrophic activity in benthic microbial communities. *Int. Symp. Environ. Biogeochem. Abstr.* **8**:48.

Bauld, J., 1988, Microbial mats in playa lakes and other saline habitats: Early Mars analog?, in: *Exobiology and Future Mars Missions*, NASA Conference Publication 10027, (C. P. McKay and W. L. Davis, eds.), pp. 7–8, National Aeronautics and Space Administration, Washington, D.C.

Bauld, J., and Brock, T. D., 1974, Algal excretion and bacterial assimilation in hot spring algal mats, *J. Phycol.* **10**:101–106.

Bauld, J., and Chambers, L. A., 1983, Carbon flow in microbial mats, *Aust. Microbiol.* **4**:92.

Bauld, J., Chambers, L. A., and Skyring, G. W., 1979, Primary productivity, sulfate reduction and sulfur isotope fractionation in algal mats and sediments of Hamelin Pool, Shark Bay, W.A., *Aust. J. Mar. Freshwater Res.* **30**:753–764.

Bauld, J., Burne, R. V., Chambers, L. A., Ferguson, J., and Skyring, G. W., 1980, Sedimentological and geobiological studies of intertidal cyanobacterial mats in north-eastern Spencer Gulf, South Australia, in: *Biogeochemistry of Ancient and Modern Environments* (P. A. Trudinger, M. R. Walter, and B. J. Ralph, eds.), pp. 157–166, Australian Academy of Science, Canberra.

Bauld, J., Favinger, J. L., Madigan, M. T., and Gest, H., 1987, Obligately halophilic *Chromatium vinosum* from Hamelin Pool, Shark Bay, Australia, *Curr. Microbiol.* **14**:335–339.

Bauld, J., Farmer, J. D., and D'Amelio, E., 1990, Modern microbial mats, in: *The Proterozoic Biosphere: A Multidisciplinary Study* (J. W. Schopf and C. Klein, eds.), Cambridge University Press, Cambridge (in press).

Bebout, B. M., Paerl, H. W., Crocker, K. M., and Prufert, L. E., 1987, Diel interactions of oxygenic photosynthesis and N_2 fixation (acetylene reduction) in a marine microbial mat community, *Appl. Environ. Microbiol.* **53**:2353–2362.

Borowitzka, M. A., Larkum, A. W. D., and Borowitzka, L. J., 1978, A preliminary study of algal turf communities of a shallow coral reef lagoon using an artificial substratum, *Aquat. Bot.* **5**:365–381.

Boynton, W. R., Hall, C. A., Falkowski, P. G., Keefe, C. W., and Kemp, W. M., 1983, Phytoplankton productivity in aquatic ecosystems, *Encyc. Plant Physiol. (New Ser.)* **12D**:305–327.

Brown, A. D., 1976, Microbial water stress, *Bacteriol. Rev.* **40**:803–846.

Bubela, B., 1980, Some aspects of the interstitial water movements in simulated sedimentary systems, *BMR J. Aust. Geol. Geophys.* **5**:257–263.

Bubela, B., Ferguson, J., and Davies, P. J., 1975, Biological and abiological processes in a simulated sedimentary system, *J. Geol. Soc. Aust.* **22**:135–143.

Burne, R. V., and Bauld, J., 1985, The origin and preservation of microbial organic facies in coastal saline lakes and Western Australia, in: *Abstr. Geol. Soc. London Symp. Lacustrine Petroleum Source Rocks*, pp. 17–18.

Burne, R. V., and Colwell, J. B., 1982, Temperate carbonate sediments of northern Spencer Gulf, South Australia: A high salinity "foramol" province, *Sedimentology* **29**:223–238.

Burne, R. V., and Ferguson, J., 1983, Contrasting marginal sediments of a seasonally flooded saline lake—Lake Eliza, South Australia: Significance for oil shale genesis, *BMR J. Aust. Geol. Geophys.* **8**:99–108.

Burne, R. V., and Moore, L. S., 1987, Microbialites: Organosedimentary deposits of benthic microbial communities, *Palaios* **2:**241–254.

Burne, R. V., Bauld, J., and De Deckker, P., 1980, Saline lake charophytes and their geological significance, *J. Sediment. Petrol.* **50:**281–293.

Burris, R. H., 1976, Nitrogen fixation by blue-green algae of Lizard Island area of the Great Barrier Reef, *Aust. J. Plant. Physiol.* **3:**41–51.

Carpenter, R. C., 1985, Relationships between primary production and irradiance in coral reef algal communities, *Limnol. Oceanogr.* **30:**784–793.

Castenholz, R. W., Bauld, J., and Pierson, B. K., 1990, Photosynthetic activity in modern mat-building communities, in: *The Proterozoic Biosphere: A Multidisciplinary Study* (J. W. Schopf and C. Klein, eds.), Cambridge University Press, Cambridge (in press).

Castenholz, R. W., Jørgensen, B. B., D'Amelio, E., and Bauld, J., 1988, *Oscillatoria boryana*, a fluctuating component of sulfide-rich microbial mats, *Int. Symp. Photosynthetic Prokaryotes Abstr.* **6:**87.

Chambers, L. A., 1982, Sulfur isotope study of a modern intertidal environment, and interpretation of ancient sulfides, *Geochim. Cosmochim. Acta* **46:**721–728.

Chambers, L. A., 1985, Biochemical aspects of the carbon metabolism of microbial mat communities, in: *Proceedings of the Fifth International Coral Reef Congress, Tahiti, 1985,* Vol. 3 (C. Gabrie, J. L. Toffart, and B. Salvat, eds.), pp. 371–376, Antenne Museum-Ephe, Moorea, French Polynesia.

Chambers, L. A., Trudinger, P. A. Smith, J. W., and Burns, M. S., 1975, Fractionation of sulfur isotopes by continuous cultures of *Desulfovibrio desulfuricans, Can. J. Microbiol.* **21:**1602–1607.

Cohen, Y., 1989, Hypersaline cyanobacterial mats: Photosynthesis in cyanobacterial mats and its relation to the sulfur cycle: a model for microbial sulfur interactions, in: *Microbial Mats: Physiological Ecology of Benthic Microbial Communities* (Y. Cohen and E. Rosenberg, eds.), pp. 22–36, American Society for Microbiology, Washington, D.C.

Cohen, Y., and Rosenberg, E. (eds.), 1989, *Microbial Mats: Physiological Ecology of Benthic Microbial Communities,* American Society for Microbiology, Washington, D.C.

Cohen, Y., Aizenshtat, Z., Stoler, A., and Jørgensen, B. B., 1980, The microbial geochemistry of Solar Lake, Sinai, in: *Biogeochemistry of Ancient and Modern Environments* (P. A. Trudinger and M. R. Walter, eds.), pp. 167–172, Australian Academy of Science, Canberra,

Cohen, Y., Castenholz, R. W., and Halvorson, H. O. (eds.), 1984, *Microbial Mats: Stromatolites,* Alan R. Liss, Inc., New York.

Cohen, Y., Jørgensen, B. B., Revsbech, N. P., and Poplawski, R., 1986, Adaptation to hydrogen sulfide of oxygenic and anoxygenic photosynthesis among cyanobacteria, *Appl. Environ. Microbiol.* **51:**398–407.

Cook, P. J., and Mayo, W., 1977, Sedimentology and Holocene history of a tropical estuary (Broad Sound, Queensland), *BMR Bull.* **170:**1–206.

Cook, P. J., and Shergold, J. H., 1986, Proterozoic and Cambrian phosphorites—nature and origins, in: *Phosphate Deposits of the World,* Vol. 1, *Proterozoic and Cambrian Phosphorites* (P. J. Cook and J. H. Shergold, eds.), pp. 369–386, Cambridge University Press, Cambridge.

Davies, G. R., 1970, Algal-laminated sediments, Gladstone Embayment, Shark Bay, Western Australia, *Am. Soc. Petrol. Geol. Mem.* **13:**169–205.

Davis, J. S., 1978, Biological communities of a nutrient enriched salina, *Aquat. Bot.* **4:**23–42.

De Deckker, P., Bauld, J., and Burne, R. V., 1982, Pillie Lake, Eyre Peninsula, South Australia: Modern environment and biota, dolomite sedimentation and Holocene history, *Trans. R. Soc. S. Aust.* **106:**169–181.

Dill, R. F., Shinn, E. A., Jones, A. T., Kelly, K., and Steinen, E. P., 1986, Giant subtidal stromatolites forming in normal salinity waters, *Nature (London)* **324:**55–58.

Dravis, J. J., 1983, Hardened subtidal stromatolites, Bahamas, *Science* **219:**385–386.

Ellis, B. K., and Stanford, J. A., 1982, Comparative photoheterotrophy, chemoheterotrophy, and photolithotrophy in a eutrophic reservoir and an oligotrophic lake. *Limnol. Oceanogr.* **27:**440–454.

Fattom, A., and Shilo, M., 1984, Hydrophobicity as an adhesion mechanism of benthic cyanobacteria, *Appl. Environ. Microbiol.* **47:**135–143.

Ferguson, J., Plumb, L. A., and Skyring, G. W., 1988, Genetic models of low temperature stratiform Cu-(Pb-Zn) deposits in semiarid permeable carbonate sabkha, in: *Mobilité et Concentration des Metaux de Base dans les Couvertures Sédimentaire; Manifestations, Mécanismes, Prospection,* Conference Abstracts, p. 87.

Fowler, M. G., and Douglas, A. G., 1987, Saturated hydrocarbon biomarkers in oils of Late Precambrian age from Eastern Siberia, *Org. Geochem.* **11:**201–213.

Fuchs, G., Stupperich, E., and Eden, G., 1980, Autotrophic CO_2 fixation in *Chlorobium limicola.* Evidence for the operation of a reductive tricarboxylic acid cycle in growing cells, *Arch. Microbiol.* **128:**64–71.

Garlick, W. G., 1981, Sabkhas, slumping and compaction at Mufubra, Zambia, *Econ. Geol.* **76:**1817–1847.

Garrett, P., 1970, Phanerozoic stromatolites: Noncompetitive ecological restriction by grazing and burrowing animals, *Science* **169:**171–173.

Gieskes, J. M., 1973, Interstitial water studies, leg 15—alkalinity, pH, Mg, Ca, Si, PO_4, and NH_4, in: *Initial Report of the Deep Sea Drilling Project* (B. C. Heezen and I. D. Macgregor, eds.), pp. 813–829, U.S. Government Printing Office, Washington, D.C.

Goertemiller, T., 1988, Prototype for a Great Barrier Reef replica, *Aust. Sci. Mag.* **3:**18–20.

Goldman, C. R., Mason, D. T., and Wood, B. J. B., 1972, Comparative study of the limnology of two small lakes on Ross Island, Antarctica, *Antarct. Res. Ser.* **20:**1–50.

Gorin, G. E., Racz, L. G., and Walter, M. R., 1982, Late Precambrian-Cambrian sediments of Huqf Group, Sultanate of Oman, *Am. Assoc. Pet. Geol. Bull.* **66:**2609–2627.

Grantham, P. J., Lijmbach, G. W. M., Posthuma, J., Hughes Clarke, M. W., and Willink, R. J., 1987, Origin of crude oils in Oman, *J. Pet. Geol.* **11:**61–80.

Grey, K., Moore, L. S., Burne, R. V., Pierson, B. K., and Bauld, J., 1990, Lake Thetis, Western Australia: An example of saline lake sedimentation dominated by benthic microbial processes, *Aust. J. Mar. Freshwater Res.* **41:**(in press).

Griffin, D. M., 1981, Water and microbial stress, in: *Advances in Microbial Ecology* Vol. 5 (M. Alexander, ed.), pp. 91–136, Plenum Press, New York.

Grotzinger, J. P., 1989, Facies and evolution of Precambrian carbonate depositional systems: Emergence of the modern platform archetype, in: *Controls of Carbonate Platforms and Basin Development,* (P. Crevello, J. F. Read, R. Sarg, and J. Wilson, eds.), Society of Economic Paleontologists and Mineralogists, Special Publication 44.

Henrichs, S. M., and Reeburgh, W. S., 1987, Anaerobic mineralization of marine sediment organic matter: Rates and role of anaerobic processes in the oceanic carbon economy, *Geomicrobiol. J.* **5:**191–237.

Holo, H., and Sirevag, R., 1986, Autotrophic growth and CO_2 fixation of *Chloroflexus aurantiacus,* *Arch. Microbiol.* **145:**173–180.

Howarth, R. W., and Marino, R., 1984, Sulfate reduction in salt marshes, with some comparisons to sulfate reduction in microbial mats, in: *Microbial Mats: Stromatolites* (Y. Cohen, R. W. Castenholz, and H. O. Halvorson, eds.), pp. 254–263. Alan R. Liss, Inc., New York.

Imbus, S. W., Engel, M. H., Elmore, R. D., and Zumberge, J. E., 1988, The origin, distribution and hydrocarbon generation potential of the organic-rich facies in the Nonesuch Formation, Central North American Rift System: A regional study, in: *Advances in Organic Geochemistry 1987* (L. Mattavelli and L. Novelli, eds.), pp. 207–219, Pergamon Press, Oxford.

Jackson, M. J., Powell, T. G., Summons, R. E., and Sweet, I. P., 1986, Hydrocarbon shows and petroleum source rocks in sediments as old as 1.7×10^9 years. *Nature (London)* **322:**727–729.

Jannasch, H. W., and Wirsen, C. O., 1979, Chemosynthetic primary production at East Pacific sea floor spreading centers, *Bioscience* **29:**592–598.

Javor, B. J., and Castenholz, R. W., 1984, Invertebrate grazers of microbial mats, Laguna Guerrero Negro, Mexico, in: *Microbial Mats: Stromatolites* (Y. Cohen, R. W. Castenholz, and H. O. Halvorson, eds.), pp. 85–94, Alan R. Liss, Inc., New York.

Johannes, R. E., and Project Symbiosis Team, 1972, The metabolism of some coral reef communities, *Bioscience* **22**:541–543.

Johns, I. A., and Jacobson, G., 1988, Hydrological monitoring for a remote area—the Curtin Springs (N.T.) data acquisition platform, in: *Hydrology and Water Resources Symposium 1988*, National Conference Publication 88/1, pp. 275–277, The Institution of Engineers, Melbourne, Australia.

Jones, A. G., Ewing, C. M., and Melvin, M. V., 1981, Biotechnology of solar saltfields, *Hydrobiologia* **82**:391–406.

Jørgensen, B. B., and Cohen, Y., 1977, Solar Lake (Sinai). 5. The sulfur cycle of the benthic microbial mats, *Limnol. Oceanogr.* **22**:657–666.

Jørgensen, B. B., Cohen, Y., and Des Marais, D. J., 1987, Photosynthetic action spectra and adaptation to spectral light distribution in a benthic cyanobacterial mat, *Appl. Environ. Microbiol.* **53**:879–886.

Kimmerer, W. J., McKinnon, A. D., Atkinson, M. J., and Kessel, J. A., 1985, Spatial distributions of plankton in Shark Bay, Western Australia, *Aust. J. Mar. Freshwater Res.* **36**:421–432.

King, G. M., 1988, Methanogenesis of methylated amines in a hypersaline algal mat. *Appl. Environ. Microbiol.* **54**:130–136.

Knoll, A. H., and Bauld, J., 1989, The evolution of ecological tolerance in prokaryotes, *Proc. R. Soc. Edinburgh: Earth Sciences* **80**:209–223.

Krumbein, W. E., Cohen, Y., and Shilo, M., 1977, Solar Lake (Sinai). 4. Stromatolitic cyanobacterial mats, *Limnol. Oceanogr.* **22**:635–656.

Lazar, B., Javor, B., and Erez, J., 1989, Total alkalinity in marine-derived brines and pore waters associated with microbial mats, in: *Microbial Mats: Physiological Ecology of Benthic Microbial Communities* (Y. Cohen and E. Rosenberg, eds.), pp. 84–94, American Society for Microbiology, Washington, D.C.

Logan, B. W., 1961, Cryptozoon and associated stromatolites from the Recent, Shark Bay, Western Australia, *J. Geol.* **69**:517–533.

Logan, B. W., Read, J. F., and Davies, G. R., 1970, History of carbonate sedimentation, Quaternary epoch, Shark Bay, Western Australia, *Am. Soc. Petrol. Geol. Mem.* **13**:38–84.

Logan, B. W., Hoffman, P., and Gebelein, C. D., 1974, Algal mats, cryptalgal fabrics, and structures, Hamelin Pool, Western Australia, *Am. Soc. Petrol. Geol. Mem.* **22**:140–194.

Lyons, W. B., Hines, M. E., and Gaudette, H. E., 1984, Major and minor element porewater geochemistry of modern marine sabkhas: The influence of cyanobacterial mats, in: *Microbial Mats: Stromatolites* (Y. Cohen, R. W. Castenholz, and H. O. Halvorson, eds.), pp. 411–424. Alan R. Liss, Inc., New York.

Marshall, J. F., and Davies, P. J., 1978, Skeletal carbonate variation on the continental shelf of eastern Australia, *BMR J. Aust. Geol. Geophys.* **3**:85–92.

Marshall, K. C., 1976, *Interfaces in Microbial Ecology*, Harvard University Press, Cambridge, Mass.

McKinley, W. R., and Wetzel, R. G., 1979, Photolithotrophy, photoheterotrophy, and chemoheterotrophy: Patterns of resource utilization on an annual and a diurnal basis within a pelagic microbial community, *Microb. Ecol.* **5**:1–15.

Mendelsohn, F., 1976, Mineral deposits associated with stromatolites, in: *Stromatolites. Developments in Sedimentology*, Vol. 20 (M. R. Walter, ed.), pp. 645–662, Elsevier, Amsterdam.

Monty, C. L., 1979, Monospecific stromatolites from the Great Barrier Reef Tract and their paleontological significance, *Ann. Soc. Geol. Belg.* **101**:163–171.

Moore, L. S., 1987, Water chemistry of the coastal saline lakes of the Clifton-Preston lakeland system, south-western Australia, and its influence on stromatolite formation, *Aust. J. Mar. Freshwater Res.* **38**:647–660.

Moore, L. S., Knott, B., and Stanley, N. F., 1983, The stromatolites of Lake Clifton, Western Australia, *Search* **14:**309–314.

Moriarty, D. J. W., 1983, Bacterial biomass and productivity in sediments, stromatolites, and water of Hamelin Pool, Shark Bay, Western Australia, *Geomicrobiol. J.* **3:**121–133.

Morrissey, J., and Jones, M., 1988, Water—clean, clear and warm, *Aust. Sci. Mag.* **3:**33–41.

Naiman, R. J., 1976, Primary production, standing stock, and export of organic matter in a Mohave Desert thermal stream, *Limnol. Oceanogr.* **21:**60–73.

Nedwell, D. B., and Abram, J. W., 1978, Bacterial reduction in relation to sulfur geochemistry in two contrasting areas of a salt marsh sediment, *Estuarine Coastal Mar. Sci.* **6:**341–351.

Oehler, D. Z., 1976, Transmission electron microscopy of organic microfossils from the late Precambrian Bitter Springs Formation of Australia: Techniques and survey of preserved ultrastructure, *J. Paleontol.* **50:**90–106.

Oehler, D. Z., 1978, Microflora of the middle Proterozoic Balbirini Dolomite (McArthur Group) of Australia. *Alcheringa* **2:**269–309.

Oehler, D. Z., Oehler, J. H., and Stewart, A. J., 1979, Algal fossils from a late Precambrian, hypersaline lagoon. *Science* **205:**388–390.

Paling, E. I., 1986, Ecological significance of blue-green algal mats in the Dampier mangrove system, Technical Series 2, 134 pp., Department of Conservation and Environment, Western Australia.

Palmisano, A. C., Summons, R. E., Cronin, S. E., and Des Marais, D. J., 1989, Lipophilic pigments from cyanobacterial (blue-green algal) and diatom mats in Hamelin Pool, Shark Bay, Western Australia, *J. Phycol.* **25:**655–662.

Pentecost, A., and Bauld, J., 1988, Nucleation of calcite on the sheaths of cyanobacteria using a simple diffusion cell, *Geomicrobiol. J.* **6:**129–135.

Pentecost, A., and Riding, R., 1986, Calcification in cyanobacteria, in: *Biomineralization in Lower Plants and Animals* (B. S. C. Leadbeater and R. Riding, eds.), pp. 73–90, Clarendon Press, Oxford.

Pierson, B. K., Oesterle, A., and Murphy, G. L., 1987, Pigments, light penetration, and photosynthetic activity in the multi-layered microbial mats of Great Sippewissett Salt Marsh, Massachusetts, *FEMS Microbiol. Ecol.* **45:**365–376.

Playford, P. E., and Cockbain, A. E., 1976, Modern algal stromatolites at Hamelin Pool, a hypersaline barred basin in Shark Bay, Western Australia, in: *Stromatolites* (M. R. Walter, ed.), pp. 389–411, Elsevier, Amsterdam.

Plumb, L. A., Bauld, J., Ho, D., and Reichstein, I., 1982, Production and fate of organic carbon in cyanobacterial mats, in: *Baas Becking Geobiological Laboratory, Annual Report, 1982*, pp. 25–31, Bureau of Mineral Resources, Canberra, Australia.

Plumb, L. A., Bauld, J., Ho, D., and Reichstein, I., 1983, Production and fate of organic carbon in cyanobacterial mats, in: *Baas Becking Geobiological Laboratory, Annual Report, 1983*, pp. 23–33, Bureau of Mineral Resources, Canberra, Australia.

Potts, M., and Bowman, M. A., 1985, Sensitivity of *Nostoc commune* UTEX 584 (Cyanobacteria) to water stress, *Arch. Microbiol.* **141:**51–56.

Renfro, A. R., 1974, Genesis of evaporative associated stratiform metalliferous deposits—a sabkha process. *Econ. Geol.* **69:**33–45.

Revsbech, N. P., and Jørgensen, B. B., 1986, Microelectrodes: Their use in microbial ecology, in: *Advances in Microbial Ecology*, Vol. 9 (K. C. Marshall, ed.), pp. 293–352, Plenum Press, New York.

Schopf, J. W., 1968, Microflora of the Bitter Springs Formation, late Precambrian, central Australia, *J. Paleontol.* **42:**651–688.

Schopf, J. W., 1983, *Earth's Earliest Biosphere: Its Origin and Evolution* (J. W. Schopf, ed.), Princeton University Press, Princeton, N.J.

Schopf, J. W., and Parker, B. M., 1987, Early Archean (3.3 to 3.5 Ga-old) fossil microorganisms from the Warrawoona Group, Western Australia, *Science* **273**:70–73.

Skyring, G. W., 1981, Sulfate reduction in modern sediments and implications for ore formation, *BMR J. Aust. Geol. Geophys.* **6**:335.

Skyring, G. W., 1984, Sulfate reduction in marine sediments associated with cyanobacterial mats in Australia, in: *Microbial Mats: Stromatolites* (Y. Cohen, R. W. Castenholz, and H. O. Halvorson, eds.), pp. 265–275, Alan R. Liss, Inc., New York.

Skyring, G. W., 1985a, Sulfate reduction in sediments associated with carbonate reefs in: *Research Review, 1985*, pp. 88–89, CSIRO, Division of Mineralogy and Geochemistry, Perth, Australia.

Skyring, G. W., 1985b, Biogeochemistry of Holocene environments: Sulfate reduction in anoxic sediments, in: *Baas Becking Geobiological Laboratory, Annual Report, 1985*, pp. 4–6, Bureau of Mineral Resources, Canberra, Australia.

Skyring, G. W., 1987, Sulfate reduction in coastal ecosystems, *Geomicrobiol. J.* **5**:295–374.

Skyring, G. W., 1988, Sulfate reducers in oxygenic cyanobacterial mats. *Aust. Microbiol.* **9**:168.

Skyring, G. W., 1989, Quantitative relationships between sulfate reduction and carbon metabolism in marine sediments, in: *Interaction of Sulfur and Carbon Cycles in Marine Sediments, Chapter 6*, SCOPE (P. Brimblecombe and A. Yu Lein, eds.), pp. 125–143, Wiley and Sons, Chichester.

Skyring, G. W., and Chambers, L. A., 1980, Sulfate reduction in intertidal sediments, in: *Sulfur in Australia* (J. R. Freney and A. J. Nicholson, eds.), pp. 88–94, Australian Academy of Science, Canberra.

Skyring, G. W., and Johns, I. A., 1980, Iron in cyanobacterial mats, *Micron* **11**:407–408.

Skyring, G. W., and Lupton, F. S., 1986, Anaerobic microbial activity in organic-rich sediments of a coastal lake, in: *Sediments Down-Under, 12th International Sedimentological Congress, Abstracts*, p. 280, Canberra, Australia.

Skyring, G. W., Oshrain, R. L., and Wiebe, W. J., 1979, Sulfate reduction rates in Georgia marshland soils, *Geomicrobiol. J.* **1**:389–400.

Skyring, G. W., Chambers, L. A., and Bauld, J., 1983, Sulfate reduction in sediments colonized by cyanobacteria, Spencer Gulf, South Australia, *Aust. J. Mar. Freshwater Res.* **34**:359–374.

Skyring, G. W., Lynch, R. M., and Smith, G. D., 1988, Acetylene reduction and hydrogen metabolism by a cyanobacterial/sulfate reducing bacterial mat ecosystem, *Geomicrobiol. J.* **6**:25–31.

Skyring, G. W., Lynch, R. M., and Smith, G. D., 1989, Quantitative relationships between carbon, hydrogen, and sulfur metabolism in cyanobacterial mats, in: *Microbial Mats: Physiological Ecology of Benthic Microbial Communities* (Y. Cohen and E. Rosenberg, eds.), pp. 170–179, American Society for Microbiology, Washington, D.C.

Smith, S. V., and Atkinson, M. J., 1983, Mass balance of carbon and phosphorus in Shark Bay, Western Australia, *Limnol. Oceanogr.* **28**:625–639.

Soudry, D., and Southgate, P. N., 1989, Ultrastructure of a Middle Cambrian primary nonpelletal phosphorite and its early transformation into phosphate vadoids: Georgina Basin, Australia. *J. Sediment. Petrol.* **59**:53–64.

Southgate, P. N., 1980, Cambrian stromatolitic phosphorites from the Georgina Basin, Australia, *Nature (London)* **285**:395–397.

Stal, L. J., Grossberger, S., and Krumbein, W. E., 1984, Nitrogen fixation associated with cyanobacterial mat of a marine laminated microbial ecosystem, *Mar. Biol.* **82**:217–224.

Summons, R. E., Powell, T. G., and Boreham, C. J., 1988, Petroleum geology and geochemistry of the Middle Proterozoic McArthur Basin, Northern Australia: III. Composition of extractable hydrocarbons. *Geochim. Cosmochim. Acta* **52**:1747–1763.

Tobschall, H. J., and Dissanayake, C. B., 1986, Precious metals in cyanobacterial mats of Mannar Lagoon, Sri Lanka, in: *Sediments Down-Under, 12th International Sedimentological Congress, Abstracts*, p. 305, Canberra, Australia.

Trudinger, P. A., 1982, Biology of sulfate reduction in intertidal sediments, in: *Baas Becking Geobiological Laboratory, Annual Report, 1982*, p. 35, Bureau of Mineral Resources, Canberra, Australia.

Trudinger, P. A., Lambert, I. B., and Skyring, G. W., 1972, Biogenic sulfide ores: A feasibility study, *Econ. Geol.* **67:**1114–1127.

Van Baalen, C., Hoare, D. S., and Brandt, E., 1971, Heterotrophic growth of blue-green algae in dim light, *J. Bacteriol.* **105:**685–689.

van Liere, L., and Walsby, A. E., 1982, Interactions of cyanobacteria with light, in: *The Biology of Cyanobacteria* (N. G. Carr and B. A. Whitton, eds.), pp. 9–45, Blackwell Scientific Publications, Oxford.

Walter, M. R., 1983, Archean stromatolites: Evidence of the Earth's earliest benthos, in: *Earth's Earliest Biosphere: Its Origin and Evolution* (J. W. Schopf, ed.), pp. 187–213, Princeton University Press, Princeton, N.J.

Walter, M. R., and Bauld, J., 1983, The association of sulphate evaporites, stromatolitic carbonates and glacial sediments: Examples from the Proterozoic of Australia and the Cainozoic of Antarctica, *Precambrian Res.* **21:**129–148.

Walter, M. R., and Bauld, J., 1986, Subtidal stromatolites of Shark Bay, in: *Sediments Down-Under: 12th International Sedimentological Congress Abstracts*, p. 315, Canberra, Australia.

Walter, M. R., Golubic, S., and Priess, W. V., 1973, Recent stromatolites from hydromagnesite and aragonite depositing lakes near the Coorong Lagoon, South Australia, *J. Sediment. Petrol.* **43:**1021–1030.

Walter, M. R., Buick, R., and Dunlop, J. S. R., 1980, Stromatolites 3,400–3,500 Myrs old from the North Pole area, Western Australia, *Nature (London)* **284:**443–445.

Ward, D. M., 1984, Decomposition of microbial mats—discussion, in: *Microbial Mats: Stromatolites* (Y. Cohen, R. W. Castenholz, and H. O. Halvorson, eds.), pp. 277–280, Alan R. Liss, Inc., New York.

Ward, D. M., Bauld, J., Castenholz, R. W., Cohen, Y., Jørgensen, B. B., Nelson, D. C., Pierson, B. K., and Summons, R. E., Modern microbial mats: Anoxygenic, transitional, thermal, chemolithotrophic, eukaryotic and terrestrial, in: *The Proterozoic Biosphere: A Multidisciplinary Study* (J. W. Schopf and C. Klein, eds.), Cambridge, University Press, Cambridge (in press).

Webb, K. L., Du Paul, W. D., Wiebe, W., Sottile, W., and Johannes, R. E., 1975, Enewetak (Eniwetok) Atoll: Aspects of the nitrogen cycle on a coral reef, *Limnol. Oceanogr.* **20:**198–210.

Wiebe, W. J., 1985, Nitrogen dynamics on coral reefs, in: *Proceedings of the Fifth International Coral Reef Congress, Tahiti, 1985*, Vol. 3 (C. Gabrie, J. L. Toffart, and B. Salvat, eds.), pp. 401–406, Antenne Museum-Ephe, Moorea, French Polynesia.

Wiebe, W. J., Johannes, R. E., and Webb, K. L., 1975, Nitrogen fixation in a coral reef community, *Science* **188:**257–259.

Wilkinson, C. R., 1979, Nitrogen fixation in coral reef sponges with symbiotic cyanobacteria, *Nature (London)* **279:**527–529.

Wilkinson, C. R., Williams, D. McB., Sammarco, P. W., Hogg, R. W., and Trott, L. A., 1984, Rates of nitrogen fixation on coral reefs across the continental shelf of the central Great Barrier Reef, *Mar. Biol.* **80:**255–262.

Williams, W. D., 1981, Inland salt lakes: An introduction, *Hydrobiologia* **81:**1–14.

Index

Abies albus spp., 407
Absidia corymbifera, 129
Acacia pulchella, 359
Acacia spp., 359
Acetobacter spp., 204
Acholeplasma spp., 244
Achromobacter spp., 204, 398, 403, 406
Acid production, 239–242, 243, 282
Acid rain, 375
Acinetobacter calcoaceticus, 197, 203, 204, 206, 209, 210
Acinetobacter spp., 173, 197, 198, 199, 200, 203, 206, 218, 219
Actinomyces spp., 244
Actinomycetes, 124, 393, 396, 406, 409; *see also specific types*
Activated sludge systems, 1
  microbiology of, 196–201
  nutrients in, 189–196, 199, 201
  phosphorus removal in: *see* Biological phosphorus removal
Adhesion, 232, 275
  to chitin, 411–415
  of lactobacilli, 151–153, 156–157
Aermonas spp., 398
Aerobacter aerogenes, 202, 203, 204, 206; *see also Klebsiella aerogenes*
Aerobiosis, 195
  in corrosion, 233–235
  in microbial mats, 481–482
  in nitrification, 265
  in nitrogen fixation, 317
  in phosphorus removal, 174, 195–196, 219
Aeromonas caviae, 393, 407
Aeromonas hydrophila, 403
Aeromonas punctata, 197

Aeromonas spp., 197, 244
Agrobacterium radiobacter, 372
Agrobacterium spp., 448
Alcaligenes spp., 129
Algae, 6, 20, 65, 86, 87, 123, 124, 463
  in activated sludge systems, 173
  in Antarctic, 73, 80, 83, 84, 85, 86, 91, 93, 96, 100, 102, 103, 105
  chitin in, 389
  cryotolerance of, 119
  in fellfield soil, 121, 122, 123
  as food, 15, 19
  in microbial mats, 113, 471, 481, 489
  nitrogen fixation and, 334
  nutrient cycling and, 126
  in sea ice microbial community, 113, 114, 115, 116, 117
  size of, 39
  sulfate-reducing, 232
  sulfur compounds and, 351, 352–353, 354, 355, 358, 361, 362, 372
  symbionts of, 23
Alteromonas putrefaciens, 246
Aluminum, 240, 243
Aluminum alloys, 232, 241
Amebae, 2, 7, 10, 11, 12, 26, 124; *see also specific types*
  chitin degradation and, 388, 393
Ammonia, 87, 113
  lactobacilli and, 154
  microbial mats and, 481
  nitrification and, 264, 265, 267, 270, 273, 276, 280, 281, 282, 284, 287, 289, 292, 294, 298
  nitrogen fixation and, 305, 326
  protozoa and, 3, 4, 6, 10, 19–20

Ammonia oxidation, 267, 269, 270, 271, 272, 273, 274, 294, *see also* Ammonia-oxidizing bacteria
Ammonia-oxidizing bacteria, 263, 264, 270
Ammonium, 86, 108
  nitrification and, 276, 277–278, 280, 282, 285, 287, 290, 291, 292, 298
  protozoa and, 3, 20
Ammonium oxidation, 291, 292, 293, 298
Amphiprora spp., 114, 115, 116
Anabaena azollae, 307
Anabaenais spp., 438
Anabaena oscillarioides, 312, 313, 314
Anabaena spp., 306, 323, 335, 433
  nitrogen fixation and, 307, 309
Anaerobiosis, 175, 203, 204
  in biological phosphorus removal, 175, 179, 182, 186, 189, 190, 192, 204, 219
  in corrosion, 235–236, 245
  in nitrogen fixation, 317, 333
  protozoa and, 25–28
Anguilla japonica, 416
Anhydrobiosis, 74
Anion/cation transport, 194–196
Antarctic microbiology, 71–132
  in deserts, 78–96
  distinctiveness of, 74–83
  environmental impact and, 127–131
  in ice-covered water bodies, 98–117
  nutrient cycling and, 125–127
  of patterned ground, 117–125
  seasonal changes and, 81–83
Antibiotic resistance, 161, 164
Antibiotics, 67, 160, 417; *see also specific types*
Antibodies, 311
Ants, 418
A/O process, 180
Aphanizomenon flos-aquae, 312
Aphanizomenon spp., 307, 309, 323
Aphanocapsa spp., 315
Aphanomyces astaci, 418
Aquatic habitats, 65; *see also specific types*
  chitin degradation in, 393–398, 411
  nitrification in, 265, 287–289
  nitrogen fixation in: *see* Nitrogen fixation
  protozoa in, 3, 4, 12
Archaebacteria, 38
Arrhenius equation, 281, 283
Arthrobacter atrocyaneus, 203, 204

Arthrobacter spp., 128, 244
  chitin degradation and, 406, 407, 408
  nitrogen fixation and, 308
  sulfur compounds and, 373
Asafetida oil, 361
Aspergillus spp., 241, 406, 407
Asymbionts, 306
Aureobasidium spp., 244
Azomonas spp., 308
Azorhisobium spp., 308
Azospirillium spp., 306, 308
Azotobacter deijerinckii, 206, 209
Azotobacter spp., 306, 308, 317
Azotobacter vinelandii, 203, 205
Azotococcus spp., 308

Bacateroides spp., 155
Bacillariophyceae, 353
Bacilli, 393; *see also specific types*
Bacillus cereus, 212, 403
Bacillus circulans, 408
Bacillus spp., 96, 124, 129, 156, 244
  chitin degradation and, 398, 403, 406, 408
  corrosion and, 250
  nitrogen fixation and, 308
Bacillus subtilis, 162, 163, 439
  diffusion in, 38, 54
Bacillus thuringiensis, 403
Bacteria, 1–2, 108, 128, 129, 130, 233, 234; *see also specific types*
  in activated sludge systems, 173, 196–198, 199–200
  ammonia-oxidizing, 263, 264, 270
  in Antarctic, 71, 72, 73, 74, 85, 94, 96, 98, 100, 102, 111
  categorization of, 127
  cellulose-degrading, 239
  chemolithoautotrophic, 308
  chemotaxis in, 8
  chitin degradation and: *see under* Chitin degradation
  corrosion and: *see under* Microbial corrosion
  denitrifying, 191
  diffusion in: *see* Diffusion
  endosymbiotic: *see* Endosymbionts
  in fellfield soil, 121, 123
  as food, 6, 8, 12, 19, 25
  hydrogen and, 245–248, 251, 252, 321
  iron and, 232, 233, 242, 244, 246

Bacteria (*cont.*)
  manganese and, 233, 242, 244
  membranes of: *see* Membranes
  metabolic control in, 213
  metabolites of, 210–213
  methanogenic: *see* Methanogens
  nitrate-reducing, 243
  nitrifying: *see* Nitrifying bacteria
  nitrite-oxidizing, 243, 263, 264, 270
  nitrogen fixing: *see* Nitrogen-fixing bacteria
  nutrient cycling and, 126
  in phosphorus removal, 174, 181, 184, 189, 199, 209–210
  plasticity of: *see* Microbial plasticity
  polyP: *see* PolyP bacteria
  in sea ice microbial community, 113, 114, 115–116, 117
  size of, 38–39
  sulfate-reducing: *see* Sulfate-reducing bacteria
  sulfur compounds and, 353, 358, 366, 369, 372, 373–374
  thermophilic, 248–250
Bacteriocin, 155–156
Bacteriorhodopsin, 55
Bacterioplankton, 111
Bacteroides fragilis, 441
Bacteroides spp., 129
Bardenpho process, 175, 179, 195
  modified, 179, 185
Barium, 202
Barley, 418
Bass, 417
Bats, 417
Bays, 72, 79
  chitin degradation in, 390, 391, 393, 398, 413
  microbial mats in, 461, 463, 464, 468, 473, 475, 476, 477, 478, 480, 481, 482, 483, 485, 486, 487–488, 490
Bdellovibrio bacteriovorus, 205
Bean root rot, 410
Beans, 411
Beauveria bassiana, 418
Beetles, 418
Beggiatoa spp., 308
Beijerinckia spp., 308
Beneckea spp., 406
Benthic communities, 287, 468, 476, 477, 479, 489

Benthic mats, 100, 106, 109, 111, 472
  nitrogen fixation in, 308
Benthic sediments, 80, 98
Bergey's Manual of Systematic Bacteriology, 149
Best equation, 57, 66
Biddulphia punctata, 117
Biddulphia spp., 115
Bifidobacterium bufidus, 417
Bifidobacterium spp., 148
Biodenipho process, 180–181
Biological phosphorus removal, 173–219
  biochemical model of, 201–218
  design and operational aspects of, 184–189
  ecological implications of, 218–219
  evolution of, 175–184
  microbiology of, 196–201
  nutrient dynamics in, 189–196
  wastewater composition and, 186
Birds, 417; *see also specific types*
Black-body radiation, 77, 81, 115
Blackman's law, 57, 61
Blastobacter spp., 96
Blastocaulis spp., 244
Blepharisma spp., 12
Bordetella bronchiseptica, 445
Bordetella parapertussis, 445
Bordetella spp., 197
Boron, 326
Borrelia hermsii, 437, 438
Botrytis spp., 129
Brachonella spp., 4, 6
Brachysira aponina, 464
Bradyrhizobium japonicum, 447, 448
Bradyrhizobium spp., 308
Branchinecta poppei, 113
Brassica spp., 359
Brevibacterium spp., 96, 204
Brownian motion, 65, 233
Brucella spp., 447
Bryum algens, 105
Budworm, 361

Cadmium, 238
Calcium, 80, 199, 202, 212, 243
Calcium carbonate, 80
Calcium chloride, 93
Calcium sulfate, 80
Callinectes sapidus, 413
Calothrix spp., 307, 308, 309, 335

Calvin cycle, 369
Campylobacter coli, 447
Campylobacter spp., 308
Candida albicans, 388, 389
Candida pintolopesii, 154, 155
Candida spp., 129
Carbon, 24, 66, 68, 178, 186, 202, 283, 294, 319, 372, 391
Carbon cycle, 315, 419
Carbon dioxide, 108, 373
  corrosion and, 243, 245, 249
  diffusion and, 45
  in lakes, 102
  nitrification and, 264
  photosynthesis and, 18
  protozoa and, 19, 20, 21, 23, 24
Carbon dioxide fixation, 20, 24, 86, 108, 119, 270, 368, 369
  microbial mats and, 473, 474, 477, 478
Carbon disulfide (CS2), 346, 349, 350, 351, 358–360, 371–372
Carbon fixation, 101
Carbon flux, 488, 489–490
Carbon metabolism, 206–210
Carbon monoxide, 211, 326
Carbon sulfide, 349–358
Carbonyl sulfide (COS), 346, 349–360, 371–372
Cattle
  chitinases in, 417
  lactobacilli and, 150, 159
Caulobacter crescentus, 447
Caulobacter spp., 96, 244
Caulococcus spp., 244
Cell aggregation, 8, 10, 12, 49, 51, 52, 310, 326, 331
Chaetomium spp., 406
Chasmoendolithic communities, 83
Chemiosmotic theory, 210
Chemolithoautotrophy, 307, 321
Chemotaxis, 8, 10, 65, 233
  defined, 8
Chemotrophy, 490
Chickens
  chitinases in, 417
  lactobacilli and, 150, 154, 155, 157, 159
Chionoecetes tanneria, 397
Chitin; *see also* Chitin degradation
  adhesion of microbes to, 411–415
  annual production of, 390

Chitin (*cont.*)
  digestion of by microbes, 393
  digestion of in animals, 416–418
  fossils of, 389–390, 391
  occurrence of, 388–389
  in soil, 409–411
  structure of, 387–388
Chitinase, 402, 407, 410, 417, 418, 419
  in soil, 408–409
Chitin degradation, 387–419
  in aquatic habitats, 393–398, 411
  bacteria and, 392–395, 405, 406–407, 408, 409, 413, 414, 418–419
  in estuaries, 398–402, 411, 413
  in freshwater, 402–405
  in pathogenesis, 418–419
  pathways of, 391–393
  rates of, 397–398, 400–402, 403–404
  in sediments, 396, 398, 401, 402, 403
  in soil, 405–415
  in symbiosis, 418–419
Chitosan, 392, 396, 408
Chitosanase, 407
Chitosan-degrading bacteria, 407
Chlamydomonas spp., 116
Chlamydomonas ulvaensis, 209
Chloramphenicol resistance, 162
Chlorella fusca, 356, 370, 372
Chlorella spp., 372
Chloride, 80
Chlorobiaceae, 321
Chlorobium spp., 308, 473
Chloroflexus spp., 469, 473
Chlorophyll, 23, 24, 117, 123, 353
Chloroplasts, 14, 24–25
Chromatiaceae, 321
Chromatium spp., 102, 308, 474
Chromatium vinosum, 478
Chromium, 211
Chromobacterium lividum: *see Janthinobacterium lividum*
Chromobacterium spp., 398, 403, 406
Chromobacterium violaceum, 392
Chromosomes, 446–447, 448
Chroococcus spp., 315
Chrysamoeba spp, 353
Chrysophyceae, 353
Ciliates, 2, 4, 6, 7, 12, 13, 14, 24, 25; *see also* specific types
  chitin degradation and, 388, 389, 418

Ciliates (*cont.*)
  cryptomonad-bearing, 21
  microaerophilic, 15–18, 19, 21
  polymorphic, 11–12
  zoochlorellae-bearing, 18–22
Circular symmetry, 41
Cirrhosis, 367
Citrobacter spp., 197, 244
Cladobotryum spp., 418
Cladosporium herbarum, 241
Cladosporium resinae: see *Hormoconis resinae*
Clams, 397
Clonothrix spp., 244
Clostridia, 160, 393
Clostridium acetobutylicum, 247
Clostridium bryantii, 251
Clostridium limosum, 247
Clostridium perfringens, 441
Clostridium ramosum, 156
Clostridium spp., 129, 156
  chitin degradation and, 392, 395, 399, 414
  corrosion and, 239, 245
  nitrogen fixation and, 308
  sulfur compounds and, 353
Coal, 361, 362, 372, 374
Cobalt, 199, 321
Cod, 416
Codium spp., 307
Coffee, 362
Coliforms, 129, 130, 155; *see also specific types*
Colimiting transport, 55–57
Collozoum spp., 23
Colobanthus quitensis, 82
Colpoda spp., 11, 12
Comeau model, 214, 215
Conduction of Heat in Solids, 50
Coniothyrium spp., 244
Conjugation, 161, 162
Conservation, 130–131
Consortia, 240, 243
  corrosion and, 251–252
  diffusion and, 68
  nitrogen fixation and, 311, 319
  sulfur compounds and, 367
Contamination, 127, 128, 130, 242; *see also* Pollution
Continental drift, 74–75
Continuous-flow systems, 192
Copper, 199, 211, 237, 250, 272, 326

Copper alloys, 232, 244
Coprinus cinereus, 389
Coral reefs, 461, 463, 467, 470, 481, 489
Corals, 23, 322, 479
Corn, 359, 409, 418
Corophium volutator, 390, 402
Corrosion: see Microbial corrosion
Corynebacterium bovis, 212
Corynebacterium diphtheriae, 441
Corynebacterium spp., 204, 308, 398
Corynebacterium xerosis, 203, 204
Coryneforms, 96; *see also specific types*
Corythion dubium, 124
COS: *see* Carbonyl sulfide
Coscinodiscus spp., 115
Cotton, 409
Crabs, 397, 413, 418
Crayfish, 403, 418
Crenothrix spp., 243, 244
Crocuta crocuta, 366
Cryotolerance, 74, 90–96, 116
  in a wet habitat, 117–119
Cryptic functions, 444–446
Cryptococcus spp., 124, 129, 244
Cryptococcus vishniacii, 93, 94
Cryptomonad-bearing ciliates, 21
Cryptomonads, 353; *see also specific types*
Cryptomonas spp., 353
CS2: *see* Carbon disulfide
Cucurbita pepo, 410
Cuttlefish, 406
Cyanobacteria, 124
  in Antarctic, 71, 73, 74, 80, 83, 84, 86, 87, 91
  cryotolerance of, 119
  in fellfield soil, 121, 122, 123
  in lakes, 100, 104, 105, 106, 109
  nitrogen fixing: *see* Nitrogen-fixing cyanobacteria
  nutrient cycling and, 126
  in sea ice microbial community, 115
  sulfur compounds and, 352, 353
Cyanobacterial mats, 107, 108, 111, 113, 461, 468–469, 488, 489, 490
  in Antarctic, 94, 98, 100, 105
  nitrogen fixation and, 322, 329
  phototrophic activity in: *see under* Phototrophy
  sulfate reduction and, 482, 483, 484, 485, 486

Cycads, 307; *see also specific types*
Cyclidium spp., 6
Cylindrospermum spp., 308, 309
Cytophaga aquatilis, 398
Cytophaga johnsonae, 392, 393, 403, 407,
    414
Cytophaga spp., 244, 394, 398, 403, 406

DCMU: *see* 3-(3,4-Dichlorophenyl)-1,1-
    dimethylurea
DEDS: *see* Diethyldisulfide
Dehydration, 7, 12
Deinococcus spp., 88, 96
Deletion, 437–438
Denitrification, 175, 177, 178, 180, 182, 185,
    186, 195, 199, 200, 278, 283, 289, 299,
    374
    oxygen and, 183–184
Dental caries, 165
Derxia spp., 308
DES: *see* Diethylsulfide
Deschampsia antarctica, 82
Deserts, 73, 74, 78–96, 104, 105, 108, 117,
    123, 128, 129, 131, 284
    endolithic communities in, 83–89
    hot, 78, 83, 86
Desulfobacter postgatei, 245
Desulfobulbus propionicus, 245
Desulfonema magnum, 245
Desulfosarcina variabilis, 245
Desulfotomaculum acetoxidans, 245
Desulfovibrio desulfuricans, 193
Desulfovibrio sapovorans, 245
Desulfovibrio spp., 102, 245, 308, 373
Desulfovibrio thermophilus, 249
Desulfovibrio vulgaris, 236
Desulfuromonas acetoxidans, 245
Devanathan cell, 247
Diatoms, 114, 115, 116, 117, 307, 322; *see
    also specific types*
    chitin degradation and, 388, 395, 412
    sulfur compounds in, 354
3-(3,4-Dichlorophenyl)-1,1-dimethylurea
    (DCMU), 20, 23, 473, 474, 476, 477,
    481
Dictyostelium discoideum, 7–8
Dictyostelium spp., 7–11, 12, 28
Didinium spp., 6, 12
Diethyldisulfide (DEDS), 369
Diethylsulfide (DES), 369

Diffusion, 37–68
    dynamics of, 46–47
    growth limitations and, 47–49
    kinetics of, 50–57
    mathematics of, 49–50
    nutrient uptake systems and, 62–67
    principle of, 43
    through gels, 52, 57–59
    through obstacles, 43–46
    within two-dimensional ecosystems, 37, 60–
        61
Dileptus spp., 6
Dimethyl disulfide (DMDS), 346, 349, 350,
    351, 355–358, 359, 361, 363, 367, 369
Dimethylphosphorodithioate (DMPT), 374
Dimethyl sulfide (DMS), 346, 349, 350, 351–
    355, 356–357, 358, 359, 360, 361, 367,
    368, 369, 374, 375
Dimethyl sulfone, 363
Dimethylsulfonium propionate (DMSP), 352–
    353, 355–358, 360, 368
Dimethyl sulfoxide (DMSO), 346, 353–354,
    358, 360, 369
Dinophyceae, 353
Disease control in plants, 410–411
Disease resistance, 148, 158, 159–160
Dissolved oxygen (DO), 176, 179, 184, 188,
    190
Disulfides, 362, 366, 367; *see also specific
    types*
DMDS: *see* Dimethyl disulfide
DMPT: *see* Dimethylphosphorodithioate
DMS: *see* Dimethyl sulfide
DMSO: *see* Dimethyl sulfoxide
DMSP: *see* Dimethylsulfonium propionate
DNA, 451, 452
    of lactobacilli, 149, 161, 162, 163, 164
    nitrogen fixation and, 307
    plasmid, 447–449
    in polyP bacteria, 202
    rearrangement of, 434, 437, 438–439, 440,
        441, 446, 447
DNA synthesis, 101, 108
Dogs, 417
Drosophila spp., 404
Drought, 73
Drug resistance, 440
Dry Valleys Drilling Project (DVDP), 127–
    129
Dunaliella spp., 103, 116
DVDP: *see* Dry Valleys Drilling Project

Earworm, 361
Ectosymbionts, 28
Ectothiorhodospira spp., 308
Eels, 416
Elodea spp., 335
Emericella spp., 406
Emphysema, acute bovine pulmonary, 159
Endocarditis, 165
Endolithic communities, 78, 83–89, 86, 128
Endolithic habitats, 113, 131
Endosymbionts, 26, 27, 28, 322, 334
Enterobacteriaceae, 237, 416, 442, 448
Enterobacter spp., 244, 308
Enterococci: *see specific types*
Enterococcus faecalis, 155, 156, 162, 164
Entomoneis spp., 114
Entophysalis major, 464, 469, 474
Entophysalis spp., 474, 477, 484, 487
Enzymatic biosynthesis and degradation, 203–206
Enzymes, 19–20, 102, 162, 213, 418, 444;
  see also specific types
  ammonia monooxygenase, 275
  catabolic, 442
  chitinolytic, 416
  corrosion and, 250
  digestive, 419
  extracellular, 153, 239
  food-processing, 417
  lytic, 410
  Monod, 57
  peripheral, 55
  periplasmic, 50, 54–55
  in phosphorus removal, 203–209, 213
  proteolytic, 156
  sulfur compounds and, 371
Erythromycin resistance, 161, 162, 163
Escherichia coli
  chitin degradation and, 413
  diffusion in, 38, 48, 66, 67
  lactobacilli and, 155, 159, 162
  in phosphorus removal, 203, 204, 205, 206, 210, 211, 212
  plasticity of, 434, 436, 438–439, 440, 441, 442, 443, 444, 447, 448, 450, 451
Escherichia spp., 156, 308
Estuaries
  chitin degradation in, 391, 398–402, 411, 413
  microbial mats in, 461
  nitrogen fixation in, 305, 316, 335

Eubacteria, 38
  nitrogen fixing: *see* Nitrogen-fixing bacteria
Euphausia superba, 390
Euplotes daidaleos, 18
Euplotes spp., 13
Eutrophication, 109, 130, 173, 265
  control of, 218
Exopolymer–metal interactions, 233, 236–239
Exopolymers, 232, 243, 250
Exosymbiosis, 322

Fellfield soil, 96, 111
  cryotolerance and, 117, 119
  defined, 82
  microbial stabilization of, 119–123
Fermentation, 247, 482–486
Ferns, 322
Ferrets, 366
Ferrobacillus spp., 244
Ferrous hydrate, 236
Ferrous hydroxide, 235
Fickian diffusion barrier, 66
Fick's diffusion law, 39–40, 44, 53, 55, 57
First-order kinetics, 278, 281, 283
Fischerella spp., 309
Fish, 18, 416–417; *see also specific types*
Fisherellia spp., 308
Flavobacterium spp., 197, 244, 373, 374
  chitin degradation and, 398, 403, 406, 407
Flow regime, 185
Food: *see* Nutrients
Food chains, 24, 50, 74, 78, 107, 109, 113
Foraminifera, 1, 22–25; *see also specific types*
Formaldehyde, 368, 369
Fossils, 1, 88, 104, 322, 461, 463, 468
  chitin, 389–390, 391
Fowl, 150, 154, 157, 158; *see also specific types*
Foxes, 364, 417
Fragilaria spp., 115
Fragilariopsis sublinearis, 116
Freshwater habitats; *see also* Aquatic habitats; *specific types*
  chitin degradation in, 402–405, 415
  nitrogen fixation in, 334–336
  protozoa in, 22, 27, 28
Freshwater lakes, 77, 98, 105, 107
  chitin degradation in, 403, 404
  ice-covered, 108–113
  nitrogen fixation in, 323
  protozoa in, 15, 25

Frontonia vernalis, 18
Fumerate, 236
Fungi, 108, 124; *see also specific types*
  in Antarctic, 71, 83, 86, 93, 94
  chitin degradation and, 388, 389, 393, 402,
    404, 405, 406, 407, 408, 409, 410, 411,
    415, 416, 418
  corrosion and, 240–242
  diffusion in, 46
  pathogenic, 128
  size of, 39
  sulfate-reducing, 232
  sulfur compounds and, 359, 374
Fusarium oxysporum spp., 408
Fusarium solani, 410
Fusarium spp., 241, 410, 411
Fusarium udum, 410
Fusobacterium spp., 155

Gadus morhua, 416
Gallionella spp., 242, 243, 244
Gamasellus racovitzai, 74, 82–83, 126
Garlic, 361
Gastrointestinal tract bacteria: *see* Lactobacilli
Gaussian distribution, 46
Gels, 52, 57–59
Genes, 445, 446
Genetic manipulation, 434, 437, 438, 440,
    444
  of lactobacilli, 163, 165
Geotaxis, 15
Glaucoma spp., 4, 6
Gliomastix spp., 407
Globigerina spp., 22
Globigerinoides saculifer, 22
Gloeocapsa spp., 87, 108, 309, 311
Gloeothece spp., 315
Gloeotrichia spp., 308, 309, 323, 335
Gompertz function, 280
Gravity, 2, 15, 16, 17, 38
Gunnera spp., 307
Gyrodinium spp., 353

Haemophilus influenzae, 439
Haldane equation, 296
Halitosis, 367–368
Halobacterium halobium, 441
Halobacterium spp., 102, 446, 449
Halobacterium volcanii, 441
Hamsters, 364

Hapalosiphon spp., 309
HDO: *see* High dissolved oxygen
Heavy metal cations, 397
Heavy metals, 394
Hedgehogs, 417
High dissolved oxygen (HDO), 98, 100, 106
Hormoconis resinae, 240, 241
Humicola fuscoatra, 410
Humicola spp., 406, 407
Hyaena brunnea, 366
Hydra spp., 20
Hydra viridis, 19, 20
Hydrogen, 210, 211, 212, 247, 251
Hydrogen bacteria, 245–248, 251, 252, 321
Hydrogenomonas spp., 202
Hydrogenosomes, 26; *see also specific types*
Hydrogen sulfide, 346, 355–358, 359, 367
  corrosion and, 236, 240
  protozoa and, 4, 6
Hyenas, 364, 366
Hymenomonas spp., 353
Hyphomicrobium spp., 243, 244

Ice habitats, 75–78
Immune system, 52
Insects, 119; *see also specific types*
  chitin degradation and, 388, 418, 419
Insertion, 438
Interorganismal exchange, 433–434
Interstitial zone, 14–15
Inversion, 434, 452
Iron, 199, 235, 245, 319
  chitin degradation and, 394
  deposition of, 242–244
  ferric, 240
  ferrous, 240, 346
  nitrogen fixation and, 324–325, 327, 328,
    329
Iron bacteria, 232, 233, 242, 244, 246
Iron sulfides, 356
Iron transport, 211
Iron tubercles, 242–243

Janthinobacterium lividum, 111, 118, 121,
    403

Karlingia asterocysta, 393
Kinetics
  of biological phosphorus removal, 200
  chemostat, 270

Kinetics (*cont.*)
  of diffusion, 50–57
  first-order, 278, 281, 283
  Michaelis–Menten, 269, 273, 278, 291
  multiplicative, 290
  of nitrification: *see* Nitrification kinetics
  reaction rate, 298
  threshold, 290
  of viral absorption, 42
  zero-order, 278, 281, 283, 285
Klebsiella aerogenes, 442, 443
Klebsiella spp., 124, 197, 308
Krebs cycle, 374
Krill, 126, 130
K strategies, 200, 201
Kuznetsovia spp., 244

Lactobacilli, 147–165; *see also specific types*
  adhesion mechanisms of, 151–153
  cell walls of, 151–153
  colonization of, 156–158
  hosts of, 149–150, 151, 154, 158–160
  interaction with other microbes, 151, 155–156
  metabolic properties of, 151, 153–155
  plasmids in, 155, 161–162, 164
Lactobacillus acidophilus, 147, 149, 154, 156, 161, 162
Lactobacillus bulgaricus, 156
Lactobacillus casei, 161
Lactobacillus confusus, 151
Lactobacillus delbrueckii, 147, 154, 156, 159, 239
Lactobacillus fermentum, 149, 152, 154, 155, 156, 161, 163
Lactobacillus gasseri, 149, 154, 161
Lactobacillus helveticus, 156, 161
Lactobacillus lactis, 156
Lactobacillus plantarum, 162, 164
Lactobacillus reuteri, 149, 154, 161, 163
Lactobacillus salivarius, 154
Lactobacillus spp., 129, 156, 445
Lactose intolerance, 158
Lakes, 47, 73, 79, 80, 109, 111, 117, 121, 127, 129, 130; *see also specific types*
  chitin degradation in, 390, 402–405, 415
  eutrophic, 109
  freshwater: *see* Freshwater lakes
  hypersaline, 78, 116, 351, 488
  ice-covered, 77, 98–107, 108–113, 131

Lakes (*cont.*)
  microbial mats in, 461, 464, 467, 469, 471, 475, 480, 483, 488, 490
  nitrogen fixation in, 305, 323–324, 325, 333, 334, 335
  protozoa in, 6, 14, 15, 22, 25
  saline: *see* Saline lakes
  sulfur compounds in, 351, 358
Lamprothamnium papulosum, 479
Lateolabrox japonicus, 417
Lembadion spp., 13
Lentinus spp., 361
Leptospira spp., 447
Leptospirillum spp., 244
Leptothrix spp., 243, 244
Lichens, 83, 84, 85, 86, 87, 101, 128
  cryotolerance of, 117, 119
Lieskeella spp., 244
Light, 87, 98, 128
  in lakes, 101, 102, 104, 106
  microbial mats and, 471–473, 477, 480–481, 490
  nitrogen fixation and, 309, 310
  protozoa and, 10, 11, 15, 17, 18, 19, 20
  sulfur compounds and, 360
Linear flow, 40, 43
Lipoic acid, 348, 362
Loligo edulis, 418
Loxodes spp., 15–18, 19
Luxury uptake, 177, 178
Lyngbya aestuarii, 311, 313, 464, 476
Lyngbya marensiana, 105
Lyngbya spp., 123–124, 309, 324, 477, 480, 487
Lysobacter spp., 393, 407

Macroalgae, 322, 336; *see also specific types*
Macrophytes, 173, 335, 336, 479; *see also specific types*
Maggots, 418
Magnesium, 243
  in activated sludge systems, 199, 202, 204, 211, 218
Magnesium chloride, 80
Malabsorption syndromes, 165
Malbranchea aurantiaca, 410
Malustella spp., 406
Manganese, 202, 211, 243
  chitin degradation and, 394, 397

Manganese (*cont.*)
  deposition of, 242–244
  nitrogen fixation and, 326
Manganese bacteria, 233, 242, 244
Marine habitats; *see also* Aquatic habitats; *specific types*
  nitrogen fixation in, 334–336
  sulfur compounds in, 355–358
Marine sands, 14–15
Mars, 79, 88–89, 131; *see also* Viking Mars program
Marshes, 328
  sulfur compounds in, 347, 352, 355–358, 359, 360
Mastigocladus laminosus, 124
Mastigocladus spp., 308, 309
Mastogloia spp., 464
Medicago spp., 359
Meloidogyne incognita, 411
Meloidogyne javanica, 411
Meloidogyne spp., 410
Membranes
  cytoplasmic, 38, 50, 55, 66, 67
  outer, 37, 38, 50, 52–54
Mephitinae, 366
Mephitis mephitis, 366
3-Mercaptopropionate, 355–358
Mesodinium rubrum: *see Myrionecta rubrum*
Metabolite transport, 210–213
Metal cations, 411
Metal–exopolymer interactions, 233, 236–239
Metallogenium spp., 244
Metal–polymer interactions, 234
Metals, 322; *see also* Heavy metals; *specific types*
Metal sulfides, 356, 463; *see also specific types*
Metarrhizium anisophiae, 418
Metazoa, 107; *see also specific types*
Methane, 28, 355–358
Methane sulfonic acid (MSA), 354, 355, 361, 372, 375
Methanethiol (MT), 346, 349, 350, 351, 355, 356, 361, 367, 368, 369
Methanobacterium formicicum, 26
Methanobacterium spp., 250
Methanogenesis, 102, 111, 125, 126, 130; *see also* Methanogens
  corrosion and, 236, 250
  in microbial mats, 486

Methanogens, 102, 246, 321; *see also* Methanogenesis
  chitin degradation and, 399
  corrosion and, 246, 251
  in lakes, 113
  microbial mats and, 486
  protozoa and, 25–28
  sulfur compounds and, 356, 357, 371
  symbiotic, 25–28
  thermophilic, 250
Methanol, 368
Methanoplanus endosymbiosus, 26
Methanosarcina barkeri, 246
Methanthermus spp., 250
Methionine, 354, 360, 361, 367
Methylamine, 368
Methylated sulfides, 346, 349–358, 367–369
Methylobacter spp., 308
Methylococcus spp., 308
Methylocystis spp., 308
Methylosinus spp., 308
Methyl sulfate, 374
Methyl sulfide, 363–364
Metopus contortus, 26
Metopus striatus, 26
Mice
  chitinases in, 417
  lactobacilli in, 150, 154, 155, 157, 159–160, 161
Michaelis–Menten relationship, 57, 61, 269, 273, 278, 280, 287, 291
Microaerophily, 13–21
Microalgae, 71, 74, 113, 336, 469, 471, 473
Microbial corrosion, 231–252
  acid production and, 239–242
  consortia in, 251–252
  hydrogen bacteria and, 245–248
  surface microbiota in, 232–239
  thermophilic processes in, 248–250
Microbial mats, 105–107, 113, 461–491; *see also specific types*
  benthic: *see* Benthic mats
  characteristics of, 463–467
  colloform, 471, 472, 481
  cyanobacterial: *see* Cyanobacterial mats
  degradative processes in, 481–486
  diatomaceous, 461, 469, 471, 472, 473, 475, 478
  diffusion within, 37, 60, 68
  environmental constraints on, 471–486

Microbial mats (*cont.*)
  ice cover, 105
  lift-off, 105, 107
  light and, 471–473, 477, 480–481, 490
  moat, 105, 107
  nitrogen fixation in, 308, 315, 322, 331, 480–481, 488
  occurrence of, 463–467
  phototrophic activity in, 471–481, 490
  pinnacle, 105, 107
  prostrate, 105, 107
  pustular, 471, 480, 481
  smooth, 464, 471, 477, 480, 481, 486
  stromatolitic, 105
  tufted, 464, 471, 476, 477, 480, 481, 486, 487
Microbial plasticity, 431–452
  evidence for, 446–449
  interorganismal exchange and, 433–434
  phenotypic change and, 449–451
  programmed rearrangement and, 434–438
  significance of, 451–452
  unprogrammed rearrangement and, 438–446
Micrococci, 398; *see also specific types*
Micrococcus denitrificans, 202
Micrococcus lysodeikticus, 205
Micrococcus spp., 88, 96, 129, 204, 244
Microcoleus chthonoplastes, 474, 477
Microcoleus spp., 309, 310, 335
  in microbial mats, 472, 480, 487
Microcystis spp., 324
Microfungi, 74, 96, 123; *see also specific types*
Microchaete spp., 309
Micromonospora spp., 403, 406
Microthrix parvicella, 188, 189
Microzones, 327–333, 334, 336
Milk, 147, 148, 156, 158
Mima spp., 174
Minerals, 85, 91, 242, 276; *see also specific types*; Trace minerals
  in activated sludge systems, 199
  microbial mats and, 487–488
Mink, 366
Mites, 82, 119, 126; *see also specific types*
Modeling, 487–488
Molds, 7–11, 124, 127, 128, 129; *see also specific types*
Moles, 417
Molluscs, 389, 395, 396, 397

Molybdate, 325
Molybdenum, 319, 321, 325, 326, 327, 328, 329
Monod equation, 57, 61, 269, 271, 281, 287, 292, 296
Moraxella spp., 174, 197, 397, 398, 403
Mortierella spp., 406, 407, 408, 410
Mosquitoes, 361, 362
Mosses, 81, 82, 117, 123, 125, 128
Moss–peat ecosystems, 73, 125, 130
Mountains, 72, 78, 123, 128
MSA: *see* Methane sulfonic acid
MT: *see* Methanethiol
Mucor spp., 129, 408
Mushrooms, 7, 39
Mussels, 417
Mustela vison, 366
Mustelidae, 366
Mutations, 55, 63, 65, 444, 450, 452
  of lactobacilli, 153, 164
Mycobacterium phlei, 205
Mycobacterium smegmatis, 204
Mycobacterium spp., 129, 202
Mycoplasma gallosepticum, 447
Mycoplasma ovipneumoniae, 447
Mycoplasma spp., 449
Myrionecta rubrum, 21, 24
Myrionecta spp., 24
Myriophyllum spp., 336
Mytilus edulis, 397
Myxobacteria, 408; *see also specific types*
Myxobacter spp., 408
Myxococcus spp., 46

Nadsoniella nigra, 119
Naumaniella spp., 244
Nautilus spp., 390
Neem oil, 361
Neisseria gonorrhoeae, 203, 437, 438, 445, 446
Neisseria meningitidis, 445
Nematodes, 82, 93, 107, 113, 359; *see also specific types*
  chitin degradation and, 389, 406, 410, 411, 418
Neurospora crassa, 389
New Yorker phenomenon, 66
Nickel, 211, 250
Nitrate, 87, 94, 108, 113, 126
  corrosion and, 236

Nitrate (*cont.*)
  microbial mats and, 481
  nitrification and, 264, 265, 267, 278, 280,
    285, 287, 289
  nitrogen fixation and, 325
  in phosphorus removal, 176, 177, 178,
    179–180, 182–184, 185, 186, 187, 189,
    190–193, 199
  protozoa and, 15, 24
Nitrate-reducing bacteria, 243
Nitrification, 263–300; *see also* Nitrifying
    bacteria
  in aquatic habitats, 287–289
  in batch culture, 267–269, 291, 297
  chitin degradation and, 409
  defined, 264
  described, 264–265
  inhibition of, 264, 266, 267, 273–275, 296
  kinetics of: *see* Nitrification kinetics
  pH and, 264, 265, 266, 267, 273–275, 276,
    280, 282, 283, 284, 287, 293, 294–296
  phosphorus removal and, 175, 177, 178,
    179, 180, 183, 185, 187, 195, 200
  pure-culture studies of, 266–277, 281, 291
  in sewage treatment, 289–299
  in soil, 265, 277–287
  temperature and, 264, 265, 267, 272–273,
    282, 283, 284, 287, 294–296
Nitrification kinetics, 264, 266, 267, 271,
    276, 283, 287, 289, 290, 299
Nitrifying bacteria, 203, 263, 281, 292, 296,
    297, 300; *see also* Nitrification; *specific*
    types
  in aquatic environment, 287
  in batch culture, 267–269, 271, 272, 275,
    276, 278
  in continuous culture, 270–272
  in lakes, 100
  pH effects on, 274
  in sewage, 290
  in soil, 286
  substrate concentration effects on, 266, 267,
    268, 269–270
  sulfur compounds and, 359
  surface growth of, 275–277
Nitrite, 95, 113
  nitrification and, 264, 267, 278, 297, 298
Nitrite oxidation, 270, 271, 272, 273, 274,
    275, 294, 298
Nitrite-oxidizing bacteria, 243, 263, 264, 270

Nitrobacter spp., 271, 274, 275, 276, 291,
    292, 293
Nitrobacter winogradskyi, 267, 274
Nitrogen, 22, 68, 81, 87, 109, 111, 126, 127,
    199, 372
  chitin degradation and, 391, 402
  deficiency of, 305, 306, 334
  in lakes, 100, 104
  in microbial mats, 106
  nitrification and, 264, 265, 284
  protozoa and, 20, 22
Nitrogenase, 310, 311, 316, 319, 320, 321,
    325, 328, 334
Nitrogen cycle, 87, 94, 125, 130, 264, 315,
    419
Nitrogen fixation, 73, 87, 106, 113, 124,
    305–337, 433; *see also* Nitrogen-fixing
    bacteria
  differences in freshwater and marine hab-
    itats, 334–336
  environmental constraints on, 323–327
  evolutionary and ecological considerations
    in, 333–334
  microbial mats and, 308, 315, 322, 331,
    480–481, 488
  microzone formation and, 33, 327–333, 334
  organic matter in, 327–333
  physiological ecology of, 319–322
Nitrogen-fixing bacteria, 306, 307–319, 321–
    322, 325, 328, 335, 336; *see also* Nitro-
    gen fixation; *specific types*
Nitrogen-fixing cyanobacteria, 306, 307–315,
    322, 323, 324, 326–327, 328, 330, 331,
    334, 335, 336; *see also* Nitrogen fixation;
    specific types
Nitrogen limitation, 353
Nitrogen removal, 175, 178, 179, 180, 185
  single-sludge systems for, 176–177
Nitrogen uptake, 285
Nitrosococcus spp., 274
Nitrosolobus spp., 268
Nitrosomonas europaea, 203, 268, 270, 271,
    272, 273, 274, 280
Nitrosomonas spp., 269, 271, 274, 276, 292,
    293
Nitrosospira spp., 268, 271
Nitzschia spp., 115
Nocardia erythropolis, 188
Nocardia minima, 205
Nocardia spp., 205, 244, 403, 406

Nocardioforms, 403; *see also specific types*
Nodularia spp., 308, 309, 323
Nostoc commune, 119
Nostoc spp., 91, 108, 126
  in microbial mats, 108
  nitrogen fixation and, 306, 307, 308, 309, 323, 335
Nosularia spp., 335
Nummulites gizehensis, 1
Nutrient cycling, 71, 72, 113, 125–127, 263, 315
Nutrient deficiency, 1–2, 24–25; *see also* Starvation
Nutrients, 82, 232, 233, 322
  in activated sludge systems, 189–196, 199, 201
  chitin as, 392, 393
  diffusion and, 47, 50
  for lactobacilli, 153, 155
  in lakes, 109
  microbial mats and, 489
  nitrogen fixation and, 336
  for protozoa, 2, 6, 7–13, 15, 18, 19, 22–23
  in sea ice microbial community, 116
Nutrient uptake systems, 62–67

Oceanospirillum spp., 244
Oceans, 74, 76, 94, 113, 114, 116, 117, 131
  chitin degradation in, 390, 394, 395
  microbial mats in, 489
  nitrogen fixation in, 305, 311, 316, 325, 326, 327, 329, 333, 335, 337
  protozoa in, 22, 23, 25
  sulfur compounds in, 347, 351, 353, 356, 360
Ochrobium spp., 244
Ochromonas spp., 353
Octopus, 418
Odors, 363–368
Ogsten theory, 58–59
Oil, 73, 130, 361, 362, 370, 372, 463
Oil production systems, 246
Oil spills, 130
Onion, 361
Organic sulfur compounds, 345–375; *see also* specific types
  aromatic, 362–363, 372–374
  effects of on ecology, 374–375
  heterocyclic, 362–363

Organic sulfur compounds (*cont.*)
  microbiological degradation of, 368–375
  in the natural environment, 348–368
Oscillatoria boryana, 477
Oscillatoria spp., 94, 307, 309, 324, 331, 335
  in microbial mats, 464
Oxygen; *see also* Dissolved oxygen; High dissolved oxygen
  corrosion and, 231, 233, 234, 243
  denitrification and, 183–184
  diffusion and, 45, 60
  lactobacilli and, 154
  in lakes, 98, 100, 105, 106, 107, 111
  nitrification and, 275, 280, 291, 297, 298
  nitrogen fixation and, 309, 311, 317, 320, 322, 328, 330, 331, 332, 333, 334, 336
  protozoa and, 2, 4, 6, 14, 15, 17, 18, 19, 20, 22, 25, 28
  in sea ice microbial community, 117
Oysters, 417

Paecilomyces lilacinus, 410, 411
Paecilomyces spp., 407
Pagrus major, 416
Panaeus setiforus, 418
Papilospora spp., 244
PAR: *see* Photosynthetically active radiation
Parabroteas sarsi, 113
Paracoccus denitrificans, 212
Paramecia, 18; *see also specific types*
Paramecium bursaria, 18, 19, 20
Pasteurella spp., 197
Pathogenesis, 418–419
Patterned ground, 117–125
Peat, 82, 130, 362; *see also* Moss–peat ecosystem
Pedomicrobium spp., 244
Pelomyxa palustris, 26
Peloploca spp., 244
Penguins, 81, 109, 125, 126
Penicillin, 155
Penicillium spp., 94, 129
  chitin degradation and, 406, 407
  corrosion and, 241
  sulfur compounds and, 374
Peptidoglycan, 37, 38, 50, 54, 151
Periplasmic space, 50
Perodicticus potto, 417
Petroleum, 487–488

pH
  chitin degradation and, 407, 410, 413
  corrosion and, 240, 243
  lactobacilli and, 154, 156, 158
  microbial mats and, 467
  nitrification and, 264, 265, 266, 267, 273–
    275, 276, 280, 282, 283, 284, 287, 293,
    294–296
  phosphorus removal and, 212, 213
  protozoa and, 3, 4, 20
Phaeocystis pouchetii, 115
Phaeocystis spp., 115, 126, 353
Phanerochaete chrysosporium, 374
PHB: *see* Polyhydroxybutyrate
Phenotypic change, 449–451
Pheromones, 363–368
Phialophora dermatidis, 128
Phialophora gougerotii, 128
Phoma spp., 129
Phoredox configuration, 179, 180
Phormidium frigidum, 105, 107
Phormidium hendersonii, 471
Phormidium spp., 91, 108, 111, 121, 123,
    309, 330
Phosphates, 108, 126
  microbial mats and, 479, 480, 488, 489
  nitrogen fixation and, 321
Phosphorus, 22, 68, 81, 109, 111, 127; *see
    also* Biological phosphorus removal
  in fellfield soil, 123
  in lakes, 104
  nitrogen fixation and, 321, 322, 324, 327,
    328, 329, 334, 335, 336, 337
  protozoa and, 22
Phosphorus cycling, 315
Phosphorus fixation, 24
Phosphorus limitation, 323
Phosphorus release, 189, 190, 194–196, 201
Phosphorus uptake, 189, 192, 194–196, 201
  in *Escherichia coli*, 201
Phostrip process, 178, 185
Photoadaptation, 101, 104, 116, 131
Photoassimilation, 476
Photoautotrophy, 121, 473–476
Photobacterium spp., 393, 397, 398, 416, 418
Photoheterotrophy, 476–477, 489–490
Photolithoautotrophy, 307
Photosynthesis, 18, 20, 23, 67–68, 108, 119
  in Antarctic, 84, 86, 88, 98, 101, 107
  in microbial mats, 463, 474, 476, 484
  nitrogen fixation and, 309, 311, 321

Photosynthetically active radiation (PAR), 77,
    80, 86, 101, 106, 107, 109, 116
Photosynthetic rates, 19, 22, 23, 24, 490
Phototaxis, 10, 12
Phototrophy, 471, 471–481, 472, 473, 474,
    475, 477, 478, 479, 480, 481, 490
Phytoplankton, 22, 71, 74, 101, 109, 111,
    113, 116, 126, 130, 353; *see also specific
    types*
  nitrogen fixation and, 324, 325
Pigs
  chitinases in, 417
  lactobacilli in, 150, 154, 157, 158, 159
Pinna nobilis, 397
Pit system, 212
Plagiopyla nasuta, 26
Planctomyces spp., 244
Plankton, 2, 98, 100, 476; *see also specific
    types*
Plants, 65, 78, 82
  chitin degradation and, 402, 406
  disease control in, 410–411
  nitrogen fixation and, 306
  sulfur compounds and, 347, 352, 359, 361,
    371
Plasmid DNA, 447–449
Plasmids, 131, 164, 431, 439
  in lactobacilli, 155, 161–162, 164
  linear, 437
Plectonema boryanum, 202
Plectonema spp., 309
pmf: *see* Proton motive force
Pollution, 4, 6, 128, 130, 131, 173, 265, 372;
    see also Contamination
Polygonum cuspidatum, 284
Polyhydroxybutyrate (PHB), 174, 201, 206–
    209, 213, 214, 218, 219
Polymer–metal interactions, 234
Polymorphism, 7–13
PolyP, 174, 201, 213, 218, 219; *see also*
    PolyP bacteria
  degradation of, 217
  metabolism of, 202–206
PolyP bacteria, 173, 190, 194, 195, 196, 198,
    199, 202, 213, 219; *see also specific
    types*
Polyphosphates, 174
Polypus vulgaris, 418
Polysaccharides, 114, 237, 402; *see also spe-
    cific types*
  extracellular, 209–210, 232, 450

Polysiphonia fastigiata, 352
Polysulfur compounds, 361–362
Ponds, 15, 18, 28, 78, 121
Potassium, 80, 199, 202, 210, 211, 218
  nitrification and, 267, 270, 271, 274, 278, 280, 290, 291, 292, 294, 296
Potassium chloride, 80
Potassium transport, 210
Poultry: *see* Fowl
Prasiola crispa, 119
Procambarus versutus, 403
Propionibacterium freudenreichii, 205
Propionibacterium shermanii, 203, 204, 205
Propyl sulfide, 363
Prorocentrum spp., 353
Prosthecobacter spp., 116
Proteus spp., 156
Proton motive force (pmf), 210
Protozoa, 1–29, 128; *see also specific types*
  in Antarctic, 82, 107, 113
  basic requirements of, 2–7
  chitin degradation and, 395, 405, 406
  interstitial, 14–15
  microaerophily in, 13–21
  nitrogen fixation and, 322
  polymorphism in, 13
  size of, 39
  in soil, 7–12, 28
Providencia spp., 441
Prymnesiophyceae, 353
Pseudeurotium ovale, 410
Pseudoboeckella gaini, 113
Pseudomonads, 232, 373, 393, 398, 403; *see also specific types*
Pseudomonas aeruginosa, 209, 312, 313, 441, 442, 443, 449, 450
Pseudomonas atlantica, 450
Pseudomonas cepacia, 441, 443, 445, 447
Pseudomonas chitinovorans, 393, 407
Pseudomonas fluorescens, 413
Pseudomonas putida, 447, 448, 451
Pseudomonas spp., 66, 96, 129, 445, 447, 449
  in activated sludge systems, 197
  chitin degradation and, 397, 398, 403, 406, 407, 413
  corrosion and, 244, 246
  lactobacilli and, 156
  sulfur compounds and, 354, 373, 374
Pseudomonas syringae, 447
Pseudomonas vesicularis, 203, 204

Pseudomonas vulgaris, 441
Pyramimonas gelidicola, 116
Pyramimonas spp., 115, 116
Pyrite, 356
Pyrodictium spp., 249

Quercus spp., 359

Rabbits, 362
Radiation, 86, 111
  black-body, 77, 81, 115
  photosynthetically active: *see* Photosynthetically active radiation
  ultraviolet, 87, 96, 107
Radiolaria, 1, 22, 23; *see also specific types*
Random Walks in Biology, 44
Raphidiopsis spp., 309
Rats
  chitinases in, 417
  lactobacilli in, 150, 155
  sulfur compounds and, 354, 371
Rearrangement
  programmed, 434–438
  unprogrammed, 438–446
Redox conditions, 17, 490
Redox potential, 179, 190, 191, 231, 289
Red sea bream, 416
Renkin theory, 52–53, 59
Rhizobium japonicum: *see Bradyrhizobium japonicum*
Rhizobium leguminosarum, 410
Rhizobium meliloti, 441
Rhizobium phaseoli, 447
Rhizobium spp., 306, 448, 449
Rhizoctonia spp., 410
Rhizopus arhizus, 129
Rhizopus spp., 241, 408
Rhizoselenia spp., 307, 322
Rhodobacter spp., 307, 308
Rhodococcus spp., 373, 374
Rhodopseudomonas sphaeroides, 448
Rhodopseudomonas spp.: *see Rhodobacter* spp.
Rhodospirillaceae, 321
Rhodospirillium rubrum, 205
Rhodospirillium spp., 102, 307, 308
Rhodotorula pallida, 129
Rhodotorula spp., 129
Richellia spp., 309, 322
Rickettsiae, 447
Ristella spp., 155

Rivers, 78, 79
  chitin degradation in, 390, 391, 400, 403,
    411
  nitrification in, 287
  nitrogen fixation in, 305, 323
  sulfur compounds in, 372
  transient, 107–108
Rivularia spp., 309, 335
RNA, 88, 131, 307, 432
Rodents, 154, 157, 158, 160; *see also specific
    types*
Rotifers, 107, 113, 389; *see also specific types*
r strategies, 200, 201

Saccharomyces cerevisiae, 129
Saline lakes, 77, 80, 98, 102, 103, 105, 107,
    115
  microbial mats in, 467, 469, 471
Salmo gairdneri, 417
Salmonella minnesota, 203
Salmonella paratyphi, 441
Salmonella spp., 156, 159
Salmonella typhimurium, 434, 441, 446, 448,
    451
Sarcina spp., 156
Schizophyllum commune, 389
Schizothrix spp., 108
Scopulariopsis brevicaulis, 241
S cycling, 315
Scytonema spp., 308, 309, 471
Sea
  protozoa in, 1, 22
Seabirds, 81
Seagrass, 328, 335
Sea ice microbial community (SIMCO), 74,
    77, 100, 106, 113–117, 131
Seals, 81, 83, 90, 130
Seas, 71, 77, 85, 102, 113, 114, 115, 117,
    359
  chitin degradation in, 390, 391, 395, 413
  microbial mats in, 467, 477, 489
  nitrogen fixation in, 316, 322, 329, 335
  protozoa in, 1, 22, 24
  sulfate in, 346
Seaweed, 307, 352, 391
Sediments, 111; *see also specific types*
  benthic, 80, 98
  chitin degradation in, 392, 394–395, 396,
    398, 401, 402, 403, 404, 412
  microbial mats and, 478–479, 483, 488,
    490

Sediments (*cont.*)
  nitrification in, 275, 288, 289
  nitrogen fixation in, 315
  protozoa in, 2, 3, 14, 15, 25, 28
  sulfur compounds in, 356, 357–358, 359,
    362
Seliberia spp., 244
Seriola quinqueradiata, 416
Serratia lignefaciens, 418
Serratia marcescens, 396, 403, 418, 450
Serratia spp., 156, 403, 418
Sewage, 130, 283; *see also* Sludge
  nitrification in, 289–299
  nitrogen fixation in, 324
  protozoa in, 4
  sulfur compounds in, 354, 366, 368
Shigella spp., 197
Shipworms, 322
Shrimp, 322, 401, 417, 418
Siderocapsa spp., 243
Siderococcus spp., 244
SIMCO: *see* Sea ice microbial community
Single-sludge nitrogen removal systems, 176–
    177
Skunks, 366
S-layer, 50, 52–54
Slime mold, 7–11
Sludge; *see also* Sewage
  management of, 187–188
Sludge age, 185–186, 187, 188, 290, 296
  defined, 185
Sludge bulking, 188–189, 193
Sludge retention time, 290, 297; *see also*
    Sludge age
Snails, 362, 418
Sodium, 80
Sodium chloride, 80
Sodium nitrate, 86
Soil, 105, 124, 125, 130, 131
  in Antarctic, 71, 72, 73, 88
  chitinase activity in, 408–409
  chitin degradation in, 405–415
  cryotolerance and, 90–96
  fellfield: *see* Fellfield soil
  microbial mats and, 490
  nitrification in, 265, 275, 277–287
  ornithogenic, 81, 126–127
  protozoa in, 7–12, 28
  sulfur compounds in, 347, 359, 368, 369,
    372
  wet, 117–119

Soil column, 285–287
  chitin degradation and, 409
Sole, 417
Solea solea, 417
Sorogena spp., 11–12
Soyabean, 409
Spartina alterniflora, 355, 359
Spartina spp., 328, 331, 352, 353
Sphaerotilus natans, 189
Sphaerotilus spp., 243, 244
Spherical symmetry, 41–42
Spiders, 418
Spirillum spp., 66, 308
Sponges, 322, 463
Sporocytophaga spp., 408
Springtails, 82
Squash, 410
Squid, 418
Staphylococcus aureus, 441, 448, 449
Staphylococcus epidermidis, 155
Staphylococcus spp., 129, 156
Starvation, 7, 10, 12–13; *see also* Nutrient deficiency
Steady-state, 203, 275
  in diffusion, 40, 47, 50, 56, 62
Steel, 236, 247
  mild, 245–246, 247
  stainless, 243
Stickstofffressers, 306
Stigonema spp., 309
Stoat, 366
Stokes' law, 52
Stratified ecosystems, 47
Streams, 78, 105, 107–108
Streptococcus faecalis, 210, 211, 441
Streptococcus lactis, 162, 163
Streptococcus pyogenes, 152
Streptococcus sanguis, 162, 441
Streptococcus spp., 156
Streptococcus thermophilus, 147
Streptomyces alivocinereus, 410
Streptomyces coelicolor, 448
Streptomyces fradiae, 441
Streptomyces rimosus, 445
Streptomyces spp., 129, 244
  chitin degradation and, 388, 391, 398, 403, 406, 407, 408
Streptomycetes, 96, 403, 407; *see also specific types*
Stromatolites, 68, 104–107, 461, 463, 468, 479, 483, 486, 488

Stryphnodendron excelsum, 359
Stryphnodendron spp., 359
Sulfate reducers, 251, 321; *see also specific types*
Sulfate-reducing algae, 232
Sulfate-reducing bacteria; *see also specific types*
  chitin degradation and, 399
  corrosion and, 231, 232, 235, 236, 240, 243, 245–246, 247, 249
  in lakes, 113
  in microbial mats, 470
  phosphorus removal and, 193
  sulfur compounds and, 356
Sulfate-reducing fungi, 232
Sulfate reduction, 463, 481, 482–486
Sulfates, 111, 346, 369, 372, 373, 374; *see also specific types*; Sulfate reducers; Sulfate reduction
  ferrous, 86
  nitrification and, 282
  nitrogen fixation and, 325
Sulfate–sulfur, 346
Sulfides, 346, 362, 366, 369–370, 372, 373; *see also specific types*
  alkyl, 361–362
  aromatic, 361–362
  corrosion and, 236, 247
  ferrous, 235–236
  microbial mats and, 484, 487
  in phosphorus removal, 193–195
Sulfite, 236, 372
Sulfolobus spp., 244, 250, 362, 374
Sulfonates, 360–361, 372; *see also specific types*
Sulfones, 372; *see also specific types*
Sulfoxides, 360–361; *see also specific types*
Sulfur, 362, 372; *see also* Organic sulfur compounds
  in activated sludge systems, 199, 202
  corrosion and, 236, 249
Sulfur cycle, 345, 346, 358
Sulfur dioxide, 372
Sulfur oxidation, 239–240
Swamps, 315, 347, 358
Symbionts, 7, 18, 19, 20, 23, 24; *see also specific types*; Symbiosis
  methanogenic, 25–28
  nitrogen fixation and, 306
  nonmethanogen, 28
  photosynthetic, 21–25

Symbiosis, 24, 26, 116; see also Symbionts
    chitin degradation in, 418–419
    nitrogen fixation and, 322
Synechococcus spp., 309, 311, 315
Synechocystis spp., 315
Syntrophomonas wolfei, 251
Syntrophus buswellii, 251

Tardigrades, 94, 107, 113; see also specific
    types
Temperature, 3, 4, 6, 10, 11, 125
    in Antarctic, 72, 73, 76, 77, 78, 80, 81–82,
        84, 86, 87, 96, 98, 102, 103, 111
    chitin degradation and, 400, 403, 406, 413
    microbial mats and, 485, 490
    nitrification and, 264, 265, 267, 272–273,
        282, 283, 284, 287, 294–296
    nitrogen fixation and, 326, 328
    phosphorus removal and, 185
    in sea ice microbial community, 114, 116,
        117
Tetracycline resistance, 161, 163
Tetrahymena pyriformis, 4
Tetrahymena spp., 4, 12
Thermodesulfobacterium commune, 249
Thermophilic bacteria, 248–250
Thermus aquaticus, 249
Thermus spp., 250
Thiamine, 348, 349, 362
Thielaria basicola, 410
Thielavia spp., 406
Thietanes, 366
Thiobacilli, 239–240, 250, 369, 482, see also
    specific types
Thiobacillus ferrooxidans, 240
Thiobacillus spp., 244, 308, 368, 374
Thiobacillus thiooxidans, 239, 240
Thiobacillus thioparus, 240, 369, 371
Thiols, 362, 366, 367
Thiopedia spp., 244
Thiophenes, 363, 372–374
Thiosulfate, 86, 236
Thiothrix spp., 189, 193
Thrombolites, 468
Tintinnidium spp., 4, 6
Titanium, 232
Toadstools, 7
Tolypothrix spp., 111, 308, 309, 335
Tomatoes, 410
Toxothrix spp., 244

Trace elements, 199
Trace metals, 326, 334, 336; see also specific
    types
Trace minerals, 68; see also specific types
Tranposition, 452
Transduction, 161, 162–163
Transformation, 161, 163–164
Transition, 438
Transport, 210–213, 233
    colimiting, 55–57
Transposition, 437–438
Transposons, 164
Transversion, 438
Trebonema spp., 111
Trichoderma spp., 241, 406, 407
Trichodesmium spp., 309, 310, 326, 327, 335
Trichosporon spp., 241
Tridacna gigas, 397
Trimyema compressum, 27
Trout, 417
Tubercle formation, 242–243
Tumor cells, 160
Two-dimensional ecosystems, 37, 60–61

UCT process, 179, 183, 200
    modified, 180, 183
Ultraviolet radiation, 87, 96, 107
Urinary tract infections, 165
UV: see Ultraviolet

Vaginitis, 165
Vanadium, 321
Verticillium lecanii, 418
Verticillium spp., 406, 407
Vibrio alginolyticus, 417
Vibrio cholerae, 413
Vibrio parahaemolyticus, 413, 414
Vibrios, 393, 398; see also specific types
Vibrio spp., 88
    chitin degradation and, 394, 396, 397, 398,
        413, 416
    corrosion and, 244
    nitrogen fixation and, 308
Viking Mars program, 73, 79, 88, 89; see also
    Mars
Viral absorption, 42
Viruses, 51, 433
Volcanoes, 75, 79, 96, 123, 124, 125, 347,
    369
von Smoluchowski formula, 51

Wagner–Traud theory, 231
Wastewater
  nitrification in, 289–299
  phosphorus removal from: *see* Biological
    phosphorus removal
  sulfur compounds in, 361, 366, 372
Water, 2, 123; *see also specific types*
  corrosion and, 234
  in deserts, 83–84, 85, 87, 91–93, 94, 96
  in diffusion, 45
  protozoa and, 7–13, 17, 18
Water column, 98, 100, 101, 113, 114, 232
  chitin degradation in, 393, 404, 413
  diffusion and, 46
Welds, 243–244
Westiellopsis spp., 309
Whales, 130, 417
Wheat, 410, 418

Xanthomonas campestris, 441, 446
Xenobiotics, 372

Xerotolerance, 90–96

Yeasts, 128, 129; *see also specific types*
  in Antarctic, 71, 73, 74, 83, 93, 94, 96
  categorization of, 127
  chitin degradation and, 389
  cryotolerance of, 119
  in fellfield soil, 123
  lactobacilli and, 154, 155, 164
  nutrient cycling and, 125, 126
Yellowtail, 416
Yersinia spp., 197

Zea mays, *see* Corn
Zinc, 199, 321, 326
Zoochlorellae-bearing ciliates, 18–22
Zooglea ramigera, 196
Zooplankton, 18, 22, 353, 393, 395, 413; *see*
  also specific types
Zostera marina, 352
Zostera spp., 328, 331, 335

2